FORTSCHRITTE DER CHEMIE ORGANISCHER NATURSTOFFE

PROGRESS IN THE CHEMISTRY OF ORGANIC NATURAL PRODUCTS

PROGRÈS DANS LA CHIMIE DES SUBSTANCES ORGANIQUES NATURELLES

HERAUSGEGEBEN VON   EDITED BY   RÉDIGÉ PAR

L. ZECHMEISTER
CALIFORNIA INSTITUTE OF TECHNOLOGY, PASADENA

NEUNTER BAND NINTH VOLUME NEUVIÈME VOLUME

VERFASSER   AUTHORS   AUTEURS

J. G. BAXTER · H. BORSOOK · N. BULMAN · D. H. CAMPBELL · F. M. DEAN
H. H. INHOFFEN · H. M. KALCKAR · W. S. McNUTT · P. MEUNIER
H. SIEMER · A. STOLL · M. TOMITA

MIT 20 ABBILDUNGEN   WITH 20 ILLUSTRATIONS   AVEC 20 ILLUSTRATIONS

WIEN · SPRINGER-VERLAG · 1952

ALLE RECHTE, INSBESONDERE DAS DER ÜBERSETZUNG
IN FREMDE SPRACHEN, VORBEHALTEN
SOFTCOVER REPRINT OF THE HARDCOVER 1ST EDITION 1952

ISBN-13:978-3-7091-7171-4   e-ISBN-13:978-3-7091-7169-1
DOI: 10.1007/978-3-7091-7169-1

Titel Nr. 8215

Inhaltsverzeichnis.
Contents. — Table des matières.

Synthetische Chemie der Carotinoide. Von H. H. INHOFFEN und
H. SIEMER, Organisch-chemisches Institut der Technischen Hochschule,
Braunschweig 1

I. Allgemeine Einleitung 1
   Nomenklatur-Vorschlag und Einteilung des Stoffes................... 2
   Zur Stereochemie der Carotinoide 4
      Die möglichen Stereoisomeren.................................... 6
      Spektrale Veränderungen durch *trans* → *cis*-Isomerisierung 8
      „*cis*-peak"-Effekt 8

II. Synthesen von C_{30}- und C_{32}-Kohlenwasserstoffen als Modellsynthesen... 9

III. Bis-nor-methyl-β-carotin und 7,7'-Dihydro-β-carotin 13
   Bis-nor-methyl-β-carotin 13
   7,7'-Dihydro-β-carotin 16

IV. β-Carotin-Synthesen 19
   β-Carotin-Synthesen $C_{16} + C_8 + C_{16}$................................... 19
   β-Carotin-Synthese $C_{19} + C_2 + C_{19}$..................................... 22
      Stereoisomerisierung von 15,15'-mono-*cis*-β-Carotin.................... 26
   β-Carotin-Synthese $C_{18} + C_4 + C_{18}$..................................... 27

V. Lycopin und ε_1-Carotin.. 28
   Lycopin 28
   ε_1-Carotin 30

VI. Synthesen von höheren Carotin-Homologen......................... 31
   16,16'-Homo-β-carotin 31
   Decapreno-β-carotin und Decapreno-ε_1-carotin...................... 34
   Dodecapreno-β-carotin 35

Literaturverzeichnis 38

Synthesis and Properties of Vitamin A and Some Related Compounds. By J. G. BAXTER, Distillation Products Industries, Rochester,
New York 41

Introduction 42

I. Synthesis of Vitamin A... 43
   1. *Via* Esters of β-Ionylideneacetic Acid 43
      Ethyl β-Ionylideneacetate and its β,γ-Unsaturated Isomer 46
      β-Ionylidene-ethanol... 50
      C_{18}-Ketone 50

Vitamin A Acid ethyl ester and its β,γ-Unsaturated Isomer 50
Vitamin A 51
Isomer of Vitamin A 51
Synthesis by Wendler, Slates, Trenner and Tishler 51
β-Ionylideneacetaldehyde 52
2. Synthesis of Vitamin A via Esters of β-Ionylidenecrotonic Acid 52
β-Ionylidenecrotonic Acid...................................... 53
C_{18}-Ketone 53
Vitamin A Aldehyde.. 56
Vitamin A Acids 56
Vitamin A 57
Synthesis by Schwarzkopf, Cahnmann, Lewis, Swidinsky and Wuest 57
3. Synthesis of Vitamin A via "C_{14}-Aldehyde" 58
C_{14}-Aldehyde 60
Allylic Rearrangement 60
4. Synthesis of Vitamin A by Other Methods 64

II. Synthesis and Biological Activity of Some Compounds Related to Vitamin A 66
Alcohols... 66
Esters and Ethers of Vitamin A 74
Aldehydes 75
Vitamin A_2 76
Hydrocarbons.. 77
Vitamin A Acids... 79
Conclusions 80

III. Relationship between Vitamin A and Carotenes 80

References 80

Les Antivitamines. Par P. Meunier, Laboratoire de chimie biologique de la Faculté des Sciences, Lyon 88
Introduction 88

I. Les antagonistes des vitamines hydrosolubles 90
   1° Les antagonistes de la thiamine 90
      a) Pyrithiamine 90
      b) Oxythiamine 91
      c) Homothiamine-glycol....................................... 91
      d) Antagonistes de la thiamine de nature enzymatique: thiaminases diverses.. 92
      e) Antithiamines des fougères 93
   2° Les antagonistes des flavines et de la vitamine B_{12} 94
   3° Les antagonistes de l'acide pantothénique 95
   4° Les antagonistes de la pyridoxine 95
      a) Activité vitaminique B_6.................................... 95
      b) Constitution chimique des antivitamines B_6 96
      c) Désoxypyridoxine.. 96
      d) Méthoxypyridoxine 97
      e) Autres antivitamines B_6.................................... 97
   5° Les antagonistes de l'acide nicotinique 98
   6° Les antagonistes de la biotine................................. 98
   7° Les antisulfamides.. 100

8° Les anti-acides foliques	100
9° L'activité antisulfamide et antisulfone de dérivés voisins de l'acide *p*-aminobenzoïque	101
II. Les antagonistes des vitamines liposolubles	102
1° L'antivitamine A	102
2° Les antivitamines E	103
3° Les antivitamines K	104
Bibliographie	107

Recent Investigations on Ergot Alkaloids. By A. STOLL, Chemische Fabrik Sandoz, Basle, Switzerland 114

I. Historical Introduction	114
II. The Structure of the Ergot Alkaloids	119
1. Introduction	119
2. The Structure of Lysergic Acid	122
3. Structure of the Peptide Portion	134
III. The Individual Alkaloids of Ergot	149
1. Ergobasine and Ergobasinine	149
2. Ergotamine and Ergotaminine	153
3. Ergosine and Ergosinine	155
4. Alkaloids of the Ergotoxine Group	156
5. Ergocristine and Ergocristinine	159
6. Ergokryptine and Ergokryptinine	161
7. Ergocornine and Ergocorninine	162
IV. Partially Synthetic and Hydrogenated Derivatives of Ergot Alkaloids	163
1. Partially Synthetic Derivatives of Lysergic Acid	164
2. The Dihydro Derivatives of the Natural Alkaloids of Ergot	167
References	170

Die Alkaloide der Menispermaceae-Pflanzen. Von M. TOMITA, Pharmazeutisches Institut der Universität Kyoto 175

I. Einleitung	176
II. Die in Menispermaceae aufgefundenen Alkaloide	177
A. Durch KONDO und seine Mitarbeiter untersuchte Pflanzen	177
B. Von nicht-japanischen Forschern untersuchte Pflanzen	177
C. Pflanzen, in denen von KONDO und Mitarbeitern das Vorkommen von Alkaloiden festgelegt wurde.	178
D. In der Literatur als alkaloidhaltig angegebene Pflanzen	178
E. Aus Rohmaterialien des chinesischen Drogenmarktes isolierte Alkaloide	178
F. Nicht zu den Menispermaceae gehörende, Biscoclaurin-Basen enthaltende Pflanzen	179
Berberidaceae 179. — Anonaceae 179. — Magnoliaceae 179. — Monimiaceae 179.	
G. In Curare enthaltene Biscoclaurin-Alkaloide	179

| VI | Inhaltsverzeichnis. — Contents. — Table des matières. |

III. Klassifizierung der Menispermaceae-Alkaloide 180
    Systematik der Biscoclaurin-Basen 180
    Gruppe I. Basen mit einem Äthersauerstoff 180
    Gruppe IIa. Basen mit zwei Äthersauerstoffen (Tetrandrin-Typus) 180
    Gruppe IIb. Basen mit zwei Äthersauerstoffen (Isochondendrin-Typus) 180
    Gruppe IIIa. Basen mit drei Äthersauerstoffen (Diphenylendioxyd-Typus) 180
    Gruppe IIIb. Basen mit drei Äthersauerstoffen (Depsidan-Typus) 180

IV. Allgemeine Untersuchungsprinzipien der Biscoclaurin-Alkaloide 181
    1. Permanganat-Oxydation des Alkaloides selbst oder seines HOFMANNschen Abbauproduktes 181
    2. Ozon-Spaltung von Methinbasen 182
    3. Aufspaltung durch Natrium in flüssigem Ammoniak 184

V. Spezieller Teil 186
    1. Benzylisochinolin-Typus 186
        Coclaurin 186. — Isococlaurin 187. — Magnocurarin 187. — Salicifolinchlorid 188.
    2. Phenanthropyridin-Typus 188
        Sinomenin 188. — Disinomenin 188. — Tuduranin 189. — Stephanin 189. — Crebanin 189. — Phanostenin 190. — Dicentrin 190.
    3. Berberin-Typus 190
        Berberin 190. — Palmatin 190. — Columbamin 190. — Jatrorrhizin 190. — Sinactin 191.
    4. Benzochinolizin-Typus 191
        Rotundin 191.
    5. Biscoclaurin-Typus 192
        Gruppe I. Dauricin 192. — Magnolin 192. — Magnolamin 192. — Aztequin 193.
        Gruppe IIa. Berbamin 193. — Isotetrandrin 194. — Tetrandrin 194. — Phaeanthin 195. — Cepharanthin 195. — Oxyacanthin 195. — Repandin 197. — Daphnandrin 198. — Daphnolin (Trilobamin) 198. — Aromolin 198. — Epistephanin 199. — Hypoepistephanin 199.
        Gruppe IIb. Isochondodendrin 199. — Cycleanin 200. — Protocuridin 200. — Neoprotocuridin 201. — Bebeerin 201. — Chondrofolin 202. — Tubocurarinchlorid 202. — Chondocurin 202.
        Gruppe IIIa. Trilobin 203. — Isotrilobin 204. — Menisarin 204. — Normenisarin 204. — Micranthin 205.
        Gruppe IIIb. Insularin 205.
    6. Strukturell ungeklärte Basen 207
    7. Optische Isomerie der Biscoclaurin-Basen 207
    8. Charakterisierung der Biscoclaurin-Basen 209

VI. Biogenetische Betrachtungen über Biscoclaurin-Basen 209
VII. Medizinische Anwendungen 213
Literaturverzeichnis 214

Inhaltsverzeichnis. — Contents. — Table des matières. VII

Naturally Occurring Coumarins. By F. M. DEAN, The University of Liverpool, Department of Organic Chemistry 225

I. General Structural Features.. 226

II. The Chemistry of the Coumarin System 229

    Conversions and Degradation................................... 229
    The Synthesis of Coumarins.................................... 235
    Theoretical Considerations..................................... 237

III. Occurrence, Isolation and Determination 239

IV. Some Biochemical Properties..................................... 240

V. Simple Coumarins... 242

    Coumarin 242. — Dihydro-coumarin 243. — Umbelliferone 243. — Herniarin 243. — Aesculetin 244. — Scopoletin 245. — Fabiatrin 245. — Ayapin 245. — Citropten 245. — Daphnetin 246. — Fraxetin 247. — Fraxidin 248. — Isofraxidin 249. — Fraxinol 249. — Eugenin 249. — 5-Geranoxy-7-methoxycoumarin 249. — Suberosin 250. — Collinin 250. — Brayleyanin 250. — Umbelliprenin 251. — Toddalolactone 251. — Aculeatin 252. — Auraptene 252. — Ostruthin 253. — Osthenol 254. — Osthol 254. — Ammoresinol 256. — Dicoumarol 257.

VI. Furanocoumarins.. 257

    Psoralene 260. — Angelicin 261. — Bergapten 262. — Bergaptol 263. — Isobergapten 264. — Xanthotoxin 264. — Xanthotoxol 265. — Sphondin 265. — Sphondylin 265. — Isopimpinellin 265. — Pimpinellin 266. — Isoimperatorin 266. — Oxypeucedanin 267. — Ostruthol 267. — Imperatorin 268. — Bergamottin 269. — Phellopterin 269. — Byakangelicol 270. — Byakangelicin 270. — Ferulin 271. — Nodakenetin 272. — Marmesin 272. — Peucedanin 273. — Athamantin 275.

VII. Chromeno-α-pyrones 276

    Xanthyletin 277. — Seselin 278. — Xanthoxyletin 278. — Luvangetin 280. — Alloxanthoxyletin 281. — Braylin 281.

VIII. 3:4-Benzcoumarins... 282

    2′:3″-Dihydroxydibenz-α-pyrone 282. — 4:6:4′:6′-Dihydroxydiphenic acid dilactone 282. — Ellagic acid 283. — 4:4′-Dihydroxy-6:6′-dimethoxydiphenic acid dilactone 284.

References.. 285

The Biosynthesis of Proteins and Peptides, including Isotopic Tracer Studies. By H. BORSOOK, California Institute of Technology, Pasadena, California 292

I. Introduction 293

   1. The Theory of Endogenous and Exogenous Protein Metabolism.... 294
   2. The Theory of Protein Metabolism as a Dynamic Steady State.... 294

    a) Indirect Evidence 294
    b) Direct Evidence 297
    c) Lability of Enzyme Proteins 298

II. The Measurement of Protein Turnover 299

III. Incorporation of Labeled Amino Acids *in vivo* 300
   1. N^{15}-labeled Amino Acids as Tracers 300
   2. C^{14}- and S^{35}-labeled Amino Acids as Tracers 303
      a) In Normal Tissues 303
      b) In Tumors 305
      c) Influence of Hormones 305
      d) Incorporation of Foreign Amino Acids 308
IV. Incorporation of Labeled Amino Acids *in vitro* 309
   1. Incorporation of Carbon Dioxide into Amino Acids 310
   2. Net Synthesis of Protein *in vitro* 312
   3. Comparison of Incorporation of Amino Acids *in vivo* and *in vitro* ... 313
   4. Amino Acid Incorporation in Different Cell Fractions 313
   5. The Nucleus, Amino Acid Incorporation, and the Maintenance of the Amino Acid Pattern in Proteins 314
   6. Nucleic Acids, Protein Synthesis and Amino Acid Incorporation into Proteins 315
   7. Normal, Foetal and Tumor Tissue 315
   8. Effect of Concentration of Labeled Amino Acid on its Rate of Incorporation 315
   9. Does Incorporation of One Amino Acid Require the Presence of Others? 316
      a) Feeding Experiments 316
      b) *In vivo* Experiments with Single Labeled Amino Acids 316
      c) *In vitro* Experiments with Labeled Amino Acids 317
V. The Biological Significance of the High Lability of the Proteins in the Cell 319
VI. Mechanism of Peptide Bond Synthesis 319
   1. Heats and Free Energies of Formation of Some Amino Acids and Peptides (Solids) 319
   2. Free Energies of Formation of Some Peptides in Aqueous Solution 320
   3. The Effect of pH on the Free Energy Change in Peptide Formation 321
   4. Peptide Synthesis by Proteases and Peptidases 322
      a) Classification of Enzymatic Peptide Syntheses According to the Sign and Magnitude of the Free Energy Change $(-\Delta F)$ ... 322
      b) Peptide Syntheses where $-\Delta F$ is Positive and Large ... 323
      c) Peptide Syntheses where $-\Delta F$ is Small and the Peptide is Relatively Insoluble 325
      d) Plastein Formation 327
      e) Peptide Synthesis in an Exchange Reaction during Hydrolysis (Transamidation and Transpeptidation) 328
      f) Peptide Synthesis from Amino Acid Esters 333
   5. Glutamo- and Asparto-Transferases 334
   6. Syntheses where $-\Delta F$ is Negative and Large, Coupled with High Energy Phosphate 336
      a) Synthesis of Glutamine 336
      b) Synthesis of Hippuric Acid 337
      c) Synthesis of *p*-Aminohippuric Acid 338
      d) Synthesis of Ornithuric Acids 339
      e) Synthesis of Glutathione 341

VII. Mechanism of Amino Acid Incorporation into Proteins 343
   1. Effect of Inhibitors 343
   2. Amino Acid Incorporation and Phosphorylation 346
   3. Heat-Stable Co-factors for Amino Acid Incorporation 346
   4. Is Amino Acid Incorporation Synthesis of Protein *de novo* or an Exchange? 347
   5. The Possibility of Peptides as Intermediates in Protein Synthesis... 348
   6. The Linkage of Incorporated Amino Acids 349
References 352

The Enzymes of Nucleoside Metabolism. By HERMAN M. KALCKAR, Cytophysiological Institute of the University, Copenhagen 363

Introduction 363
 I. The Preparation of Nucleosides 364
II. The Enzymes of Nucleoside Metabolism 365
   1. Purine Nucleoside Phosphorylase 367
   2. Pyrimidine Nucleoside Phosphorylase 372
   3. Trans-N-Glycosidase 372
   4. Ribosidase 374
   5. Phosphoribomutase 375
   6. Degradation and Synthesis of Ribose-Phosphoric Esters 376
   7. Nucleoside Deaminases 378
III. Phospho-Ribosides 381
   1. Preparation and Properties of Ribose-1-phosphate 381
   2. Enzymatic Synthesis of Ribosides 382
   3. Preparation and Properties of Deoxyribose-1-phosphate 385
   4. Enzymatic Synthesis of Hypoxanthine Deoxyriboside 386
IV. Trans-N-Glycosidic Reactions 387
   1. Non-participation of Deoxyribose-1-Phosphate in Trans-N-glycosidic Reactions 387
   2. Trans-N-Glycosidic Reactions in the Deoxyribose Nucleoside Series.. 387
   3. Enzymatic Formation of New Deoxyribosides 388
V. Phosphorylation of Nucleosides 390
VI. Incorporation of Purines and Pyrimidines into Nucleic Acids 391
   In vivo Studies with Labelled Purines 391. — *In vivo* Studies with Labelled Pyrimidines 392. — *In vitro* Studies with Labelled Purines 393. — Studies on the Amphibian and Echinoderm Egg 394. — Studies on Micro-organisms 394.
References 395

Nucleosides and Nucleotides as Growth Substances for Microorganisms. By W. S. McNUTT, Vanderbilt University, School of Medicine, Department of Biochemistry, Nashville, Tennessee 401

Introduction 402
 I. Nucleosides and Nucleotides of Ribose 405
   1. Coenzyme I, "Desamino-codehydrogenase I," Coenzyme II and Nicotinamide Riboside 405

2. Purine-Nucleosides and Nucleotides		405
a) Growth-promoting Activity		405
b) Growth-inhibiting Activity and the Ability to Reverse Growth-inhibition		409
3. Nucleotides in the Nutrition of *Lactobacillus gayonii*		410
4. Pyrimidine-Nucleosides and Nucleotides		411
a) Growth-promoting Activity		411
b) Growth-inhibiting Activity		412
5. The Biosynthesis of Ribosides and Ribonucleotides		413
A Comparison between Microorganisms and Higher Animals with Regard to Purine Precursors in Nucleic Acid Biosynthesis		413
6. Vitamin B_{12}		417
Microbiological Functions of Vitamin B_{12}		418
Different Forms of Vitamin B_{12}		419
II. Nucleosides and Nucleotides of Desoxyribose		420
1. The Biosynthesis of Desoxyribosides		421
Considerations of the Mode of Formation of the Desoxyribosidic Linkage		422
2. The Growth-promoting Activity of Desoxyribosides and Desoxyribonucleotides		424
a) The Specificity of Certain Desoxyribosides in Eliciting the Growth-response of Bacteria		424
b) The Non-specificity of the Natural Desoxyribosides in Promoting the Growth of Certain Bacteria		424
3. The Relationship of the Desoxyribosides, Vitamin B_{12}, Reducing Agents, and the "Citrovorum-Factor" in Supporting the Growth of Various Microorganisms		426
Relationship between Certain Reducing Agents and Vitamin B_{12} Requirement		427
The "Citrovorum Factor"		431
References		433

Some Current Concepts of the Chemical Nature of Antigens and Antibodies By DAN H. CAMPBELL and N. BULMAN, California Institute of Technology, Pasadena, California ... 443

I. Introduction		443
II. Antigens and Haptens		445
1. Antigens		446
2. Haptens		449
III. Antibodies		451
1. Chemical Composition of Antibodies		451
2. Electrophoretic Properties of Antibodies		452
3. Shape and Size of Antibodies		453
4. Nature of Combining Sites		455
5. Purification of Antibodies		461

IV. The Physical Nature of Antigen-Antibody Reactions 463
  1. The Properties of Specific Precipitates.......................... 463
    a) Composition 463
    b) Formation and Specificity 465
    c) 'Ageing'.. 466
  2. Thermodynamic Properties of Antigen-Antibody Reactions 466
    a) The Free Energy and Heat Changes in Antigen-Antibody Reactions 467
    b) Differences in Free Energies of Combination 468
  3. Nature of the Forces Involved................................. 471
  4. Mathematical Interpretations of the Precipitin Reaction............ 475
  5. A Note on the Use of Polyvalent Haptens 476

V. Conclusions 477

References 478

Namenverzeichnis. Index of Names. Index des Auteurs 485

Sachverzeichnis. Index of Subjects. Index des Matieres 502

Synthetische Chemie der Carotinoide.

Von H. H. INHOFFEN und H. SIEMER, Braunschweig.

Mit 4 Abbildungen.

Inhaltsübersicht.

	Seite
I. Allgemeine Einleitung	1
Nomenklatur-Vorschlag und Einteilung des Stoffes	2
Zur Stereochemie der Carotinoide	4
Die möglichen Stereoisomeren	6
Spektrale Veränderungen durch trans → cis-Isomerisierung	8
„cis-peak"-Effekt	8
II. Synthesen von C_{30}- und C_{32}-Kohlenwasserstoffen als Modellsynthesen	9
III. Bis-nor-methyl-β-carotin und 7,7'-Dihydro-β-carotin	13
Bis-nor-methyl-β-carotin	13
7,7'-Dihydro-β-carotin	16
IV. β-Carotin-Synthesen	19
β-Carotin-Synthesen $C_{16} + C_8 + C_{16}$	19
β-Carotin-Synthese $C_{19} + C_2 + C_{19}$	22
Stereoisomerisierung von 15,15'-mono-cis-β-Carotin	26
β-Carotin-Synthese $C_{18} + C_4 + C_{18}$	27
V. Lycopin und ε_1-Carotin	28
Lycopin	28
ε_1-Carotin	30
VI. Synthesen von höheren Carotin-Homologen	31
16,16'-Homo-β-carotin	31
Decapreno-β-carotin und Decapreno-ε_1-carotin	34
Dodecapreno-β-carotin	35
Literaturverzeichnis	38

I. Allgemeine Einleitung.

Im folgenden soll vor allem die synthetische Chemie der Carotinoide beschrieben werden, die etwa um das Jahr 1942 mit ersten Arbeiten von HEILBRON und seiner Schule beginnt und sich bis 1951 stetig entwickelt hat. Vorkommen, Isolierung und Konstitutionsaufklärung der natürlichen Carotinoide sowie deren Abbau- und Umwandlungsprodukte sind in der im Jahre 1948 erschienenen Monographie von KARRER und JUCKER (34) hinreichend behandelt, so daß hierauf verwiesen werden kann. Auch die Carotinoid-epoxyde und furanoiden Oxyde von Carotinoid-farbstoffen sind von KARRER in diesen „Fortschritten" ausführlich beschrieben (25). Betr. Vitamin A, s. BAXTER (20) auf S. 41 dieses Bandes.

Nomenklatur-Vorschlag und Einteilung des Stoffes.

Hinsichtlich des Begriffes „Carotinoide" möchten wir eine erweiterte Nomenklatur zur Diskussion stellen. Hiernach werden alle Verbindungen, die einen Teil des typischen Kohlenstoffgerüstes vom γ-Carotin bis hinunter zu Isopren-Abkömmlingen mit 10 C-Atomen enthalten, grundsätzlich zu den Carotinoiden gezählt, gleichgültig, ob sie partiell hydriert oder dehydriert sind. In diese Nomenklatur läßt sich jetzt zwanglos die große Zahl der synthetischen Stoffe, die für den Aufbau der „klassischen Carotinoide" (C_{40}) von Bedeutung sind, sinngemäß einordnen.

Es wird jedoch für übersichtlicher gehalten, diejenigen Carotinoid-Verbindungen, die einen Mehr- oder Mindergehalt an Methylgruppen aufweisen, wie z. B. das Iron und das nor-Methyl-Vitamin A, sowie Verbindungen mit „versetzten" Methylgruppen, wie das Jonyliden-aceton, ferner Carotinoide mit mehr als vierzig Kohlenstoffatomen, wie z. B. 16,16′-Homo-β-carotin, in einem gesonderten Kapitel „Verbindungen mit Carotinoid-Charakter" zusammenzufassen. Hierzu gehören insbesondere die nor-Vitamin-A-Verbindungen von HEILBRON.

Die Unterkapitel werden nach der Kohlenstoffzahl gegliedert, doch erscheint hier eine schärfere Einteilung zweckmäßig, damit die Kohlenstoffzahlen auch weiterhin sogleich die Grundstruktur sowie strukturelle Zusammenhänge erkennen lassen:

1. Unsymmetrische Carotinoide mit weniger als 40 C-Atomen.
2. Carotinoide mit 40 C-Atomen.
3. Symmetrische Carotinoide mit weniger als 40 C-Atomen.
4. Verbindungen mit Carotinoid-Charakter.

1. Bei den unsymmetrischen Carotinoiden mit weniger als 40 C-Atomen kann man an der Kohlenstoffzahl sofort ablesen, wie weit die betreffende Verbindung „aufgebaut" ist, d. h. wie weit der Aufbau des Carotinoidgerüstes ab C_{10} bis C_{40} fortgeschritten ist. Im Falle des β-Jonons liegt z. B. ein C_{13}-Keton vor. Wird die Kette um die vier C-Atome der Crotonsäure verlängert, so resultiert eine C_{17}-Säure, und nach weiterem Aufbau kommt man beispielsweise zum C_{20}-Alkohol Vitamin A. Die hier aufgeführten Verbindungen stellen also hinsichtlich ihres vollständigen Kohlenstoffgerüstes exakte Teilstücke des C_{40}-Bezugskohlenwasserstoffes γ-Carotin dar.

Ferner zeigt der griechische Buchstabe α bzw. β das am „linken" Ende stehende Ringsystem an, während das Fehlen eines solchen Buchstaben unmittelbar den rein aliphatischen Aufbau kundgibt. Das Präfix α- bzw. β- sollte daher immer verwendet werden. Die Wortenden -Alkohol, -Aldehyd, -Keton, -Säure, -Äthinyl-Verbindung, -Kohlenwasserstoff belegen zugleich die praktisch entscheidende reaktive Gruppe am Ende der aliphatischen Kette.

2. Bezüglich der Carotinoide mit 40 C-Atomen ist nichts weiter hinzuzufügen, da deren Aufbau-Prinzip hinlänglich bekannt ist.

3. Die Grundstruktur der symmetrischen Carotinoide mit weniger als 40 C-Atomen erhellt gleichfalls unmittelbar aus ihrer Kohlenstoffzahl, wie z. B. beim Crocetin C_{20} und Norbixin C_{24}. Hier treten an den Enden zwei reaktive Gruppen symmetrisch in Funktion.

4. Die hier zusammengefaßten Verbindungen sind nur mit einem Teil ihres Kohlenstoffgerüstes (definitionsgemäß mindestens mit einem C_{10}-Teilstück) mit γ-Carotin strukturidentisch.

Es gibt naturgemäß auch Übergangsglieder, die sowohl zu den Carotinoiden als auch zu den Terpenen gerechnet werden können, wie z. B. Geraniol und Farnesol.

Es wird also der Versuch unternommen, die unmöglich langen und kaum zu handhabenden Namen durch kurze und typische Bezeichnungen zu ersetzen, die dazu noch den Vorteil des unmittelbaren Zusammenhanges mit der Grundstruktur aufweisen. Ein vollständiges Erfassen aller Möglichkeiten ist mit diesem Schema nicht gewährleistet, denn man kann sich aus dem Carotin-Molekül Teilstücke herausgeschnitten denken, die sich hier nicht einordnen lassen.

Eine *Ausnahme* hinsichtlich der Systematik mittels der Kohlenstoffzahl erscheint zur Wahrung des dargelegten Zusammenhangs mit der Grundstruktur unerläßlich. Alle Verbindungen mit Carotinoid-Charakter unter 4., vor allem die nor-Methyl-Verbindungen, werden *nicht* nach ihrer Kohlenstoffzahl eingeordnet, da sie sonst in eine irreführende Beziehung zu den eigentlichen Carotinoiden geraten. Zum Beispiel, das Iron mit 14 C-Atomen hat nichts mit dem C_{14}-Aldehyd zu tun. Fernerhin sollte man die nor-Methyl-Verbindungen auf die „vollmethylierten" Isopren-Abkömmlinge beziehen, wodurch sie in eine klare Beziehung zu den entsprechenden Carotinoiden treten. Das 13-nor-Methyl-Vitamin A mit 19 C-Atomen gehört also seiner Ketten-Struktur nach zum C_{20}-Vitamin A. Diejenigen C-Atome, an denen Methyl-Gruppen fehlen, werden zweckmäßig mit ihrer Ziffer genannt.

Hinsichtlich der *Bezifferung der Kohlenstoffatome* und der Festlegung der Doppelbindungen wird der Vorschlag von KARRER übernommen. Dieses System sei an den Beispielen des β- und γ-Carotins nachstehend erläutert. Es werden also weder die endständigen noch die Seitenketten-Methylgruppen beziffert, und ferner wird durchwegs vom linken Ende der Molekel bis zur Mitte bzw. nach rechts gezählt, so daß hierdurch die höchste Kohlenstoffziffer 15 beträgt. Die andere Molekülhälfte wird

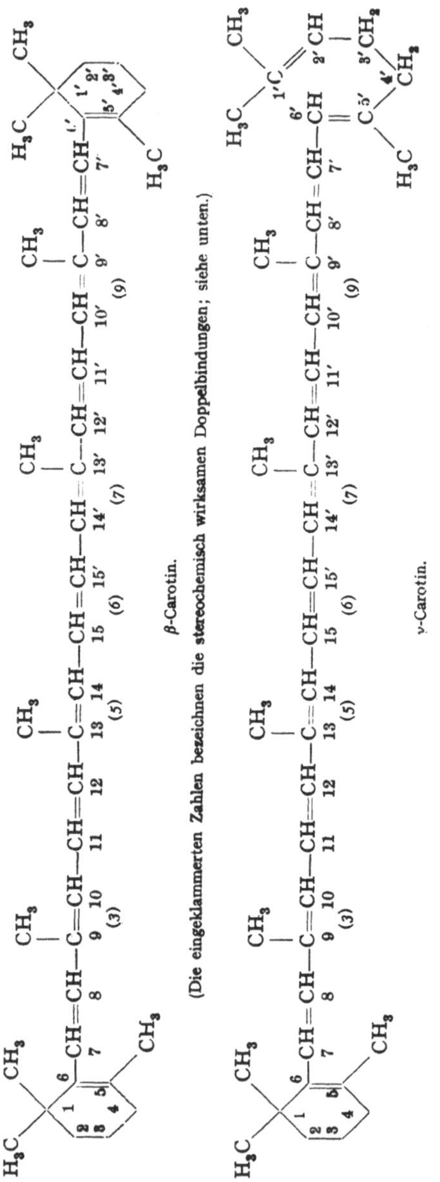

β-Carotin.

(Die eingeklammerten Zahlen bezeichnen die stereochemisch wirksamen Doppelbindungen; siehe unten.)

γ-Carotin.

mit einem Index zurückgezählt. Die Lage einer jeden Doppelbindung ist somit eindeutig durch die Ziffern der beiden diesbezüglichen C-Atome definiert. Liegt an einer bestimmten Stelle bereits eine Doppelbindung vor, so zeigt die Angabe *x,x*-Dehydro an diesem Ort das Vorhandensein einer Dreifachbindung an.

Auch bei Molekülen mit einer kleineren Zahl von C-Atomen als 20, z. B. beim C_{19}-Aldehyd, soll konsequent die obige Numerierung der C-Atome beibehalten werden, und nicht, wie dies häufig der Fall ist, von der sauerstoffhaltigen Gruppe der aliphatischen Kette anfangend zum anderen Ende, d. h. nach „links" gezählt werden.

Die künstlich dargestellten *cis-trans*-isomeren Carotinoide erhalten das Präfix „Neo" und werden nach ihrer Lage im Chromatogramm durch einen anschließenden Buchstaben näher gekennzeichnet. Beim β-Carotin z. B. sieht ein Stereoisomeren-Chromatogramm etwa folgendermaßen aus (*41*):

Neo-β-carotin V (oben)
Neo-β-carotin U
all-*trans*-β-Carotin
Neo-β-carotin A
Neo-β-carotin B (unten)

Oberhalb des β-Carotins liegen in der Tswett-Säule die „Neo-isomeren", beginnend mit dem Buchstaben U (von „ultra"); unterhalb des β-Carotins befinden sich die „Neo-isomeren", beginnend mit dem Buchstaben A.

Zur Stereochemie der Carotinoide.

Die Stereochemie der Carotinoide hat in den letzten zehn Jahren eine außerordentliche Entwicklung erfahren, die wir vor allem den umfangreichen Experimentalarbeiten von Zechmeister und seiner Schule verdanken. In Zusammenarbeit mit PAULING konnten die spektralen Erscheinungen theoretisch gedeutet werden, so daß aus einer sich stetig vermehrenden Zahl von Einzelbefunden klare Zusammenhänge zwischen der Lichtabsorption und der sterischen Konfiguration der Carotinoide abgeleitet werden konnten, die unter Einbeziehung des übrigen Verhaltens schließlich ihren Ausdruck in definierten Vorstellungen von der räumlichen Feinstruktur von zahlreichen *cis-trans*-Isomeren fanden. Was die präparative Seite des Problems anbetraf, so hatte TSWETTS Methode der chromatographischen Stofftrennung auch hier wieder unersätzliche Dienste geleistet.

Die Carotine mit 40 Kohlenstoffatomen waren lange Zeit nur in jeweils einer einzigen sterischen Form bekannt — wie wir heute wissen, der langgestreckten, sogenannten ,,all-*trans*"-Form — und man hatte sogar an der Möglichkeit der Existenz von Stereoisomeren vorübergehend gezweifelt, obgleich eine sehr große Anzahl von *cis-trans*-isomeren Carotinen theoretisch abzuleiten waren. Vor allem an Hand der eingehenden Untersuchungen an den Kohlenwasserstoffen α-, β- und γ-Carotin sowie Lycopin, von denen hier hauptsächlich die Rede sein soll, hat sich dann aus Einzelbefunden und gestützt auf frühzeitige Beobachtungen von GILLAM und EL RIDI, durch ZECHMEISTER und seine Mitarbeiter eine Systematik der *cis-trans*-Umlagerung entwickelt, die seit 1944 in ihren wesentlichen Grundzügen als geklärt anzusehen ist. Mit der Auffindung von Prolycopin und Pro-γ-carotin sowie einer größeren Anzahl von kristallisierten Poly-*cis*-lycopinen (*42, 38, 44, 45, 49*) in der Natur wurden wichtige *cis*-Carotine bekannt, die als erste Vertreter dieser neuen Reihe natürlicher Pigmente in die Untersuchungen mit einbezogen werden konnten. Nachdem PAULING (*40*) in dieser Schriftenreihe die Isomerie und Struktur der Carotine im Licht der Elektronentheorie sowie der Bindungs-Abstände und -Winkel behandelt hatte, konnte ZECHMEISTER (*42*) in einem umfassenden Bericht das ganze Gebiet im Zusammenhang darstellen.

Folgende Methoden der *cis-trans*-Isomerisierung von Carotinoiden wurden entwickelt:

1. Thermische *cis-trans*-Umlagerung in Lösung.
2. *cis-trans*-Umlagerung durch Schmelzen der Kristalle.
3. *cis-trans*-Umlagerung durch Jod-Katalyse und Belichtung bei Zimmertemperatur in Lösung.
4. *cis-trans*-Umlagerung durch Säure-Katalyse in Lösung.
5. Photochemische *cis-trans*-Isomerisierung in Lösung.

1. Die spontane Isomerisierung der ,,all-*trans*"-Carotine bei Zimmertemperatur ist ein langsamer Prozeß, der z. B. beim β-Carotin in Benzol- oder Benzin-Lösung innerhalb eines Tages nur zu etwa 1% vor sich geht. In siedendem Benzol oder Benzin (Kp. 60—80°) ist der Gleichgewichtszustand der Isomerisierung nach 15 bis

60 Minuten erreicht. Wird z. B. β-Carotin in Benzinlösung (Kp. 60—80°) 60 Minuten zum Sieden erhitzt, so liegt danach folgendes Isomerengemisch vor: All-*trans* : NeoU : : NeoB : NeoE : labile Isomere = 86 : 4 : 8 : 1 : 1.

2. Werden Kristalle von Carotinoiden einige Grade oberhalb ihres Schmelzpunktes erhitzt, vollziehen sich reversible und irreversible Veränderungen, deren Ausmaß naturgemäß von der Höhe der Schmelztemperatur und der Erhitzungsdauer abhängt. Beim β-Carotin liegen z. B. die Verhältnisse folgendermaßen, nach 15 Minuten Erhitzen (Schmelzen) auf 190°: All-*trans* : NeoU : NeoV : NeoA : NeoB : : NeoE : labile Isomere = 33 : 19 : 4 : 8 : 24 : 8 : 4.

3. Jod ist bekanntlich ein Katalysator, der die räumliche Struktur von Äthylenderivaten sehr stark beeinflußt. Wird Jod zu einer Carotin-Lösung in Mengen von 1—2% (des Farbstoffes) bei Zimmertemperatur zugesetzt, so bewirkt es rasch weitgehende sterische Veränderungen. In dem erhaltenen Gleichgewichtsgemisch sollten theoretisch alle Isomeren enthalten sein, jedoch findet man in praxi meistens nur zwischen 2 und 12 Vertreter. Die Zusammensetzung des so erhaltenen Gleichgewichtsgemisches ist gewöhnlich verschieden von dem durch Kochen in Lösung erhaltenen, und zwar nicht nur im Verhältnis zueinander, sondern auch in bezug auf die Art. Das wesentliche bei der Stereoisomerisierung mit Jod ist die Umkehrbarkeit des Prozesses. Man gelangt praktisch zu demselben Gleichgewicht, wenn man von jedem einzelnen der erhaltenen Isomeren ausgeht. Wie auch in vielen anderen Fällen einer Jod-Katalyse, wird diese *cis-trans*-Umlagerung der Carotinoide durch Licht begünstigt, z. B.: All-*trans* : NeoU : NeoB : NeoE : labile Isomere = = 47 : 24 : 24 : 3 : 2.

4. Die Säure-Katalyse wird gewöhnlich so ausgeführt, daß eine Benzin- oder Benzol-Lösung des Polyens mit einer mehr oder weniger konzentrierten wäßrigen Säure unter CO_2 geschüttelt wird. Am Beispiel des β-Carotins wird mit konz. Salzsäure folgendes Gemisch zurückerhalten: All-*trans* : NeoU : NeoB : NeoE : labile Isomere = 50 : 23 : 23 : 3 : 1. (Unter Umständen werden auch irreversible Produkte gebildet.)

5. Sämtliche Carotinoid-Lösungen, die bisher untersucht wurden, erwiesen sich als lichtempfindlich, jedoch das Ausmaß der Veränderungen variiert stark und hängt sehr von der räumlichen Struktur ab. Sichtbares Licht ist wirksamer als ultraviolettes. Intensive Sonnenbestrahlung in Quarzgefäßen unter CO_2 erwies sich als eine zweckmäßige Versuchsanordnung, wobei die Temperatur möglichst niedrig zu halten ist, um die thermische Reaktion zurückzudrängen. Werden ,,all-*trans*"-Carotinoide dem Licht ausgesetzt, so finden sowohl *cis-trans*-Umlagerungen als auch irreversible Zerstörungen statt; beides setzt die Farbintensität herab. Wenn dagegen Poly-*cis*-Verbindungen belichtet werden, so überwiegt die durch sterische Umlagerung herbeigeführte, deutlich erkennbare Farbvertiefung alle anderen Effekte, z. B.: All-*trans* : NeoU : NeoB : NeoC + D usw. = 55 : 36,5 : 6 : 2,5.

Zu den allgemeinen Eigenschaften der *cis*-Carotinoide im Vergleich zu denen der ,,all-*trans*"-Verbindungen sind zu rechnen: Niedriger Schmelzpunkt, größere Löslichkeit und Ultraviolett-Verschiebung der Lichtabsorption (siehe unten).

Die möglichen Stereoisomeren.

Bezüglich der theoretisch möglichen *cis-trans*-Isomeren eines gegebenen konjugierten Systems muß zunächst betont werden, daß nicht alle in Frage stehenden (aliphatischen) Doppelbindungen einer räumlichen Umlagerung zugänglich sind. Die Doppelbindungen der Carotinoide lassen sich nach PAULING (40) in „stereochemisch wirksame" und „stereochemisch unwirksame" einteilen, wobei an den zuletzt Genannten infolge sterischer Hinderung die Wahrscheinlichkeit einer *cis*-Konfiguration relativ sehr gering ist. In einem C_{40}-Carotinoid vermag sich also meist nur an *einer* Doppelbindung eines jeden Isoprenteils der aliphatischen Kette *cis*-Stellung auszubilden. Und zwar sind dies nach PAULING jene Doppelbindungen, die ein Seitenketten-Methyl tragen bzw. die von zwei =CH-Gruppen benachbart sind. Hinzu kommt als einzige Ausnahme die eine zentrale Doppelbindung, die infolge ihrer Lage strukturell und stereochemisch bevorzugt erscheint. Wie aus dem nachstehenden Formelbild zu ersehen ist, besitzt in einer ungesättigten Kette die *cis*-Form folgende Struktur:

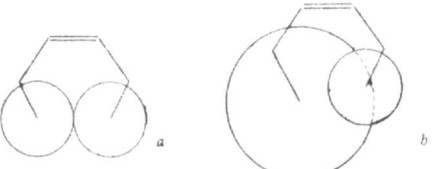

Wenn sowohl X als auch X' Wasserstoff-Atome sind, wird eine *cis*-Stellung durch die

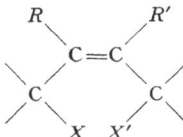

Abb. 1. Gegenseitige sterische Beeinflussung *a*) von zwei H-Atomen und *b*) von einem H-Atom und einer Methyl-Gruppe in den *cis*-Gruppierungen: *a*) —CH—CH=CH—CH— und *b*) —C—CH=CH—CH—
            |
           CH₃

räumlichen Abstände noch gestattet, bei $X =$ Methyl ist eine *cis*-Gruppierung sterisch jedoch nicht mehr möglich (Abb. 1). Hieraus hat PAULING (40) geschlossen, daß *cis*-Konfiguration in einer aliphatischen Carotin-Kette nur von solchen Doppelbindungen eingenommen werden kann, die vom Typus =CH—CH=CR'—CH= sind, d. h. von solchen, die von zwei =CH-Gruppen benachbart sind. Die stereochemisch unwirksamen Doppelbindungen anderseits sind dagegen die, welche im Ring liegen oder einer =C—CH_3-Gruppe benachbart sind. Dieser letztere Doppelbindungstyp ist unwirksam, weil sich die Wirkungssphären des Wasserstoffs der =CH-Gruppe und des Methyls der =C—CH_3-Gruppe räumlich überschneiden (Abb. 1).

So ergeben sich nach den vorstehenden Darlegungen folgende Strukturformeln für β-Carotin (S. 4), γ-Carotin (S. 4) sowie Lycopin (S. 7), in denen die stereochemisch wirksamen Doppelbindungen numeriert sind.

Wir haben es also in den genannten Fällen nur mit fünf bis sieben stereochemisch aktiven Doppelbindungen zu tun. Demnach lassen sich jetzt die möglichen Stereoisomeren wie folgt berechnen: β-Carotin 20, Lycopin 72, α-Carotin 32, γ-Carotin 64.

Spektrale Veränderungen durch trans → cis-Isomerisierung.

Wird die Lösung eines „all-*trans*"-Carotins stereoisomerisiert, so findet in allen Fällen eine Verschiebung der Hauptabsorptionsbanden nach kürzeren Wellenlängen statt, unter gleichzeitigem Absinken der Extinktion. Bei genügend hoher Konzentration sind diese Veränderungen auch an der Farbaufhellung der Lösung zu erkennen. Veränderungen des Spektrums im entgegengesetzten Sinne vollziehen sich, wenn ein Pigment mit *cis*-Bindungen mit Jod behandelt wird.

„cis-peak"-Effekt. Beim Studium der spektralen Veränderungen auch im langwelligen Ultraviolett wurde von ZECHMEISTER und POLGÁR (*46—48*) ein bemerkenswertes neues Maximum entdeckt, das unter der Bezeichnung „*cis*-peak" in die Literatur eingegangen ist. Während die quantitative Lichtabsorptionskurve der „all-*trans*"-Carotinoide zwischen 320 und 340 mμ höchstens eine leichte Erhebung aufweist, zeigen gewisse *cis*-Carotine in diesem Bereich ein ausgeprägtes Maximum. Die Differenz zwischen seiner Wellenlänge und der des langwelligsten Maximums der „all-*trans*"-Form ist praktisch konstant und beträgt 142 mμ ($\pm$ 2 mμ). Die relative Stellung dieses Maximums wird nicht durch die Anwesenheit von funktionellen Gruppen beeinflußt, die an den Enden der Molekeln ihren Sitz haben. Es ist eines der wesentlichsten Ergebnisse der Arbeit von ZECHMEISTER, LE ROSEN, SCHROEDER, POLGÁR und PAULING (*44*), erkannt zu haben, daß das Auftreten dieses „*cis*-peaks" mit einer ganz bestimmten *cis*-Struktur des betreffenden Pigments im Zusammenhang steht. Auf Grund der verfügbaren experimentellen Daten wird gefolgert, daß die spektrale Erscheinung eines hohen „*cis*-peaks" für solche Carotinoid-Stereoisomeren charakteristisch ist, deren Moleküle eine gebogene, z. B. V-förmige Gestalt besitzen, wie sie beim Vorliegen einer *cis*-Bindung im Zentrum oder in der Nähe des Zentrums des chromophoren Systems an-

genommen werden muß. Es wird hiernach verständlich, warum sowohl die „all-*trans*-" als auch die „poly-*cis*"-Konfigurationen spektroskopisch in der „cis-peak"-Region unwirksam sind.

Es ist weiterhin zu folgern, daß die Intensität des „*cis*-peaks" nicht einfach von der Zahl der *cis*-Doppelbindungen abhängt, sondern von der Gesamtform des Moleküls.

Bei den Extinktions-Kurven der Carotinoide sind drei Hauptbanden zu unterscheiden: 1. Die Bande der sehr hohen Lichtabsorption im sichtbaren Bereich, wo die Hauptabsorption liegt („Hauptbande"); 2. die „*cis*-peak"-Bande im nahen Ultraviolett; 3. eine Bande im ferneren Ultraviolett (1. Oberton). Diese Absorptionsbereiche entsprechen den Übergängen vom normalen Elektronenzustand in bestimmte angeregte Zustände.

Das Molekül mit einer zentralen *cis*-Bindung besitzt, wegen seiner V-förmigen Struktur, unter allen Raumformen das stärkste Dipol-Moment, das senkrecht gerichtet ist zur geraden Linie, die von einem Ende des konjugierten Systems zum anderen gezogen wird; dieses Isomere besitzt daher den höchsten „*cis*-peak". Die Modelle zeigen, daß nur eine kleine Minderzahl der *cis*-Formen eines gegebenen Carotinoids einen bedeutenden „*cis*-peak" aufweisen kann.

II. Synthesen von C_{30}- und C_{32}-Kohlenwasserstoffen als Modellsynthesen.

Als Modellreaktion für Synthesen von Carotinoiden diente der Aufbau des C_{30}-Kohlenwasserstoffs, der nach den folgenden Schemata gewonnen werden konnte:

a) $C_{14} + C_2 + C_{14} \rightarrow C_{30}$;
b) $C_{16} + C_{14} \rightarrow C_{30}$;
c) $C_{13} + C_4 + C_{13} \rightarrow C_{30}$.

Die Aufbauprinzipien a) und c) sind später für zwei Carotin-Synthesen von Bedeutung geworden.

Die Kondensation von β-C_{14}-Aldehyd (I) mit Acetylen-dimagnesiumbromid führte zu dem symmetrischen C_{30}-In-diol (II) (*11*). Dieses Diol (II) stellte ein Isomerengemisch dar, das weitgehend aufgetrennt werden konnte. Nach Wasserabspaltung mit p-Toluolsulfosäure wurde der Kohlenwasserstoff $C_{30}H_{42}$ (III) mit zentraler Dreifachbindung erhalten, wobei sämtliche isomeren Diole den gleichen Kohlenwasserstoff lieferten.

Das Diol (II) konnte weiterhin an der Dreifachbindung selektiv hydriert und hierauf durch Allylumlagerung und Wasserabspaltung in den Kohlenwasserstoff (IV) übergeführt werden. Dieser bildet gelbe Nadeln vom Schmp. 136° und weist ein Lichtabsorptions-Maximum bei

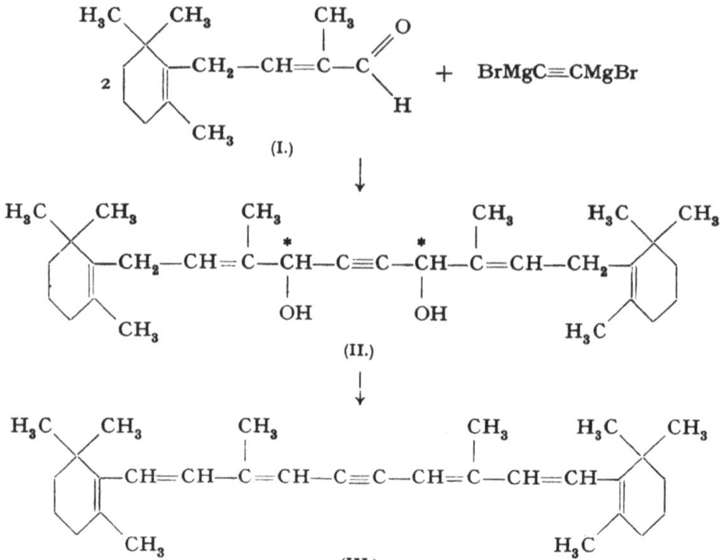

$$2 \underset{\text{(I.)}}{\overset{\displaystyle H_3C\diagdown\!/CH_3}{\underset{CH_3}{\bigcirc}}\!\!-CH_2-CH=C(CH_3)-CHO} + BrMgC\equiv CMgBr$$

$$\downarrow$$

$$\underset{\text{(II.)}}{\overset{H_3C\diagdown\!/CH_3}{\underset{CH_3}{\bigcirc}}\!\!-CH_2-CH=C(CH_3)-\underset{OH}{\overset{*}{C}H}-C\equiv C-\underset{OH}{\overset{*}{C}H}-C(CH_3)=CH-CH_2-\!\!\overset{H_3C\diagdown\!/CH_3}{\underset{H_3C}{\bigcirc}}}$$

$$\downarrow$$

$$\underset{\text{(III.)}}{\overset{H_3C\diagdown\!/CH_3}{\underset{CH_3}{\bigcirc}}\!\!-CH=CH-C(CH_3)=CH-C\equiv C-CH=C(CH_3)-CH=CH-\!\!\overset{H_3C\diagdown\!/CH_3}{\underset{H_3C}{\bigcirc}}}$$

370 mμ auf. Der Äthylenkohlenwasserstoff (IV) absorbiert also zirka 15 mμ langwelliger als das Acetylen-Derivat (355 mμ), was schon mehrfach beobachtet wurde (4). Dieses 7fach ungesättigte Polyen ist aus 6 Isopren-Resten aufgebaut und unterscheidet sich vom β-Carotin dadurch, daß zwischen den beiden Cyclocitral-Resten und der aliphatischen C_{10}-Kette je ein Isoprenrest fehlt.

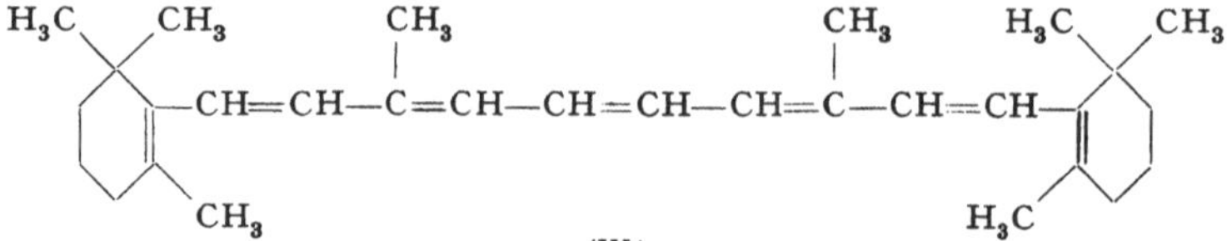

$$\underset{\text{(IV.)}}{\overset{H_3C\diagdown\!/CH_3}{\underset{CH_3}{\bigcirc}}\!\!-CH=CH-C(CH_3)=CH-CH=CH-CH=C(CH_3)-CH=CH-\!\!\overset{H_3C\diagdown\!/CH_3}{\underset{H_3C}{\bigcirc}}}$$

Für die zweite Synthese wurde das sogenannte β-C_{16}-Äthinylcarbinol (V) benutzt, das aus β-Jonon und Propargylbromid dargestellt werden kann (5).

In dieser Verbindung läßt sich das Acetylen-Wasserstoffatom gut durch Lithium ersetzen und die entstandene metallorganische Verbindung mit β-C_{14}-Aldehyd (I, s. oben) kondensieren. Das so gewonnene unsymmetrische C_{30}-In-diol (VI) wurde einer einseitigen Allylumlagerung und doppelten Wasserabspaltung unterworfen, was wiederum durch Behandlung mit p-Toluolsulfosäure in siedendem Toluol gelang. Nach chromatographischer Auftrennung konnte der gleiche Acetylenkohlenwasserstoff $C_{30}H_{42}$ (III) erhalten werden, der bereits nach Schema a)

(S. 10) beschrieben wurde. Bei Verwendung der GRIGNARD-Verbindung war die Ausbeute wesentlich geringer.

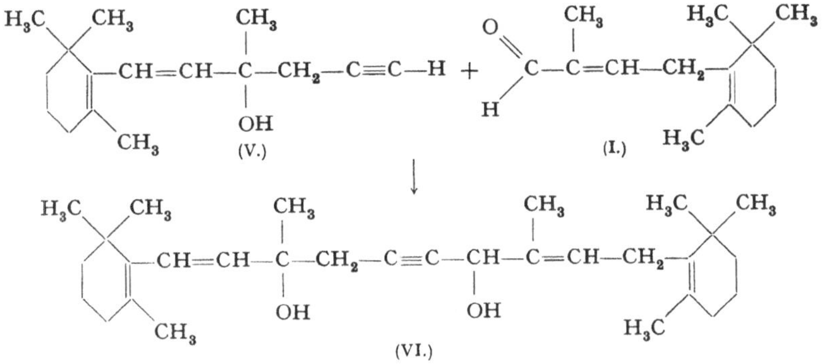

$$\text{(V.)} + \text{(I.)}$$

$$\downarrow$$

$$\text{(VI.)}$$

Eine Variation dieses Syntheseweges erreicht man dadurch, daß an Stelle des β-C_{16}-Äthinyl-carbinols (V) der β-C_{16}-Kohlenwasserstoff (VII) eingesetzt wird. Dieser Acetylenkohlenwasserstoff $C_{16}H_{22}$ (VII) entsteht durch Wasserabspaltung aus (V). Zur Wasserabspaltung kann entweder Chlorwasserstoffsäure (21) oder wasserfreie Oxalsäure (15) verwendet werden; Toluolsulfosäure erwies sich als unbrauchbar.

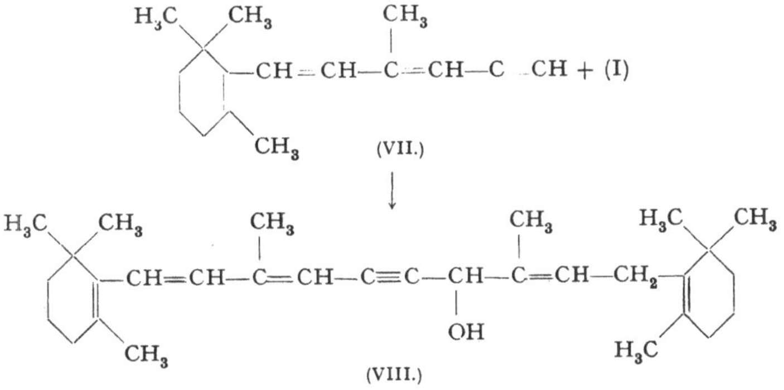

$$\text{(VII.)} + \text{(I)}$$

$$\downarrow$$

$$\text{(VIII.)}$$

Durch Kondensation mit β-C_{14}-Aldehyd, wobei sich die Lithiumverbindung wiederum als besser geeignet erwies als die GRIGNARD-Verbindung, wurde der Alkohol (VIII) erhalten. Durch Allylumlagerung und Wasserabspaltung entstand ebenfalls der vorstehend beschriebene β-C_{30}-Acetylenkohlenwasserstoff (III), jedoch in schlechter Ausbeute.

Ähnlich schlechte Ausbeuten waren schon von ISLER (21) beobachtet worden, als er den Kohlenwasserstoff (VII) mittels einer metallorganischen Synthese zum Vitamin A umsetzen wollte. In (VII) liegt ein konjugiertes System von drei Doppelbindungen und einer Acetylengruppe vor. Man darf annehmen, daß nach Metallierung des Acetylenwasserstoffs sich das metallorganisch bindende Elektronenpaar an der

Resonanz des konjugierten Systems beteiligt, wodurch offenbar die Reaktivität herabgesetzt wird.

Vor kurzem wurde auch das Diacetylen für Carotinoid-Synthesen mit Erfolg herangezogen (6). Es reagiert als Dimagnesiumbromid-Verbindung mit ungesättigten Aldehyden und Ketonen in normaler Weise, unter Bildung von Diin-diolen, ohne gegenüber der analogen Kondensation mit Acetylen Besonderheiten zu zeigen.

Kondensiert man nach dem dritten Schema (S. 9) β-Ionon mit Diacetylen-dimagnesiumbromid, so erhält man ein C_{30}-Diin-Diol der Struktur (IX), das als kristallines Isomerengemisch erhalten wurde (6). Bemerkenswert ist, daß das β-Jonon praktisch unverändert zurückgewonnen wird, wenn man es mit dem Dinatriumsalz des Diacetylens in flüssigem Ammoniak behandelt (1).

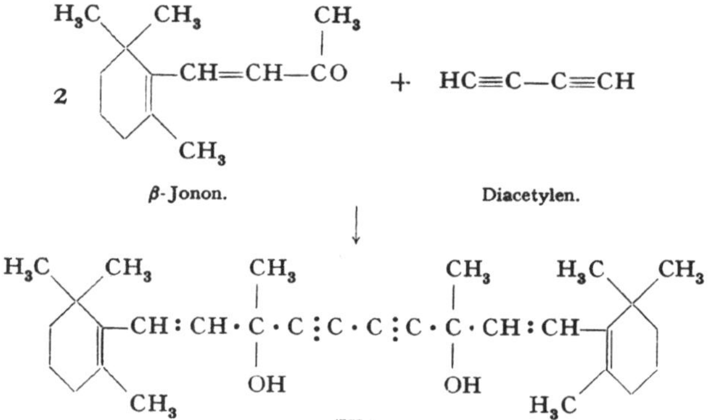

Auch bei dem Diin-diol (IX) gelang die partielle Hydrierung der beiden Acetylengruppen glatt und man erhielt das entsprechende C_{30}-Diendiol (X).

Die Herausnahme der beiden Hydroxylgruppen nach KUHN und WALLENFELS (37) mittels Phosphordijodid war die Methode der Wahl, um unter Hinzutritt einer weiteren Doppelbindung die durchlaufende Konjugation zu erzeugen, wobei das intermediär entstehende Dijodid unter Jodabgabe zerfällt. Der resultierende Kohlenwasserstoff erwies sich als identisch mit der vorstehend beschriebenen Verbindung $C_{30}H_{44}$ (IV, S. 10).

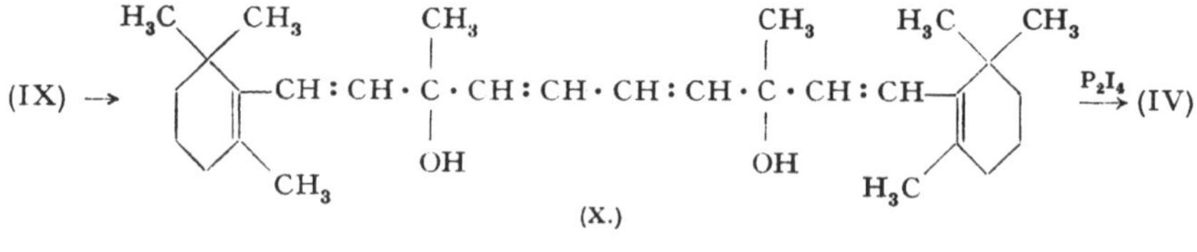

Die Kondensation des β-C_{14}-Aldehyds mit Diacetylen führte in analoger Weise zu einem Isomerengemisch von C_{32}-Diin-diolen (XI). Nach zweifacher Wasserabspaltung mit p-Toluolsulfosäure entstand der Kohlenwasserstoff (XII), der nach chromatographischer Auftrennung in gelben Kristallen vom Schmp. 88° erhalten wurde. Partielle Hydrierung mit Palladium/Calciumcarbonat lieferte die entsprechende Äthylenverbindung. Das rohe Hydrierungsprodukt, das wahrscheinlich im wesentlichen die Di-*cis*-Konfiguration besaß, konnte nach Isomerisierung mit Jod im Tageslicht zur „all-*trans*"-Verbindung (XIII) umgelagert werden; die intensiv gelben Kristalle schmelzen bei 152—153°. Das Abs.-Max. von (XIII) liegt bei 398 mμ und ist somit um 30 mμ langwelliger als das des Diins mit einem Max. bei 368 mμ. Die Diacetylengruppe bewirkt also eine Verschiebung des Maximums in das Kurzwellige, die dem doppelten Betrag entspricht, der bei der vergleichbaren Monoacetylen-Verbindung beobachtet wurde.

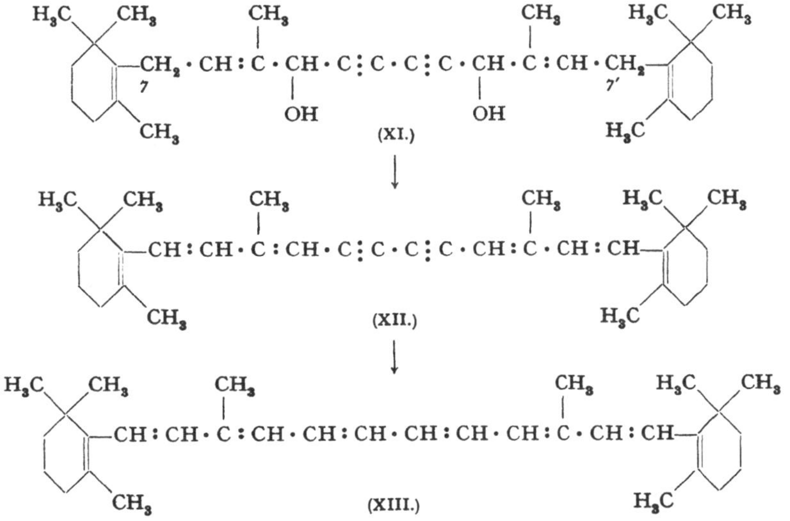

$$\text{—CH}_2 \cdot \text{CH} \vdots \text{C} \cdot \text{CH} \cdot \text{C} \vdots \text{C} \cdot \text{C} \vdots \text{C} \cdot \text{CH} \cdot \text{C} \vdots \text{CH} \cdot \text{CH}_2\text{—}$$
(XI.)

$$\text{—CH} \vdots \text{CH} \cdot \text{C} \vdots \text{CH} \cdot \text{C} \vdots \text{C} \cdot \text{C} \vdots \text{C} \cdot \text{CH} \cdot \text{C} \vdots \text{CH} \cdot \text{CH} \vdots \text{CH—}$$
(XII.)

$$\text{—CH} \vdots \text{CH} \cdot \text{C} \vdots \text{CH} \cdot \text{CH} \vdots \text{CH} \cdot \text{CH} \vdots \text{CH} \cdot \text{CH} \vdots \text{C} \cdot \text{CH} \vdots \text{CH—}$$
(XIII.)

III. Bis-nor-methyl-β-carotin und 7,7'-Dihydro-β-carotin.

Bis-nor-methyl-β-carotin.

Ausgehend vom β-C_{14}-Aldehyd (I) ließ sich auf folgendem Wege ein C_{18}-Aldehyd aufbauen (*10*): β-C_{14}-Aldehyd (I) wurde mit der Lithiumverbindung des Äthoxyvinyl-acetylens (*24*) kondensiert und hierbei der C_{18}-Oxy-enoläther erhalten (XIV). Dieser C_{18}-Oxy-enoläther (XIV) ließ sich glatt zum Oxy-aldehyd hydrolysieren (XV), der durch ein Dinitrophenylhydrazon charakterisiert werden konnte.

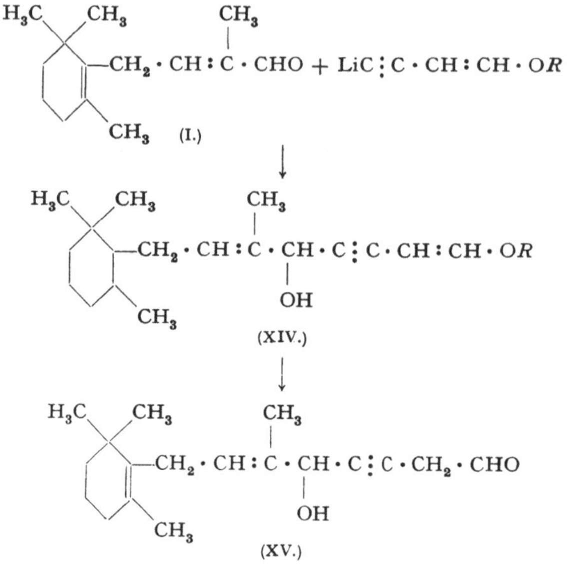

$$\text{[cyclohexene ring with } H_3C, CH_3, CH_3 \text{ substituents]} - CH_2 \cdot CH \vdots C \cdot CHO + LiC \vdots C \cdot CH \vdots CH \cdot OR$$

(I.)

↓

$$\text{[ring]} - CH_2 \cdot CH \vdots C \cdot CH \cdot C \vdots C \cdot CH \vdots CH \cdot OR$$
$$\overset{|}{OH}$$

(XIV.)

↓

$$\text{[ring]} - CH_2 \cdot CH \vdots C \cdot CH \cdot C \vdots C \cdot CH_2 \cdot CHO$$
$$\overset{|}{OH}$$

(XV.)

Der Oxy-enoläther (XIV) konnte weiterhin nach der Vorschrift von ISLER und Mitarbeitern (22) mit vergiftetem Palladium-Kohle-Katalysator zum Dihydroderivat (XVI) selektiv hydriert werden. Läßt man diesen Dihydro-enoläther (XVI) in alkoholischer Lösung unter Zusatz von 5% 20proz. Schwefelsäure stehen, so gelangt man direkt zum freien Aldehyd

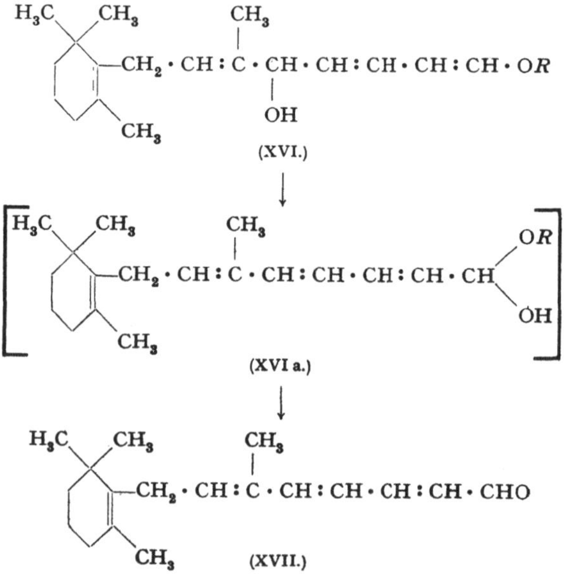

$$\text{[ring]} - CH_2 \cdot CH \vdots C \cdot CH \cdot CH \vdots CH \cdot CH \vdots CH \cdot OR$$
$$\overset{|}{OH}$$

(XVI.)

↓

$$\left[\text{[ring]} - CH_2 \cdot CH \vdots C \cdot CH \vdots CH \cdot CH \vdots CH \cdot CH \underset{OH}{\overset{OR}{<}} \right]$$

(XVI a.)

↓

$$\text{[ring]} - CH_2 \cdot CH \vdots C \cdot CH \vdots CH \cdot CH \vdots CH \cdot CHO$$

(XVII.)

(XVII). Unter den angegebenen Bedingungen wird offenbar nach Allylumlagerung zum Halbacetal (XVI a) gleich der Anhydro-aldehyd gebildet.

Das Phenylsemicarbazon zeigte beim spektralen Vergleich mit dem entsprechenden Derivat des β-C_{14}-Aldehyds, daß durch das Hinzukommen von zwei weiteren Doppelbindungen in Konjugation zur Carbazongruppierung die Lichtabsorption um etwa 50 mμ ins Langwellige verschoben und die Extinktion entsprechend auf etwa das Doppelte erhöht wird.

Der β-C_{18}-Aldehyd (XVII) wurde nun mit Acetylen-dimagnesiumbromid bei Zimmertemperatur umgesetzt, wobei man das β-C_{38}-Acetylendiol (XVIII) erhielt.

(XVII)
↓

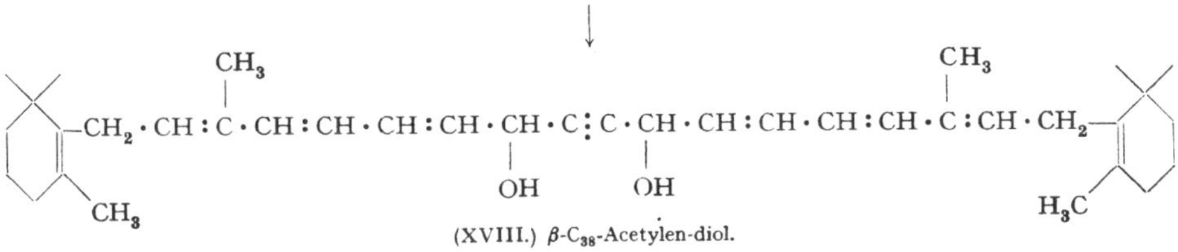

$-CH_2 \cdot CH : C(CH_3) \cdot CH : CH \cdot CH : CH \cdot CH \cdot C \vdots C \cdot CH \cdot CH : CH \cdot CH : CH \cdot C(CH_3) : CH \cdot CH_2-$
             | | |
                                                    OH   OH

(XVIII.) β-C_{38}-Acetylen-diol.

Das C_{38}-Isomerengemisch konnte ohne weitere Reinigung direkt der Allylumlagerung und Wasserabspaltung mit p-Toluolsulfosäure unterworfen werden, und zwar entweder in siedendem Toluol oder Benzol; im ersteren Fall genügten 15 Minuten, während im zweiten Fall 5 Stunden gekocht wurde. Durch Chromatographie konnte das 13,13′-Bis-nor-methyl-15,15′-dehydro-β-carotin $C_{38}H_{50}$ in roten Kristallnadeln vom Schmp. 174° erhalten werden (XIX).

(XVIII)
↓

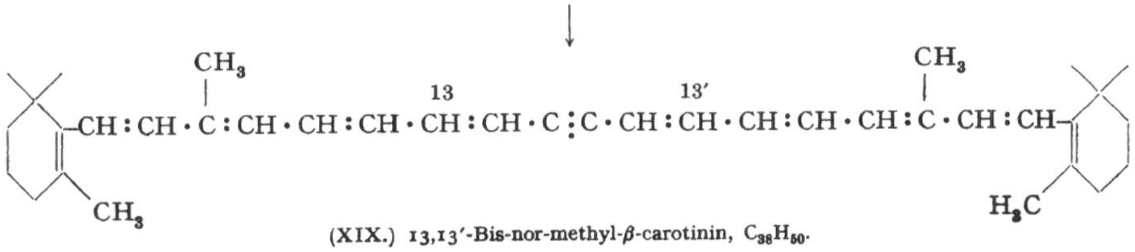

$-CH : CH \cdot C(CH_3) : CH \cdot CH : CH \cdot \overset{13}{CH} : CH \cdot C \vdots C \cdot \overset{13'}{CH} : CH \cdot CH : CH \cdot CH : C(CH_3) \cdot CH : CH-$

(XIX.) 13,13′-Bis-nor-methyl-β-carotinin, $C_{38}H_{50}$.

Die Gewinnung des Carotinoid-Farbstoffes mit zentraler Doppelbindung gelang auf zwei Wegen: Einmal durch partielle Hydrierung der

(XIX)
↓

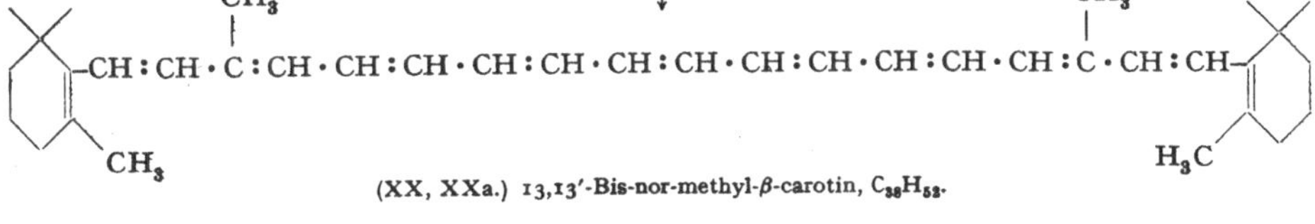

$-CH : CH \cdot C(CH_3) : CH \cdot CH : CH \cdot CH : CH \cdot CH : CH \cdot CH : CH \cdot CH : CH \cdot CH : C(CH_3) \cdot CH : CH-$

(XX, XXa.) 13,13′-Bis-nor-methyl-β-carotin, $C_{38}H_{52}$.

Dreifachbindung im C_{38}-Acetylen-diol (XVIII) mit anschließender Wasserabspaltung und zum anderen durch eine Umkehrung der Reaktionsfolge, indem die Hydrierung der Dreifachbindung erst nach der Wasserabspaltung vorgenommen wurde.

Auf beiden Wegen wurden zwei isomere 13,13'-Bis-nor-methyl-β-carotine mit der Summenformel $C_{38}H_{52}$ (XX, XXa) erhalten. Das bei 157° schmelzende Isomere kann als die 15,15'-Mono-*cis*-Verbindung angesprochen werden, denn sie zeigt deutlich den sogenannten „*cis*-peak", der den von ZECHMEISTER und POLGAR (46) geforderten 140 mμ-Abstand vom langwelligsten Maximum aufweist. Die „all-*trans*"-Verbindung schmilzt bei 177°, besitzt keinen „*cis*-peak" und weist im Vergleich zum β-Carotin bezüglich der Lage der beiden Hauptmaxima eine UV-Verschiebung um 10 mμ auf, die durch das Fehlen von zwei seitenständigen Methylgruppen bedingt ist.

Das Bis-nor-methyl-β-carotin ist im Wachstumstest an der Ratte nach GRIDGEMAN noch mit Tagesdosen von 20 γ physiologisch unwirksam.

7,7'-Dihydro-β-carotin.

Der erstmalige Anschluß an die natürlichen Carotine auf synthetischem Wege wurde durch den Aufbau des 7,7'-Dihydro-β-carotins erhalten. Seine Synthese gelang auf folgendem Wege (*16, 17*).

Für das Schema $C_{14} + C_{12} + C_{14} \rightarrow C_{40}$ mußte zunächst der C_{12}-Mittelteil aufgebaut werden. Dies gelang durch eine WURTZsche Reaktion zwischen zwei Molekülen 1-Brom-3-methyl-penten-2-in-4 (XXI) („C_6-Bromid" genannt) (*19*). Als Reaktionsprodukt entstand hierbei ein Kohlenwasserstoffgemisch folgender Zusammensetzung (XXII, XXIIa, XXIIb):

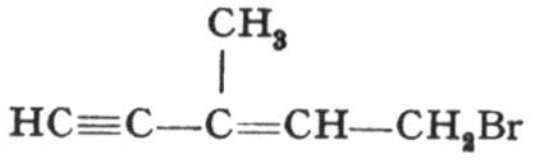

$$\underset{(XXI.)}{HC\equiv C-\underset{\underset{CH_3}{|}}{C}=CH-CH_2Br}$$

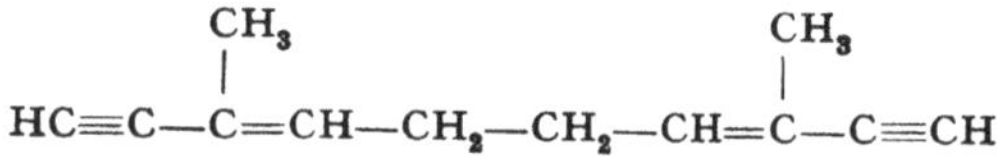

$$\underset{(XXII.)}{HC\equiv C-\underset{\underset{CH_3}{|}}{C}=CH-CH_2-CH_2-CH=\underset{\underset{CH_3}{|}}{C}-C\equiv CH}$$

$$\underset{(XXIIa.)}{HC\equiv C-\underset{\underset{CH_2}{\|}}{C}-CH_2-CH_2-CH_2-CH=\underset{\underset{CH_3}{|}}{C}-C\equiv CH}$$

$$\underset{(XXIIb.)}{HC\equiv C-\underset{\underset{CH_2}{\|}}{C}-CH_2-CH_2-CH_2-CH_2-\underset{\underset{CH_2}{\|}}{C}-C\equiv CH}$$

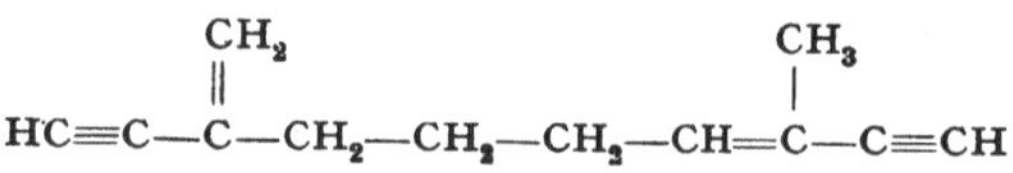

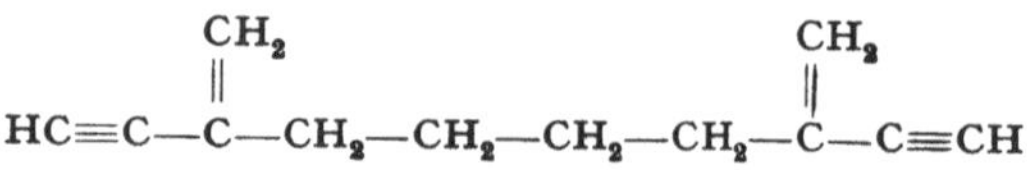

Zur Darstellung des C_6-Bromids diente der entsprechende C_6-Alkohol, das 1-Oxy-3-methyl-penten-2-in-4 als Ausgangsmaterial, das erstmals von CYMERMAN, HEILBRON und JONES (3) aus Vinylmethylketon und Natriumacetylid mit nachfolgender Allylumlagerung dargestellt werden konnte, und das von ISLER, HUBER, RONCO und KOFLER (23) zur Synthese von kristallisiertem Vitamin A mit so gutem Erfolg verwendet wurde.

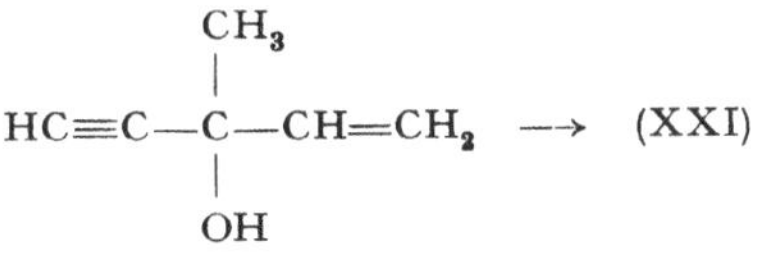

$$HC\equiv C-\underset{\underset{OH}{|}}{\overset{\overset{CH_3}{|}}{C}}-CH=CH_2 \quad \rightarrow \quad (XXI)$$

(XXI a.)

Der Ersatz der primären Hydroxylgruppe durch Brom erfolgte mittels Phosphortribromid, jedoch kann das Bromid (XXI) aus dem Carbinol (XXI a) auch direkt erhalten werden, wobei gleichzeitig bei der Behandlung mit Phosphortribromid Allylumlagerung erfolgt. Das sehr unbeständige C_6-Bromid wurde zweckmäßigerweise im Rohzustand eingesetzt. Eine Änderung der Zusammensetzung des Isomeren-Gemisches der drei Kohlenwasserstoffe (XXII, XXIIa und b), d. h. die Verhinderung einer Doppelbindungsverschiebung in die Methylseitenketten konnte durch eine Variation der Bedingungen der WURTZschen Reaktion erreicht werden.

Bei Verwendung von Zinkspänen in Äther mußte nämlich die Reaktion durch Zugabe von Jod in Gang gebracht werden; unter diesen Bedin-

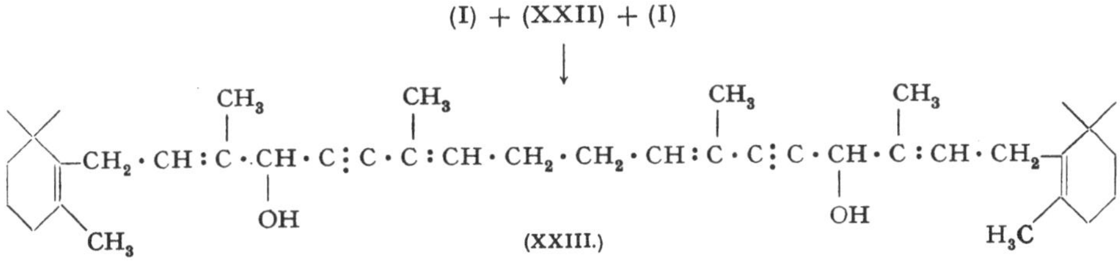

$$(I) + (XXII) + (I)$$
$$\downarrow$$

(XXIII.)

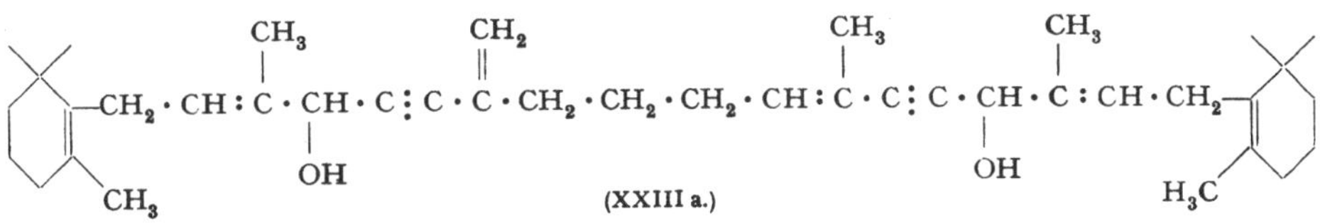

(XXIII a.)

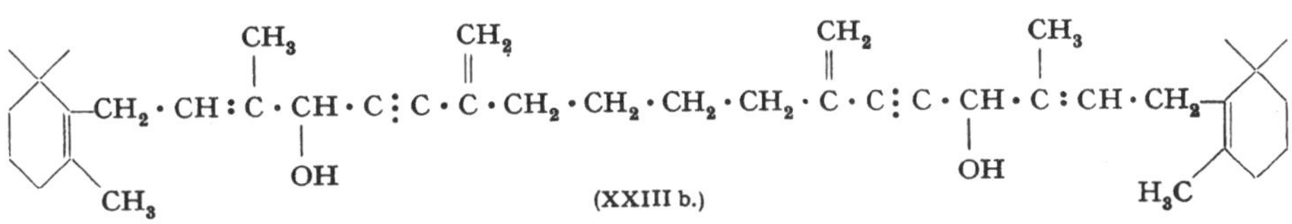

(XXIII b.)

gungen entstehen erhebliche Mengen der Isomeren (XXIIa) und (XXIIb), wie durch Ozonabbau bewiesen werden konnte. Durch fraktionierte Destillation kann bis zu einem gewissen Grade eine Auftrennung erzielt werden, da sich das Mischungsverhältnis zwischen den drei Isomeren mit steigendem Siedepunkt zugunsten von (XXII) verschiebt. Führt man die WURTZsche Reaktion unter Ausschaltung von Jod als Katalysator mittels Zink-Kupfer-Paar durch, so enthält das entstandene Produkt nur noch geringe Mengen an den C_{12}-Kohlenwasserstoffen (XXIIa und b), während die Hauptmenge aus dem gewünschten Kohlenwasserstoff (XXII) besteht.

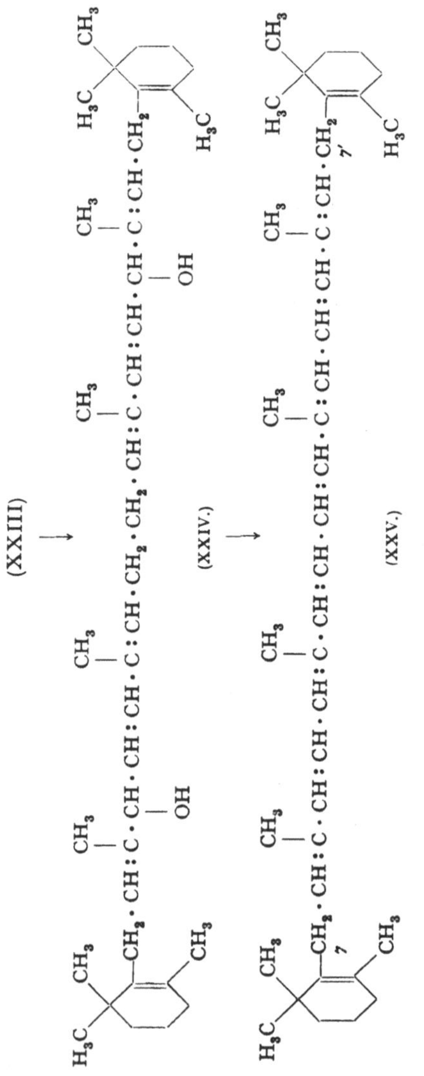

Mit Hilfe der beiden reaktionsfähigen Acetylenwasserstoffatome des Di-äthinyl-Kohlenwasserstoffes $C_{12}H_{14}$ wurde nun nach dem oben angegebenen Aufbauschema eine Kondensation mit β-C_{14}-Aldehyd durchgeführt. Da das C_{12}-Produkt nicht völlig einheitlich vorlag, war ein Gemisch von drei isomeren Diolen mit 40 C-Atomen und dem Kohlenstoffgerüst des natürlichen Carotins zu erwarten (XXIII, XXIIIa und XXIIIb).

Die Lithiumverbindung war hierfür besser geeignet als die magnesiumorganische Verbindung. Die Reinigung des nicht kristallisierenden Kondensationsproduktes gestaltete sich schwierig, jedoch deuteten die analytischen Befunde (Diacetylderivat, Molekulargewicht, ZEREWITINOFF-Bestimmung, UV-Absorption, Perhydrierung und vor allem der Ozonabbau) darauf hin, daß ein Teil der C_{40}-Diin-diol-Isomeren in der gewünschten Form (XXIII) vorlag.

Mit dem so erhaltenen C_{40}-Diindiol-Isomerengemisch wurde zunächst die partielle Hydrierung der beiden Acetylenbindungen zu Doppelbindungen durchgeführt, was wiederum mittels eines Palladium-

Kohle-Katalysators gelang, der durch Adsorption von Chinolin so weit desaktiviert worden war, daß nach Absättigung der Acetylenbindungen zu Doppelbindungen keine weitere Wasserstoffaufnahme mehr erfolgte. Das Hydrierungsprodukt (XXIV) wurde unmittelbar einer doppelten Allylumlagerung und anschließenden Wasserabspaltung unterworfen und das Reaktionsgemisch sorgfältig chromatographisch aufgetrennt. Aus den orangerot gefärbten Zonen konnte hierbei ein kristallisierter Kohlenwasserstoff vom Schmp. 181—182° isoliert werden, der die Zusammensetzung $C_{40}H_{58}$ besaß und sich als identisch erwies mit dem von KARRER und RÜEGGER durch Reduktion des β-Carotins mit Aluminiumamalgam erhaltenen Dihydro-β-carotin (XXV). Diese Reaktionen können daher wie folgt formuliert werden (XXIII—XXV).

IV. β-Carotin-Synthesen.

Die Totalsynthese des β-Carotins wurde gleichzeitig von KARRER und EUGSTER (26, 27) in Zürich und von INHOFFEN, BOHLMANN, BARTRAM und POMMER (7, 8) in Braunschweig durchgeführt, wobei die letzteren Autoren auf vier verschiedenen Wegen zum Ziele gelangten. Etwas später erschien die kurze Ankündigung der Synthese von MILAS (39).

Eine Chronologie der einschlägigen Veröffentlichungen wurde von KARRER und EUGSTER (30) gegeben.

β-Carotin-Synthesen $C_{16} + C_8 + C_{16}$.

Für das Aufbauschema $C_{16} + C_8 + C_{16} \rightarrow C_{40}$ diente wiederum das β-C_{16}-Äthinyl-carbinol (V, S. 11) als der eine Baustein, der schon bei der Synthese des β-C_{30}-Kohlenwasserstoffs Verwendung gefunden und hiermit seine Brauchbarkeit für Carotinoid-Synthesen bereits unter Beweis gestellt hatte (14). Der zweite, mittlere Baustein ist das Octendion, das zuerst von KARRER und EUGSTER (28) beschrieben worden war und für das noch eine zweite Synthese angegeben wurde (20). Durch Kondensation des C_{16}-Äthinyl-carbinols mit Octendion konnte ein β-C_{40}-Diin-tetrol (XXVI) erhalten werden (27, 18). Die beiden Acetylenbindungen ließen sich wiederum mit Hilfe des desaktivierten Palladium-Katalysators selektiv zu Doppelbindungen hydrieren. Das so erhaltene β-C_{40}-Tetrol (XXVII) wurde zur vierfachen Wasserabspaltung in siedendem Toluol mit p-Toluolsulfosäure behandelt. Dabei erwies es sich als wesentlich, daß die Konzentration der p-Toluolsulfosäure im Moment des Einsetzens der Reaktion genügend groß war. Aus dem Reaktionsprodukt konnte schließlich nach sorgfältiger Chromatographie reines β-Carotin erhalten werden (XXVIII).

Eine Variation dieses Syntheseweges benutzt dasselbe Aufbauschema, jedoch wird an Stelle des β-C_{16}-Äthinyl-carbinols der β-C_{16}-Kohlenwasser-

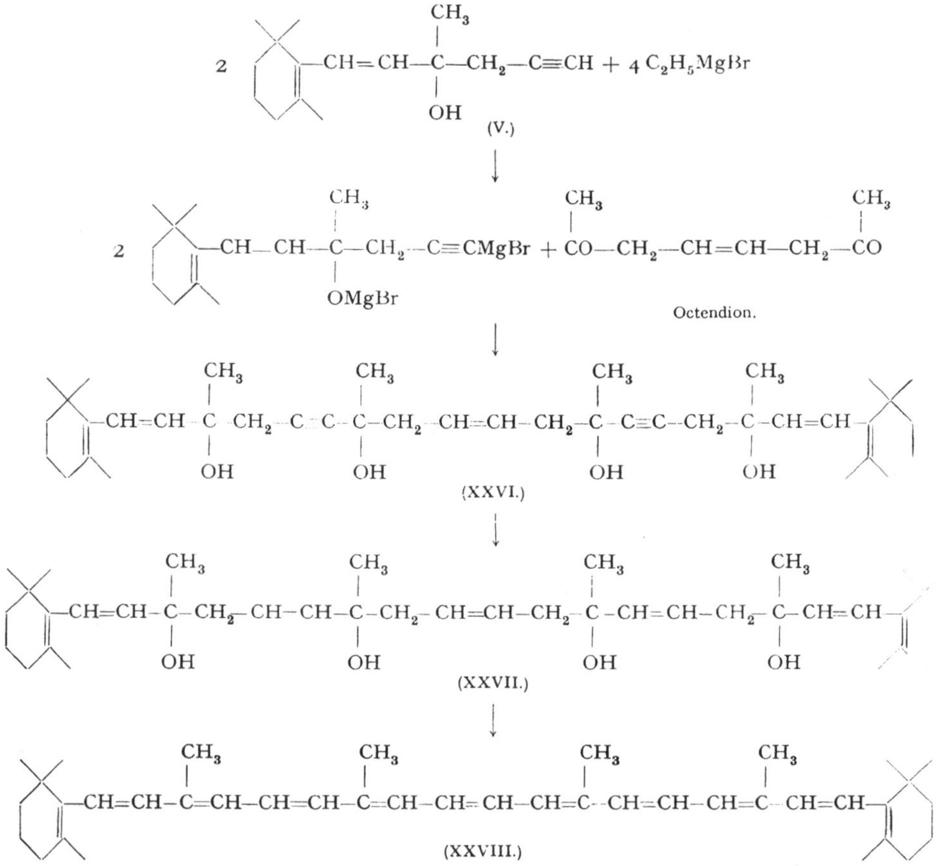

$$2 \; \bigg\langle\!\!\!-CH\!=\!CH-\underset{\underset{OH}{|}}{\overset{\overset{CH_3}{|}}{C}}-CH_2-C\!\equiv\!CH + 4\,C_2H_5MgBr$$

(V.)

↓

$$2 \; \bigg\langle\!\!\!-CH-CH-\underset{\underset{OMgBr}{|}}{\overset{\overset{CH_3}{|}}{C}}-CH_2-C\!\equiv\!CMgBr + CO-CH_2-\overset{\overset{CH_3}{|}}{C}H=CH-CH_2-\overset{\overset{CH_3}{|}}{C}O$$

Octendion.

↓

$$\bigg\langle\!\!\!-CH\!=\!CH-\underset{\underset{OH}{|}}{\overset{\overset{CH_3}{|}}{C}}-CH_2-C-C-\underset{\underset{OH}{|}}{\overset{\overset{CH_3}{|}}{C}}-CH_2-CH\!=\!CH-CH_2-\underset{\underset{OH}{|}}{\overset{\overset{CH_3}{|}}{C}}-C\!\equiv\!C-CH_2-\underset{\underset{OH}{|}}{\overset{\overset{CH_3}{|}}{C}}-CH\!=\!CH-\!\!\!\bigg\rangle$$

(XXVI.)

↓

$$\bigg\langle\!\!\!-CH\!=\!CH-\underset{\underset{OH}{|}}{\overset{\overset{CH_3}{|}}{C}}-CH_2-CH-CH-\underset{\underset{OH}{|}}{\overset{\overset{CH_3}{|}}{C}}-CH_2-CH\!=\!CH-CH_2-\underset{\underset{OH}{|}}{\overset{\overset{CH_3}{|}}{C}}-CH\!=\!CH-CH_2-\underset{\underset{OH}{|}}{\overset{\overset{CH_3}{|}}{C}}-CH\!=\!CH-\!\!\!\bigg\rangle$$

(XXVII.)

↓

$$\bigg\langle\!\!\!-CH\!=\!CH-\overset{\overset{CH_3}{|}}{C}\!=\!CH-CH\!=\!CH-\overset{\overset{CH_3}{|}}{C}\!=\!CH-CH\!=\!CH-CH\!=\!C-CH\!=\!CH-CH\!=\!\overset{\overset{CH_3}{|}}{C}-CH\!=\!CH-\!\!\!\bigg\rangle$$

(XXVIII.)

stoff (VII) (*12, 13, 15*) angewandt. Dabei zeigte sich Lithium gegenüber Magnesium als die günstigere Metallkomponente. Nach Umsetzung mit Octendion gelangte man in diesem Falle zu einem Gemisch von Isomeren β-C$_{40}$-Diin-diolen nach (XXIX).

(VII) + Octendion + (VII)

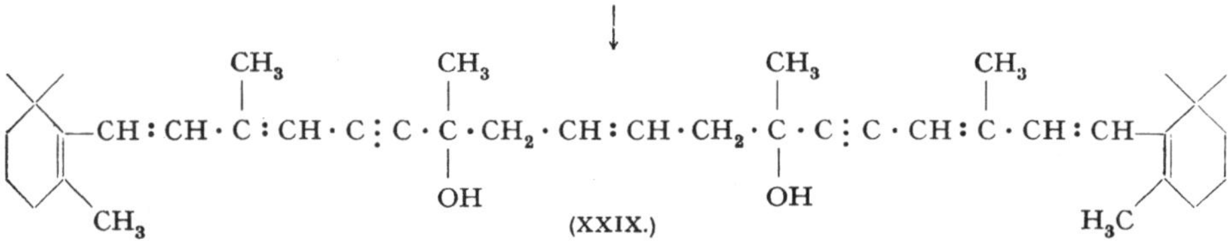

↓

$$\bigg\langle\!\!\!-CH\!:\!CH\cdot\overset{\overset{CH_3}{|}}{C}\!:\!CH\cdot\overset{\overset{CH_3}{|}}{C}\!:\!C\cdot\underset{\underset{OH}{|}}{C}\cdot CH_2\cdot CH\!:\!CH\cdot CH_2\cdot\underset{\underset{OH}{|}}{\overset{\overset{CH_3}{|}}{C}}\cdot C\!:\!C\cdot CH\!:\!\overset{\overset{CH_3}{|}}{C}\cdot CH\!:\!CH-\!\!\!\bigg\rangle$$

CH$_3$ (XXIX.) H$_3$C

Zur partiellen Anlagerung von Wasserstoff an die beiden Acetylenbindungen war es zweckmäßig, alle Katalysatorgifte sorgfältig zu entfernen, was durch Schütteln mit Platinkohle in absolutem Alkohol gelang. Aus dem nach der partiellen Hydrierung erhaltenen ditertiären β-C$_{40}$-Diol (XXX) konnten danach in üblicher Weise 2 Mole Wasser abgespalten werden. Das aus dem Reaktionsprodukt isolierte β-Carotin schmolz bei 178—179°; es zeigte einen etwas höheren Schmelzpunkt als das vorangehend beschriebene Carotin-Präparat. Misch-schmelzpunkt und Ab-

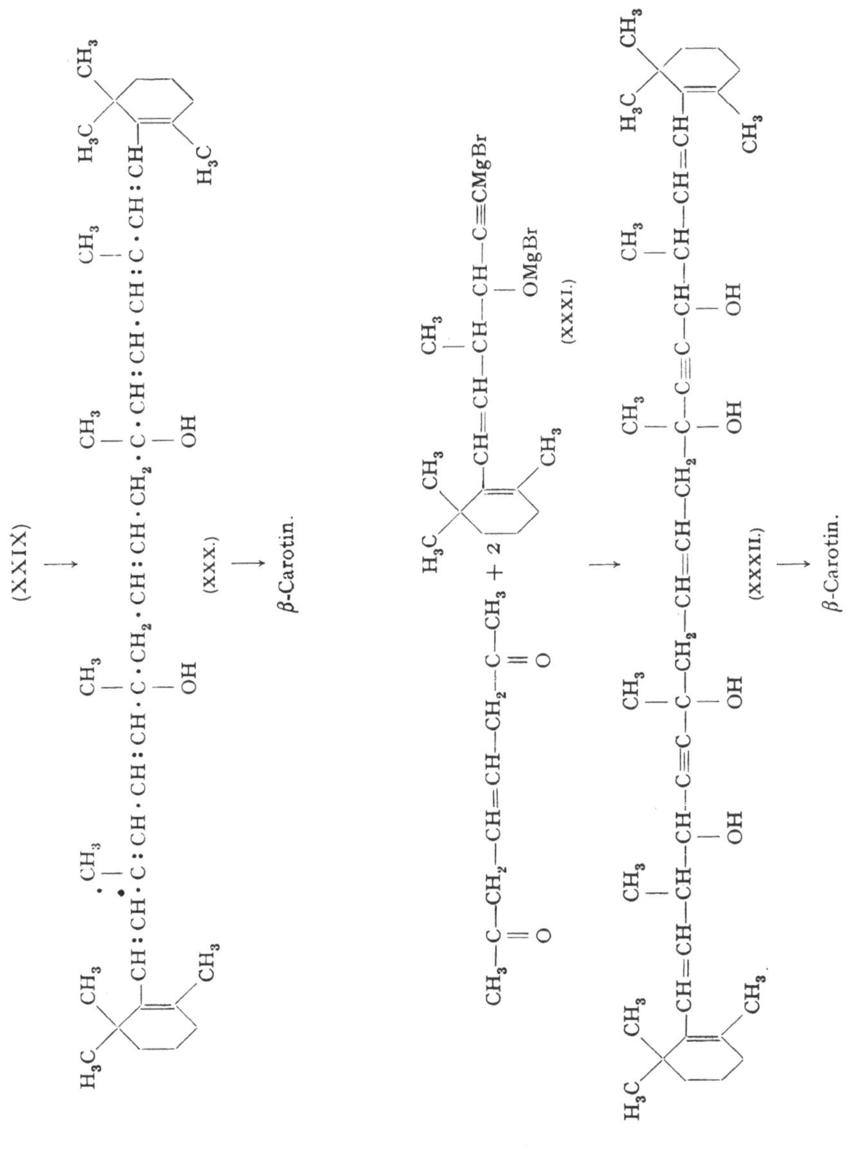

sorptionsspektrum bewiesen die völlige Identität mit natürlichem β-Carotin. Auch war die Ausbeute auf diesem Wege nicht ganz so schlecht wie bei dem ersten Verfahren.

Weiter wurde von MILAS (*39*), allerdings nur in einer kurzen Notiz, angegeben, daß er Octendion mit dem Acetylen-carbinol (XXXI) zum Tetrol (XXXII) kondensiert und dieses einer partiellen Hydrierung und Wasserabspaltung unterworfen habe.

β-Carotin-Synthese $C_{19} + C_2 + C_{19}$.

Seit der im technischen Ausmaß durchgeführten Vitamin A-Synthese von ISLER, HUBER, RONCO und KOFLER (*23*) ist der C_{14}-Aldehyd für Carotinoid-Synthesen in steigendem Maße in den Vordergrund des Interesses gerückt. Die gegenüber dem β-Jonon noch geeignetere Reaktionsbereitschaft und Reaktionsweise der Aldehydgruppe sowie das Vorhandensein der Methylengruppe am $C_{(7)}$, die das für eine Wasserabspaltung notwendige Wasserstoffatom enthält, machen diesen C_{14}-Aldehyd zum wichtigsten Ausgangsprodukt vieler Syntheseversuche. Wenn man ihn für eine Carotin-Synthese direkt einsetzen wollte, sähe das Aufbauschema folgendermaßen aus: $C_{14} + C_{12} + C_{14} \rightarrow C_{40}$.

Dieser Aufbauweg war bereits schon einmal beschritten worden und hatte erstmalig zum totalsynthetischen Dihydro-β-carotin geführt. Das C_{12}-Mittelstück mußte daher neu aufgebaut werden. Führt man in die C_{12}-Kette Acetylen als zentrales Bindeglied ein, gelangt man schließlich zu folgender Unterteilung (*7, 8, 9*):

β-C_{14}-Aldehyd + C_5 + CH≡CH + C_5 + β-C_{14}-Aldehyd,

entsprechend: $C_{14} + C_5 + C_2 + C_5 + C_{14} = C_{40}$.

Das zunächst noch unbekannte C_5-Molekül mußte folglich ein Isoprenbaustein mit zwei reaktiven Gruppen sein, so daß also folgender C_5-Äthinylaldehyd bzw. ein Derivat desselben darzustellen war. Hierzu brauchte man nur Acetoläther mit Lithiumacetylid zum C_5-Oxyäther (XXXIII) zu kondensieren:

$$H_3C \cdot O \cdot CH_2 \cdot \overset{\overset{\displaystyle CH_3}{|}}{C} : O + LiC \vdots CH \rightarrow H_3C \cdot O \cdot CH_2 \cdot \overset{\overset{\displaystyle CH_3}{|}}{\underset{\underset{\displaystyle OH}{|}}{C}} \cdot C \vdots CH$$

(XXXIII.)

Eine nachfolgende Wasserabspaltung mittels Aluminiumphosphats oder Magnesiumsulfat in der Dampfphase ergab zwar den Vinyläther (XXXIIIa),

$$HC\equiv C-\overset{\overset{\displaystyle CH_3}{|}}{C}=CH \cdot OCH_3$$

(XXXIII a.)

jedoch in so schlechter Ausbeute, daß mit dem Oxyäther (XXXIII) weitergearbeitet werden mußte.

Der Zusammenbau der Einzelteile gemäß dem obigen Schema wurde nun in der Weise in Angriff genommen, daß der neue C_5-Oxyäther (XXXIII) zunächst mit β-C_{14}-Aldehyd kondensiert wurde, um den für die letzte Kondensationsreaktion günstig erscheinenden β-C_{19}-Aldehyd zu gewinnen. Die Kondensation beider Komponenten lieferte den bishin ebenfalls unbekannten C_{19}-Dioläther (XXXIV) als Isomerengemisch, das sich in 30—40% Ausbeute zur Kristallisation bringen ließ. Aber auch das restliche ölige Isomerengemisch erwies sich für die weiteren Umsetzungen als geeignet.

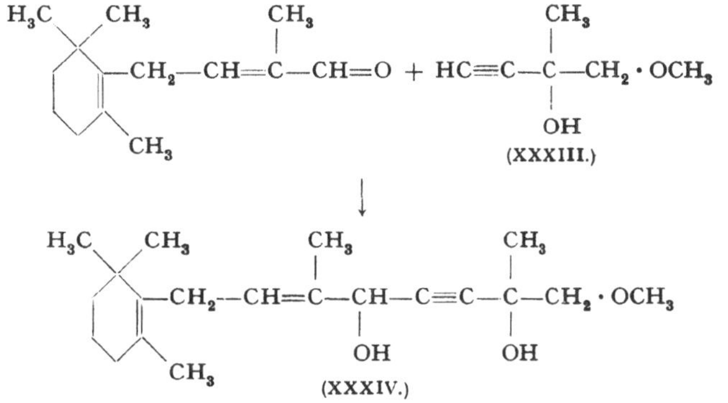

$$\text{Ring}-CH_2-CH=C(CH_3)-CH=O + HC\equiv C-C(CH_3)(OH)-CH_2 \cdot OCH_3 \quad \text{(XXXIII.)}$$

$$\downarrow$$

$$\text{Ring}-CH_2-CH=C(CH_3)-CH(OH)-C\equiv C-C(CH_3)(OH)-CH_2 \cdot OCH_3 \quad \text{(XXXIV.)}$$

Für den weiteren Aufbau wurde die Acetylengruppe zunächst partiell hydriert. Hierzu mußte der C_{19}-Dioläther (XXXIV) erst einer Allylumlagerung unterworfen werden, da die von den beiden OH-Gruppen flankierte Acetylengruppe in diesem Fall nur schwierig selektiv zu hydrieren war. Die Allylumlagerung konnte leicht mittels eines BECKMAN-Spektrophotometers verfolgt werden; das sich ausbildende En-in-System (XXXV) wies eine Absorptionsbande bei 233 mμ auf, so daß nur der Maximalwert der Extinktion abgewartet zu werden brauchte. Die partielle Hydrierung gelang danach wiederum mit Palladium-Kohle, an die vor Gebrauch Chinolin adsorbiert war.

Das erhaltene Dien (XXXVI) zeigte im Spektrum die durch das Verschwinden der Dreifachbindung zu erwartende Rotverschiebung von etwa 9 mμ (λ_{max} bei 242 mμ).

Für die nunmehr auszuführende Dehydratisierung erwies sich kurzes Kochen mit p-Toluolsulfosäure in Benzol als am besten geeignet. Es werden etwa 25% des freien Aldehyds direkt erhalten, so daß die Dehydratisierung zum Teil folgenden Verlauf nimmt: Nach Eliminierung von 1 Mol Wasser aus (XXXVI) zum Oxyvinyläther (XXXVII) findet Allylumlagerung zum Halbacetal (XXXVIII) statt, das spontan unter Bildung des Aldehyds zerfällt (XXXIX).

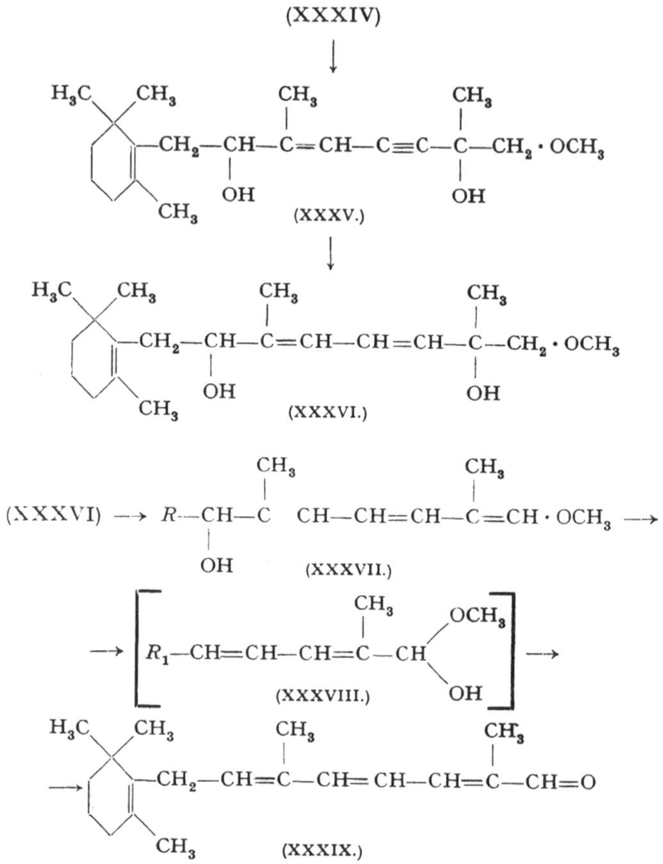

(XXXIV)
↓

H_3C \ CH_3 CH_3 CH_3
[cyclohexene ring]—CH_2—CH—C=CH—C≡C—C—$CH_2 \cdot OCH_3$
            | |
            OH   OH
  CH_3 (XXXV.)

↓

H_3C \ CH_3 CH_3 CH_3
[cyclohexene ring]—CH_2—CH—C=CH—CH=CH—C—$CH_2 \cdot OCH_3$
            | |
            OH   OH
  CH_3 (XXXVI.)

(XXXVI) → R—CH—C   CH—CH=CH—C=CH·OCH_3 →
              | |
              OH CH_3 CH_3
                   (XXXVII.)

$$\rightarrow \left[R_1\text{—CH=CH—CH=C—CH} \begin{array}{c} OCH_3 \\ OH \end{array} \right] \rightarrow$$
                         |
                        CH_3
              (XXXVIII.)

H_3C \ CH_3 CH_3 CH_3
→ [cyclohexene ring]—CH_2—CH=C—CH=CH—CH=C—CH=O
  CH_3 (XXXIX.)

Dieser synthetische Carotinoid-Aldehyd, $C_{19}H_{28}O$, konnte nach chromatographischer Auftrennung in kristallisiertem Zustand gewonnen werden. Er bildet unregelmäßige, honiggelbe Blöcke und schmilzt bei 64,5°. Mit Hydrazinhydrat liefert er das Azin (XL), das sich vom 7,7'-Dihydro-β-carotin dadurch unterscheidet, daß es an Stelle der beiden mittelständigen =CH—CH=-Gruppen eine =N—N=-Gruppierung trägt:

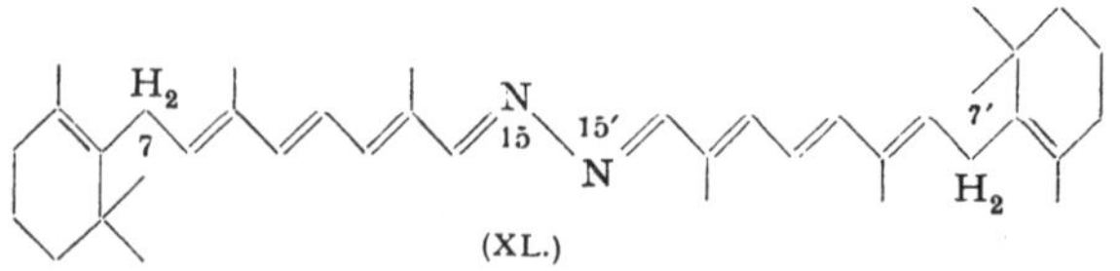

[structure of XL showing symmetric carotenoid azine with H_2 at positions 7 and 7', and N=N linkage at positions 15, 15']

(XL.)

Die doppelte Kondensation dieses C_{19}-Aldehyds mit Acetylen-dimagnesiumbromid führte zum C_{40}-In-diol (XLI), das sich gleichfalls in kristallinem Zustand abtrennen ließ. Dieses Diol fällt naturgemäß infolge der beiden neuentstandenen asymmetrischen C-Atome 14 und 14' als Isomerengemisch an.

$$2 \text{ (XXXIX)} + \text{BrMgC} \equiv \text{CMgBr}$$

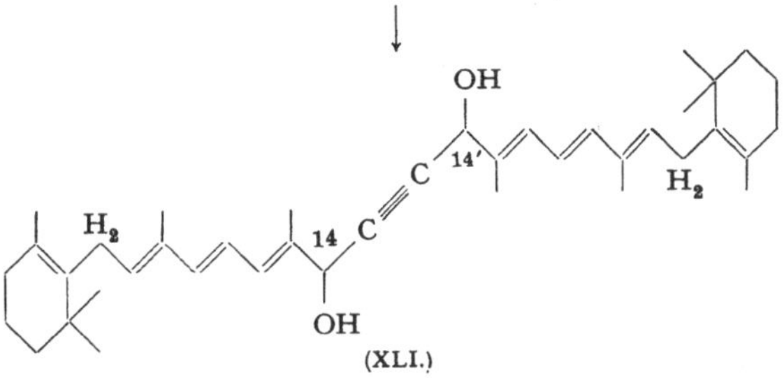

(XLI.)

Nach beiderseitiger Allylumlagerung und zweifacher Wasserabspaltung mit p-Toluolsulfosäure in Benzol wurde das 15,15'-Dehydro-β-carotin bzw. das β-Carotin-15,15'-in (XLII) erhalten. Dieser synthetische Kohlenwasserstoff unterscheidet sich vom β-Carotin also nur noch dadurch, daß er in der Mitte an Stelle der Doppelbindung eine Dreifachbindung enthält.

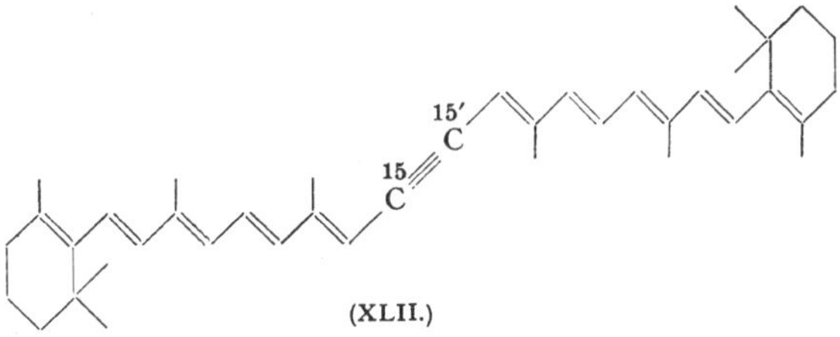

(XLII.)

Die physiologische Untersuchung nach GRIDGEMAN hat ergeben, daß das β-Carotinin im Wachstumstest an der Ratte noch mit einer Tagesdosis von 10 γ unwirksam ist.

Der letzte noch ausstehende Schritt zur Synthese des β-Carotins bestand in der partiellen Hydrierung der mittleren Dreifachbindung. Dies gelang mit dem von LINDLAR beschriebenen, mit Blei vergifteten Katalysator der Firma Hoffmann-La Roche. Obwohl also die zentrale Dreifachbindung von zwei fünffach konjugierten Systemen benachbart ist, bleibt die Hydrierung nach Aufnahme von etwa 1 Mol Wasserstoff praktisch stehen.

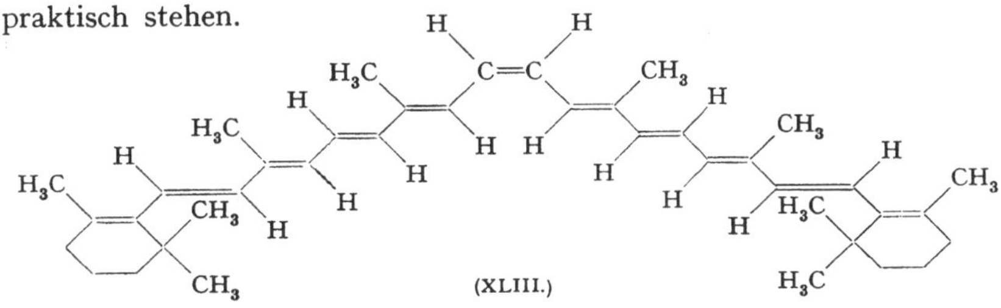

(XLIII.)

Da auch hier wieder mit einer *cis*-Hydrierung der Acetylen-Bindung zu rechnen war, konnte die Isolierung von 15,15'-mono-*cis*-β-Carotin (XLIII) nicht überraschen. Dieses spezielle β-Carotin-Isomere war bisher noch nicht bekannt. Von POLGÁR und ZECHMEISTER (*41*) war unter anderem das Neo-β-carotin U in kristallisierter Form durch Isomerisierung von natürlichem β-Carotin erhalten worden. Im Gegensatz dazu zeigt das neue mono-*cis*-Carotin im Absorptionsspektrum in charakteristischer Weise einen außerordentlich hohen *cis*-peak (S. 8). Mit der Synthese dieses eindeutig definierten 15,15'-mono-*cis*-β-Carotins, das im Spektrum genau den von ZECHMEISTER und POLGÁR (*46*) postulierten 140-mμ-Abstand vom langwelligsten Maximum aufweist, findet die auf S. 8 erwähnte Theorie des *cis*-peak-Effektes ihre Bestätigung.

Die Provitamin-A-Aktivität des synthetischen *cis*-β-Carotins beträgt für niedrige Tagesdosen die Hälfte derjenigen der all-*trans* Form (*43*).

Stereoisomerisierung von 15,15'-mono-cis-β-Carotin. Es konnte von vornherein erwartet werden, daß das neue mono-*cis*-β-Carotin außerordentlich stark zur Isomerisierung neigen würde, da nur mit einer verhältnismäßig geringen Übergangsenergie von *cis* in *trans* zu rechnen war, bedingt durch das Bestreben des Moleküls, in die gestreckte Lage überzugehen. Bei der Darstellung und Isolierung von (XLIII) mußten daher sorgfältig alle isomerisierenden Einflüsse ausgeschaltet werden. Vor allem mußten alle Operationen unter vollständigem Ausschluß von Licht bzw. nur bei Rotlicht und bei möglichst tiefer Temperatur durchgeführt werden.

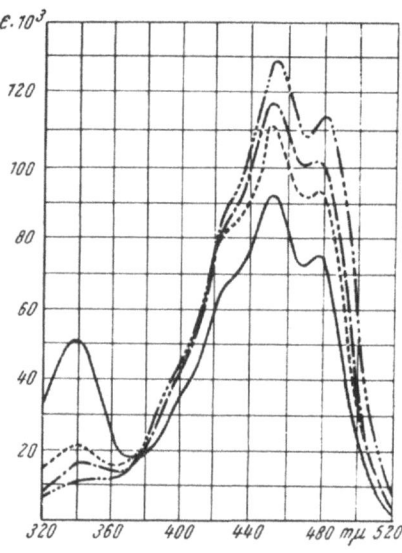

Abb. 2. Stereoisomerisierung von synthetischem 15,15'-mono-*cis*-β-Carotin: ——, Spektralkurve der all-*trans*-Form (in Hexan); - - - -, nach 2-stündiger Belichtung mit Tageslicht, - · - ·, nach Jod-Katalyse im Tageslicht (3 Minuten); und - - - -, nach 2-stündigem Kochen im Dunkeln.

Wurde das 15,15'-mono-*cis*-Carotin dem diffusen Tageslicht ausgesetzt, so hatte sich bereits nach etwa 3 Stunden eine nahezu restlose Umlagerung in all-*trans*-β-Carotin vollzogen. Die spektralen Veränderungen, die hierbei beobachtet wurden, sind eindeutig. Während der „*cis*-peak" bei 338 mμ abgebaut wird, steigt gleichzeitig das Hauptmaximum stark an unter geringer Verschiebung bei 476 mμ ins Langwellige. Führte man die Umlagerung mit Jod als Katalysator im Tageslicht durch, so war bereits nach etwa drei Minuten das Gleichgewicht erreicht. Jedoch erhält man auf diese Weise ein anderes Isomerengemisch als durch Belichtung allein. Bei der Jod-Isomerisierung ist gleichfalls Licht erforderlich, denn eine mit Jod versetzte Lösung blieb im Dunkeln 20 Stunden praktisch unverändert.

Bezüglich der Lichtisomerisierung wurde ferner beobachtet, daß Ultraviolettlicht weniger wirksam ist, was auch schon ZECHMEISTER erwähnt hat. Die Umlagerung mit Tageslicht geht wesentlich schneller vor sich als mit einer Quecksilberlampe. Man darf daher annehmen, daß für die cis → trans-Umlagerung der mittelständigen cis-Bindung Licht mit der Wellenlänge des Hauptmaximums verantwortlich zu machen ist. Licht mit Wellenlängen oberhalb 550 mμ erwies sich während der Versuchsdauer praktisch als wirkungslos. Auch durch Erhitzen einer Hexanlösung der mono-cis-Verbindung unter Kohlensäure und unter völligem Lichtausschluß auf 70° war nach $1^1/_2$ Stunden das Isomerisierungsgleichgewicht erreicht.

Diese Versuche zeigen, daß die Lichtisomerisierung des 15,15′-mono-cis-β-Carotins für die präparative Darstellung von all-trans-β-Carotin den übrigen Methoden vorzuziehen ist. Die chromatographische Auftrennung einer solchen Lichtisomerisierungslösung ergab, daß praktisch nur all-trans-β-Carotin neben wenig irreversiblen Umwandlungsprodukten vorlag. Die Einheitlichkeit des Umwandlungsproduktes ist auf Grund der vorerwähnten Tatsachen verständlich. Der erste Schritt bei der Umlagerung ist das Umklappen an der zentralen —C=C—Bindung. Diese Reaktion verläuft allem Anschein nach schneller als die weitere Isomerisierung der entstandenen trans-Verbindung.

β-Carotin-Synthese $C_{18} + C_4 + C_{18}$.

Das schon einmal mit Erfolg verwendete Diacetylen ließ sich auch für eine weitere Synthese des β-Carotins benutzen (6). Die doppelseitige Kondensation der GRIGNARD-Verbindung des Diacetylens mit 2 Mol β-C_{18}-Keton (XLIV) führte zu einem C_{40}-Diin-diol (XLV). Die partielle Hydrierung der beiden Acetylengruppen der Diacetylen-Verbindung ließ sich ohne Schwierigkeiten durchführen; die gegenüber einem Monoacetylen-diol weiter auseinanderstehenden OH-Gruppen bewirken anscheinend keine sterische Hinderung mehr (XLVI).

Nach der Behandlung dieses Diols (XLVI) mit P_2J_4 konnte das entstandene β-Carotin einwandfrei abgetrennt und charakterisiert werden, obgleich die Ausbeute bei der letzten Stufe nur etwa 1% betrug. Auch hier ist wieder interessanterweise das starke Absinken der Ausbeute mit steigender Zahl von konjugierten C=C-Bindungen zu beobachten; beim C_{30}-Kohlenwasserstoff lag die Ausbeute noch über 10%. Führt man mit dem C_{40}-Dien-diol (XLVI) an Stelle der Herausnahme der beiden OH-Gruppen mit P_2J_4 eine reguläre Wasserabspaltung durch, so gelangt man infolge der Mitnahme von 2 H-Atomen zu einem bisher unbekannten Dehydro-β-carotin $C_{40}H_{54}$ (XLVII). Es handelt sich also hiernach um

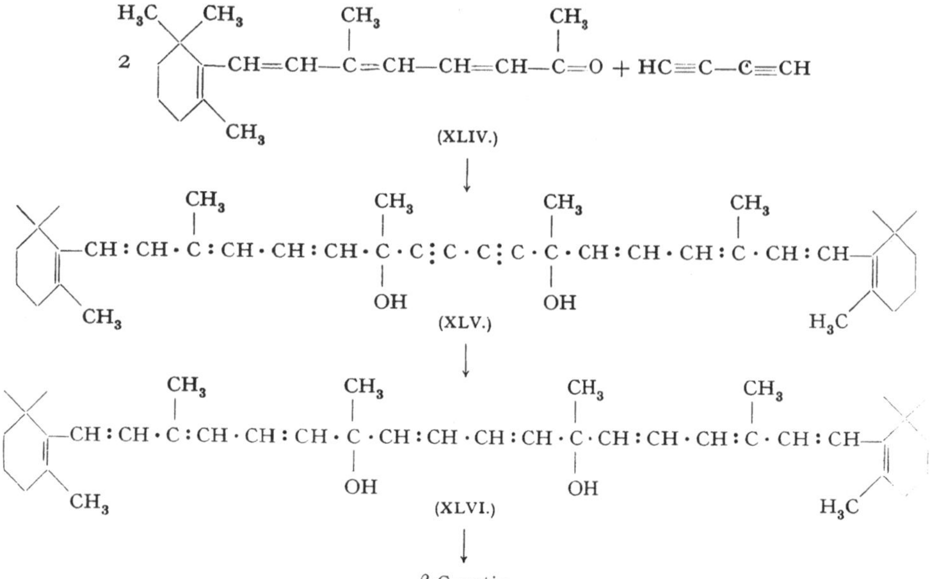

$$2 \quad \text{[ring]}-CH=CH-\underset{CH_3}{C}=CH-CH=CH-\underset{CH_3}{C}=O + HC\equiv C-C\equiv CH$$

(XLIV.)

$$\text{[ring]}-CH{:}CH\cdot\underset{CH_3}{C}{:}CH\cdot CH{:}CH\cdot\underset{OH}{C}\cdot C{:}C\cdot C{:}C\cdot\underset{OH}{C}\cdot CH{:}CH\cdot CH{:}\underset{CH_3}{C}\cdot CH{:}CH-\text{[ring]}$$

(XLV.)

$$\text{[ring]}-CH{:}CH\cdot\underset{CH_3}{C}{:}CH\cdot CH{:}CH\cdot\underset{OH}{C}\cdot CH{:}CH\cdot CH{:}CH\cdot\underset{OH}{C}\cdot CH{:}CH\cdot CH{:}\underset{CH_3}{C}\cdot CH{:}CH-\text{[ring]}$$

(XLVI.)

↓

β-Carotin.

eine Bis-methylenverbindung mit 12 Doppelbindungen, deren durchlaufende Konjugation durch die beiden Methylengruppen an den C-Atomen 13 und 13′ zweimal unterbrochen ist.

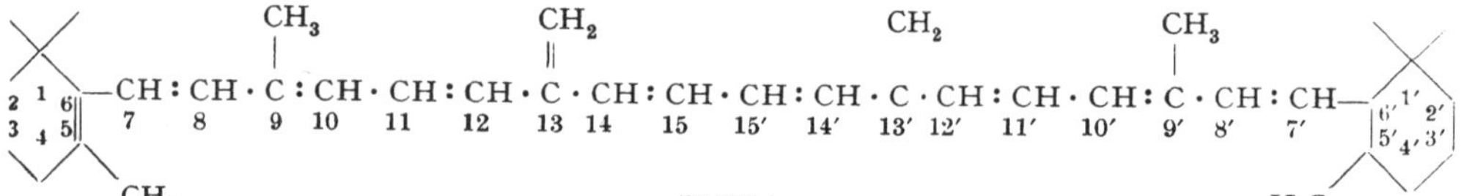

$$\underset{7}{-}CH{:}\underset{8}{CH}\cdot\underset{9}{\underset{|}{\overset{CH_3}{C}}}{:}\underset{10}{CH}\cdot\underset{11}{CH}{:}\underset{12}{CH}\cdot\underset{13}{\underset{\|}{\overset{CH_2}{C}}}\cdot\underset{14}{CH}{:}\underset{15}{CH}\cdot\underset{15'}{CH}{:}\underset{14'}{CH}\cdot\underset{13'}{\underset{\|}{\overset{CH_2}{C}}}\cdot\underset{12'}{CH}{:}\underset{11'}{CH}\cdot\underset{10'}{CH}{:}\underset{9'}{\overset{CH_3}{C}}\cdot\underset{8'}{CH}{:}\underset{7'}{CH}-$$

(XLVII.)

V. Lycopin und ε_1-Carotin.

Lycopin.

In Analogie zur Darstellung des β-Carotins gelang KARRER, EUGSTER und TOBLER (33) die Synthese des Lycopins.

Durch Kondensation von Pseudo-jonon mit Propargylbromid und Zink erhielten die Autoren das Acetylen-carbinol (XLVIII) in nicht ganz reiner Form, dessen Dimagnesiumsalz mit Octendion umgesetzt wurde. Nach einer chromatographischen Reinigung des entstandenen Tetrols (XLIX) wurde dieses selektiv an den Dreifachbindungen hydriert (L) und anschließend aus dem Dien mit p-Toluolsulfosäure Wasser abgespalten. Nach einer Verteilung des Rohproduktes der Wasserabspaltung in Methanol und Petroläther wurde der epiphasische Anteil an Calciumhydroxyd chromatographiert. Die ebenfalls entstandenen *cis*-Isomeren wurden durch eine Behandlung mit Jod zum Teil in die *trans*-Formen

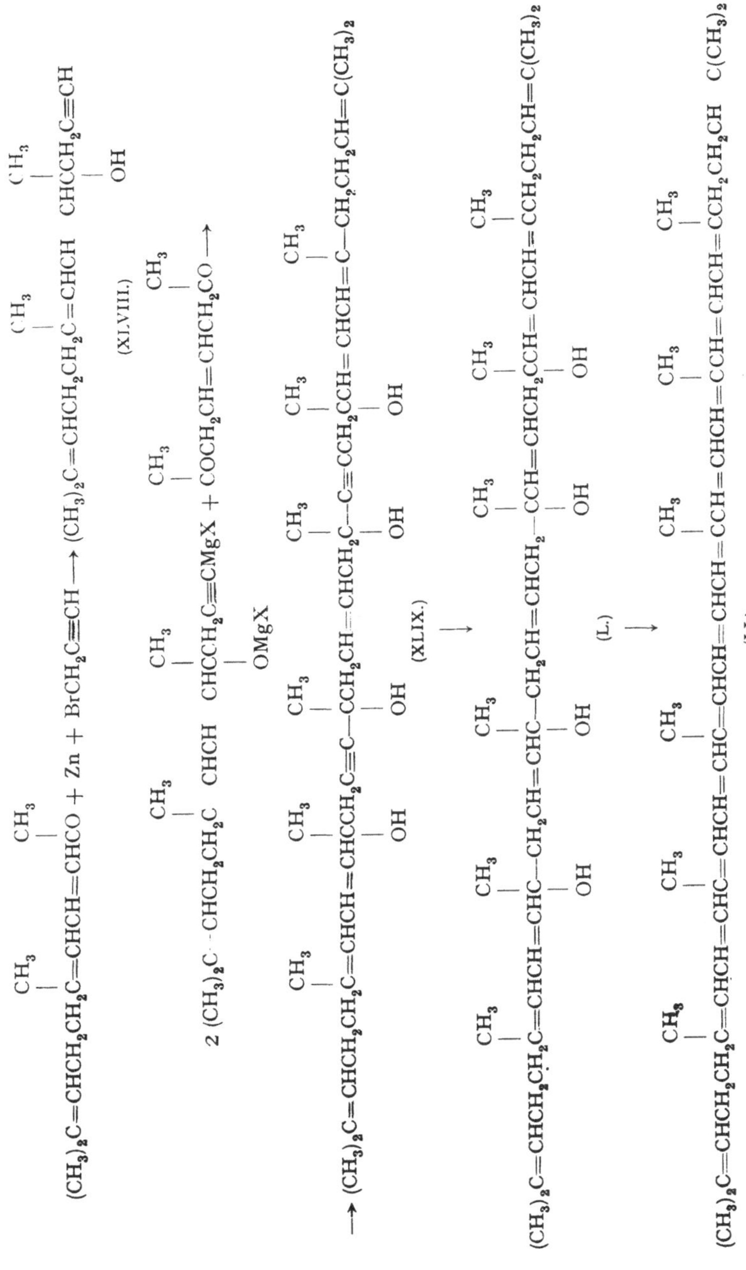

$$(CH_3)_2C=CHCH_2CH_2\underset{\underset{CH_3}{|}}{C}=CHCH=CHCO + Zn + BrCH_2C\equiv CH \longrightarrow (CH_3)_2C=CHCH_2CH_2\underset{\underset{CH_3}{|}}{C}=CHCH=CHCH\underset{\underset{OH}{|}}{\underset{|}{C}H}CH_2C\equiv CH$$

(XLVIII.)

$$2\ (CH_3)_2C\cdot CHCH_2CH_2\underset{\underset{CH_3}{|}}{C}=CHCH\ CHCH_2C\equiv CMgX + COCH_2CH=CHCH_2CO \longrightarrow$$

$$\underset{OMgX}{|}$$

$$\longrightarrow (CH_3)_2C=CHCH_2CH_2\underset{\underset{CH_3}{|}}{C}=CHCH=CHCCH_2C\equiv C-\underset{\underset{OH}{|}}{\underset{\underset{CH_3}{|}}{C}}-CCH_2CH=CHCH=CHC\equiv CCH_2\underset{\underset{OH}{|}}{\underset{\underset{CH_3}{|}}{C}}CH=CHCH=C-CH_2CH_2CH=C(CH_3)_2$$

(XLIX.) →

$$(CH_3)_2C=CHCH_2CH_2\underset{\underset{CH_3}{|}}{C}=CHCH=CHC\underset{\underset{OH}{|}}{\underset{\underset{CH_3}{|}}{-}}CH_2CH=CH-\underset{\underset{OH}{|}}{\underset{\underset{CH_3}{|}}{C}}-CCH_2-CCH=CHCH=CCH_2CH_2CH=C(CH_3)_2$$

(L.) →

$$(CH_3)_2C=CHCH_2CH_2\underset{\underset{CH_3}{|}}{C}=CHCH=CHC=CHCH=CH-\underset{\underset{CH_3}{|}}{C}=CH-\underset{\underset{CH_3}{|}}{C}=CHCH=CCH_2CH_2CH=C(CH_3)_2$$

(LI.)

überführt und diese nochmals chromatographiert. Die vereinigten Lycopin-Zonen ergaben schließlich aus Petroläther charakteristische rote Nadeln und Plättchen, deren Schmelzpunkt und optische Daten mit denen des natürlichen Lycopins identisch waren (LI).

ε_1-Carotin.

ε_1-Carotin (LII) (*29*) ist bisher noch nicht in der Natur aufgefunden worden und stellt ein Isomeres des β-Carotins dar, von dem es sich durch seine beiden α-Jononringe unterscheidet. α-C_{16}-Äthinyl-carbinol (LIII) wurde in das Dimagnesiumsalz überführt und mit Octendion zum Tetrol

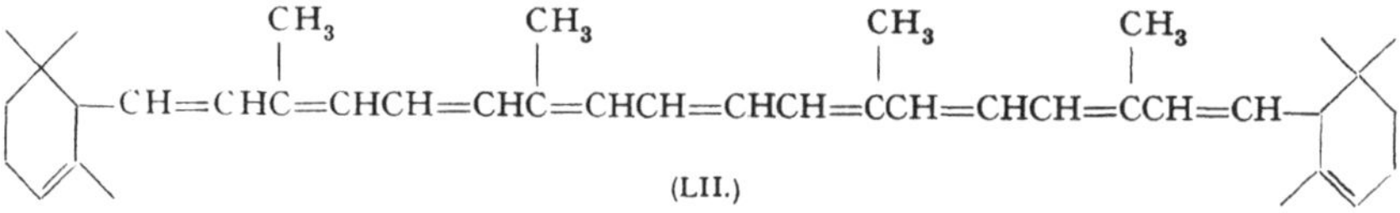

$$\underset{(LII.)}{\text{—CH=CHC(CH}_3\text{)=CHCH=CHC(CH}_3\text{)=CHCH=CHCH=CC(CH}_3\text{)=CHCH=CC(CH}_3\text{)=CH—}}$$

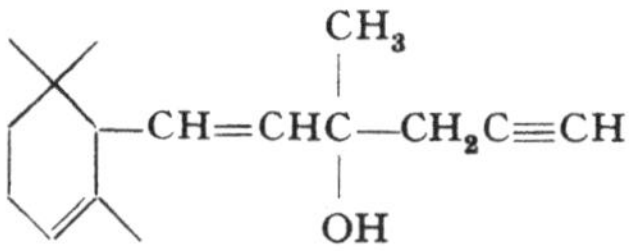

$$\underset{(LIII.)}{\text{—CH=CHC(CH}_3\text{)(OH)—CH}_2\text{C}\equiv\text{CH}}$$

(LIV) kondensiert. Dieses konnte nach einer chromatographischen Reinigung mit einem Pd-BaSO$_4$-Katalysator partiell hydriert und aus dem entstandenen Tetrol (LV) mit p-Toluolsulfosäure Wasser abgespalten werden. Das Produkt wurde in Epi- und Hypophase getrennt und der epiphasische Anteil an einer Säule von Calciumhydroxyd chromatographiert, wobei sich ergab, daß *cis*- und *trans*-Isomere des ε_1-Carotins vorlagen. Das Isomerengemisch wurde mit Jod behandelt und nochmals chromatographiert. Aus dem erhaltenen roten Öl kristallisierte nach

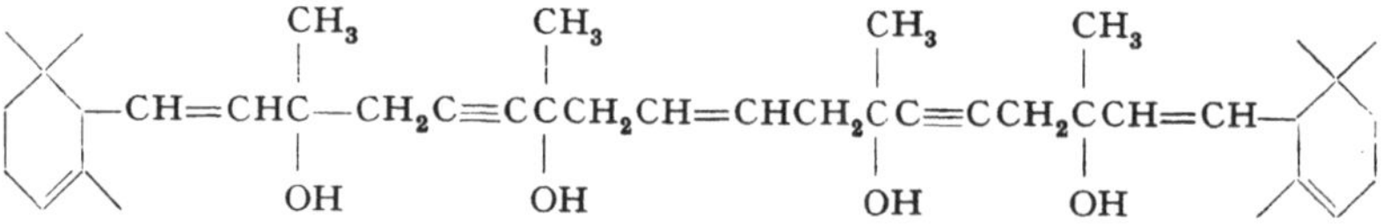

$$\underset{(LIV.)}{\text{—CH=CHC(CH}_3\text{)(OH)—CH}_2\text{C}\equiv\text{CC(OH)(CH}_3\text{)CH}_2\text{CH=CHCH}_2\text{C(OH)(CH}_3\text{)C}\equiv\text{CC(OH)(CH}_3\text{)CH}_2\text{CCH=CH—}}$$

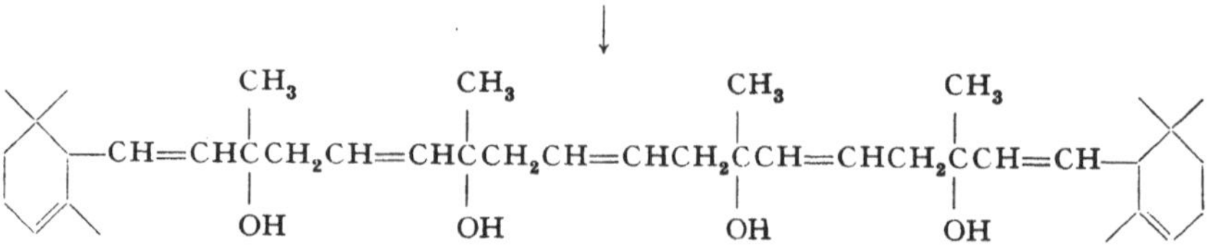

↓

$$\underset{(LV.)}{\text{—CH=CHC(CH}_3\text{)(OH)CH}_2\text{CH=CHC(CH}_3\text{)(OH)CH}_2\text{CH=CHCH}_2\text{C(CH}_3\text{)(OH)CH=CHCH}_2\text{C(CH}_3\text{)(OH)CH=CH—}}$$

↓

(LII)

einer Extraktion mit Methanol aus einem Gemisch von Benzol/Methanol das ε_1-Carotin in großen, gelbroten Blättchen aus. Die Verbindung schmilzt bei 190° (korr.); ihre CARR-PRICE-Reaktion ergibt eine reine blaue Färbung, die sich nach Blaugrün verfärbt. Das Hauptabsorptionsmaximum des ε_1-Carotins liegt in Petroläther bei 470 mμ und in CS_2 bei 501 mμ [KARRER und EUGSTER (29)].

Eine gleichzeitige synthetische Bildung von ε_1-Carotin, β-Carotin und DL-α-Carotin wurde von KARRER und EUGSTER (29a) beobachtet.

VI. Synthesen von höheren Carotin-Homologen.
16,16-Homo-β-carotin (6).

Durch Kondensation des Diacetylens mit 2 Molen des β-C_{19}-Aldehyds (XXXIX, S. 24) wurde ein C_{42}-Diin-diol (LVI) erhalten. Allylumlagerung und Wasserabspaltung konnten in normaler Weise durchgeführt werden, wobei das C_{42}-Diin (LVII) der Summenformel $C_{42}H_{54}$ in blattgoldähnlichen Blättchen erhalten wurde. Bei diesen C_{42}-Kohlenwasserstoffen liegt unterschiedlich zu den C_{40}-Carotinen zentral keine Doppelbindung sondern eine Einfachbindung.

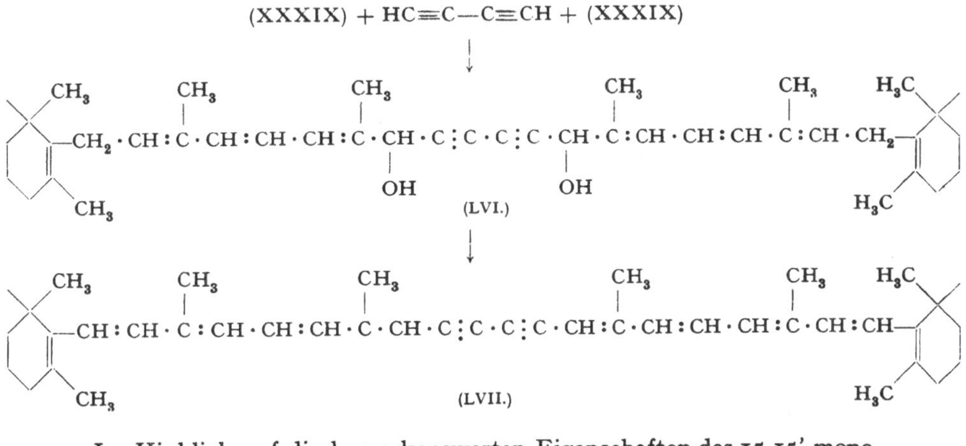

$$(XXXIX) + HC\equiv C-C\equiv CH + (XXXIX)$$

Im Hinblick auf die bemerkenswerten Eigenschaften des 15,15'-mono-cis-β-Carotins wurde auch bei den homologen $C_{42}H_{58}$-Kohlenwasserstoffen nach cis-Formen gesucht und ihr spektrales Verhalten sowie die Umlagerung zur trans-Konfiguration näher untersucht.

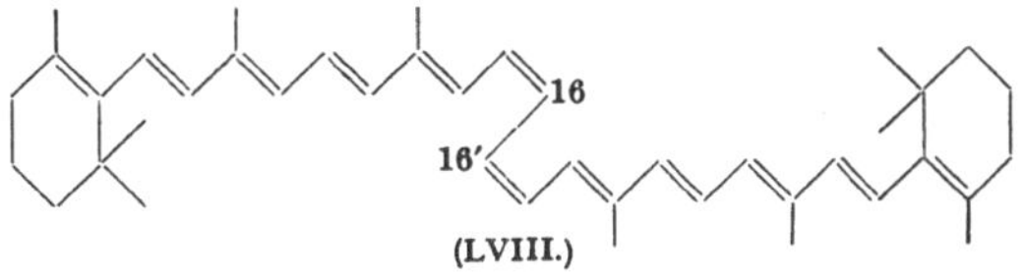

Die partielle Hydrierung der beiden Acetylengruppen wurde mit Rücksicht auf die beim mono-*cis*-β-Carotin gemachten Erfahrungen von vornherein unter Ausschluß von Tageslicht und ohne Erwärmung durchgeführt.

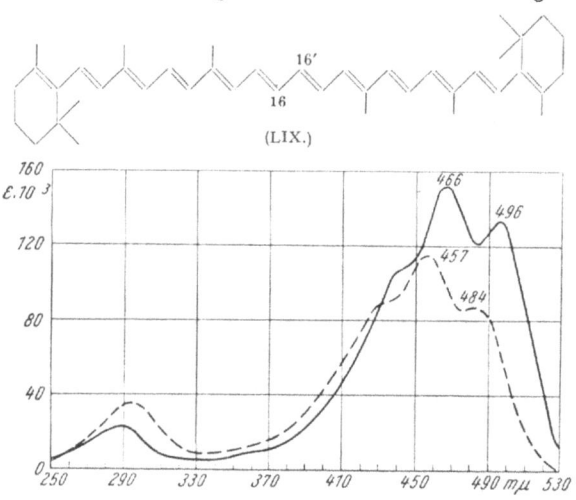

Abb. 3. Spektralkurve von 16,16'-Homo-β-carotin (in Hexan): ———, all-*trans*-Form; und - - - -, di-*cis*-Form.

Der in dunkelroten Kristallen anfallende Kohlenwasserstoff $C_{42}H_{58}$ (LVIII) mit 12facher Konjugation schmilzt bei 142°, um kurz danach noch einmal zu kristallisieren und dann erst wieder bei 190° zu schmelzen. Schon dieses Verhalten deutete darauf hin, daß eine thermische Stereoisomerisierung stattgefunden hat. Der bei 142° schmelzende Kohlenwasserstoff, der also unter allen Kautelen bei der Hydrierung des $C_{42}H_{54}$-Diins erhalten worden war, ließ sich nicht nur durch Erhitzen in kristallisiertem Zustand bzw. in Lösung, sondern auch durch Behandlung mit Jod im Licht, sowie vor allem durch

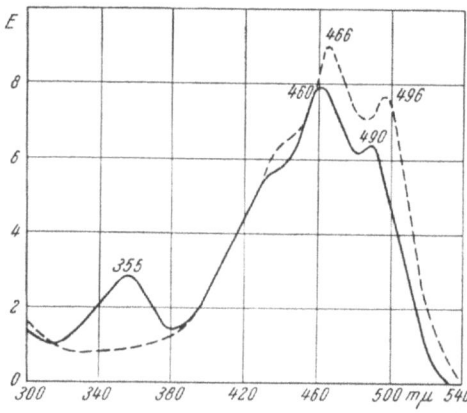

Abb. 4. Spektralkurve von 16,16'-Homo-β-carotin: ———, mono-*cis*-Form (in Hexan); und - - - -, nach 5-stündigem Verweilen in diffusem Tageslicht.

Belichtung allein in die isomere, bei 190° schmelzende all-*trans*-Verbindung umwandeln (LIX). Bei dieser *cis* → *trans*-Isomerisierung findet eine Rotverschiebung des Hauptmaximums um etwa 10 mμ statt. Diese weist etwa den doppelten Wert auf, der nach den bisherigen Erfahrungen für die cis-trans-Umlagerung an einer Doppelbindung gefunden wurde (4—5 mμ). In (LVIII) kann daher die di-cis-Konfiguration angenommen werden. Die beiden Isomeren unterscheiden sich auch deutlich in der Farbe; die all-trans-Verbindung ist wesentlich dunkler gefärbt — tiefdunkelrot bis violett — als das cis-Isomere. Beide zeigen eine namhafte Provitamin A-Aktivität (43).

Betrachtungen am STUART-Modell lassen erkennen, daß durch zweifaches Umklappen an den beiden mittelständigen =CH—CH=-Gruppen (16 und 16') — auch im Kristall — eine einfache Streckung der Moleküle zustande kommen kann, die mit Farbvertiefung und Schmelzpunktserhöhung verbunden ist.

Im Chromatogramm lassen sich beide C_{42}-Isomere leicht voneinander trennen. Während die di-cis-Verbindung rasch durchläuft, bleibt das all-trans-Isomere fest haften, ein Verhalten, das ebenfalls in Übereinstimmung mit den Befunden von ZECHMEISTER und Mitarbeitern steht (42).

Ferner wurde noch versucht, mit dem C_{42}-di-cis-Carotinoid eine stufenweise Isomerisierung zu erreichen und die eventuell erhaltene mono-cis-Verbindung auf ihr spektrales Verhalten zu untersuchen. Dieses Ziel konnte in der Tat erreicht werden. Wurde das di-cis-Isomere im kristallinen Zustand kurz zum Schmelzen erhitzt und die Schmelze rasch abgekühlt, so ließ sich durch chromatographische Reinigung der mono-cis-C_{42}-Kohlenwasserstoff (LX) isolieren.

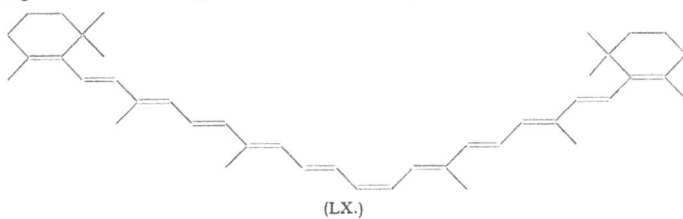

(LX.)

Dieser zeigte deutlich die Ausbildung eines cis-Gipfels bei 350 mμ, der 141 mμ vom langwelligsten Maximum der all-trans-Verbindung (496 mμ) entfernt lag. Trotzdem also in diesem Fall die cis-Bindung nicht mehr ganz genau zentral liegt, verhält sie sich noch so, wie es die Anschauungen von ZECHMEISTER, LEROSEN, SCHROEDER, POLGAR und PAULING (44) verlangen.

Wird das ölige, jedoch chromatographisch gereinigte, d. h. von di-cis und all-trans befreite mono-cis-Isomere noch einmal, jedoch dieses Mal länger erhitzt, so erhält man glatt die all-trans-Verbindung vom Schmelzpunkt 190°. Der cis-peak ist nunmehr wieder verschwunden und dafür das Spektrum von (LIX) erschienen. Auch durch einfaches Belichten läßt sich das mono-cis-Isomere in die all-trans-Verbindung umlagern. Daß die di-cis-Form keinen cis-peak aufweist, obgleich zwei zentrale cis-Bindungen vorliegen, wird damit zusammenhängen, daß in der allgemein gerade gebliebenen Molekülform die beiden cis-Effekte einander kompensieren.

Decapreno-β-carotin und Decapreno-ε₁-carotin.

Decapreno-β-carotin ist als ein Isoprenhomologes des β-Carotins zu betrachten. Es ist aus 10 Isopreneinheiten mit 15 konjugierten Doppelbindungen aufgebaut. Von derartigen Polyenen mit 15 konjugierten Doppelbindungen waren bisher das Diphenyl-triacontapentadecaen von KUHN (36) und das Dehydrolycopin von KARRER (35) bekannt.

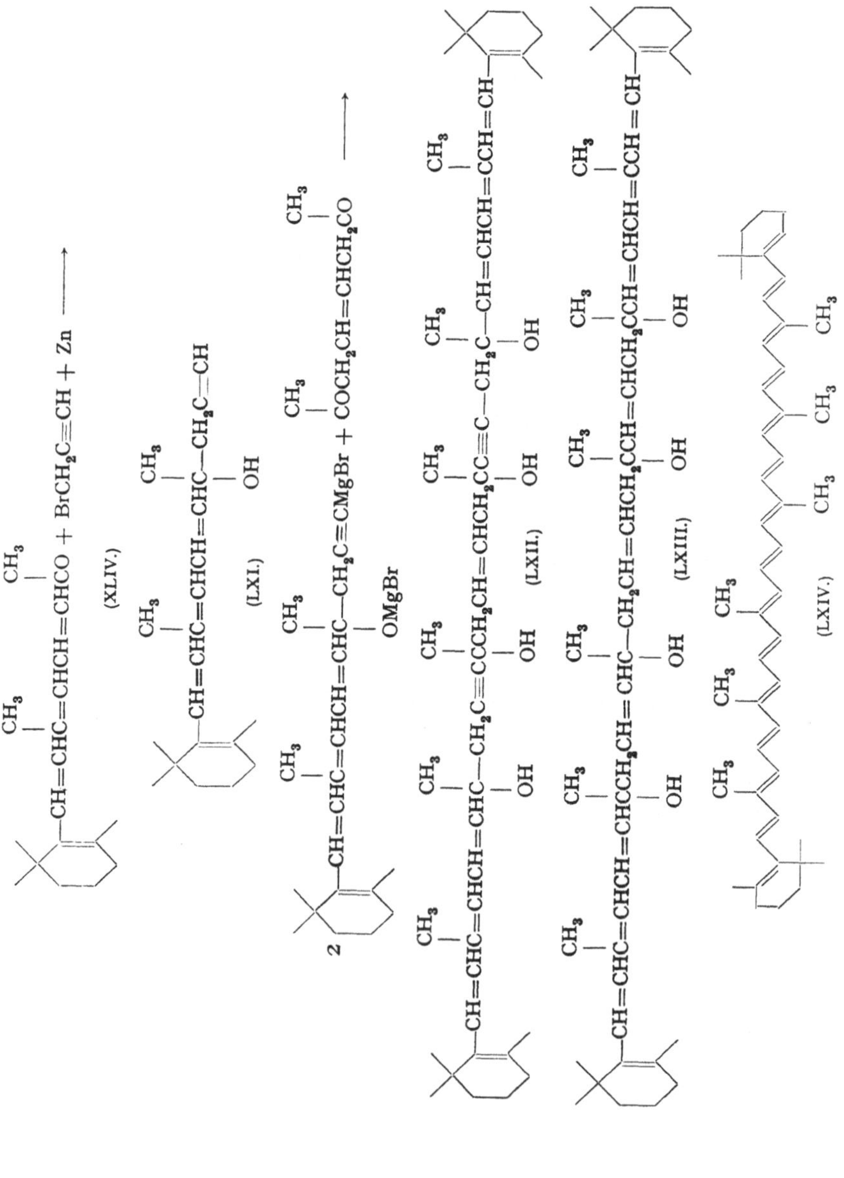

KARRER und EUGSTER (*30*) führten die Synthese des Decapreno-β-carotins in ähnlicher Weise wie beim β-Carotin von C_{18}-Keton (XLIV) ausgehend durch. Sie kondensierten (XLIV) mit Propargylbromid und Zink zum C_{21}-Äthinyl-carbinol (LXI), dessen Dimagnesiumsalz mit Octendion zu einem Gemisch von Stereoisomeren des Tetrols (LXII) führte. Nach partieller Hydrierung von (LXII) wurde das Tetrol (LXIII) erhalten, aus dem mittels p-Toluolsulfosäure in siedendem Toluol Wasser abgespalten wurde, unter Bildung des *trans*-Decapreno-β-carotins (LXIV), das durch eine Chromatographie an Zinkcarbonat isoliert wurde. Die Verbindung kristallisiert in schwarzvioletten Spießen (Schmelzp. 192°, korr.), deren CARR-PRICE-Reaktion eine blaue Färbung liefert und bei 578 mμ absorbiert.

Die Darstellung des Decapreno-$ε_1$-carotins als höheres Isoprenhomologes des $ε_1$-Carotins gelang KARRER, EUGSTER und FAUST (*32*) in gleicher Reaktionsfolge wie beim Decapreno-β-carotin, nur mit dem Unterschied, daß die Autoren vom α-C_{18}-Keton (LXV) ausgehend über die entsprechenden Zwischenprodukte mit α-Jononringen (LXVI—LXVIII) zum Decapreno-$ε_1$-carotin gelangten (LXIX). Dieses konnte mittels Chromatographie an einer Magnesiumoxyd-Celitsäule gereinigt und aus einem Benzol/Methanol-Gemisch in feinen, schwarzroten Kristallen (Schmelzp. 216,5—217°) erhalten werden. Die CARR-PRICE-Reaktion führte zu einer blaugrünen Lösung.

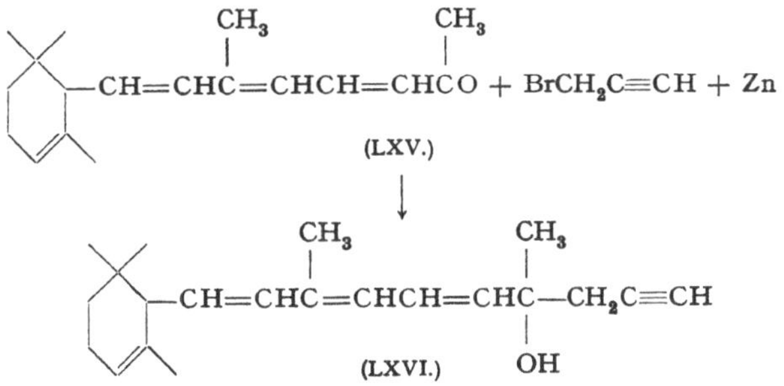

$$\text{[Struktur]}-CH=CHC(CH_3)=CHCH=CHCO + BrCH_2C\equiv CH + Zn$$

(LXV.)

↓

$$\text{[Struktur]}-CH=CHC(CH_3)=CHCH=CHC(CH_3)(OH)-CH_2C\equiv CH$$

(LXVI.)

Dodecapreno-β-carotin.

Durch die Synthese des Dodecapreno-β-carotins wurde ein β-Carotin-Homologes mit 12 Isopreneinheiten und 19 konjugierten Doppelbindungen dargestellt. KARRER und EUGSTER (*31*); [vgl. auch BATTY und Mitarbeiter (*2*)] schlugen prinzipiell denselben Weg ein wie zur Darstellung des β- und $ε_1$-Carotins sowie seiner Isopren-Homologen.

Kristallisiertes Vitamin-A-Acetat wurde verseift und der entstandene Vitamin-A-Alkohol nach OPPENAUER mit tertiärem Aluminiumbutylat in Gegenwart von Aceton zum C_{23}-Keton (LXX) oxydiert, wobei gleichzeitig eine Kondensation mit Aceton stattfand. Auf gleichem Wege war bereits von HEILBRON und Mitarbeitern (2) das gleiche Produkt synthetisiert worden. Das Keton lieferte mit Zink und Propargylbromid das tertiäre C_{20}-Äthinyl-carbinol (LXXI), dessen Dimagnesiumsalz sich mit Octendion zum Tetrol (LXXII) kondensieren ließ. Das chromatographisch gereinigte Tetrol wurde nunmehr partiell hydriert (LXXIII) und einer Dehydrati-

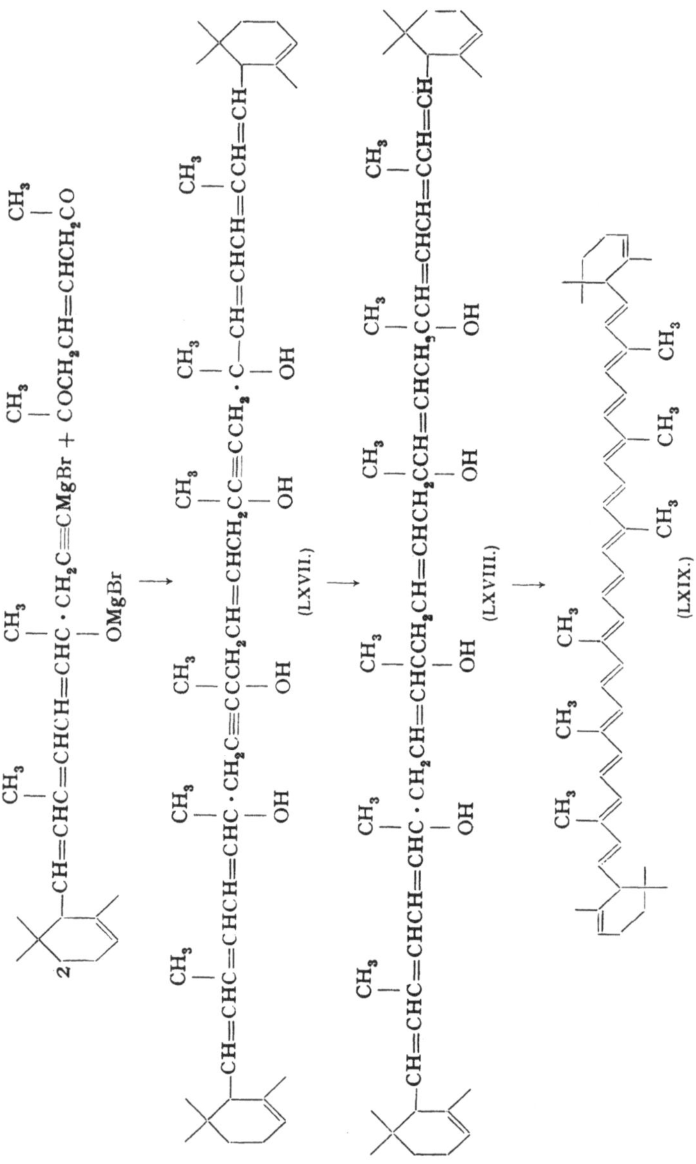

Synthetische Chemie der Carotinoide.

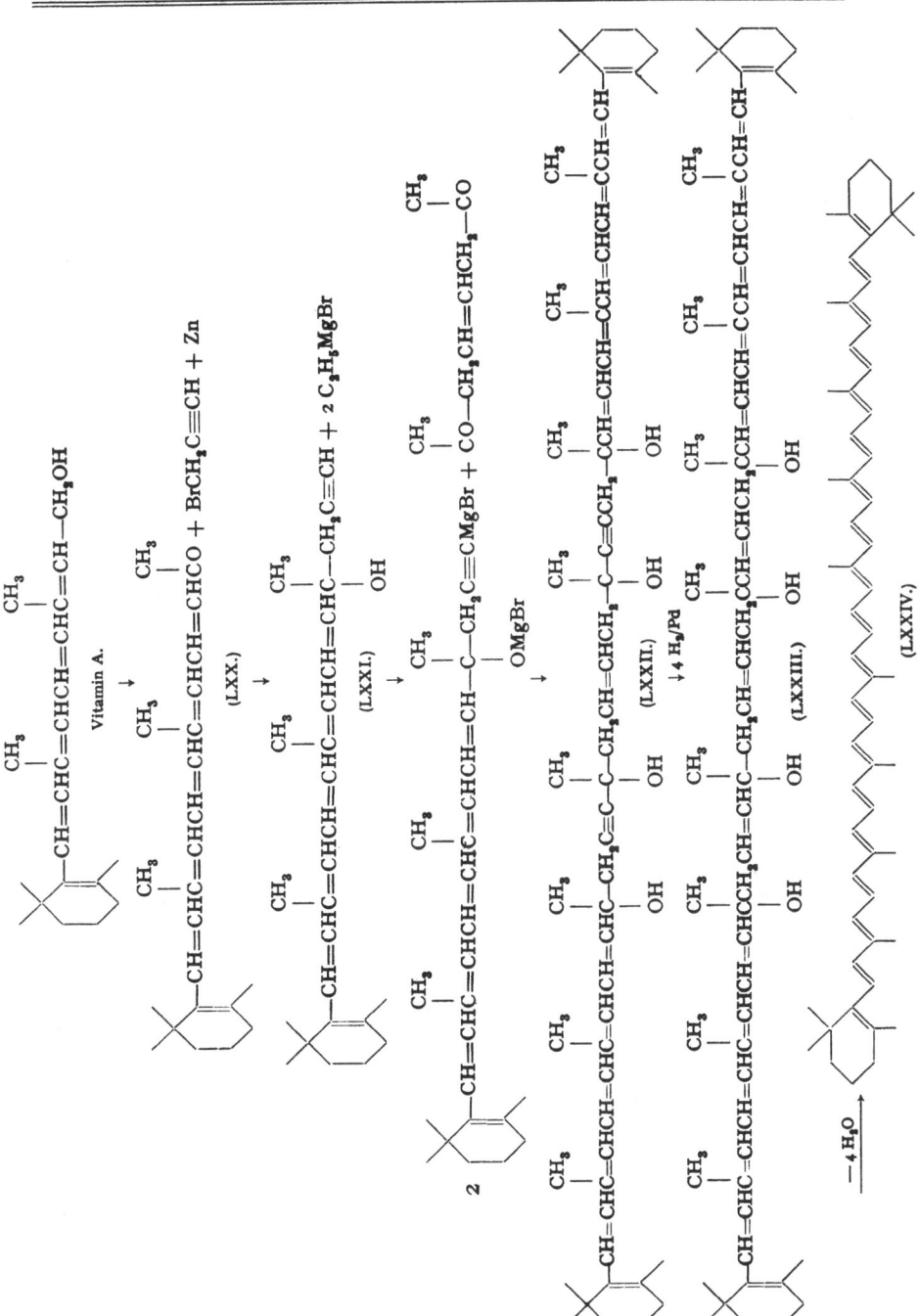

sierungs-Reaktion unterworfen. Nach mehrmaliger Chromatographie an Magnesiumoxyd und Calciumcarbonat konnte eine geringe Menge des Dodecapreno-β-carotins (LXXII) als schwarzes, mikrokristallines Pulver erhalten werden. Hauptabsorptions-Maximum in Cyclohexan bei 537 mμ, in CS_2 bei 577 mμ.

KARRER nimmt an, daß eine Darstellung noch höherer Carotinhomologer nach seiner Methode infolge der kritischen Wasserabspaltungs-Stufe aus dem Tetrol nicht möglich sei.

Literaturverzeichnis.

1. ARMITAGE, J. B., E. R. H. JONES and M. C. WHITING: Researches on Acetylenic Compounds. XXVIII. A New Route to Diacetylene and its Symmetrical Derivatives. J. chem. Soc. (London) 1951, 44.
2. BATTY, J. W., A. BURAWOY, S. H. HARPER, I. M. HEILBRON and W. E. JONES: The Action of the Oppenauer Reagent on Primary Alcohols Including Vitamin A. J. chem. Soc. (London) 1938, 175.
2a. BAXTER, J. G.: Synthesis and Properties of Vitamin A and Some Related Compounds. Fortschr. Chem. organ. Naturstoffe 9, 41 (1952).
3. CYMERMAN, J., I. M. HEILBRON and E. R. H. JONES: Studies in the Polyene Series. XXI. Ethynylcarbinols from α,β-Unsaturated Ketones: Their Anionotropic Rearrangements and other Reactions. J. chem. Soc. (London) 1945, 90.
4. HEILBRON, I. M., E. R. H. JONES, J. T. McCOMBIE and B. C. L. WEEDON: Studies in the Polyene Series. XIX. The Semihydrogenation of Ethynylcarbinols derived from α,β-Unsaturated Aldehydes and the Anionotropic Rearrangements of the Resulting Vinylcarbinols. J. chem. Soc. (London) 1945, 84.
5. HOFFMANN-LA ROCHE, Basel: Schweiz. Patent 258514, 16. Mai 1949.
6. INHOFFEN, H. H., F. BOHLMANN, H. J. ALDAG, S. BORK und G. LEIBNER: Synthesen in der Carotinoid-Reihe. XXI. Kondensation von Carotinoidketonen und -aldehyden mit Diacetylen; zugleich eine weitere Synthese des β-Carotins. Liebigs Ann. Chem. 573, 1 (1951).
7. INHOFFEN, H. H., F. BOHLMANN, K. BARTRAM und H. POMMER: Synthesen in der Carotinoid-Reihe. XI. Totalsynthese des β-Carotins. Chemiker-Ztg. 74, 285 (1950).
8. — — — — Synthesen in der Carotinoid-Reihe. XIII. Totalsynthese des β-Carotins. Abh. Braunschweig. Wiss. Ges. 2, 75 (1950).
9. INHOFFEN, H. H., F. BOHLMANN, K. BARTRAM, G. RUMMERT und H. POMMER: Synthesen in der Carotinoid-Reihe. XV. Über die Darstellung von trans- und von 15,15'-mono-cis-β-Carotin. Liebigs Ann. Chem. 570, 54 (1950).
10. INHOFFEN, H. H., F. BOHLMANN und G. RUMMERT: Synthesen in der Carotinoid-Reihe. X. Synthese des Bis-nor-methyl-β-carotins. Liebigs Ann. Chem. 569, 226 (1950).
11. INHOFFEN, H. H., H. POMMER und F. BOHLMANN: Synthesen in der Carotinoid-Reihe. II. Über die Synthese eines Kohlenwasserstoffs $C_{30}H_{42}$. Liebigs Ann. Chem. 561, 26 (1948).
12. — — — Synthesen in der Carotinoid-Reihe. XII. Zweite Synthese des β-Carotins. Chemiker-Ztg. 74, 309 (1950).
13. — — — Synthesen in der Carotinoid-Reihe. XIV. Aufbau des β-Carotins. Liebigs Ann. Chem. 569, 237 (1950).

14. INHOFFEN, H. H., H. POMMER und E. G. METH: Synthesen in der Carotinoid-Reihe. V. Eine neue Methode zur Darstellung des Kohlenwasserstoffs $C_{30}H_{42}$. Liebigs Ann. Chem. 565, 45 (1949).
15. — — — Synthesen in der Carotinoid-Reihe. IX. Über einen Kohlenwasserstoff $C_{16}H_{22}$. Liebigs Ann. Chem. 569, 74 (1950).
16. — — — Synthesen in der Carotinoid-Reihe. VIII. Totalsynthese des β-Dihydro-carotins. Chemiker-Ztg. 74, 211 (1950).
17. — — — Synthesen in der Carotinoid-Reihe. XIX. Totalsynthese des 7,7'-Dihydro-β-carotins. Liebigs Ann. Chem. 572, 151 (1950).
18. INHOFFEN, H. H., H. POMMER und F. WESTPHAL: Synthesen in der Carotinoid-Reihe. XVI. Eine weitere Synthese des β-Carotins. Liebigs Ann. Chem. 570, 69 (1950).
19. INHOFFEN, H. H., H. POMMER und K. WINKELMANN: Synthesen in der Carotinoid-Reihe. VII. Über einen Kohlenwasserstoff $C_{12}H_{14}$. Liebigs Ann. Chem. 586, 174 (1950).
20. INHOFFEN, H. H., H. POMMER, K. WINKELMANN und H. J. ALDAG: Synthese des Octadien-(3,5)-dions-(2,7). Ber. dtsch. chem. Ges. 84, 87 (1951).
21. ISLER, O.: Über Synthesen in der Vitamin-A-Reihe. Chimia 3, 150 (1949); sowie Privatmitteilung von O. ISLER.
22. ISLER, O., W. HUBER, A. RONCO und M. KOFLER: Synthese von Vitamin-A-Methyl-äther. EMIL BARELL-Festschrift 1946, 31 [Chem. Zbl. 118 II, 519 (1947)].
23. — — — — Synthese des Vitamin A. Helv. chim. Acta 30, 1911 (1947).
24. JOHNSON, A. W.: 2-Butyne-1:4-diol. I. Reactions of the Hydroxyl Groups. J. chem. Soc. (London) 1946, 1009.
25. KARRER, P.: Carotinoid-epoxyde und furanoide Oxyde von Carotinoidfarbstoffen. Fortschr. Chem. organ. Naturstoffe 5, 1 (1948).
26. KARRER, P. et C. H. EUGSTER: Synthèse totale du β-carotène. C. R. hebd. Séances Acad. Sci. 230, 1920 (1950).
27. — — Synthese von Carotinoiden. II. Totalsynthese des β-Carotins. I. Helv. chim. Acta 33, 1172 (1950).
28. — — Darstellung und einige Umsetzungen des Octadien-(3,5)-dions-(2,7). Helv. chim. Acta 32, 1934 (1949).
29. — — Synthese von Carotinoiden. IV. Synthese eines ε_1-Carotins. Helv. chim. Acta 33, 1433 (1950).
29 a. — — Synthese von Carotinoiden V. Gleichzeitige synthetische Bildung von ε-Carotin, β-Carotin und d,l-α-Carotin. Helv. chim. Acta 33, 1952 (1950).
30. — — Synthese von Carotinoiden. VI. Synthese eines Homologen des β-Carotins mit 15 konjugierten Doppelbindungen: Decapreno-β-carotin. Helv. chim. Acta 34, 28 (1951).
31. — — Carotinoidsynthesen. VIII. Synthese des Dodecapreno-β-carotins. Helv. chim. Acta 34, 1805 (1951).
32. KARRER, P., C. H. EUGSTER und M. FAUST: Synthesen von Carotinoiden. VII. Synthese des Decapreno-ε_1-carotins. Helv. chim. Acta 34, 823 (1951).
33. KARRER, P., C. H. EUGSTER und E. TOBLER: Synthese von Carotinoidfarbstoffen. III. Totalsynthese des Lycopins. Helv. chim. Acta 33, 1349 (1950).
34. KARRER, P. und E. JUCKER: Carotinoide. Basel: Birkhäuser. 1948.
35. KARRER, P. und J. RUTSCHMANN: Dehydro-lycopin, ein Carotinoidfarbstoff mit 15 konjugierten Doppelbindungen. Helv. chim. Acta 28, 793 (1945).
36. KUHN, R.: Über die Synthese höherer Polyene. Angew. Chem. 50, 703 (1937).
37. KUHN, R. und K. WALLENFELS: Synthese von Polyenen mit Hilfe von Acetylen und Diacetylen. Ber. dtsch. chem. Ges. 71, 1889 (1938).

38. LeRosen, A. L. and L. Zechmeister: Prolycopene. J. Amer. chem. Soc. 64, 1075 (1942).
39. Milas, N. A., P. Davis, I. Belič and D. A. Fleš: Synthesis of β-Carotene. J. Amer. chem. Soc. 72, 4844 (1950).
40. Pauling, L.: Recent Work on the Configuration and Electronic Structure of Molecules; with some Applications to Natural Products. Fortschr. Chem. organ. Naturstoffe 3, 203 (1939).
41. Polgar, A. and L. Zechmeister: Isomerization of β-Carotene. Isolation of a Stereoisomer with Increased Adsorption Affinity. J. Amer. chem. Soc. 64, 1856 (1942).
42. Zechmeister, L.: cis-trans-Isomerization and Stereochemistry of Carotenoids and Diphenylpolyenes. Chem. Reviews 34, 267 (1944).
43. Zechmeister, L., H. J. Deuel, Jr., H. H. Inhoffen, J. Leemann, S. M. Greenberg and J. Ganguly: Stereochemical Configuration and Provitamin A Activity. X. A Comparison of Synthetic 15,15'-Mono-cis-β-carotene (Central Mono-cis-β-carotene) with All-trans-β-carotene in the Rat and Chick. Arch. Biochem. Biophys. 36, 80 (1952).
44. Zechmeister, L., A. L. LeRosen, W. A. Schroeder, A. Polgár and L. Pauling: Spectral Characteristics and Configuration of Some Stereoisomeric Carotenoids Including Prolycopene and Pro-γ-carotene. J. Amer. chem. Soc. 65, 1940 (1943).
45. Zechmeister, L. and J. H. Pinckard: Some Poly-cis-lycopenes Occurring in the Fruit of *Pyracantha.* J. Amer. chem. Soc. 69, 1930 (1947).
46. Zechmeister, L. and A. Polgár: cis-trans-Isomerization and Spectral Characteristics of Carotenoids and Some Related Compounds. J. Amer. chem. Soc. 65, 1522 (1943).
47. — — cis-trans-Isomerization and cis-Peak-Effect in the α-Carotene Set and in Some Other Stereoisomeric Sets. J. Amer. chem. Soc. 66, 137 (1944).
48. — — Contribution to the Stereochemistry of γ-Carotene. J. Amer. chem. Soc. 67, 108 (1945).
49. Zechmeister, L. and W. A. Schroeder: Pro-γ-carotene. J. Amer. chem. Soc. 64, 1173 (1942).

(Eingelaufen am 17. April 1952.)

Synthesis and Properties of Vitamin A and Some Related Compounds.

By J. G. BAXTER, Rochester, New York*.

With 2 Figures.

Contents.	Page
Introduction	42
I. Synthesis of Vitamin A	43
1. *Via* Esters of β-Ionylideneacetic Acid	43
Ethyl β-Ionylideneacetate and its β,γ-Unsaturated Isomer	46
β-Ionylidene-ethanol	50
C_{18}-Ketone	50
Vitamin A Acid ethyl ester and its β,γ-Unsaturated Isomer	50
Vitamin A	51
Isomer of Vitamin A	51
Synthesis by WENDLER, SLATES, TRENNER and TISHLER	51
β-Ionylideneacetaldehyde	52
2. Synthesis of Vitamin A *via* Esters of β-Ionylidenecrotonic Acid	52
β-Ionylidenecrotonic Acid	53
C_{18}-Ketone	53
Vitamin A Aldehyde	56
Vitamin A Acids	56
Vitamin A	57
Synthesis by SCHWARZKOPF, CAHNMANN, LEWIS, SWIDINSKY and WUEST	57
3. Synthesis of Vitamin A *via* "C_{14}-Aldehyde"	58
C_{14}-Aldehyde	60
Allylic Rearrangement	60
4. Synthesis of Vitamin A by Other Methods	64

* Communication No. 185 from the Laboratories of Distillation Products Industries, Rochester, New York.

The work on synthetic vitamin A in this Laboratory has been cooperative in nature and in certain references to unpublished work in the review [Dist. Prod. Ind. (*23*)] the author acts as spokesman for a group in which J. D. CAWLEY, C. D. ROBESON, L. WEISLER and E. M. SHANTZ have made major contributions and A. J. CHECHAK, F. B. CLOUGH, C. C. EDDINGER, N. D. EMBREE, G. R. SEIDEL and M. H. STERN have given valuable assistance. The help of the Ultraviolet and the Infrared Spectroscopy Depts. and the Biochemistry Dept. supervised, respectively, by A. P. BESANCON, W. P. BLUM and P. L. HARRIS, has been indispensable.

	Page
II. Synthesis and Biological Activity of Some Compounds Related to Vitamin A	66
Alcohols	66
Esters and Ethers of Vitamin A	74
Aldehydes	75
Vitamin A_2	76
Hydrocarbons	77
Vitamin A Acids	79
Conclusions	80
III. Relationship between Vitamin A and Carotenes	80
References	80

Introduction.

Following the preparation of rich concentrates of vitamin A (vitamin A alcohol, vitamin A_1, axerophthol) from the unsaponifiable matter of fish liver oils, the following formula was established by KARRER, MORF and SCHÖPP (*89*):

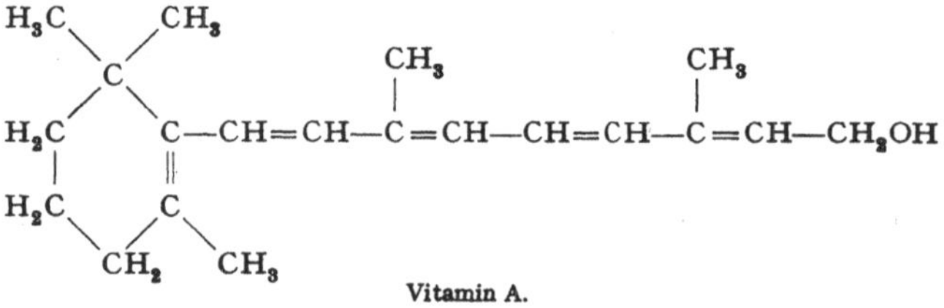

$$\text{Vitamin A structure: trimethylcyclohexenyl—C—CH=CH—C(CH}_3\text{)=CH—CH=CH—C(CH}_3\text{)=CH—CH}_2\text{OH}$$

Vitamin A.

Other concentrates were made [HEILBRON and co-workers (*46*)] and crystals were soon prepared of certain esters, such as the β-naphthoate, anthraquinone-β-carboxylate, acetate and succinate [HAMANO (*41*); MEAD (*97*); BAXTER and ROBESON (*10*)]. The free vitamin was also crystallized, first as the addition compound with methyl alcohol [HOLMES and CORBET (*57*)] and later in solvent-free form [BAXTER and ROBESON (*11*)]. Much of this work in the U. S. A. was facilitated by the development of the molecular still by HICKMAN and associates (*55*).

Certain facts about the vitamin were discovered. It was found to occur in fish liver oils almost exclusively as an ester of the higher fatty acids [BACHARACH and SMITH (*6*); HICKMAN (*56*)]. It was shown that the vitamin occurs in a number of fish liver oils as a mixture of geometrical isomers, the all-*trans* form being called vitamin A and a *cis* isomer, neo-vitamin A. The ratio of vitamin A to neovitamin A in these oils was found to be approximately 65 : 35 [ROBESON and BAXTER (*129*)].

An important new development was the identification of retinene$_1$, the prosthetic group of the visual pigment rhodopsin, with vitamin A aldehyde [MORTON (*120*)]. This followed the studies of WALD on the chemistry of the visual process, first reported on in 1935 and later reviewed [WALD (*145*)]. Concentration and crystallization of vitamin A aldehyde followed [BALL, GOODWIN and MORTON (*7*)]. The retinene forming rhodopsin was found to be a *cis* isomer, not the Δ^2-*cis*-Δ^6-*trans*, or neo-isomer [HUBBARD and WALD (*57a*)].

Certain close relatives of vitamin A, such as anhydrovitamin A, isoanhydrovitamin A, subvitamin A and kitol were identified in fish liver oils [SHANTZ, CAWLEY and EMBREE (*138*); EMBREE and SHANTZ (*26, 27, 28*)]. A relative of vitamin A, called vitamin A_2, was discovered in the liver oils of fresh water fish [GILLAM and co-workers (*33*)] and its purification through a crystalline ester followed (SHANTZ (*134*)].

The synthesis, reported in 1937, of a biologically active vitamin A preparation by KUHN and MORRIS (*94*) stimulated research in this field. The feasibility of a more practical synthesis became evident in 1946 with publications on the synthesis of vitamin A methyl ether [ISLER and co-workers (*75*); MILAS (*111*)], and vitamin A acid [ARENS and VAN DORP (*1*)]. In 1947 the synthesis of vitamin A itself was announced by VAN DORP and ARENS (*143*); ISLER and co-workers (*74*); and by CAWLEY et al. (*17*). Many other contributions have since been made which are discussed below.

A number of reviews on vitamin A synthesis have appeared [SOBOTKA and BLOCH (*140*); EMBREE (*25*); MILAS (*112*); HEILBRON (*45*); JOHNSON (*77*); KARRER (*78*); INHOFFEN and BOHLMANN (*61*); HUNTER (*59*); ISLER (*72*)]. This has made it possible to give chief attention in the present review to those syntheses in which the intermediates and the vitamin A were highly purified and characterized, particularly by their ultraviolet absorption spectra. The increasing amount of such data and their usefulness to the organic chemist working in this field makes it desirable that it be reviewed at this time.

The biological potency of certain close relatives of the vitamin will also be considered.

I. Synthesis of Vitamin A.

1. *Via* Esters of β-Ionylideneacetic Acid.

The esters of β-ionylideneacetic acid (e. g., Formula III, Scheme 1, p. 47) became key compounds in a number of methods of vitamin A synthesis [HARPER and OUGHTON (*43*); CAWLEY et al. (*16*); WENDLER et al. (*153*)], following the discovery of the selective reducing agent, lithium aluminumhydride [FINHOLT, BOND and SCHLESINGER (*32*)]. Partial completion of a proposed synthesis, using this reagent, has also been reported [MILAS and HARRINGTON (*116*)].

The synthesis as described by CAWLEY and co-workers proceeded according to the symbols in Scheme 1. β-Ionone was reacted with

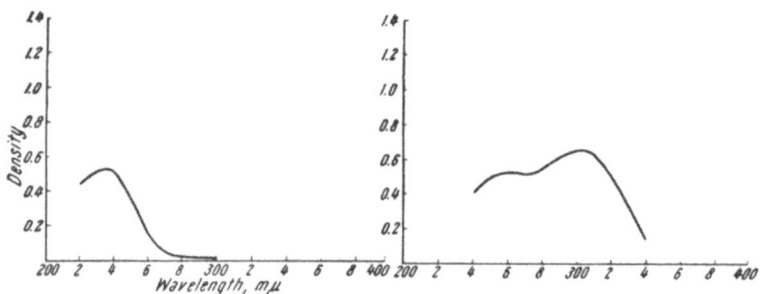

Ethyl Ester of β-Ionolacetic Acid
E (1%, 1 cm.) (234 mμ) = 214 (Ethanol).

Ethyl β-Ionylideneacetate; E (1%, 1 cm.)
(304 mμ) = 548; E (1%, 1 cm.) (256 mμ) =
= 434 (Cyclohexane).

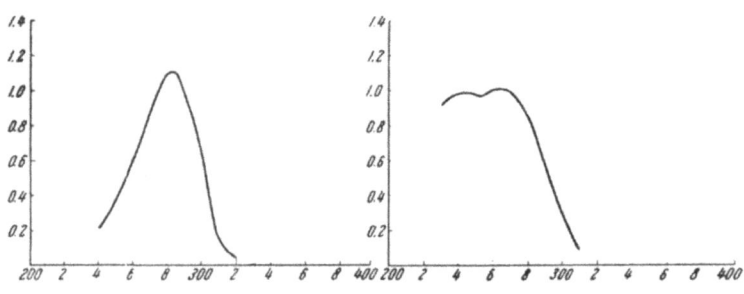

Isomer of Ethyl β-Ionylideneacetate (IV)
E (1%, 1 cm.) (284 mμ) = 1190
(Cyclohexane).

β-Ionylidene Ethanol; E (1%, 1 cm.)
(264 mμ) = 588; E (1%, 1 cm.) (242 mμ) =
= 570 (Ethanol).

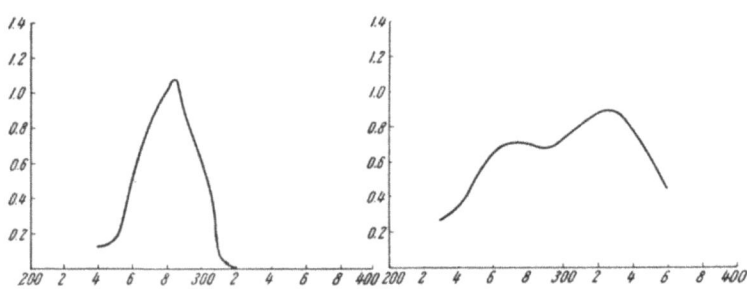

Isomer of β-Ionylidene Ethanol (VI)
E (1%, 1 cm.) (284 mμ) = 1450 (Ethanol).

β-Ionylideneacetaldehyde; E (1%, 1 cm.)
(326 mμ) = 676; E (1%, 1 cm.) (272 mμ) =
= 540 (Ethanol).

Fig. 1. Ultraviolet absorption spectra for intermediates in vitamin A synthesis (Dist. Prod. Ind. (23)).

Synthesis of Vitamin A and Some Related Compounds. 45

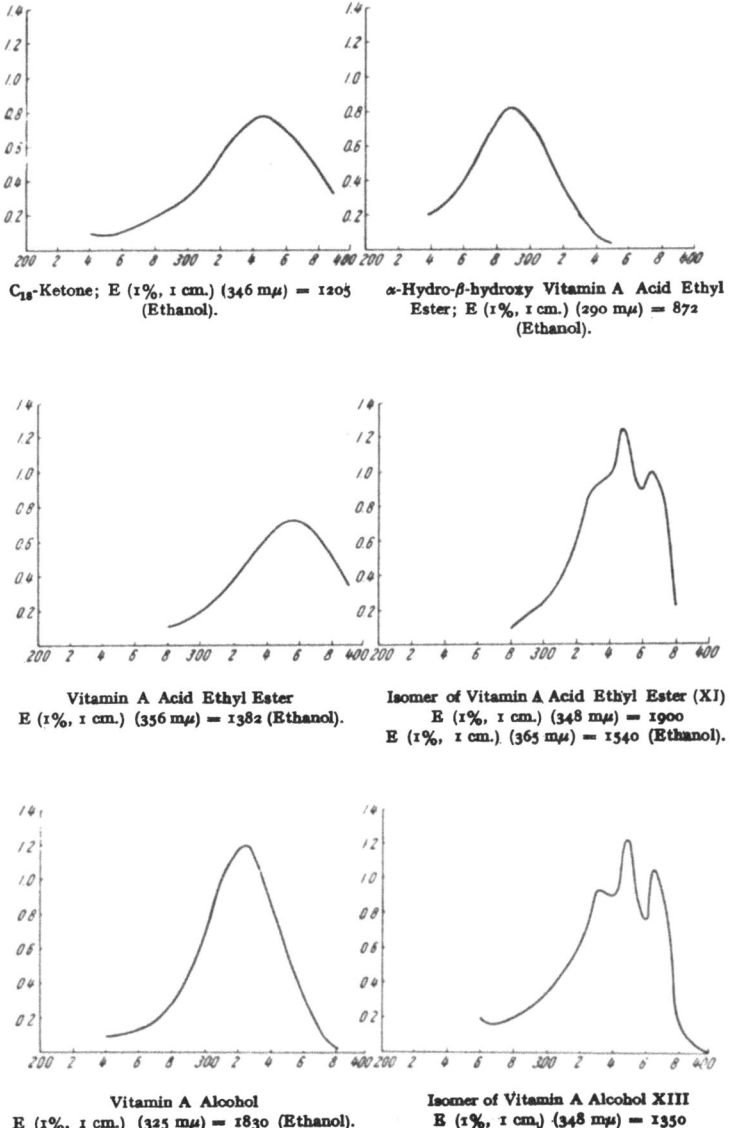

Fig. 1. Ultraviolet absorption spectra for intermediates in vitamin A synthesis (Dist. Prod. Ind. (23)).

ethyl bromoacetate and zinc in benzene solution, by the Reformatsky technique, to give the C_{15}-hydroxy ester (II, ethylester of β-ionolacetic acid). By treatment with iodine or acid-type reagents, e. g., phosphorus oxychloride, dehydration was effected to yield ethyl β-ionylideneacetate (III). This, by reduction with lithium aluminumhydride, gave β-ionylideneethanol (V), with little or no concomitant attack on the conjugated polyene system. Oxidation of the α,β-unsaturated alcohol and condensation with acetone was achieved by treatment with an acetone-benzene mixture in the presence of aluminum isopropoxide, according to the Oppenauer method, to give the C_{18}-ketone (VIII). Repetition of the initial steps of the synthesis followed. The C_{18}-ketone was reacted with ethyl bromoacetate and zinc. The intermediate C_{20}-hydroxy ester (IX, α-hydro-β-hydroxy-vitamin A acid ethyl ester) was dehydrated with an acid-type reagent to give vitamin A acid ethyl ester (X). This, by reduction with lithium aluminum hydride, yielded vitamin A.

The ultraviolet absorption spectra for the intermediates are given in Fig. 1 and the extinction values in Table 1.

The product of the synthesis had a potency of 2–2.5 million u./g. (60–75% purity). From this the free vitamin and a number of its esters were crystallized [CAWLEY and co-workers (*18*)]. It was further shown by reaction with maleic anhydride, according to the procedure of ROBESON and BAXTER (*129*), that the synthetic concentrate contained vitamin A and the *cis* isomer neovitamin A in the approximate ratio, 65 : 35. Thus the synthetic concentrate contained the two isomers previously found in fish liver oils in about the same proportions.

Certain features of the chemistry of the intermediates are of interest.

Ethyl β-ionylideneacetate and its β, γ-unsaturated isomer. Dehydration of the C_{15}-hydroxy ester (II) by acidic reagents gives a product which was assigned the formula of ethyl β-ionylideneacetate [KARRER and co-workers (*90*)], but the reaction had to be studied extensively [SOBOTKA, BLOCH and GLICK (*141*); YOUNG, ANDREWS and CRISTOL (*154*)] because it was recognized that it was complex in nature. The anomalously low wave length position of the ultraviolet absorption maximum of the reaction product (at 284 mμ) was observed and attributed, in part, to the presence of geometrical isomers.

Studies in these laboratories begun by SHANTZ, CAWLEY and CLOUGH (*137*) and continued by SHANTZ, ROBESON and KASCHER (*139*) showed that a mixture of esters results from the dehydration reaction. Besides ethyl β-ionylideneacetate, an isomeric ester absorbing at 284 mμ is formed. If the time of contact with the acidic reagent is short, this constitutes 90% or more of the reaction product, thus explaining the maximum at 284 mμ previously referred to.

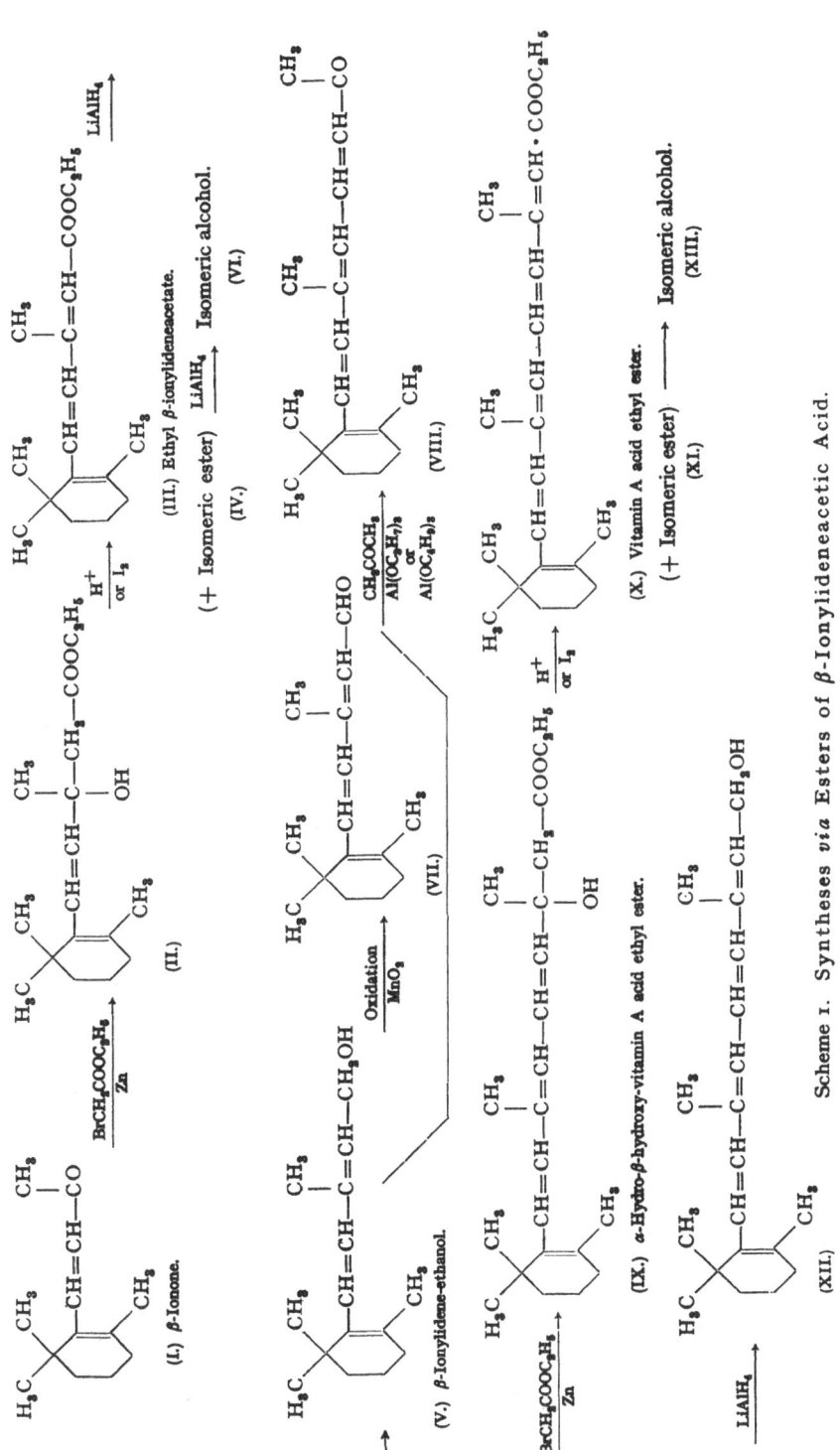

Scheme 1. Syntheses *via* Esters of β-Ionylideneacetic Acid.

Table 1. Ultraviolet Absorption Values for Intermediates in Syntheses *via* Esters of β-Ionylideneacetic Acid.

Formula in Scheme 1 (p. 47)	Compound	M. p.	λ_{max} and E (1%, 1 cm.)	Solvent	Reference
(II)	Ethyl ester of β-ionolacetic acid	—	E (234 mμ) = 214	Ethanol	Dist. Prod. i. (23)
(III)	Ethyl β-ionylideneacetate	—	E (256, 304 mμ) = 434, 548	Cyclohexane	Dist. Prod. i. (23)
	β-Ionylideneacetic acid	124°	E (260, 294 mμ) = 550, 585	Ethanol	Young (154)
		127°	E (255, 300 mμ) = 575, 660	Ethanol	Wendler (153)
		126°	E (258, 304 mμ) = 518, 650	Cyclohexane	Dist. Prod. i. (23)
(IV)	Isomer of ethyl β-ionylideneacetate	—	E (284 mμ) = 1190	Cyclohexane	Dist. Prod. i. (23)
(V)	β-Ionylidene ethanol	—	E (242, 264 mμ) = 570, 588	Ethanol	Dist. Prod. i. (23)
(VI)	Isomer of β-ionylidene ethanol	—	E (284 mμ) = 1450	Ethanol	Dist. Prod. i. (23)
(VII)	β-Ionylidene-acetaldehyde	—	E (267, 310 mμ) = 560, 670	Cyclohexane	Huisman (58)
			E (272, 326 mμ) = 540, 676	Ethanol	Dist. Prod. i. (23)
			E (320 mμ) = 710	?	Harper (43)
	β-Ionylidene-acetaldehyde, *trans*-isomer	—	E (265, 315 mμ) = 567, 760)	Isooctane	Wendler (153)
	β-Ionylidene-acetaldehyde, *trans*-isomer, semicarbazone	196°	E (323 mμ) = 1330	Chloroform	Wendler (153)
	β-Ionylidene-acetaldehyde, *cis*-isomer	193°	E (320 mμ) = 1302	Ethanol	Huisman (58)
	β-Ionylidene-acetaldehyde, *cis*-isomer, semicarbazone	—	E (318 mμ) = 904	Isooctane	Wendler (153)
(VIII)	C_{19}-Ketone	176°	E (317.5 mμ) = 1000	Chloroform	Wendler (153)
(IX)	α-Hydro-β-hydroxy-vitamin A acid ethyl ester	—	See Table 2, p. 54		
			E (292.5 mμ) = 587	Ethanol	Wendler (153)
			E (290 mμ) = 872	Ethanol	Dist. Prod. i. (23)
			E (291 mμ) = 810	Hexane	Schwarzkopf (133)

(X)	Vitamin A acid ethyl ester (from crystalline acid)	—	E (356 mμ) = 1382	Ethanol	Dist. Prod. i. (*23*)
	Same, methyl ester	—	E (354 mμ) = 1385	Hexane	SCHWARZKOPF (*133*)
(XI)	Isomer of vitamin A acid ethyl ester	—	E (348, 365 mμ) = 1900, 1540	Ethanol	Dist. Prod. i. (*23*)
(XII)	Vitamin A	64°	E (325 mμ) = 1830	Ethanol	Dist. Prod. i. (*23*)
	Vitamin A acetate	58°	E (328 mμ) = 1560	Ethanol	Dist. Prod. i. (*23*)
	Vitamin A anthraquinone-β-carboxylate	122°	E (330 mμ) = 1065	Ethanol	ROBESON (*129*)
	Vitamin A *p*-phenylazobenzoate	80°	E (330 mμ) = 1540	Ethanol	ROBESON (*129*) Dist. Prod. i. (*23*)
	Neovitamin A	60°	E (328 mμ) = 1690	Ethanol	ROBESON (*129*) Dist. Prod. i. (*23*)
	Neovitamin A anthraquinone-β-carboxylate	136°	E (333 mμ) = 1020	Ethanol	ROBESON (*129*)
	Neovitamin A *p*-phenylazobenzoate	96°	E (330 mμ) = 1460	Ethanol	ROBESON (*129*)
(XIII)	Isomer of vitamin A	—	E (348, 365 mμ) = 1350, 1160 (conc.)	Ethanol	Dist. Prod. i. (*23*)

When the time of contact with the acidic reagent is increased, this latter ester isomerizes, in part, to ethyl β-ionylideneacetate, the equilibrium mixture containing about 50% of each. The two isomers can be separated, e. g., by chromatography, solvent extraction or distillation, and the isomeric ester thus recovered can be isomerized to give a further yield of ethyl β-ionylideneacetate (SHANTZ, ROBESON and KASCHER (*139*)].

Ethyl β-ionylideneacetate has absorption maxima at 256 mμ and 304 mμ. The crystalline β-ionylideneacetic acid earlier prepared [KARRER et al. (*90*)] had nearly the same maxima and similar extinction coefficients (on an equivalent basis). Thus the absorption spectra of the products obtained by dehydration of the C_{15}-hydroxyester (II) are explained.

Preliminary studies on the isomeric ester (formula IV), carried out by SHANTZ, ROBESON and KASCHER (*139*), indicated that it had a β,γ-unsaturated structure. Later work by CAWLEY and SEIDEL (*23*) supported the formula (IV).

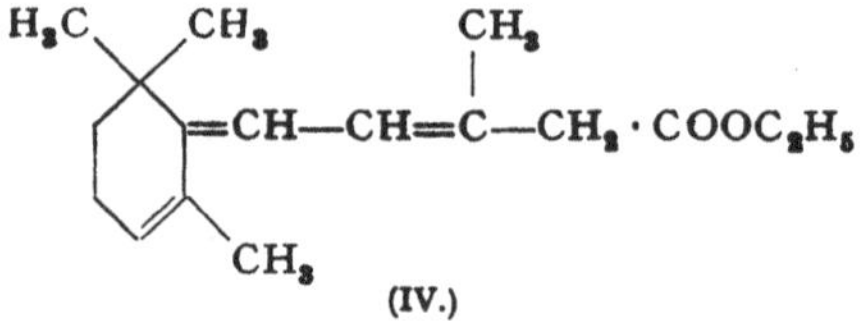

$$\underset{(IV.)}{\underset{CH_3}{\underset{|}{\bigcirc}}\!\!=\!CH\!-\!CH\!=\!\underset{\underset{CH_3}{|}}{C}\!-\!CH_2\cdot COOC_2H_5}$$
(with gem-dimethyl H_3C, CH_3 on the ring)

β-Ionylidene-ethanol. Attempts in this laboratory to synthesize β-ionylidene-ethanol by the method of GOULD and THOMPSON (*37*) were largely unsuccessful. As prepared by reduction of ethyl β-ionylideneacetate, the compound shows two absorption maxima in ethanol, 242 mμ and 264 mμ. The maximum is often reported at 284 mμ because of the presence of the isomeric alcohol (VI, p. 48) which has a strong absorption band at this wave length.

C_{18}-*Ketone.* The spectral properties of this compound are indicated on p. 45.

Vitamin A acid ethyl ester and its β,γ-unsaturated isomer. Vitamin A acid ethyl ester was prepared by dehydration of the C_{20}-hydroxy ester (IX, p. 47) with iodine or with acidic reagents such as were employed in the dehydration of the ester (II). Dehydration under mild conditions produces little vitamin A acid ethyl ester [SCHWARZKOPF and co-workers (*132*); SHANTZ, ROBESON and KASCHER (*139*)] but mainly an isomeric β,γ-unsaturated ester. By extended treatment, this latter compound isomerizes, in part, to vitamin A acid ethylester, the equilibrium mixture containing approximately 50% of each. The two esters can be separated by solvent extraction, adsorption or distillation, and the β,γ-compound can be isomerized and separated again to give a further yield of the vitamin A acid ester. This ester (X) has an absorption maximum at 354–356 mµ.

The isomeric ester (XI) is characterized by two peaks, at 348 mµ and 365 mµ, and has had the following formula suggested for it by CAWLEY and SEIDEL (23) [cf. also ROBESON (128a)].

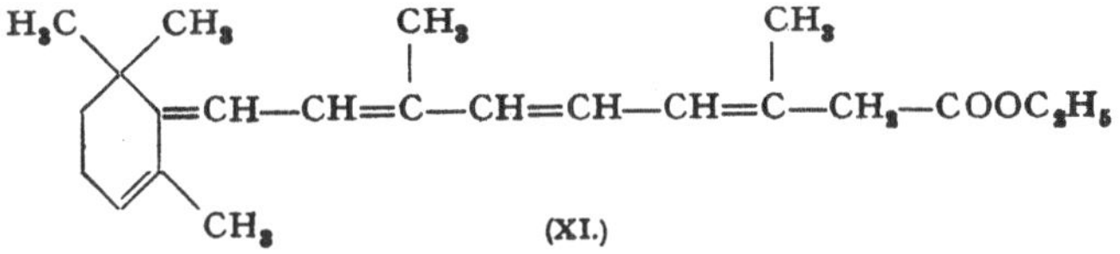

$$\text{(ring)}=CH-CH=C(CH_3)-CH=CH-CH=C(CH_3)-CH_2-COOC_2H_5$$

(XI.)

Vitamin A. The product obtained by the reduction of vitamin A acid ethylester with lithium aluminumhydride contains two geometrical isomers, vitamin A and neovitamin A. They may be further identified, respectively, as Δ^2-*trans*-Δ^6-*trans* and Δ^2-*cis*-Δ^6-*trans* vitamin A according to the following numbering system:

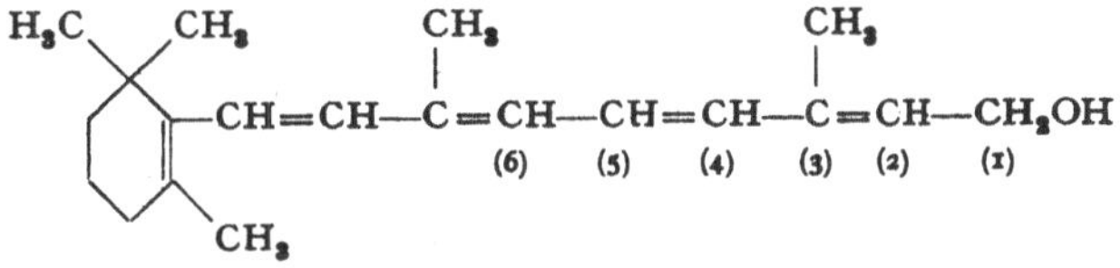

$$\text{(ring)}-CH=CH-C(CH_3)=CH-CH=CH-C(CH_3)=CH-CH_2OH$$
$$(6)(5)(4)(3)(2)(1)$$

Cis-trans isomerism in the vitamin A molecule has been previously discussed by ROBESON and BAXTER (129), based on the work of PAULING and ZECHMEISTER (cf. 155).

Isomer of vitamin A. A C_{20}-alcohol (formula XIII), isomeric with vitamin A, was prepared by reduction of the ester (XI) with lithium aluminumhydride. It was characterized by absorption maxima in ethanol at 348 mµ and 365 mµ and has had the following possible structure suggested for it by CAWLEY and SEIDEL (23),

$$\text{(ring)}=CH-CH=C(CH_3)-CH=CH-CH=C(CH_3)-CH_2-CH_2OH$$

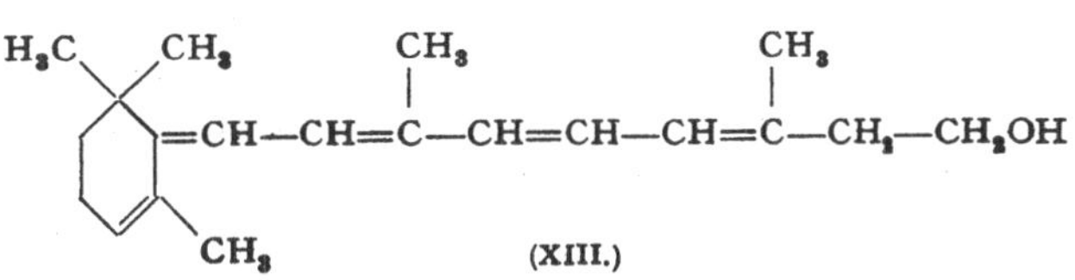

(XIII.)

Synthesis by WENDLER, SLATES, TRENNER *and* TISHLER (152, 153). This group has reported a synthesis of vitamin A which differs, in certain respects, from the scheme described. Ethyl β-ionylideneacetate was prepared in the usual manner, although the ultraviolet absorption maximum reported (λ_{max} at 285 mµ, in ethanol) suggests that the isomeric ester was not separated. β-Ionylidene-ethanol was then obtained by reduction with lithium aluminumhydride and oxidized by manganese dioxide according to the procedure earlier developed for vitamin A by BALL, GOODWIN and MORTON (7).

The β-ionylideneacetaldehyde thus produced (VII, p. 47) was resolved by chromatography into *trans*- and *cis*-isomers. These were each condensed with acetone in the presence of aluminum isobutoxide to give the corresponding C_{18}-ketones (VIII). By the Reformatsky reaction with ethyl bromoacetate, the C_{20}-hydroxy esters (IX) were made and dehydrated with iodine. After saponification, the crystalline acids were prepared which appeared to be identical with *trans*-vitamin A acid (Table 2, p. 54). The acid was reduced with lithium aluminumhydride to vitamin A, the product having E (1%, 1 cm.) (326 mμ) = 1340 (in isooctane).

β-Ionylideneacetaldehyde. The usefulness of β-ionylideneacetaldehyde in the synthesis of vitamin A has long been recognized [(DAVIES, HEILBRON, JONES and LOWE (*22a*)] and a method of preparation was reported by KUHN and MORRIS (*94*). Others could not repeat this synthesis [cf. MILAS (*112*)].

The preparation has since been reported by a number of authors. HARPER and OUGHTON (*43*), like WENDLER and co-workers, oxidized β-ionylidene-ethanol with manganese dioxide by the method of BALL, GOODWIN and MORTON (7). ROBESON and EDDINGER (*130*) oxidized β-ionylidene-ethanol with acetone and aluminum isopropoxide in the presence of such amines as aniline. The aldehyde so formed was protected as an anil and prevented from condensation with acetone to give C_{18}-ketone. The reaction of β-ionone with ethoxyacetylene magnesium bromide, $C_2H_5OC{\equiv}CMgBr$, was the basis of the procedure of ARENS et al. (*5*), HEILBRON, JONES and WEEDON (*48*) and of PREOBRAZHENSKII and RUBSTOV (*128*). VAN DORP and ARENS (*144*) converted β-ionone to the anil of β-ionylidenepyruvic acid by a reaction involving the pyridine salt of hydroxymaleic anhydride. This, upon heating, gave β-ionylideneacetaldehyde, identified by its semicarbazone.

β-Ionylideneacetaldehyde has a twin-peaked absorption spectrum with maxima at 265—272 mμ and 310—326 mμ, depending on the solvent (Table 1, Fig. 1). The *trans* isomer absorbs at a lower wave length (315 mμ) than the *cis* isomer (318 mμ) in isooctane [WENDLER and co-workers (*153*)]. The latter compound was further identified by conversion to the corresponding C_{18}-ketone, earlier prepared by ARENS and VAN DORP (*2*).

2. Synthesis of Vitamin A *via* Esters of β-Ionylidenecrotonic Acid.

The synthesis of vitamin A *via* esters of β-ionylidenecrotonic acid [ARENS and VAN DORP (*1, 3*); VAN DORP and ARENS (*143*)] is related to those already described, in that the initial step was a Reformatsky reaction with the vinylog of ethyl β-ionylideneacetate. A feature of the synthesis is that in an earlier form it lead to the preparation of vitamin A acid.

The synthesis was accomplished according to Scheme 2 (p. 55). β-Ionone was condensed with methyl-γ-bromocrotonate in the presence of zinc to give the C_{17}-hydroxy ester (XV). By treatment with oxalic acid, this was dehydrated. After saponification, β-ionylidenecrotonic acid (XVI) was crystallized which, with methyl-lithium, followed by hydrolysis, gave the C_{18}-ketone (XVII). By reaction with ethoxyacetylene-magnesiumbromide, a hydroxy-acetylenic ether was prepared which was selectively reduced with hydrogen in the presence of a palladium-barium sulfate to give the ethoxyvinyl-carbinol (XVIII). This was rearranged with oxalic acid to vitamin A aldehyde (XIX). By reduction with lithium aluminumhydride or by the Meerwein-Ponndorf method, a concentrate (XX) was obtained containing 50% vitamin A. From this, the crystalline β-naphthoate and anthraquinone-β-carboxylate esters were prepared (Table 2).

In earlier reports [ARENS and VAN DORP (*1*); VAN DORP and ARENS (*143*)], the C_{18}-ketone was converted to the C_{20}-hydroxy ester (XXI) (α-hydro-β-hydroxy-vitamin A acid methyl ester) by a Reformatsky reaction with methyl bromoacetate and zinc. This ester was dehydrated, saponified and the acid crystallized to give the vitamin A acid (XXII).

β-Ionylidenecrotonic acid. This compound was early recognized as being of importance in vitamin A synthesis. A German patent application by ZIEGLER, on the methyl ester, is referred to by VAN DORP and ARENS (*142a*), and preparations by HEILBRON, JONES and O'SULLIVAN (*50*) as well as CAWLEY (*17*) have been described. It is of interest that the compound was isolated by the Dutch workers not only in a crystalline *trans* but also in a *cis* form of lower melting point [ARENS and VAN DORP (*2*); Table 2]. For a third acid, of m. p. 162.5°, the formula (XIIIa) was suggested by ARENS and VAN DORP (*1*). This is of interest in view of the previous discussion on similar isomers of ethyl β-ionylideneacetate.

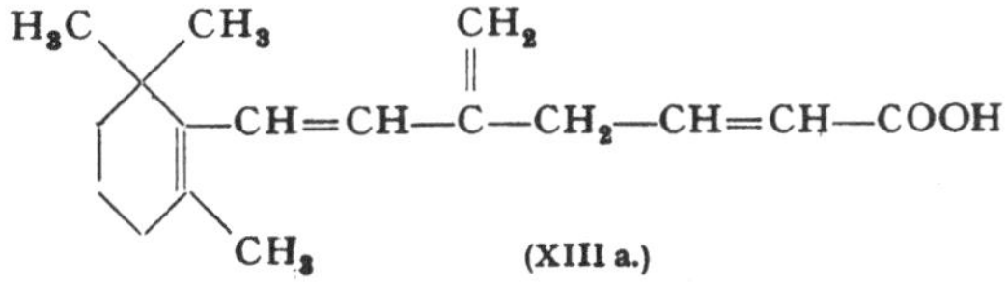

$$\text{(XIIIa)}$$

H₃C CH₃   CH₂
   \ /   ‖
    X—CH=CH—C—CH₂—CH=CH—COOH
   / \
      CH₃   (XIIIa.)

C_{18}-Ketone. This compound has also been extensively investigated [HEILBRON et al. (*53, 50*); CAWLEY et al. (*15, 16*)]. The position of the ultraviolet absorption maximum has varied, however, in different reports, as shown in Table 2 (next page).

Table 2. Ultraviolet Extinction Values for Intermediates in Syntheses via Esters of β-Ionylidenecrotonic Acid.

Formula in Scheme 2 (p. 55)	Compound	M. p.	λ_{max} and E (1%, 1 cm.)	Solvent	Reference
(XVI)	β-Ionylidenecrotonic acid, trans	157.5°	E (323 mμ) = 1400	Ether	VAN DORP (142a)
	β-Ionylidenecrotonic acid, trans	160°	E (324 mμ) = 1290	Ethanol	INHOFFEN (62)
	β-Ionylidenecrotonic acid, cis	143°	E (323 mμ) = 910	Ethanol	ARENS (2)
(XIIIa)	Isomer of β-ionylidenecrotonic acid	162.5°	—	—	INHOFFEN (62)
(XVII)	C_{18}-Ketone	—	E (335 mμ) = 1470	Ethanol	ARENS (1)
		—	E (336 mμ) = 839	Ethanol	HEILBRON (50)
		189.6°	E (346 mμ) = 1205	Ethanol	WENDLER (153)
	C_{18}-Ketone, trans, semicarbazone		E (349 mμ) = 1900	Chloroform	Dist. Prod. i. (23)
	C_{18}-Ketone, cis	143°	E (334—337 mμ) = 1100	Isooctane	VAN DORP (143)
	C_{18}-Ketone, cis, semicarbazone	64°	E (340 mμ) = 1500	Ethanol	WENDLER (153)
(XIX)	Vitamin A aldehyde, all-trans	62°	E (375 mμ) = 1090	Ethanol	WENDLER (153)
		—	E (385.5 mμ) = 1400	Petroleum ether	ARENS (3)
		—	E (369.5 mμ) = 1685	Cyclohexane	BALL (7)
		—	E (373 mμ) = 1548	Chloroform	BALL (7)
		57°	E (389 mμ) = 1303	Ethanol	BALL (7)
		195°;	E (381 mμ) = 1530	Chloroform	BALL (7)
	Vitamin A aldehyde, trans, semicarbazone	164°	E (385 mμ) = 2062; 1742	Chloroform	Dist. Prod. i. (23)
		190°;	E (385 mμ) = 1540; 1860	Chloroform	BALL (7)
		201°			
(XIX)	Vitamin A aldehyde, Δ²-trans-Δ⁶-cis isomer	209°	E (375 mμ) = 2180	Chloroform	WENDLER (151)
	Vitamin A aldehyde, semicarbazone	60°	E (372 mμ) = 930	Ethanol	VAN DORP (143)
(XXII)	Vitamin A acid, all-trans	195°	E (371 mμ) = 1610	Chloroform	GRAHAM (38)
		181.5°	E (353 mμ) = 1510	Ethanol	GRAHAM (38)
		181.5°	E (347 mμ) = 1460	Ethanol	WENDLER (153)
		180°	E (350 mμ) = 1507	Ethanol	VAN DORP (142a)
(XX)	Vitamin A acid, Δ²-cis-Δ⁶-trans isomer	146°	E (347 mμ) = 1600	?	Dist. Prod. i. (23)
	Vitamin A β-naphthoate	76.5°	—	—	INHOFFEN (64)
	Vitamin A anthraquinone-β-carboxylate	120°	—	—	ARENS (3)
	Vitamin A, Δ²-trans-Δ⁶-cis isomer	85.3°	—	—	ARENS (3)
					GRAHAM (38)

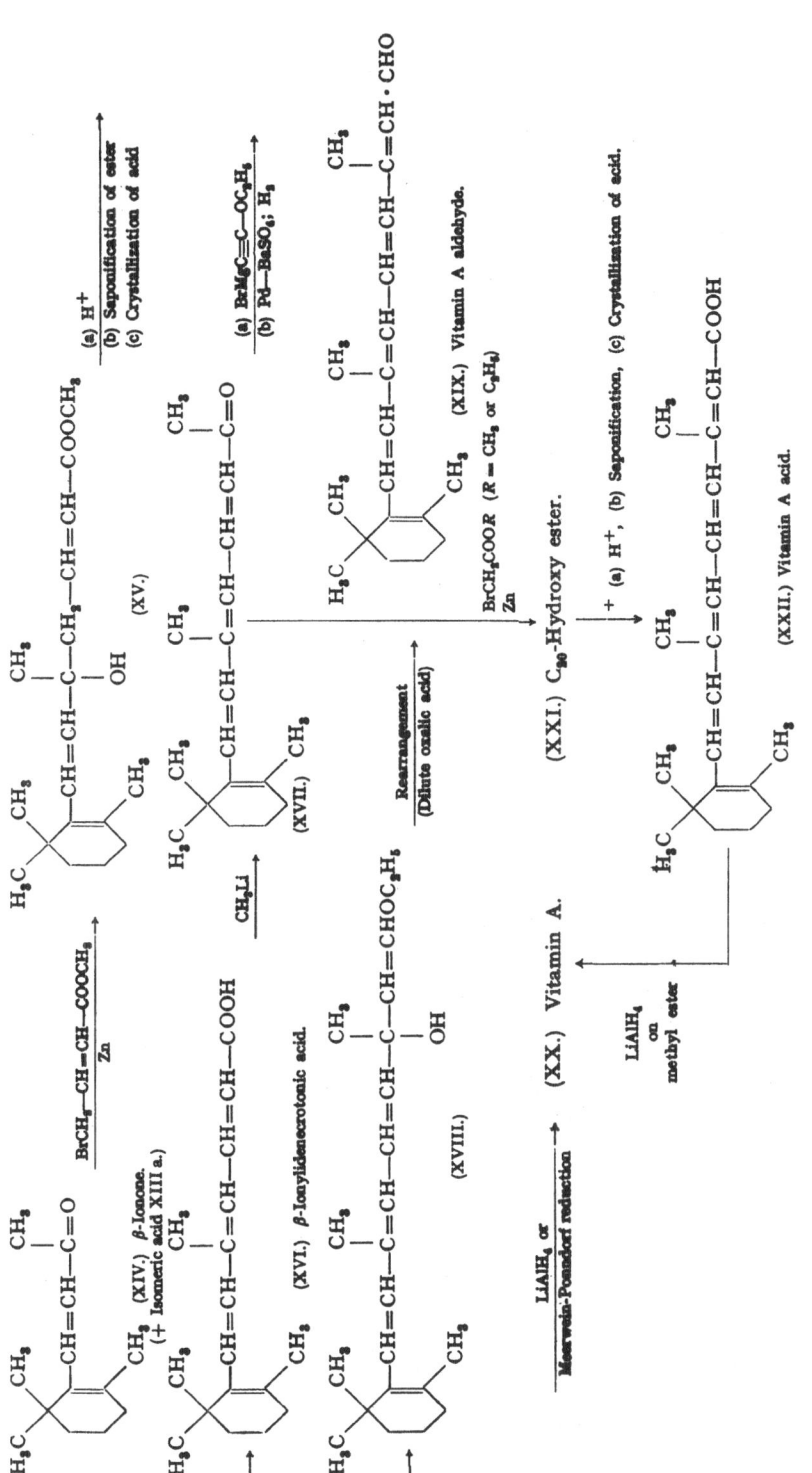

Scheme 2. Syntheses via Esters of β-Ionylidenecrotonic Acid.

ARENS and VAN DORP (2) described the preparation of the crystalline *trans* ketone but gave absorption data only for the semicarbazone ($\lambda_{max} = 349$ mμ, in CHCl$_3$).

As prepared from ethyl β-ionylideneacetate [CAWLEY and co-workers (15)], the compound had in partially purified form E (1%, 1 cm.) (343 mμ) = 660 (C$_2$H$_5$OH) and formed a semicarbazone having m. p. 189° and E (1%, 1 cm.) (344 mμ) = 1630 (C$_2$H$_5$OH). Further purification of the ketone changed the absorption maximum to 346 mμ [Dist. Prod. i. (23)]. WENDLER and co-workers (153) as well as HEILBRON, JONES and O'SULLIVAN (50), however, reported the maximum in ethanol to be at 335—336 mμ. An explanation of this discrepancy has not yet been available.

The *cis* C$_{18}$-ketone has been prepared by ARENS and VAN DORP (2) from *cis*-β-ionylidenecrotonic acid and crystallized in the form of the semicarbazone (Table 2). What appears to be the same compound was obtained by WENDLER and co-workers (153) from *cis*-β-ionylideneacetaldehyde by condensation with acetone.

Vitamin A aldehyde. This compound has been studied as an intermediate in vitamin A synthesis and also because of its identity as retinene$_1$, a polyene liberated by the bleaching of rhodopsin by light. Rhodopsin is the conjugated protein pigment in the retina which is concerned with vision in dim light [WALD (146—149); GLOVER, GOODWIN and MORTON (34); BALL and MORTON (8)].

Synthetic vitamin A aldehyde was crystallized (m. p. 64°) by ARENS and VAN DORP (3) and appeared to be identical with all-*trans*-vitamin A aldehyde prepared from natural vitamin A by the manganese dioxide procedure [BALL, GOODWIN and MORTON (7)], although the ultraviolet absorption maximum was reported to be at a lower wave length and the extinction coefficient was lower (Table 2, p. 54).

From *cis* C$_{18}$-ketone, by the method indicated in Scheme 2 (p. 55), GRAHAM, VAN DORP and ARENS (38) prepared Δ^2-*trans*-Δ^6-*cis* vitamin A aldehyde. This had a lower melting point than its all-*trans* isomer.

Other syntheses of vitamin A aldehyde by the oxidation of vitamin A with potassium permanganate [MEUNIER and JOUANNETEAU (103)] and by the oxidation of β-carotene have been reported [MEUNIER, JOUANNETEAU and ZWINGELSTEIN (106)].

Vitamin A Acids. The vitamin A acid (XXII, p. 55) prepared by ARENS and VAN DORP (1) was the first close relative of vitamin A obtained in crystalline form, by total synthesis. Its preparation has been studied by HEILBRON et al. (53, 50); KARRER, JUCKER and SCHICK (88); MILAS (113); SCHWARZKOPF et al. (133); INHOFFEN, BOHLMANN and BOHLMANN (64); and by WENDLER et al. (153). This all-*trans* acid appears to be well characterized as far as its extinction coefficient is concerned, although the positions reported for the maximum differ slightly (Table 2, p. 54).

An isomeric C_{20}-acid, m. p. 146°, has been isolated by INHOFFEN, BOHLMANN and BOHLMANN (*64*) from the reaction product of *trans* C_{18}-ketone with ethyl bromoacetate, followed by dehydration and saponification. The compound was identified as the Δ^2-*cis*-Δ^6-*trans* acid. It is of special interest because it corresponds in structure to neovitamin A.

Vitamin A. The characteristics of the vitamin A prepared according to Scheme 2 (p. 55) and purified through two crystalline derivatives have already been discussed.

Of interest is the isomeric Δ^2-*trans*-Δ^6-*cis* form, m. p. 85.3°, obtained by reducing the corresponding aldehyde [GRAHAM, VAN DORP and ARENS (*38*)]. The spectrum and other properties of this compound have not been reported as yet.

Synthesis by SCHWARZKOPF, CAHNMANN, LEWIS, SWIDINSKY *and* WÜST (*133*). This group synthesized vitamin A, starting from the C_{20}-hydroxy methyl ester (Scheme 2, XXI), prepared by a modification of the ARENS and VAN DORP procedure which gave a higher yield. The hydroxy-ester was dehydrated with *p*-toluene-sulfonic acid and the product purified chromatographically. It was then reduced to vitamin A with lithium aluminumhydride to give a concentrate of about 65% purity.

In an alternative procedure the dehydrated ester was saponified and the vitamin A acid crystallized. The latter was then esterified with diazomethane and the pure ester reduced with lithium aluminumhydride to yield vitamin A of 95% purity [E (1%, 1 cm.) (325 mμ) = 1646]. This was converted into the crystalline acetate which was identical with all-*trans*-vitamin A acetate.

In a later publication [SCHWARZKOPF and co-workers (*132*)], an ester isomeric with vitamin A acid ethyl ester was prepared from the C_{20}-hydroxy ester by a short treatment with acids or iodine. The new ester was called vitamin A_2 acid ester and was assigned the formula (XXIII). By reduction

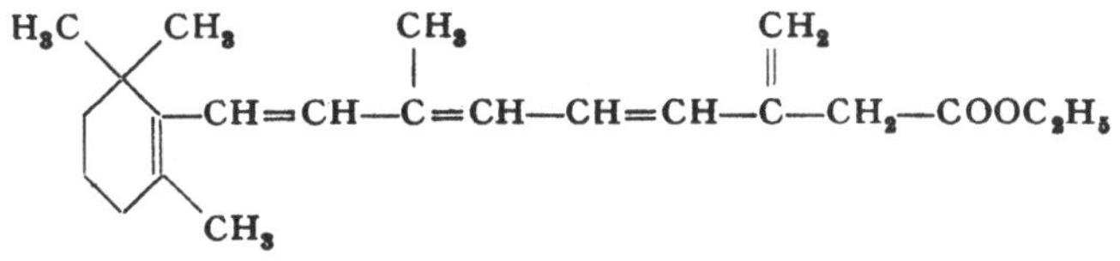

$$H_3C\diagdown\diagup CH_3 \qquad\qquad CH_3 \qquad\qquad\qquad CH_2$$
$$-CH=CH-C=CH-CH=CH-C-CH_2-COOC_2H_5$$
$$CH_3$$

(XXIII.) Vitamin A_2 acid ester.

$$H_3C\diagdown\diagup CH_3 \qquad\qquad CH_3 \qquad\qquad\qquad CH_2$$
$$-CH=CH-C=CH-CH=CH-C-CH_2-CH_2OH$$
$$CH_3$$

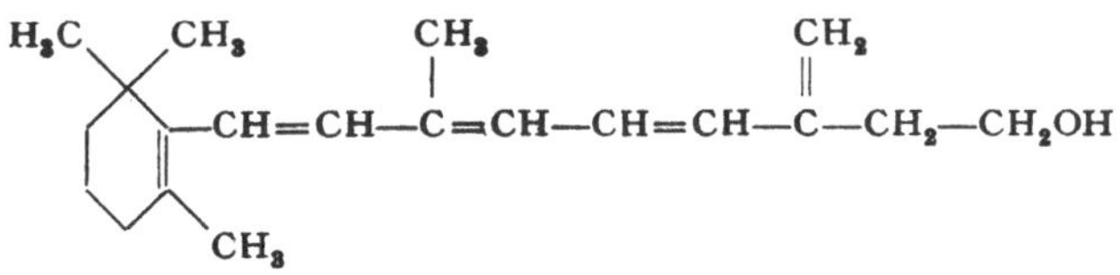

(XXIII a.) Vitamin A_2.

with lithium aluminumhydride a C_{20}-alcohol resulted which was isomeric with vitamin A and was termed vitamin A_2 (XXIIIa).

From the method of preparation, vitamin A_2 acid ester and vitamin A_2 appear to be similar, if not identical, to the "isomer of vitamin A acid ethyl ester (XI)" and "isomer of vitamin A (XIII)" in Scheme 1 (p. 47). Preliminary assays on (XIII), however, indicated that its biological potency was substantially lower than the value of 1000000 u./g. reported for vitamin A_2. This discrepancy requires further study.

3. Synthesis of Vitamin A *via* "C_{14}-Aldehyde".

The usefulness of the C_{14}-aldehyde (XXV, Scheme 3, p. 59) in the synthesis of vitamin A and its derivatives was recognized by MILAS and co-workers in the U. S. A., HEILBRON and co-workers in England, and ISLER and co-workers in Switzerland. The MILAS group pioneered in the use of this and other acetylenic compounds in the synthesis of vitamin A ethers and esters (*110, 111, 119*), homovitamin A ethers (*118*), vitamin A acid (*113*), and other derivatives such as dimethylamino vitamin A (*114*). A number of patents describe this work (*117*).

HEILBRON et al. contributed an early disclosure on the projected use of C_{14}-aldehyde in vitamin A synthesis (*47*) and a series of papers on allylic rearrangements, beginning in 1943. ISLER and his colleagues developed a method for the preparation of C_{14}-aldehyde in improved yield, and synthesized vitamin A as well as a number of ethers and esters (*74–76*). The synthetic plan was in principle similar to that employed by MILAS but the vitamin A products were higher in biological potency. Further discussions of the contributions of each group of workers have been published by HEILBRON (*45*); MILAS et al. (*117*); and by ISLER (*72*).

The synthesis of vitamin A reported by ISLER and co-workers (*74*) gave rise to crystallizable products and is shown in Scheme 3 (p. 59). β-Ionone (XXIV) was reacted with ethyl chloroacetate and sodium ethylate, according to the Darzen procedure, to produce an intermediate glycide ester. This was decomposed with alkali, at low temperatures, to give C_{14}-aldehyde (XXV) in a yield of approximately 80%. Earlier procedures for this synthesis, following the work of ISHIKAWA and MATSUURA (*68*), had given yields of only 10–20% [HEILBRON (*45*)].

The C_{14}-aldehyde was next condensed with 1-hydroxy-3-methyl-2-pentene-4-yne (XXVI) in the normal Grignard manner to give the acetylenic carbinol (XXVII). Compound (XXVI) was prepared by the condensation of acetylene and methylvinyl-ketone with sodium in liquid ammonia, followed by rearrangement of the resulting carbinol (cf. formula) with H_2SO_4. By selective hydrogenation of the triple bond in (XXVII) with a poisoned palla-

$$HC \equiv C - \underset{\underset{OH}{|}}{\overset{\overset{CH_3}{|}}{C}} - CH = CH_2$$

Synthesis of Vitamin A and Some Related Compounds.

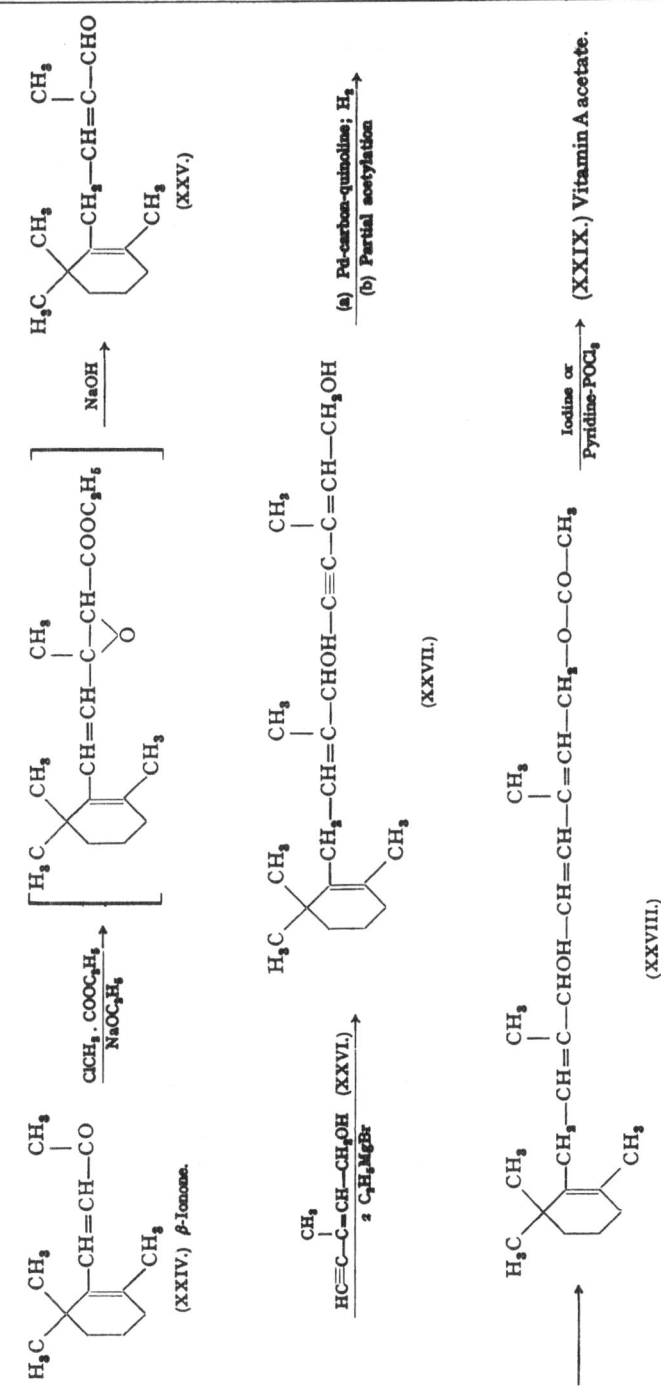

Scheme 3. Synthesis *via* C_{16}-Aldehyde.

dium catalyst, followed by acetylation, the polyene glycol acetate (XXVIII) was obtained. By treatment with iodine in petroleum ether solution, allylic rearrangement accompanied by dehydration occurred, giving a concentrate of vitamin A acetate. This was converted to the crystalline acetate, to crystalline vitamin A, and a series of esters (Table 3, next page).

In similar fashion, the methyl ether of (XXVI, p. 59) was used to prepare crystalline vitamin A methyl ether. A phenyl- and tetrahydropyranyl-ether were also prepared [ISLER (72)].

C_{14}-*Aldehyde.* The structure of the C_{14}-aldehyde obtained from β-ionone by DARZEN's procedure has been disputed. ISHIKAWA and MATSUURA (68) considered the compound to be the β,γ-unsaturated aldehyde (XXIXa). MILAS and co-workers (117) supported this formula

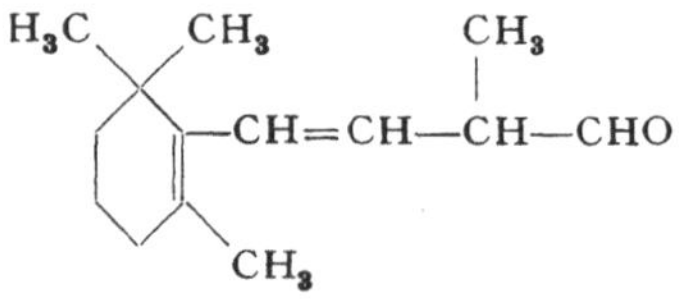

(XXIX a.) 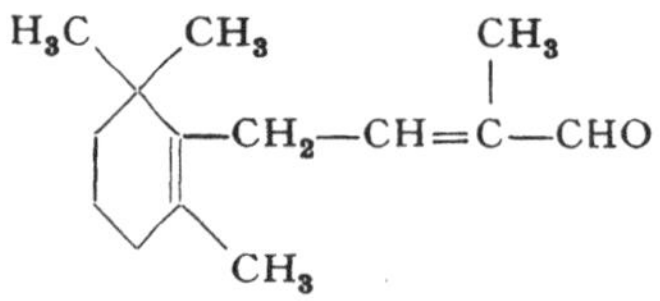 (XXX.)

on the basis of structural determinations involving, in part, ozonolysis. HEILBRON and co-workers (47) assigned formula (XXX) to the compound, principally on the basis of its spectrum and that of certain of its derivatives. In a later publication [CHEESEMAN et al. (20)], the homogeneity of the C_{14}-aldehyde preparation used as the basis of the MILAS studies was questioned and further results supporting (XXX) were reported. The C_{14}-aldehyde used was synthesized by the ISLER method. ISLER and co-workers (74, 72) adopted formula (XXX) for the aldehyde on the basis of its Raman spectrum (absorption lines at 1527 and 1681 cm^{-1}) and because of the ultraviolet spectrum of the compound formed by condensation with acetylene.

A study by BLUM in this laboratory [Dist. Prod. Ind. (23)] on the infrared spectrum of the C_{14}-aldehyde prepared by saponification of the intermediate glycide ester at low temperatures suggested that the compound is an α,β-unsaturated aldehyde with the probable formula (XXX) (C=O band, 1687 cm^{-1}; C=C band, 1636 cm^{-1}).

Allylic Rearrangement. The allylic rearrangement and dehydration of compound (XXVIII) to form vitamin A acetate has been discussed at length by ISLER and associates (74, 76). In earlier work, iodine, in such solvents as toluene or petroleum ether, was used to effect the reaction. In a preferred method, phosphorus oxychloride and pyridine were used (Scheme 3, p. 59).

Table 3. Ultraviolet Extinction Values for Products of the Synthesis via C_{16}-Aldehyde.

Formula in Scheme 3 (p. 59)	Compound	M. p.	λ_{max} and E (1%, 1 cm.)	Solvent	Reference
(XXV)	C_{16}-Aldehyde	2°	E (230 mμ) = 850	Ethanol	Cheeseman (20) Isler (72)
(XXVI)	1-Hydroxy-3-methyl-2-pentene-4-yne	−14°	—	—	Isler (73)
(XXVII)	C_{20}-Acetylenic diol	59°	E (229 mμ) = 455	Ethanol	Isler (76)
(XXVIII)	Acetate of C_{20} diol	74°	—	—	Isler (76)
(XXIX)	Vitamin A acetate	60.4°	E (325–328 mμ) = 1525*	Ethanol	Isler (76, 73)
	Vitamin A	62.2°	E (328 mμ) = 1720*	Ethanol	Isler (74, 73)
	Vitamin A palmitate	29°	E (325–328 mμ) = 975	Ethanol	Isler (76)
	Vitamin A anthraquinone-β-carboxylate	123°	E (255, 326 mμ) = 1030, 1080	Ethanol	Isler (74)
	Vitamin A β-naphthoate	75°	E (325–328 mμ) = 1210	Ethanol	Isler (74)
	Vitamin A p-phenylazobenzoate	80°	E (325–328 mμ) = 1650	Ethanol	Isler (74)
	Vitamin A methylether	35°	E (327 mμ) = 1730	Ethanol	Isler (76)
	Vitamin A phenylether	92°	E (327 mμ) = 1470	Ethanol	Isler (76)
	Anhydrovitamin A	70°	E (352, 370, 391 mμ) = 1900, 2910, 2620	Ethanol	Isler (74)
	Isoanhydrovitamin A	—	E (330, 347, 367 mμ) = 1100, 1360, 1100	Ethanol	Isler (74)

* A recent paper by Boldingh and co-workers (12) gives absorption values for trans-vitamin A and its acetate supplied by Dr. Isler. Mean values in ethanol were, respectively, E (1%, 1 cm.) (324–325 mμ) = 1835; and E (1%, 1 cm.) (325–326 mμ) = 1561

Table 4. Key Intermediates for Vitamin A Syntheses by Other Methods.

Method	Starting Materials	Key Intermediate	Product, Vitamin A	Relative Biopotency (Vitamin A = 1)	Reference
(a)	C_{14}-Aldehyde, HC≡CH, methyl-vinyl-ketone	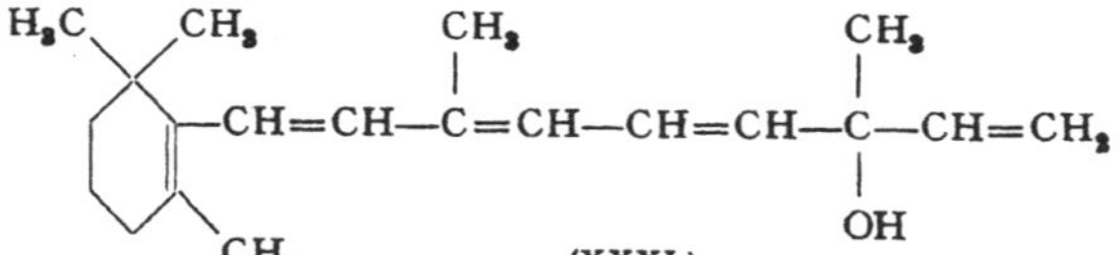 (XXXI.)	Acetate		MILAS (110)
(b)	β-Ionone, HC≡CH, $ClCH_2-C(CH_3)=CH-CH_2OCH_3$	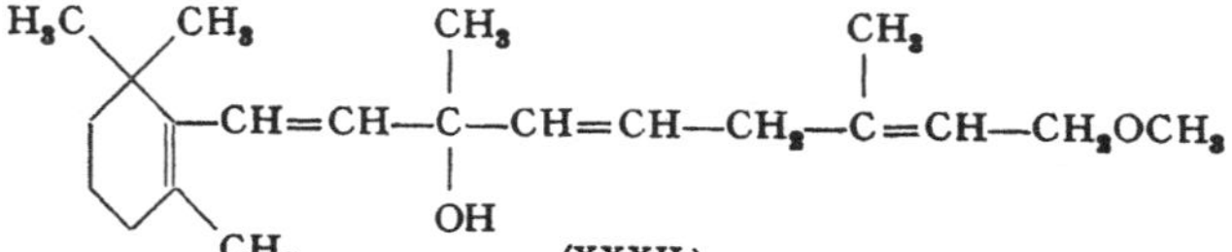 (XXXII.)	Ether	0.01–0.03	OROSHNIK (122, 123a) Ortho (125, 126)
(c)	C_{14}-Aldehyde, HC≡CH, $CH_3 \cdot CO \cdot CH_2 \cdot CH_2OR$ or C_{14}-Aldehyde, $HC≡C-C(CH_3)=CH-CH_2OR$ (R = alkyl or acyl)	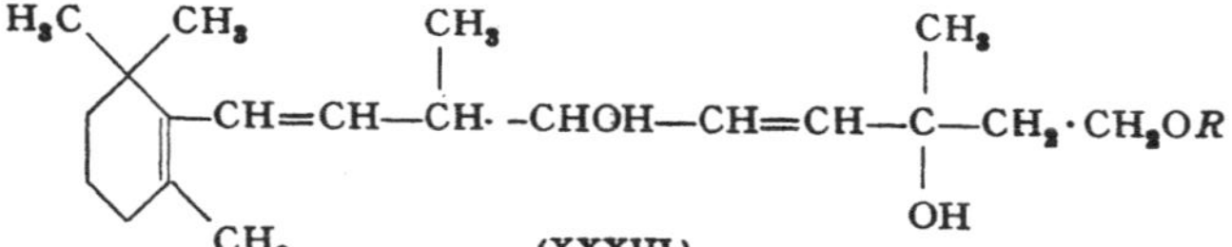 (XXXIII.)	Ethers or Esters	0.02–0.03	MILAS (111, 119)
(d)	β-Ionone, propargyl bromide, methyl-vinyl-ketone	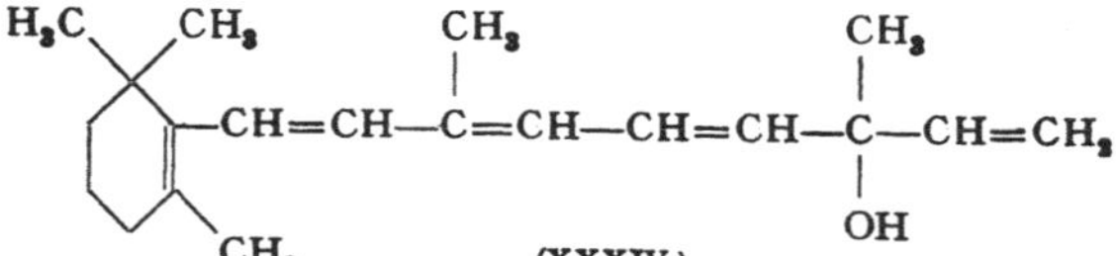 (XXXIV.)	Alcohol	0.02–0.05	ISLER (69, 72)
(e)	β-Ionone + propargyl bromide + $CH_3 \cdot CO \cdot CH_2 \cdot CH_2OCH_3$	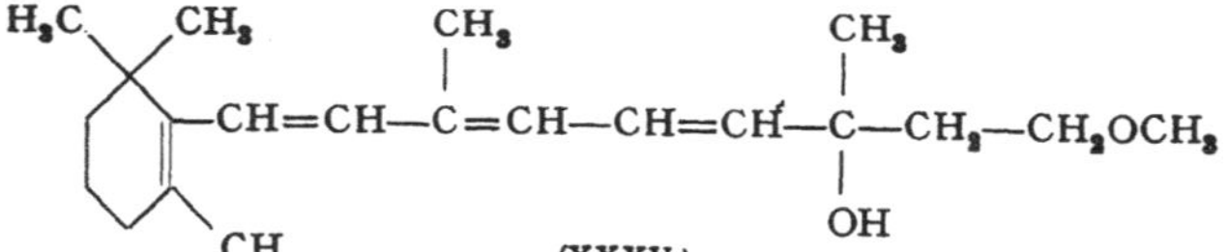 (XXXV.)	Ether	0.02–0.05	ISLER (69, 72)

	Reactants	Structure	Solvent	Conc.	Reference
(f)	β-Ionone + + BrCH$_2$–C≡C–C(CH$_3$)=CH– –CH$_2$OCH$_3$	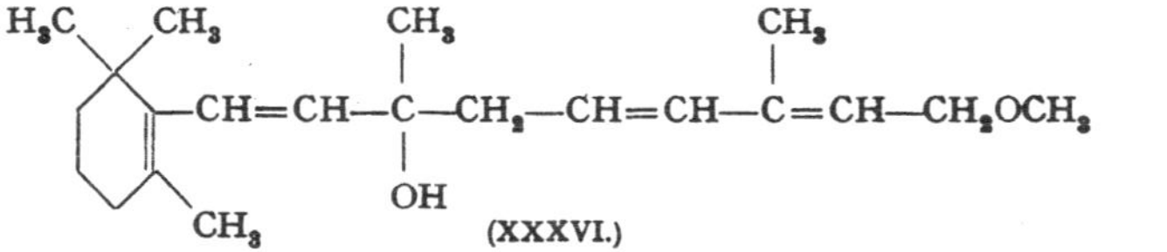 (XXXVI) ring-CH=CH–C(CH$_3$)–C–CH$_2$–CH=CH–C(CH$_3$)=CH–CH$_2$OCH$_3$ with OH	Ether	0.02–0.05	ISLER (71, 72)
(g)	β-Ionone + + HC≡C–CH$_2$–C(CH$_3$)=CH– –CH$_2$OCH$_3$	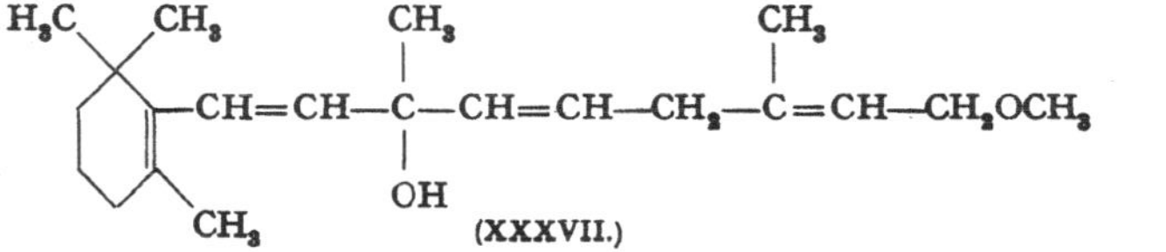 (XXXVII) ring-CH=CH–C(CH$_3$)–C–CH$_2$–CH=CH–C(CH$_3$)=CH–CH$_2$OCH$_3$ with OH	Ether	0.02–0.05	ISLER (70, 72)
(h)	β-Ionylideneacetaldehyde, ethyl β-methylglutaconate	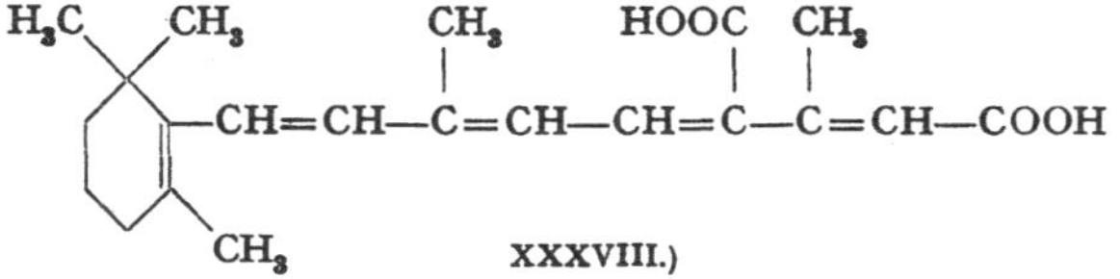 (XXXVIII) ring-CH=CH–C(CH$_3$)=CH–CH=C(HOOC)–C(CH$_3$)=CH–COOH	Alcohol	—	Dist. Prod. i. (23)
(i)	β-Ionone, propargyl bromide, CH$_3$·CO·CH$_2$·CH$_2$OCH$_3$	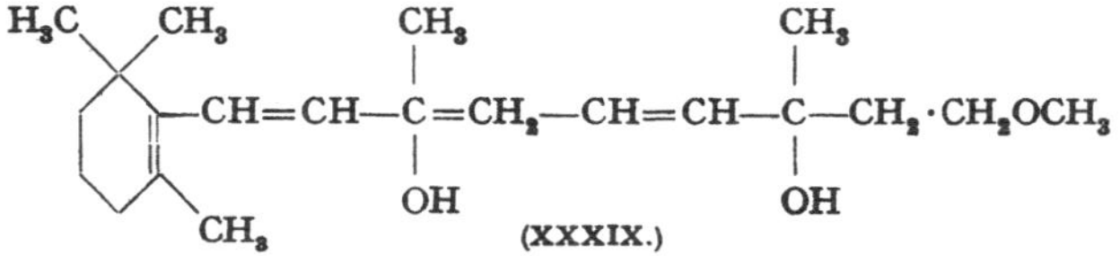 (XXXIX) ring-CH=CH–C(CH$_3$)=CH–CH$_2$–C(CH$_3$)(OH)–CH$_2$OCH$_3$	Ether	—	GOLSE (35)
(j)	β-Ionylideneacetaldehyde, ethyl γ-bromosenecioate BrCH$_2$–C(CH$_3$)=CH–COOC$_2$H$_5$	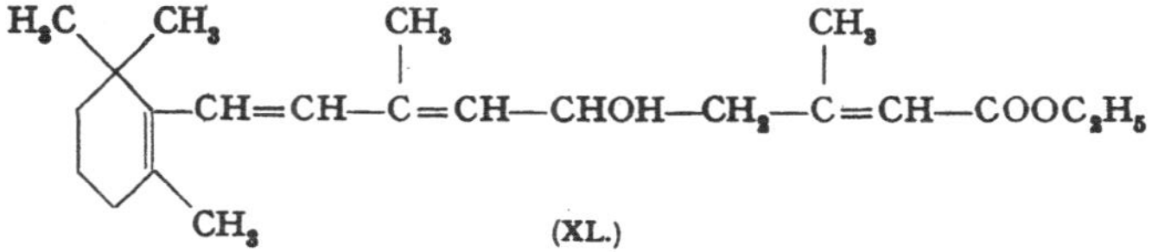 (XL) ring-CH=CH–C(CH$_3$)=CH–CHOH–CH$_2$–C(CH$_3$)=CH–COOC$_2$H$_5$	Alcohol	—	HUISMAN (58)
(k)	β-Cyclocitral, methyl-γ-bromosenecioate	Methylester of vitamin A acid	Alcohol	—	HARPER (43)

Experiments on the formation of vitamin A methyl ether [ISLER et al. (*75*)] indicated that replacement of the hydroxyl group by halogen and rearrangement could be effected by phosphorus tribromide. Acid splitting then followed by treatment with alkali. Rearrangement and dehydration could be carried out in one step by treatment with iodine, hydriodic acid, phosphorus di-iodide or iodine chloride, in toluene or petroleum ether. Rearrangement and dehydration could also be carried out by heating with acetic anhydride in the presence of potassium acetate, or with organic acids such as oxalic, glycolic, phthalic and malonic, or with anhydrides such as phthalic anhydride.

The best yields of vitamin A esters were obtained by processing the acetate (XXVIII). The butyrate, benzoate and palmitate of vitamin A were obtained in lower yields.

4. Synthesis of Vitamin A by Other Methods.

The starting materials and key intermediates for a number of syntheses of vitamin A or its esters and ethers are summarized in Table 4 (p. 62). Their chemistry cannot be discussed effectively at this time, either because the biological potency or the chemical properties of the product indicated it to be impure, or because insufficient evidence has yet been reported to make it possible to properly assess the synthesis.

Methods (*a*) and (*c*) (Table 4, p. 62), described by MILAS and co-workers, were among the early vitamin A syntheses but the potencies of the preparations were comparatively low (50 000—100 000 u./g.). The principal intermediates were the same or similar to those of Methods (*d*) and (*e*).

Method (*b*), reported by OROSHNIK, employed α-ethynyl-β-ionol which was prepared from β-ionone and acetylene by the Nef reaction [OROSHNIK and MEBANE (*123, 124*)]. The other component in the synthesis was 1-chloro-2-methyl-4-methoxy-2-butene (Table 4) from the reaction of isoprene with t-butyl hypochlorite in methanol. By catalytic reduction of the triple bond in the condensation product, the carbinol (XXXII) (Table 4, p. 62) was formed. This, on dehydration, gave a small yield of vitamin A methyl ether mixed with isomers (biopotency of product, 20 000—100 000 u./g.) [OROSHNIK, KARMAS and MEBANE (*123a*)]. From degradative studies, the principal isomer had the structure (XLI). By isomerization with alcoholic alkali the isolated double bond was brought

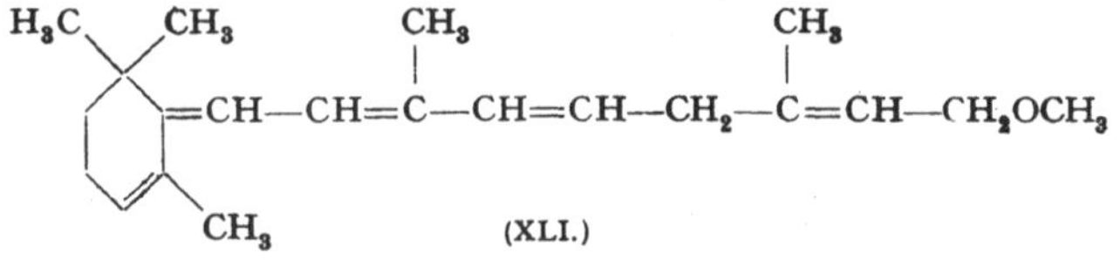

$$\text{[ring with } H_3C, CH_3, CH_3\text{]}=CH-CH=C(CH_3)-CH=CH-CH_2-C(CH_3)=CH-CH_2OCH_3$$

(XLI.)

into conjugation to form a substance called retrovitamin A methyl ether. It had only "a trace" of growth promoting activity.

The "retro-ionylidene" structure will be recognized in other formulas given in the text (e. g., IV, XI, XIII, XLIII, LII). Such structures have not previously been supported in the literature by degradative studies.

Methods (*d*)–(*g*) by ISLER and co-workers as well as Method (*i*) by GOLSE and co-workers make use of the ability of propargyl bromide (or substituted propargyl bromides) to undergo Reformatsky reactions with ketones, a reaction first described by ZEILE and MEYER (*156*). In this fundamental way, the carbon chain of vitamin A was built up by a number of methods. The triple bond was then selectively reduced and vitamin A ethers formed by allylic rearrangement. According to ISLER (*72*) the biological potency of the preparations amounted to only 60000 to 150000 u./g., and crystalline products could not be obtained from them. No bioassay data were reported in the GOLSE paper.

A key intermediate in Method (*f*) had been described earlier by KIPPING and WILD (*93*); a key intermediate in Method (*g*) was the same as in Method (*b*).

PETROW and STEPHENSON (*127*) reported on the synthesis of *trans*-4-carboxy-vitamin A acid (XLII), from β-ionylideneacetaldehyde, obtained as previously described by ARENS et al. (*5*). The aldehyde was condensed

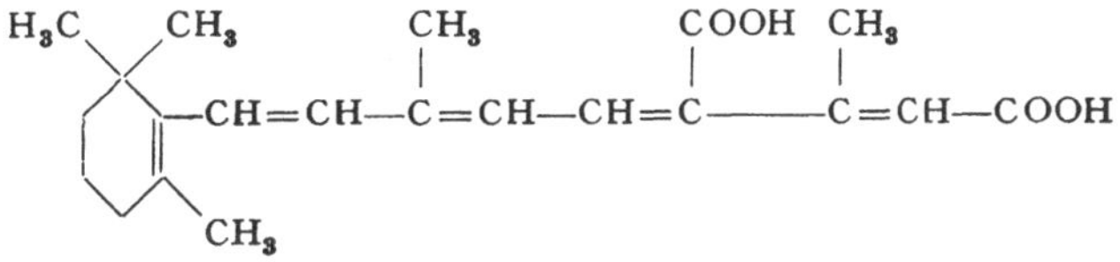

$$\text{(XLII.) 4-Carboxy-vitamin A acid.}$$

with ethyl (*cis*- + *trans*-) β-methylglutaconate, $C_2H_5OOC-CH_2-C=CH-COOC_2H_5$, in the presence of methanolic potassium hydroxide. The free acid formed yellow needles, m. p. 205° [$E(1\%, 1\text{ cm.})$ (320 mμ) = = 883 in isopropyl alcohol]. From β-ionylideneacetaldehyde and β-methylglutaconic anhydride, in pyridine, 4-carboxy-vitamin A acid anhydride, m. p. 126° [E (435 mμ) = 1040 in cyclohexane] was prepared. From the anhydride on hydrolysis the *cis* diacid [E (1%, 1 cm.) (327 mμ) = 571 (in water)] was obtained. Neither the *trans* diacid nor the anhydride gave vitamin A acid when preferential decarboxylation was attempted with alkali hydroxides, with acids or with pyridine or quinoline.

Earlier, the synthesis of 4-carboxy-vitamin A acid [Method (*h*) in Table 4] and its selective decarboxylation was accomplished in this laboratory. The vitamin A acid so formed was then reduced to vitamin A [Dist. Prod. Ind. (*23*)].

Methods (j) and (k) have points of similarity. HARPER and OUGHTON (43) reacted β-cyclocitral with methyl-γ-bromosenecioate by the Reformatsky procedure to obtain a product which, after dehydration, followed by reduction with lithium aluminumhydride, gave β-ionylidene-ethanol. The latter was converted to β-ionylideneacetaldehyde by oxidation with manganese dioxide. The sequence was then repeated to give a vitamin A concentrate which appeared to contain 20% vitamin A on the basis of its extinction value [E (1%, 1 cm.) (319–325 mμ) = 545]. No biological assay data were reported.

HUISMAN (58) reacted β-ionylideneacetaldehyde, prepared by an undisclosed method, with ethyl γ-bromosenecioate. After dehydration, vitamin A acid ethylester was reportedly obtained and reduced to vitamin A. The ultraviolet spectral curve and the biological potency of the vitamin A concentrate thus prepared were not given.

II. Synthesis and Biological Activity of Some Compounds Related to Vitamin A.

A number of compounds related to vitamin A can be synthesized from it or from intermediates used in its synthesis. Other vitamin A relatives are also known which so far have been found only in fish liver oils. In the present section and in Table 5 (p. 70) some data on these compounds are reviewed, to illustrate how structural changes influence physiological activity, as measured by the rat growth test. Photomicrographs of certain of the compounds mentioned in Table 5 and elsewhere in this review are shown in Fig. 2, next pages.

Alcohols.

The relative *biological potencies of vitamin A and neovitamin A* have recently been determined by HARRIS, AMES and BRINKMAN (44). It was found that neovitamin A had a biological potency of 2690000 U. S. P. or I. U./g. On a molar basis, neovitamin A was 80.7% as potent as vitamin A in the rat growth test; on an E value basis it was 85.3% as potent. By the liver storage assay method, neovitamin A was found to have 71.5% and 75.6% of the activity of vitamin A on a molar and E value basis, respectively.

The differences in values for the two test methods were not considered significant, and it was concluded that the two geometrical isomers are probably as well utilized by the rat for storage as for growth.

It was earlier shown that ingested vitamin A or neovitamin A acetate is stored in the rat liver as an equilibrium mixture of the two isomers, the proportion, vitamin A : neovitamin A being approximately 85 : 15 [ROBESON and BAXTER (129)]. Thus, there was a higher proportion of

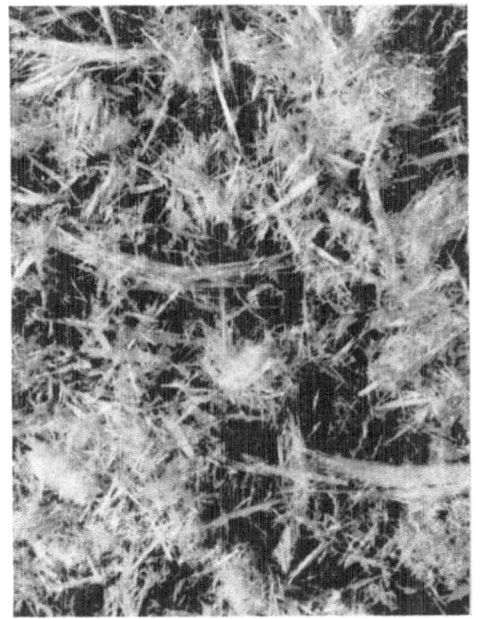

Neovitamin A (7.5 ×).

Vitamin A (15 ×).

Fig. 2. Photomicrographs of crystalline, synthetic vitamin A and some related compounds (Courtesy of Mr. R. P. Loveland, Eastman Kodak Company, Research Laboratories).

Continued p. 68

68 J. G. BAXTER:

Continued Fig. 2

trans-Vitamin A acid (10 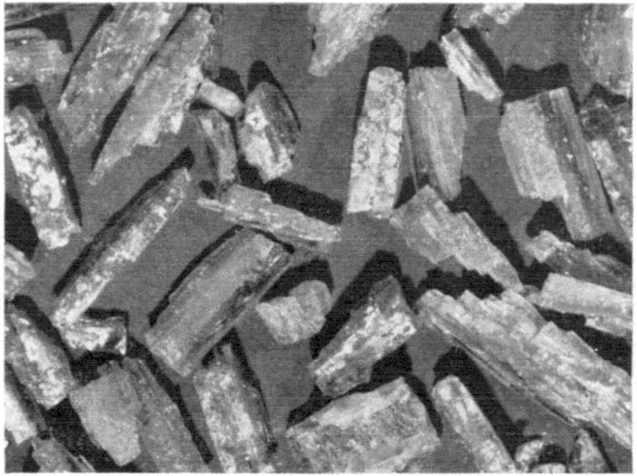 ×). Neovitamin A acid (5 ×). (Δ^2-*cis*-Δ^6-*trans*-vitamin A acid.)

Fig. 2. Photomicrographs of crystalline, synthetic vitamin A and some related compounds (Courtesy of Mr. R. P. Loveland, Eastman Kodak Company, Research Laboratories).

Synthesis of Vitamin A and Some Related Compounds. 69

trans-Vitamin A aldehyde (7.5 ×).

4-Carboxyvitamin A acid (7.5 ×).

Fig. 2. Photomicrographs of crystalline, synthetic vitamin A and some related compounds (Courtesy of Mr. R. P. Loveland, Eastman Kodak Company, Research Laboratories).

Table 5. Properties of Some Compounds Related to Vitamin A.

Compound	M. p.	λ_{max} and E (1%, 1 cm.)	Solvent	λ_{max} and E (1%, 1 cm.) of $SbCl_3$, blue complex	Relative Biopotency (Vitamin A = 1)	Reference
A. Alcohols.						
Vitamin A	64°	E (325 mμ) = 1830	Ethanol	E (620 mμ) = 4400	1	Dist. Prod. i. (23); Cawley (18)
Neovitamin A	60°	E (328 mμ) = 1690	Ethanol	E (620 mμ) = 4400	0.7—0.85	Dist. Prod. i. (23); Harris (44); Robeson (129)
Rehydrovitamin A	—	E (330, 351, 369 mμ) = —, 10.2, — (conc.)	Ethanol	612 mμ	0.1	Shantz (135)
Subvitamin A	—	E (290 mμ) = 150 (conc.)	Ethanol	E (617 mμ) = 310	0	Embree (27)
Kitol	90°	E (290 mμ) = 707	Ethanol	E (428, 505, 580 mμ) = 228, 162, 104 (conc.)	0	Embree (26); Clough (22)
Kitol diacetate	100° (Kofler block)	E (293 mμ) = 698	Isopropanol	430 mμ	—	Chatain (19)
Norvitamin A acetate	150°	E (293 mμ) = 565	Isopropanol	590—600 mμ	—	Chatain (19)
	—	E (229, 243, 281, 324 mμ) = 270, 270, 420, 400 (25% conc.)	Ethanol	—	—	Cheeseman (20)
C_{17}-Alcohol	76°	E (320, 337 mμ) = 1740, 1170	Ethanol	—	0.001	Cheeseman (21)
Benzene analogue of vitamin A	137°	E (330, 343, 362 mμ) = 1940, 2100, 1870	Ethanol	—	0	Weedon (150)

Isovitamin A acetate............	—	E (223, 326 mμ) = 350, 270 (15% conc.)	Ethanol	—	0	CHEESEMAN (20)
B. *Ethers*.						
Vitamin A methyl ether............	33.5° 35°	E (328 mμ) = 1800 E (327 mμ) = 1730	Ethanol Ethanol	E (620 mμ) = 4400 E (620 mμ) = ca. 4000	ca. 1 ca. 1	CAWLEY (14) ISLER (76)
Vitamin A phenyl ether............	92°	E (327 mμ) = 1470	Ethanol	—	ca. 0.05	ISLER (76)
Norvitamin A methyl ether......	—	E (323 mμ) = 720 (45% conc.)	—	—	ca. 0.03	CHEESEMAN (20)
Isovitamin A methyl ether......	—	E (328, 383 mμ) = 630, 290 (40% conc.)	—	—	0	CHEESEMAN (20)
C. *Aldehydes*.						
Vitamin A aldehyde		See Table 2 (p. 54)				BALL (7)
Δ^2-*trans*-Δ^6-*cis* vitamin A aldehyde...	60°	E (372 mμ) = 930	Ethanol	—	1	GRAHAM (38)
Compound Y (dihydroxy-5,6-vitamin A aldehyde) .	—	345 mμ	Chloroform	560 mμ	0.04	MEUNIER (101, 103, 108)
Compound Z (hydroxylated vitamin A aldehyde) .	—	255, 290, 340 mμ	—	485, 545 mμ	0	MEUNIER (105, 109)
D. *Vitamin A_2*.						
Alcohol.............	—	E (287, 351 mμ) = 820, 1460	Ethanol	E (693 mμ) = 4100	0.4	SHANTZ (134, 136)
Aldehyde............	61°, 77°	E (385 mμ) = 1460	Cyclohexane	E (705 mμ) = 3720	—	FARRAR (30); SALAH (131)

Continued p. 72

Continued Table 5

Compound	M. p.	λ_{max} and E (1%, 1 cm.)	Solvent	λ_{max} and E (1%, 1 cm.) of SbCl$_3$, blue complex	Relative Biopotency (Vitamin A = 1)	Reference
E. Hydrocarbons.						
Anhydrovitamin A	77°	E (351, 371, 392 mμ) = 2540, 3680, 3200	Ethanol	E (620 mμ) = 5500	0.005	Shantz (138, 135)
Anhydrovitamin A$_2$	89.5°	E (352, 370, 391 mμ) = 2040, 2980, 2620	Ethanol	E (693 mμ) = 4400	—	Shantz (134)
Axerophthene	—	E (331, 346, 364 mμ) = 1080, 1260, 952	Ethanol	E (474, 577 mμ)	0.1	Euler (29); Karrer (79, 81)
Desmethyl-axerophthene	—	E (332, 346, 364 mμ) = 1490, 1920, 1090	Ethanol	—	0	Karrer (80)
Isoanhydrovitamin A	—	E (330, 350, 370 mμ) = —, 1320, —	Ethanol	E (620 mμ) = 3200	ca. 0.005	Meunier (102); Shantz (138)
F. Acids.						
Vitamin A acid		See Table 2 (p. 54)			0.1–1.0	van Dorp (142, 142a)
C$_{16}$-Acid	168°	E (338 mμ) = 1980	Ethanol	—	0.001	Heilbron (52)
C$_{17}$-Acid	179°	E (257, 260, 339 mμ) = 430, 470, 1075	Ethanol	—	0.001	Heilbron (51, 49)
C$_{18}$-Acid	167°	E (338 mμ) = 1150	Ethanol	—	0	Heilbron (49)
C$_{19}$-Acid	171°	E (258, 351 mμ) = 530, 1250	Ethanol	—	0.001	Heilbron (49)

the *trans* isomer stored in the rat liver than in the fish liver where the ratio was found to be 65 : 35. It is of interest that in a sample of Spanish red tuna liver oil the percentage of neo- in the mixed isomers was found to be 55% [MEUNIER and JOUANNETEAU (*104*)].

This significant difference in the growth-promoting properties of two isomers of vitamin A is further evident in the relatively low potencies found as earlier mentioned for vitamin A_3 or "isomer of vitamin A" (XIII, Scheme 1, p. 47). It is evident that even the slight change in structure accompanying the change of vitamin A to a β,γ-unsaturated isomer is accompanied by a marked drop in biological activity. The rat appears to be unable to effect re-isomerization to vitamin A in anything approaching a quantitative yield.

No biological assay data has been reported on the Δ^2-*trans*-Δ^6-*cis*-vitamin A of GRAHAM, VAN DORP and ARENS (*38*).

Another interesting vitamin A isomer, called *rehydrovitamin A*, was isolated from rat livers, in unpurified form, after feeding anhydrovitamin A [SHANTZ (*135*)]. The preliminary assay estimated the biological potency to be approximately 300 000 u./g. Again a sharp drop in potency, compared to vitamin A itself, was observed. The compound was tentatively assigned the β,γ-unsaturated structure (XLIII).

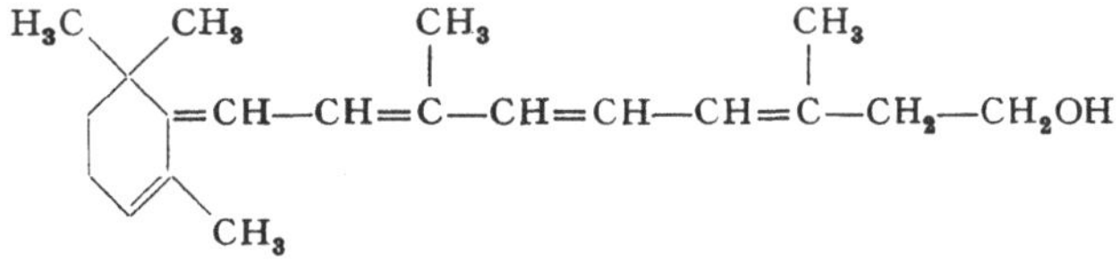

$$H_3C\underset{CH_3}{\overset{CH_3}{\diagdown\diagup}}=CH-CH=\underset{|}{\overset{CH_3}{C}}-CH=CH-CH=\underset{|}{\overset{CH_3}{C}}-CH_2-CH_2OH$$

(XLIII.) Rehydrovitamin A (?).

Subvitamin A and *kitol* are relatives of vitamin A which so far have been found only in fish liver oils. Subvitamin A is an alcohol which was separated in concentrated form from the unsaponifiable fraction of shark liver oil [EMBREE and SHANTZ (*27*)]. It may be identical with, or related to, vitamin A-epoxide or hepaxanthin [KARRER and JUCKER (*86, 87*)]. It has no biological potency.

Kitol is a dihydric alcohol, occurring in esterified form in whale, dogfish and shark liver oils [EMBREE and SHANTZ (*26*); CLOUGH et al. (*22*); BARUA and MORTON (*9*); CHATAIN and DEBODARD (*19*)]. It is a bimolecular form of vitamin A which has no biological potency but is pyrolyzed to yield vitamin A. No structure has as yet been proposed for the compound.

CHATAIN and DEBODARD (*19*) have made the interesting observation that pure kitol diacetate (Table 5, p. 70) gives a violet color with antimony trichloride, whereas free kitol gives a rose color, having only about one-tenth the absorption at 620 mμ. It was concluded that the presence of kitol esters would cause a serious error in the estimation of vitamin A by the Carr-Price reaction while free kitol would cause much less interference.

Extensive work has been done by HEILBRON and co-workers on the synthesis of modified vitamin A structures. A *norvitamin A* acetate, an *isovitamin A* acetate and an acetylenic, C_{17}-alcohol (XLIV–XLVI) were synthesized (cf. Table 5, p. 70). Their biopotencies were small, demonstrating the important part played by methyl groups and the conjugated double bond system in determining growth-promoting activity.

$$\text{—CH=CH—C(CH}_3\text{)=CH—CH=CH—CH=CH—CH}_2\text{O·CO·CH}_3$$

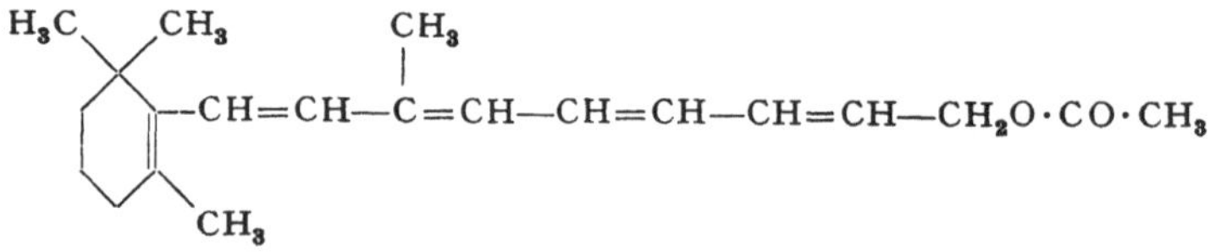

(XLIV.) Norvitamin A acetate.

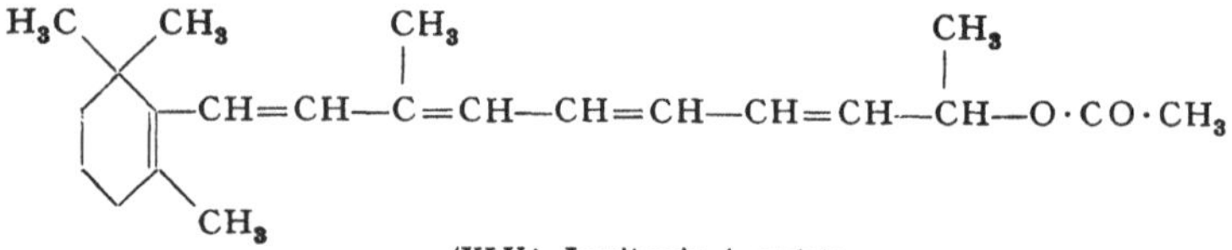

$$\text{—CH=CH—C(CH}_3\text{)=CH—CH=CH—CH=CH—CH(CH}_3\text{)—O·CO·CH}_3$$

(XLV.) Isovitamin A acetate.

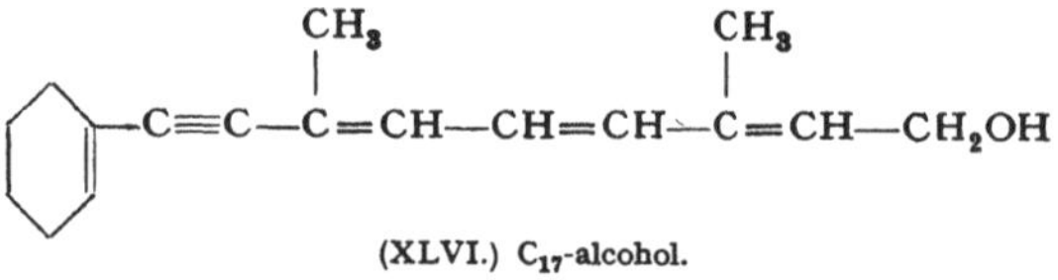

$$\text{—C≡C—C(CH}_3\text{)=CH—CH=CH—C(CH}_3\text{)=CH—CH}_2\text{OH}$$

(XLVI.) C_{17}-alcohol.

In related experiments the benzene analog of vitamin A (XLVII) was prepared [LINNELL and SHEN (*96*); WEEDON and WOODS (*150*)]. It had no growth-promoting action.

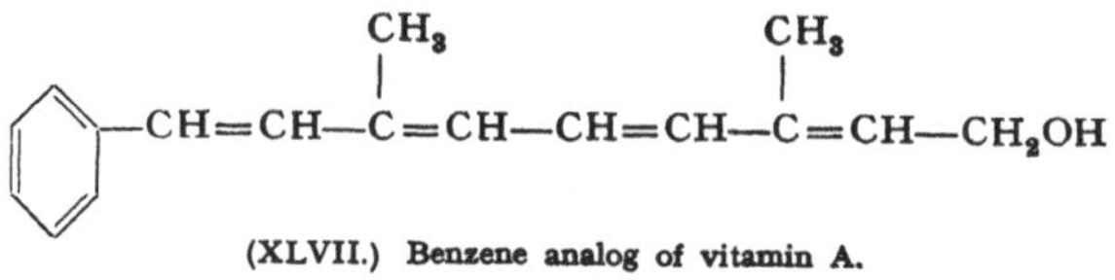

$$\text{—CH=CH—C(CH}_3\text{)=CH—CH=CH—C(CH}_3\text{)=CH—CH}_2\text{OH}$$

(XLVII.) Benzene analog of vitamin A.

Esters and Ethers of Vitamin A.

A number of esters of vitamin A have been synthesized. Their growth-promoting activity, on a chemically equivalent basis, appears to be substantially the same as that of vitamin A [BRAUDE et al. (*13*); BAXTER and ROBESON (*10*); ISLER et al. (*76*)]. The esters appear to be saponified in the body and the vitamin A made available. Cf. also (*12*).

A few ethers of vitamin A have been synthesized by ISLER and coworkers (76). The methyl ether was found to have a potency of 2.7 million u./g., and was identical in properties with the crystalline, natural *vitamin A methyl ether* earlier obtained [CAWLEY (14); HANZE et al. (42)]. The *phenyl ether* had a potency of 180000 u./g. or about one-tenth that of vitamin A, on an equivalent basis. The tetrahydropyranyl ether was reported by ISLER (72) as having the full activity of vitamin A, also on an equivalent basis. The potency of the ethers thus appears to depend on the extent to which they are hydrolyzed to vitamin A in the body.

A *nor- and iso-vitamin A methyl ether* (45% and 40% pure, respectively), analogous to the nor- and iso-vitamin A acetates previously described, have also been prepared. The former compound showed some growth-promoting activity, the latter none (cf. Table 5, p. 70).

Aldehydes.

Identification of the visual pigment *retinene*$_1$ with vitamin A aldehyde, prompted study of the oxidation products of vitamin A. Oxidation with manganese dioxide gave vitamin A aldehyde in excellent yield [BALL, GOODWIN and MORTON (7)], and its properties are well defined (Table 2, p. 54). The aldehyde is substantially as active as vitamin A in the growth test [WENDLER, ROSENBLUM and TISHLER (151)], and is stored as vitamin A in the liver [GLOVER, GOODWIN and MORTON (34)].

Other products have been obtained by oxidation with manganese dioxide. One of these was a compound having absorption maxima at 376 mμ and 290 mμ (in ethanol) and giving with antimony trichloride a wine-red colored product with a maximum at 545 mμ [WALD (146)]. This colored product proved to be highly photosensitive. It appeared to be a derivative of vitamin A containing hydroxyl and carbonyl groups.

The Δ^2-*trans*-Δ^6-*cis*- or "*cis*"-vitamin A aldehyde of GRAHAM, VAN DORP and ARENS was bioassayed and reported to be as active as vitamin A (38).

Oxidation of vitamin A alcohol with vanadium tetroxide gave a "Compound Z", characterized by MEUNIER and co-workers (105, 109) as a mixture of two derivatives of vitamin A aldehyde, each hydroxylated on the side chain and in equilibrium with each other. One of the aldehydic compounds appeared to be similar or identical to the "545 chromogen" of WALD (146). Neither compound had growth-promoting activity.

Oxidation of Vitamin A with potassium permanganate gave a "Compound Y" which was characterized as 5,6-dihydroxy-vitamin A aldehyde (XLVIII) by MEUNIER and co-workers (103, 108). It was found to have $1/20$–$1/25$ of the growth-promoting activity of vitamin A [MEUNIER and

FERRANDO (*101*)]. The introduction of hydroxyl groups into the ring or into the side chain of vitamin A thus resulted in a marked drop in biopotency.

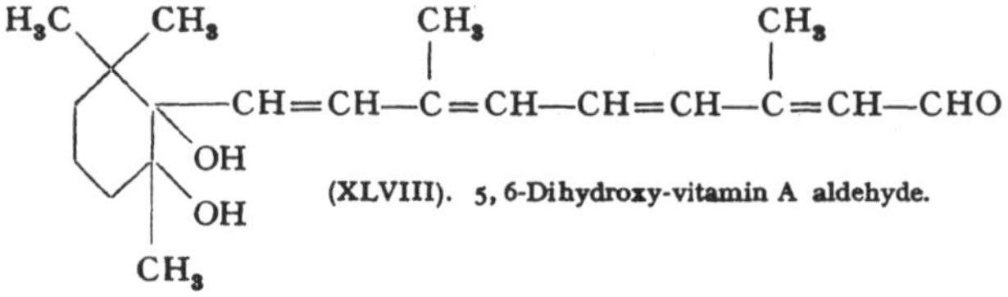

$$\text{(ring with } H_3C, CH_3, OH, OH, CH_3\text{)}-CH=CH-C(CH_3)=CH-CH=CH-C(CH_3)=CH-CHO$$

(XLVIII). 5, 6-Dihydroxy-vitamin A aldehyde.

Vitamin A_2.

The discovery of a vitamin, called vitamin A_2, which accompanies or replaces vitamin A in the liver oils of fresh water fish (LEDERER and ROZANOVA (*95*); GILLAM et al. (*33*)] stimulated research in many laboratories. The pure vitamin was isolated from fish liver oils by SHANTZ (*134*) and found to have 40% of the biopotency of vitamin A [SHANTZ and BRINKMAN (*136*)]. A number of publications on its structure appeared [GRAY (*39*); GRAY and CAWLEY (*40*); MORTON, SALAH and STUBBS (*121*); FIESER (*31*); KARRER and SCHNEIDER (*91*)]. The role of vitamin A_2 and its aldehyde in vision was also investigated [WALD (*147, 148*)].

The studies on its structure will not be reviewed here in detail, except to mention two of the proposed formulas which bear on the configuration required for vitamin A activity. The first of these (XLIX) [MORTON, SALAH and STUBBS (*121*)] pictured vitamin A_2 as a dehydrovitamin A_1; the second, (L) [KARRER, GEIGER and BRETSCHER (*85*)] assigned it an open chain structure. At present, it appears that a decision has not been made between the two formulas although KARRER and SCHNEIDER (*91*) have withdrawn support for (L).

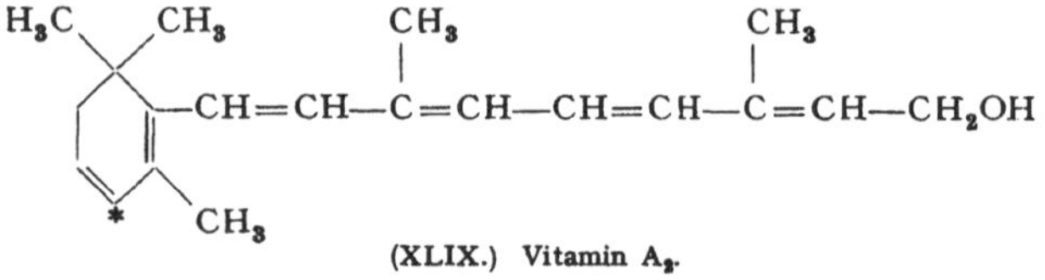

$$\text{(ring with } H_3C, CH_3, CH_3, *\text{)}-CH=CH-C(CH_3)=CH-CH=CH-C(CH_3)=CH-CH_2OH$$

(XLIX.) Vitamin A_2.

$$\text{(open chain with } H_3C, CH_3, CH_3\text{)}-CH=CH-C(CH_3)=CH-CH=CH-C(CH_3)=CH-CH_2OH$$

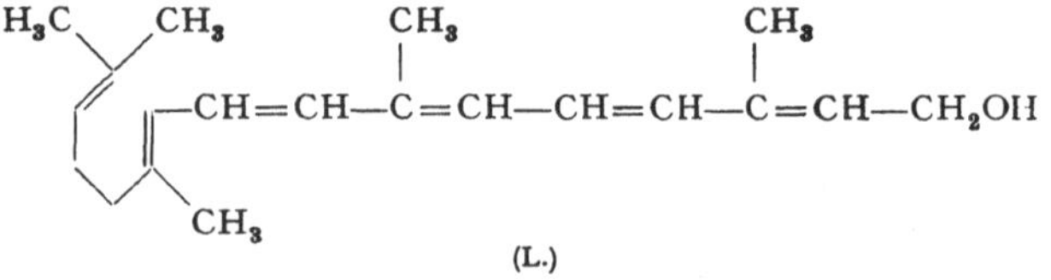

(L.)

FARRAR and co-workers (*30*) supported the structure (XLIX) by synthetic studies. Vitamin A_1 acid was converted into its crystalline methyl ester (m. p. 56°) and brominated with N-bromosuccinimide, presumably at the position starred in (XLIX). Dehydrohalogenation with N-phenylmorpholine gave a compound characterized as vitamin A_2 acid methyl ester. This was saponified to the corresponding acid (m. p. 77°) and re-esterified to give a pure methyl ester (m. p. 47°). By reduction with lithium aluminumhydride a product designated as vitamin A_2 was obtained as a deep yellow oil, with ultraviolet absorption properties similar to the preparation of SHANTZ from natural sources. The anhydro compound was similar in properties to natural anhydrovitamin A_2. The corresponding aldehyde was prepared by oxidation of the alcohol with manganese dioxide and was similar to retinene$_2$. Further studies to identify the synthetic with the natural vitamin are in progress. No bioassay data have been given.

The recent synthesis of 3-dehydro-β-ionone (LI) [HENBEST (54)] should offer another means of synthesizing (XLIX) and provide further support for the proposed formula for vitamin A_2.

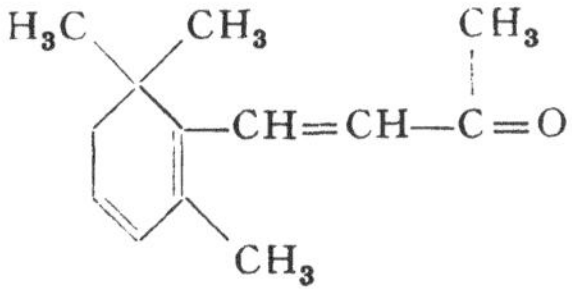

(LI.) 3-Dehydro-β-ionone.

MEUNIER, JOUANNETEAU and ZWINGELSTEIN (*107, 99*), however, supported the open-chain formula (L) from oxidation experiments on lycopene. The latter was oxidized with manganese dioxide and a product was isolated which was characterized as vitamin A_2 aldehyde, partly from spectral measurements on the aldehyde and on the blue color formed from it with antimony trichloride. While lycopene has no vitamin A activity, the aldehyde preparation was reported to have a biological activity equal to that of vitamin A_2. By reduction with lithium aluminumhydride, an alcohol was also obtained which was characterized as vitamin A_2 (λ_{max} at 352 mμ; with SbCl$_3$, at 690 mμ).

This leaves the structure of vitamin A_2 and the specificity of vitamin A activity still in doubt. Formula (XLIX) is consistent with the high degree of specificity so far evidenced. High biological activity on the part of a substance of formula (L) would not be so consistent.

Vitamin A_2 aldehyde has been prepared in crystalline form by SALAH and MORTON (*131*) but bioassay data were not reported.

Hydrocarbons.

By the action of mineral acids on vitamin A a hydrocarbon is formed which was termed *anhydrovitamin A* [SHANTZ, CAWLEY and EMBREE (*138*);

Karrer and Schwyzer (*92*)] and axerophthene [Meunier, Dulou and Vinet (*98*, *100*)]. In the latter reference (*100*), the compound was assigned the formula (LII).

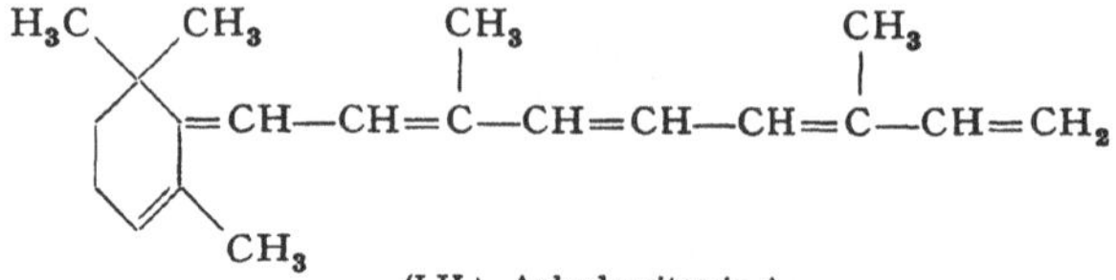
H3C
CH3
CH3
CH3
=CH—CH=C—CH=CH—CH=C—CH=CH2
CH3

(LII.) Anhydrovitamin A.

Anhydrovitamin A_2 is similarly prepared from vitamin A_2 by the action of mineral acids. Its ultraviolet spectrum is similar to that of anhydrovitamin A, but the absorption maximum of the blue complex with antimony trichloride is located at 693 mμ rather than 620 mμ. Shantz (*134*) considered that this last fact might be explained on the basis of an open-chain formula for vitamin A_2 and anhydrovitamin A_2 (possibly, LIII) by assuming that antimony trichloride causes the external double bond to fall into conjugation. Farrar and co-workers (*30*), in supporting the ring structure for vitamin A_2, assigned formula (LIV) to the anhydro compound.

Anhydrovitamin A_1 has a biological potency of about 17000 u./g. [Shantz, Cawley and Embree (*138*)]. The biological activity of anhydrovitamin A_2 has not yet been reported.

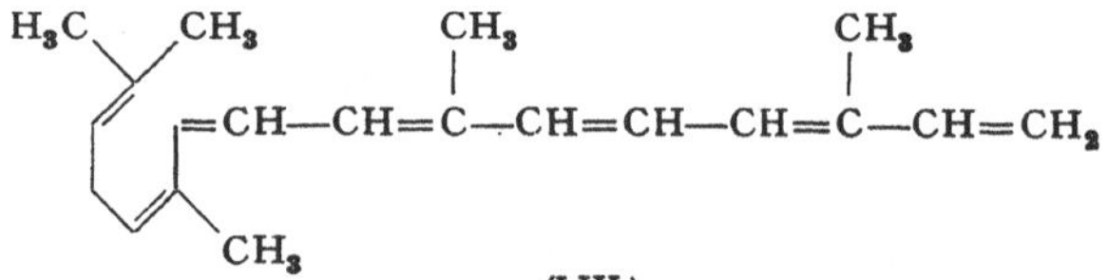
H3C
CH3
CH3
CH3
=CH—CH=C—CH=CH—CH=C—CH=CH2
CH3

(LIII.)

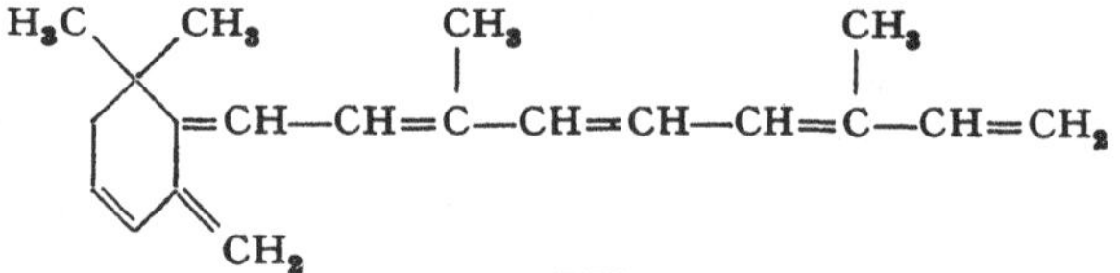
H3C
CH3
CH3
CH3
=CH—CH=C—CH=CH—CH=C—CH=CH2
CH2

(LIV.)

A hydrocarbon, likewise called "axerophthene", has been prepared [Karrer and Benz (*79*)] by the reaction of C_{18}-ketone with ethyl magnesium bromide followed by dehydration. Meunier (*98*) objected to this nomenclature. The hydrocarbon had a triple peak in its ultraviolet spectrum (Table 5, p. 70) and was about one-tenth as active as vitamin A in the rat growth test [Euler and Karrer (*29*)]. It was assigned formula (LV) corresponding to vitamin A hydrocarbon:

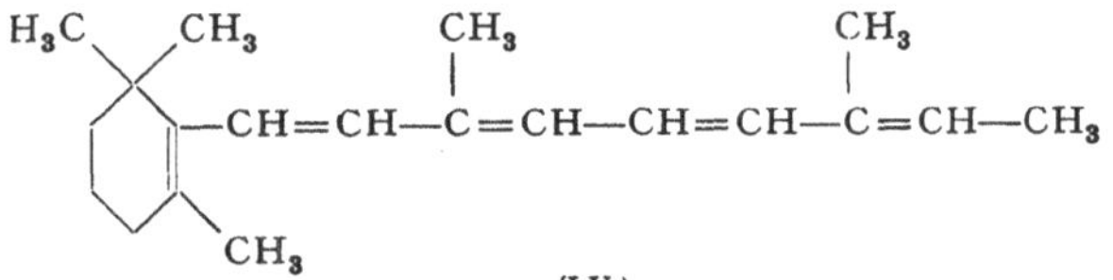

Synthesis of Vitamin A and Some Related Compounds.

$$\text{cyclohexenyl}(H_3C, CH_3, CH_3)\text{—}CH=CH\text{—}C(CH_3)=CH\text{—}CH=CH\text{—}C(CH_3)=CH\text{—}CH_3$$

(LV.)

The corresponding desmethyl compound (LVI) had no A-activity.

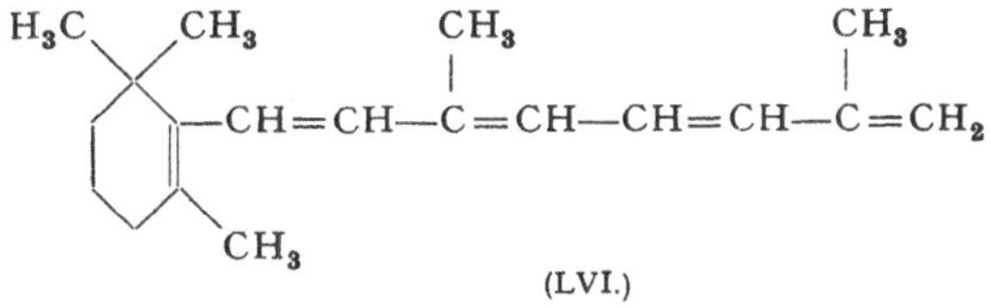

$$\text{cyclohexenyl}(H_3C, CH_3, CH_3)\text{—}CH=CH\text{—}C(CH_3)=CH\text{—}CH=CH\text{—}C(CH_3)=CH_2$$

(LVI.)

Extended treatment of vitamin A with hydrochloric acid gave a substance known as *isoanhydrovitamin A* [SHANTZ, CAWLEY and EMBREE (*138*)], showing approximately the same biological activity as anhydrovitamin A [MEUNIER et al. (*102*)]. By further treatment with acid a compound containing chlorine was produced, having no biological potency (*102*).

Vitamin A Acids.

The vitamin A acid synthesized by VAN DORP and ARENS (*143*) was reported to have one-tenth the biological activity of vitamin A, when given orally. As the sodium salt, injected subcutaneously in aqueous solution, it was one-half as active as vitamin A. The sodium salt, buffered to p_H 10 in aqueous solution and given orally, was later reported to be as active as vitamin A itself [VAN DORP and ARENS (*142*)]. Neither oral nor subcutaneous administration of the salt mentioned gave vitamin A in the liver [ARENS and VAN DORP (*4*)].

A number of other acids related to vitamin A have been synthesized by HEILBRON and co-workers (*45*), including the C_{16}-, C_{17}-, C_{18}- and C_{19}-acids (LVII). These acids had an acetylenic bond in the side chain and like the C_{17}-alcohol, earlier mentioned, showed little biopotency (Table 5, p. 70).

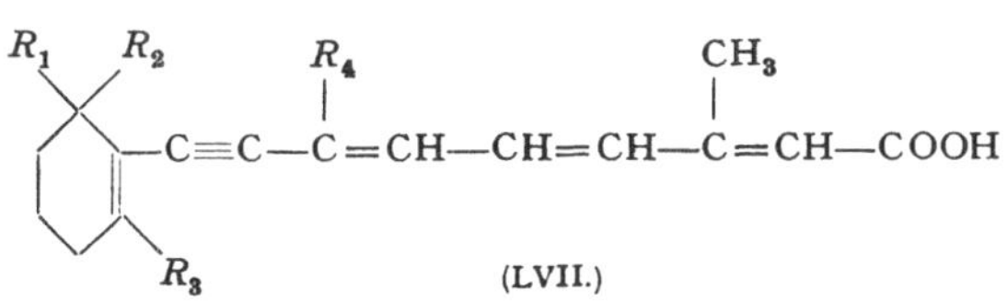

$$\text{cyclohexenyl}(R_1, R_2, R_3)\text{—}C{\equiv}C\text{—}C(R_4)=CH\text{—}CH=CH\text{—}C(CH_3)=CH\text{—}COOH$$

(LVII.)

	R_1	R_2	R_3	R_4
C_{16}-acid	H	H	H	H
C_{17}-acid	H	H	H	CH_3
C_{18}-acid	H	H	CH_3	CH_3
C_{19}-acid	CH_3	CH_3	H	CH_3

Conclusions.

It is evident from what has been said that high growth-promoting activity is characteristic only of vitamin A and certain of its derivatives, such as the esters and a few ethers, which are readily convertible to the vitamin in the body. The one apparent exception to this generalization, vitamin A acid, is unlike vitamin A since it is not stored in the liver. Whether it is converted to vitamin A at some other site in the body remains to be determined.

The available evidence does not permit the further conclusion that vitamin A-like substances owe their activity to metabolic conversion to vitamin A. In one case, that of anhydrovitamin A, there is evidence of conversion to an isomer of vitamin A, rehydrovitamin A, which may further undergo partial isomerization to vitamin A. It appears not improbable that hydrocarbons like isoanhydrovitamin A and axerophthene may owe their activity to a similar series of metabolic changes. However, it would be difficult to account for the apparent growth-promoting action of substances such as "C_{17}-alcohol", norvitamin A methylether or the acetylenic relatives of vitamin A acid on such a basis. It would appear, at present, that the rat growth assay is not specific for vitamin A and that other substances having the vitamin A skeleton can be utilized to a degree which depends on their structure and their ability to be absorbed and carried to the proper body tissues.

III. Relationship between Vitamin A and Carotenes.

As well known, the relationship between β-carotene (p. 4) and vitamin A is close. Carotene has long been known to have growth-promoting activity and to function as a provitamin A. Also, vitamin A aldehyde has been prepared from β-carotene by oxidation with such agents as hydrogen peroxide [HUNTER and WILLIAMS (60)], hydrogen peroxide-osmium tetroxide [GOSS and MCFARLANE (36); WENDLER, ROSENBLUM and TISHLER (151)], and manganese dioxide [MEUNIER, JOUANNETEAU and ZWINGELSTEIN (106)]. These reactions suggest that β-carotene is converted to vitamin A in the body by oxidation and not by a hydrolytic process, as earlier supposed.

Syntheses of β-carotene have been reported by KARRER (82–84), INHOFFEN (63, 65–67), and MILAS (115), together with their associates. A general survey of this field written by INHOFFEN and SIEMER (67a) appears on p. 1 of the present volume.

References.

1. ARENS, J. F. and D. A. VAN DORP: Synthesis of Some Compounds Possessing Vitamin A Activity. Nature (London) **157**, 190 (1946).

2. ARENS, J. F. and D. A. VAN DORF: A Cis C_{18}-Ketone, Related to Vitamin A. Recueil Trav. chim. Pays-Bas 66, 759 (1947).
3. — — Synthesis of Vitamin A. Recueil Trav. chim. Pays-Bas 68, 604 (1949).
4. — — Activity of "Vitamin A Acid" in the Rat. Nature (London) 158, 622 (1946).
5. ARENS, J. F., D. A. VAN DORP, G. VAN DIJK and B. J. BRANDT: A New Method for the Synthesis of α,β-Unsaturated Aldehydes. Preparation of β-Methylcinnamaldehyde, Citral, and β-Ionylideneacetaldehyde. Recueil Trav. chim. Pays-Bas 67, 973 (1948).
6. BACHARACH, A. L. and E. L. SMITH: Notes on the Chemistry of the Fat-Soluble Vitamins in Cod Liver Oil. Quart. J. Pharmac. Pharmacol. 1, 539 (1928).
7. BALL, S., T. W. GOODWIN and R. A. MORTON: Studies in Vitamin A. V. Preparation of Retinene$_1$-Vitamin A Aldehyde. Biochemic. J. 42, 516 (1948).
8. BALL, S. and R. A. MORTON: Studies in Vitamin A. X. Vitamin A_1 and Retinene$_1$ in Relation to Photopic Vision. Biochemic. J. 45, 298 (1949).
9. BARUA, R. K. and R. A. MORTON: Studies in Vitamin A. XII. Whale Liver Oil Analysis: Preparation of Kitol Esters. Biochemic. J. 45, 308 (1949).
10. BAXTER, J. G. and C. D. ROBESON: Crystalline Aliphatic Esters of Vitamin A. J. Amer. chem. Soc. 64, 2407 (1942).
11. — — Crystalline Vitamin A. J. Amer. chem. Soc. 64, 2411 (1942).
12. BOLDINGH, J., H. R. CAMA, F. D. COLLINS, R. A. Morton, N. T. GRIDGEMAN, O. ISLER, M. KOFLER, R. J. TAYLOR, A. S. WELLAND and T. BRADBURY: Pure All-Trans Vitamin A Acetate and the Assessment of Vitamin A Potency by Spectrophotometry. Nature (London) 168, 598 (1951).
13. BRAUDE, R., A. S. FOOT, K. M. HENRY, S. K. KON, S. Y. THOMPSON and T. H. MEAD: Vitamin A Studies with Rats and Pigs. Biochemic. J. 35, 693 (1941).
14. CAWLEY, J. D.: Improved Vitamin A Products and Methods of Preparation Thereof. Brit. 579449, June 2 (1943).
15. — Vitamin A Intermediate. Brit. 636935, May 10 (1950).
16. CAWLEY, J. D., C. D. ROBESON, E. M. SHANTZ, L. WEISLER and J. G. BAXTER: Synthesis of Vitamin A Active Compounds Containing Repeated Isoprene Units. U. S. 2576103, November 27 (1951).
17. CAWLEY, J. D., C. D. ROBESON, L. WEISLER, E. M. SHANTZ, N. D. EMBREE and J. G. BAXTER: Crystalline Synthetic Vitamin A. Abstracts, 112th Meeting American Chem. Soc., New York City, N. Y., September 15–19, 26 C (1947).
18. — — — — — — Crystalline Synthetic Vitamin A and Neovitamin A. Science (New York) 107, 346 (1948).
19. CHATAIN, H. et M. DEBODARD: A propos du dosage de la vitamine A: spectre ultraviolet et réaction de Carr-Price du kitol alcool et du kitol ester. C. R. hebd. Séances Acad. Sci. 233, 105 (1951).
20. CHEESEMAN, G. W. H., I. HEILBRON, E. R. H. JONES, F. SONDHEIMER and B. C. L. WEEDON: Studies in the Polyene Series. XXVIII. The Structure of the C_{14}-Aldehyde Derived from β-Ionone and its Use for the Synthesis of Norvitamin A and Isovitamin A Derivatives. J. chem. Soc. (London) 1949, 1516.
21. CHEESEMAN, G. W. H., I. HEILBRON, E. R. H. JONES and B. C. L. WEEDON: Studies in the Polyene Series. XXXIV. The Synthesis of a C_{17}-Alcohol Related to Vitamin A. J. chem. Soc. (London) 1949, 3120.
22. CLOUGH, F. B., H. M. KASCHER, C. D. ROBESON and J. G. BAXTER: Crystalline Kitol. Science (New York) 105, 436 (1947).
22a. DAVIES, W. H., I. HEILBRON, W. E. JONES and A. LOWE: Studies in the Synthesis of Vitamin A. Part I. J. chem. Soc. (London) 1935, 584.

23. Distillation Products industries: Unpublished work.
23a. DEUEL, H. J., Jr.: The Lipids. Their Chemistry and Biochemistry. I. Chemistry. New York: Interscience. 1951.
24. DEVINE, J., R. F. HUNTER and N. E. WILLIAMS: The Preparation of β-Carotene of a High Degree of Purity. Biochemic. J. **39**, 5 (1945).
25. EMBREE, N. D.: Fat-Soluble Vitamins. Annu. Rev. Biochem. **1947**, 323.
26. EMBREE, N. D. and E. M. SHANTZ: Kitol, a New Provitamin A. J. Amer. chem. Soc. **65**, 910 (1943).
27. — — A Possible New Member of the Vitamins A_1 and A_2 Group. J. Amer. chem. Soc. **65**, 906 (1943).
28. — — Vitamin A Active Esters. U. S. 2434687, January 20 (1948).
29. EULER, H. v. u. P. KARRER: Über die biologische Wirksamkeit des Axerophtens und verwandter Kohlenwasserstoffe. Helv. chim. Acta **32**, 461 (1949).
30. FARRAR, K. R., J. C. HAMLET, H. B. HENBEST and E. R. H. JONES: Elucidation of the Structure of Vitamin A_2 by Total Synthesis. Chem. and Ind. **1951**, 49.
31. FIESER, L. F.: Absorption Spectra of Carotenoids; Structure of Vitamin A_2. J. org. Chemistry **15**, 930 (1950).
32. FINHOLT, A. E., A. C. BOND, Jr. and H. I. SCHLESINGER: Lithium Aluminum Hydride, Aluminum Hydride and Lithium Gallium Hydride, and Some of Their Applications in Organic Chemistry. J. Amer. chem. Soc. **69**, 1199 (1947).
33. GILLAM, A. E., I. HEILBRON, W. E. JONES and E. LEDERER: The Occurrence and Constitution of the 693 mμ Chromogen (Vitamin A_2?) of Fish Liver Oils. Biochemic. J. **32**, 405 (1938).
34. GLOVER, J., T. W. GOODWIN and R. A. MORTON: Studies in Vitamin A. VI. Conversion *in vivo* of Vitamin A Aldehyde... Biochemic. J. **43**, 109 (1948).
35. GOLSE, R., J. GAVARRET et G. DEMANGE: Synthèse de l'éther-oxyde méthylique de la vitamine A. Bull. Soc. chim. France **1950**, 285.
36. GOSS, G. C. and W. D. McFARLANE: Oxidation of β-Carotene with Osmium Tetroxide. Science (New York) **106**, 375 (1947).
37. GOULD, R. G., Jr. and A. F. THOMPSON, Jr.: The Synthesis of Certain Unsaturated Compounds from β-Ionone and Tetrahydroquinone. J. Amer. chem. Soc. **57**, 340 (1935).
38. GRAHAM, W., D. A. VAN DORP and J. F. ARENS: Synthesis of a Cis Isomer of Vitamin A Aldehyde. Recueil Trav. chim. Pays-Bas **68**, 609 (1949).
39. GRAY, E. LEB.: Comparison of Vitamins A and A_2 by Distillation. J. biol. Chemistry **131**, 317 (1939).
40. GRAY, E. LEB. and J. D. CAWLEY: The Influence of the Structure on the Elimination Maximum. I. The Structure of Vitamin A_2. J. biol. Chemistry **134**, 397 (1940).
41. HAMANO, S.: Physiologically Active Crystalline Esters of Vitamin A. Sci. Pap. Inst. physic. chem. Res., Tokyo **28**, 69 (1935).
42. HANZE, A. R., T. W. CONGER, E. C. WISE and D. I. WEISBLAT: Crystalline Vitamin A Methyl Ether. J. Amer. chem. Soc. **70**, 1253 (1948).
43. HARPER, S. H. and J. F. OUGHTON: Reformatsky Reactions with Methyl γ-Bromo-β,β-Dimethylacrylate. Chem. and Ind. **1950**, 574.
44. HARRIS, P. L., S. R. AMES and J. H. BRINKMAN: Biochemical Studies on Vitamin A. IX. Biopotency of Neovitamin A in the Rat. J. Amer. chem. Soc. **73**, 1252 (1951).
45. HEILBRON, I.: Recent Developments in the Vitamin A Field. J. chem. Soc. (London) **1948**, 386.
46. HEILBRON, I., R. N. HESLOP, R. A. MORTON, E. T. WEBSTER, J. L. REA and J. C. DRUMMOND: Characteristics of Highly Active Vitamin A Preparations. Biochemic. J. **26**, 1178 (1932).

47. HEILBRON, I., A. W. JOHNSON, E. R. H. JONES and A. SPINKS: Studies in the Polyene Series. V. The Employment of 3-(2':6':6'-Trimethylcyclohexenyl)-1-methylcrotonaldehyde for the Synthesis of Vitamin A and Analogues. J. chem. Soc. (London) **1942**, 727.
48. HEILBRON, I., E. R. H. JONES and B. C. L. WEEDON: Studies in the Polyene Series. XXIX. Ethoxyacetylenic Carbinols and Their Conversion to α,β-Unsaturated Aldehydes and Acids. J. chem. Soc. (London) **1949**, 1823.
49. HEILBRON, I., E. R. H. JONES, D. G. LEWIS and B. C. L. WEEDON: Studies in the Polyene Series. XXXI. The Synthesis of C_{18}- and C_{19}-Acids Related to Vitamin A. J. chem. Soc. (London) **1949**, 2023.
50. HEILBRON, I., E. R. H. JONES and D. G. O'SULLIVAN: Studies in the Polyene Series. XXIV. The C_{17}-Acid and C_{18}-Ketone Related to Vitamin A. J. chem. Soc. (London) **1946**, 866.
51. HEILBRON, I., E. R. H. JONES and R. W. RICHARDSON: Studies in the Polyene Series. XXV. Molecular Structure and Vitamin A Activity. Synthesis of a Biologically Active C_{17}-Acid. J. chem. Soc. (London) **1949**, 287.
52. HEILBRON, I., E. R. H. JONES, D. G. LEWIS, R. W. RICHARDSON and B. C. L. WEEDON: Studies in the Polyene Series. XXVII. The Synthesis of a Biologically Active C_{16}-Acid. J. chem. Soc. (London) **1949**, 742.
53. HEILBRON, I., W. E. JONES, A. LOWE and H. R. WRIGHT: Studies in the Synthesis of Vitamin A. Part II. J. chem. Soc. (London) **1936**, 561.
54. HENBEST, H. B.: Studies in the Polyene Series. XXXVII. Preparation of 3-Dehydro-β-Ionone and Some 3-Substituted β-Ionones. J. chem. Soc. (London) **1951**, 1074.
55. HICKMAN, K. C. D.: Molecular Distillation. Apparatus and Methods. Ind. Engng. Chem. **29**, 968 (1937).
56. — State of the Vitamins in Certain Fish Liver Oils. Ind. Engng. Chem. **29**, 1107 (1937).
57. HOLMES, H. N. and R. E. CORBET: The Isolation of Crystalline Vitamin A. J. Amer. chem. Soc. **59**, 2042 (1937).
57a. HUBBARD, R. and G. WALD: Cis-trans Isomers of Vitamin A and Retinene in Vision. Science (New York) **115**, 60 (1952).
58. HUISMAN, H. O.: Investigations in the Vitamin A Series. I. A New Synthesis of Vitamin A Acid and Vitamin A. Recueil Trav. chim. Pays-Bas **69**, 851 (1950).
59. HUNTER, R. F.: Carotenoids and Vitamin A. Research **3**, 453 (1950).
60. HUNTER, R. F. and N. E. WILLIAMS: Chemical Conversion of β-Carotene into Vitamin A. J. chem. Soc. (London) **1945**, 554.
61. INHOFFEN, H. H. u. F. BOHLMANN: Synthesen in der Carotinoid-Chemie seit 1939. Fortschr. chem. Forschg. **1**, 175 (1949).
62. INHOFFEN, H. H., F. BOHLMANN u. K. BARTRAM: Synthesen in der Carotinoid-Reihe. I. Kondensationen von β-Jonon und homologen Ketonen mit Oxalester. Liebigs Ann. Chem. **561**, 13 (1948).
63. INHOFFEN, H. H., F. BOHLMANN, K. BARTRAM u. H. POMMER: XI. Totalsynthese des β-Carotins. Chemiker-Ztg. **74**, 285 (1950).
64. INHOFFEN, H. H., F. BOHLMANN u. M. BOHLMANN: Synthesen in der Carotinoid-Reihe. VI. Darstellung der Vitamin-A-Säure und einer isomeren C_{20}-Säure. Liebigs Ann. Chem. **568**, 47 (1950).
65. INHOFFEN, H. H., H. POMMER u. F. BOHLMANN: XII. Zweite Synthese des β-Carotins. Chemiker-Ztg. **74**, 309 (1950).
66. — — — Synthesen in der Carotinoid-Reihe. XIV. Aufbau des β-Carotins. Liebigs Ann. Chem. **569**, 237 (1950).

67. INHOFFEN, H. H., H. POMMER u. F. WESTPHAL: Synthesen in der Carotinoid-Reihe. XVI. Eine weitere Synthese des β-Carotins. Liebigs Ann. Chem. 570, 69 (1950).
67a. INHOFFEN, H. H. u. H. SIEMER: Synthetische Chemie der Carotinoide. Fortschr. Chem. Organ. Naturstoffe 9, 1 (1952).
68. ISHIKAWA, S. and T. MATSUURA: Synthesis of Aldehydic Isomers of Methylionone. Sci. Rep. Tokyo Bunrika Daigaku, Sect. A 3, No. 60, 173 (1937).
69. ISLER, O.: Über Synthesen in der Vitamin A-Reihe. Chimia 3, 150 (1949).
70. — Swiss 254948.
71. — Swiss 261888.
72. — Die Chemie des Vitamin A. Chimia 4, 103 (1950).
73. — Leaflet distributed, American Chemical Society Meeting, New York City, N. Y., September (1951).
74. ISLER, O., W. HUBER, A. RONCO u. M. KOFLER: Synthese des Vitamin A. Helv. chim. Acta 30, 1911 (1947).
75. ISLER, O., M. KOFLER, W. HUBER u. A. RONCO: Synthese von Vitamin A-Methyläther. Experientia 2, 31 (1946).
76. ISLER, O., A. RONCO, W. GUEX, N. C. HINDLEY, W. HUBER, K. DIALER u. M. KOFLER: Über die Ester und Äther des synthetischen Vitamins A. Helv. chim. Acta 32, 489 (1949).
77. JOHNSON, A. W.: Recent Advances in Science; Organic Chemistry, Science Progress 36, 112, 496 (1948).
78. KARRER, P.: Neuere Fortschritte auf dem Gebiete der Carotinoide und des Vitamins A. Österr. Chemiker-Ztg. 49, 215 (1948).
79. KARRER, P. u. J. BENZ: Axerophten, der dem Vitamin A zugrunde liegende Kohlenwasserstoff. Helv. chim. Acta 31, 1048 (1948).
80. — — Eine neue Synthese des Desmethylaxerophtens. Helv. chim. Acta 31, 1607 (1948).
81. — — Zur Synthese des Axerophtens. II. Mitteilung. Helv. chim. Acta 32, 232 (1949).
82. KARRER, P. u. C. H. EUGSTER: Darstellung und einige Umsetzungen des Octadien-(3,5)-dions-(2,7). Helv. chim. Acta 32, 1934 (1949).
83. — — II. Totalsynthese des β-Carotins. Helv. chim. Acta 33, 1172 (1950).
84. — — Synthèse totale du β-carotène. C. R. hebd. Séances Acad. Sci. 230, 1920 (1950).
85. KARRER, P., A. GEIGER u. E. BRETSCHER: Über das sog. Vitamin A_2. Helv. chim. Acta 24, 161 E (1941).
86. KARRER, P. u. E. JUCKER: Die Konstitution des „574 Chromogens" oder Hepaxanthins aus Lebertran. Helv. chim. Acta 28, 717 (1945).
87. — — Über Vitamin A-Epoxyd (Hepaxanthin). II. Helv. chim. Acta 30, 559 (1947).
88. KARRER, P., E. JUCKER u. E. SCHICK: Synthese der Vitamin-A-Säure und der entsprechenden, den α-Jononkohlenstoffring enthaltenden Verbindung (3,7-Dimethyl-9-[1',1',3'-trimethyl-c.-hexen-3'-yl-2]-nonatetraen-[2,4,6,8]-carbonsäure). Helv. chim. Acta 29, 704 (1946).
89. KARRER, P., R. MORF u. K. SCHÖPP: Zur Kenntnis des Vitamins A aus Fischtranen. Helv. chim. Acta 14, 1036, 1431 (1931).
90. KARRER, P., H. SALOMON, R. MORF u. O. WALKER: Über mehrfach ungesättigte, den β- oder α-Jonon-Kohlenstoffring enthaltende Verbindungen. Helv. chim. Acta 15, 878 (1932).
91. KARRER, P. u. P. SCHNEIDER: Beitrag zur Konstitution des Vitamins A_2. Helv. chim. Acta 33, 38 (1950).

92. KARRER, P. u. R. SCHWYZER: Überführung von Vitamin A in Anhydrovitamin A und ein Carotinoid, wahrscheinlich identisch mit β-Carotin. Helv. chim. Acta 31, 1055 (1948).
93. KIPPING, F. B. and F. WILD: Synthesis of Vitamin A Methyl Ether. Chem. and Ind. 1939, 802.
94. KUHN, R. u. C. J. O. R. MORRIS: Synthese von Vitamin A. Ber. dtsch. chem. Ges. 70, 853 (1937).
95. LEDERER, E. A. and V. A. ROZANOVA: Studies on Vitamin A of Fish-Liver Oils. I. An Abnormal Reaction of CARR and PRICE. Biokimiya (U. S. S. R.) 2, 293 (1937).
96. LINNELL, W. H. and C. C. SHEN: Synthesis of the Benzene Analog of Vitamin A. J. Pharmac. Pharmacol. 1, 971 (1949).
97. MEAD, T. H.: Crystalline Esters of Vitamin A. I. Preparation and Properties. Biochemic. J. 33, 589 (1939).
98. MEUNIER, P.: Sur la véritable formule de l'axérophtène. C. R. hebd. Séances Acad. Sci. 227, 206 (1948).
99. — Des carotènes aux rétinènes. Revue internationale de Vitaminologie 23, 31 (1951).
100. MEUNIER, P., R. DULOU et A. VINET: Sur les conditions de formation et la constitution de la vitamine A dite « cyclisée ». Bull. Soc. Chim. biol. 25, 371 (1943).
101. MEUNIER, P. et R. FERRANDO: Activité biologique du « chromogène à 560 mμ » obtenu dans l'oxydation permanganique de l'axérophtol. Bull. Soc. Chim. biol. 31, 227 (1949).
102. MEUNIER, P., A. GUÉRILLOT-VINET, J. JOUANNETEAU et M. GOUREVITCH: Sur les corps obtenus par action progressive de l'acide chlorohydrique sur la vitamine A et leur activité biologique. C. R. hebd. Séances Acad. Sci. 226, 128 (1948).
103. MEUNIER, P. et J. JOUANNETEAU: Sur un congénère de l'axérophtal (rétin-aldéhyde). Bull. Soc. Chim. biol. 30, 185 (1948).
104. — Recherches sur l'isomérie cis-trans dans la série de la vitamine A (axérophtal). Bull. Soc. Chim. biol. 30, 260 (1948).
105. MEUNIER, P., J. JOUANNETEAU et R. FERRANDO: Sur la toxicité pour le Rat blanc d'un nouveau dérivé d'oxydation de la vitamine A. C. R. hebd. Séances Acad. Sci. 230, 140 (1950).
106. MEUNIER, P., J. JOUANNETEAU et G. ZWINGELSTEIN: Sur la coupure oxydante du β-carotène en rétinène (axérophtal) par MnO_2. C. R. hebd. Séances Acad. Sci. 231, 1170 (1950).
107. — — — La coupure oxydante du lycopène par MnO_2 et l'activité biologique du rétinène$_2$ ainsi obtenu. C. R. hebd. Séances Acad. Sci. 231, 1570 (1950).
108. MEUNIER, P., R. MALLEIN et J. JOUANNETEAU: Recherches chimiques dans la série du rétinène (axérophtal). I. Identification du congénère du rétinène responsable de l'halochromie à 560 mμ. Bull. Soc. Chim. biol. 31, 965 (1949).
109. MEUNIER, P., G. ZWINGELSTEIN, J. JOUANNETEAU et R. MALLEIN: Sur la structure exacte du dérivé à action antivitaminique A obtenu par oxydation de l'axérophtol. C. R. hebd. Séances Acad. Sci. 230, 1323 (1950).
110. MILAS, N. A.: Vitamin A. U. S. 2369156, Feb. 13 (1945).
111. — Synthesis of Biologically Active Vitamin A Substances. Science (New York) 103, 581 (1946).
112. — The Synthesis of Vitamin A and Related Products. Vitamins and Hormones 5, 1 (1947).
113. — Vitamin A Acid and Esters. U. S. 2424994, August 5 (1947).
114. — Dimethylamino Vitamin A. U. S. 2507802, May 16 (1950).

115. MILAS, N. A., P. DAVIS, I. BELIČ and D. A. FLEŠ: Synthesis of β-Carotene. J. Amer. chem. Soc. **72**, 4844 (1950).
116. MILAS, N. A. and T. M. HARRINGTON: New Synthesis of 1-(2',6',6'-Trimethylcyclohexen-1'-yl)-3-methyl-1,3,5-trien-7-one (C_{18}-ketone). J. Amer. chem. Soc. **69**, 2247 (1947); cf. **70**, 4275 (1948).
117. MILAS, N. A., S. W. LEE, E. SAKAL, H. C. WOHLERS, N. S. MCDONALD, F. X. GROSSI and H. F. WRIGHT: Synthesis of Products Related to Vitamin A. IV. The Application of the Darzens Reaction to β-Ionone. J. Amer. chem. Soc. **70**, 1584 (1948).
118. MILAS, N. A., S. W. LEE, C. SCHUERCH, R. O. EDGERTON, J. T. PLATI, F. X. GROSSI, Z. WEISS and M. A. CAMPBELL: Synthesis of Products Related to Vitamin A. V. The Synthesis of [1-(2',6',6-trimethylcyclohexen-1'-yl)-3,7-dimethyldeca-1,3,5,7-tetraenyl]-10-ethyl Ether. J. Amer. chem. Soc. **70**, 1591 (1948).
119. MILAS, N. A., E. SAKAL, J. T. PLATI, J. T. RIVERS, J. K. GLADDING, F. X. GROSSI, Z. WEISS, M. A. CAMPBELL and H. F. WRIGHT: Synthesis of Products Related to Vitamin A. VI. The Synthesis of Biologically Active Vitamin A Ethers. J. Amer. chem. Soc. **70**, 1597 (1948).
120. MORTON, R. A.: Chemical Aspects of the Visual Process. Nature (London) **153**, 69 (1944).
121. MORTON, R. A., M. K. SALAH and A. L. STUBBS: Retinene$_2$ and Vitamin A_2. Nature (London) **159**, 744 (1947).
122. OROSHNIK, W.: A Synthesis of Vitamin A Methyl Ether—Preliminary Report. J. Amer. chem. Soc. **67**, 1627 (1945).
123. — Ethynyl Carbinols. U. S. 2425201, August 5 (1947).
123a. OROSHNIK, W., G. KARMAS and A. D. MEBANE: Synthesis of Polyenes. I. *Retro*vitamin A Methyl Ether, Spectral Relationships between the β-Ionolidene and *Retro*ionolidene Series. J. Amer. chem. Soc. **74**, 295 (1952).
124. OROSHNIK, W. and A. D. MEBANE: The Nef Reaction with α,β-Unsaturated Ketones. J. Amer. chem. Soc. **71**, 2062 (1949).
125. Ortho Pharmaceutical Company: Brit. 630865.
126. — Brit. 653679.
127. PETROW, V. and O. STEPHENSON: The Synthesis of 4-Carboxyvitamin A Acid. J. chem. Soc. (London) **1950**, 1310.
128. PREOBRAZHENSKII, N. A. and I. A. RUBSTOV: Syntheses in the Field of Vitamin A. III. Synthesis of β-Ionylideneacetaldehyde. J. gen. Chem. (USSR.) **18**, 1719 (1948).
128a. ROBESON, C. D.: Esterification of Polyene Acids. U. S. 2583594 (January 29, 1952).
129. ROBESON, C. D. and J. G. BAXTER: Neovitamin A. J. Amer. chem. Soc. **69**, 136 (1947).
130. ROBESON, C. D. and C. C. EDDINGER: Unsaturated Aldehydes. U. S. 2507647, May 16 (1950).
131. SALAH, M. K. and R. A. MORTON: Crystalline Retinene$_2$. Biochemic. J. **43**, LVI (1948).
132. SCHWARZKOPF, O., H. J. CAHNMANN, A. D. LEWIS, J. SWIDINSKY and H. M. WÜEST: Synthesis of Vitamin A. A Synthesis of Vitamin A_3 Acid Ester, Its Conversion to Vitamin A Acid Ester and Its Reduction to Vitamin A_3. Abstracts, 115th Meeting American Chem. Soc., San Francisco, California. March 27–April 1, 11 C (1949).
133. — — — — — Zur Synthese des Vitamins A. I. Eine neue Methode zur direkten Darstellung von Vitamin A von hoher biologischer Wirksamkeit. Helv. chim. Acta **32**, 443 (1949).

134. SHANTZ, E. M.: Isolation of Pure Vitamin A₂. Science (New York) **108**, 417 (1948).
135. — Rehydrovitamin A, the Compound Formed from Anhydrovitamin A *in vivo*. J. biol. Chemistry **182**, 515 (1950).
136. SHANTZ, E. M. and J. H. BRINKMAN: Biological Activity of Pure Vitamin A₂. J. biol. Chemistry **183**, 467 (1950).
137. SHANTZ, E. M., J. D. CAWLEY and F. B. CLOUGH: The Structure of the Ionylideneacetic Acids. Unpublished report. (Distillation Products industries.)
138. SHANTZ, E. M., J. D. CAWLEY and N. D. EMBREE: Anhydro ("Cyclized") Vitamin A. J. Amer. chem. Soc. **65**, 901 (1943).
139. SHANTZ, E. M., C. D. ROBESON and H. M. KASCHER: Dehydration and Isomerization of Straight-Chain Conjugated Polyene Esters and Acids. U. S. 2 576 104, Nov. 27 (1951).
140. SOBOTKA, H. and E. BLOCH: Polyene Syntheses. Chem. Reviews **34**, 435 (1941).
141. SOBOTKA, H., E. BLOCH and D. GLICK: Studies on Ionone. I. Cleavage of Ethyl Ionylidene Acetate. J. Amer. chem. Soc. **65**, 1961 (1943).
142. VAN DORP, D. A. and J. F. ARENS: Biological Activity of Vitamin A Acid. Nature (London) **158**, 60 (1946).
142a. — — The Synthesis of "Vitamin A Acid", A Biologically Active Substance. Recueil Trav. chim. Pays-Bas **65**, 338 (1946).
143. — — Synthesis of Vitamin A Aldehyde. Nature (London) **160**, 189 (1947).
144. — — Condensation Reaction of Ketones with Hydroxymaleic Anhydride. Transformation of the Condensation Products into α,β-Unsaturated Aldehydes. Recueil Trav. chim. Pays-Bas **67**, 459 (1948).
145. WALD, G.: The Chemical Evolution of Vision. The Harvey Lectures **41**, 117 (1945–1946).
146. — Synthesis from Vitamin A₁ of Retinene and of a New 545 mμ Chromogen Yielding Light-Sensitive Products. J. gen. Physiol. **31**, 489 (1947–1948).
147. — The Enzymic Reduction of the Retinenes to the Vitamins A. Science (New York) **109**, 482 (1949).
148. — Interconversion of the Retinenes and Vitamins A *in vitro*. Biochem. et Biophys. Acta **4**, 215 (1950).
149. WALD, G. and R. HUBBARD: Reduction of Retinene₁ to Vitamin A₁ *in vitro*. J. gen. Physiol. **32**, 367 (1948–1949).
150. WEEDON, B. C. L. and R. J. WOODS: Studies in the Polyene Series. XXXVII. The Synthesis of Phenyl Analogues of Vitamin A Acids and Vitamin A. J. chem. Soc. (London) **1951**, 2687.
151. WENDLER, N. L., C. ROSENBLUM and M. TISHLER: The Oxidation of β-Carotene. J. Amer. chem. Soc. **72**, 234 (1950).
152. WENDLER, N. L., H. L. SLATES and M. TISHLER: Synthesis of Vitamin A. J. Amer. chem. Soc. **71**, 3267 (1949).
153. WENDLER, N. L., H. L. SLATES, N. R. TRENNER and M. TISHLER: Synthesis of Vitamin A. J. Amer. chem. Soc. **73**, 719 (1951).
154. YOUNG, W. G., L. J. ANDREWS and S. J. CRISTOL: Polyenes I. The Synthesis and Absorption Spectra of the Ionylideneacetones and Related Compounds. J. Amer. chem. Soc. **66**, 520 (1944).
155. ZECHMEISTER, L.: Stereoisomeric Provitamins A. Vitamins and Hormones **7**, 57 (1949).
156. ZEILE, K. u. H. MEYER: Über einige Umsetzungen von Propargylderivaten. Ber. dtsch. chem. Ges. **75**, 356 (1942).

(Received, March 26, 1952.)

Les Antivitamines.
Par P. Meunier, Lyon.
Sommaire.

	Page
Introduction	88
I. Les antagonistes des vitamines hydrosolubles	90
1° Les antagonistes de la thiamine	90
a) Pyrithiamine	90
b) Oxythiamine	91
c) Homothiamine-glycol	91
d) Antagonistes de la thiamine de nature enzymatique: thiaminases diverses	92
e) Antithiamines des fougères	93
2° Les antagonistes des flavines et de la vitamine B_{12}	94
3° Les antagonistes de l'acide pantothénique	95
4° Les antagonistes de la pyridoxine	95
a) Activité vitaminique B_6	95
b) Constitution chimique des antivitamines B_6	96
c) Désoxypyridoxine	96
d) Méthoxypyridoxine	97
e) Autres antivitamines B_6	97
5° Les antagonistes de l'acide nicotinique	98
6° Les antagonistes de la biotine	98
7° Les antisulfamides	100
8° Les anti-acides foliques	100
9° L'activité antisulfamide et antisulfone de dérivés voisins de l'acide *p*-aminobenzoïque	101
II. Les antagonistes des vitamines liposolubles	102
1° L'antivitamine A	102
2° Les antivitamines E	103
3° Les antivitamines K	104
Bibliographie	107

Introduction.

Au cours de ces dix dernières années, la notion d'antivitamines n'a cessé d'apporter à la biochimie de nouvelles données expérimentales d'une importance fondamentale, non seulement sur le plan théorique mais également pratique. Parmi les antagonistes des vitamines, plusieurs peuvent être directement utilisés pour l'étude du métabolisme intermédiaire de certains constituants des êtres vivants. Il est en effet plus simple de provoquer une avitaminose par administration d'une antivitamine qu'en soumettant l'animal à un régime carencé durant une période plus ou moins longue. La plupart du temps, les troubles provoqués par une telle administration sont bien localisés et concernent certaines réactions diastasiques bien délimitées.

Envisageons par exemple une chaîne de réactions $A \to B \to C$ etc. $\to Z$ dans laquelle une substance A donne naissance à Z, en passant par les intermédiaires B, C, D, etc. . . . Nous savons que la plupart des diastases responsables de ces transformations successives nécessitent la présence de coferments de nature vitaminique. Si nous bloquons maintenant l'un de ces coferments par l'administration de l'antivitamine correspondante, l'un des intermédiaires de la chaîne ne sera plus transformé et s'accumulera dans les tissus ou les humeurs de l'organisme envisagé.

Il a été ainsi démontré que la pyridoxine doit jouer le rôle de coferment au cours de la transformation de la cynurénine en alanine et acide anthranilique (*12*). D'autres problèmes pourront sans doute être résolus d'une façon analogue.

Sur le plan pratique, plusieurs antivitamines en tant qu'inhibiteurs de certains mécanismes biochimiques commencent à être appliquées au traitement de quelques maladies. L'emploi des antivitamines dans la lutte contre les thromboses est déjà très répandu en clinique humaine. Les antagonistes de l'acide folique présentent un certain intérêt dans le traitement du cancer expérimental. Là encore, d'intéressantes possibilités se présentent et il est fort probable que les divers antibiotiques actuellement connus agissent en inhibant des facteurs de croissance normalement nécessaires aux microorganismes. L'étude de la chloromycétine, analogue structural de la sérine ou de la phénylalanine (*55, 55a, 67, 67a*) est particulièrement intéressante sous cet angle et a fait l'objet, dans ce laboratoire, de plusieurs investigations expérimentales.

Les recherches dont il est question dans l'exposé qui va suivre concernent plus particulièrement les travaux des cinq dernières années, mais même limitée à cette période relativement courte, notre mise au point ne peut avoir la prétention d'être complète, en raison de l'ampleur du sujet et du nombre considérable de publications qui ont trait à ce problème. Nous nous excusons donc auprès de tous les auteurs dont les travaux auraient pu échapper à nos investigations.

Durant la rédaction de notre manuscrit, un ouvrage sur l'ensemble des antimétabolites actuellement connus a été publié par WOOLLEY (*122*). En ce qui concerne les travaux antérieurs à 1948, nous mentionnons plus particulièrement les rapports du Colloque de Lyon en septembre 1948 sur les « Antivitamines » (*9*) et les mises au point de MENTZER (*52*) et de ROBLIN (*83*).

I. Les antagonistes des vitamines hydrosolubles.
I° Les antagonistes de la thiamine.
a) Pyrithiamine.

Au Colloque de Lyon en 1948, Schopfer (*90*) a fait une mise au point sur la pyrithiamine que l'on peut considérer comme la première étudiée parmi les antivitamines B_1, celle qui avait fait l'objet d'une note préliminaire de Schmelkes et Joiner (*87*). Woolley et White (*123*) avaient, dès 1943, signalé que cette pyrithiamine déterminait une avitaminose B_1 chez la souris et agissait comme inhibiteur de certains microorganismes. C'est sur ce même produit, synthétisé par Baumgarten et Dornow (*3*) et appelé « hétérovitamine B_1 » que Schopfer (*90*) avait fait ses premières recherches sur les carences en thiamine produites dans les espèces végétales par la pyrithiamine précédemment indiquée. Dans l'esprit des auteurs ci-dessus, l'analogue essayé était un produit dans lequel le groupe thiazol était remplacé par une pyridine substituée de façon identique à celle de la vitamine. Mais le doute ne tarda pas à se faire jour quand Baumgarten et Dornow (*4*) purent préciser que, dans l'hétérovitamine, c'était un groupe α-oxyéthyl qui était substitué et non le groupe β-oxyéthyl, comme dans la vitamine elle-même. Pourtant, dès 1941, Tracy et Elderfield (*102*) avaient réussi à synthétiser le véritable isostère pyridinique de la thiamine (I) qu'ils appelèrent pyrithiamine (II). Robbins (*81*) qui essaya ce corps nouveau au point de vue biologique ne semble pas pouvoir interpréter les résultats obtenus faute d'une notion très claire de l'antivitamine. En 1949, Wilson et Harris (*109*) pressentent que l'ancienne pyrithiamine n'était qu'un mélange et font la synthèse d'une néopyrithiamine dont l'analyse élémentaire est conforme aux calculs du véritable isostère et qui se révèle d'emblée beaucoup plus active que la soi-disant pyrithiamine du début. Raffauf (*80*), du laboratoire de Reichstein, a également effectué la synthèse de la néopyrithiamine, sur la demande de Schopfer, lequel trouve cette néopyrithiamine 15 à 300 fois plus active que la première pyrithiamine selon les organismes essayés (Schopfer, sous presse). Woolley (*121*) a récemment confirmé l'activité antivitaminique de la néopyrithiamine chez la souris, sans indiquer expressément dans son étude enzymatique que l'antivitamine étudiée n'est plus tout à fait la pyrithiamine de ses premiers travaux, bien qu'il emploie l'expression de « néopyrithiamine » dans ce dernier article.

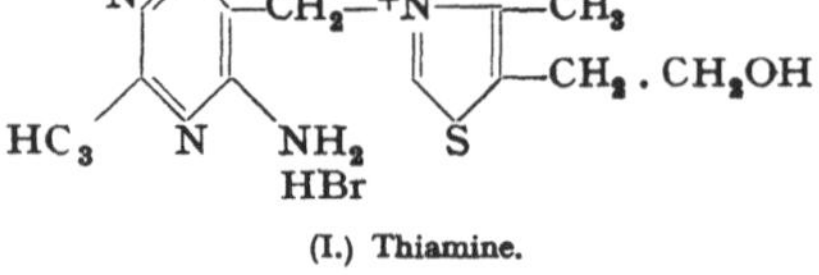

(I.) Thiamine.

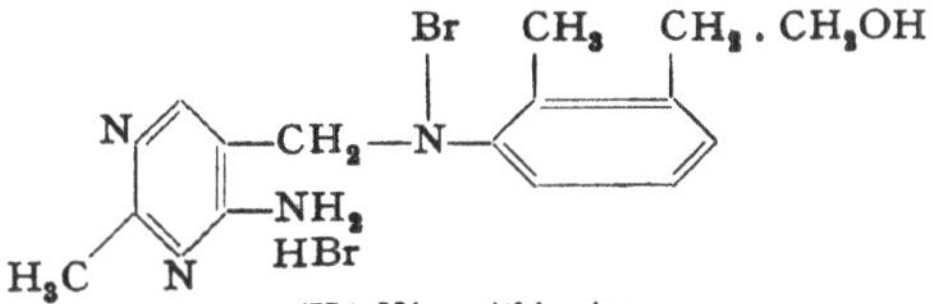

Les Antivitamines.

$$\text{H}_3\text{C}-\underset{N}{\overset{N}{\diagup\!\!\!\diagdown}}-\text{CH}_2-\underset{\underset{\text{HBr}}{\text{NH}_2}}{\overset{\overset{\text{Br}}{|}}{N}}-\underset{}{\overset{\text{CH}_3\ \ \text{CH}_2.\text{CH}_2\text{OH}}{\diagup\!\!\!\diagdown}}$$

(II.) Néopyrithiamine.

b) Oxythiamine.

L'oxythiamine de SOODAC et CERECEDO (96) a été signalée comme conférant une protection significative de la souris contre la souche de Lansing de virus de la poliomélyte (36). RYDON (85) a indiqué, plus récemment, après le Colloque de Lyon, une méthode éprouvée pour la préparation de l'oxythiamine sans thiamine avec 80% de rendement en chauffant à reflux la thiamine avec l'acide chlorhydrique 5 N pendant six heures. Dès 1949, FROHMAN et DAY (25) avaient déjà étudié l'oxythiamine (III) chez le rat et montré que c'était un antagoniste puissant de la thiamine.

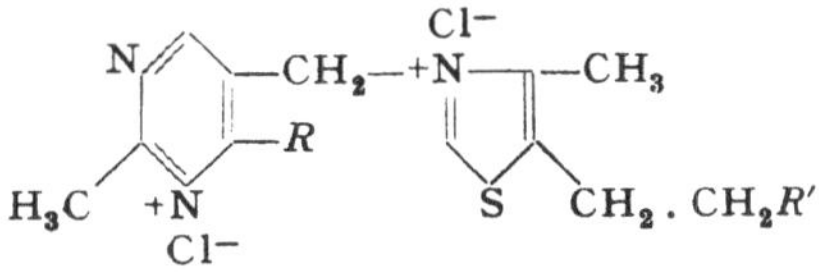

$$\text{H}_3\text{C}-\underset{\underset{\text{Cl}^-}{+\text{N}}}{\overset{N}{\diagup\!\!\!\diagdown}}-\underset{R}{\text{CH}_2}-\overset{\overset{\text{Cl}^-}{|}}{+\text{N}}\underset{S}{\diagup\!\!\!\diagdown}\begin{matrix}-\text{CH}_3\\ \text{CH}_2.\text{CH}_2R'\end{matrix}$$

(III.) Oxythiamine. ($R = R' = $ OH.)

Dans ce travail, le déplacement de la vitamine B_1 par l'oxythiamine est démontré d'une façon directe par l'excès de pyruvate dans le sang et par l'augmentation de l'excrétion de thiamine suivant l'administration de l'antagoniste. Dans cette expérience, l'oxythiamine et la thiamine sont administrées en solution saline dans le péritoine et les animaux sont décapités une ou quatre heures après cette administration; le sang est recueilli pour l'analyse des acides lactique et pyruvique. Une heure après l'administration d'oxythiamine, un rapport normal pyruvate/lactate est encore obtenu, mais quatre heures après l'administration de l'antagoniste, il y a un grand excès de pyruvate et de lactate. Ces résultats semblent bien prouver que l'oxythiamine interfère directement avec le fonctionnement de la thiamine dans le métabolisme pyruvique, soit que le diphosphate de la thiamine soit déplacé de son transporteur protéique par l'oxythiamine, soit qu'un diphosphate d'oxythiamine inerte au point de vue physiologique ait été formé.

c) Homothiamine-glycol.

L'homothiamine-glycol (IV), homologue de la thiamine préparé récemment par KARRER et SCHOELLER (39), a été étudié tout dernièrement pour ses propriétés antivitaminiques B_1 par SCHOPFER et BEIN (91) qui ont prouvé qu'il s'agissait bien là d'une antivitamine B_1 à antagonisme compétitif vis-à-vis de la thiamine, le rapport d'inhibition homothiamine/thiamine étant voisin de 100. Dans cette même publication, ces auteurs ont vérifié que l'antagonisme de l'homothiamine et de la thiamine s'exerçait aussi vis-à-vis de la culture d'un tissu de racine (méristème radiculaire de PISUM) cultivé en milieu synthétique. Cette dernière application représente certainement une méthode assez générale pour l'étude des antivitamines dans le règne végétal.

$$\underset{H_3C}{\overset{N}{\diagdown}}\underset{N}{\diagup}\overset{-CH_2-}{\underset{-NH_2 \quad HCl}{}}\overset{Cl^-}{\underset{S}{+N\diagdown\diagup}}\overset{-CH_3}{\underset{-CH_2 . CHOH . CH_2OH}{}}$$

(IV.) Homothiamine-glycol.

d) Antagonistes de la thiamine de nature enzymatique: thiaminases diverses.

Il y a une vingtaine d'années, l'attention des vétérinaires d'Amérique du Nord était attirée par une maladie d'allure épizootique qui affectait les élevages de renards. GREEN et EVANS (*30*) constataient que l'animal atteint de la paralysie dite de CHASTEK présente des symptômes analogues à ceux observés chez l'homme au cours de la polyencéphalite hémorragique dite paralysie de WERNICKE.

Les signes cliniques sont les suivants: anorexie, faiblesse générale suivie d'ataxie progressive et de paraplégie entrecoupée de crises nerveuses; les malades gémissent continuellement dans la dernière période de la maladie. La mort survient entre 48 et 72 heures après l'apparition des premiers symptômes. L'autopsie montre la dégénérescence du foie et des lésions vasculaires des centres nerveux confirmant ainsi la ressemblance avec la maladie de WERNICKE. Quand les femelles en gestation sont atteintes, les foetus meurent quelque temps avant leur mère puisque, à l'autopsie, on les découvre en décomposition dans l'utérus. Les avortements sont aussi fréquents et certaines femelles, sans paraître malades, ne peuvent nourrir convenablement leur portée qui meurt en cours d'allaitement. La maladie est aiguë, subaiguë ou chronique.

GREEN, CARLSON et EVANS (*28, 29*) s'aperçurent, après enquête, que, dans tous les élevages atteints, la ration des renards contenait au moins 10% de poisson cru. Ils rapprochèrent cette constatation de leur diagnostic d'avitaminose B_1. GREEN et ses collaborateurs furent alors amenés à faire, sur plusieurs lots comprenant au total 300 sujets, des séries d'expériences mettant en évidence et le rôle déterminant du poisson cru dans l'étiologie de la maladie et l'action empêchante de la coprophagie. Continuant leurs recherches, ils purent indiquer:

— que l'action d'une livre (anglaise) de carpe entière est neutralisée par un supplément à la ration de 10 mg. de chlorhydrate d'aneurine;

— que les filets du poisson sont sans danger mais que les autres parties, et notamment les viscères, provoquent la maladie.

La carpe n'est pas seule à mettre en cause. Le merlan, le maquereau, le brochet, de nombreux poissons des lacs américains et canadiens auraient la même action. On ne peut donc faire de différence entre les poissons d'eau douce et les poissons de mer; certains mollusques et même le caviar sont aussi à incriminer. Ni le milieu ni l'espèce n'entrent en ligne de compte.

Cette action du poisson cru fut également étudiée chez l'homme (*49, 49a*). Elle peut entraîner la destruction de la moitié de la thiamine de son régime.

Tous ces faits permettent d'affirmer que certains poissons et certains mollusques contiennent des thiaminases ou antivitamines B_1. Quelle est la nature de cette substance? En 1941, WOOLLEY (*114*) et de leur côté SPITZER et ses collaborateurs (*97*) préparèrent un extrait hydro-alcoolique de carpe dont l'action antithiamine *in vitro* est très nette. Les propriétés physiques et chimiques de la substance permettent de l'assimiler à un enzyme de nature protéidique, d'où son nom de thiaminase.

La destruction de la thiamine *in vivo* n'est pas aussi grande que celle produite *in vitro*. Pourtant MELNICK (*49, 49a*) a montré que cette destruction peut être importante dans les mélanges d'aliments et dans l'estomac où elle peut atteindre 50% de la thiamine de la ration. En 1949, FERRANDO (*21*) a fait une revue de cette question.

e) Antithiamines des fougères.

Des accidents semblables, quant à leur étiologie, à la paralysie de CHASTEK ont été étudiés par WESWIG et ses collaborateurs (*107*) chez des animaux de l'espèce bovine consommant des fougères du genre *Pteris aquilina* et qui succombèrent. On étudie l'action de ces fougères sur le rat. Les groupes de rats reçoivent tous une ration de base contenant 40% de fougères séchées à l'air libre et à la température du laboratoire. Ce régime contient au total de 0,2 à 0,6 mg. de thiamine, mais on donne *per os* à chacun des animaux d'un groupe un supplément quotidien de 5 mg. de thiamine. Les sujets de ce dernier groupe ne présentent aucun symptôme. Les autres meurent au bout de 20 jours après avoir montré tous les signes d'une carence en vitamine B_1.

Quand on administre un supplément de thiamine aux sujets présentant les premiers signes de carence, tout rentre dans l'ordre à une exception près. Différents lots de fougères provenant de divers endroits et recueillis à des moments variables de leur cycle de végétation se sont montrés toxiques. Le chauffage à 105° ne diminue pas la toxicité. Le principe insoluble dans l'éther et dans l'acétone est légèrement soluble dans l'éthanol à 92%.

Doit-on invoquer une action enzymatique ? Mais alors, dans ce cas, les bactéries des réservoirs digestifs des ruminants ne risquent-elles pas de pouvoir utiliser à nouveau les deux portions de la thiamine ? Existe-t-il une action semblable à celle qui se produit pour la biotine et l'avidine ou bien l'action est-elle plus profonde encore ? On comprend que devant la complexité de la nutrition chez les ruminants, Weswig et ses collaborateurs demeurent très prudents. Pour l'instant, aucun autre travail n'a été, à notre connaissance, effectué sur cette question et l'on en est réduit à des hypothèses.

2° **Les antagonistes des flavines et de la vitamine B_{12}.**

La dichloroflavine (VI) semble être la seule antivitamine correspondant à la flavine ayant fait l'objet de recherches jusqu'à ce jour.

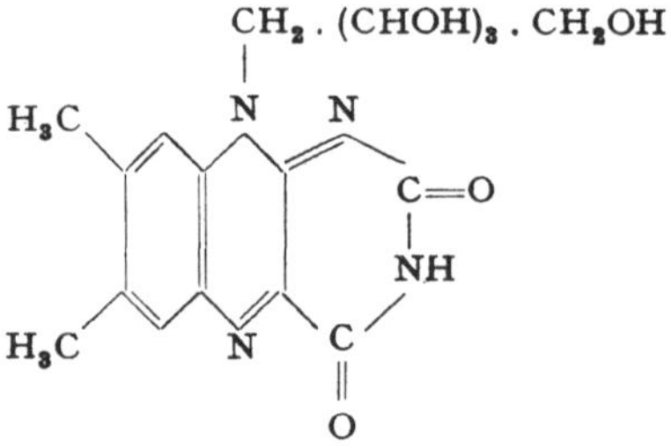

(V.) Lactoflavine. 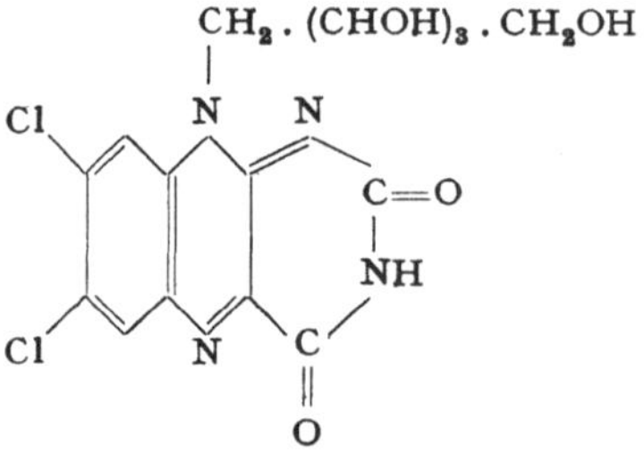   (VI.) Dichloroflavine.

Cette dichloroflavine a été synthétisée pour la première fois par Kuhn, Weygand et Möller (*41*) qui ont étudié son action inhibitrice vis-à-vis de *Bacillus lactis acidi*, de *Staphylococcus aureus* et de *Streptobacterium plantarum*. Depuis cette date, la théorie du déplacement a été réfutée dans ce cas par Karrer (*38*) qui a étudié *in vitro* l'action de la dichloroflavine sur la réaction fermentaire due à l'action de la xanthine déhydrase, de l'aldéhydodéhydrase du lait et d'oxydases de D-aminoacides qui n'ont en aucune façon été inhibés par la dichloroflavine. Schopfer (*88*) a repris cette antivitamine pour montrer qu'elle inhibait la croissance de *Eremothecium Ashbyii* qui est pourtant producteur de lactoflavine.

Une autre substance très simple est récemment apparue à Woolley (*119*) comme un antagoniste de la riboflavine et de la vitamine B_{12}. Il s'agit du 1,2-dichloro-4,5-diaminobenzène (VII).

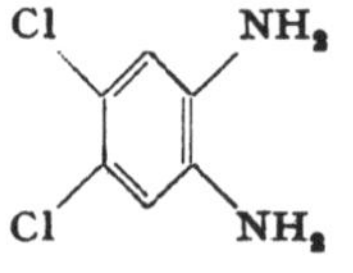

(VII.) 1,2-Dichloro-4,5-diaminobenzène. 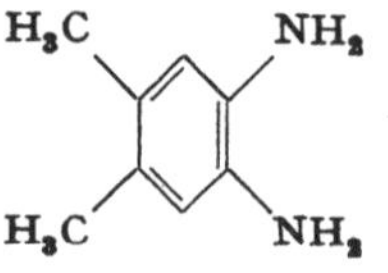   (VIII.) 1,2-Diméthyl-4,5-diaminobenzène.

Dans les cultures de *Bacillus megatherium*, l'inhibition se produit avec 7 γ de dérivé dichloré. WOOLLEY admet que, dans ses expériences, la diméthyl-diamine correspondante (VIII) est un précurseur à la fois de la riboflavine et de la vitamine B_{12} (*120*). L'analogue dichloré déterminerait l'arrêt des processus biosynthétiques dans la formation de ces vitamines.

3° Les antagonistes de l'acide pantothénique.

La pantoyltaurine (IX) et ses dérivés ont certainement été, parmi les antivitamines en général, celles qui avaient laissé entrevoir le plus d'espoir pratique dans la lutte contre les parasites de la Malaria en particulier, depuis l'observation de TRAEGER (*103*), d'après laquelle le *Plasmodium lophurrae* requérait l'acide pantothénique pour sa croissance. Mais depuis ces premiers espoirs, il semble bien que les fonctions de l'acide pantothénique et de ses analogues aient été limitées à l'étude du métabolisme du radical acétyle (*72*).

$$HOCH_2 . \underset{\underset{CH_3}{|}}{\overset{\overset{CH_3}{|}}{C}} . CHOH . CO . NH . CH_2 . CH_2SO_3H$$

(IX.) Pantoyltaurine.

4° Les antagonistes de la pyridoxine.

a) Activité vitaminique B_6.

La pyridoxine (X) partage avec le pyridoxal (XI) et la pyridoxamine (XII) l'activité vitaminique B_6 à des degrés variables; suivant le test considéré pour les animaux: rats et poulets en particulier, les trois dérivés sont équivalents. Pour certains microorganismes, *Streptococcus lactis*, le pyridoxal et la pyridoxamine sont 1000 à 10000 fois plus actifs sur la croissance. L'activité de la « pseudopyridoxine » est due au pyridoxal et à la pyridoxine.

Divers analogues structuraux ont été essayés pour leur action vitaminique B_6 (*106*), puis pour une activité anti-B_6.

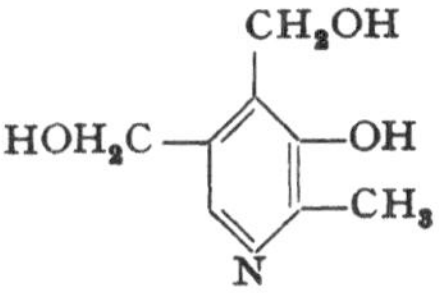

(X.) Pyridoxine. 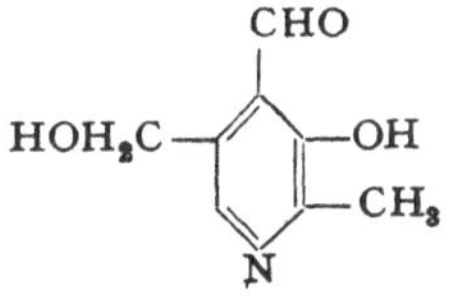   (XI.) Pyridoxal. 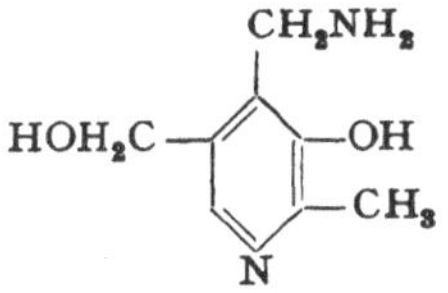   (XII.) Pyridoxamine.

b) *Constitution chimique des antivitamines B_6.*

La première substance à activité antivitaminique B_6 a été décrite par OTT en 1946 (*77*); c'est la 2,4-diméthyl-3-hydroxy-5-hydroxyméthyl-pyridine ou «désoxypyridoxine» (XIII). Elle inhibe la croissance du poulet. Puis OTT (*78*) a montré que la 2-méthyl-3-hydroxy-4-méthoxyméthyl-5-hydroxyméthyl-pyridine ou «méthoxypyridoxine» (XIV) était également anti-B_6.

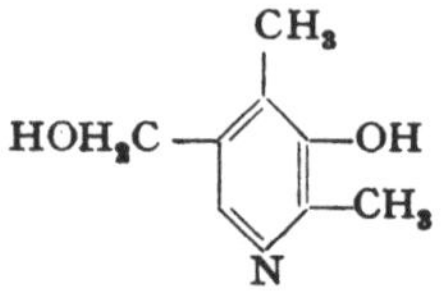

(XIII.) Désoxypyridoxine. 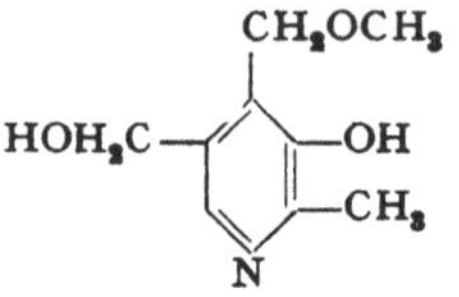   (XIV.) Méthoxypyridoxine.

MARTIN, AVAKIAN et MOSS (*47*) ont trouvé que la 2-méthyl-3-hydroxy-4-hydroxyméthyl-pyridine était déjà légèrement inhibitrice pour *Saccharomyces cerevisiae*, bien que ne différant de la B_6 que par l'absence d'un groupement hydroxyméthyl en 5. La 2-éthyl-3-amino-4-éthoxyméthyl-5-aminométhyl-pyridine a une action inhibitrice plus marquée que la désoxypyridoxine pour ce même *Saccharomyces*. La 2-acétoxy-3,5-diacétyl-méthyltoluène et la 2-méthyl-3-hydroxy-4-diméthylamino-méthyl-pyridine n'ont pas d'activité anti-B_6.

Les deux substances antivitamines B_6 les plus étudiées sont la désoxy- et la méthoxypyridoxine (XIII, XIV).

c) *Désoxypyridoxine.*

L'activité antivitaminique B_6 de cette substance se retrouve sur les différentes actions biologiques de la vitamine B_6: croissance des animaux, des microorganismes, métabolisme du tryptophane, activité enzymatique, etc.... Elle est annulée par la pyridoxine, le pyridoxal ou la pyridoxamine à des degrés divers suivant le test considéré. Voici les principaux résultats obtenus:

La désoxypyridoxine aggrave l'acrodynie des rats carencés en vitamine B_6 (*19*), augmente également l'excrétion d'acide xanthurénique après ingestion de tryptophane (*79*) et inhibe la croissance du poulet (*77*).

Elle provoque une atrophie du tissu lymphoïde (*99*) et diminue le pourcentage des lymphosarcomes de la souris (*100*). On a essayé d'utiliser chez l'homme cette action dans le lymphosarcome étendu ou la leucémie lymphoïde aiguë (*26*); la désoxypyridoxine n'a pas donné de résultats cliniques positifs; par ailleurs les auteurs n'ont pas constaté de modifications dans le métabolisme du tryptophane des sujets traités.

La désoxypyridoxine diminue également la sensibilité des souris au virus de la pneumonie (*43*). Elle inhibe la multiplication du T 2-bactériophage du Colibacille (*113*). Cette action est annulée par la vitamine B_6 et aussi par quelques acides gras.

Elle entraîne d'autre part des troubles de la reproduction chez les rats (*71*).

L'injection précoce de désoxypyridoxine à des oeufs entraîne la mort de l'embryon en trois à six jours (*11*); l'injection simultanée de pyridoxal, pyridoxamine ou de pyridoxine prévient cette action nocive; le rapport vitamine/inhibiteur est variable:

> 1/10 pyridoxal,
> 1/50 pyridoxamine,
> 1/100 pyridoxine.

Il faut remarquer que l'injection de désoxypyridoxine au cours de l'incubation (quatre ou cinq jours) n'est plus toxique. La vitamine B_6 est indispensable aux tous premiers stades du développement embryonnaire.

La désoxypyridoxine est inactive sur la tyrosine décarboxylase de *Streptococcus foecalis* (*5*) mais, fait curieux, elle devient inhibitrice après phosphorylation. C'est le premier exemple connu d'antivitamine qui doit être d'abord (et peut l'être) intégrée à un coenzyme, phosphate de désoxypyridoxine pour avoir une action inhibitrice (*105*).

Par ailleurs, l'activité de la décarboxylase de l'acide glutamique ne diminue pas plus rapidement dans le cerveau de rats carencés recevant de la désoxypyridoxine (*82*).

L'addition de 100 mg. par jour au régime de carence B_6, à base de caséine, a permis de constater chez l'homme normal l'apparition en trois semaines de symptômes de carence: troubles cutanés, faiblesse, nausées qui ont disparu en deux à trois jours après l'injection de pyridoxine (*70*).

d) *Méthoxypyridoxine*.

Les propriétés biologiques de la méthoxypyridoxine ont été moins largement étudiées que celles de la désoxypyridoxine.

La méthoxypyridoxine, outre son action inhibitrice sur la croissance du poulet (*78*), est très toxique pour l'embryon de poulet (*37*).

La pyridoxine, le pyridoxal et la pyridoxamine protègent l'embryon (ordre d'activité décroissante, la pyridoxamine étant la moins active); le phosphate de pyridoxal est encore moins efficace.

Certains auteurs ont trouvé que la méthoxypyridoxine avait une faible action vitaminique B_6 sur le métabolisme du tryptophane chez les rats carencés (*79*).

e) *Autres antivitamines B_6*.

L'activité antivitaminique B_6 de la 2-méthyl-3-hydroxy-4-hydroxy-méthylpyridine et celle de la 2-éthyl-3-amino-4-éthoxyméthyl-5-amino-méthylpyridine (*47*) n'a été étudiée, à notre connaissance, que sur la croissance de *Saccharomyces cerevisiae*.

L'opinion de Martin, Avakian et Moss. (*47*), selon laquelle le nombre encore trop restreint de substances analogues à la vitamine B_6 étudiées ne permet pas de définir avec précision les caractéristiques structurales indispensables à l'activité antivitaminique B_6, nous servira de conclusion provisoire dans ce domaine.

5° Les antagonistes de l'acide nicotinique.

L'acétyl-pyridine (XVI), analogue de la niacine (XV), a été étudié pour la première fois par Woolley (*117*) qui montra qu'il produisait des symptômes apparentés à ceux de la pellagre chez la souris et que ces symptômes étaient annulés soit par la niacine elle-même, soit par son précurseur, le tryptophane. Plus récemment, une investigation a été faite à l'aide de ce même analogue durant l'incubation de l'oeuf de poule (*73*). L'injection de doses graduelles d'antagoniste produit une mortalité croissante des embryons de poulets et cet effet est annulé par la nicotinamide elle-même.

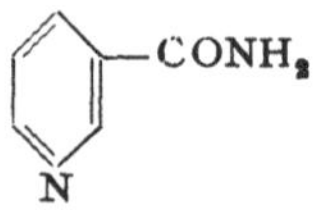

(XV.) Nicotinamide. 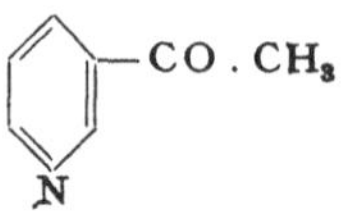   (XVI.) 3-Acétyl-pyridine.

Dans un ordre d'idées un peu inattendu, Schopfer a, en 1948, envisagé que l'action antibiotique de la méthyl-2-naphtoquinone chez la plante était due, dans certains cas bien déterminés, à la perturbation du métabolisme de la nicotinamide (*92*).

6° Les antagonistes de la biotine.

La structure de la biotine a été établie par Melville, Hofmann et du Vigneaud (*51*) en 1941, par synthèse. De même, c'est l'Ecole de du Vigneaud qui a étudié les premières antibiotines.

En 1943, Melville, Dittmer, Brown et du Vigneaud ont décrit une substance sans soufre, une desthiobiotine (XVIII) se conduisant comme antagoniste de la biotine (XVII) vis-à-vis de la croissance de *Lactobacillus casei* (*50, 16*).

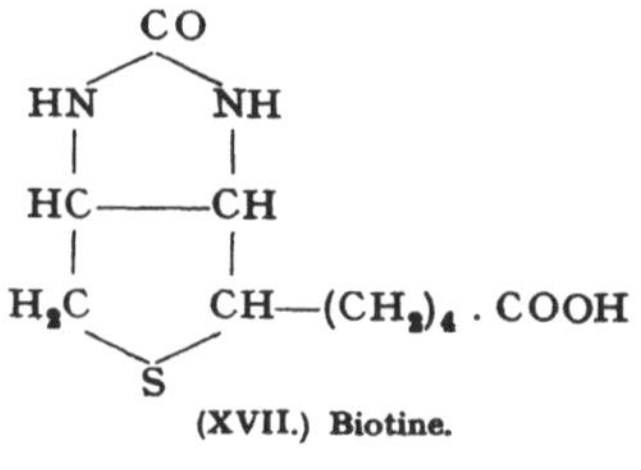

(XVII.) Biotine. 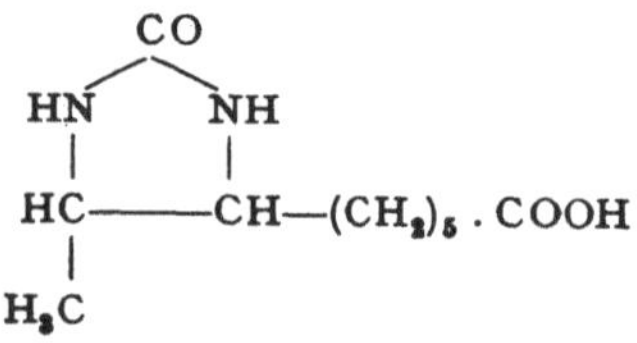   (XVIII.) Desthiobiotine.

En 1947, Rogers et Shive (*84*) d'une part et Dittmer et du Vigneaud (*16*) d'autre part suppriment le méthyle angulaire de la desthiobiotine et obtiennent des dérivés du type (XIX); pour $n = 5$, ils observent le maximum d'activité antibiotine.

$$\text{HN}\underset{\underset{\text{H}_2\text{C}\text{———}\text{CH}-(\text{CH}_2)_n\cdot\text{COOH}}{}}{\overset{\text{CO}}{\frown}}\text{NH}$$

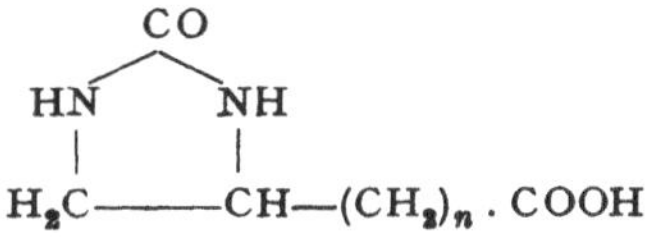

(XIX.)

En 1945, English et ses collaborateurs (*20*) avaient décrit un composé de structure (XX) avec le maximum d'activité antibiotine pour *Lactobacillus casei*.

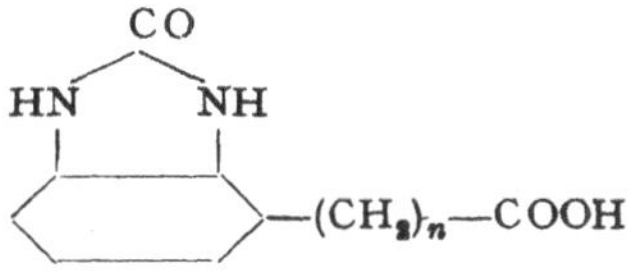

(XX.)

Dès 1944, Dittmer et collaborateurs (*17*) avaient préparé la biotine-sulfone (XXI), antagoniste pour *Lactobacillus casei* et au contraire probiotine pour *Saccharomyces cerevisiae*.

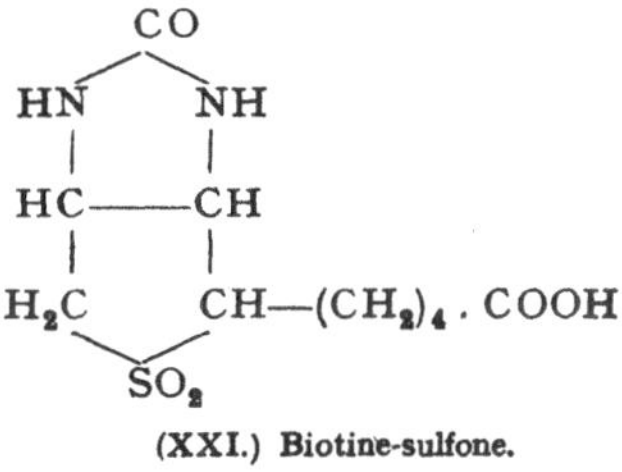

(XXI.) Biotine-sulfone.

Le remplacement du soufre dans la molécule de biotine par un atome d'oxygène conduisit à des dérivés d'activité similaire (oxybiotine) (*34*).

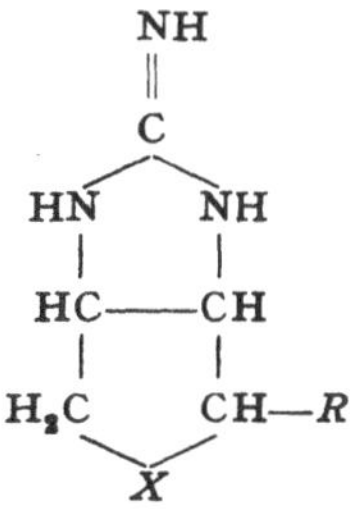

(XXII.) R variant de $-\text{CH}_3$ à $-(\text{CH}_2)_4\cdot\text{COOH}$.
$X = \text{O}$ ou S.

Citons enfin les corps de formule (XXII) qui se sont révélés inactifs à la fois comme biotines et comme antibiotines (*33*) vis-à-vis de la croissance des levures *S. cerevisiae* ou *L. casei* et qui sont des analogues guanidinés de la biotine ou de l'oxybiotine.

7° Les antisulfamides.

On sait que « l'acide *p*-aminobenzoïque occupe, parmi les facteurs de croissance pour les microorganismes, une place unique, par le fait que son importance biologique fut d'abord reconnue non par une action directe sur la croissance, mais par sa capacité de surmonter l'inhibition de la croissance bactérienne par les sulfamidés » (Woods, 1940) (*111, 112*). On se rappelle en effet que le nom de vitamine H' a été donné à ce corps (*40*) bien après la découverte de l'action thérapeutique du *prontosil rubrum* par Domagk (*18*) et la découverte de la *p*-amino-phénylsulfamide comme agent thérapeutique par Tréfouel, Mme Tréfouel, Nitti et Bovet (*104*). A ce moment, c'était le déplacement de l'acide *p*-aminobenzoïque par les sulfamidés qui paraissait le seul phénomène important.

Plus tard seulement, l'arrêt de la synthèse des substances liées à cet acide *p*-aminobenzoïque, des purines en particulier, est apparu comme non moins important (*94*); ainsi s'explique l'influence des purines et même des acides nucléiques comme antagonistes des sulfamidés [Schopfer (*89*)], indépendamment du travail des Américains Shive et Collaborateurs (*95, 95a*). C'est peu de temps après qu'il est devenu évident que les sulfamidés arrêtaient la synthèse de systèmes enzymatiques liés à l'acide folique et aux produits de même série (ptéroyl-glutamate) (*110*). Depuis cette date (1948), c'est à un développement considérable des études chimiques, physiologiques et des applications en clinique des ptéroyl-glutamates que nous assistons (*74*).

Rappelons à ce sujet qu'au Colloque de Lyon, en 1948, Woolley (*118*) avait déjà insisté sur l'importance de l'aminoptérine dans les essais de traitement de cancers.

8° Les antiacides foliques.

Des antagonistes puissants de l'acide folique (XXIII) sont obtenus en remplaçant dans la molécule de ce composé la fonction OH du noyau pyrimidine par un groupement NH_2.

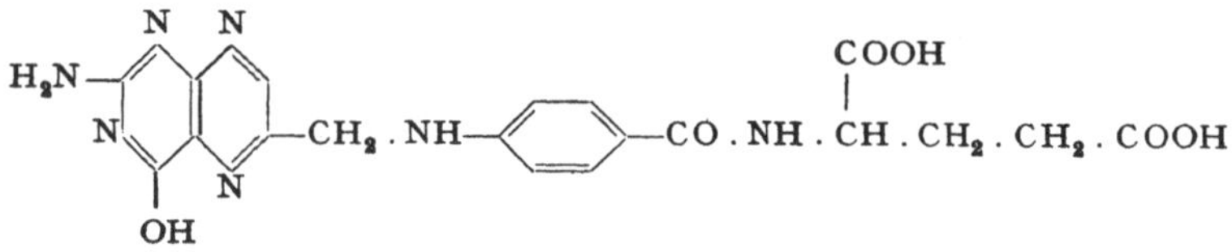

(XXIII.) Acide folique.

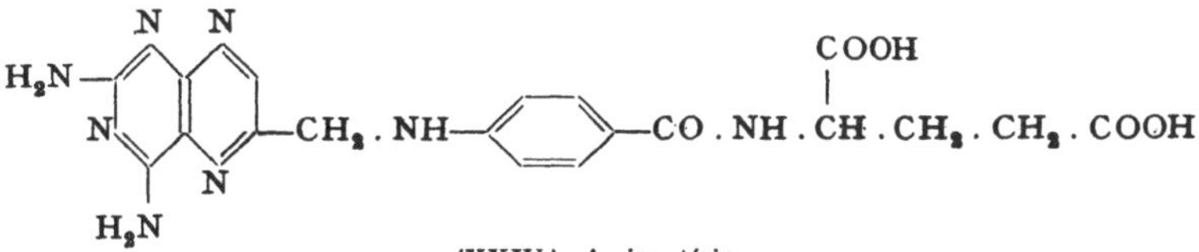

$$H_2N-\underset{\underset{H_2N}{N}}{\overset{N\;N}{\diagdown}}-CH_2.NH-\underset{}{\bigcirc}-CO.NH.\underset{\underset{COOH}{|}}{CH}.CH_2.CH_2.COOH$$

(XXIV.) Aminoptérine.

La plupart des corps de cette série ont été essayés soit comme anticancéreux, soit surtout comme antileucémiques. A la Conférence de Boston en Janvier 1949, des hématologues ont donné leurs impressions sur les rémissions spontanées de cette maladie, rémissions plus fréquentes chez les enfants que chez les adultes. Bien que de véritables cures n'aient pu, sans doute, être obtenues, des rémissions durables sont déjà constatées, mais les antagonistes foliques sont trop toxiques. C'est pourquoi une suggestion semble à retenir: la découverte par SAUBERLICH (*86*) dans le *Leuconostoc citrovorum* d'un facteur capable de renverser les effets de l'aminoptérine (XXIV) et qui vient d'être synthétisé par BROCKMANN et collaborateurs (*7*). Mais il existe d'autres procédés également fondés sur la notion d'analogie structurale pour l'attaque du problème de la thérapeutique des cancers (*54*).

9° L'activité antisulfamide et antisulfone de dérivés voisins de l'acide *p*-aminobenzoïque.

L'effet antisulfamide de l'acide *p*-aminobenzoïque a été longtemps à la base même de la notion d'antivitamine selon la terminologie de KUHN (*40*), les antisulfamides étant des antivitamines *H'*. Aussi, est-ce l'antagonisme entre l'acide *p*-aminobenzoïque et les sulfamidés qui a incité d'autres esprits à rechercher des antagonismes analogues entre des composés légèrement différents. C'est ainsi que dès 1942, LEVADITI, MENTZER et PERRAULT (*44*) ont décrit les propriétés de l'acide *p*-oxybenzoïque (XXV) comme antagoniste spécifique à l'égard de la 4,4'-*p*-oxydiphénylsulfone (XXVI).

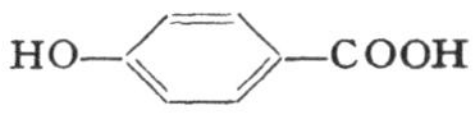

$$HO-\bigcirc-COOH \qquad HO-\bigcirc-SO_2-\bigcirc-OH$$

(XXV.) Acide *p*-oxybenzoïque. 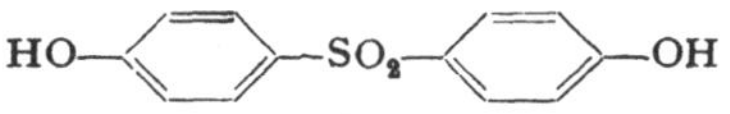   (XXVI.) 4,4'-*p*-Oxydiphénylsulfone.

Il est intéressant de noter que l'acide *p*-hydroxybenzoïque a été décrit comme un nouveau facteur essentiel pour la croissance bactérienne en 1950 seulement par DAVIS (*15*), travail qui a été analysé peu de temps après dans Nutrition Reviews sous le titre: « Identification d'un nouveau facteur essentiel à la croissance bactérienne » (*75*).

En 1951, un antagonisme analogue a été étudié par les Français LAVOLLAY et NEUMAN sur l'*Aspergillus niger* (*42*), celui de l'acide *p*-oxybenzoïque avec la cétone correspondante.

Dans plusieurs cas d'ailleurs, le remplacement d'un groupement —COOH par un groupement —COCH$_3$ s'accompagne d'un renversement de l'action physiologique. Ainsi, l'acétyl-3-pyridine est douée de propriétés anti-vitaminiques PP (*115*) et l'amino-4-acétophénone est capable de neutraliser les effets de l'acide *p*-aminobenzoïque vis-à-vis de *Streptobacterium plantarum* (*1*).

Ces vitamines bactériennes ont été découvertes uniquement à la faveur de la connaissance de leurs antagonistes, comme cela s'est passé d'ailleurs pour l'acide *p*-aminobenzoïque lui-même, connu depuis sous le nom de vitamine H' (*112*). En 1951, Mlle MARNAY confirma l'action favorable pour *E. coli* de l'acide *p*-hydroxybenzoïque dont l'intensité d'action est intermédiaire entre celle de l'acide *p*-aminobenzoïque et de l'acide nicotinique ou du tryptophane (*46*).

II. Les antagonistes des vitamines liposolubles.

Il est curieux de noter que les antivitamines s'appliquant aux vitamines liposolubles sont d'apparition beaucoup plus récente que les antivitamines hydrosolubles et que, sauf peut-être pour les antivitamines K, elles sont limitées à des travaux de ce laboratoire, antivitamine A, antivitamines E.

I° L'antivitamine A.

C'est en 1950, au cours d'études sur l'oxydation directe du β-carotène en rétinène par différents oxydes métalliques que MEUNIER, JOUANNETEAU et FERRANDO (*59*) ont trouvé que, parmi les produits d'oxydation obtenus, se trouve une substance qui doit être regardée comme la première antivitamine A. Cette substance s'obtient facilement par action oxydante directe de l'oxyde de vanadium (V$_2$O$_4$) sur la vitamine A elle-même. Elle est essentiellement caractérisée par l'apparition d'une bande d'absorption à 340 mμ accompagnée d'une bosse à 290 mμ, cependant que

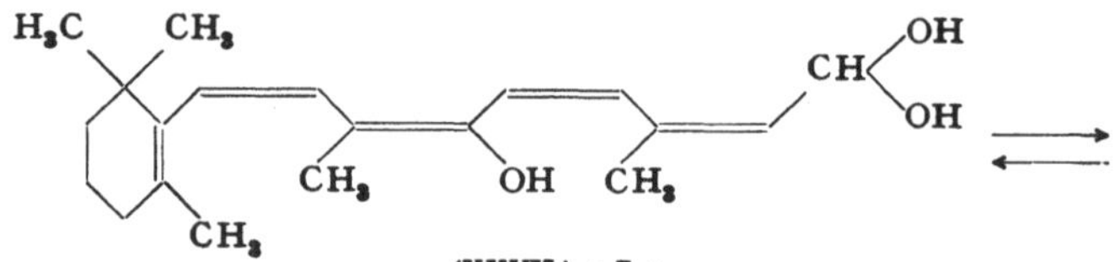

(XXVII.) « Z$_1$ ».

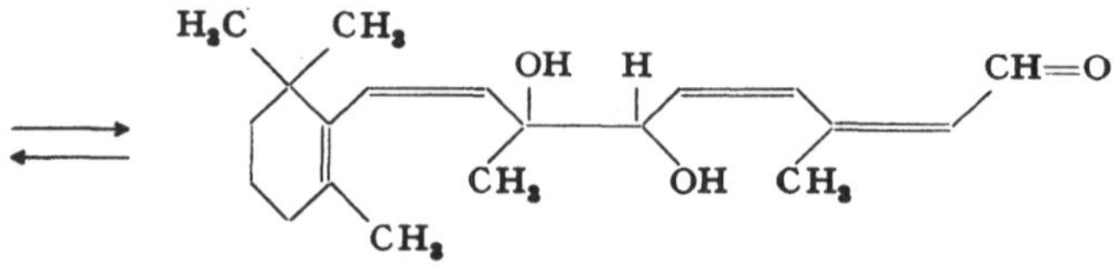

(XXVIII.) « Z$_2$ ». (On ne doit pas attacher de signification stéréochimique précise à cette écriture qui ne correspond pas nécessairement à la forme all-*cis*.)

sa réaction avec le réactif au trichlorure d'antimoine produit une coloration violet-rouge à 545 mµ. Plus tard, nous avons reconnu que ce corps désigné sous le nom de « corps Z » était très probablement un mélange en équilibre entre un hydrate d'aldéhyde brut (XXVII) et l'aldéhyde (XXVIII) résultant de la migration de —H—OH au sein de la molécule de l'hydrate de rétinène oxydé en son milieu (65, 76).

Administrée à la dose de 12 γ par jour à des rats, cette substance accélère considérablement la carence en vitamine A, en même temps que l'on observe des symptômes hémorragiques ressemblant à ceux du scorbut (48). Cependant, les foies des rats ainsi traités ne renferment plus de vitamine A mais seulement le corps « Z » reconnaissable par sa réaction avec $SbCl_3$ à 545 mµ.

Dans une dernière expérimentation, nous avons pu obtenir une perte de poids de rats pesant plus de 150 g. avec 7 γ par jour de ce même corps « Z » (64) mais ces quantités ne permettent plus la mise en réserve du corps « Z » dans le foie du rat. Disons enfin que l'énorme perte de poids déterminée par l'ingestion de ce corps « Z » peut être dans une faible mesure compensée en partie par l'acide ascorbique lui-même (64). De là, cette idée que la carence en vitamine A pourrait déterminer l'arrêt de la synthèse de la vitamine C; en outre cette carence détermine l'arrêt de la glycurono-conjugaison chez le rat intoxiqué au benzoate de sodium, comme cela a été établi par FERRANDO (22).

2° Les antivitamines E.

Au Colloque de Lyon en 1948, CORMIER avait présenté un rapport sur les « Actions antivitaminiques des huiles de foies de poissons » (10). Dans la discussion qui avait suivi la lecture de ce rapport, FERRANDO, en collaboration avec Mlle CHENAVIER, avaient indiqué que le traitement de jeunes moutons en voie de croissance par 5 à 10 cm³ par jour d'huile de fois de morue amène une baisse sensible du tocophérol sanguin rapidement arrêtée par l'administration ultérieure d'huile de germe de maïs (23, 58). Cette action a également été mise en évidence chez la vache laitière par FERRANDO et collaborateurs (24), en même temps que par LOOSLI et Collaborateurs (108).

On se souvient que dès 1947, MEUNIER, Mlle VINET et JOUANNETEAU (63) avaient montré que 5% d'huile de foie de morue dans le régime des lapins amenaient une baisse sensible du tocophérol du sang en même temps qu'une perte de poids des animaux, suivie d'intenses phénomènes de dystrophie musculaire. Le tocophérol, administré à temps, sauve de la mort les lapins ou les moutons. Dans ce même travail (63), il était confirmé que le contre-poison efficace du phosphate de tri-o-crésyle était bien le tocophérol lui-même, comme l'avaient déjà indiqué BLOCH et HOTTINGER (6). De la sorte, il apparaissait, par un contrôle parallèle du taux du tocophérol sanguin, que le phosphate de tri-*o*-crésyle pouvait bien être considéré comme une antivitamine E. En 1949, MEUNIER et Mlle CHENAVIER (*57*) ont signalé un nouvel antagoniste de la vitamine E, moins toxique que le phosphate de tri-*o*-crésyle, qui n'était autre que le succinate de di-*o*-crésyle. Cette constatation avait l'avantage de montrer que l'action antivitaminique découverte pour le phosphate de tri-*o*-crésol était indépendante de la présence de l'ion phosphorique dans cette molécule. Le succinate de di-*o*-crésyle détermine également une perte de poids du lapin en voie de croissance qui est parfaitement annulée par administration de tocophérol, à la dose de 50 mg. pour 250 mg. de succinate de di-*o*-crésyle, expérience faite sur un lapin de 1,500 g.

Dès que DAM, PRANGE et SÖNDERGAARD (*13*) eurent montré que la réserve en vitamine A des poulets carencés en vitamine E pouvait être indifféremment rétablie par l'acétate de tocophérol ou le bleu de méthylène, de même que la croissance pondérale de ces animaux, nous n'eûmes pas de peine à montrer que le bleu de méthylène se comportait comme antagoniste, c'est à dire comme la vitamine E elle-même, vis-à-vis de l'intoxication par le phosphate de tri-*o*-crésyle (XXVII) ou même par le succinate de di-*o*-crésyle (XXVIII) chez de jeunes lapins (*8*).

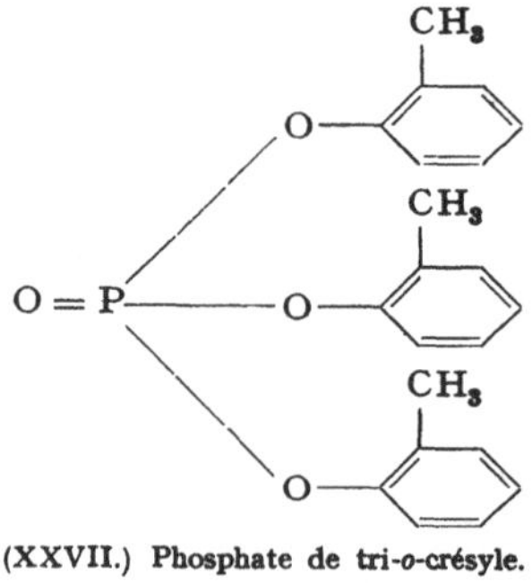

(XXVII.) Phosphate de tri-*o*-crésyle. 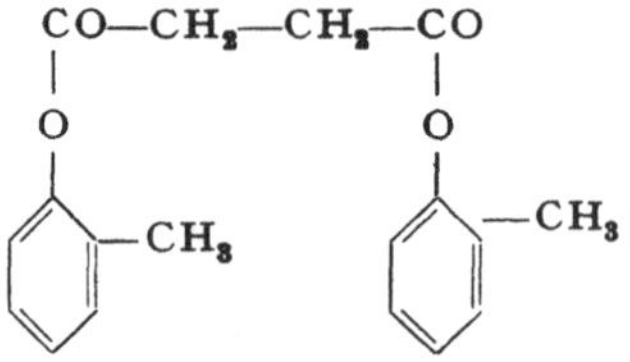   (XXVIII.) Succinate de di-*o*-crésyle.

L'analogie structurale entre la vitamine et ses antagonistes n'existe donc pas dans ce cas particulier et les substances dites « antivitamines E » exercent très probablement une action toxique complexe qui se manifeste entre autres par un abaissement de la teneur en tocophérol du plasma.

3° Les antivitamines K.

Au Colloque de Lyon sur les « Antivitamines » en 1948, deux mémoires avaient été consacrés au problème des antivitamines K chez les animaux (*53*, *56*), c'est à dire aux substances qui s'opposent à la synthèse de la prothrombine par les animaux, substances dont le premier type connu était le dicoumarol (XXIX), découvert par STAHMANN, HUEBNER et LINK (*98*) et qui est en même temps le plus actif.

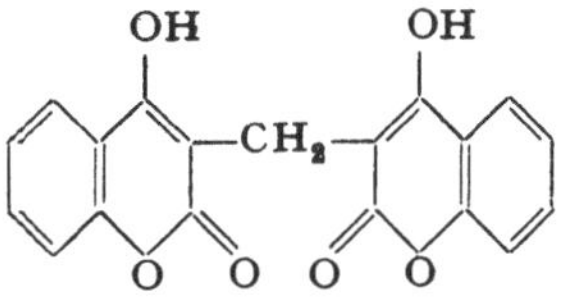

(XXIX.) Dicoumarol = méthylène-3:3'-*bis*[hydroxy-4-coumarine].

A côté de cette antivitamine K, dite artificielle, MEUNIER (*56*) avait rappelé les conditions dans lesquelles il avait été amené à fabriquer de toutes pièces, sur le modèle du dicoumarol, une substance qui abaissait également le taux de la prothrombine du lapin par exemple et qu'avec MENTZER, BUU-HOÏ et CAGNIANT (*61*) ils avaient appelée le diphtiocol (XXXI) dont la parenté avec le phtiocol (XXX) était absolument évidente puisqu'il s'agissait de la méthylène-3 : 3'-bis-oxy-2-naphtoquinone :

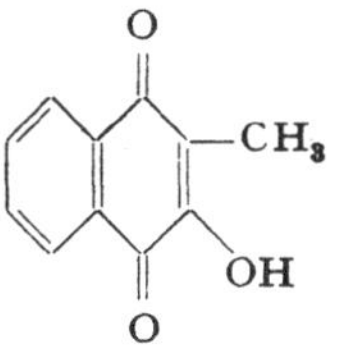

(XXX.) Phtiocol. 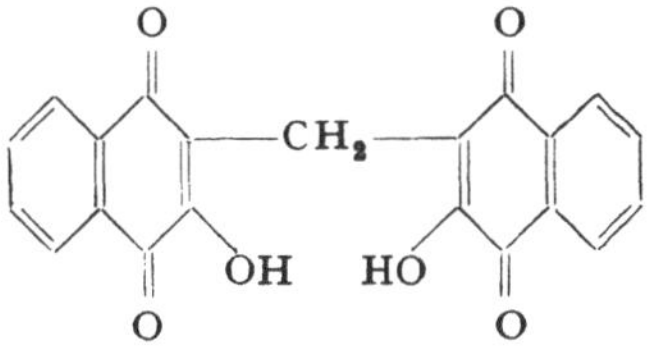   (XXXI.) Diphtiocol (méthylène-3: 3'-*bis*-oxy-2-naphtoquinone).

Depuis cette date, MOLHO, MORAUX et MEUNIER (*68*) ont reconnu qu'un second produit naturel autre que le dicoumarol et très voisin de la méthyl-2-naphtoquinone (XXXII) primitivement isolé de la Balsamine des jardins comme fongicide notable par LITTLE, SPROSTON et FOOTE (*45*), la méthoxy-2-naphtoquinone-1,4 (XXXIII), se comportait également comme un produit hypoprothrombinémiant chez le lapin, son action étant annulée par la méthyl-2-naphtoquinone-1,4 elle-même.

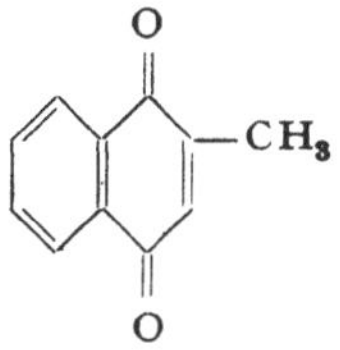

(XXXII.) Méthyl-2-naphtoquinone-1,4. 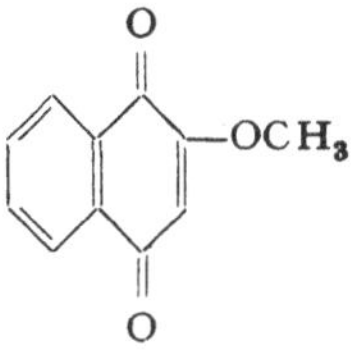   (XXXIII.) Méthoxy-2-naphtoquinone-1,4.

Le même année, à une conférence de Londres, MEUNIER reprit le problème des antivitamines K, dites artificielles, et rattacha à la plus ou moins facile élimination de ces corps par l'organisme leur activité plus ou moins grande comme hypoprothrombinémiant (*56a, 27*). Il indiqua en outre que le représentant le plus intéressant était la phényl-indanedione, découverte comme agent hypoprothrombinémiant par MEUNIER, MENTZER et Molho, en 1947 (*62*) et dont la structure (XXXIV) est pourtant déjà assez éloignée de celle du noyau naphtoquinonique. Cette phényl-in-

(XXXIV.) Phényl-indanedione (P. I. D.).

danedione a été reprise pour le traitement des thromboses par Jaques et Collaborateurs (*35*) et par Badin et collaborateurs en France (*2, 44a*).

Peu de temps après le Colloque de Lyon, Dam et Söndergaard (*14*) établirent que, par l'étude des mélanges de plasma de poulets carencés en vitamine K et de poulets intoxiqués au dicoumarol, non seulement les plasmas dicoumarolés manquaient de prothrombine, mais encore d'un autre facteur protéique encore indéterminé capable d'activer la coagulation plasmatique. Très récemment dans ce laboratoire, Moraux (*69*) s'est efforcé de préciser chez des poulets carencés en vitamine K ou dicoumarolés, ou encore traités par la phényl-indanedione ou même par du diphtiocol, si le plasma des animaux ainsi traités était déficient en prothrombine ou encore s'il était altéré sous le rapport d'un autre constituant du plasma normal, facteur de nature protéique appelé facteur δ ou encore facteur χ ou encore différemment dénommé, selon les auteurs.

Les conclusions de Moraux indiquent très nettement que le dicoumarol agit non seulement sur la prothrombine, comme Dam l'avait montré, mais en plus sur un autre facteur (*69*) et qu'il en est de même pour les animaux traités par la phényl-indanedione ou même par le diphtiocol. Il s'agirait d'un seul et même facteur altéré en même temps par la prothrombine par les trois anticoagulants essayés, mais facteur non encore identifiable, étant donné la grande divergence des opinions sur sa nature. Quoi qu'il en soit, cette conclusion provisoire semble indiquer que les substances que nous avons appelées, avec Mentzer, des « anti-vitamines K » dès 1943 (*60*) ont en plus de leur action sur le taux de prothrombine des propriétés inhibitrices par rapport à des mécanismes biochimiques dont nous ne connaissons pas encore la nature exacte. Parmi ces propriétés, figurent également les effets antibiotiques de certaines de nos antivitamines K vis-à-vis des champignons inférieurs.

C'est en 1943 que Ter Horst et Felix (*101*) ont découvert la puissante activité antifongique de la dichloro-2,3-naphtoquinone et c'est en 1945 que Woolley (*116*), frappé par la ressemblance de la structure de ce corps avec la vitamine K, réussit à neutraliser l'action néfaste de ce composé sur les levures par addition simultanée de méthyl-2-phytyl-3-naphtoquinone-1,4. Au Colloque de Lyon en 1948, c'est Mme Guérillot-Vinet (*31*) qui avait exposé le rapport « Vitamines K et antivitamines K en microbiologie ». Dans ce rapport, Mme GUÉRILLOT-VINET insista sur la diversité des substances dont l'action antibiotique était capable d'être annulée par une véritable vitamine K d'une part et d'autre part sur le fait que les quinones elles-mêmes étaient très facilement antibiotiques. La même existence de couples antagonistes a été étudiée peu de temps après par MOLHO et MOLHO-LACROIX (*66, 66a*) vis-à-vis du protozoaire *Glaucoma piriformis* et de l'*Aspergillus niger*: il s'agissait du couple méthyl-2-naphtoquinone-1,4 et méthoxy-2-naphtoquinone-1,4.

Plus généralement, l'action des quinones sur la croissance de la levure a été étudiée par HOFFMANN-OSTENHOF, WERTHEIMER et GRATZL (*32, 32a*).

La démonstration la plus nette et la plus intéressante de ces couples antagonistes est constituée par celle décrite par SCHOPFER (*93*) sur l'action de la chloro-2-naphtoquinone-1,4 (XXXV) et de la méthyl-2-naphtoquinone-1,4, toutes deux cristallisées sur l'uréase également cristallisée. C'est à l'occasion de recherches sur l'action de la méthyl-2-naphtoquinone-1,4 chez les plantes qu'un mécanisme possible de l'effet antibiotique de cette substance a été envisagé par SCHOPFER et Mlle Boss. Il s'agit de la perturbation de la synthèse de la nicotinamide chez la plante (*92*) (p. 98).

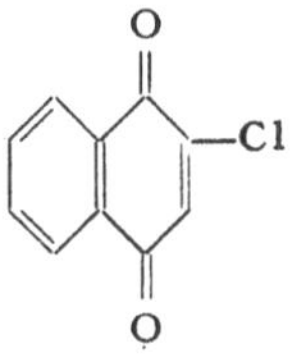

(XXXV.) Chloro-2-naphtoquinone-1,4.

Nous adressons ici nos très vifs remerciements à Monsieur le Professeur W. H. SCHOPFER de Berne, à Monsieur le Professeur C. MENTZER et à Mademoiselle Ch. MARNAY du C.N.R.S. à Paris pour l'aide qu'ils nous ont apportée dans la rédaction de ce travail.

Bibliographie.

1. AUHAGEN, E.: *p*-Aminophenylketone als Antagonisten der *p*-Aminobenzoesäure. Hoppe-Seyler's Z. physiol. Chem. **274**, 48 (1942).
2. BADIN, J., C. MENTZER, J. MORAUX et P. MEUNIER: Sur une certaine exaltation de l'action hypoprothrombinémiante de quelques antivitamines K par ingestion simultanée d'esculoside. C. R. Séances Soc. Biol. Filiales Associées **144**, 871 (1950).
3. BAUMGARTEN, P. und A. DORNOW: Über das 2-Methyl-3-oxyäthyl-*N*-[(2-methyl-4-amino-pyrimidyl-(5))-methyl]-pyridinium-bromid, ein Heterovitamin B_1. Ber. dtsch. chem. Ges., Ser. B **73**, 44 (1940).
4. — — Zur Kenntnis zweier Heterovitamine B_1. Ber. dtsch. chem. Ges., Ser. B **73**, 353 (1940).
5. BEILER, J. M. and G. J. MARTIN: Inhibition of the Action of Tyrosine Decarboxylase by Phosphorylated Desoxypyridoxine. J. biol. Chemistry **169**, 345 (1947).

6. BLOCH, H. und A. HOTTINGER: Über eine bei der o-Trikresylphosphatvergiftung auftretende Kreatinurie und deren Beeinflussung durch Vitamin E. Z. Vitaminforsch. **13**, 9 (1943).
7. BROCKMAN, J. A., Jr., B. ROTH, H. P. BROQUIST, M. E. HULTQUIST, J. M. SMITH, Jr., M. J. FAHRENBACH, D. B. COSULICH, R. P. PARKER, E. L. R. STOKSTAD and T. H. JUKES: Synthesis and Isolation of a Cristalline Substance with the Properties of a New B Vitamin. J. Amer. chem. Soc. **72**, 4325 (1950).
8. CARRÉ-CHENAVIER, (Mme) P. et P. MEUNIER: Comparaison de l'effet du bleu de méthylène et du tocophérol chez des lapins recevant des corps à allure antagoniste de la vitamine E (Sous presse).
9. Colloque de Lyon sur les « Antivitamines » — Septembre 1948. Bull. Soc. Chim. biol. **30**, 725—960 (1948).
10. CORMIER, M.: Les actions antivitaminiques des huiles de foie de Poisson. Bull. Soc. Chim. biol. **30**, 921 (1948).
11. CRAVENS, W. W. and E. E. SNELL: Effects of Desoxypyridoxine and Vitamine B_6 on Development of the Chick Embryo. Proc. Soc. exp. Biol. Med. **71**, 73 (1949).
12. DALGLIESH, C. E., W. E. KNOX and A. NEUBERGER: Intermediary Metabolism of Tryptophan. Nature (London) **168**, 20 (1951).
13. DAM, H., I. PRANGE and E. SÖNDERGAARD: Similar Effects of Methylene Blue and of Vitamin E on Liver Storage of Vitamin A in Chicks. Experientia **7**, 184 (1951).
14. DAM, H. and E. SÖNDERGAARD: Observations on the Coagulation Anomaly in Vitamin K-Deficiency and Dicumarol Poisoning. Biochim. Biophys. Acta **2**, 409 (1948).
15. DAVIS, B. D.: p-Hydroxybenzoic Acid: A New Bacterial Vitamin. Nature (London) **166**, 1120 (1950).
16. DITTMER, K. and V. DU VIGNEAUD: Antibiotin Activity of Imidazolidone Aliphatic Acids. J. biol. Chemistry **169**, 63 (1947).
17. DITTMER, K., V. DU VIGNEAUD, P. GYÖRGY and C. S. ROSE: A Study of Biotin Sulfone. Arch. Biochemistry **4**, 229 (1944).
18. DOMAGK, G.: Chemotherapy of Bacterial Infections. Dtsch. Med. Wschr. **61**, 250 (1935).
19. EMERSON, G. A.: The Antivitamin B_6 Activity of Desoxypyridoxine in the Rat. Federat. Proc. (Amer. Soc. exp. Biol.) **6**, 406 (1947).
20. ENGLISH, J. P., R. C. CLAPP, Q. P. COLE, I. F. HALVERSTADT, J. O. LAMPEN and R. O. ROBLIN, Jr.: Studies in Chemotherapy. IX. Ureylenebenzene and Cyclohexane Derivatives as Biotin Antagonists. J. Amer. chem. Soc. **67**, 295 (1945).
21. FERRANDO, R.: Antivitamines et nutrition des animaux domestiques. Rapport et conséquences. Rev. Pathol. comp. et Hygiène gén. **49**, Nos 609—610, 579 (1949).
22. — Influence de la vitamine A sur la glycuronoconjugaison chez le rat intoxiqué au bonzoate de sodium. C. R. Séances hebd. Acad. Sci. **231**, 1264 (1950).
23. FERRANDO, R. et (Mlle) P. CHENAVIER: Discussion qui suivit le rapport de M. CORMIER au Colloque de Lyon en 1948. Bull. Soc. Chim. biol. **30**, 937—939 (1948).
24. FERRANDO, R., (Mlle) P. CHENAVIER et M. CORMIER: Action de l'huile de foie de morue sur le taux du tocophérol sanguin de la vache laitière et le taux butyreux de son lait. Bull. Soc. Chim. biol. **31**, 810 (1949).
25. FROHMAN, C. E. and H. G. DAY: Effect of Oxythiamine on Blood Pyruvate-Lactate Relationships and the Excretion of Thiamine in Rats. J. biol. Chemistry **180**, 93 (1949).

26. GELLHORN, A. and L. O. JONES: Pyridoxine Deficient Diet and Desoxypyridoxine in the Therapy of Lymphsarcoma and Acute Leukemia in Man. Blood 4, 60 (1949).
27. GLEY, P. et (Mlle) J. DELOR: Action antivitaminique P de la phényl-indanedione. Bull. Soc. Chim. biol. 30, 891 (1948).
28. GREEN, R. G., W. E. CARLSON and C. A. EVANS: A Deficiency Disease of Foxes Produced by Feeding Fish. J. Nutrit. 21, 243 (1941).
29. — — — The Inactivation of Vitamin B_1 in Diets Containing Whole Fish. J. Nutrit. 23, 165 (1942).
30. GREEN, R. G. and C. A. EVANS: A Deficiency Disease of Foxes. Science (New York) 92, 154 (1940).
31. GUÉRILLOT-VINET, (Mme) A.: Vitamines K et antivitamines K en microbiologie. Bull. Soc. Chim. biol. 30, 863 (1948).
32. HOFFMANN-OSTENHOF, O.: Vorkommen und biochemisches Verhalten der Chinone. Fortschr. Chem. organ. Naturstoffe 6, 154 (1950).
32a. HOFFMANN-OSTENHOF, O., P. WERTHEIMER und K. GRATZL: Die Wirkung von Chinonen auf das Hefewachstum. Experientia 3, 327 (1947).
33. HOFMANN, K. and A. E. AXELROD: Microbiological Activity of the Guanidino Analogues of Biotin and Oxybiotin. J. biol. Chemistry 187, 29 (1950).
34. HOFMANN, K., T. WINNICK and A. E. AXELROD: The Use of Raney's Nickel in a Differential Assay for Oxybiotin and Biotin. J. biol. Chemistry 169, 191 (1947).
35. JAQUES, L. B., E. GORDON and E. LEPP: A New Prothrombopenic Drug: Phenylindanedione. Canad. med. Assoc. J. 62, 465 (1950).
36. JONES, J. H., C. FOSTER and W. HENLE: Effect of Oxythiamine on Infection of Mice with the Lansing Strain of *Poliomyelitis* Virus. Proc. Soc. exp. Biol. Med. 69, 454 (1948).
37. KARNOFSKY, D. A., C. C. STOCK, L. P. RIDGWAY and L. P. PATTERSON: The Toxicity of Vitamin B_6, 4-Desoxypyridoxine, and 4-Methoxymethyl-pyridoxine, Alone and in Combination, to the Chick Embryo. J. biol. Chemistry 182, 471 (1950).
38. KARRER, P. und H. RUCKSTUHL: Zur Frage der Ursache der antagonistischen Wirkung von Vitaminen und Antivitaminen. Bull. Schweiz. Akad. Med. Wissensch. 1, 236 (1945).
39. KARRER, P. und M. SCHOELLER: Synthesen Thiamin-ähnlicher Verbindungen: Homothiaminglykol und 3-[4'-Amino-2'-methyl-pyrimidyl-(5')-methyl]-4-methyl-5-allyl-thiazoliumchlorid-hydrochlorid. Helv. chim. Acta 34, 826 (1951).
40. KUHN, R.: Vitamine und Arzneimittel. Die Chemie 55, 1 (1942).
41. KUHN, R., F. WEYGAND und E. F. MÖLLER: Über einen Antagonisten des Lactoflavins. Ber. dtsch. chem. Ges. 76, 1044 (1943).
42. LAVOLLAY, J. et J. NEUMAN: Toxicité des *p*-amino et *p*-hydroxy-acétophénones pour *Aspergillus Niger*. Spécificité d'action antitoxue des acides *p*-amino et *p*-hydroxybenzoïques. C. R. Séances hebd. Acad. Sci. 232, 758 (1951).
43. LEFWICH, W. B., G. S. MIRICK and E. I. CORDDRY: The Effect of Diet on the Susceptibility of the Mouse to Pneumonia Virus of Mice (PVM). I. Influence of Pyridoxine in the Period After the Inoculation of Virus. J. exp. Medicine 89, 155 (1949).
44. LEVADITI, C., C. MENTZER et R. PERRAULT: Vitamine H': antisulfamide et antisulfone. C. R. Séances Soc. Biol. Filiales Associées 136, 769 (1942).
44a. LEVY-SOLAL, E., J. BADIN et J. CHOUKROUN: Le traitement de la maladie thrombo-embolique par l'association phényl-indanedione et esculoside. Gynécologie et Obstétrique 50, No. 1, 1 (1951).

45. LITTLE, J. E., T. J. SPROSTON and M. W. FOOTE: Isolation and Antifungal Action of Naturally Occuring 2-Methoxy-1,4-naphtoquinone. J. biol. Chemistry 174, 335 (1948).
46. MARNAY, CH.: Recherches sur l'action de l'acide p-hydroxy-benzoïque sur le colibacille. Bull. Soc. Chim. Biol. 33, 1304 (1951).
47. MARTIN, G. J., S. AVAKIAN and J. MOSS: Studies of Pyridoxine Displacement. J. biol. Chemistry 174, 495 (1948).
48. MAYER, J. and W. A. KREHL: Scorbutic Symptoms in Vitamin A-Deficient Rats. Arch. Biochemistry 16, 313 (1948).
49. MELNICK, D., M. HOCHBERG and B. L. OSER: Physiological Availability of the Vitamins. I. The Human Bioassay Technic. J. Nutrit. 30, 67 (1945).
49a. — — — Physiological Availability of the Vitamins. II. The Effect of Dietary Thiaminase in Fish Products. J. Nutrit. 30, 81 (1945).
50. MELVILLE, D. B., K. DITTMER, G. B. BROWN and V. DU VIGNEAUD: Desthiobiotin. Science (New York) 98, 497 (1943).
51. MELVILLE, D. B., K. HOFMANN and V. DU VIGNEAUD: Resynthesis of Biotin from a Degradation Product. Science (New York) 94, 308 (1941).
52. MENTZER, C.: Etat actuel de nos connaissances sur les antivitamines. Ann. Nutrit. et Aliment. 1, 339 (1947).
53. — Les divers groupes de substances synthétiques douées d'une activité antivitaminique K et la signification biologique des résultats obtenus. Bull. Soc. Chim. biol. 30, 872 (1948).
54. MENTZER, C. et P. MEUNIER: Comment appliquer la notion d'analogie structurale au problème de la thérapeutique des cancers. Thérapie 5, n° 4, 192 (1950).
55. MENTZER, C., P. MEUNIER et L. MOLHO-LACROIX: Faits de synergie et d'antagonisme entre la chloromycétine et divers acides aminés vis-à-vis de cultures de *E. Coli*. C. R. Séances hebd. Acad. Sci. 230, 241 (1950).
55a. MENTZER, C., P. MEUNIER, L. MOLHO-LACROIX et D. BILLET: Recherches d'analogues structuraux simplifiés de la chloromycétine. I. Sur l'analogie de l'action *in vitro* de la sérine, de la phényl-sérine et de la chloromycétine sur *E. Coli*. Bull. Soc. Chim. biol. 32, 55 (1950).
56. MEUNIER, P.: Le mécanisme d'action des antivitamines K chez les animaux. Bull. Soc. Chim. biol. 30, 884 (1948).
56a. — Du dicoumarol aux antivitamines K artificielles. Brit. J. Nutrit. 2, 397 (1949).
57. MEUNIER, P. et (Mlle) P. CHENAVIER: Un nouvel antagoniste de la vitamine E: le succinate de di-ortho-crésyle. C. R. Séances Soc. Biol. 143, 1046 (1949).
58. MEUNIER, P., R. FERRANDO et (Mlle) P. CHENAVIER: Le taux de tocophérol (vitamine E) sanguin du mouton. Action de l'huile de foie de morue. C. R. Séances Soc. Biol. Filiales Associées 142, 525 (1948).
59. MEUNIER, P., J. JOUANNETEAU et R. FERRANDO: Sur la toxicité pour le rat blanc d'un nouveau dérivé d'oxydation de la vitamine A. C. R. Séances hebd. Acad. Sci. 230, 140 (1950).
60. MEUNIER, P. et C. MENTZER: Sur l'activité antihémorragique de certains dérivés du chromane et la notion d'antivitamine K. Bull. Soc. Chim. biol. 25, 80 (1943).
61. MEUNIER, P., C. MENTZER, BUU-HOÏ et P. CAGNIANT: Contribution au problème des antivitamines K. II. Obtention d'un dérivé hémorragique naphtalénique. Bull. Soc. Chim. biol. 25, 384 (1943).
62. MEUNIER, P., C. MENTZER et D. MOLHO: Sur l'activité antivitaminique K (hémorragique) d'une indanedione. C. R. Séances hebd. Acad. Sci. 224, 1666 (1947).

63. MEUNIER, P., (Mlle) A. VINET et J. JOUANNETEAU: Sur les actions antagonistes de la vitamine E et des huiles de foie de poissons sur la croissance du lapin. Bull. Soc. Chim. biol. 24, 507 (1947).
64. MEUNIER, P., G. ZWINGELSTEIN et J. JOUANNETEAU: Activité biologique sur des rats adultes du dérivé à action antivitaminique A obtenu par oxydation de l'axérophtol et effet de l'acide ascorbique (Sous presse).
65. MEUNIER, P., G. ZWINGELSTEIN, J. JOUANNETEAU et R. MALLEIN: Sur la structure exacte du dérivé à action antivitaminique A obtenu par oxydation de l'axérophtol. C. R. Séances hebd. Acad. Sci. 230, 1323 (1950).
66. MOLHO, D. et L. MOLHO-LACROIX: Effets antibiotiques de quelques quinones. I. Comportement de *Glaucoma Piriformis* en milieu peptoné vis-à-vis du couple vitamine K — antivitamine K. Bull. Soc. Chim. biol. 31, 1341 (1949).
66a. — — Effets antibiotiques de quelques quinones. III. Influence de la méthoxy-2-naphtoquinone-1,4 sur la croissance d'*Aspergillus Niger*. Bull. Soc. Chim. biol. 31, 1357 (1949).
67. — — Méthodes des antagonistes et des synergiques dans l'étude comparée des modes d'action de la cystéine et de la β_2-thiénylalanine sur *E. Coli* en milieu synthétique. Bull. Soc. Chim. biol. 32, 680 (1950).
67a. — — Etude comparée de l'antagonisme entre quelques dérivés de la phénylalanine et la chloromycétine, la β_2-thiénylalanine et la β-phénylsérine. Bull. Soc. Chim. biol. 34, 99 (1952).
68. MOLHO, D., J. MORAUX et P. MEUNIER: Contribution au problème des antivitamines K. V. Sur l'activité hémorragique pour le lapin de la méthoxy-2-naphtoquinone-1,4 et de composés voisins. Bull. Soc. Chim. biol. 30, 637 (1948).
69. MORAUX, J.: Observations sur le mécanisme d'action de quelques antivitamines K (dicoumarol, phénylindanedione, diphtiocol). C. R. Séances hebd. Acad. Sci. 233, 711 (1951).
70. MUELLER, J. F. and R. W. VILTER: Pyridoxine Deficiency in Human Beings Induced With Desoxypyridoxine. J. clin. Invest. 29, 193 (1950).
71. NELSON, M. M. and H. M. EVANS: Effect of Desoxypyridoxine on Reproduction in the Rat. Proc. Soc. exp. Biol. Med. 68, 274 (1948).
72. Nutrition Reviews: Pantothenic Acid Antagonists. 6, 153 (1948).
73. Nutrition Reviews: Use of Metabolic Inhibitors in Studies with Developing Chick Embryos. 8, 42 (1950).
74. Nutrition Reviews: Present Knowledge of Pteroylglutamates ("Folic Acid"). 8, 260 (1950).
75. Nutrition Reviews: Identification of a New Factor Essential for Bacterial Growth. 9, 156 (1951).
76. Nutrition Reviews: An Antivitamin A. 9, 140 (1951).
77. OTT, W. H.: Antipyridoxine Activity of 2,4-Dimethyl-3-hydroxy-5-hydroxymethyl-pyridine in the Chick. Proc. Soc. exp. Biol. Med. 61, 125 (1946).
78. — Antipyridoxine Activity of Methoxy-pyridoxine in the Chick. Proc. Soc. exp. Biol. Med. 66, 216 (1947).
79. PORTER, C. C., I. CLARK and R. H. SILBER: The Effect of Pyridoxine Analogues on Tryptophane Metabolism in the Rat. J. biol. Chemistry 167, 573 (1947).
80. RAFFAUF, R. F.: Neopyrithiamin. Helv. chim. Acta 33, 102 (1950).
81. ROBBINS, W. J.: The Pyridine Analog of Thiamin and the Growth of Fungi. Proc. nat. Acad. Sci. USA 27, 419 (1941).
82. ROBERTS, E., F. YOUNGER and S. FRANKEL: Influence of Dietary Pyridoxine on Glutamic Decarboxylase Activity of Brain. J. biol. Chemistry 191, 277 (1951).
83. ROBLIN, R. O., Jr.: Metabolite Antagonists. Chem. Reviews 38, n° 2, 255 (1946).

84. Rogers, L. L. and W. Shive: Biological Transformation as Determined by Competitive Analogue-Metabolite Growth Inhibitions. VI. Preventing of Biotin Synthesis by 2-Oxo-4-imidazolidine-caproic Acid. J. biol. Chemistry **169**, 57 (1947).
85. Rydon, H. N.: Note of an Improved Method for the Preparation of Oxythiamin. Biochemic. J. **48**, 383 (1951).
86. Sauberlich, H. E.: The Relationship of Folic Acid, Vitamin B_{12} and Thymidine in the Nutrition of *Leuconostoc citrovorum*. Arch. Biochemistry **24**, 224 (1949).
87. Schmelkes, F. C. and R. R. Joiner: Synthesis of 2-Methyl-3-β-hydroxyethyl-N-[(2-methyl-6-aminopyrimidyl-(5))-Methyl]pyridinium Bromide Hydrobromide. J. Amer. chem. Soc. **61**, 2562 (1939).
88. Schopfer, W. H.: Recherches sur l'action antivitamine de la dichloro-flavine sur un microorganisme producteur de lactoflavine (*Eremothecium Ashbyii*). Z. Vitaminforsch. **20**, 116 (1948).
89. — Sulfamidés, acides nucléiques et purines. Bull. Soc. Chim. biol. **30**, 748 (1948).
90. — La pyrithiamine, antivitamine B_1. Bull. Soc. Chim. biol. **30**, 940 (1948).
91. Schopfer, W. H. et M. L. Bein: L'homothiamine-glycol, antivitamine B_1. Z. Vitaminforsch. **23**, 47 (1951).
92. Schopfer, W. H. et (Mlle) Boss: Recherches sur le rôle de la vitamine K et de diverses quinones chez les plantes. Un mécanisme possible de l'effet antibiotique de la vitamine K. Arch. Sci. Phys. Hist. Nat. Genève **1**, 521 (1948).
93. Schopfer, W. H. et E. C. Grob: Recherches sur l'action de diverses quinones sur l'activité enzymatique de l'uréase cristallisée. Arch. Sci. Phys. Hist. Nat. Genève **2**, 575 (1949).
94. Schopfer, W. H. et M. Guilloud: L'action antisulfamide des purines chez un microorganisme. Helv. physiol. Acta **4**, C 24 (1946).
95. Shive, W., W. W. Ackermann, M. Gordon, M. E. Getzendaner and R. E. Eakin: 5(4)-Amino-4(5)-imidazolcarboxinamide, a Precursor of Purines. J. Amer. chem. Soc. **69**, 725 (1947).
95a. Shive, W. and E. C. Roberts: Biochemical Transformations as Determined by Competitive Analogue-Metabolite Growth Inhibitions. II. Some Transformations Involving p-Aminobenzoic Acid. J. biol. Chemistry **162**, 463 (1946).
96. Soodac, M. and L. R. Cerecedo: Studies of Oxythiamine. J. Amer. chem. Soc. **66**, 1988 (1944).
97. Spitzer, E. H., A. I. Coombes, C. A. Elvehjem and W. Wisnicky: Inactivation of Vitamin B_1 by Raw Fish. Proc. Soc. exp. Biol. Med. **48**, 376 (1941).
98. Stahmann, M. A., C. F. Huebner and K. P. Link: Studies on the Hemorrhagic Sweet Clover Disease. V. Identification and Synthesis of the Hemorrhagic Agent. J. biol. Chemistry **138**, 513 (1941).
99. Stoerk, H. C.: The Regression of Lymphosarcoma Implants in Pyridoxin-Deficient Mice. J. biol. Chemistry **171**, 437 (1947).
100. — Desoxypyridoxine. Morphologic and Functional Changes in Acute Pyridoxine Deficiency. Federat. Proc. (Amer. Soc. exp. Biol.) **7**, 281 (1948).
101. Ter Horst, W. P. and E. L. Felix: 2,3-Dichloro-1,4-naphtoquinone. A Potent Organic Fungicide. Ind. Engng. Chem. **35**, 1255 (1943).
102. Tracy, A. H. and R. C. Elderfield: Studies in the Pyridine Series. II. Synthesis of 2-Methyl-3(β-hydroxyethyl)-pyridine and of the Pyridine Analog of Thiamine (Vitamin B_1). J. org. Chemistry **6**, 54 (1941).
103. Traeger, W.: Further Studies on the Survival and Development *in vitro* of Malarial Parasite. J. exp. Medicine **77**, 411 (1943).
104. Tréfouel, J., (Mme) Tréfouel, F. Nitti et D. Bovet: Activité du p-aminophénylsulfamide sur les infections streptococciques expérimentales de la souris et du lapin. C. R. Séances Soc. biol. Filiales Associées **120**, 756 (1935).

105. UMBREIT, W. W. and J. G. WADDELL: Mode of Action of Desoxy-pyridoxine. Proc. Soc. exp. Biol. Med. 70, 293 (1949).
106. UNNA, K. and W. ANTOPOL: Toxicity of Vitamin B_6. Proc. Soc. exp. Biol. Med. 43, 116 (1940).
107. WESWIG, P. H., A. M. FREED and J. R. HAAG: Antithiamine Activity of Plant Materials. J. biol. Chemistry 165, 737 (1946).
108. WHITING, F., J. K. LOOSLI, V. N. KRUKOVSKY and K. L. TURK: The Influence of Tocopherols and Codliver Oil on Milk and Fat Production. J. Dairy Sci. 32, 133 (1949).
109. WILSON, A. N. and S. A. HARRIS: Synthesis and Properties of Neopyrithiamine Salts. J. Amer. chem. Soc. 71, 2231 (1949).
110. WINKLER, K. C. and P. G. DE HAAN: On the Action of Sulfanilamide. XII. A Set of Non Competitive Sulfanilamide Antagonists for *Escherichia coli*. Arch. Biochemistry 18, 97 (1948).
111. WOODS, D. D.: The Relation of p-Aminobenzoic Acid to the Mecanism of the Action of Sulfanilamide. Brit. J. exp. Pathol. 21, 74 (1940).
112. — Les sulfamides en tant qu'antagonistes de l'acide p-aminobenzoïque. Bull. Soc. Chim. biol. 30, 730 (1948).
113. WOOLEY, J. G. and M. K. MURPHY: Metabolic Studies on T_2 *Escherichia coli* Bacteriophage. I. A Study of Desoxypyridoxine Inhibition and its Reversal. J. biol. Chemistry 178, 869 (1949).
114. WOOLLEY, D. W.: Destruction of Thiamine by Substance in Certain Fish. J. biol. Chemistry 141, 997 (1941).
115. — Production of Nicotinic Acid Deficiency with 3-Acetylpyridine, the Ketone Analogue of Nicotinic Acid. J. biol. Chemistry 157, 455 (1945).
116. — Observations on Antimicrobial Action of 2,3-Dichloro-1,4-Naphtoquinone, and its Reversal by Vitamins K. Proc. Soc. exp. Biol. Med. 60, 225 (1945).
117. — Reversal by Tryptophan of the Biological Effects of 3-Acetylpyridine. J. biol. Chemistry 162, 179 (1946).
118. — Some Structural Analogs Antagonistic to Pteroyl Glutamic Acid (Folic Acid). Bull. Soc. Chim. biol. 30, 805 (1948).
119. — Inhibition of Synthesis of Vitamin B_{12} and of Riboflavin by 1,2-Dichloro-4,5-diaminobenzene in Bacterial Cultures. Proc. Soc. exp. Biol. Med. 75, 745 (1950).
120. — Selective Toxicity of 1,2-Dichloro-4,5-diaminobenzene, its Relations to Requirements for Riboflavin and Vitamin B_{12}. J. exp. Medicine 93, 13 (1951).
121. — An Enzymatic Study of the Mode of Action of Pyrithiamine (Neopyrithiamine). J. biol. Chemistry 191, 43 (1951).
122. — A Study of Antimetabolites. Vol. 1. New York: Wiley & Sons. 1952.
123. WOOLLEY, D. W. and A. G. C. WHITE: Production of Thiamine Deficiency Disease by the Feeding of Pyridine Analogue of Thiamine. J. biol. Chemistry 149, 285 (1943).

(Reçu le 20 février 1952.)

Recent Investigations on Ergot Alkaloids.

By A. STOLL, Basle, Switzerland.

With 10 Figures.

Contents.

	Page
I. Historical Introduction	114
II. The Structure of the Ergot Alkaloids	119
1. Introduction	119
2. The Structure of Lysergic Acid	122
3. Structure of the Peptide Portion	134
III. The Individual Alkaloids of Ergot	149
1. Ergobasine and Ergobasinine	149
2. Ergotamine and Ergotaminine	153
3. Ergosine and Ergosinine	155
4. Alkaloids of the Ergotoxine Group	156
5. Ergocristine and Ergocristinine	159
6. Ergokryptine and Ergokryptinine	161
7. Ergocornine and Ergocorninine	162
IV. Partially Synthetic and Hydrogenated Derivatives of Ergot Alkaloids	163
1. Partially Synthetic Derivatives of Lysergic Acid	164
2. The Dihydro-Derivatives of the Natural Alkaloids of Ergot	167
References	170

I. Historical Introduction.

Ergot occupies a special position among the drugs of our therapeutic armamentarium, not only on account of its unusual classification in the vegetable kingdom but also because of its interesting biological characteristics and the remarkable nature of its active principles. Known botanically as *Claviceps purpurea*, ergot is a parasitic filamentous fungus which grows on the ears of plants of the Gramineae family. It is found principally on cereals, and thrives best on the ears of rye. Ergot of rye, or *Secale cornutum*, is the officinal form of the pharmacopoeias and the starting material for pharmaceutical preparations.

Ergot is strictly speaking the form in which the fungus passes the winter and consists of brownish-violet, horn-shaped sclerotia which project from the ripe ears of rye in place of the rye grains (Fig. 3, p. 150).

From the histological point of view, ergot may be considered as a pseudo-parenchyma and biologically as a rhizomorph. The sclerotium, which readily becomes detached from the rye before the harvest and falls to the ground, contains a rich store of foodstuffs and remains in a resting stage throughout the winter. Moisture causes the sclerotia to swell and, with the onset of warm weather in the spring, they germinate, first putting out bundles of hyphae and later long stromata with spherical heads (Fig. 4, p. 150).

On the surface of these stroma heads appear numerous perithecia, flask-shaped cavities from which filiform ascospores are expelled. These are blown away by the wind or carried upwards by rising currents of warm air, afterwards settling on open rye flowers. In this way a primary infection of the rye fields occurs. The spores then germinate and the resulting mycelium destroys the ovaries of the rye. Very soon the mycelium abstricts enormous numbers of conidia surrounded by a saccharine fluid, the so-called honey-dew, which is secreted simultaneously. The infectious secretion is transferred to other rye flowers either by insects or when neighboring ears are brought into contact by the wind. This results in the so-called secondary infection. After a few weeks the mycelium solidifies to the dark-colored pseudo-parenchyma which consists of the thickly matted hyphae and forms the sclerotium known as "ergot". This is either collected from the ears of rye immediately before or during the harvest, or is separated during the thrashing process.

Although the large size and dark color of the sclerotia render them very conspicuous, the injurious effect of ergot on health remained for a long time unrecognized. The number of persons who became ill through eating contaminated bread was sometimes so great that the outbreaks are often referred to as "ergot epidemics". The term "ergotism" is preferable, however, in view of the non-infectious nature of the disease.

Two characteristic forms of ergotism can be distinguished: gangrenous ergotism and convulsive ergotism.

Gangrenous ergotism is described as beginning with tingling and a furry sensation in the fingers, and with vomiting and diarrhoea. After a few days, visible evidence of gangrene appears, usually beginning on the fingers and toes. The skin turns bluish-black, probably as a result of injury to the peripheral blood vessels, and later the epidermis peels off. In cases of severe poisoning, dry gangrene of entire limbs can occur, beginning with violent, burning pain, followed by loss of sensation. Finally, the affected limbs may become completely detached from the body, sometimes without loss of blood.

The convulsive form of ergotism begins with symptoms similar to the gangrenous form. These are followed, however, by very painful muscle contractions, particularly in the extremities. Finally the convulsions assume an epileptic-like character, although they may persist for several hours. Severe disturbances of the central nervous system, accompanied by histological changes, may result. Since contamination of the bread with ergot was particularly likely to occur following bad harvests, when undernourishment and avitaminosis were common, ergotism had devastating effects. Severe ergot intoxication frequently proved fatal.

According to KOBERT (49), who collected a vast amount of information on the subject of ergotism, the epidemic of the year 994 in Aquitaine and Limoges in France killed about 40000 people, while the epidemic of 1129 in the region of Cambrai claimed at least 12000 victims. Of course, in dealing with data handed down over so many centuries, we can never be certain whether ergotism alone was responsible for the tremendous number of deaths or whether other factors such as avitaminosis or infectious diseases also played a rôle. One often finds this doubt expressed in the literature, especially with regard to the assumption that ergot was the main cause of many of the epidemics of former times.

Many of the victims of gangrenous ergotism regarded the burning sensation in their limbs as a divine punishment, a fact which accounts for such descriptions as Holy Fire, St. Anthony's Fire, feu sacré and ignis sacer, a name which goes back to the writings of Virgil.

At this point it may be mentioned that the quantities of ergot needed to produce ergotism are very large in comparison with the doses employed therapeutically, although patients already suffering from vascular diseases show a hypersensitivity to ergot. A similar hypersensitivity is seen under certain septic conditions.

While the cause of the widespread epidemics of ergotism escaped recognition, the ability of ergot to excite contractions of the uterus had long been recognized and put to use in popular medicine. In the 1582 edition of the "Kreuterbuch" by ADAM LONITZER, an interesting account is to be found on p. 285. Not only the description of the drug but also the application and even the dosage leave little doubt that he was referring to ergot. He recognized the constrictory action on the uterus, but not the toxicity of ergot in cereals, since it is expressly stated that the "corn pegs" are quite harmless.

The drug also finds frequent mention in later herbal books, and an illustration of it appeared for the first time in BAUHIN's "Theatrum botanicum" which was published in Basle in 1658. Nevertheless, it was not until well into the 18th century that ergot was officially recognized by the medical profession. Scientific interest in the application of ergot in midwifery was first aroused by the publication in America in 1808 of JOHN STEARNS' "Account of the Pulvis Parturiens, a Remedy for Quickening Child-birth" (70). Publication after publication followed, so that by 1827 as many as 90 papers on the medicinal use of ergot had appeared. This period also witnessed the publication of an account of the first chemical investigations by VAUQUELIN (*101*) in 1816.

However, these studies, like many others which followed them in the subsequent 100 years, were unable to provide any convincing information regarding the chemical nature of the active principles in ergot. Until a few decades ago, opinions on this question changed frequently, even after the isolation of a crystalline alkaloidal preparation, "ergotinine cristallisée", by the French pharmacist TANRET (*93*) in 1875 and of ergotoxine by BARGER and CARR (*5*) in England in 1906. The latter substance was also obtained simultaneously in Switzerland by KRAFT (*54*) who gave it the name "hydroergotinine". These alkaloidal preparations, which represent landmarks on the road to our present knowledge of the active principles of ergot, will be discussed in more detail later.

A new impetus was given to research on ergot by the isolation of *ergotamine* (*71, 74*) in 1918 and by the investigation of its pharmacological properties, especially when it was shown that ergotamine possesses the uterotonic action of ergot (*73*), till then the most striking property. Since that time it has been possible to isolate a whole series of ergot alkaloids and to establish their constitutions. Further investigations by STOLL and his associates have led to the preparation of a number of partially synthetic compounds, and both these and the purified natural alkaloids have found a wide variety of unexpected clinical applications.

Not only has the pharmacological investigation of the ergot alkaloids made great progress during the past few decades, but the study of their chemical constitution has also been pursued with success. An achievement of considerable practical importance was the isolation of the alkaloids in a pure state, for it was a well-known fact that the alkaloid content of ergot varies greatly according to the age and source of the sample (*73*). Consequently, the clinician can only be certain of reliable dosage when he is working with the pure alkaloids.

By chemical analysis of the pure compounds and by their degradation to known substances of simpler structure, it was possible to gain an insight into the very complicated constitution and configuration of the ergot alkaloids. In this way the foundation was laid for an approach to their synthesis. Very recently, it has been possible to clear up the last remaining uncertainties regarding the structures of the individual bases (*89*), so that the chemical investigation of the ergot alkaloids has now been brought more or less to a conclusion.

For a more detailed account of the early research on ergot, the reader is referred to the excellent monograph "Ergot and Ergotism" by BARGER (*2*), and to a more recent publication by the writer (*73*). To conclude this short historical survey of ergot research, it will suffice here to summarize the most important dates in tabular form.

Table 1. Historical Development of Ergot Research.

1582. First written description of the application of ergot for healing purposes by the Frankfort physician ADAM LONITZER (Lonicerus) in his "Kreuterbuch".

1808. Publication by the American physician STEARNS (70) of his "Account of the Pulvis Parturiens, a Remedy for Quickening Child-birth" through which the use of ergot in medicine was first placed on a scientific basis.

1816. VAUQUELIN (101) published the first noteworthy pharmaceutical and chemical investigations on ergot.

1875. TANRET (93) succeeded in preparing the first crystalline ergot alkaloid preparation, termed "ergotinine".

1906. A further alkaloidal preparation, "ergotoxine" was isolated from ergot by BARGER and CARR (5) and simultaneously by KRAFT (54), who gave it the name "hydroergotinine".

1918. STOLL (72, 74, 75) succeeded in isolating the first homogeneous alkaloid of ergot, ergotamine. Shortly afterwards it was proved conclusively that the oxytocic action is a property of the alkaloidal components of ergot and is not due to the compounds of low molecular weight, such as histamine or tyramine.

1932. CHASSAR MOIR (17) demonstrated that aqueous extracts of ergot exert a powerful constrictory action on the uterus, even when they contain no ergotamine or ergotoxine.

1934. SMITH and TIMMIS (65–68) succeeded in isolating a homogeneous derivative of *iso*lysergic acid. By degradation of natural alkaloids of ergot they had obtained *ergine* which was later recognized as *iso*lysergic acid amide.

1934. JACOBS and CRAIG (34) obtained lysergic acid for the first time by degradation of ergot alkaloids.

1935. DUDLEY and CHASSAR MOIR (21), STOLL and BURCKHARDT (76, 77), KHARASCH and LEGAULT (47, 48) as well as THOMPSON (97) isolated simultaneously a previously unknown alkaloid of ergot, *ergometrine* (21) or *ergobasine* (76, 77) (*ergonovine*).

1936. SMITH and TIMMIS (69) obtained a further alkaloid of ergot, termed *ergosine*.

1937. STOLL and BURCKHARDT succeeded in isolating a previously unknown ergot alkaloid, *ergocristine* (78).

1938. JACOBS (19) and his associates proposed a formula for lysergic acid which, with the sole exception of the position of the easily hydrogenated double bond, has subsequently proved correct.

1938. STOLL and HOFMANN (80, 82) prepared *ergobasine* (*ergometrine*, *ergonovine*) by partial synthesis. This was the first synthesis of a natural ergot alkaloid.

1943. STOLL and HOFMANN (83) proved that ergotoxine is not a homogeneous alkaloid but possesses a complex and variable composition; they succeeded in separating ergotoxine into its components *ergocristine*, *ergocornine* and *ergokryptine*.

1943. STOLL and HOFMANN (84) prepared the dihydroderivatives of the natural alkaloids of ergot. On the basis of thorough pharmacological investigations by ROTHLIN et al., these compounds have been introduced into therapy as dihydroergotamine and as a combination of dihydroergocristine, dihydroergokryptine and dihydroergocornine under the name "Hydergine".

1945. UHLE and JACOBS (98) succeeded in accomplishing the total synthesis of a mixture of racemic dihydrolysergic acid and racemic dihydro*iso*lysergic acid.

1949. STOLL, HOFMANN and TROXLER (90) proposed the now accepted formula for lysergic acid. On the basis of experimental data the steric relationships between lysergic acid and *iso*lysergic acid were clarified.

1950. STOLL, RUTSCHMANN and SCHLIENTZ (*92*) carried out the first total synthesis of the optically active dihydrolysergic acids.
1951. STOLL, HOFMANN and PETRZILKA (*89*) succeeded in settling all remaining questions regarding the constitution of the ergot alkaloids of the peptide type.

II. The Structure of the Ergot Alkaloids.

1. Introduction.

During the past 50 years, the search for the true active principles of ergot has led to the isolation of a large number of compounds, an impressive testimony to the powers of biosynthesis of the fungus *Claviceps purpurea*. In addition to the alkaloids, a number of other interesting compounds have been discovered, some of which are listed in Table 2. In contrast to the alkaloids which have so far been found only in ergot, the compounds contained in the Table are ubiquitously distributed in nature; most of them are of simpler composition than the alkaloids.

Although the ergot alkaloids have attracted the interest of chemists for many decades, it is only in the course of the last 30 years that appreciable progress has been made. As already mentioned, these alkaloids are nowadays generally accepted as being the specific active principles of

Table 2. Non-specific Compounds Found in Ergot.

Compound	Formula	References
Tyramine	$(p)HO \cdot C_6H_4 \cdot CH_2 \cdot CH_2NH_2$	(*3, 11*)
Histamine	$(C_3H_3N_2)-CH_2 \cdot CH_2NH_2$	(*8, 57, 102, 103, 63*)
Agmatine = δ-Guanidyl-butylamine	$HN=C\underset{NH(CH_2)_4NH_2}{\overset{NH_2}{\diagup}}$	(*24, 25, 52, 53*)
Putrescine	$H_2N(CH_2)_4 \cdot NH_2$	(*7*)
Cadaverine	$H_2N(CH_2)_5 \cdot NH_2$	(*7*)
Isoamylamine	$(CH_3)_2CH \cdot CH_2 \cdot CH_2NH_2$	(*7*)
Trimethylamine	$(CH_3)_3N$	(*26, 27*)
Choline	$HO \cdot CH_2 \cdot CH_2\overset{+}{N}(CH_3)_3OH^-$	(*26, 27*)
Betaine	$(CH_3)_3\overset{+}{N} \cdot CH_2 \cdot COO^-$	(*99, 100*)
Clavine		(*99, 100, 6*)
Tyrosine	$(p)HO \cdot C_6H_4 \cdot CH_2 \cdot CH(NH_2)COOH$	(*28*)
Histidine	$(C_3H_3N_2) \cdot CH_2 \cdot CH(NH_2)COOH$	(*28*)
Tryptophane	$(C_8H_6N) \cdot CH_2 \cdot CH(NH_2)COOH$	(*28*)
Ergothioneine		(*96, 10, 62, 30, 1, 59, 22, 23, 32, 12*)
Ergotinic acid	$H_2N \cdot C_{15}H_{26}O_{15} \cdot SO_3H$	(*54, 50, 100*)
Ergosterine		(*94, 95*)
Vitamin D		(*58*)

ergot, but with few other natural products has the elucidation of the constitution presented such great difficulties. On the one hand, these bases possess a very unusual structure and, on the other, they are extremely sensitive to chemical agents, light, air etc. As will be shown later, most of these compounds have relatively high molecular weights and are made up of a complicated heterocyclic portion (lysergic acid) combined with a peptide portion*. Consequently, while the fact that these compounds readily undergo decomposition and isomerization had to be kept constantly in mind, chemists investigating their structure had three main questions to settle:

a) The structure of the lysergic acid portion of the molecule.
b) The structure of the peptide portion.
c) The nature of the linkage between the two portions.

Table 3 includes all the natural alkaloids of ergot so far known. They have been divided into three groups, on the basis of their chemical structures.

Table 3. The Natural Alkaloids of Ergot and their Dextrorotatory Isomers.

Name	Formula	$[\alpha]_D^{20}$ in Chloroform	Discoverer
1. *Ergotamine group*			
Ergotamine	$C_{33}H_{35}O_5N_5$	$-155°$	STOLL (1918)
Ergotaminine		$+385°$	
Ergosine	$C_{30}H_{37}O_5N_5$	$-179°$	SMITH and TIMMIS (1936)
Ergosinine		$+420°$	
2. *Ergotoxine group*			
Ergocristine	$C_{35}H_{39}O_5N_5$	$-183°$	STOLL and BURCKHARDT (1937)
Ergocristinine		$+366°$	
Ergokryptine	$C_{32}H_{41}O_5N_5$	$-187°$	STOLL and HOFMANN (1943)
Ergokryptinine		$+408°$	
Ergocornine	$C_{31}H_{39}O_5N_5$	$-188°$	STOLL and HOFMANN (1943)
Ergocorninine		$+409°$	
3. *Ergobasine group*			
Ergobasine	$C_{19}H_{23}O_2N_3$	$-44°$	DUDLEY and MOIR
Ergobasinine		$+414°$	KHARASCH and LEGAULT
			STOLL and BURCKHARDT
			THOMPSON (1935)

As Table 3 shows, the natural ergot alkaloids occur in pairs. The two alkaloids of a pair are stereoisomers, each of which readily undergoes

---

* Of the six natural alkaloids of ergot so far known, five have a peptide nature; only one, ergobasine (ergometrine, ergonovine) has as its basic component an aminoalcohol group.

reversible rearrangement to the other. This property depends upon the ease of isomerization exhibited by lysergic acid, a characteristic constituent of all the alkaloids mentioned. This compound, either as such or built into the alkaloid molecule, readily undergoes reversible rearrangement into the isomeric *iso*lysergic acid. The natural, levorotatory ergot alkaloids are derived from lysergic acid, while the isomeric, dextrorotatory members of each pair are derivatives of *iso*lysergic acid. This isomerism is closely connected with a number of interesting problems to be discussed later. It may be mentioned at this point, that only the naturally occurring, levorotatory alkaloids are pharmacologically effective. The corresponding dextrorotatory compounds possess only a fraction of the activity of their levorotatory isomers.

The first five pairs of alkaloids shown in Table 3 are also in the chronological sequence of their discovery. The first pair consists of ergotamine and ergotaminine. Ergotamine was isolated in crystalline form and analyzed as far back as 1918 (*72*); its rearrangement to the strongly dextrorotatory, pharmacologically less active ergotaminine was accomplished shortly afterwards.

Preliminary pharmacological tests with pure ergotamine were performed first by the writer and then by SPIRO, while more comprehensive investigations were carried out somewhat later, mainly by ROTHLIN. From these studies and from preliminary clinical tests it became obvious that very small doses of ergotamine exert the full action of ergot. In obstetrics and gynecology, which were the only fields of medicine in which ergot was employed at that time, ergotamine gave complete satisfaction. From this alone it was evident that the specific activity of ergot is due to its content of alkaloids, a view which was of decisive importance for the subsequent development of the chemistry and pharmacology of ergot. The very numerous pharmacological and clinical investigations carried out with ergotamine fulfilled a pioneer function, but it was not until after the discovery of ergometrine (ergobasine, ergonovine) in the middle of the 1930's that the view that the specific active principles of ergot are alkaloids found universal acceptance.

The second pair of alkaloids in Table 3, ergosine and its isomer ergosinine, first isolated by SMITH and TIMMIS (*69*) in 1936, have not so far found any application in medicine.

The three pairs of alkaloids belonging to the ergotoxine group—ergocristine–ergocristinine, ergokryptine–ergokryptinine and ergocornine–ergocorninine—(Table 3) exhibit a number of special features, but only a brief account of their discovery and characterization will be given here. It may be recalled (Chapter I) that in 1906, an apparently homogeneous but nevertheless amorphous alkaloidal preparation was isolated simultaneously by BARGER and CARR (*5*) in England and by the Swiss pharmacist

KRAFT (*54*). The latter called this new preparation "hydroergotinine", but the name *ergotoxine* is generally employed in the literature. On the basis of various chemical and pharmacological investigations, we were able to show in 1943 that, although ergotoxine had meanwhile been obtained in crystalline form, it is not a homogeneous substance but a mixture of three isomorphous alkaloids (*83*). It consists of ergocristine, which had been isolated in 1937, and two other, previously unknown alkaloids which were named ergokryptine and ergocornine.

As mentioned, lysergic acid, or its isomer *iso*lysergic acid, is a characteristic constituent of all the alkaloids of ergot, including ergobasine. The alkaloids of the ergotamine and ergotoxine groups are polypeptides, formed by the combination of lysergic acid or *iso*lysergic acid with further aminoacids. Their polypeptide structure secures them a special place among the plant alkaloids.

The last pair of alkaloids listed in Table 3, ergobasine and ergobasinine, have a more simple structure (*21, 76, 77, 47, 48, 97*). They consist of lysergic acid and *iso*lysergic acid respectively, combined with an aminoalcohol. The structure of ergobasine (known in England as "ergometrine" and in America as "ergonovine") was established by JACOBS, and shortly afterwards STOLL and his associates (*80, 82*) succeeded in accomplishing its partial synthesis (see later).

2. The Structure of Lysergic acid.

This acid was first investigated by JACOBS and CRAIG (*34*), who isolated it from the degradation products obtained by energetic alkaline hydrolysis of ergot alkaloids. SMITH and TIMMIS (*68*) converted lysergic acid into an isomeric compound, *iso*lysergic acid, which exhibited stronger dextrorotation. Shortly afterwards, STOLL and HOFMANN (*79*) accomplished the cleavage of ergot alkaloids with hydrazine hydrate, a method which provides a good yield of the racemic hydrazide of *iso*lysergic acid, without appreciable resinification. The racemic *iso*lysergic acid hydrazide shows the characteristic color reactions of the ergot alkaloids which will be discussed later. If this racemic hydrazide is treated with strong caustic potash, not only is hydrazine split off, but at the same time, re-isomerization to the racemic form takes place.

Racemic *iso*lysergic acid hydrazide, racemic lysergic acid and racemic *iso*lysergic acid form salts only with difficulty. STOLL and HOFMANN (*79*) were able to obtain the optically active components by converting racemic *iso*lysergic acid hydrazide, for example, to the azide and condensing this with optically active bases, such as L-*nor*-ephedrine. The mere addition of ether to an ethanol solution of the *nor*-ephedride caused the levorotatory component to crystallize out in the form of a sparingly soluble ether addition compound, while the antipode remained in the mother liquors. Energetic alkaline hydrolysis of the dextrorotatory compound yielded a lysergic acid with weak dextrorotation which could be identified as natural D-lysergic acid. In an analogous manner, levorotatory L-lysergic acid was obtained crystalline from the levorotatory nor-ephedride. Both D- and L-lysergic acid readily undergo racemization; it is sufficient to dissolve them in 400 parts of hot water.

Some of the properties of the optically active forms of the hydrazides of lysergic and isolysergic acids (81) are presented in Table 4.

Table 4. Properties of Lysergic and Isolysergic Acid Hydrazides.

Substance	M. P.*	$[\alpha]_D^{20}$ (in pyridine)	Solubility and typical crystalline form
D-lysergic acid hydrazide ...	218°	+ 11°	From 50 volumes of methanol in long, thin prisms.
L-lysergic acid hydrazide ...	218°	− 11°	
D-isolysergic acid hydrazide .	204°	+ 452°	Readily soluble in methyl and ethyl alcohol, from which it crystallizes in massive prisms.
L-isolysergic acid hydrazide .	204°	− 454°	
rac. lysergic acid hydrazide..	220°	—	From 100 parts of hot alcohol in long needles.
rac. isolysergic acid hydrazide	240°	—	From 300 parts of hot alcohol in hexagonal plates.

JACOBS and CRAIG (34) succeeded relatively quickly in establishing in lysergic acid the presence of a carboxyl group, an NCH_3 group and a double bond. This double bond can be readily saturated by catalytic hydrogenation (84, 40), giving rise to a mixture of isomeric dihydrolysergic acids. These are more stable than lysergic acid itself, and are therefore more suitable for carrying out degradation reactions directed at establishing the constitution.

If dihydrolysergic acid is subjected to alkaline fusion, the products obtained include methylamine, propionic acid, 1-methyl-5-amino-naphthalene, and an indole carboxylic acid which is converted by decarboxylation into 3:4-dimethylindole (40, 37, 39, 44). By oxidation of "ergotinine" with nitric acid, JACOBS and his associates (40, 39, 33) obtained a tri-

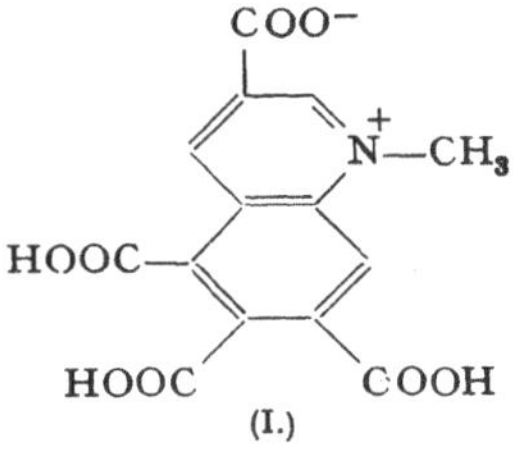

$$\text{(I.)}$$

---

* Indistinct with decomposition. All melting points are corrected.

carboxylic acid having the composition $C_{14}H_9O_8N$ which yielded quinoline on degradation with soda lime. JACOBS ascribed to this acid the structure of an N-methylquinoline-betaine-tricarboxylic acid (I).

From the results of these investigations and further degradation experiments, JACOBS et al. deduced the formulas (II) and (III), respectively, for lysergic and *iso*lysergic acids (*19*):

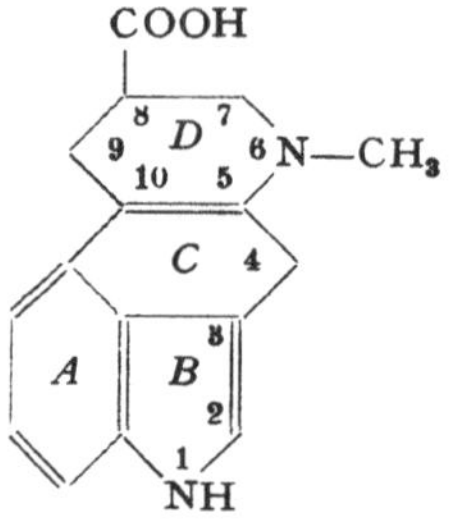

(II.) Lysergic acid (JACOBS). 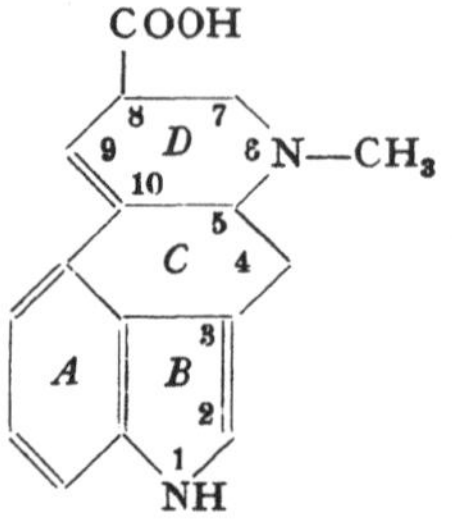 (III.) *Iso*lysergic acid (JACOBS).

Examination of these formulas shows that the following groupings are present: an indole system (rings *A* and *B*), a naphthalene system (rings *A* and *C*), and an N-methylquinoline system (rings *C* and *D*). JACOBS considered that lysergic acid differed from *iso*lysergic acid in the position of the readily hydrogenated double bond in ring *D*. Since the dihydro-acids exhibit asymmetric centers at positions 5, 8 and 10, saturation of the double bond with hydrogen must lead to complicated racemic mixtures, particularly if the starting material is racemic lysergic or *iso*lysergic acid. These complications had to be taken into account by UHLE and JACOBS (*98*) in 1945 when they carried out the total synthesis of a mixture of racemic dihydrolysergic acids. By comparing the product with a mixture of racemic dihydrolysergic acids of natural origin, they were able to obtain proof that the skeleton of the lysergic acid formula given above is correct. The synthesis was accomplished by the following route:

$(C_2H_5O)_2CH-CH_2-Br$ $\xrightarrow{KCN}$ $(C_2H_5O)_2CH-CH_2-CN$ $\xrightarrow{HCOOC_2H_5}{Na}$

Bromoacetal. Cyanoacetal.

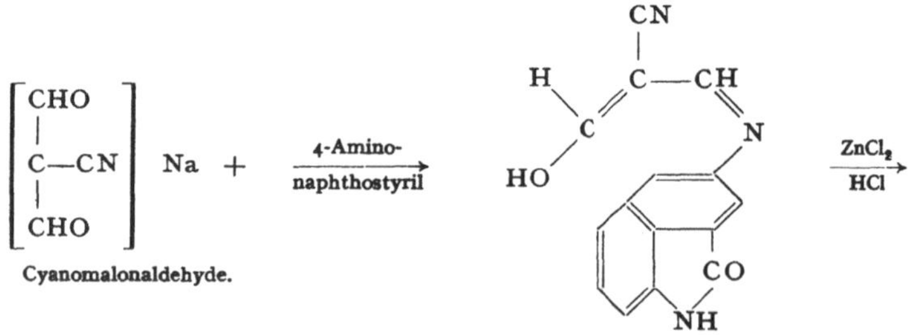

$\begin{bmatrix} CHO \\ | \\ C-CN \\ | \\ CHO \end{bmatrix}$ Na + $\xrightarrow{\text{4-Amino-}}{\text{naphthostyril}}$

Cyanomalonaldehyde.

2-Cyano-2-formylethylidene-4-aminonaphthostyril.

$\xrightarrow{ZnCl_2}{HCl}$

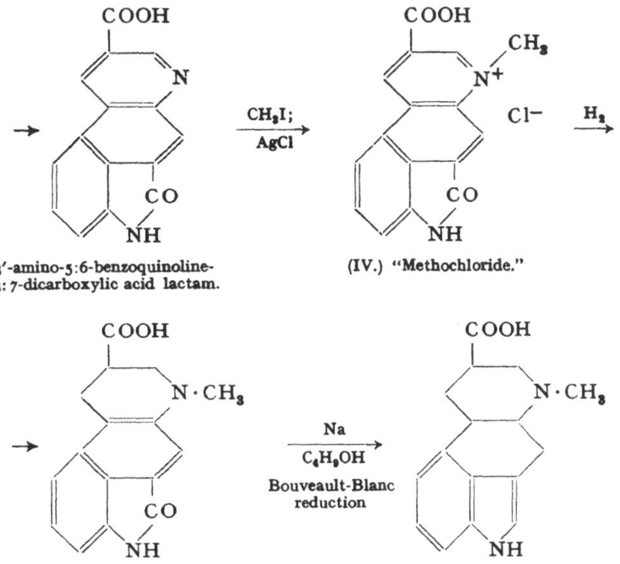

3'-amino-5:6-benzoquinoline-3:7-dicarboxylic acid lactam.

(IV.) "Methochloride."

(V.) Dihydrolysergic acid.

Bromoacetal was converted by means of potassium cyanide into cyanoacetal which was then treated with ethyl formate and sodium to give the sodium salt of cyanomalondialdehyde. On allowing this salt to react with 4-amino-naphthostyril, UHLE and JACOBS obtained 2-cyano-2-formylethylidene-4-amino-naphthostyril which, on treatment with zinc chloride and hydrochloric acid, underwent ring closure to yield 3'-amino-5:6-benzoquinoline-3:7-dicarboxylic acid lactam. The latter was converted by methyl iodide and silver chloride into the corresponding chlormethylate (IV), catalytic hydrogenation of which yielded 3'-amino-1-methyl-1:2:3:4-tetrahydro-5:6-benzoquinoline-3:7-dicarboxylic acid lactam. By further reduction of this lactam with sodium in boiling butanol, a mixture of racemic dihydrolysergic acid (V) and racemic dihydro*iso*lysergic acids was obtained.

This mixture of racemates, isolated in very small yield, proved to be identical with a product prepared by catalytic hydrogenation of racemic lysergic acid of natural origin. Like the synthetic product, the preparation obtained from natural lysergic acid was a mixture of racemates.

This total synthesis of a mixture of racemic, isomeric dihydrolysergic acids confirmed that the formula proposed by JACOBS for lysergic acid was correct, not only with regard to the carbon skeleton, but also with regard to the nature and position of the substituents. Still to be settled remained the position of the one relatively easily hydrogenated double bond lying outside the indole system as well as the mechanism of isomerization. The position of the double bond mentioned derived particular importance from the fact that JACOBS and his associates explained the peculiar isomerism between lysergic acid and *iso*lysergic acid as due to a change in the position of the double bond.

Recent investigations by STOLL and his associates (*90*), however, have demonstrated that both in lysergic acid and in *iso*lysergic acid the double bond in ring *D* is in the same position, namely $\Delta^{9:10}$.

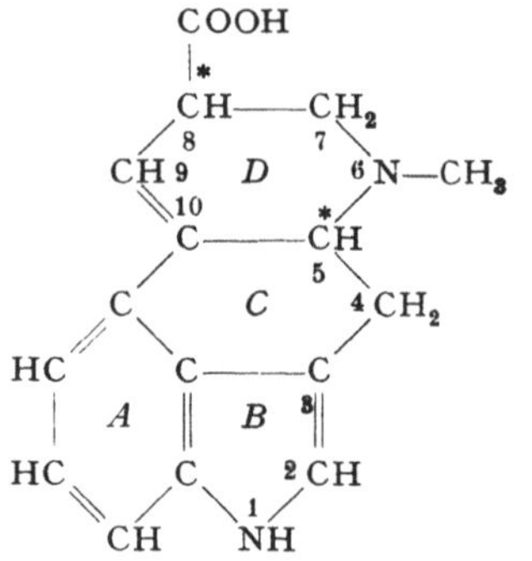

(VI.) Lysergic acid, *Iso*lysergic acid.

On heating the isomeric lysergic acids and the isomeric dihydrolysergic acids with acetic anhydride, STOLL *et al.* obtained lactams, the formation of which was due to opening of the ring in the manner of a β-amino carboxylic acid. After the opening of ring *D*, a new ring closure can take place between the secondary amino group and the carboxyl to give a lactam ring. This sequence of reactions is shown in the following diagram:

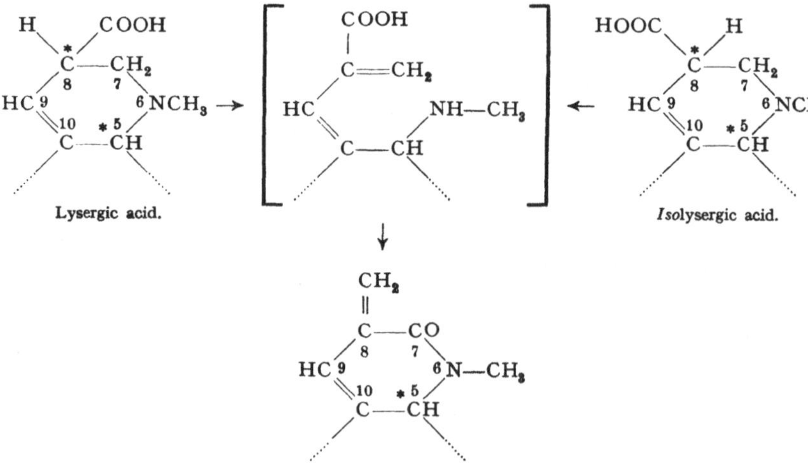

Lysergic acid.   *Iso*lysergic acid.

"Lactam".

As a result of the introduction of the new double bond, the asymmetric center at $C_{(8)}$ is abolished and, what is especially important, both lysergic acid and *iso*lysergic acid give rise to the same optically active lactam in equally good yield.

From this result it may be concluded that

(a) The steric configuration at the asymmetric $C_{(8)}$ is the sole difference between lysergic and *iso*lysergic acids. The remainder of the molecule is identical in both isomers, especially with regard to the arrangement

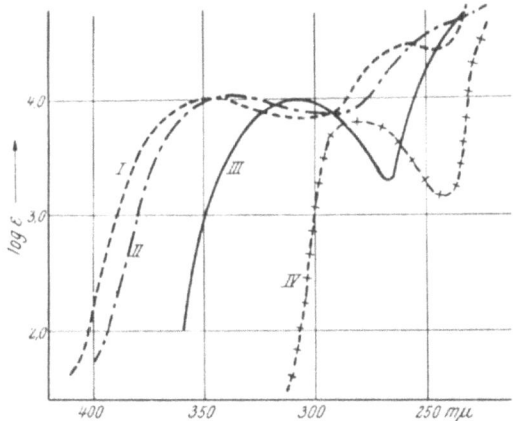

Fig. 1. Ultraviolet spectra. I = Lactam obtained from lysergic or *iso*lysergic acid; II = product obtained by decarboxylation of lysergic or *iso*lysergic acid; III = lysergic or *iso*lysergic acid; IV = dihydro-lysergic acid.

of the double bonds between the carbon atoms, a finding which is in harmony with the fact that the ultraviolet spectra of the two acids are identical.

(b) The readily hydrogenated double bond, which is conjugated with the indole system, can only be in position 4:5 or in 9:10. The position 5:10 as proposed by JACOBS for the double bond in lysergic acid must be ruled out, since this would allow only one asymmetric center, viz. that at $C_{(8)}$. Abolition of this asymmetric center would then lead to an optically inactive lactam which, however, is not the case.

Whether the readily hydrogenated double bond is located at 4:5 or at 9:10, was decided by comparing the ultraviolet spectrum of the lactam with that of the acids. The marked displacement of the lactam spectrum in the direction of longer wavelengths (Fig. 1) indicates that the new double bond in position 7:8 in the lactam must be conjugated with the unsaturated system already present in lysergic acid and *iso*lysergic acid. Hence, the reducible double bond in both acids occupies the position 9:10, since in position 4:5 it would not be conjugated with the new 7:8 double bond of the lactam.

From these experimental results alone, it may be concluded that lysergic acid and *iso*lysergic acid are diastereomers and not structural isomers, as was formerly assumed. They differ in the spatial arrangement of the substituents at the asymmetric carbon $C_{(8)}$. In both isomers, the readily reducible double bond is located in 9:10.

In the following formulas (VII) and (VIII) based on the above findings, the arrangement of the substituents at the atoms $C_{(5)}$ and $C_{(8)}$ has been chosen arbitrarily.

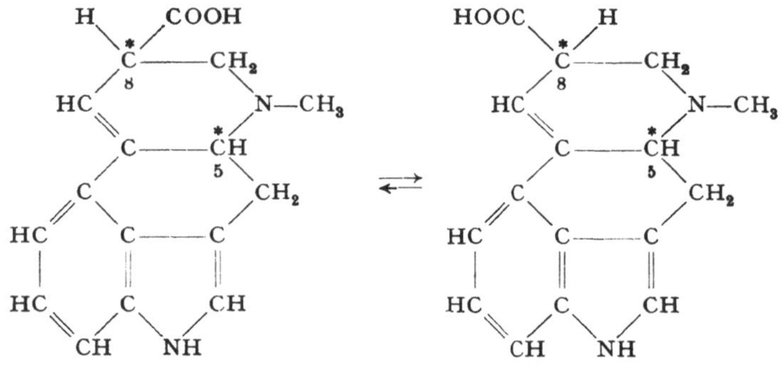

(VII.) Lysergic acid. (VIII.) *Iso*lysergic acid.

These new formulas are also in agreement with the results of decarboxylation experiments carried out by STOLL and his associates (*90*), both acids yielding the same decarboxylation product, as shown below. This decarboxylation results not only in the removal of the carboxyl group but also in the breaking of the linkage between the nitrogen atom 6 and the carbon atom 7. The ultraviolet spectrum of the decarboxylation product (Fig. 1) demonstrates clearly that the newly formed double

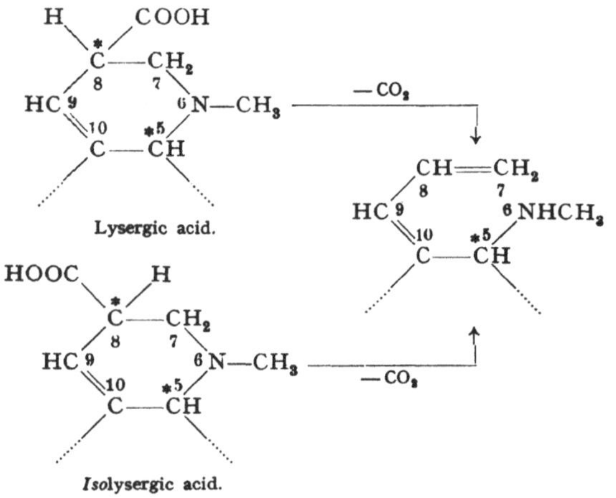

Lysergic acid.

*Iso*lysergic acid.

bond between $C_{(7)}$ and $C_{(8)}$ is conjugated with the double bond $\Delta^{9\cdot 10}$ already present. The chromophore system of this decarboxylation product is identical with that of the lactam, and this is confirmed by the similarity between the spectra of the two compounds.

If the double bond in lysergic acid is placed at $\Delta^{9:10}$, two asymmetric centers result, namely at $C_{(5)}$ and $C_{(8)}$. That a second asymmetric center must be present both in lysergic acid and in *iso*lysergic acid (in addition to that at $C_{(8)}$) is proved by the fact that the lactam discussed is optically active, although it has no asymmetric center at $C_{(8)}$. This observation further indicates that the two acids have the same configuration at $C_{(5)}$; and since the double bond in ring D is in the same $\Delta^{9:10}$ position in both acids, they can differ only in the spatial arrangement of the substituents at $C_{(8)}$. Evidence has been obtained that the carboxyl group is nearer the nitrogen atom 6 in lysergic acid than it is in *iso*lysergic acid.

Theoretically, either a D or an L configuration may be present at each of the two asymmetric carbon atoms 5 and 8. Represented purely schematically the following stereoisomers are possible:

$$C_{(5)} \qquad\qquad C_{(8)}$$
$$D_{(5)} \quad L_{(5)} \qquad D_{(8)} \quad L_{(8)}$$

By combining these four possibilities, the following four optically active compounds are obtained:

$$\begin{array}{cccc} D_{(5)}D_{(8)} & D_{(5)}L_{(8)} & L_{(5)}L_{(8)} & L_{(5)}D_{(8)} \\ \text{I} & \text{II} & \text{III} & \text{IV} \end{array}$$

I and III are optical antipodes and so are II and IV, while II is the diastereomer of I; and IV is the diastereomer of III. Consequently, I and II agree in the configuration at $C_{(5)}$ and differ only at $C_{(8)}$; and III and IV are similarly related. Thus, either I and III are the two optical antipodes of lysergic acid, while II and IV are the optical antipodes of *iso*lysergic acid, or vice versa, the absolute configurations at $C_{(5)}$ and $C_{(8)}$ being unknown.

The experimental results obtained with lysergic and *iso*lysergic acids are in complete harmony with these theoretical considerations. Besides natural D-lysergic acid and its optical antipode L-lysergic acid, both D-*iso*lysergic acid and L-*iso*lysergic acid have been prepared. The two corresponding racemates are also known.

Having elucidated these steric relationships, it was possible to give a simple explanation of the interconvertibility of lysergic acid and *iso*lysergic acid. Enolization of the carboxyl group at the position 8 gives rise to the formation of an intermediate acid enolate, which subsequently rearranges back to the acid form, a process which may be accompanied by a change in the configuration of the substituents at $C_{(8)}$:

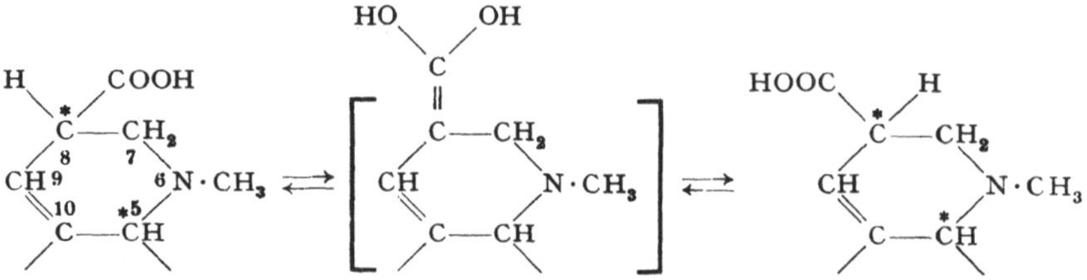

[Lysergic acid.] ⇌ ["Enol Form".] ⇌ [Isolysergic acid.]

In the light of these formulas a ready explanation can also be given for the behavior exhibited by the two acids and by the natural ergot

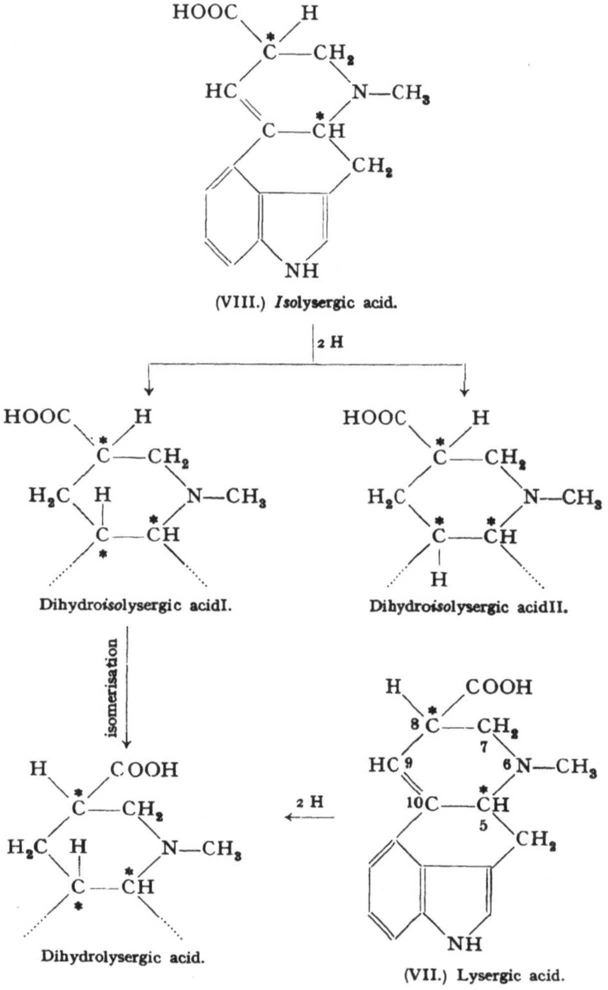

(VIII.) *Iso*lysergic acid.

| 2 H

Dihydro*iso*lysergic acid I.   Dihydro*iso*lysergic acid II.

isomerisation

Dihydrolysergic acid.   ← 2 H   (VII.) Lysergic acid.

alkaloids and their dextrorotatory isomers on catalytic hydrogenation (*88*). Saturation of the double bond in ring *D* leads to the formation of a new asymmetric center at $C_{(10)}$, so that lysergic acid and *iso*lysergic acid would each be expected *a priori* to give rise to two partial racemates. This was, in fact, found to be the case when *iso*lysergic acid was hydrogenated; it was then possible to isolate two well-defined dihydro*iso*lysergic acids, which were distinguished by the Roman figures I and II. As can be seen from the formulas on p. 130, the isomerism between the dihydro*iso*lysergic acidsI and II is due to a difference in the spatial arrangement of the hydrogen atom at $C_{(10)}$. (The configurations shown in the formulas are arbitrary.)

In contrast to *iso*lysergic acid, lysergic acid, when hydrogenated, yields only a single dihydro derivative. This must have the same configuration at $C_{(10)}$ as dihydro*iso*lysergic acidI, since it can be obtained from the latter by isomerization. The two acids differ merely in the configuration at $C_{(8)}$. In dihydro*iso*lysergic acidII, a different configuration must be present at $C_{(10)}$, since it cannot be converted into dihydrolysergic acid.

The structural relationships having been settled, it became possible to work out methods of resolving the racemic preparations obtained by synthesis into optically homogeneous lysergic and *iso*lysergic acids or their dihydro derivatives. Final proof that the lysergic acid and dihydrolysergic acid formulas were correct was not obtained, however, until the synthetic product had been shown to agree in all its properties, including optical rotation, with the dihydro derivative of natural origin.

To obtain the compounds needed for this comparison, and in order to work out suitable methods for separating the isomers, STOLL, RUTSCHMANN and SCHLIENTZ (*92*) converted lysergic acid of natural origin via the hydrazide to the racemates of lysergic and *iso*lysergic acids, catalytic hydrogenation of which yielded racemic dihydrolysergic acid and the racemic dihydro*iso*lysergic acidsI and II. With the aid of these compounds, it was possible to work out chromatographic separation methods which could subsequently be employed to resolve the mixtures of isomers obtained by synthesis.

In order to make the separation of the isomers a practical proposition, it was first necessary to increase the yield which was accomplished by employing a somewhat different route from that given by JACOBS. First, a mixture of isomeric racemates of dihydro-*nor*-lysergic acids was prepared and these were then separated into their racemic components by chromatographic analysis.

Starting from 4-amino-naphthostyril (IX) and ethoxy-methylene-malonic acid diethyl ester, a condensation product was obtained which, on ring closure as described by GOULD and JACOBS (*29*), yielded the

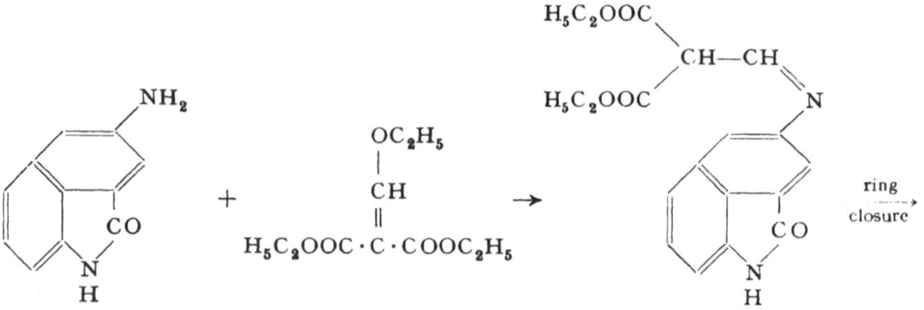

(IX.) 4-Amino-naphthostyril.   Diethyl-ethoxymethylene-malonate.

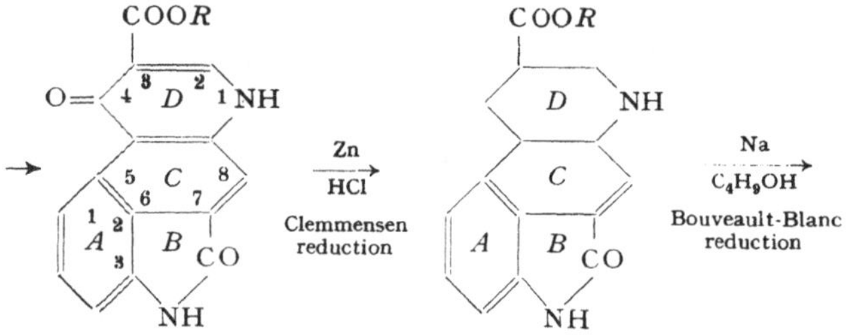

(X.) 3′-Amino-5:6-benzoquinolone(4)-3:7-dicarboxylic acid lactam ethyl ester.

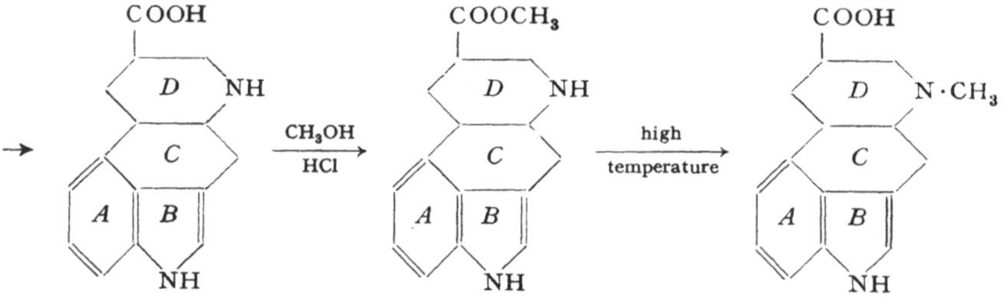

(XI.) Nor-dihydrolysergic acid.

(XII.) Nor-dihydrolysergic acid methyl ester.

(XIII.) Dihydrolysergic acid.

lactam of 3′-amino-5:6-benzoquinolone(4)-3:7-dicarboxylic acid ethyl ester (X). By means of the Clemmensen reduction, the carbonyl group in ring D was converted to a methylene group; and further reduction of the product with sodium in boiling butanol gave a good yield of a mixture of stereoisomeric dihydro-*nor*-lysergic acids (XI). This mixture was separated chromatographically into the racemates of dihydro-*nor*-lysergic acid, dihydro-*nor-iso*lysergic acidI and dihydro-*nor-iso*lysergic acidII. The homogeneous dihydro-*nor*-lysergic acid was esterified with methanol in the presence of gaseous hydrogen chloride. On merely heating the dihydro-*nor*-lysergic acid methyl ester (XII) to a high temperature a novel type of rearrangement took place, the methyl group migrating from the carboxyl to the nitrogen atom 6. In this way, a satisfactory yield of racemic dihydrolysergic acid (XIII) was obtained (92).

The next step was the resolution of racemic dihydrolysergic acid via its *L-nor*-ephedride into the optical antipodes, *D*-dihydrolysergic acid and *L*-dihydrolysergic acid; and it could then be shown that the former compound was identical with *D*-dihydrolysergic acid obtained from the dihydro derivatives of natural ergot alkaloids (92). The constitution was thus established beyond doubt and a considerable contribution made towards the total synthesis of the therapeutically important dihydro alkaloids.

Table 5. Properties of Dihydro*iso*lysergic Acids I and II, Dihydrolysergic Acid and Some Simple Derivatives.

	Dihydro*iso*lysergic acid I	Dihydro*iso*lysergic acid II	Dihydrolysergic acid
Acid			
Empirical formula .	$C_{16}H_{18}O_2N_2 \cdot H_2O$	$C_{16}H_{18}O_2N_2$	$C_{16}H_{18}O_2N_2$
Melting point	280° (block)	310° (block)	318° (block)
$[\alpha]_D^{20}$ (in pyridine) .	− 86°	+ 17°	− 122°
Crystal form (from water)	irregular platelets	massive polyhedra	hexagonal platelets
Hydrazide			
Empirical formula .	$C_{16}H_{20}ON_4$	$C_{16}H_{20}ON_4$	$C_{16}H_{20}ON_4$
Melting point	227°	260°	247°
$[\alpha]_D^{20}$ (in pyridine) .	− 23°	+ 56°	− 123°
Crystal form (from water)	needles	needles	needles
Azide			
$[\alpha]_D^{20}$ (in pyridine) .	− 48°		− 79°
Crystal form (from ether)	prisms	(amorphous)	prisms
Amide			
Empirical formula .	$C_{16}H_{19}ON_3$	$C_{16}H_{19}ON_3$	$C_{16}H_{19}ON_3$
Melting point	275° (block)	307° (block)	276° (block)
$[\alpha]_D^{20}$ (in pyridine) .	0°	+ 17°	− 131°
Crystal form (from methanol)	rectangular or hexagonal plates	prisms	prisms and plates
Methyl ester			
Empirical formula .	$C_{17}H_{20}O_2N_2$	$C_{17}H_{20}O_2N_2$	$C_{17}H_{20}O_2N_2$
Melting point	190°		187°
$[\alpha]_D^{20}$ (in pyridine) .	− 82°		− 96°
Crystal form (from aqueous methanol)	long prisms	(amorphous)	long prisms

The properties of the different dihydrolysergic acids and some simple derivatives are summarized in Table 5 (*88*).

The synthesis of lysergic acid itself, with its readily hydrogenated double bond at $\Delta^{9:10}$ in ring D, presents a much more difficult problem than the synthesis of the dihydro compound. Despite strenuous efforts made in laboratories in various parts of the world, the synthesis of natural lysergic acid has not yet been achieved.

3. Structure of the Peptide Portion.

Until very recently, our knowledge of the structure of the peptide portion of the ergot alkaloids was not as well advanced as that of lysergic acid. As far back as 1911, however, BARGER and EWINS (*9*) were able to show that thermal decomposition of ergotoxine preparations yielded the amide of dimethylpyruvic acid; about 15 years ago, JACOBS and CRAIG (*36, 41*) investigated the alkaline cleavage of ergotinine and ergotamine and isolated two other amino acids in addition to lysergic acid. Irrespective of the alkaloid, one of the amino acids obtained is always proline, while the other varies from alkaloid to alkaloid. A further study of these reactions, which was the starting point of the investigations of the peptide portion, has yielded considerable information regarding the structural units present in the alkaloids of the ergotamine and ergotoxine types. The cleavage products obtained from the ergot alkaloids of the peptide type are listed in Table 6.

Table 6. Cleavage Products Obtained from Peptide Type Ergot Alkaloids.

Ergotamine		*Ergokryptine*	
Lysergic acid amide	$C_{16}H_{17}ON_3$	Lysergic acid	$C_{16}H_{16}O_2N_2$
Pyruvic acid	$C_3H_4O_3$	Dimethylpyruvic acid	$C_5H_8O_3$
L-Phenylalanine	$C_9H_{11}O_2N$	L-Leucine	$C_6H_{13}O_2N$
D-Proline	$C_5H_9O_2N$	D-Proline	$C_5H_9O_2N$
		Ammonia	H_3N
$-3\,H_2O$	$C_{33}H_{35}O_5N_5$	$-4\,H_2O$	$C_{32}H_{41}O_5N_5$

Ergocristine		*Ergocornine*	
Lysergic acid	$C_{16}H_{16}O_2N_2$	Lysergic acid	$C_{16}H_{16}O_2N_2$
Dimethylpyruvic acid	$C_5H_8O_3$	Dimethylpyruvic acid	$C_5H_8O_3$
L-Phenylalanine	$C_9H_{11}O_2N$	L-Valine	$C_5H_{11}O_2N$
D-Proline	$C_5H_9O_2N$	D-Proline	$C_5H_9O_2N$
Ammonia	H_3N	Ammonia	H_3N
$-4\,H_2O$	$C_{35}H_{39}O_5N_5$	$-4\,H_2O$	$C_{31}H_{39}O_5N_5$

That the ketoacid is not present as such in the ergotamine molecule but is formed from a precursor during the degradation process, had already been shown by JACOBS and CRAIG (*43*) in 1938. A few years later, STOLL and his co-workers (74) isolated L-phenylalanyl-D-proline lactam from ergotamine, this being the first occasion on which a large fragment of the peptide portion had been obtained intact from this alkaloid molecule. In most recent investigations (91), which have been carried out using the more stable dihydroalkaloids, it was possible to split off the peptide group as a whole from practically all the ergot alkaloids of the peptide type. Finally, STOLL et al. have succeeded in establishing the structural formula, which was confirmed by synthesis (91). Since the cleavage was performed with anhydrous hydrazine, the peptide residue contained the precursor of the ketoacid in a reduced form. The cleavage products obtained in this way are shown in Table 7.

Table 7. Cleavage Products Obtained from Ergot Alkaloids.

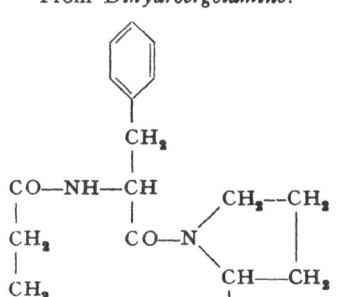

From *Dihydroergotamine:*

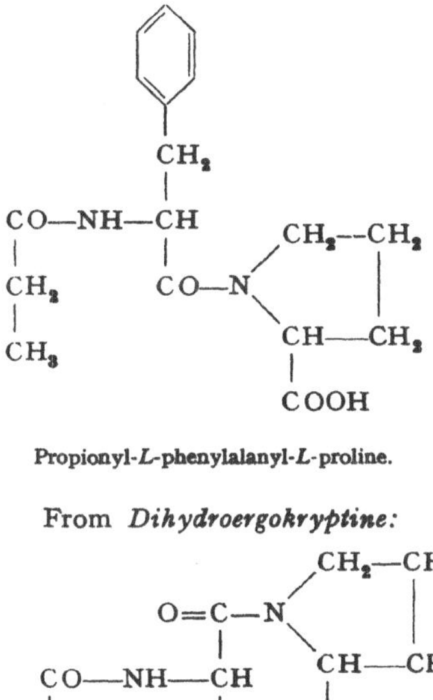

Propionyl-L-phenylalanyl-L-proline.

From *Dihydroergocristine:*

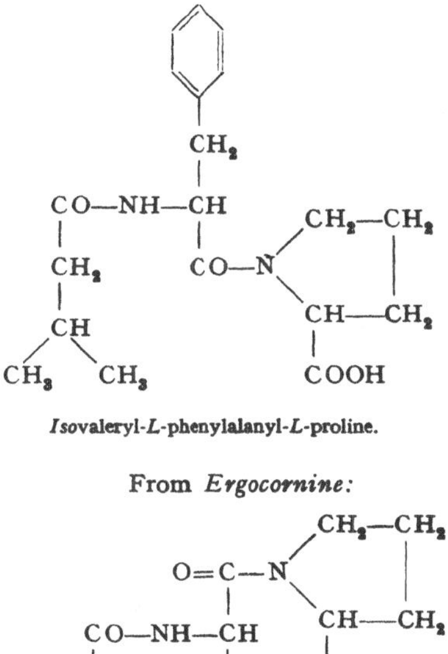

Isovaleryl-L-phenylalanyl-L-proline.

From *Dihydroergokryptine:*

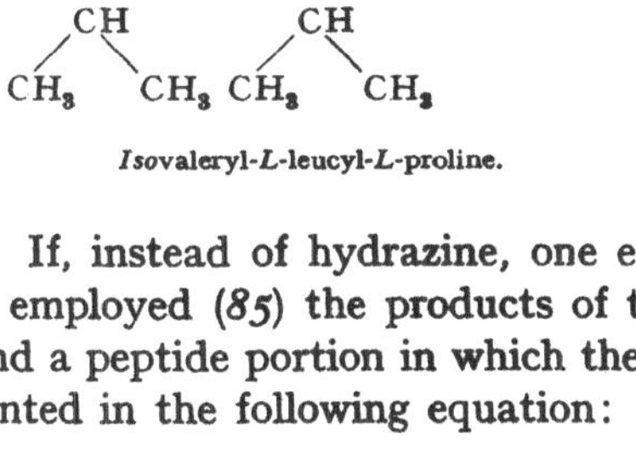

Isovaleryl-L-leucyl-L-proline.

From *Ergocornine:*

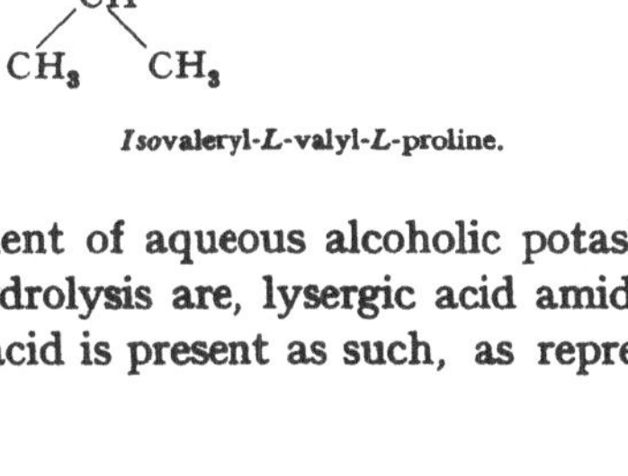

Isovaleryl-L-valyl-L-proline.

If, instead of hydrazine, one equivalent of aqueous alcoholic potash is employed (85) the products of the hydrolysis are, lysergic acid amide and a peptide portion in which the ketoacid is present as such, as represented in the following equation:

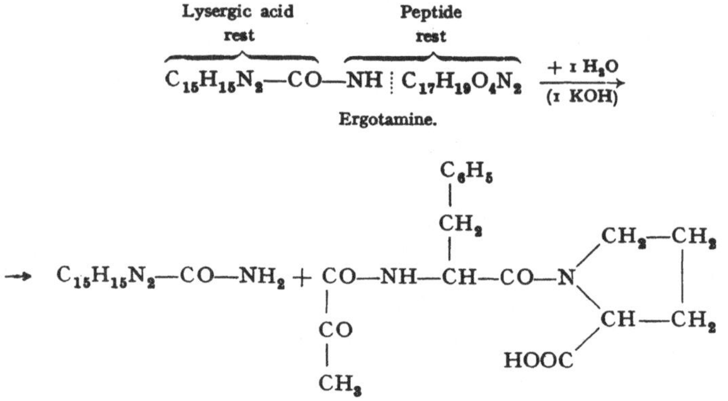

$$\underbrace{C_{15}H_{15}N_2\text{—CO}}_{\text{Lysergic acid rest}}\text{—NH} \mid \underbrace{C_{17}H_{19}O_4N_2}_{\text{Peptide rest}} \xrightarrow[\text{(1 KOH)}]{+\ 1\ H_2O}$$

Ergotamine.

$$\rightarrow\ C_{15}H_{15}N_2\text{—CO—NH}_2\ +\ \begin{array}{c}\text{CO—NH—CH—CO—N}\\ \mid\\ \text{CO}\\ \mid\\ \text{CH}_3\end{array}\begin{array}{c}\text{C}_6\text{H}_5\\ \mid\\ \text{CH}_2\end{array}\quad\begin{array}{c}\text{CH}_2\text{—CH}_2\\ \diagup\qquad\qquad\mid\\ \diagdown\text{CH—CH}_2\\ \text{HOOC}\end{array}$$

Lysergic acid amide.   Pyruvyl-*L*-phenylalanyl-*L*-proline.

From these investigations it was apparent that the carboxyl group of the precursor of the ketoacid is bound by an amide-like linkage to the amino group of a second amino acid. The carboxyl group of the latter which varies from alkaloid to alkaloid, is again connected by an amide-like linkage with the amino group of a third amino acid which is proline in all ergot alkaloids of the polypeptide type. The carboxyl group of proline must form a linkage of the lactone type with a hydroxyl group in the precursor of the ketoacid, since there is no free carboxyl group present in the alkaloids.

These findings were in agreement with the provisional structural formula proposed by JACOBS and CRAIG and modified by BARGER (*4*) but two important structural features still remained uncertain:

(a) The nature of the linkage between the lysergic acid residue and the labile connecting link, and

(b) the structure of this connecting link.

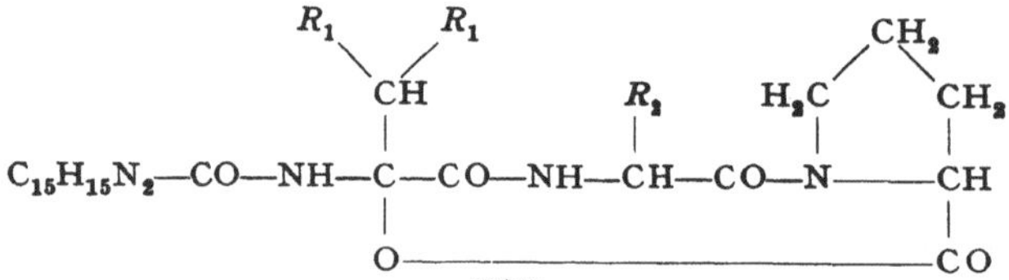

$$C_{15}H_{15}N_2\text{—CO—NH—C(R_1)(R_1)—CO—NH—CH(R_2)—CO—N—CH(CH_2CH_2CH_2)—CO}$$
$$\text{O———————————————————————}$$
(XIV.)

$R_1 = $ H or CH$_3$

$R_2 = $ CH(CH$_3$)$_2$ or CH$_2$—CH(CH$_3$)$_2$ or CH$_2$—C$_6$H$_5$

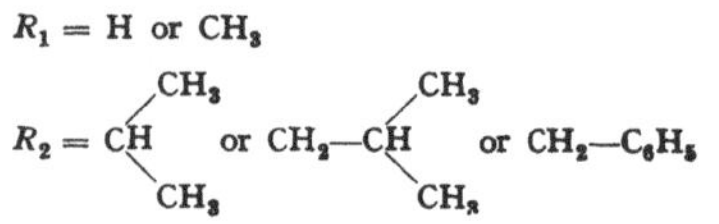

The suggestion had been put forward at one time by JACOBS and CRAIG (*43*) that this unidentified fragment of the molecule might have the structure of an α-hydroxy-α-aminoacid, which would have explained the formation of α-ketoacids during the hydrolysis of ergot alkaloids. Since, however, pyruvic acid may also be formed on hydrolytic cleavage of serin derivatives (*13, 14*), all the previous findings could have been explained equally well by assuming an α,β-structure for the hydroxyamino acid. Furthermore, there were certain observations which could not be accounted for by a structure containing a nine-membered lactone ring such as that present in formula (XIV). The most important of these was the observation that the two aminoacids which remain unchanged on cleavage frequently appear as a mixture of diketopiperazines, and that this occurs under conditions which favor the opening rather than the closing of a lactam ring, such as alkaline hydrolysis (*85*) or cleavage with hydrazine (*91*). The simple lactone structure of formula (XIV) also made it impossible to account for the formation of the neutral, saturated product which was obtained on thermal cleavage and which still contained the same number of carbon atoms as the original peptide residue.

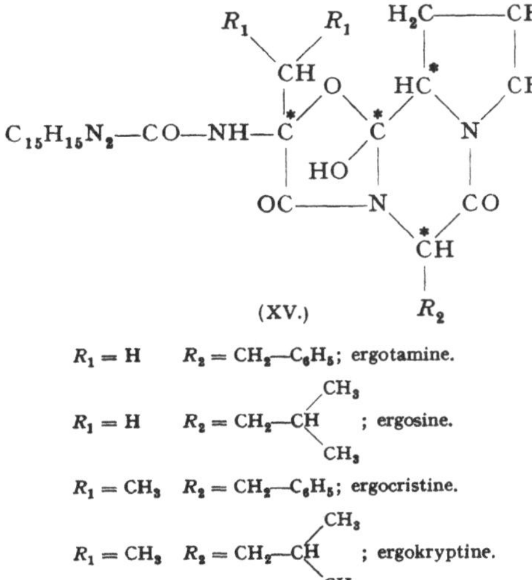

$$C_{15}H_{15}N_2\text{—CO—NH—}\overset{R_1}{\underset{*}{C}}H\text{—}\overset{R_1}{\underset{HO}{C}}\cdots$$

(XV.)

$R_1 = H \quad R_2 = CH_2\text{—}C_6H_5;$ ergotamine.

$R_1 = H \quad R_2 = CH_2\text{—}CH\underset{CH_3}{\overset{CH_3}{<}}$; ergosine.

$R_1 = CH_3 \quad R_2 = CH_2\text{—}C_6H_5;$ ergocristine.

$R_1 = CH_3 \quad R_2 = CH_2\text{—}CH\underset{CH_3}{\overset{CH_3}{<}}$; ergokryptine.

$R_1 = CH_3 \quad R_2 = CH\underset{CH_3}{\overset{CH_3}{<}}$; ergocornine.

These considerations prompted STOLL, HOFMANN and PETRZILKA (*89*) to investigate the possibility that the six-membered ring found in the degradation products, either as a diketopiperazine or as the mixture of piperazines obtained by JACOBS and CRAIG (*35, 36*), might be already preformed in the alkaloid, in the manner represented in formula (XV).

As can be seen from (XV), the nine-membered lactone ring is here subdivided into a six-membered and a five-membered ring by an additional bond between the nitrogen atom of the variable aminoacid and the carbon atom of the proline carboxyl group, with the result that this carbon atom carries a tertiary hydroxyl.

A linkage between amide groups of neighboring peptide chains, corresponding to that present in this formulation, had previously been suggested by WRINCH (*104, 105*) in connection with her cyclol theory of the structure of peptides. She held the view that certain proteins are built up from cyclized peptides in such a manner that several amino acid residues, for example 2, 6, 18 or more, combine to form a system of six-membered rings. According to this hypothesis, condensation within such a system is brought about by the transference of an atom of hydrogen from an NH-group to a CO-group giving rise to a N—C-linkage, the carbonyl group being at the same time converted into a tertiary alcohol group. In the case of a protein derived from six amino acid residues, the resulting formula would be (XVI).

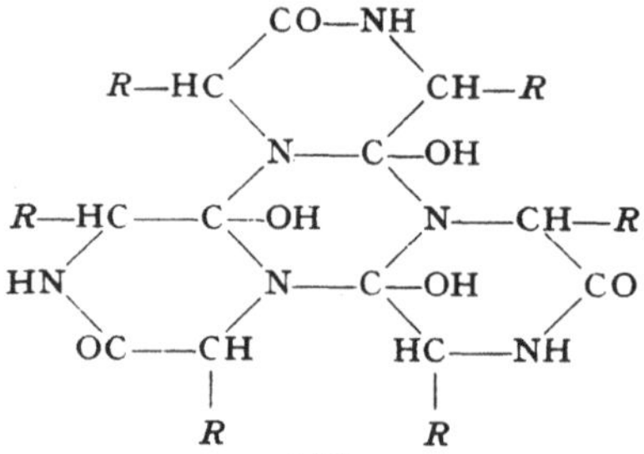

$$
\begin{array}{c}
\diagup\text{CO—NH}\diagdown\\
R\text{—HC}\text{CH—}R\\
\diagdown\text{N——C—OH}\diagup\\
R\text{—HC——C—OH}\quad\diagdown\text{N——CH—}R\\
\text{HN}\diagup\diagdown\text{N——C—OH}\diagup\diagdown\text{CO}\\
\diagdown\text{OC——CH}\diagup\text{HC——NH}\diagup\\
||\\
RR\\
\text{(XVI.)}
\end{array}
$$

Subsequently, a number of objections were raised to this cyclol theory (*15, 31, 61, 16*). Nevertheless, at least in this simple example of the peptide residue in ergot alkaloids, experimental confirmation of the theory of WRINCH seems to have been obtained; and as research enables us to penetrate further into the constitution of the peptides, it may well be that analogous structures will also be found in compounds of low molecular weight belonging to this class.

Proof of the correctness of the cyclol formula (XV) was provided in the following two ways:

(a) Identification by synthesis of the reduction products obtained on treatment with LiAlH$_4$ (*89*).

(b) Identification of the products obtained on thermal cleavage of the alkaloids in high vacuum.

The use of the more stable dihydroalkaloids, in which the double bond $\Delta^{9:10}$ in ring D of the lysergic acid residue is saturated with hydrogen, was found to be an advantage, particularly in the case of the thermal cleavage experiment.

All previous experience would indicate that compounds having structure (XVII) should yield open polyamino alcohols of type (XVIII) on reduction with LiAlH$_4$.

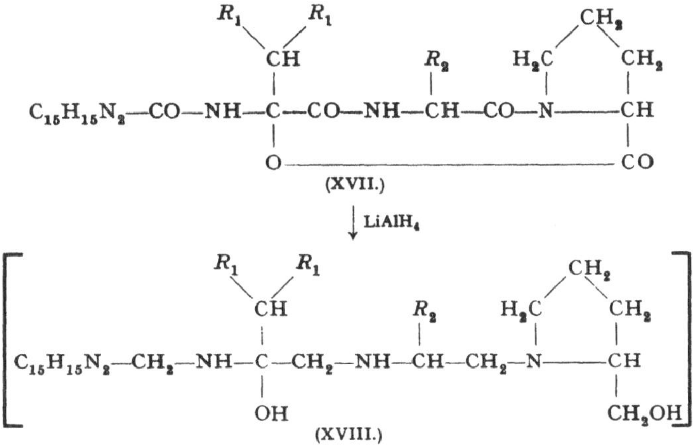

$$C_{15}H_{15}N_2-CO-NH-\underset{\underset{O\underline{\hspace{4em}}}{|}}{\overset{\overset{R_1}{\diagdown}\overset{R_1}{\diagup}}{\underset{|}{CH}}}-CO-NH-\overset{R_2}{\underset{|}{CH}}-CO-N\underline{\hspace{3em}}\overset{\overset{CH_2}{\diagup}\overset{CH_2}{\diagdown}}{\underset{\underset{CO}{|}}{\overset{H_2C\diagdown\diagup CH_2}{CH}}}$$

(XVII.)

↓ LiAlH$_4$

$$\left[C_{15}H_{15}N_2-CH_2-NH-\overset{\overset{R_1}{\diagdown}\overset{R_1}{\diagup}}{\underset{\underset{OH}{|}}{CH}}-CH_2-NH-\overset{R_2}{\underset{|}{CH}}-CH_2-N\underline{\hspace{3em}}\overset{\overset{CH_2}{\diagup}\overset{CH_2}{\diagdown}}{\underset{\underset{CH_2OH}{|}}{\overset{H_2C\diagdown\diagup CH_2}{CH}}}\right]$$

(XVIII.)

$R_1 = $ H or CH$_3$

$R_2 = $ CH$\diagup$CH$_3$$\diagdown$CH$_3$ or CH$_2$—CH$\diagup$CH$_3$$\diagdown$CH$_3$ or CH$_2$—C$_6$H$_5$

$R_1 = $ H $R_2 = $ CH$_2$—C$_6$H$_5$; ergotamine.

$R_1 = $ H $R_2 = $ CH$_2$—CH$\diagup$CH$_3$$\diagdown$CH$_3$; ergosine.

$R_1 = $ CH$_3$ $R_2 = $ CH$_2$—C$_6$H$_5$; ergocristine.

$R_1 = $ CH$_3$ $R_2 = $ CH$_2$—CH$\diagup$CH$_3$$\diagdown$CH$_3$; ergokryptine.

$R_1 = $ CH$_3$ $R_2 = $ CH$\diagup$CH$_3$$\diagdown$CH$_3$; ergocornine.

However, this is not the case. Instead, STOLL and his associates (89) obtained the cleavage products (XX), (XXI) and (XXII), the structures and configurations of which could be confirmed by synthesis.

$R_1 = $ H $R_2 = $ CH$_2$—C$_6$H$_5$; ergotamine.

$R_1 = $ H $R_2 = $ CH$_2$—CH$\diagup$CH$_3$$\diagdown$CH$_3$; ergosine.

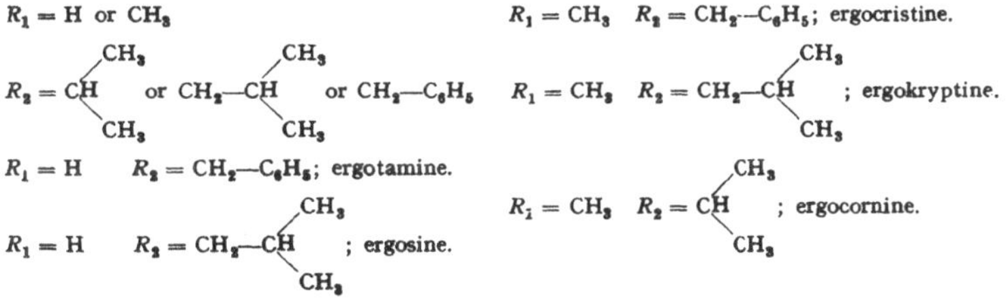

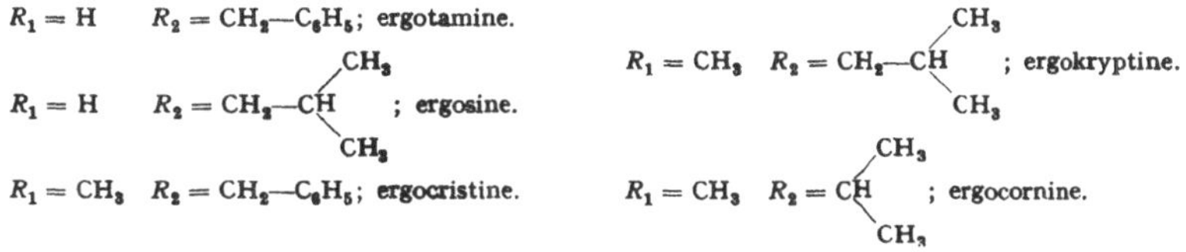

$R_1 = $ CH$_3$ $R_2 = $ CH$_2$—C$_6$H$_5$; ergocristine.

$R_1 = $ CH$_3$ $R_2 = $ CH$_2$—CH$\diagup$CH$_3$$\diagdown$CH$_3$; ergokryptine.

$R_1 = $ CH$_3$ $R_2 = $ CH$\diagup$CH$_3$$\diagdown$CH$_3$; ergocornine.

The polyamines of the general formula (XX) are obtained as crystalline compounds containing no oxygen but still possessing the same number of carbon and nitrogen atoms as the original alkaloid molecule. Since the amino acids from which the peptide portion of the ergot alkaloids is composed already contain asymmetric centers, it was to be expected that reduction with LiAlH$_4$ would lead to partial racemization. Consequently, a mixture of reduction products was obtained which, for identification purposes, had to be separated into its components. Thus, on reduction of dihydroergosine, two isomeric polyamines I and II, differing in optical rotation and melting point, were obtained.

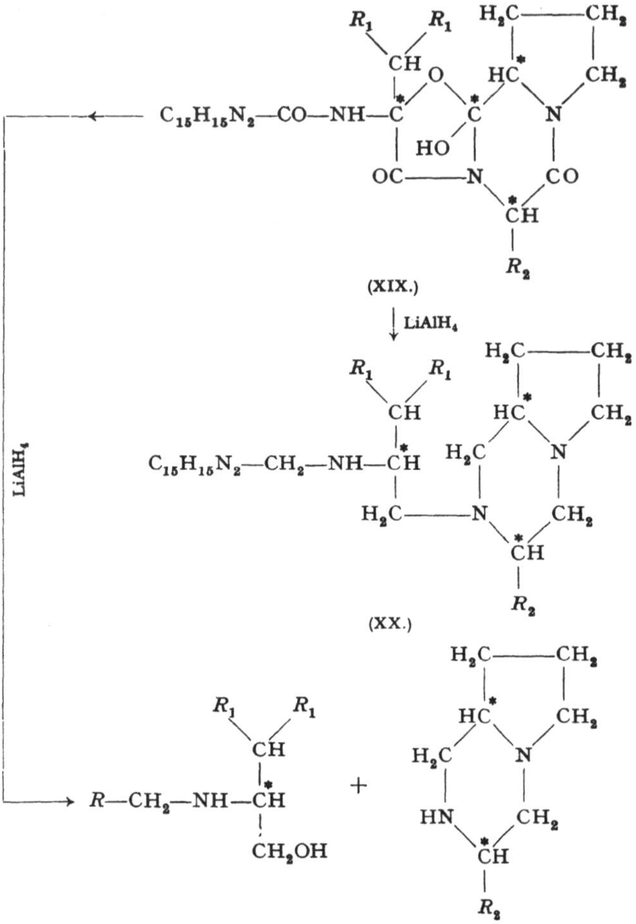

Polyamines having the general formula (XX) exhibit three asymmetric carbon atoms in the reduced peptide portion. While it was certain that the asymmetric center derived from the variable amino acid had the L-configuration, the configurations of the other two asymmetric centers derived from the labile hydroxy-amino acid and from proline remained in doubt. In order to make a comparison with the reduction products obtained from the alkaloids we thus needed four stereoisomers, each of which was synthesized.

The following route was employed for the synthesis of the various polyamines.

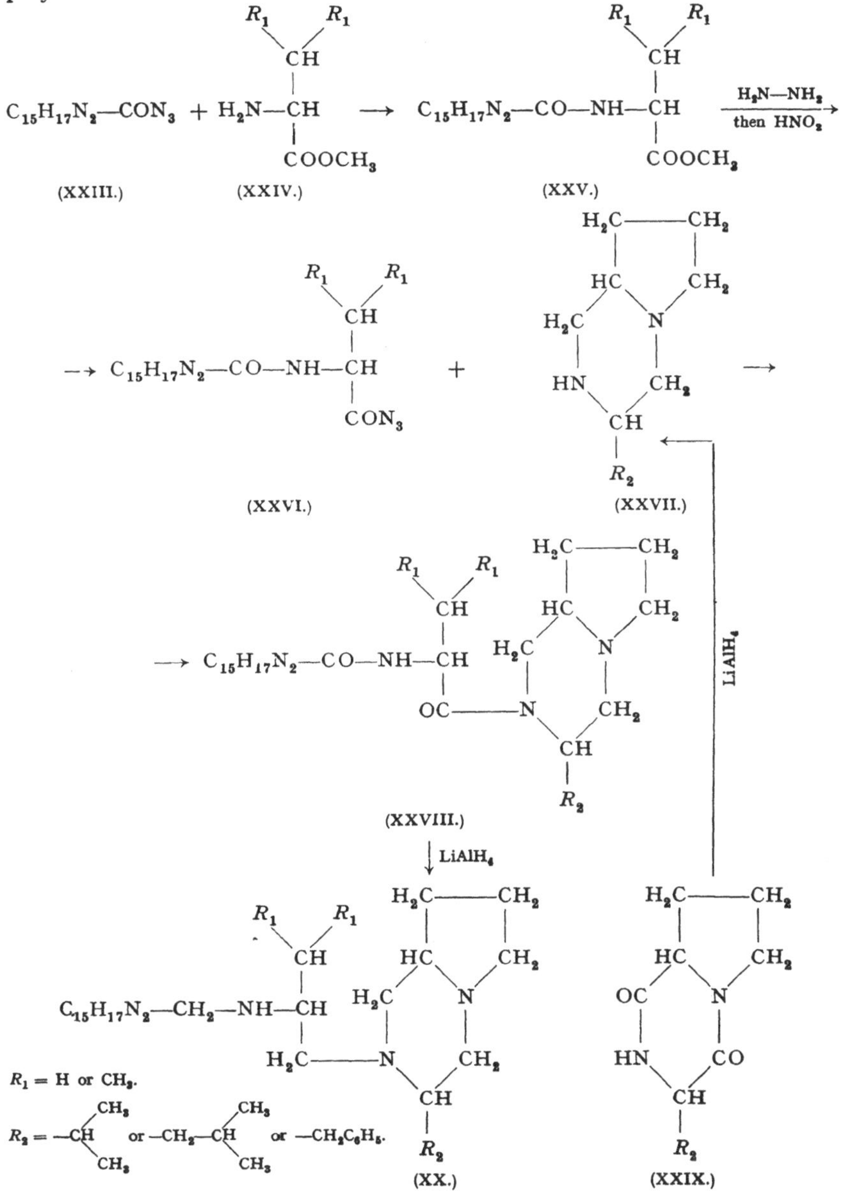

$$C_{15}H_{17}N_2\text{—}CON_3 + H_2N\text{—}\underset{\underset{COOCH_3}{|}}{\overset{\overset{R_1\diagdown\diagup R_1}{CH}}{CH}} \longrightarrow C_{15}H_{17}N_2\text{—}CO\text{—}NH\text{—}\underset{\underset{COOCH_3}{|}}{\overset{\overset{R_1\diagdown\diagup R_1}{CH}}{CH}} \xrightarrow[\text{then } HNO_2]{H_2N\text{—}NH_2}$$

(XXIII.)   (XXIV.)   (XXV.)

$$\longrightarrow C_{15}H_{17}N_2\text{—}CO\text{—}NH\text{—}\underset{\underset{CON_3}{|}}{\overset{\overset{R_1\diagdown\diagup R_1}{CH}}{CH}} \quad + \quad \text{(XXVII.)} \longrightarrow$$

(XXVI.)

$$\longrightarrow C_{15}H_{17}N_2\text{—}CO\text{—}NH\text{—}\underset{\underset{OC\text{—}\cdots\text{—}N}{|}}{\overset{\overset{R_1\diagdown\diagup R_1}{CH}}{CH}}$$

(XXVIII.)

↓ LiAlH₄

$C_{15}H_{17}N_2\text{—}CH_2\text{—}NH\text{—}CH$ … (XX.) (XXIX.)

$R_1 = H$ or CH_3.

$R_2 = \text{—}CH\!\!\begin{smallmatrix}CH_3\\CH_3\end{smallmatrix}$ or $\text{—}CH_2\text{—}CH\!\!\begin{smallmatrix}CH_3\\CH_3\end{smallmatrix}$ or $\text{—}CH_2C_6H_5$.

Dihydrolysergic acid azide (XXIII) was condensed with alanine or valine methyl ester (XXIV), and the resulting dihydrolysergyl-amino acid ester (XXV) converted via the hydrazide to the azide (XXVI). This product was allowed to react with the piperazine (XXVII) prepared by LiAlH$_4$ reduction of the corresponding diketopiperazine (XXIX), to give the piperazide of the dihydrolysergyl-amino acid (XXVIII), reduction of which with LiAlH$_4$ resulted in the removal of the two oxygen atoms and the formation of polyamines of type (XX).

By comparing the synthetic polyamines of known configuration with the polyamines obtained by LiAlH$_4$ reduction of the alkaloids, it was possible to establish the configuration at all asymmetric centers. PolyaminesI and II from dihydroergosine differ from one another in the configuration at the asymmetric carbon atom derived from the hydroxy-amino acid. PolyamineI has the D-alanine configuration, polyamineII derives from L-alanine.

The formation of polyaminesI and II, in other words, the racemization of the α-carbon atom of the residue obtained from the hydroxy-amino acid, indicates that a change of substituent has taken place at this asymmetric center in the course of the reduction.

Comparison with the synthetic compounds shows that at the asymmetric center derived from the proline residue all the polyamines have the L-configuration. This provides further proof that in the peptide residue itself the proline component is also present in the L-configuration.

The LiAlH$_4$ reduction of the alkaloids leaves the ring system of dihydrolysergic acid intact. Since the reduction leads to the formation of three alkylated nitrogen atoms in the peptide portion of the molecule, and since one is already present in the dihydrolysergic acid portion, the polyamine molecule would be expected to contain 4 basic atoms. In fact, however, the polyamines behave as tri-acid bases when titrated with hydrochloric acid; it appears that one of the nitrogen atoms is so weakly basic that it cannot be titrated with this acid. It is significant that a similar observation has been made with the partially synthetic polyamines.

A Zerewitinoff determination on the polyamines showed the presence of two active hydrogen atoms, one of which is accounted for by the imino group of the indole ring. From this and from the fact that treatment with acetic anhydride yields a monoacetyl derivative, it may be concluded that one of the three alkylated nitrogen atoms derived from the peptide portion is secondary and the other two are tertiary. All these requirements are fulfilled by structure (XX) (p. 140).

An investigation of the polyamine fraction obtained from dihydroergokryptine has been carried out by Stoll and his associates (89) on similar lines. Since it is known for certain that all ergot alkaloids of the polypeptide type have analogous structures, the knowledge gained from the study of these two bases can be applied to the other alkaloids.

Among the other principal products obtained on reduction of the alkaloids with LiAlH$_4$ was the amino alcohol (XXI) (p. 140).

Analysis showed that the two terminal amino acids of the peptide residue had been removed by reductive cleavage, leaving only the reduced residue of the hydroxy-amino acid attached to dihydrolysergic acid.

A Zerewitinoff estimation indicated the presence of three active hydrogen atoms, one of which is accounted for by the imino group of the indole system, one by the hydroxyl group and the third by the secondary amino group. Potentiometric titration with hydrochloric acid revealed two basic groups; and on treatment with acetic anhydride in pyridine it was possible to introduce two acetyl groups.

These properties are in accordance with the assumption that the amino alcohol linked to dihydrolysergic acid is either alaninol or valinol. This was confirmed by reduction of the appropriate synthetic preparations in the manner shown below in the formulas.

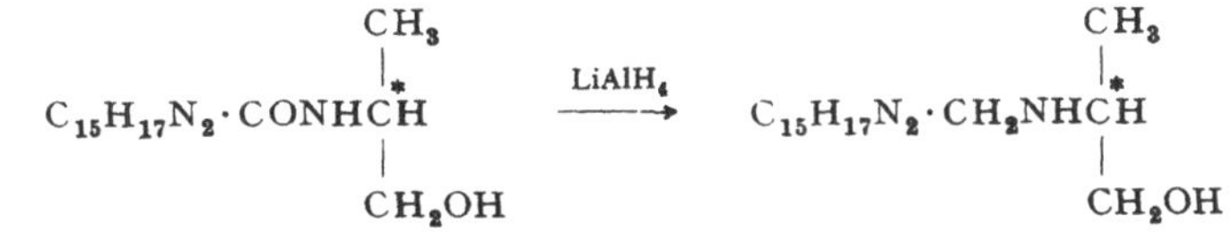

$$C_{15}H_{17}N_2 \cdot CONHCH(CH_3)CH_2OH \xrightarrow{LiAlH_4} C_{15}H_{17}N_2 \cdot CH_2NHCH(CH_3)CH_2OH$$

(XXX.) Dihydrolysergic acid-*L-isopropanolamide* =
= Dihydroergobasine = Dihydroergonovine.
Dihydrolysergic acid-*D-isopropanolamide*.

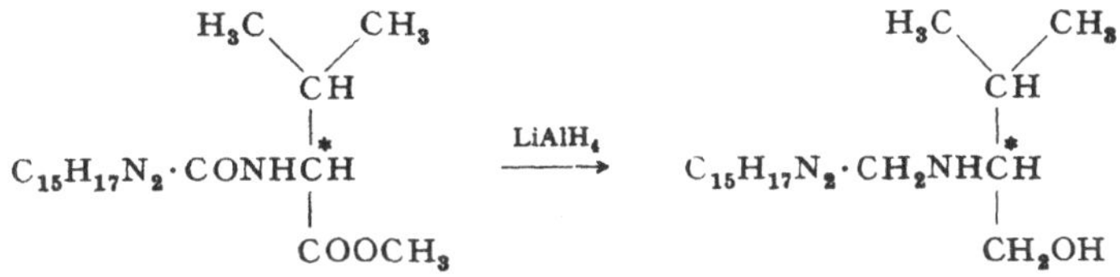

$$C_{15}H_{17}N_2 \cdot CONHCH(CH(CH_3)_2)COOCH_3 \xrightarrow{LiAlH_4} C_{15}H_{17}N_2 \cdot CH_2NHCH(CH(CH_3)_2)CH_2OH$$

(XXXI.) Dihydrolysergic acid-*L*-valine methylester.
Dihydrolysergic acid-*D*-valine methylester.

This reduction is accompanied by racemization due to a change of substituent on the asymmetric α-carbon atom of the amino alcohol. A complication which had to be overcome was that the optical rotations and melting points of the synthetic amino alcohols prepared from optically homogeneous components did not agree with those of the amino alcohols obtained by reduction of the alkaloids. Agreement with regard to these physical constants could, however, also be reached when we prepared mixtures containing equal amounts of the synthetic *L*- and *D*-forms of the amino alcohols; and it is therefore certain that the amino alcohols obtained by reduction of the peptide alkaloids are identical with those prepared synthetically.

The reduction of the ergot alkaloids with LiAlH$_4$ yielded a third important product which contained no lysergic acid component and did not give a color reaction with Keller's reagent. Its constitution proved relatively easy to determine.

From dihydroergokryptine, for example, a piperazine was obtained which was identical with the compound prepared by reduction of L-leucyl-L-proline lactam, and could therefore be identified as L:L-1:2-trimethylene-5-isobutyl-piperazine. In a similar way, the piperazine obtained from dihydroergocristine could be shown to be L:L-1:2-trimethylene-5-benzyl-piperazine (XXII) (p. 140).

The two terminal amino acids had already been isolated previously in the form of their diketopiperazines. The piperazines isolated on reduction with $LiAlH_4$ furnished new proof that in the alkaloids themselves the two terminal amino acids are linked up to form a ring system.

Considerable insight into the structure of the ergot alkaloids of the polypeptide type was gained by their *thermal cleavage*. As long ago as 1910, BARGER and EWINS (9) had obtained dimethylpyruvic acid amide by thermal decomposition of ergotinine. From the products of thermal cleavage of ergosine, SMITH and TIMMIS (69) were able to isolate L-leucyl-D-proline lactam, together with a little pyruvic acid amide. STOLL and his associates have subjected ergotamine (74) as well as ergokryptine and ergocornine (86) to thermal cleavage and obtained the corresponding diketopiperazines together with a little α-ketoacid amide.

On heating the dihydro alkaloids in high vacuum to 200 to 220°, they decompose into dihydrolysergic acid amide and a cleavage product which is of particular interest because it contains all the carbon atoms of the peptide portion (89, 86). An appreciable quantity of diketopiperazine is also formed. For the greater part, however, the thermal cleavage takes place according to the following equation, the example given here being dihydroergotamine:

$$C_{15}H_{17}N_2 \cdot CONH \cdot C_{17}H_{19}O_4N_2 = C_{15}H_{17}N_2 \cdot CONH_2 + C_{17}H_{18}O_4N_2.$$

It will be seen that none of the atoms is lost from the alkaloid molecule.

Thermal decomposition thus results in a type of cleavage different from hydrolytic methods and does not yield either lysergic acid or the intact peptide portion, since the amino group which connects these two fragments remains attached to the lysergic acid portion of the molecule. The question of the origin of the hydrogen required for the formation of the NH_2-group was not very easy to answer, but an explanation for this will be given later.

Let us turn now to the properties of the peptide fragment obtained on thermal cleavage. On Zerewitinoff determination no active hydrogen atoms could be detected, so that neither OH nor NH groups are present. The fragment is completely saturated. It does not take up hydrogen either with palladium or with platinum catalyst in glacial acetic acid solution. This excludes the presence not only of double bonds, but also of carbonyl and epoxide groups. Since an ebullioscopic determination of the molecular weight gave a value corresponding to the simple empirical

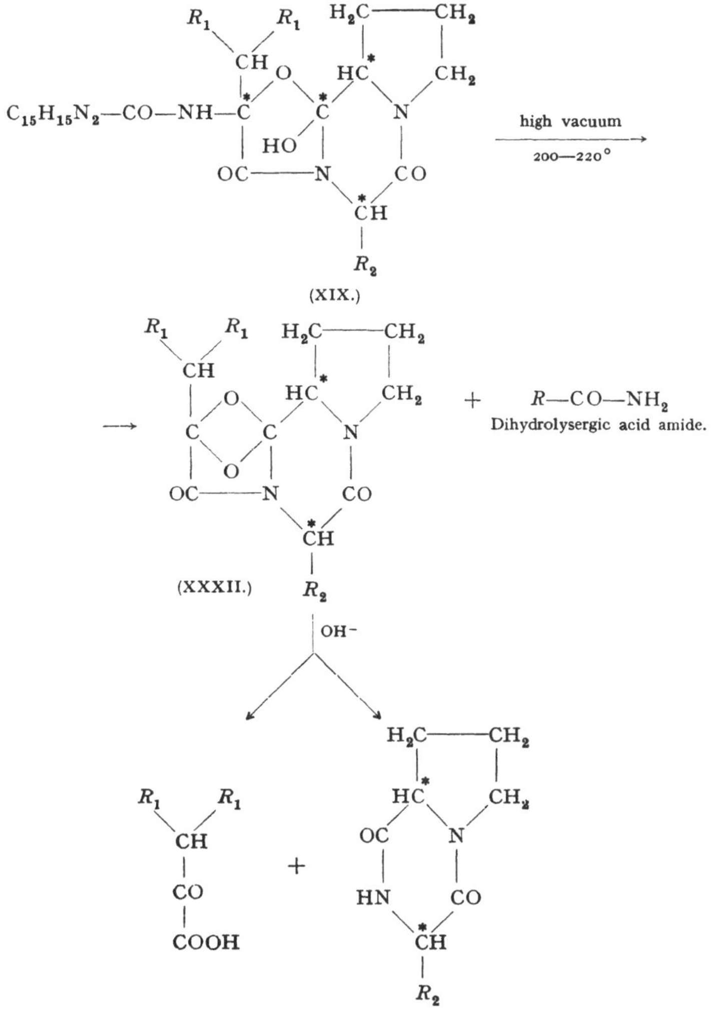

(XIX.)

$\xrightarrow{\text{high vacuum}}_{200-220°}$

(XXXII.) + $R-CO-NH_2$ Dihydrolysergic acid amide.

$\Big| OH^-$

(XXXIII.) + (XXXIV.)

$R_1 = H \quad R_2 = CH_2-C_6H_5$; ergotamine.

$R_1 = H \quad R_2 = CH_2-CH\begin{smallmatrix}CH_3\\CH_3\end{smallmatrix}$; ergosine.

$R_1 = CH_3 \quad R_2 = CH_2-C_6H_5$; ergocristine.

$R_1 = CH_3 \quad R_2 = CH_2-CH\begin{smallmatrix}CH_3\\CH_3\end{smallmatrix}$; ergokryptine.

$R_1 = CH_3 \quad R_2 = CH\begin{smallmatrix}CH_3\\CH_3\end{smallmatrix}$; ergocornine.

Fortschritte d. Chem. org. Naturstoffe. IX.

formula, and since neither double bonds nor keto groups are present, a tetracyclic structure can be deduced for the peptide residue obtained on thermal cleavage.

The peptide residue is neutral in reaction. It is fairly stable towards dilute acids, but readily decomposed even by very mild alkalies into the α-ketoacid (pyruvic acid or dimethylpyruvic acid) and the corresponding diketopiperazine. Incidentally, it may be noted that proline is present in the *D*-form in this diketopiperazine.

Taking into consideration the fact that the carboxyl group of the α-ketoacid must be connected to the nitrogen of the variable amino acid by an amide-like linkage, it is possible to deduce the above formula (XXXII) for the cleavage product (see p. 145).

The formation of a cleavage product having the structure (XXXII) can scarcely be accounted for if the peptide portion is formulated as containing a nine-membered lactone ring. On the other hand, (XXXII) can readily be derived from the cyclol formula. The mobile H-atom of the hydroxyl attached to the carbon atom of the proline carboxyl group will be split off with the amino group of the hydroxy-amino acid in the form of lysergic acid amide, and the resulting free valencies on the oxygen and the α-carbon atom will then unite to form an oxide ring.

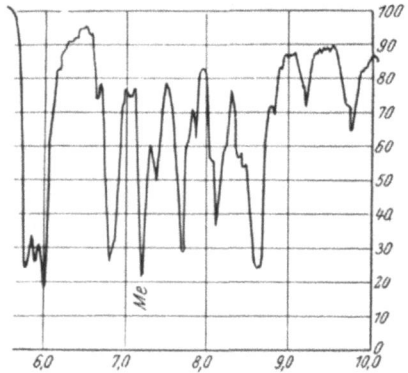

Fig. 2. Infra-red spectrum of the thermal cleavage product obtained from ergotamine.

The structure of the thermal cleavage product (XXXII) not only confirms the cyclol structure of the peptide portion, but also enables strict proof to be provided of the presence of an α-hydroxy-α-amino grouping in the labile connecting link.

(XXXV.)   (XXXVI.)

If the hydroxy-amino acid possessed an α,β or serin structure, the thermal cleavage product would have to have the constitution (XXXVI).

Evidence against this formula was provided by a study of the thermal cleavage product obtained from ergotamine, in which a C-methyl group could be detected by the method of KUHN and ROTH (55), as exhibited by (XXXV). Furthermore, the infra-red spectrum (Fig. 2) shows a decided methyl band at 1385 cm^{-1}, which is also in agreement with formula (XXXV).

This appears to be the first proof of the presence of an α-hydroxy-α-amino grouping in the labile amino acid of the peptide portion.

A novel feature of structure (XXXV) is the combination of the oxazolidone ring with the ring system of 1:3-di-oxa-cyclo-butane (60). The four-membered ring containing two ether oxygen atoms must be considered as a double acetal; and examples of this type of grouping are already to be found in the literature. Thus, DIELS and PILLOW (20), for example, described a compound having this type of structure and which was likewise stable towards acids but was split by alkalies into a ketoacid and an amide.

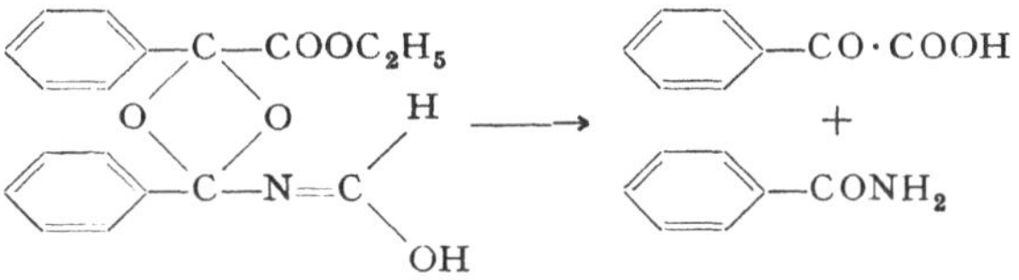

(XXXVII).

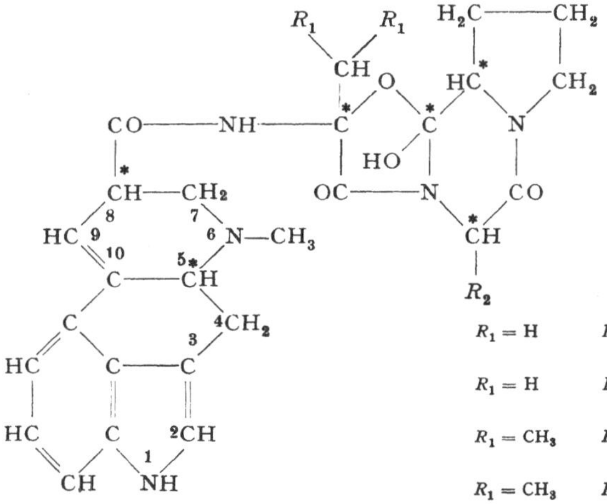

(XXXVIII.) General formula of ergot alkaloids.

$R_1 = H$ $R_2 = CH_2-C_6H_5$; ergotamine.

$R_1 = H$ $R_2 = CH_2-CH\begin{smallmatrix}CH_3\\CH_3\end{smallmatrix}$; ergosine.

$R_1 = CH_3$ $R_2 = CH_2-C_6H_5$; ergocristine.

$R_1 = CH_3$ $R_2 = CH_2-CH\begin{smallmatrix}CH_3\\CH_3\end{smallmatrix}$; ergokryptine.

$R_1 = CH_3$ $R_2 = CH\begin{smallmatrix}CH_3\\CH_3\end{smallmatrix}$; ergocornine.

Let us now take another look at the general structural formula (XXXVIII) and summarize briefly the conclusions which have been drawn from the reduction of the ergot alkaloids with LiAlH$_4$ and their thermal cleavage.

The formation of polyamines, the structures of which STOLL and his associates (89) have established by partial synthesis, indicates that the ketopiperazine ring is preformed in the peptide portion. In other words, the nine-membered lactone ring is subdivided into a six-membered and a five-membered ring by a bond connecting the nitrogen atom of the variable amino acid with the carbon atom of the proline carboxyl group. This leads to the inevitable formation of a tertiary hydroxyl group on this carbon atom. For the resulting structure, produced by a cyclization involving the transference of a hydrogen atom from the nitrogen of an amide group to the oxygen atom of a neighboring lactone carbonyl group, the description *"cyclol grouping"* first proposed by WRINCH (*104, 105*) has been adopted. The existence of such a structure has now been proved for the first time by chemical means. It seems very possible that, with the aid of the LiAlH$_4$ reduction method, it will be possible in the future to demonstrate that cyclol structures are also present in other peptides of low molecular weight.

The cyclol grouping in the peptide portion is based upon the orthocarboxylic acid form of proline; the carbon atom of the proline carboxyl carries three further substituents. This grouping exhibits the reactions of a potential carbonyl group, in consequence of which LiAlH$_4$ reduction results in the removal of all the oxygen atoms, as in the case of a true CO group.

Evidence in favor of this structural formula is likewise provided by the two other cleavage products, the amino alcohol and the piperazine. It is known that tertiary amides can be split by LiAlH$_4$ with the formation of aldehydes and that, in the presence of excess reagent, these are further reduced to alcohols. It is to this reductive cleavage that the amino alcohol and the piperazine owe their formation. Since secondary amides do not react in this way, the nitrogen atom of the variable amino acid must be tertiary in character, as required by the formula.

The polyamines and the amino alcohols are the first cleavage products to be obtained which provide direct evidence of an amide-like linkage between lysergic acid and the α-amino group of the labile hydroxy-amino acid.

The α-hydroxy-α-amino structure of this labile connecting link, previously thought to be highly probable, although not proved, is now rendered certain by the structure of the thermal cleavage products. Thus, in the case of the alkaloids of the ergotamine group, the connecting link between the lysergic acid residue and the two other amino acids is α-hydroxy-alanine, while in the case of the ergotoxine group the bridge is formed by α-hydroxy-valine.

The proposed formula also provides a ready explanation for the formation of the cleavage products obtained on thermal decomposition of the ergot alkaloids. The tertiary hydroxyl group furnishes a mobile hydrogen atom which facilitates the detachment of the lysergic acid residue in the form of lysergic acid amide.

By comparing the amines obtained by reduction of natural ergot alkaloids with synthetic amines, indisputable proof has been obtained that the amino acid proline, contained in the peptide residue of all the ergot alkaloids, is present in the L-configuration. The question of the configuration of the proline fraction had previously remained undecided because it was found that hydrolysis of the ergot alkaloids sometimes yielded L-proline and sometimes D-proline, depending upon the method employed.

III. The Individual Alkaloids of Ergot.

1. Ergobasine and Ergobasinine $C_{19}H_{23}O_2N_3$.
(Ergometrine and Ergometrinine.)

In 1932, the English gynecologist CHASSAR MOIR (*17*) reported that aqueous extracts of ergot induce powerful contractions of the uterus, even when they contain neither ergotamine nor ergotoxine. He therefore came to the conclusion that such extracts must contain a previously unknown, water-soluble substance with a powerful uterotonic action. Two years later a new water-soluble alkaloid was isolated simultaneously in four different laboratories (*21, 76, 77, 47, 48, 97*). The purest preparation was that obtained by the chemists of the Basle laboratory (*76, 77*), who were also the first to arrive at the correct empirical formula. Following a proposal made by Sir HENRY DALE, the workers in the different laboratories exchanged the preparations which they had isolated, and thus it became possible to establish beyond doubt that all four preparations were identical (*46*). However, four different names had been chosen for the new alkaloid by its discoverers, and two of these, ergometrine (*21*) and ergobasine (*76, 77*), are still in use in Europe, while in the U. S. A. ergonovine has since been adopted as the official name.

Ergobasine (ergometrine, ergonovine) differs in many respects from the other alkaloids of ergot. As mentioned, it is readily soluble in water and is characterized by a very rapid and, at the same time, powerful action on the uterus. This action is, however, of shorter duration than that of ergotamine. By consecutive application of the two alkaloids or by their simultaneous administration ("Neo-Gynergen"), the prolonged action of ergotamine may be combined with the rapid effect of ergobasine.

Fig. 3. Ripe ears of rye infected by ergot.

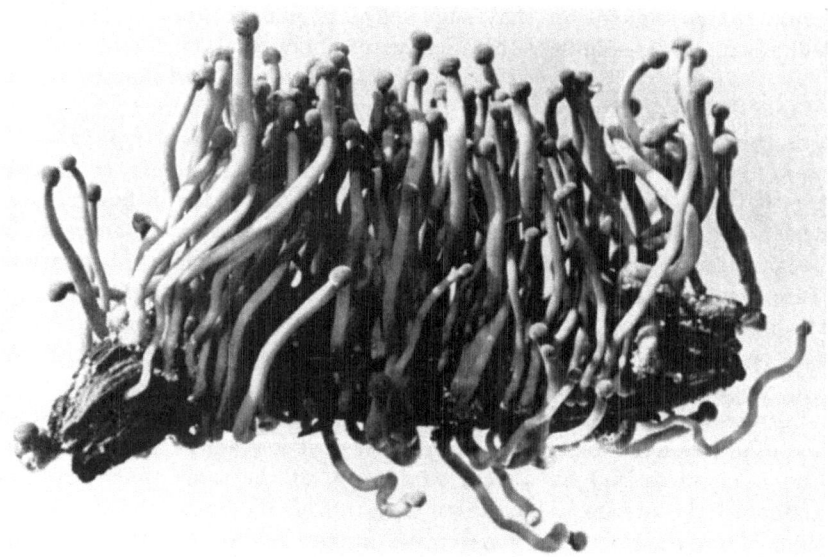

Fig. 4. Ergot sclerotia.

Table 6. Condensation Products of Lysergic and *Iso*lysergic Acids with 2-Aminopropanol-(1).

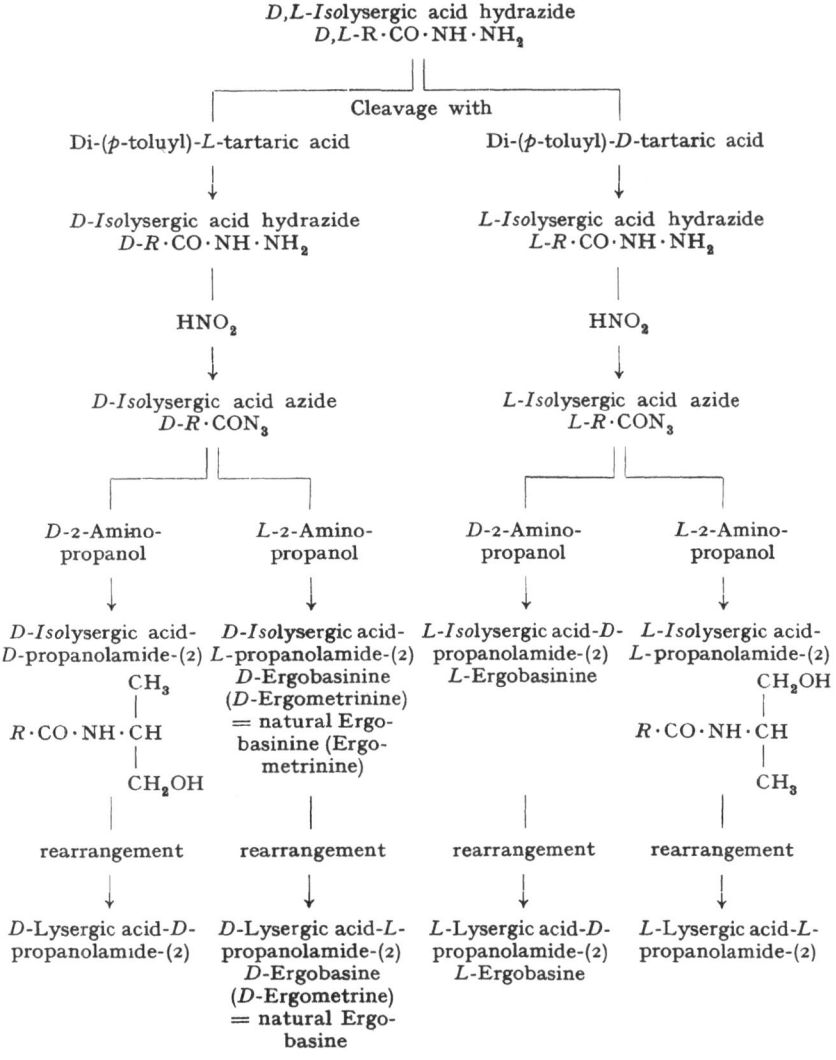

Within a short time of the discovery of ergobasine, JACOBS and CRAIG (*38*) were able to elucidate its structure (XXXIX). As was already evident from the empirical formula established by STOLL and BURCKHARDT (*76,77*), ergobasine was found to possess the simplest structure of the ergot alkaloid series, being composed of *D*-lysergic acid and *L*-2-aminopropanol-(1).

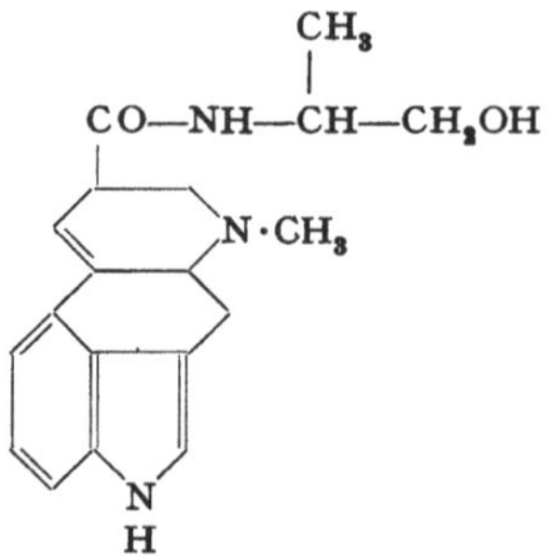

$$\text{CO—NH—CH(CH}_3\text{)—CH}_2\text{OH}$$

(XXXIX.) Ergobasine.

This formula was confirmed by STOLL and HOFMANN (80) in 1938 by the partial synthesis of ergobasine, the only synthesis of a natural alkaloid of ergot which has so far been accomplished. They employed the following route: Racemic lysergic acid hydrazide was converted by means of nitrous acid into *iso*lysergic acid azide (80, 79) which was condensed with L-2-aminopropanol-(1). Thus a mixture of different condensation products was obtained which was eventually resolved by a method specially devised for this purpose (80).

This partial synthesis of ergobasine, besides being of practical significance, also raised a number of interesting stereochemical

Table 7. The Eight Isomers of Ergobasine, $C_{19}H_{23}O_2N_3$.

Name	Decomp. point	$[\alpha]_D^{20}$	Typical crystal form
D-Lysergic acid-L-propanol-amide (2) (D-ergobasine)	162°	+90° (in water)	Well-formed tetrahedra from ethyl acetate; soft needles from benzene; very sparingly soluble molecular compound with 1 $CHCl_3$
L-Lysergic acid-D-propanolamide-(2) (L-ergobasine)	162°	−89° (in water)	
D-*Iso*lysergic acid-L-propanol-amide-(2) (D-ergobasinine)	196°	+414° (in $CHCl_3$)	From acetone; large transparent prisms
L-*Iso*lysergic acid-D-propanol-amide-(2) (L-ergobasinine)	196°	−415° (in $CHCl_3$)	
D-Lysergic acid-D-propanol-amide-(2)	220°	−11° (in pyridine)	From methanol on dilution with benzene, elongated flat prisms
L-Lysergic acid-L-propanol-amide-(2)	220°	+10° (in pyridine)	
D-*Iso*lysergic acid-D-propanol-amide-(2)	195°	+353° (in $CHCl_3$)	From acetone, massive prisms. Perchlorate from 95% alcohol, short prisms and polyhedra
L-*Iso*lysergic acid-L-propanol-amide-(2)	195°	−351° (in $CHCl_3$)	

questions. As has been shown, lysergic acid possesses two asymmetric carbon atoms, and this allows the existence of four optically active isomers, viz.: D-lysergic acid and L-lysergic acid, together with D-*iso*lysergic acid and L-*iso*lysergic acid. If one of these compounds is condensed with an optically active compound such as D-2-aminopropanol-(1) or its isomer, two optical antipodes are obtained, so that altogether a total of eight optically active condensation products, each isomeric with natural ergometrine, were to be expected. STOLL and HOFMANN (*82*) carried out this condensation and were, in fact, able to isolate all eight optically active condensation products in a pure state. In Table 6 (p. 151) the reactions by which the various isomers are obtained are shown diagrammatically.

Some of the properties of the eight isomers of ergobasine obtained in this way are listed in Table 7.

The photomicrographs show the typical crystal forms of these eight isomers (Fig. 5).

Using the same principle, STOLL and HOFMANN have prepared a whole series of homologs and analogs of ergobasine which will be discussed later.

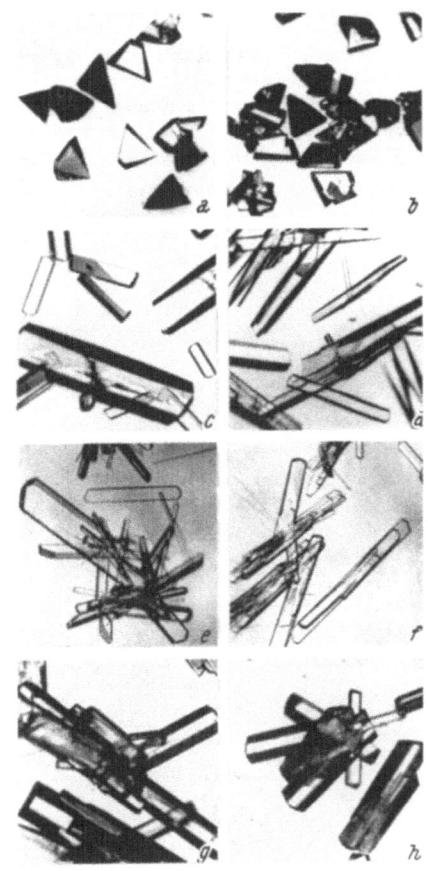

Fig. 5.
a D-Lysergic acid-L-propanolamide-(2), D-ergobasine (D-ergonovine); *b* L-Lysergic acid-D-propanolamide-(2), L-ergobasine (L-ergonovine); *c* D-*Iso*lysergic acid-L-propanolamide-(2), D-ergobasinine; *d* L-*Iso*lysergic acid-D-propanolamide-(2), L-ergobasinine; *e* D-Lysergic acid-D-propanolamide-(2); *f* L-Lysergic acid-L-propanolamide-(2); *g* D-*Iso*lysergic acid-D-propanolamide-(2); *h* L-*Iso*lysergic acid-L-propanolamide-(2).

2. Ergotamine and Ergotaminine, $C_{33}H_{35}O_5N_5$.

As already mentioned, ergotamine was isolated by STOLL (*72, 75, 74*) in 1918 and was the first homogeneous alkaloid isolated from ergot. It was also the first ergot alkaloid to find wide application in medicine, even though it was discovered at a time when many investigators still maintained that the specific uterotonic action of ergot is not due to its alkaloidal components but to compounds of simpler structure.

Ergotamine crystallizes from aqueous acetone in large, strongly refracting prisms which contain two mols of crystal acetone and two mols of crystal water. The solvent-containing crystals melt at 180° C. From 800 volumes of benzene, ergotamine is obtained in long, slender prisms, m. p. 212–214° (corr., decomp.); $[\alpha]_D^{20} = -160°$ (in chloroform). Ergotamine tartrate crystallizes in thick rhombic plates, m. p. 203° (decomp.).

 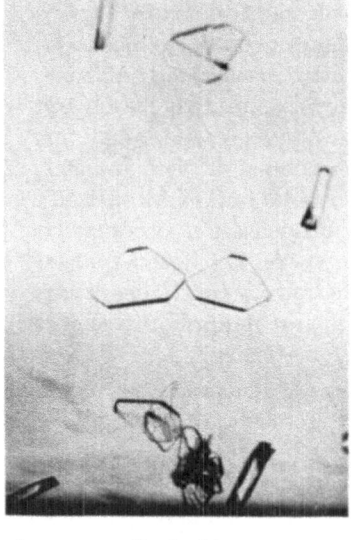

Ergotamine.　　　Fig. 6.　　　Ergotaminine.

The two photomicrographs given above (Fig. 6) show crystals of ergotamine obtained from 90% acetone and crystals of ergotaminine from methanol.

Like all other ergot alkaloids, ergotamine is very sensitive towards chemical reagents, light, atmospheric oxygen and high temperatures. It was only after making allowance for these properties that the isolation of ergotamine could be successfully accomplished (72, 75, 74).

In common with all ergot alkaloids whose lysergic acid portion contains a double bond at $\Delta^{9:10}$, ergotamine exhibits a vivid, pale-violet fluorescence in ultraviolet light.

A particularly striking property is the ease with which ergotamine rearranges to the isomeric ergotaminine, derived from *iso*lysergic acid. This isomerization takes place especially readily in alcoholic solution and is markedly accelerated by traces of acid. Ergotaminine has practically no pharmacological activity.

It is much more sparingly soluble than the natural alkaloid in organic solvents and is moreover distinguished from ergotamine by a different behavior on crystallization. Whereas ergotamine always crystallizes with solvent which it retains tenaciously, ergotaminine always separates without solvent.

The general formula for the alkaloids of the peptide type has been given above. The formula for ergotamine itself follows (XL).

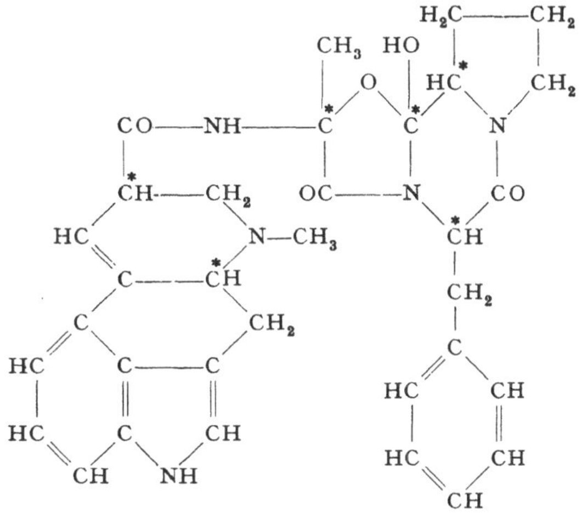

$$\text{(XL.) Ergotamine.}$$

3. Ergosine and Ergosinine, $C_{30}H_{37}O_5N_5$.

Ergosine also belongs to the ergotamine group of peptide-like ergot alkaloids. It was isolated by SMITH and TIMMIS (69) in 1937. Apparently because ergosine occurs in ergot only in rather small quantities, it has not so far attained any practical importance.

The base crystallizes from ethyl acetate in prisms which melt at 228° (decomp.); $[\alpha]_D^{20} = -179°$ (in chloroform). Ergosine hydrochloride crystallizes from acetone with 1 mol of solvent. It forms plates, m. p. 235° (decomp.). Ergosinine, which is formed by the isomerization of ergosine, crystallizes in prisms, m. p. 228° (decomp.); $[\alpha]_D^{20} = +420°$ (in chloroform).

The two photographs in Fig. 7 (p. 156) show crystals of ergosine from ethyl acetate and of ergosinine from 90% acetone.

The final stages in the elucidation of the constitution of ergosine (XLI) were carried out by STOLL and his associates (89) as described above.

 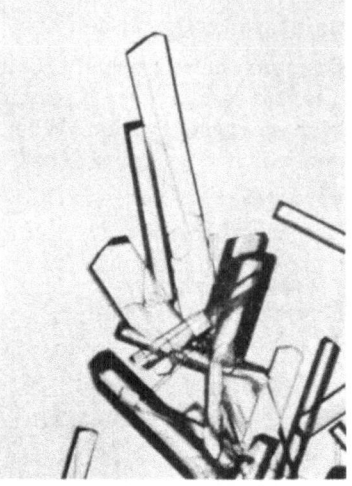

Ergosine.   Fig. 7.   Ergosinine.

(XLI.) Ergosine and Ergosinine.

4. Alkaloids of the Ergotoxine Group.

In 1906, as mentioned, an amorphous but apparently homogeneous alkaloidal preparation was isolated from ergot by BARGER and CARR (5) in England and, simultaneously, by the pharmacist KRAFT (54) in Switzerland. This previously unknown substance was given the name ergotoxine by BARGER and CARR and hydroergotinine by KRAFT. Their investigations showed that by boiling the substance in methanol it could be converted into "ergotinine cristallisée" which had been discovered by TANRET (93) in 1875. However, no reliable explanation for this conversion could be put forward at that time. KRAFT believed that his hydroergotinine was the hydrate of ergotinine and lost water during the conversion, an opinion to which BARGER and CARR likewise subscribed. Remarkably enough, it was not until 1930 that SMITH and TIMMIS (64) succeeded in obtaining ergotoxine in crystalline form. They wrote: "The sharp definition of the crystals and the constant specific rotation leave no doubt concerning the purity of the substance."

However, the more research on the ergot alkaloids progressed, the more striking became the differences between the analytical data and physical constants determined on specimens of ergotoxine originating from various sources. Furthermore, ROTHLIN reported that the pharmacological activity of ergotoxine preparations of different manufacture or origin was subject to qualitative as well as quantitative variations.

All these findings gave rise to doubts concerning the homogeneous nature of ergotoxine. STOLL and HOFMANN (83) therefore decided to examine the problem afresh and were, in fact, able to show that even crystalline ergotoxine preparations are mixtures consisting of the three isomorphous alkaloids, ergocristine, ergokryptine, and ergocornine.

Owing to the isomorphism exhibited by these alkaloids, ergotoxine preparations could not be separated into the three components by fractional crystallization, and a special method of separation had to be worked out. STOLL and HOFMANN (83) found that if they first converted the ergotoxine preparations to well-defined crystalline salts, they were then able to separate them into the individual components by fractional crystallization. Di-(p-toluyl)-L-tartaric acid was found to be a particularly suitable acid for this purpose, since it gave salts of remarkable stability.

It was possible to develop this method of separation to such an extent that it also gave approximate information regarding the quantitative composition of ergotoxine preparations obtained from different sources. Table 7 contains a few examples of assays performed in this manner.

Table 7. Composition of Ergotoxine Preparations of Various Origin.

(The figures represent percentages of the total quantity of pure alkaloids obtained.)

Origin	Ergocristine	Ergokryptine	Ergocornine
Hungarian ergot	31	19	50
Ergotoxine ethane sulphonate from The British Drug Houses Ltd., London	75	8	17
Ergotoxine ethane sulphonate from Burroughs Wellcome & Co., London	10	—	90

Investigations on the chemical and physical properties of these three homogeneous alkaloids have shown that they resemble one another very closely in many of their properties. Thus, for example, they behave so similarly on crystallization from benzene that they may reasonably be regarded as isomorphous. This accounts for the fact that ergotoxine was for so long thought to be homogeneous. It is a striking fact, however, that the free bases show marked differences in behavior when crystallized from other solvents, such as acetone or alcohol. The empirical formulas and some characteristic properties of the three alkaloids are presented in Table 8.

Table 8. The Alkaloids of the Ergotoxine Group.

	Ergocristine	Ergokryptine	Ergocornine
Empirical formula	$C_{35}H_{39}O_5N_5$	$C_{32}H_{41}O_5N_5$	$C_{31}H_{39}O_5N_5$
M. P. (with decomposition)	165–170°	212–214°	182–184°
$[\alpha]_D^{20}$ ($c = 1$ in $CHCl_3$)	−183°	−187°	−188°
$[\alpha]_{5461}^{20}$ ($c = 1$ in $CHCl_3$)	−217°	−226°	−226°
$[\alpha]_D^{20}$ ($c = 1$ in pyridine)	−93°	−112°	−105°
$[\alpha]_{5461}^{20}$ ($c = 1$ in pyridine)	−107°	−133°	−122°

Like all other alkaloids of ergot, the alkaloids of the ergotoxine group isomerize easily, the natural levorotatory alkaloids rearranging

Table 9.
The Dextrorotatory Isomers of the Alkaloids of the Ergotoxine Group.

	Ergocristinine	Ergokryptinine	Ergocorninine
Empirical Formula	$C_{35}H_{39}O_5N_5$	$C_{32}H_{41}O_5N_5$	$C_{31}H_{39}O_5N_5$
M. P. (with decomposition)	226°	240–242°	228°
$[\alpha]_D^{20}$ ($c = 1$ in $CHCl_3$)	+366°	+408°	+409°
$[\alpha]_{5461}^{20}$ ($c = 1$ in $CHCl_3$)	+460°	+508°	+512°
$[\alpha]_D^{20}$ ($c = 1$ in pyridine)	+462°	+479°	+500°
$[\alpha]_{5461}^{20}$ ($c = 1$ in pyridine)	+576°	+596°	+624°
$[\alpha]_D^{20}$ ($c = 1$ in acetone)	+383°	+396°	+414°
$[\alpha]_{5461}^{20}$ ($c = 1$ in acetone)	+479°	+493°	+517°
Solubility			
in boiling methanol	in 100 parts	in 50 parts	in 25 parts
in boiling ethanol	in 100 parts	in 50 parts	in 15 parts

to give the corresponding, strongly dextrorotatory isomers derived from *iso*lysergic acid. Table 9 gives the empirical formulas and some physical constants of the three isomers.

To conclude this condensed account of the alkaloids belonging to the ergotoxine group, it may be noted that this instance of isomorphism is not the only one found among the ergot alkaloids. Several double compounds, which were at first mistaken for homogeneous alkaloids, had been previously described. Some details of these are given in Table 10.

Table 10. Mixed Alkaloids of Ergot.

Composition	Author	Optical rotation in chloroform
Ergotamine-Ergotaminine "Sensibamine"	WOLF (*18*) STOLL and BURCKHARDT (*78*)	$[\alpha]_D = +125°$
Ergosine Ergotamine Ergosinine-Ergocristine "Ergotoxine"? "Ergoclavine"	KÜSSNER (*56*) SMITH and TIMMIS (*69*) STOLL and BURCKHARDT (*78*) KOFLER and KOFLER (*51*)	$[\alpha]_D^{22} = +124°$
Ergosinine-Ergosine	SMITH and TIMMIS (*69*)	$[\alpha]_D^{20} = +128°\ (c=0.5)$
Ergosinine-Ergotamine	KOFLER and KOFLER (*51*) STOLL and BURCKHARDT (*78*)	$[\alpha]_D^{20} = +116°\ (c=1.0)$
Ergosinine-Ergocristine	STOLL and BURCKHARDT (*78*) KOFLER and KOFLER (*51*)	$[\alpha]_D^{20} = +108°\ (c=0.5)$
Ergosinine-Ergobasine	KOFLER and KOFLER (*51*)	

5. Ergocristine and Ergocristinine, $C_{35}H_{39}O_5N_5$.

In 1937, STOLL and BURCKHARDT (*78*) succeeded in isolating in a pure state a new alkaloid of ergot to which they gave the name "ergocristine", suggested by the magnificent manner in which it crystallizes. Its most important physical properties were described and its empirical formula was determined. STOLL and BURCKHARDT also reported that the natural, levorotatory alkaloid, which is characterized by a powerful pharmacological action, readily and reversibly rearranges to the dextrorotatory isomer, ergocristinine, which is pharmacologically almost inactive. Ergocristine is present in varying amounts in all ergotoxine preparations, and, as mentioned, was subsequently isolated from this source via the di-(*p*-toluyl)-*L*-tartrate.

Ergocristine dissolves in 40 parts of boiling benzene from which it separates in long, rectangular plates. On long standing, these crystals become transformed into massive, oblique prisms. Characteristic is the manner in which ergocristine crystallizes from acetone; the crystals are many-sided, oblique prisms, containing 1 mol acetone. Ergocristine melts at 165–170° (corr., decomp.); $[\alpha]_D^{20} = -183°$ (in chloroform). The hydrochloride crystallizes from a mixture of ethanol and ether in oblong plates, with obliquely cut ends.

Ergocristine.   Fig. 8.   Ergocristinine.

(XLII.) Ergocristine.

Ergocristinine crystallizes from ethanol in long, slender prisms, m. p. 226° (corr., decomp.); $[\alpha]_D^{20} = +366°$ (in chloroform). The photographs in Fig. 8 (p. 160) show crystals of ergocristine from acetone and of ergocristinine from ethanol.

The elucidation of the structure of ergocristine (*89*) has already been described. In accordance with the experimental findings of STOLL et al., the structural formula of ergocristine is (XLII).

6. Ergokryptine and Ergokryptinine, $C_{32}H_{41}O_5N_5$.

As mentioned, ergokryptine was discovered by STOLL and HOFMANN (*83*) when investigating ergotoxine preparations of various origin.

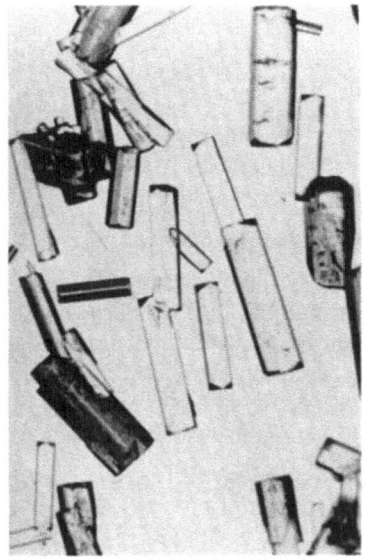
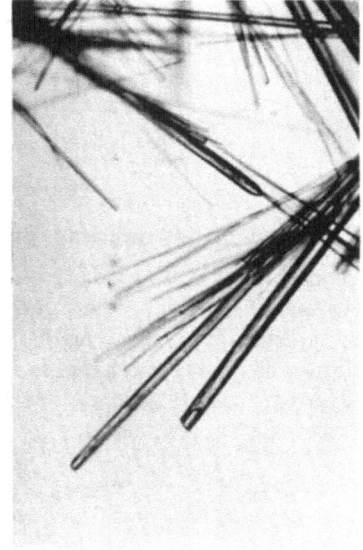

Ergokryptine. Fig. 9. Ergokryptinine.

It crystallizes from 20 volumes of benzene in long, rectangular plates, which are transformed on standing into thick, oblique prisms, containing benzene of crystallization. Ergokryptine dissolves readily in methanol, from which it crystallizes in rectangular prisms, m. p. 212–214° (corr., decomp.); $[\alpha]_D^{20} = -187°$ (in chloroform). The hydrochloride crystallizes in bundles of fine needles, m. p. 208°.

On boiling a methyl alcoholic solution of ergokryptine, the isomeric ergokryptinine (*83*) is obtained. This dissolves in 50 volumes of boiling methanol, from which it crystallizes in needles, m. p. 240–242° (corr., decomp.); $[\alpha]_D^{20} = +408°$ (in chloroform). The photographs reproduced in Fig. 9 (p. 161) show crystals of ergokryptine from methanol and of ergokryptinine from ethanol.

The structural formula (XLIII) for ergokryptine is based on the investigations reported above (89).

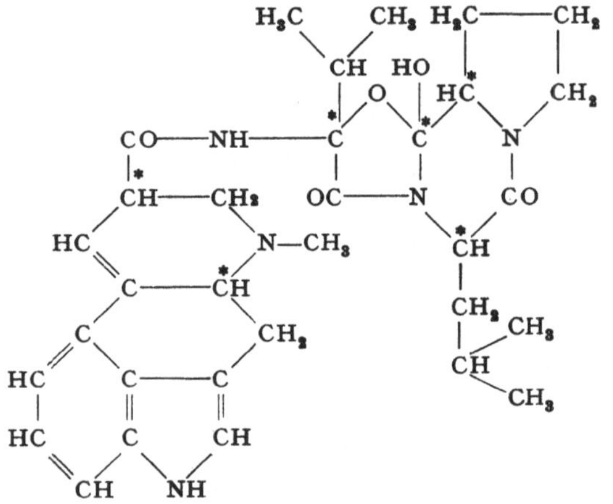

(XLIII.) Ergokryptine.

7. Ergocornine and Ergocorninine, $C_{31}H_{39}O_5N_5$.

Ergocornine was first isolated from ergotoxine preparations by STOLL and HOFMANN (83) who determined its physical and chemical characteristics. It crystallizes from methanol in glistening polyhedra, and from 20 volumes of boiling methanol in long, rectangular plates. On standing, the latter are transformed into massive, many-sided prisms and plates, m. p. 182–184° (corr., decomp.); $[\alpha]_D^{20} = -188°$ (in chloroform). The

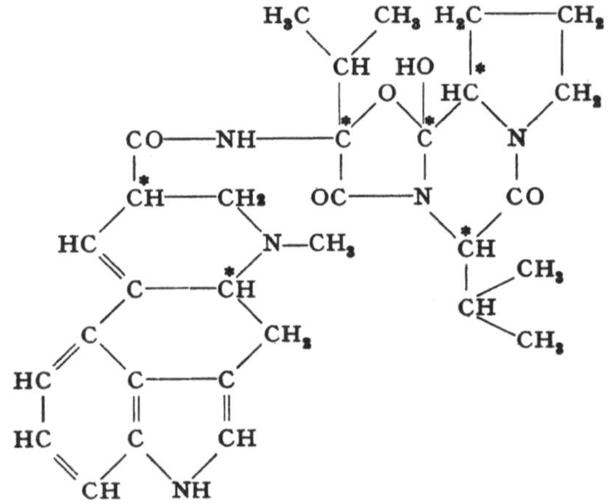

(XLIV.) Ergocornine.

hydrochloride crystallizes from acetone in well-defined prisms with obliquely cut ends, m. p. 223° (decomp.).

Like all the natural alkaloids of ergot, ergocornine readily isomerizes, giving ergocorninine, which crystallizes from ethanol in long, slender prisms, m. p. 228° (corr., decomp.); $[\alpha]_D^{20} = +409°$ (in chloroform). Photomicrographs of ergocornine crystals obtained from methanol and of ergocorninine crystals from ethanol are reproduced in Fig. 10.

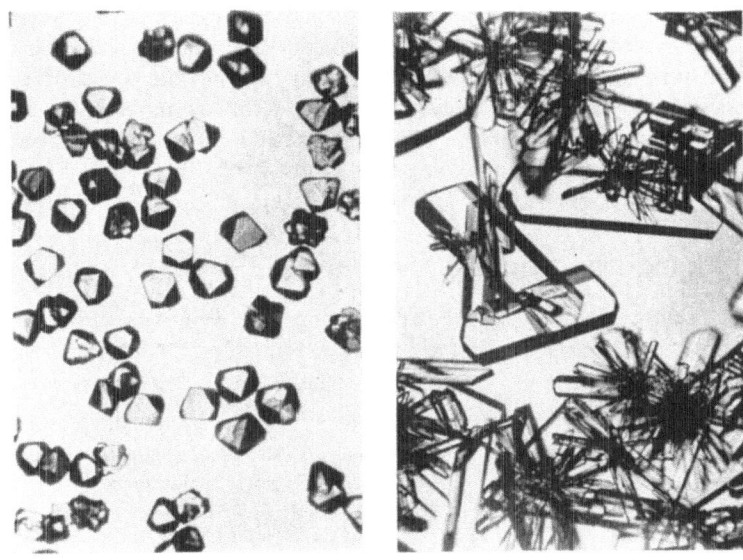

Ergocornine.   Fig. 10.   Ergocorninine.

By analogy with the other ergot alkaloids of the peptide type, ergocornine has the following constitution (XLIV) (*89*).

IV. Partially Synthetic and Hydrogenated Derivatives of the Ergot Alkaloids.

Although, at the time of writing this review, the total synthesis of both lysergic acid and the peptide portion, as well as the union of the two fragments, still remain to be accomplished, it is nevertheless possible to report a number of partial syntheses. Some of these are of practical importance and have led to compounds with unexpected and interesting pharmacological properties. The variations in composition made possible by partial synthesis are important for comparative studies on structure, configuration and pharmacological activity; and the knowledge thus obtained may be applied to the synthesis of highly active compounds.

The partially synthetic compounds described below are obtained either by varying the basic fragment which is connected by the amide-like linkage to lysergic acid or by changes in the lysergic acid nucleus itself. For syntheses of the former type, lysergic acid was first converted into the highly reactive azide, which was then treated with amines of various composition.

Racemic *iso*lysergic acid hydrazide was first prepared by STOLL and HOFMANN (79) in 1937; the action of hydrazine hydrate on natural alkaloids of ergot gave a good yield of the racemate which was then resolved into its optical antipodes (82). On treating the optically active hydrazides of lysergic and *iso*lysergic acids with nitrous acid, the azides of the two acids were obtained (79). They react in a similar manner to acid halides and enabled a whole series of amides to be prepared. A brief account of these derivatives follows.

1. Partially Synthetic Derivatives of Lysergic Acid.

Table 11. Homologs of Ergobasine and Ergobasinine.

Compound	M. P. (decomp.)	$[\alpha]_D^{20}$	Typical crystal form
D-lysergic acid-ethanolamide (*nor*-ergobasine)	95°	−10° (in pyridine)	Pentagonal plates from chloroform
D-*iso*lysergic acid-ethanolamide (*nor*-ergobasinine)	206°	+448° (in pyridine)	Obliquely truncated prisms and polyhedra from methanol
D-lysergic acid-*L*(+)-propanol-amide-(2) (ergobasine)	162°	−16° (in pyridine)	Tetrahedra from ethyl acetate
D-*iso*lysergic acid-*L*(+)-propanolamide-(2) (ergobasinine)	196°	+414° (in CHCl$_3$)	Well-formed prisms from acetone
D-lysergic acid-(+)-butanol-amide-(2) (methyl-ergobasine)	172°	−45° (in pyridine)	Horizontally truncated prisms from methanol-acetone
D-*iso*lysergic acid-(+)-butanol-amide-(2) (methyl-ergobasinine)	194°	+386° (in CHCl$_3$)	Polyhedra from methanol
D-lysergic acid-*L*(+)-4-methyl-pentanolamide-(2) (*iso*propyl-ergobasine)	130°	−38° (in pyridine)	Horizontally truncated, elongated prisms from alcohol-benzene
D-*iso*lysergic acid-*L*(+)-4-methyl-pentanolamide-(2) (*iso*propyl-ergobasinine)	160°	+330° (in CHCl$_3$)	Rosette-shaped clusters of blunt prisms from acetone

As mentioned, lysergic acid azide reacts with amines, amino alcohols or amino acids to give amides of lysergic acid. The formulas below illustrate the reaction of lysergic acid azide with diethylamine.

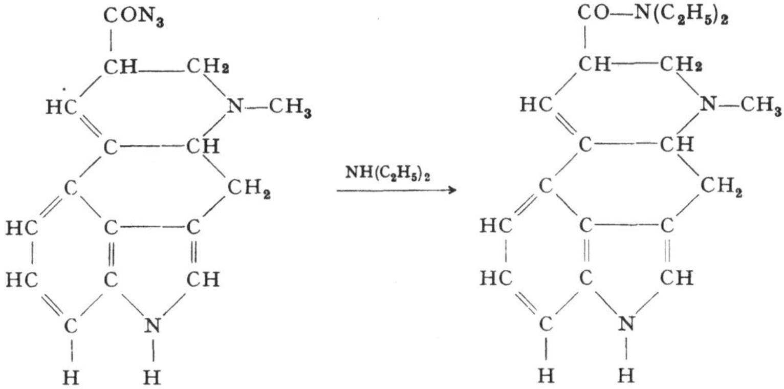

(XLV.) Lysergic acid diethylamide.

Table 12. Various Derivatives of Ergobasine.

Compound	M. P. (decomp.)	$[\alpha]_D^{20}$	Typical crystal form
D-lysergic acid-1:3-dihydroxypropaneamide-(2) (hydroxy-ergobasine)	125°	+55° (in water)	Acid oxalate from 90% alcohol in clusters of needles
D-isolysergic acid-1:3-dihydroxypropane-amide-(2) (hydroxy-ergobasinine)	231°	+445° (in pyridine)	Base from alcohol in short, obliquely truncated prisms
D-lysergic acid-L-N-benzyl-propanol-amide-(2) (N-benzyl-ergobasine)	230°	−17° (in pyridine)	Base from alcohol in rounded aggregates of short prisms
D-lysergic acid-L-ephedride (N-methyl-phenyl-ergobasine)	258°	−21° (in pyridine)	Base from alcohol in rectangular plates and polyhedra
D-lysergic acid-2-diethylamino-ethyl-amide	200°	−16° (in pyridine)	Acid oxalate from 95% alcohol in fine needles
D-isolysergic acid-2-diethylamino-ethyl-amide	163°	+396° (in pyridine)	Base from acetone in thick, shining plates and double pyramids
D-lysergic acid-diethyl-amide	83°	+30° (in pyridine)	Base from benzene in pointed prisms
D-isolysergic acid-diethyl-amide	182°	+217° (in pyridine)	Base from acetone in large prisms

D-lysergic acid diethylamide obtained in this way is a compound with extremely interesting pharmacological properties. Doses as small as 20–30 μg given orally to healthy individuals cause psychic disturbances combined with color visions and hallucinations, the effects being similar to those produced by mescaline in a dosage 300 times as high.

Using the same principle, STOLL and HOFMANN (82) prepared a considerable number of other amides of lysergic and *iso*lysergic acid; some properties of these are listed in Tables 11–13.

Table 13. Stereoisomers of Phenylergobasine and Phenylergobasinine, $C_{25}H_{27}O_2N_3$.

Compound	M. P. (decomp.)	$[\alpha]_D^{20}$ (in acetone)	Typical crystal form
D-lysergic acid-D-*nor*-ephedride	230°	+14°	Base amorphous, hydrochloride in thick prisms from alcohol-ether
L-lysergic acid-L-*nor*-ephedride	230°	−16°	
D-*iso*lysergic acid-D-*nor*-ephedride	128°	+267°	Well-shaped prisms from alcohol on diluting with ether
L-*iso*lysergic acid-L-*nor*-ephedride	128°	−267°	
D-lysergic acid-L-*nor*-ephedride	130°	−17°	From benzene in coarse aggregates. Tartrate from methanol in fine needles
D-*iso*lysergic acid-L-*nor*-ephedride	130°	+296°	Amorphous
D-lysergic acid-D-*nor*-ψ-ephedride	128°	+27°	From benzene in long needles
D-*iso*lysergic acid-D-*nor*-ψ-ephedride	—	+370°	Amorphous

Table 14. Peptide-like, Partially Synthetic Derivatives of Lysergic acid, *Iso*lysergic acid, and Dihydrolysergic acid.

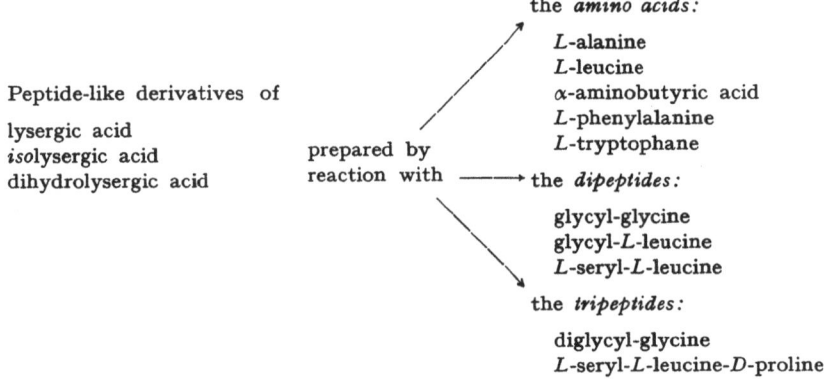

The pharmacological examination of these derivatives has in some cases yielded very interesting results, since some of the partially synthetic compounds are even more active pharmacologically than the natural products, and therefore of practical importance.

During the last few years, STOLL and his associates (87) have also accomplished the partial synthesis of a number of peptide-like derivatives of lysergic acid, as shown schematically in Table 14.

Unfortunately, these synthetic mono-, di- and tripeptides lack almost completely the sympathicolytic action so characteristic of the ergot alkaloids. Apparently this action is exhibited only when the amino acids are arranged in a ring system like that present in the natural bases.

2. Dihydro-Derivatives of the Natural Alkaloids of Ergot.

As explained in the section on the structure of lysergic acid (p. 122), JACOBS and his associates demonstrated the presence in lysergic acid of a double bond which could be relatively easily hydrogenated. STOLL et al. (90) have shown that this double bond must be situated at $\Delta^{9,10}$ [see formula (VI) on p. 126].

The first attempts at hydrogenating lysergic acid and its derivatives were carried out by JACOBS and CRAIG (40). By dissolving ergobasine

Table 15. Dihydro-derivatives of the Natural, Levorotatory Alkaloids of Ergot.

Compound	M. P. (decomp.)	$[\alpha]_D^{20}$ (pyridine)	Typical crystal form
Ergotamine group:			
Dihydro-ergotamine $C_{33}H_{37}O_5N_5$	239°	−64°	From 90% aqueous acetone in horizontally truncated polyhedral prisms
Dihydro-ergosine $C_{30}H_{39}O_5N_5$	212°	−52°	From ethyl acetate in pointed prisms and polyhedra
Ergotoxine-group:			
Dihydro-ergocristine $C_{35}H_{41}O_5N_5$	180°	−56°	From acetone in massive, hexagonal plates
Dihydro-ergokryptine $C_{32}H_{43}O_5N_5$	235°	−41°	From ethyl alcohol in massive plates and polyhedra
Dihydro-ergocornine $C_{31}H_{41}O_5N_5$	187°	−48°	From ethyl alcohol in massive plates, usually hexagonal
Ergobasine group:			
Dihydro-ergobasine $C_{19}H_{25}O_2N_3$	230°	−145°	From benzene in needles

in glacial acetic acid and treating the solution with hydrogen in the presence of a platinum oxide catalyst, they succeeded in obtaining a well-defined dihydroergobasine. In a similar manner, they also prepared a dihydrolysergic acid and a dihydro*iso*lysergic acid. The same authors had shown a little earlier (*35*) that both ergotinine and lysergic acid methyl ester yield a mixture of the two isomeric alcohols, α- and β-dihydrolysergol, on BOUVEAULT-BLANC reduction with sodium in boiling butanol. No definite results could be obtained, however, when attempts were made to reduce the alkaloids of the ergotamine and ergotoxine groups (*42*) by catalytic hydrogenation. A similar failure was also reported by KHARASCH (*45*) who could not obtain crystalline dihydro-derivatives except from ergobasine.

The explanation for these unexpected difficulties was furnished by STOLL and HOFMANN (*84*) who showed that the hydrogenation of the

Table 16. Dihydro-derivatives of the Dextrorotatory Alkaloids of Ergot.

Compound	M. P. (decomp.)	$[\alpha]_D^{20}$ (pyridine)	Typical crystal form
Dihydro-ergotaminineI $C_{33}H_{37}O_5N_5$	236°	+97°	From alcohol in hexagonal leaflets
Dihydro-ergotaminineII $C_{33}H_{37}O_5N_5$	206°	−7°	From acetone in rectangular plates
Dihydro-ergosinineI $C_{30}H_{39}O_5N_5$	234°	+108°	From alcohol in long needles
Dihydro-ergosinineII $C_{30}H_{39}O_5N_5$	223°	+3°	From acetone in fine needles
Dihydro-ergocristinineI $C_{35}H_{41}O_5N_5$	248°	+109°	From alcohol in polyhedra
Dihydro-ergocristinineII $C_{35}H_{41}O_5N_5$	175°	+13°	From acetone in hexagonal or octagonal plates
Dihydro-ergokryptinineI $C_{32}H_{43}O_5N_5$	268°	+126°	From methylene chloride-methanol in long prisms
Dihydro-ergokryptinineII $C_{32}H_{43}O_5N_5$	226°	+26°	From methylene chloride-petroleum ether in spear-shaped needles
Dihydro-ergocorninineI $C_{31}H_{41}O_5N_5$	264°	+147°	From chloroform-alcohol in long needles
Dihydro-ergocorninineII $C_{31}H_{41}O_5N_5$	180°	+32°	From methylene chloride in massive, hexagonal plates
Dihydro-ergobasinineI $C_{19}H_{25}O_2N_3$	211°	+8°	From ethyl acetate-methanol in leaflets
Dihydro-ergobasinineII $C_{19}H_{25}O_2N_3$	212°	+45°	From acetone in massive, obliquely truncated prisms

natural ergot alkaloids derived from lysergic acid takes place much more readily than that of the dextrorotatory isomers. In a relatively short time, they succeeded in preparing the dihydro-derivatives of *all* the natural levorotatory ergot alkaloids. Their main physical properties are summarized in Table 15.

In the course of subsequent investigations of STOLL, HOFMANN and PETRZILKA (*88*), the catalytic hydrogenation of the dextrorotatory alkaloids derived from *iso*lysergic acid was likewise accomplished. Instead of a single dihydro-derivative, however, each alkaloid yielded two hydrogenation products. Ergotamine, for example, gave a dihydroergotamininеI and a dihydroergotamininеII. The various dihydro-derivatives of the dextrorotatory alkaloids of ergot, and some of their properties, are listed in Table 16.

Thus, in the *iso*lysergic acid series two dihydro-derivatives are invariably obtained, whereas each alkaloid belonging to the lysergic acid series yields only a single dihydro-compound.

Apart from their theoretical interest, dihydro-derivatives of the ergot alkaloids are of practical significance in the fields of internal medicine and neurology. On saturating the double bond at $\Delta^{9:10}$, the action on organs with smooth musculature, such as the uterus, is lost, while the sympathicolytic action on organs under the control of the autonomic nervous system is enhanced. At the same time the toxicity is considerably decreased. As examples of the application of the dihydroalkaloids in medicine may be mentioned the use of dihydroergotamine in the treatment of migraine, and that of "Hydergine" (a combination of the three dihydro alkaloids of the ergotoxine group in equal proportions) in the treatment of hypertension, peripheral vascular disorders and angina pectoris.

At the present time, the problem of the total synthesis of the ergot alkaloids of the polypeptide type is once more in the foreground, the basis having been laid by the elucidation of the complicated structure of these substances. As we have seen, the total synthesis of dihydro-*D*-lysergic acid, the parent compound of the therapeutically important dihydro alkaloids, has already been accomplished. The synthesis of natural *D*-lysergic acid, with the double bond at position $\Delta^{9:10}$ in ring *D*, should not now be long delayed. There would then remain only the synthesis of the peptide residue and its linkage to lysergic or dihydro-lysergic acid. Although the nature of this linkage has now been clarified, we still do not know how such a linkage can be brought about artificially.

For the time being, therefore, we cannot say whether it will be possible eventually to synthesize the ergot alkaloids of the polypeptide type, with their structure and configuration correct in all details; and whether such a synthesis on a laboratory or technical scale would prove to be cheaper than that which the ergot fungus performs with such remarkable skill. How little we still know of the ways and means by which the living cell, even the cell of a lowly fungus, so readily accomplishes the most complicated biosyntheses, starting from simple materials. Here is one more case where we still have much to learn from Nature.

References.

1. AKABORI, S.: Synthese von Imidazol-Derivaten aus α-Aminosäuren. I. Mitt.: Eine neue Synthese von Desaminohistidin und ein Beitrag zur Kenntnis der Konstitution des Ergothioneins. Ber. dtsch. chem. Ges. **66**, 151 (1933).
2. BARGER, G.: Ergot and Ergotism. London: Gurney & Jackson. 1931.
3. — Isolation and Synthesis of p-Hydroxyphenyl-ethylamine, an Active Principle of Ergot Soluble in Water. J. chem. Soc. (London), Trans. **95**, 1125 (1909).
4. — The Alkaloids of Ergot. In: Handbuch der experimentellen Pharmakologie, Ergänz.-Werk, Bd. VI, S. 84, 222. 1938.
5. BARGER, G. and F. H. CARR: The Alkaloids of Ergot. J. chem. Soc. (London), Trans. **91**, 337 (1907).
6. BARGER, G. and H. H. DALE: Ergotoxine and some other Constituents of Ergot. Biochemic. J. **2**, 240 (1907).
7. — — The Water-Soluble Active Principle of Ergot. J. Physiol., Proc. Physiol. Soc. **38**, LXXVII (1909).
8. — — 4-β-Aminoethylglyoxaline (β-Iminazolylethylamine) and the other Active Principles of Ergot. J. chem. Soc. (London), Trans. **97**, 2592 (1910).
9. BARGER, G. and A. J. EWINS: The Alkaloids of Ergot. II. J. chem. Soc. (London), Trans. **97**, 284 (1910).
10. — — The Constitution of Ergothioneine: a Betaine Related to Histidine. J. chem. Soc. (London), Trans. **99**, 2336 (1911).
11. BARGER, G. and G. S. WALPOLE: Further Syntheses of p-Hydroxyphenyl-ethylamine. J. chem. Soc. (London), Trans. **95**, 1720 (1909).
12. BEHRE, J. A.: A Note on the Determination of Ergothioneine in Blood Filtrates. Biochemic. J. **26**, 458 (1932).
13. BERGMANN, M. u. D. DELIS: Umwandlung von α-Amino-β-oxysäuren in α-Ketosäuren. Verwandlung ihrer Hydantoine in Ketosäuren und Harnstoffe. Liebigs Ann. Chem. **458**, 76 (1927).
14. BERGMANN, M. u. A. MIEKELEY: Neue desmotrope Aminosäureanhydride vom Piperazintypus. Zur Kenntnis des Abbaues der Aminosäuren. Serin als Dehydrierungsmittel. Liebigs Ann. Chem. **458**, 40 (1927).
15. BERGMANN, M. and C. NIEMANN: The Chemistry of Amino Acids and Proteins. Annu. Rev. Biochem. **7**, 99 (1938).
16. BRAGG, W. L.: Patterson Diagrams in Crystal Analysis. Nature (London) **143**, 73 (1939).
17. CHASSAR MOIR, J.: The Action of Ergot Preparations on the Puerperal Uterus. Brit. Med. J. **1932**, 1119, part I.
18. "Chinoin" Act. Ges. u. E. WOLF: Swiss Patent 160898.
19. CRAIG, L. C., T. SHEDLOVSKY, R. G. GOULD, Jr. and W. A. JACOBS: The Ergot Alkaloids. XIV. The Positions of the Double Bond and the Carboxyl Group in Lysergic Acid and its Isomer. The Structure of the Alkaloids. J. biol. Chemistry **125**, 289 (1938).
20. DIELS, O. u. A. PILLOW: Über Bis-benzoylcyanid. Ber. dtsch. chem. Ges. **41**, 1893 (1908).

21. DUDLEY, H. W. and J. CHASSAR MOIR: The Substance Responsible for the Traditional Clinical Effect of Ergot. Brit. Med. J. **1935**, 520, part I.
22. EAGLES, B. A.: Biochemistry of Sulfur. II. The Isolation of Ergothioneine from Ergot of Rye. J. Amer. chem. Soc. **50**, 1386 (1928).
23. EAGLES, B. A. and T. B. JOHNSON: The Biochemistry of Sulfur. I. The Identity of Ergothioneine from Ergot of Rye with Sympectothion and Thiasine from Pigs' Blood. J. Amer. chem. Soc. **49**, 575 (1927).
24. ENGELAND, R. u. FR. KUTSCHER: Über eine zweite wirksame Secalebase. Zbl. Physiol. **24**, 479 (1910).
25. — — Über einige Bestandteile des *Extractum Secalis cornuti*. Zbl. Physiol. **24**, 589 (1910).
26. EWINS, A. J.: Acetylcholine, a New Active Principle of Ergot. Biochemic. J. **8**, 44 (1914).
27. — Some New Physiologically Active Derivatives of Choline. Biochemic. J. **8**, 366 (1914).
28. FRÄNKEL, S. u. J. RAINER: Über das Vorkommen von cyklischen Aminosäuren im *Secale cornutum*. Biochem. Z. **74**, 167 (1916).
29. GOULD, R. G., Jr. and W. A. JACOBS: The Synthesis of Certain Substituted Quinolines and 5,6-Benzoquinolines. J. Amer. chem. Soc. **61**, 2890 (1939).
30. HARINGTON, C. R. and J. OVERHOFF: A New Synthesis of 2-Thiolhistidine together with Experiments towards the Synthesis of Ergothioneine. Biochemic. J. **27**, 338 (1933).
31. HAUROWITZ, F.: Die Anordnung der Peptidketten in Sphäroprotein-Molekülen. Hoppe-Seyler's Z. physiol. Chem. **256**, 28 (1938).
32. HUNTER, G.: A New Test for Ergothioneine upon which is Based a Method for its Estimation in Simple Solution and in Blood-Filtrates. Biochemic. J. **22**, 4 (1928).
33. JACOBS, W. A.: The Ergot Alkaloids. I. The Oxidation of Ergotinine. J. biol. Chemistry **97**, 739 (1932).
34. JACOBS, W. A. and L. C. CRAIG: The Ergot Alkaloids. II. The Degradation of Ergotinine with Alkali. Lysergic Acid. J. biol. Chemistry **104**, 547 (1934).
35. — — The Ergot Alkaloids. IV. The Cleavage of Ergotinine with Sodium and Butyl Alcohol. J. biol. Chemistry **108**, 595 (1935).
36. — — The Ergot Alkaloids. V. The Hydrolysis of Ergotinine. J. biol. Chemistry **110**, 521 (1935).
37. — — The Ergot Alkaloids. VI. Lysergic Acid. J. biol. Chemistry **111**, 455 (1935).
38. — — On an Alkaloid from Ergot. Science (New York) **82**, 16 (1935).
39. — — The Ergot Alkaloids. The Structure of Lysergic Acid. Science (New York) **83**, 38 (1936).
40. — — The Ergot Alkaloids. IX. The Structure of Lysergic Acid. J. biol. Chemistry **113**, 767 (1936).
41. — — The Ergot Alkaloids. X. On Ergotamine and Ergoclavine. J. org. Chemistry **1**, 245 (1936/37).
42. — — The Ergot Alkaloids. XI. Isomeric Dihydrolysergic Acids and the Structure of Lysergic Acid. J. biol. Chemistry **115**, 227 (1936).
43. — — The Ergot Alkaloids. XIII. The Precursors of Pyruvic and Isobutyrylformic Acids. J. biol. Chemistry **122**, 419 (1937/38).
44. — — The Ergot Alkaloids. XVII. The Dimethylindole from Dihydrolysergic Acid. J. biol. Chemistry **128**, 715 (1939).
45. KHARASCH, M. S.: U. S. Patent 2 086 559.
46. KHARASCH, M. S., H. KING, A. STOLL u. M. R. THOMPSON: Das neue Mutterkornalkaloid. Schweiz. med. Wschr. **66**, 261 (1936).

47. KHARASCH, M. S. and R. R. LEGAULT: Ergotocin. Science (New York) 81, 388 (1935).
48. — — The New Active Principle(s) of Ergot. Science (New York) 81, 614 (1935).
49. KOBERT, R.: Zur Geschichte des Mutterkorns. In: Hist. Stud. pharmak. Inst. Kaiserl. Univ. Dorpat 1, 1 (1889).
50. — Über die Bestandteile und Wirkungen des Mutterkorns. Naunyn-Schmiedebergs Arch. exp. Pathol. Pharmakol. 18, 316 (1884).
51. KOFLER, A. u. L. KOFLER: Über zusammengesetzte Mutterkornalkaloide. Z. angew. Chem. 50, 620 (1937).
52. KOSSEL, A.: Über das Agmatin. Hoppe-Seyler's Z. physiol. Chem. 66, 257 (1910).
53. — Synthese des Agmatins. Hoppe-Seyler's Z. physiol. Chem. 68, 170 (1910).
54. KRAFT, F.: Über das Mutterkorn. Arch. Pharmaz. Ber. dtsch. pharmaz. Ges. 244, 336 (1906).
55. KUHN, R. u. H. ROTH: Mikrobestimmung von Acetyl-, Benzoyl- und C-Methylgruppen. Ber. dtsch. chem. Ges. 66, 1274 (1933).
56. KÜSSNER, W.: Über ein neues spezifisches Alkaloid des Mutterkorns. Mercks Jahresber. 47, 5 (1933).
57. KUTSCHER, FR.: Die physiologische Wirkung einer Secalebase und des Imidazolyl-äthylamins. Zbl. Physiol. 24, 163 (1910).
58. MELLANBY, E., E. SURIE and D. C. HARRISON: Vitamin D in Ergot of Rye. Biochemic. J. 23, 710 (1929).
59. NEWTON, E. B., S. R. BENEDICT and H. D. DAKIN: On Thiasine, its Structure and Identification with Ergothioneine. J. biol. Chemistry 72, 367 (1927).
60. PATTERSON, A. M. and L. T. CAPELL: The Ring Index. New York: Reinhold Publ. Corp. 1940.
61. PAULING, L. and C. NIEMANN: The Structure of Proteins. J. Amer. chem. Soc. 61, 1860 (1939).
62. PIRIE, N. W.: Improved Methods for the Isolation of Methionine and Ergothioneine. Biochemic. J. 27, 202 (1933).
63. PYMAN, F. L.: A New Synthesis of 4- (or 5-)-β-Aminoethyl-glyoxaline, one of the Active Principles of Ergot. J. chem. Soc. (London), Trans. 99, 668 (1911).
64. SMITH, S. and G. M. TIMMIS: The Alkaloids of Ergot. II. Ergotinine and ψ-Ergotinine. J. chem. Soc. (London) 1931, 1888.
65. — — The Alkaloids of Ergot. III. Ergine, a New Base Obtained by the Degradation of Ergotoxine and Ergotinine. J. chem. Soc. (London) 1932, 763.
66. — — The Alkaloids of Ergot. IV. A Complex Group Common to Ergotoxine and Ergotamine. J. chem. Soc. (London) 1932, 1543.
67. — — The Alkaloids of Ergot. V. The Nature of Ergine. J. chem. Soc. (London) 1934, 674.
68. — — The Alkaloids of Ergot. VII. *Iso*-Ergine and *iso*-Lysergic Acids. J. chem. Soc. (London) 1936, 1440.
69. — — The Alkaloids of Ergot. VIII. New Alkaloids of Ergot: Ergosine and Ergosinine. J. chem. Soc. (London) 1937, 396.
70. STEARNS, J.: Account of the *Pulvis Parturiens*, a Remedy for Quickening Childbirth. Med. Reposit. New York 5, 308 (1808).
71. STOLL, A.: Zur Kenntnis der Mutterkornalkaloide. Compt. rend. Soc. Suisse Sci. nat., Neuchâtel 1920, 190.
72. — Über die wirksamen Substanzen des Mutterkorns. Schweiz. med. Wschr. 51, 525 (1921).

73. STOLL, A.: Altes und Neues über Mutterkorn. Mitt. Naturforsch.-Ges. Bern **1942**, 45, 53.
74. — Über Ergotamin. Helv. chim. Acta **28**, 1283 (1945).
75. — Swiss Patent 79879 (1918); German Patent 357272 (1922).
76. STOLL, A. et E. BURCKHARDT: L'ergobasine, un nouvel alcaloïde de l'ergot de Seigle, soluble dans l'eau. C. R. hebd. Séances Acad. Sci. **200**, 1680 (1935).
77. — — L'ergobasine, un nouvel alcaloïde de l'ergot de seigle, soluble dans l'eau. Bull. Sci. pharmacol. **42**, 257 (1935).
78. — — Ergocristin und Ergocristinin, ein neues Alkaloidpaar aus Mutterkorn. (Vorl. Mitt.) Hoppe-Seyler's Z. physiol. Chem. **250**, 1 (1937).
79. STOLL, A. u. A. HOFMANN: Racemische Lysergsäure und ihre Auflösung in die optischen Antipoden. 2. vorl. Mitt. über Mutterkornalkaloide. Hoppe-Seyler's Z. physiol. Chem. **250**, 7 (1937).
80. — — Partialsynthese des Ergobasins, eines natürlichen Mutterkornalkaloids sowie seines optischen Antipoden. 3. Mitt. über Mutterkornalkaloide. Hoppe-Seyler's Z. physiol. Chem. **251**, 155 (1938).
81. — — Die optisch aktiven Hydrazide der Lysergsäure und der Isolysergsäure. 4. Mitt. über Mutterkornalkaloide. Helv. chim. Acta **26**, 922 (1943).
82. — — Partialsynthese von Alkaloiden vom Typus des Ergobasins. 6. Mitt. über Mutterkornalkaloide. Helv. chim Acta **26**, 944 (1943).
83. — — Die Alkaloide der Ergotoxingruppe: Ergocristin, Ergokryptin und Ergocornin. 7. Mitt. über Mutterkornalkaloide. Helv. chim. Acta **26**, 1570 (1943).
84. — — Die Dihydroderivate der natürlichen linksdrehenden Mutterkornalkaloide. 9. Mitt. über Mutterkornalkaloide. Helv. chim. Acta **26**, 2070 (1943).
85. — — Zur Kenntnis des Polypeptidteils der Mutterkornalkaloide. II. (Partielle alkalische Hydrolyse der Mutterkornalkaloide.) 20. Mitt. über Mutterkornalkaloide. Helv. chim. Acta **33**, 1705 (1950).
86. STOLL, A., A. HOFMANN u. B. BECKER: Die Spaltstücke von Ergocristin, Ergokryptin und Ergocornin. 8. Mitt. über Mutterkornalkaloide. Helv. chim. Acta **26**, 1602 (1943).
87. STOLL, A., A. HOFMANN, E. JUCKER, T. PETRZILKA, J. RUTSCHMANN u. F. TROXLER: Peptide der isomeren Lysergsäuren und Dihydro-lysergsäuren. 18. Mitt. über Mutterkornalkaloide. Helv. chim. Acta **33**, 108 (1950).
88. STOLL, A., A. HOFMANN u. T. PETRZILKA: Die Dihydroderivate der rechtsdrehenden Mutterkornalkaloide. 11. Mitt. über Mutterkornalkaloide. Helv. chim. Acta **29**, 635 (1946).
89. — — — Die Konstitution der Mutterkornalkaloide. Struktur des Peptidteils. III. 24. Mitt. über Mutterkornalkaloide. Helv. chim. Acta **34**, 1544 (1951).
90. STOLL, A., A. HOFMANN u. F. TROXLER: Über die Isomerie von Lysergsäure und Isolysergsäure. 14. Mitt. über Mutterkornalkaloide. Helv. chim. Acta **32**, 506 (1949).
91. STOLL, A., T. PETRZILKA u. B. BECKER: Beitrag zur Kenntnis des Polypeptidteils von Mutterkornalkaloiden. (Spaltung der Mutterkornalkaloide mit Hydrazin.) 16. Mitt. über Mutterkornalkaloide. Helv. chim. Acta **33**, 57 (1950).
92. STOLL, A., J. RUTSCHMANN u. W. SCHLIENTZ: Synthese der optisch-aktiven Dihydro-lysergsäuren. 19. Mitt. über Mutterkornalkaloide. Helv. chim. Acta **33**, 375 (1950).
93. TANRET, C.: Sur la présence d'un nouvel alcaloïde, l'ergotinine, dans le seigle ergoté. C. R. hebd. Séances Acad. Sci. **81**, 896 (1875).

94. TANRET, C.: Sur un nouveau principe immédiat de l'ergot de seigle, l'ergostérine. J. Pharmac. Chim. (V) **19**, 225 (1889).
95. — Sur l'ergostérine et la fongistérine. C. R. hebd. Séances Acad. Sci. **147**, 75 (1908).
96. — Sur une base nouvelle retirée du seigle ergoté, l'ergothionéine. C. R. hebd. Séances Acad. Sci. **149**, 222 (1909).
97. THOMPSON, M. R.: A New Active Principle of Ergot. Science (New York) **81**, 636 (1935).
98. UHLE, F. C. and W. A. JACOBS: The Ergot Alkaloids. XX. The Synthesis of Dihydro-*dl*-lysergic Acid. A New Synthesis of 3- Substituted Quinolines. J. org. Chemistry **10**, 76 (1945).
99. VAHLEN, E.: Clavin, ein neuer Mutterkornbestandteil. Naunyn-Schmiedebergs Arch. exp. Pathol. Pharmakol. **55**, 131 (1906).
100. — Über Mutterkorn. Naunyn-Schmiedebergs Arch. exp. Pathol. Pharmakol. **60**, 42 (1909).
101. VAUQUELIN, L. N.: Analyse du seigle ergoté du Bois de Boulogne près Paris. Ann. Chim. Phys. **3**, 337 (1816).
102. WINDAUS, A. u. H. OPITZ: Synthese einiger Imidazol-Derivate. Ber. dtsch. chem. Ges. **44**, 1721 (1911).
103. WINDAUS, A. u. W. VOGT: Synthese des Imidazolyl-äthylamins. Ber. dtsch. chem. Ges. **40**, 3691 (1907).
104. WRINCH, D. M.: The Pattern of Proteins. Nature (London) **137**, 411 (1936).
105. — Energy of Formation of "Cyclol" Molecules. Nature (London) **138**, 241 (1936).

(Received, January 22, 1952.)

Die Alkaloide der Menispermaceae-Pflanzen.
Von M. Tomita, Kyoto.

Inhaltsübersicht.
   Seite
I. Einleitung 176
II. Die in Menispermaceae aufgefundenen Alkaloide 177
   A. Durch Kondo und seine Mitarbeiter untersuchte Pflanzen 177
   B. Von nicht-japanischen Forschern untersuchte Pflanzen 177
   C. Pflanzen, in denen von Kondo und Mitarbeitern das Vorkommen
      von Alkaloiden festgelegt wurde. 178
   D. In der Literatur als alkaloidhaltig angegebene Pflanzen........ 178
   E. Aus Rohmaterialien des chinesischen Drogenmarktes isolierte Alkaloide 178
   F. Nicht zu den Menispermaceae gehörende, Biscoclaurin-Basen enthaltende Pflanzen 179
      Berberidaceae 179. — Anonaceae 179. — Magnoliaceae 179. — Monimiaceae 179.
   G. In Curare enthaltene Biscoclaurin-Alkaloide..................... 179
III. Klassifizierung der Menispermaceae-Alkaloide 180
   Systematik der Biscoclaurin-Basen 180
   Gruppe I. Basen mit einem Äthersauerstoff 180
   Gruppe II a. Basen mit zwei Äthersauerstoffen (Tetrandrin-Typus) 180
   Gruppe II b. Basen mit zwei Äthersauerstoffen (Isochondendrin-Typus) 180
   Gruppe III a. Basen mit drei Äthersauerstoffen (Diphenylendioxyd-Typus).. 180
   Gruppe III b. Basen mit drei Äthersauerstoffen (Depsidan-Typus) 180
IV. Allgemeine Untersuchungsprinzipien der Biscoclaurin-Alkaloide........ 181
   1. Permanganat-Oxydation des Alkaloides selbst oder eines Hofmannschen Abbauproduktes 181
   2. Ozon-Spaltung von Methinbasen 182
   3. Aufspaltung durch Natrium in flüssigem Ammoniak 184
V. Spezieller Teil 186
   1. Benzylisochinolin-Typus 186
      Coclaurin 186. — Isococlaurin 187. — Magnocurarin 187. — Salicifolinchlorid 188.
   2. Phenanthropyridin-Typus 188
      Sinomenin 188. — Disinomenin 188. — Tuduranin 189. — Stephanin 189. — Crebanin 189. — Phanostenin 190. — Dicentrin 190.

3. Berberin-Typus 190
    Berberin 190. — Palmatin 190. — Columbamin 190. — Jatrorrhizin 190. — Sinactin 191.
4. Benzochinolizin-Typus 191
    Rotundin 191.
5. Bisoclaurin-Typus 192
    Gruppe I. Dauricin 192. — Magnolin 192. — Magnolamin 192. — Aztequin 193.
    Gruppe IIa. Berbamin 193. — Isotetrandrin 194. — Tetrandrin 194. — Phaeanthin 195. — Cepharanthin 195. — Oxyacanthin 195. — Repandin 197. — Daphnandrin 198. — Daphnolin (Trilobamin) 198. — Aromolin 198. — Epistephanin 199. — Hypoepistephanin 199.
    Gruppe IIb. Isochondodendrin 199. — Cycleanin 200. — Protocuridin 200. — Neoprotocuridin 201. — Bebeerin 201. — Chondrofolin 202. — Tubocurarinchlorid 202. — Chondocurin 202.
    Gruppe IIIa. Trilobin 203. — Isotrilobin 204. — Menisarin 204. — Normenisarin 204. — Micranthin 205.
    Gruppe IIIb. Insularin 205.
6. Strukturell ungeklärte Basen 207
7. Optische Isomerie der Bisoclaurin-Basen 207
8. Charakterisierung der Bisoclaurin-Basen 209
VI. Biogenetische Betrachtungen über Bisoclaurin-Basen 209
VII. Medizinische Anwendungen 213
Literaturverzeichnis 214

I. Einleitung.

Die den Menispermaceae angehörigen Pflanzen umfassen derzeit mehr als 260 Arten. Da sie jedoch größtenteils in den tropischen Zonen einheimisch sind, wurden sie während längerer Zeit von den meisten europäischen Forschern außer acht gelassen.

Die ältere Literatur erwähnt lediglich die Columbo-Wurzel der afrikanischen Ostküste sowie die Pareira-Wurzel aus Brasilien und Peru. Diese wurden trotz der schwierigen Materialbeschaffung von SCHOLTZ und FALTIS untersucht. Im tropischen Asien wurden nur wenige javanische Pflanzen von holländischen Forschern in Hinsicht auf ihre basischen Bestandteile aufgearbeitet. Demgegenüber wurde in Japan bereits im Jahre 1920 das Sinomenin aus *Sinomenium acutum* REHDER et WILSON als erster Vertreter der Menispermaceae-Alkaloide durch ISHIWARI erkannt, und dessen chemische Konstitution 1922 von KONDO und OCHIAI aufgeklärt. Ferner hat GOTO umfangreiche Untersuchungen über das Sinomenin und seine Derivate ausgeführt. Gleichzeitig haben KONDO und KONDO diese Studie auf das Coclaurin, eine Base aus *Cocculus laurifolius* DC., ausgedehnt und mit zahlreichen Mitarbeitern festgestellt, daß alle in Japan einheimischen Menispermaceae alkaloidhaltig sind und besonders durch das Vorkommen des Bisoclaurin-Typus gekennzeichnet werden können. Im Jahre 1935 haben KONDO und TOMITA (*107*) die chemische Konstitution und die Biosynthese der Biscoclaurin-Alkaloide besprochen, gestützt auf das damals zugängliche Versuchsmaterial. Seitdem haben europäische Forscher, wie SPÄTH, FALTIS, BRUCHHAUSEN, KING und TODD, durch ihre hervorragenden Arbeiten eine wertvolle Vervollständigung der Chemie der Biscoclaurin-Basen ermöglicht. Vor kurzem hat nun TOMITA die Öffnung der Äther-Brücke dieser Alkaloide mit Hilfe von Natrium in flüssigem Ammoniak erzielt, wodurch das bisherige Isomerie-Problem dieser Basen klargelegt werden konnte.

Die Niederschrift der vorliegenden Zusammenfassung wurde mit der Zustimmung von Dr. KONDO unternommen.

II. Die in Menispermaceae aufgefundenen Alkaloide.

A. *Durch* KONDO *und seine Mitarbeiter untersuchte Pflanzen:*

1. *Sinomenium acutum* REHDER et WILSON (*S. diversifolium* DIELS) (Japan): Sinomenin, Diversin, Disinomenin, Tuduranin, Sinactin, Acutumin (die vier letzteren von GOTO untersucht).
2. *Cocculus laurifolius* DE CANDOLLE (Japan): Coclaurin.
3. *C. trilobus* DE CANDOLLE (Japan): Trilobin, Isotrilobin, Trilobamin, Normenisarin.
4. *C. sarmentosus* DIELS (Formosa): Trilobin, Isotrilobin, Menisarin.
5. *Menispermum dauricum* DE CANDOLLE (Japan, Mandschurei): Dauricin, Tetrandrin.
6. *M. canadense* L. (Japan, Nordamerika): Dauricin.
7. *Stephania japonica* MIERS (Japan, Formosa): Stephanin, Protostephanin, Epistephanin, Metaphanin, Stephanolin, Homostephanolin, Hypoepistephanin, ,,VIII-Base", Hasubanonin, Hasubanin.
8. *St. cepharantha* HAYATA (Japan, Formosa): Cepharanthin, Isotetrandrin, Berbamin, Cycleanin.
9. *St. Sasakii* HAYATA (Formosa): Cepharanthin, Berbamin, Crebanin, Phanostenin.
10. *St. tetrandra* S. MOORE (Formosa): Tetrandrin.
11. *St. rotunda* LOUREIRO (Französisch-Indien): Rotundin.
12. *St. capitata* SPRENGEL (Nord-Borneo): Cycleanin, Dicentrin, Stephanin, Crebanin.
13. *Cyclea insularis* (MAKINO) DIELS (*Cissampelos insularis* MAKINO) (Japan, Liu-Kiu): Insularin, Cycleanin.
14. *Cissampelos ochiaiana* YAMAMOTO (Formosa): Insularin.
15. *Coscinium blumeanum* MIERS (Nord-Borneo): Berberin, Jatrorrhizin, Palmatin.
16. *Fibraurea chloroleuca* MIERS (*F. tinctoria* LOUREIRO) (Nord-Borneo): Palmatin, Jatrorrhizin.
17. *Parabaena hirsuta* (BECCARI) DIELS (Nord-Borneo): Palmatin.
18. *Archangelicia flava* MERRILL (Philippinen, Nord-Borneo): Berberin, Palmatin, Columbamin, Jatrorrhizin.

B. *Von nicht-japanischen Forschern untersuchte Pflanzen:*

1. *Cyclea peltata* HOOKER et THOMSON (*Cocculus peltata* DE CANDOLLE) (Tropen): Cyclein.
2. *Coscinium fenestratum* COLEBROOKE (Indien, Ceylon): Berberin.

3. *Jatrorrhiza columba* MIERS [*J. palmata* (LAMARCK) MIERS; *J. columba* MIERS], Columbo-Wurzel (Ost-Afrika, Brasilien): Palmatin, Columbamin, Jatrorrhizin.
4. *Anamirta paniculata* COLEBROOKE (*A. Cocculus* WIGHT et ARNOTT; *Menispermum cocculus* L.) (Ceylon, Indien, Malaya): Menispermin, Paramenispermin.
5. *Chondrodendron platyphyllum* (ST. HILAIRE) MIERS, Pareira-Wurzel (Brasilien, Peru): Isochondodendrin (Isobebeerin), L-Bebeerin, Chondrofolin, Isoclaurin.
6. *Ch. microphyllum* (Süd-Amerika): D-Isochondodendrin, D-Bebeerin.
7. *Ch. candicans* RICHARD ex DE CANDOLLE (Süd-Amerika): D-Isochondodendrin, D-Bebeerin.
8. *Ch. tomentosum* RUIZ et PAVON (Brasilien, Peru): D-Isochondodendrin, D-Dimethyl-isochondodendrin, D-Tubocurarinchlorid, L-Curin; Chondocurin (DUTCHER); L-Tubocurarinchlorid (KING); D-Tomentocurin (KING).
9. *Tinospora Rumphii* BOERLAGE (Philippinen, Java): Berberin.
10. *T. bakis* MIERS (*Cocculus bakis* RICHARD) (Tropen): Sangolin, Pelosin.
11. *T. crispa* MIERS (*Cocculus crispus* DE CANDOLLE): Berberin.
12. *Archangelicia lemniscata* BECCARI (Tropen): Berberin.
13. *Pycnarrhena manillensis* VIDAL (Philippinen): Ambalin, Ambalinin.
14. *Tiliacora racemosa* COLEBROOKE (*T. acuminata* MIERS) (Tropen): Tiliacorin.
15. *Stephania glabra* MIERS (Indien): Gindarin, Gindaricin, Gindarinin.

C. *Pflanzen, in denen von* KONDO *und Mitarbeitern das Vorkommen von Alkaloiden festgelegt wurde (70).*

1. *Diploclisia Kunstleri* (KING) DIELS (Nord-Borneo).
2. *Pericampylus glaucus* MERRILL (Nord-Borneo).
3. *Limacia veltina* MIERS (Nord-Borneo).
4. *L. obrenga* MIERS (Nord-Borneo).
5. *L. cuspidata* HOOKER (Nord-Borneo).
6. *Stephania venosa* DIELS (Nord-Borneo).
7. *Albertisia papuana* BECCARI (Nord-Borneo).
8. *Cyclea barbata* MIERS (Nord-Borneo).

D. *In der Literatur als alkaloidhaltig angegebene Pflanzen (220).*

1. *Cocculus ovalifolius* DE CANDOLLE (Tropen).
2. *C. umbellatus* STENDEL (Tropen).
3. *Hypserpa cuspidata* MIERS (Tropen).
4. *Pericampylus incanus* MIERS (Ost-Indien, Australien).
5. *Sarcopetalum Harveyanum* MUELLER (Australien).
6. *Triclisia Gilletii* STANER.

E. *Aus Rohmaterialien des chinesischen Drogenmarktes isolierte Alkaloide.*

HENRY (56) verzeichnet: Tetrandrin, Hanfangchin A und B, Menisin, Menisidin, Mufangchin, Fangchinin oder Fangchinolin u. a. Jedoch erscheinen die Rohmaterialien dort unter dem Namen Han-fang-chi oder Mu-fang-chi, welcher eine der folgenden Drogen oder ihr Gemisch bedeutet: *Menispermum dauricum* DC., *Sinomenium acutum* RED. et WIL., *Cocculus trilobus* DC., *Stephania japonica* MIERS oder *St. tetrandra* S. MOORE u. a. Demzufolge kann man leider diesen Angaben keinen wissenschaftlichen Wert beilegen.

F. Nicht zu den Menispermaceae gehörende, Biscoclaurin-Basen enthaltende Pflanzen.

Die nachstehend aufgezählten Pflanzen gehören den Menispermaceae nahestehenden Gattungen an:

Berberidaceae.
1. *Berberis vulgaris* L. (Berberitze): Oxyacanthin, Berbamin.
 B. Thunbergii DE CANDOLLE; *B. heteropoda* SCHLENK; *B. repens* DON; *B. aquifolium* PURSCH u. a. enthalten die eben genannten Alkaloide.
2. *Mahonia borealis* TAKEDA (Indien): Oxyacanthin, Berbamin.
3. *M. Griffithii* TAKEDA (Indien): Oxyacanthin, Berbamin.
4. *M. japonica* DE CANDOLLE (Japan): Isotetrandrin, Berbamin.

M. acanthifolia DON, *M. Simonsii* TAKEDA, *M. leschenaultii* TAKEDA, *M. manipurensis* TAKEDA und *M. sikkimensis* TAKEDA enthalten Oxyacanthin. In *M. Swaseyi* FEDDE wurde Berbamin festgestellt.

Anonaceae.
1. *Phaeanthus ebracteoratus* (PRESL) MERRILL (Philippinen): Phaeanthin.

Magnoliaceae.
1. *Magnolia fuscata* ANDREUS (Kaukasien): Magnolin, Magnolamin.
2. *Talauma mexicana* DON, ,,Yoloxochitl" (Mexiko): Aztequin.
 In japanischen Magnolia-Pflanzen kommen Basen vom Coclaurin-Typus vor.
3. *Magnolia obovata* THUNBERG (JAPAN): Magnocurarin.
4. *M. salicifolia* MAXIMOWICZII (Japan): Magnocurarin, Salicifolin.
5. *M. Kobus* DE CANDOLLE (Japan): Salicifolin.

Monimiaceae.
1. *Daphnandra micrantha* BENTHAM (New South Wales, Queensland): Micranthin, Daphnandrin, Daphnolin, Aromolin.
2. *D. aromatica* BAILLON (Australien); Daphnolin, Aromolin.
3. *D. repandula* BANCROFT (Australien): Repandulin, Repandin.
4. *D. Dielsii* PERKINS (Australien): Repandulin.

G. In Curare enthaltene Biscoclaurin-Alkaloide.

Unter den Curare-Alkaloiden befinden sich viele, mit den Pareira-Basen in naher Beziehung stehende Biscoclaurin-Basen. In der Tat wird jetzt die in Brasilien einheimische Menispermacea *Chondrodendron tomentosum* RUIZ et PAVON (oder eine nahestehende Spezies) als die Mutterpflanze der ,,Tubocurare" angesehen. Aus der ,,Topf-curare" ist Protocuridin und Neoprotocuridin isoliert worden.

Vor kurzem hat YUNUSOV (*222, 223*) aus den Blättern von *Cocculus laurifolius* aus Uzbek zwei Alkaloide, Cocculidin und Cocculin, isoliert, welche nach seiner Auffassung eine als Lycoris-Base eigentümliche Phenanthridin-Base darstellen sollen. Es stellt sich aber die Frage, ob sein Ausgangsmaterial mit dem japanischen *Cocculus laurifolius* tatsächlich identisch war, da solche Alkaloide von TOMITA und Mitarbeitern weder in den Blättern noch in den Stämmen oder Wurzeln des japanischen Pflanzenmaterials vorgefunden wurden.

III. Klassifizierung der Menispermaceae-Alkaloide.

Wie erwähnt, enthalten die meisten Menispermaceae-Pflanzen Alkaloide, die fast alle dem Isochinolin-Typus angehören. Die Haupttypen dieser Basen sind:

Benzylisochinolin-Typus, Phenanthropyridin-Typus, Berberin-Typus, Benzochinolizin-Typus, Biscoclaurin-Typus.

Der Berberin-Typus wird hauptsächlich in den Berberidaceae aufgefunden, in welchen hingegen die Benzylisochinolin-, Phenanthropyridin- und Aporphin-Typen recht selten vorkommen, obgleich sie in anderen Pflanzenfamilien weit verbreitet sind; Coclaurin, Sinomenin, Crebanin usw. sind die wichtigsten Vertreter in Menispermaceae. Nur ein einziger Vertreter des Benzochinolizin-Typus, Rotundin, wurde bisher isoliert. Das häufige Vorkommen des Biscoclaurin-Typus in dieser Pflanzenfamilie ist bemerkenswert, da dies in anderen Familien keineswegs der Fall ist.

Systematik der Biscoclaurin-Basen.

Der Name „Biscoclaurin- (oder Bisbenzylisochinolin-) Alkaloid" umfaßt jene Pflanzenbasen, welche zwei Benzylisochinolin-Gruppen enthalten, die durch 1, 2 oder 3 Brückensauerstoffe miteinander verknüpft sind. Auf Grund der Zahl und Verkettungsart solcher Brücken werden sie weiter wie folgt unterteilt:

Gruppe I. Basen mit einem Äthersauerstoff: Dauricin, Magnolin Magnolamin, Aztequin.

Gruppe IIa. Basen mit zwei Äthersauerstoffen [sogenannter Tetrandrin-Typus von KING (*62*), mit beinahe planarer Struktur]: Oxyacanthin, Repandin, Aromolin, Daphnolin (Trilobamin), Daphnandrin, Epistephanin, Hypoepistephanin, Berbamin, Isotetrandrin, Tetrandrin, Phaeanthin und Cepharanthin.

Gruppe IIb. Basen mit zwei Äthersauerstoffen [„Isochondodendrin-Typus" von KING (*62*), mit beinahe symmetrisch liegender Molekül-Achse]: Isochondodendrin (Isobebeerin), Cycleanin, Protocuridin, Neoprotocuridin, Chondrofolin, *D*-Bebeerin, Curin, (*L*-Bebeerin), Tubocurarinchlorid, *D*-Chondocurin.

Gruppe IIIa. Basen mit drei Äthersauerstoffen (Diphenylen-dioxyd-Kern enthaltend): Trilobin, Isotrilobin, Menisarin, Normenisarin, Micranthin.

Gruppe IIIb. Basen mit drei Äthersauerstoffen (Depsidan-Kern enthaltend): Insularin.

IV. Allgemeine Untersuchungsprinzipien der Biscoclaurin-Alkaloide.

Wie bekannt, besitzen die Moleküle der um das Jahr 1830 aufgefundenen Alkaloide aus der Berberitze (Oxyacanthin und Berbamin) sowie aus der Curare- und Pareira-Wurzel einen oder mehrere Sauerstoffatome, die keine chemischen Funktionen zeigen. Vor etwa 30 Jahren haben KONDO und Mitarbeiter im Verlaufe ihrer Untersuchungen betreffs Menispermaceae wiederholt dasselbe beobachtet. Unter dem Einfluß von gleichzeitigen schönen Forschungen, die in Europa von SPÄTH, FALTIS, BRUCHHAUSEN, KING u. a. durchgeführt wurden, hat KONDO jedoch das Vorliegen einer Biscoclaurin-Gruppe in Menispermaceae anerkannt.

Einige *Abbaumethoden*, die sich auf diesem Gebiete als nützlich erwiesen haben, seien nun kurz besprochen.

1. Permanganat-Oxydation des Alkaloides selbst oder seines Hofmannschen Abbau-Produktes.

Diese Reaktion wurde fast gleichzeitig von SPÄTH und PIKL (*181*), BRUCHHAUSEN und SCHULTZE (*12*) sowie von KONDO und YANO (*117, 221*) angegeben. Die den Gruppen I, IIa und IIIa zugehörenden Basen und deren O-Methyläther wurden dadurch in 2-Methoxy-diphenyläther-5,4'-dicarbonsäure (I) übergeführt. Dagegen ist dies in den Gruppen IIb und IIIb nicht der Fall. Diese Reaktion gilt auch für Methinbasen oder Des-Basen. Bei den Basen der Gruppe I, z. B. Magnolin (*154*), wird auch Corydaldin (II) neben der Dicarbonsäure (I) isoliert. Durch HOF-

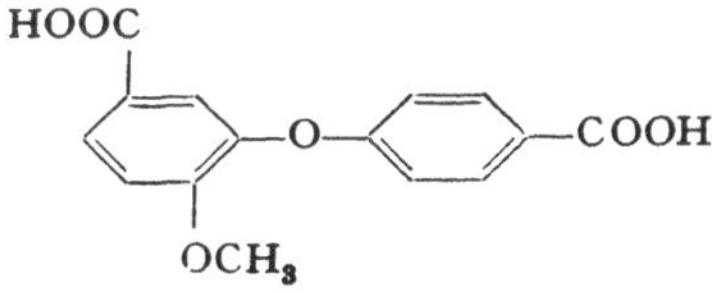

(I.) 2-Methoxy-diphenyläther-5,4'-dicarbonsäure. 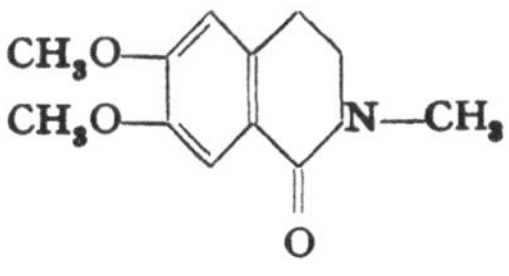   (II.) Corydaldin.

MANNschen Abbau und Oxydation der zu Gruppe IIb gehörenden Basen werden stets zwei Mole derselben Dimethoxy-diphenyläther-tricarbonsäure gewonnen, während die zum Bebeerin-Typus derselben Gruppe sowie zu Gruppe IIIb gehörenden Basen je ein Mol von zwei verschiedenen Dimethoxy-diphenyläther-tricarbonsäuren ergeben. So wird z. B. die Des-Base (IV) des 0,0-Dimethyl-isochondodendrins (III) zur 2,3-Dimethoxy-diphenyläther-5,6,4'-tricarbonsäure (V) abgebaut. Diese Reaktion wurde zuerst von FALTIS und Mitarbeitern (*27, 28, 30, 23*) ausgeführt und ist zur strukturellen Aufklärung dieser Basen-Typen sehr geeignet.

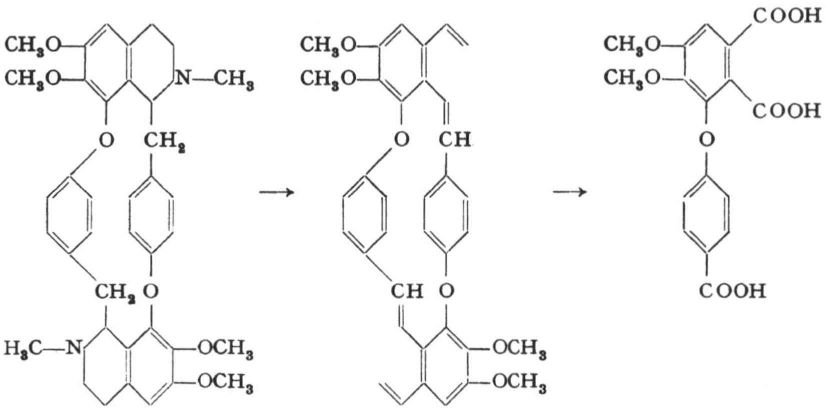

(III.) 0,0-Dimethyl-isochondodendrin. (IV.) Des-Base von (III.) (V.) 2,3-Dimethoxy-diphenyläther-5,6,4'-tricarbonsäure.

2. Ozon-Spaltung von Methinbasen.

Diese Reaktion wurde zum ersten Male 1931 von BRUCHHAUSEN und Mitarbeitern (*10, 11*) durchgeführt und leistete wichtige Dienste bei der Aufklärung des Isochinolin-Kerns der Basen in den Gruppen IIa und IIIa. In BRUCHHAUSENS Oxyacanthin-Abbau wurden z. B. 0-Methyl-oxyacanthin-methylmethin (VI) oder (VII) durch Ozon-Spaltung in den 2-Methoxy-diphenyläther-5,4'-dialdehyd (IX) und den Aminoaldehyd (VIII) übergeführt. Die Verbindung (IX) wurde dann durch Permanganat-Oxydation zur 2-Methoxy-diphenyläther-5,4'-dicarbonsäure (I) abgebaut, welche sich auch durch direkte Oxydation des o-Methyl-oxyacanthins oder dessen Methinbase darstellen ließ. Das Jodmethylat des Aminoaldehyds (VIII) spaltet beim Erhitzen mit Lauge unter Freiwerden von Trimethylamin das Vinylaldehyd (X) ab, welches in Gegenwart von Palladium-Baryumsulfat zu Trimethoxy-diäthyl-diphenyläther-dialdehyd (XI) hydriert und darauf durch eine Reduktion nach CLEMMENSEN in den Trimethoxy-diäthyl-dimethyl-diphenyläther (XII) übergeführt wurde. Die letztere Verbindung erwies sich als mit dem synthetischen Produkt identisch.

Wie erwähnt, wurde durch die schönen Abbaureaktionen (1) und (2) ein entscheidender Fortschritt in der Chemie der Biscoclaurin-Basen erzielt; ja, die Struktur der Basen der Gruppen I und IIb konnte auf diesem Wege vollständig klargelegt werden. Demgegenüber waren die den Gruppen IIa und IIIa zugehörigen Alkaloide noch einer tiefergehenden Erforschung bedürftig. Über zehn Alkaloide, wie Oxyacanthin, Berbamin (XIII) (XIV) und auch Trilobin, Isotrilobin usw. hatte man auch betr. optischer und struktureller Isomerie aufzuklären.

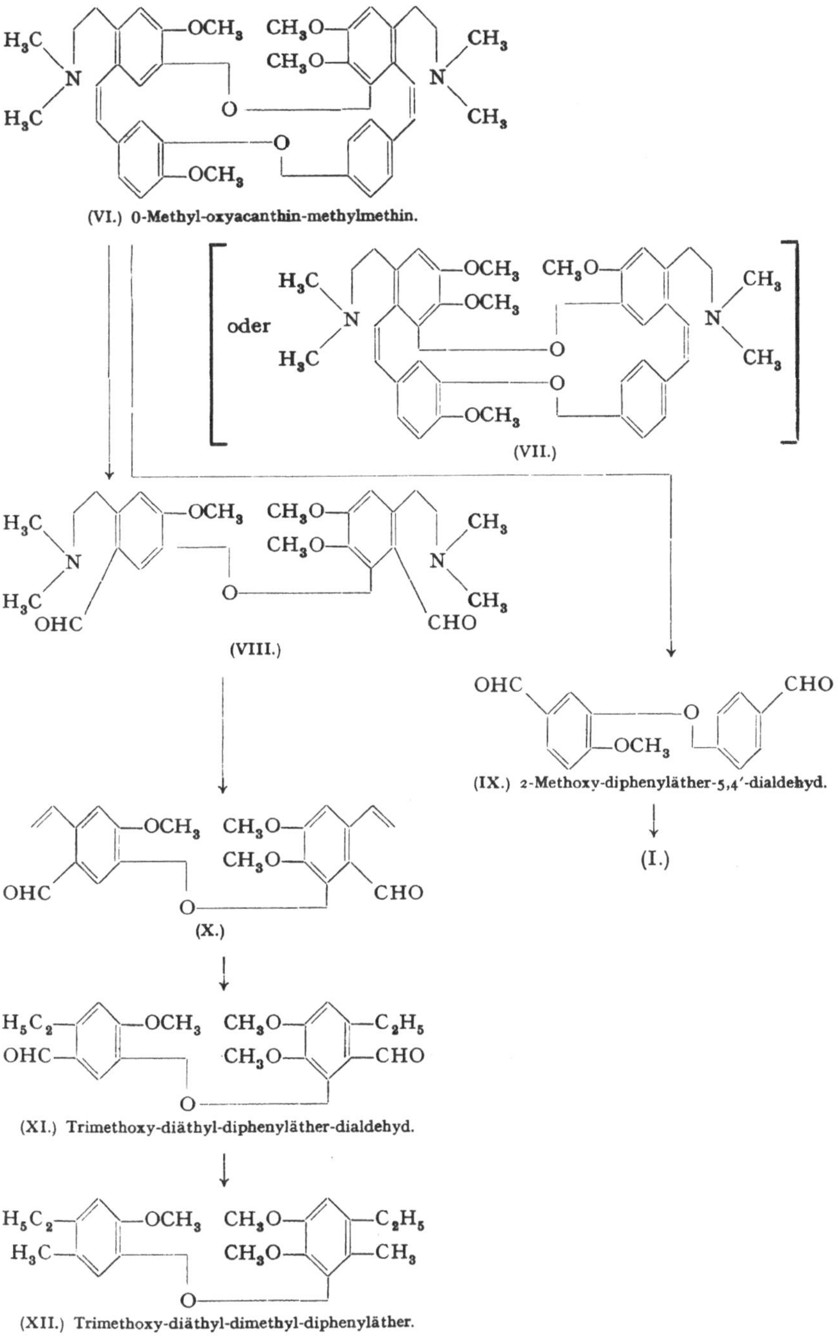

(VI.) O-Methyl-oxyacanthin-methylmethin.

oder

(VII.)

(VIII.)

(IX.) 2-Methoxy-diphenyläther-5,4'-dialdehyd.

(I.)

(X.)

(XI.) Trimethoxy-diäthyl-diphenyläther-dialdehyd.

(XII.) Trimethoxy-diäthyl-dimethyl-diphenyläther.

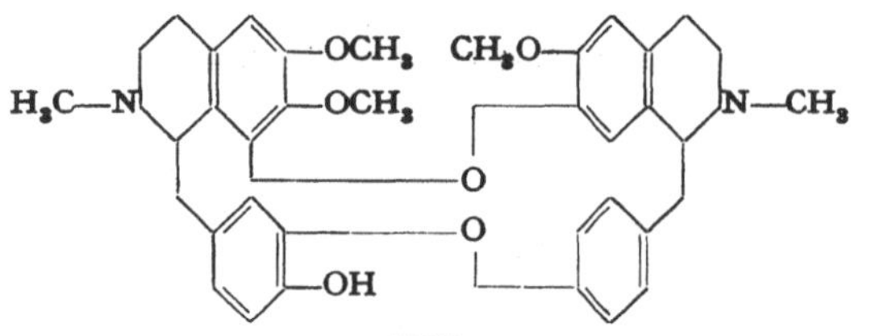

(XIII.)

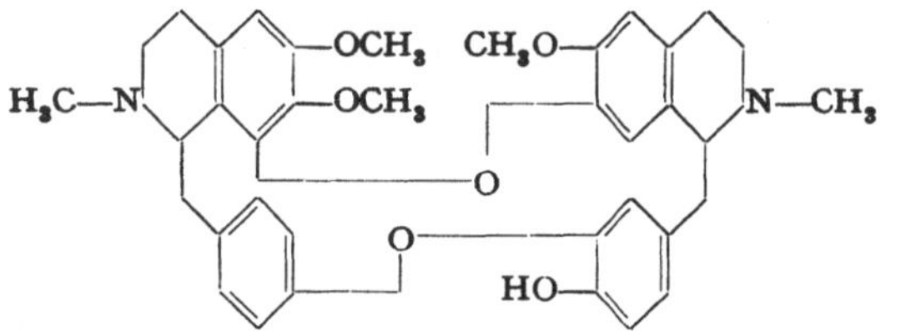

(XIV.)

Die Lösung dieser Fragen wurde durch Anwendung einer speziellen, milden Methode ermöglicht, durch welche die in den erwähnten Basenklassen eigenartig gebundene Diphenyläther-Brücke sich bequem öffnen ließ, ohne die Benzylisochinolin-Körper irgendwie anzugreifen. Dies ist vor kurzem TOMITA und Mitarbeitern geglückt, nämlich durch die Anwendung von Natrium in flüssigem Ammoniak.

3. Aufspaltung durch Natrium in flüssigem Ammoniak.

Vor einiger Zeit hatte MANSKE (*126*) diese Methode beim Cularin, einer Benzylisochinolin-Base angewandt. 1951 haben dann TOMITA und Mitarbeiter die Reaktion auf Biscoclaurin-Basen übertragen und konnten zwei Spaltstücke vom Coclaurin-Typus isolieren. Dies bedeutet, daß eine Biscoclaurin-Base, welche wohl aus zwei Molen des Coclaurin-Typus in der Pflanze aufgebaut worden ist, durch Behandlung mit Natrium und flüssigem Ammoniak, also durch Hydrieren, in die natürliche Vorstufe rückverwandelt wurde. Zum Beispiel konnte man, ausgehend von Isotetrandrin (0-Methyl-berbamin) (XV) (*191—193*) mit Hilfe der genannten Reaktion einerseits L-1-(4'-Methoxybenzyl)-6,7-dimethoxy-N-methyl-1,2,3,4-tetrahydroisochinolin (XVI) und anderseits D-1-(4'-Oxybenzyl)-6-methoxy-7-oxy-N-methyl-1,2,3,4-tetrahydroisochinolin (XVII) fast quantitativ fassen. Zwei asymmetrische Zentren wurden auf diesem Wege ermittelt.*

Die Reaktion wurde auch auf die den Gruppen I und II b angehörigen Basen übertragen (*35*), und ferner konnte auch Kalium statt Natrium angewandt werden (*197*).

---

* Ein innerhalb eines Ringes stehendes +- oder —-Zeichen gibt die Drehungsrichtung der betreffenden Molekülhälfte an, wie sie nach der Öffnung der Äthersauerstoff-Brücke beobachtet wird.

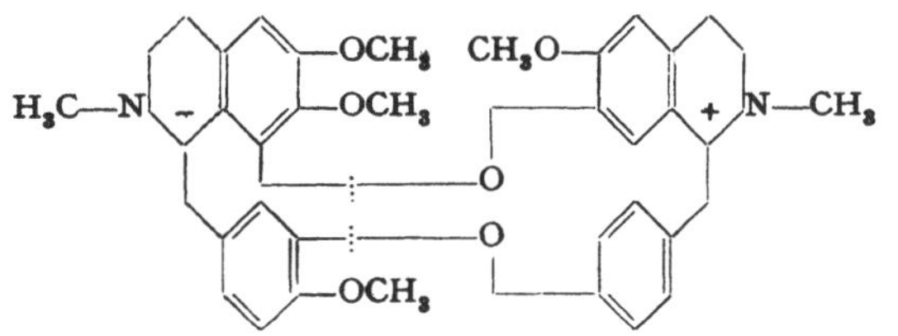

(XV.) Isotetrandrin. (Die punktierten Linien zeigen die Spaltungsstellen an.)*

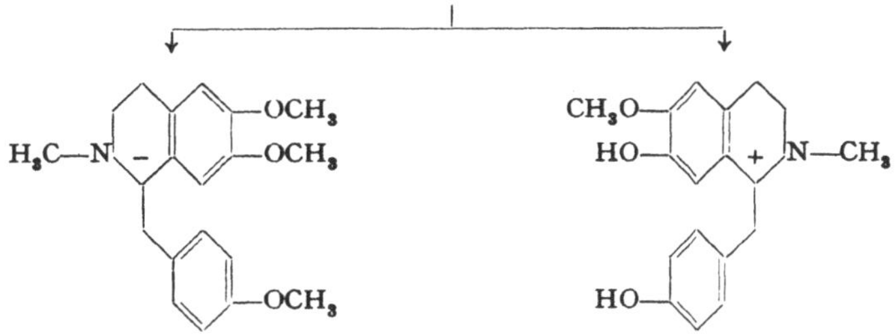

(XVI.) L-1-(4'-Methoxybenzyl)-6,7-dimethoxy-N-methyl-1,2,3,4-tetrahydroisochinolin.

(XVII.) D-1-(4'-Oxybenzyl)-6-methoxy-7-oxy-N-methyl-1,2,3,4-tetrahydroisochinolin.

Nach TOMITA und Mitarbeitern (*196*) ist diese Reaktion bei einer Phenolbase, wie Berbamin selbst, schwer ausführbar.

SOWA und Mitarbeiter (*123, 163, 218, 219*) haben diese Reaktion auf viele Diphenyläther-Derivate übertragen und beobachtet, daß sie im Falle des Oxy-diphenyläthers gehindert ist. Nach TOMITA (*196*) wird Diphenylendioxyd (XVIII) nur bis zur 2-Oxy-diphenyläther-Stufe (XIX) verwandelt. Der Umsatz kommt dann zum Stillstand, und eine weitere Zufuhr von Natrium führt lediglich zu einer Verharzung.

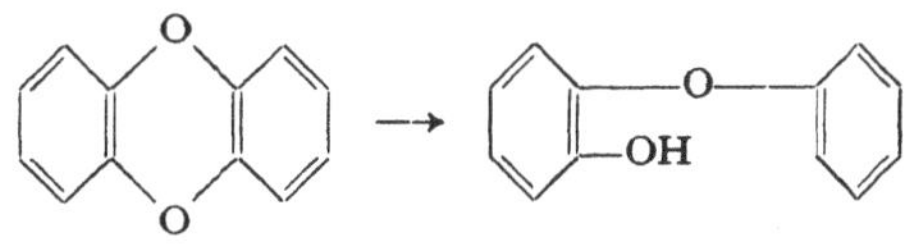

(XVIII.) Diphenylendioxyd.   (XIX.) 2-Oxy-diphenyläther.

Der letztgenannte Autor hat mit INUBUSHI (*59*) eine weitere Anwendung dieser Reaktion, nämlich die zweimalige Wiederholung derselben bei den zum Oxy-diphenyläther-Typus gehörigen Basen (z. B. Berbamin) beschrieben. Als die aus Berbamin (XX) gewonnene, dem Dauricin-Typus entsprechende Base (XXI) sogleich methyliert (XXII) und der Reaktion wieder unterworfen wurde, ergab sie L-1-(4'-Methoxybenzyl)-6,7-dimethoxy-N-methyl-1,2,3,4-tetrahydroisochinolin (XXIII) und D-1-(4'-Oxybenzyl)-6,7-dimethoxy-N-methyl-1,2,3,4-tetrahydroisochinolin (XXIV).

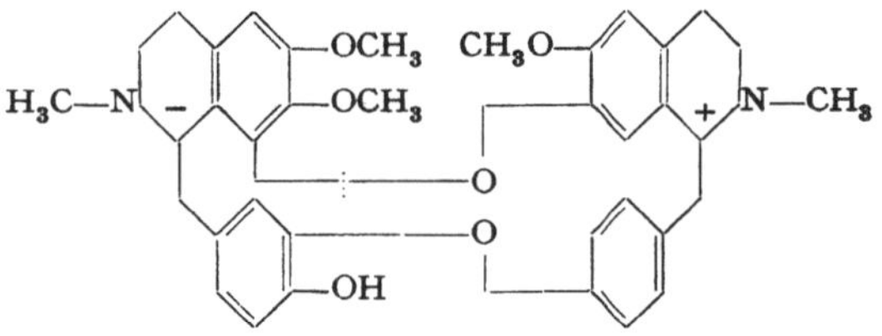

(XX.) Berbamin.

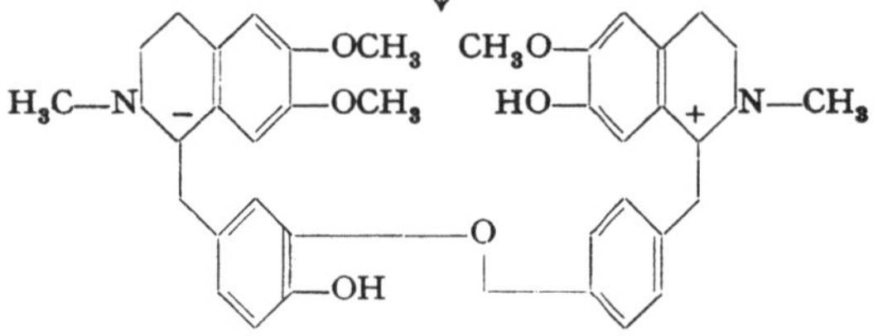

(XXI.)

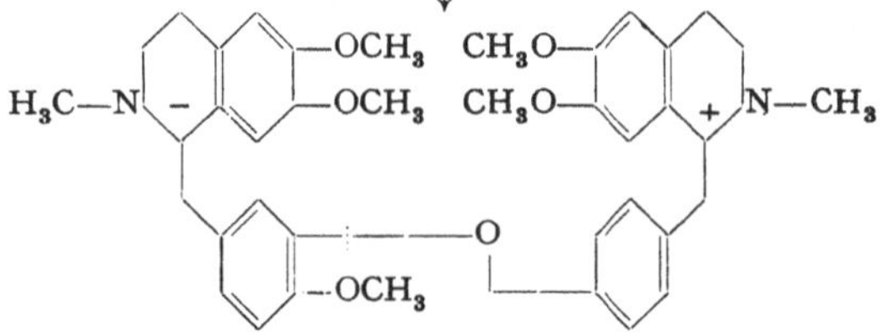

(XXII.)

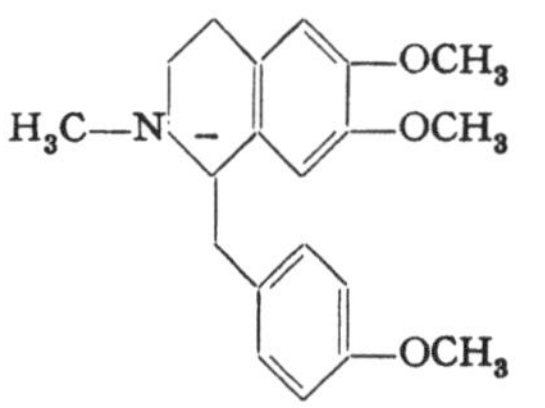

(XXIII.) L-1-(4'-Methoxybenzyl)-6,7-dimethoxy-N-methyl-1,2,3,4-tetrahydroisochinolin.

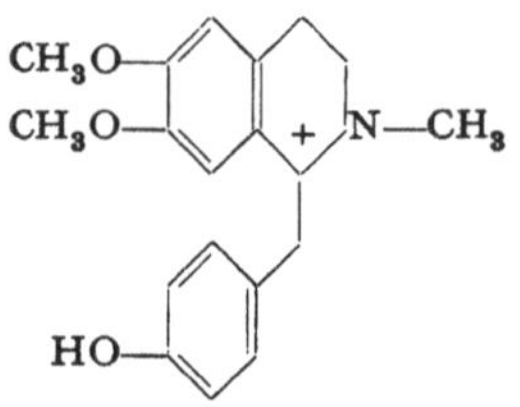

(XXIV.) D-1-(4'-Oxybenzyl)-6,7-dimethoxy-N-methyl-1,2,3,4-tetrahydroisochinolin.

V. Spezieller Teil.

In dieses Kapitel wurden nur diejenigen Alkaloide aufgenommen, deren Konstitution bereits vollständig klargelegt ist.

1. Benzylisochinolin-Typus.

Coclaurin, $C_{17}H_{19}O_3N$, tafelförmige Kristalle, Fp. 221°.

Das Coclaurin ist als die Muttersubstanz des Biscoclaurin-Typus eine der biogenetisch wichtigsten Basen. Sie wurde zuerst von KONDO und KONDO (*77, 78, 119—122*) aus *Cocculus laurifolius* DC. isoliert und ihre Konstitution (XXV) durch Abbau bestimmt. Vor kurzem haben FINKELSTEIN (*31*) sowie TOMITA, NAKAGUCHI und TAKAGI (*201*) diesen Befund auf synthetischem Wege bestätigt. Die von KONDO als der Naturstoff beschriebene Substanz ist die Razemform [TOMITA und KUSUDA (*200*)].

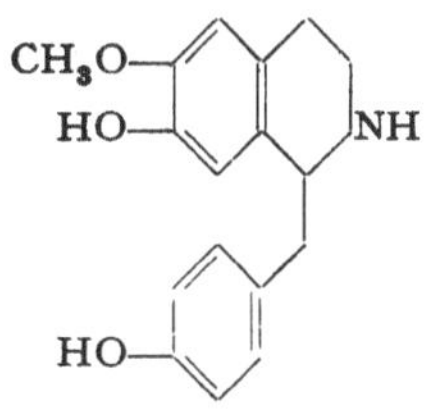

(XXV.) Coclaurin.

Isococlaurin, $C_{17}H_{19}O_3N$, Tafeln, Fp. 216—217°, $[\alpha]^{20}_{5461} = +23,9°$ (Chlorhydrat in Wasser).

Das Isococlaurin wurde von KING (*66*) aus *Radix Pareira brava* isoliert, die im englischen Drogenmarkt erhältlich ist.

Dieser Base soll die Formel (XXVI) zukommen.

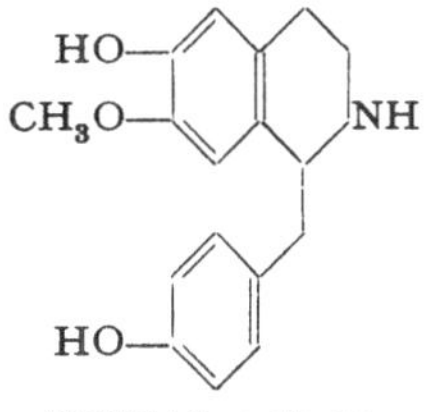

(XXVI.) Isococlaurin.

Magnocurarin. $C_{19}H_{25}O_4N$, Prismen, Fp. 200° (Zers.), $[\alpha]^{29}_D = -91,0°$ (in Wasser).

Das Magnocurarin wurde von SASAKI (*164*) aus der Rinde der *Magnolia obovata* THUNB. als ein Träger der Curare-Wirkung in harzigem Zustand erhalten. TOMITA, INUBUSHI und YAMAGATA (*198*) haben es kristallisiert und in der Form des Methylhydroxydes (XXVII) charakterisiert.

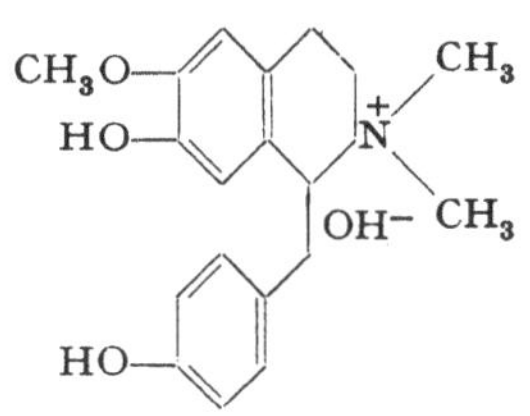

(XXVII.) Magnocurarin (Methylhydroxyd.)

Salicifolinchlorid, $C_{12}H_{20}O_2NCl$, Kristalle, Fp. 260—261° (Zers.).

Das Salicifolinchlorid wurde von TOMITA und NAKANO (*202, 203*) aus der Rinde von *Magnolia salicifolia* MAXIM. neben Magnocurarin isoliert und wurde kürzlich auch in *Magnolia Kobus* DC. aufgefunden (unveröffentlicht). Die Verbindung ist ein Ammoniumchlorid, für welches durch Abbau und Synthese das Formelbild (XXVIII) bewiesen wurde. Diese Substanz stellt eigentlich eine Kakteen-Base dar. Sie ist von biogenetischem Interesse, schon deshalb, weil sie wahrscheinlich vom Tyrosin herstammt und als eine Protobase des Coclaurins und ähnlicher Basen aufgefaßt werden darf.

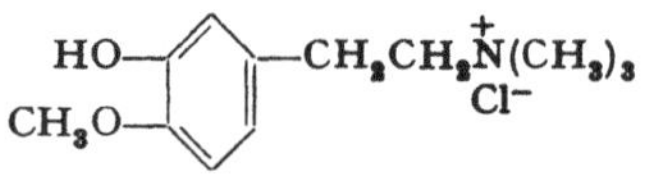

$$HO-\underset{CH_3O-}{\underset{|}{\bigcirc}}-CH_2CH_2\overset{+}{N}(CH_3)_3 \quad Cl^-$$

(XXVIII.) Salicifolinchlorid.

2. Phenanthropyridin-Typus.

Sinomenin, $C_{19}H_{23}O_4N$, Nadeln, Fp. 161° und Fp. 182°, $[\alpha]_D^{26} = -70,76°$ (in Alkohol).

Diese Verbindung wurde 1920 von ISHIWARI (*60*) aus *Sinomenium acutum* R. et W. isoliert und von KONDO und OCHIAI (*90—92, 135—141*) sowie von GOTO (*37—40, 50, 51*) strukturell vollständig geklärt (XXIX). Sie ist bis jetzt der einzige Vertreter der Morphinoid-Basen. GOTO und Mitarbeiter haben über Sinomenin- und Disinomenin-Derivate umfangreiche Arbeiten veröffentlicht (*42, 43*).

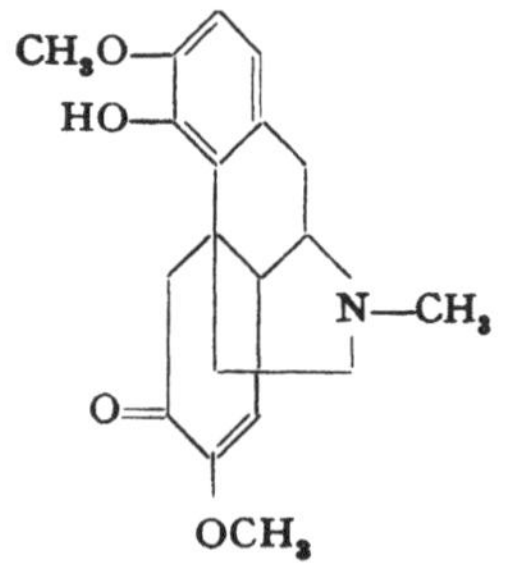

(XXIX.) Sinomenin.

Disinomenin. $(C_{19}H_{22}O_4N)_2$, Kristalle, Fp. 222°.

Das Disinomenin (XXX) begleitet das Sinomenin in der Pflanze. Es wird synthetisch durch milde Oxydation von Sinomenin zusammen mit ψ-Disinomenin erhalten [GOTO (*47, 48*)].

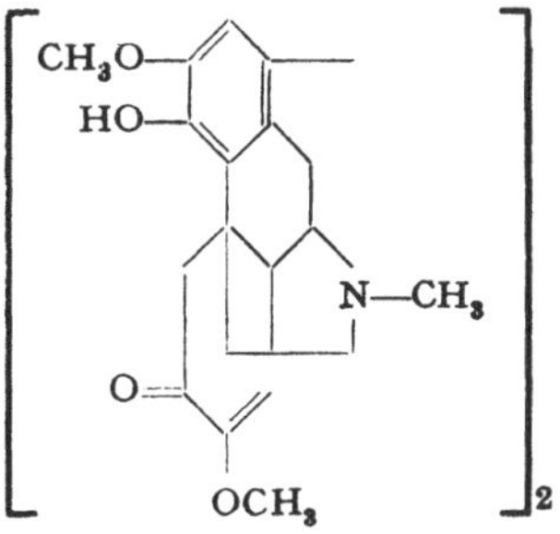

(XXX.) Disinomenin.

Tuduranin, $C_{20}H_{21}O_3N$, Kristalle, Fp. 125°, $[\alpha]_D^{14,5} = -148°$ (in verd. Methylalkohol).

Tuduranin wurde von GOTO (*41, 44, 46*) und Mitarbeitern aus *Sinomenium acutum* isoliert. Seine nachstehende Formel (XXXI) stellt ein 3-Oxy-5,6-dimethoxy-N-nor-apomorphin dar.

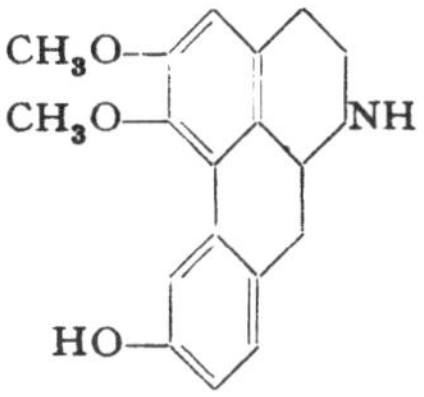

(XXXI.) Tuduranin.

Stephanin, $C_{19}H_{19}O_3N$, Rhomben, Fp. 157°, $[\alpha]_D^{10} = -93,25°$ (in Chloroform).

Diese Base wurde von KONDO und SANADA (*94*) aus *Stephania japonica* MIERS isoliert und später von TOMITA und SHIRAI (*204, 176*) auch aus *Stephania capitata* SPRENG. erhalten. Die letztgenannten Autoren haben sie als ein 1-Methoxy-5,6-methylendioxy-apomorphin erkannt (XXXII).

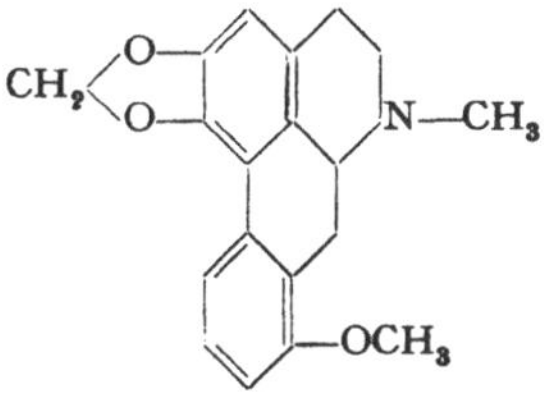

(XXXII.) Stephanin.

Crebanin, $C_{20}H_{21}O_4N$, Nadeln, Fp. 126°, $[\alpha]_D^9 = -56°$ (in Chloroform).

Nebst Stephanin haben TOMITA und SHIRAI (*204, 205*) diese Base aus *St. capitata* sowie aus *St. Sasakii* HAYATA isoliert und als 1,2-Dimethoxy-5,6-methylendioxy-apomorphin (XXXIII) identifiziert (*174, 175*).

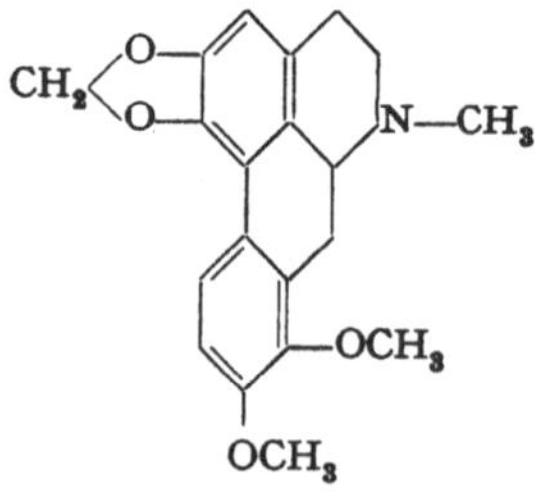

(XXXIII.) Crebanin.

Phanostenin, $C_{19}H_{19}O_4N$, Würfel, Fp. 210°, $[\alpha]_D^{20} = -36,7°$ (in Chloroform).

Das Phanostenin wurde von KONDO und TOMITA (108) aus *St. Sasakii* HAYATA in kleinen Mengen gewonnen. Nach TOMITA und SHIRAI (205) liefert es beim Methylieren L-Dicentrin. Infolgedessen wurde dem Phanostenin die Formel (XXXIV) zugeteilt ($R_1 = H$, $R_2 = CH_3$ oder umgekehrt).

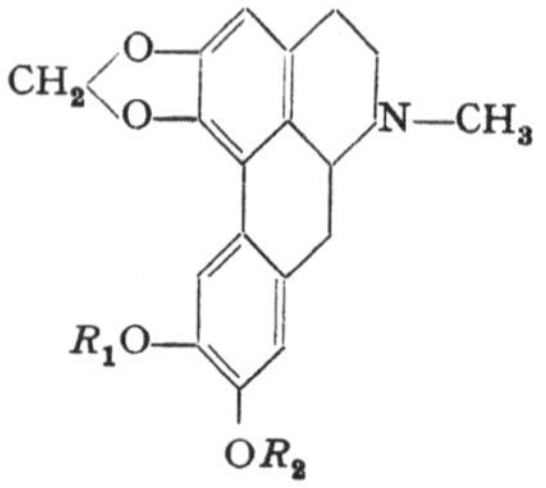

(XXXIV.) Phanostenin.

Dicentrin, $C_{20}H_{21}O_4N$, Kristalle, Fp. 168°, $[\alpha]_D^9 = +61,99°$ (in Chloroform).

TOMITA und SHIRAI (204) haben das D-Dicentrin (XXXIV, $R_1 = R_2 = CH_3$) in *Stephania capitata* SPRENG. (neben Crebanin und Phanostenin) nachgewiesen.

3. Berberin-Typus.

In tropischen Menispermaceae-Alkaloiden findet man meistens Vertreter des Berberin-Typus, wie Berberin (XXXV), Palmatin (XXXVI), Columbamin (XXXVII), Jatrorrhizin (XXXVIII) u. a. So haben TOMITA, TANI und ASADA (189, 206, 207) aus *Coscinium blumeanum* MIERS, *Fibraurea chloroleuca* MIERS, *Parabaena hirsuta* (BECC.) DIELS aus Nord-Borneo die genannten Basen isoliert. Diese wohlbekannten Alkaloide sollen an dieser Stelle nicht näher beschrieben werden.

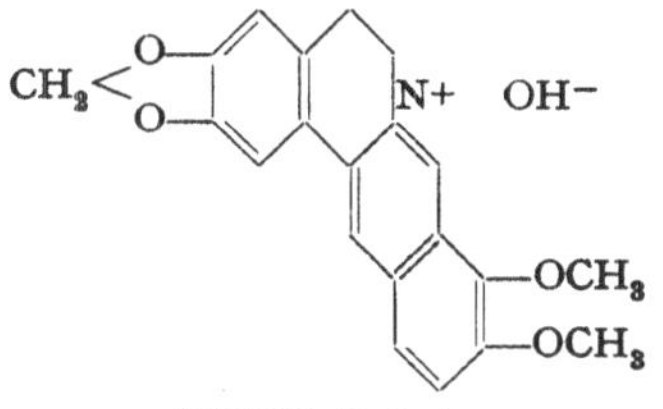

(XXXV.) Berberin. 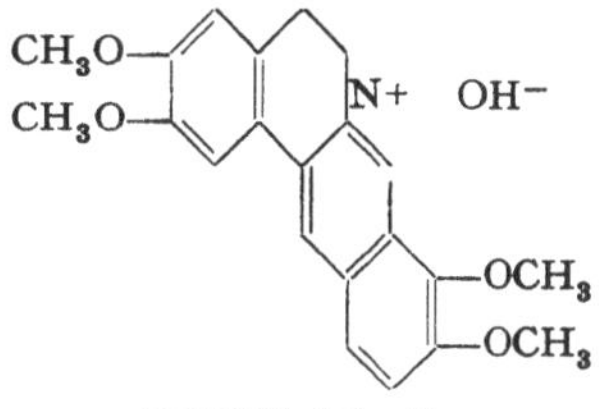   (XXXVI.) Palmatin.

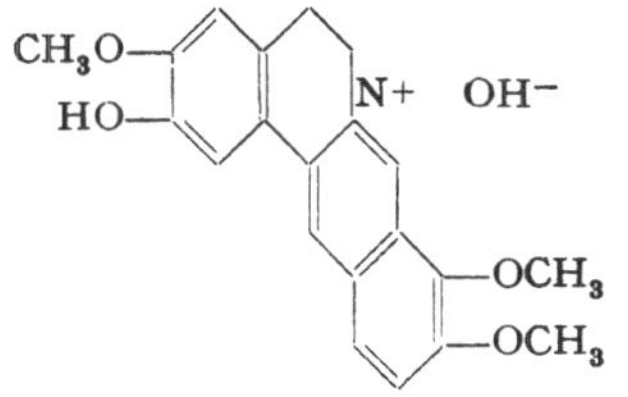

(XXXVII.) Columbamin. 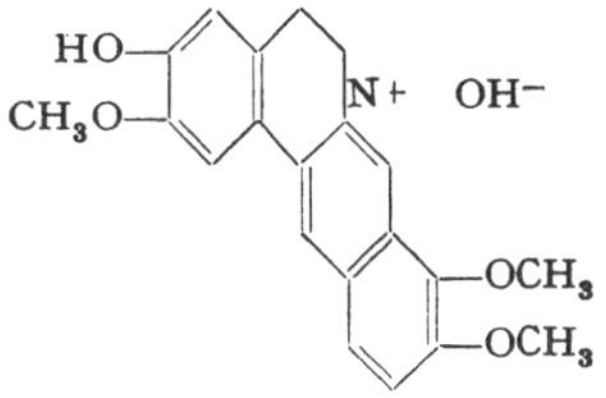   (XXXVIII.) Jatrorrhizin.

Sinactin, $C_{20}H_{21}O_4N$, Kristalle, Fp. 174°, $[\alpha]_D = -312°$ (in Chloroform).

Das Sinactin wurde von GOTO und SUDZUKI (*49*) aus *Sinomenium acutum* isoliert und seine Konstitution wurde von GOTO und KITASATO (*45*) als *L*-Tetrahydroepiberberin (XXXIX) erkannt. Nach SPÄTH und MOSETTIG (*180*) soll Sinactin von Cryptopin sich ableiten. Die Konfiguration entspricht nach LEITHE (*124*) dem (—)-Alanin.

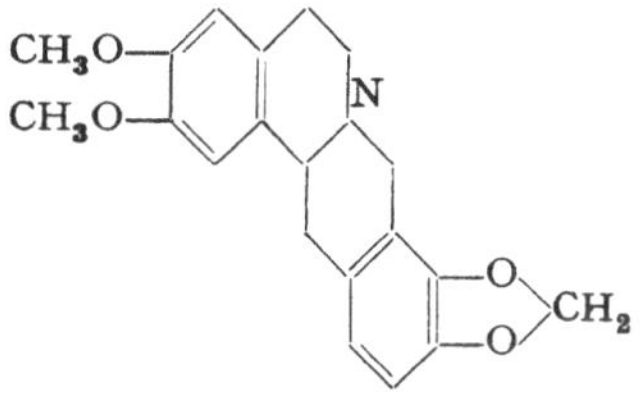

(XXXIX.) Sinactin.

4. Benzochinolizin-Typus.

Rotundin, $C_{17}H_{21}O_3N$, hexagonale Tafeln, Fp. 140—141°, $[\alpha]_D^{23} = -262,1°$ (in Alkohol).

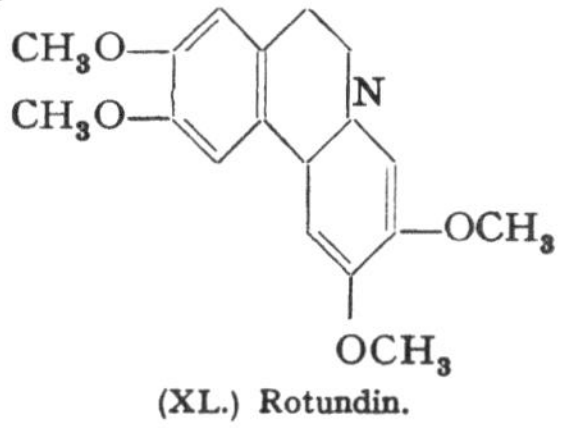

(XL.) Rotundin.

Diese Base wurde von KONDO und MATSUNO (*79, 131*) aus *Stephania rotunda* LOUREIRO gewonnen und seine Konstitution als (XL) ermittelt. Die letztere ist für ein Menispermaceae-Alkaloid einzigartig.

5. Biscoclaurin-Typus.

Gruppe I.

Dauricin, $C_{36}H_{44}O_6N_2$, amorphes Pulver, Fp. 115°, $[\alpha]_D^{11} = -139°$ (in Methanol). Jodmethylat: Nadeln, Fp. 204°.

Das Dauricin wurde von KONDO und NARITA *(83)* aus *Menispermum dauricum* DC. und von denselben Autoren mit MURAKAMI *(87)* sowie von MANSKE *(125)* auch aus *M. canadense* L. isoliert. Die in Formel (XLI) verzeichnete Struktur wurde durch KONDO, NARITA und UYEO *(84—86, 88; 10, 23)* auf synthetischem Wege gesichert. Diese interessante Verbindung könnte als ein Zwischenprodukt auf dem Wege von Coclaurin zu Oxyacanthin und Berbamin angesehen werden.

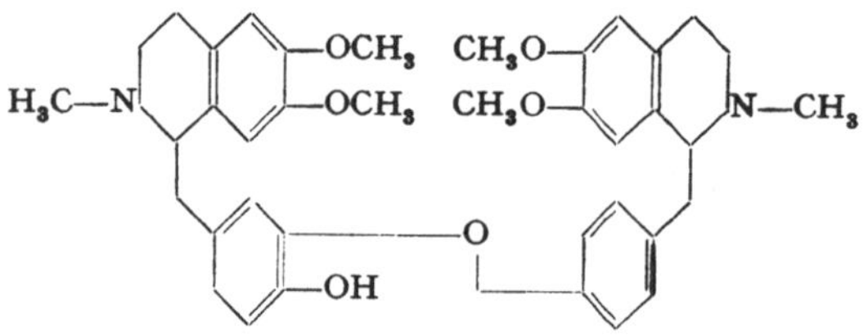

(XLI.) Dauricin.

Magnolin, $C_{36}H_{40}O_6N_2$, Kristalle, Fp. 178—179°, $[\alpha]_D = -9,6°$ (in Pyridin).

Das Magnolin wurde von PROSKOURNINA und ORECHOV *(153, 154)* aus *Magnolia fuscata* zusammen mit Magnolamin (siehe unten) isoliert und gemäß Formel (XLII) strukturell geklärt.

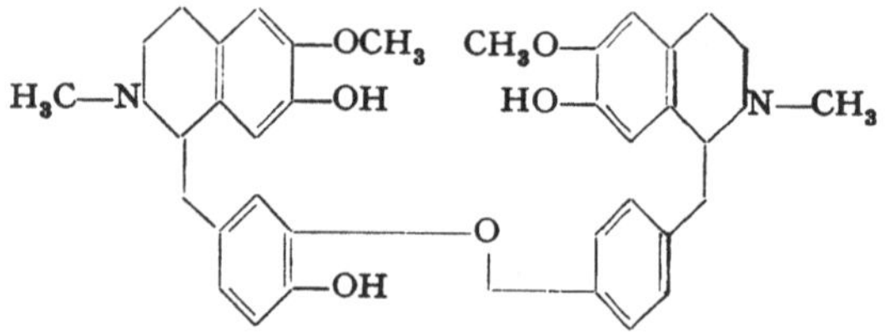

(XLII.) Magnolin.

Magnolamin, $C_{36}H_{40}O_7N_2$, Kristalle, Fp. 117—119°, $[\alpha]_D = +111,6°$ (in Alkohol).

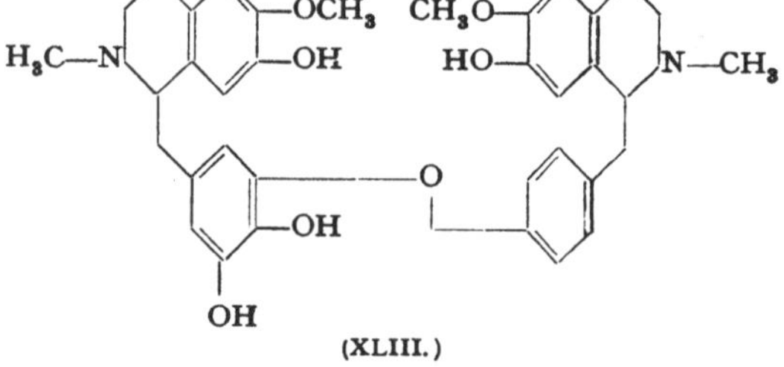

(XLIII.)

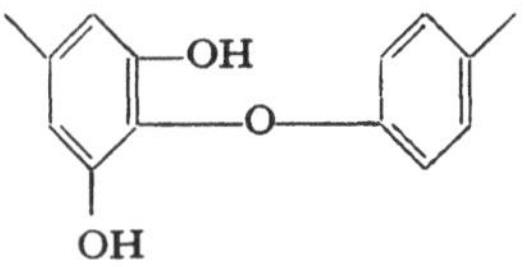

Die Alkaloide der Menispermaceae-Pflanzen.

(XLIV.)
(Sonst wie XLIII.)

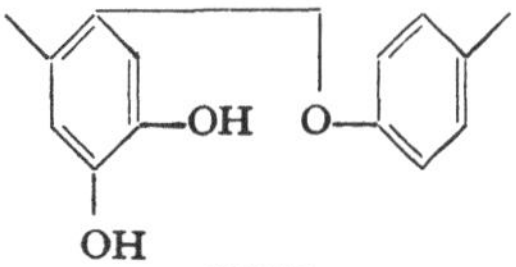

(XLV.)
(Sonst wie XLIII.)

PROSKOURNINA (*152, 153*) hat dieser Base auf Grund von Abbaureaktionen die Konstitutionen (XLIII) oder (XLIV) zugeschrieben. Demgegenüber haben TOMITA, FUJITA und NAKAMURA (*190, 194*) die wahrscheinlichere Formel (XLV) zur Diskussion gestellt. Die Anwesenheit von sieben O-Atomen ist auffallend in der Biscoclaurin-Klasse.

Aztequin, $C_{36}H_{40}O_7N_2$, Kristalle, Fp. 176°.

Das Aztequin wurde von PALLARES und GARZA (*150*) aus *Talauma mexicana* DON isoliert und gemäß dem Symbol (XLVI) als eine O_7-Base aufgefaßt.

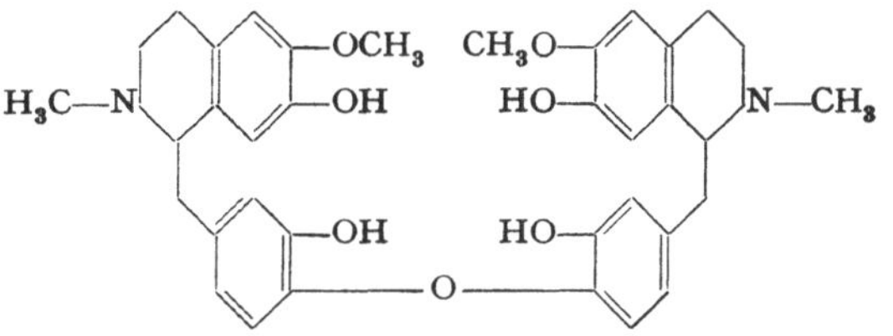

(XLVI.) Aztequin.

Gruppe II a.

Berbamin, $C_{37}H_{40}O_6N_2$. Tafeln, Fp. 156° (Zers.) (4 H_2O-haltig); Fp. 170—172° (wasserfrei); Säulen, Fp. 127° (Zers.) (Benzol-Addukt), $[\alpha]_D^{10} = = +103,1°$ (in Chloroform).

HESSE (*57*) erhielt dieses Alkaloid aus *Berberis vulgaris* L., zusammen mit Oxyacanthin. Seitdem wurde bekannt, daß Berbamin in der Berberis-Familie weit verbreitet ist. In Menispermaceae wurde es von KONDO und TOMITA (*109, 199*) in Stephania-Arten öfters vorgefunden. An seiner Konstitutionsaufklärung haben sich RUEDEL (*156*), POMMEREHNE (*151*), SANTOS (*158*) und schließlich BRUCHHAUSEN und Mitarbeiter (*10, 11*) beteiligt. Die letzteren Forscher haben im Verlaufe von schönen Abbaureaktionen gezeigt, daß die Berbamin-Struktur (XLVII) oder (XLVIII) sein muß. Jüngst haben TOMITA, INUBUSHI, FUJITA und MURAI (*59, 192, 193*) durch Aufspaltung mit Natrium in flüssigem Ammoniak die Richtigkeit der Formel (XLVII) endgültig bewiesen. Gleichzeitig konnte gezeigt werden, daß die beiden asymmetrischen Zentren gemäß den ——-Zeichen und ++-Zeichen in (XLVII) angeordnet sind.

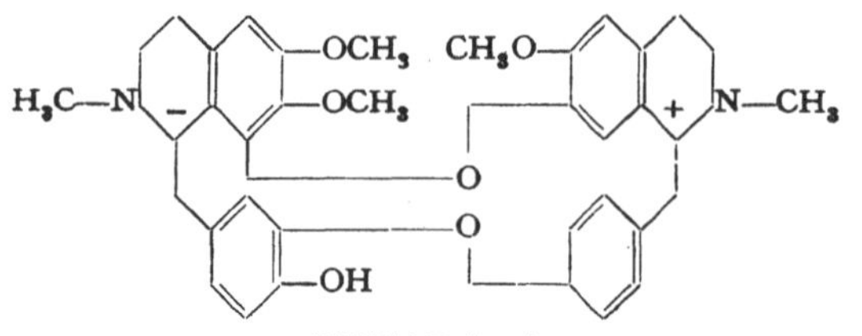

(XLVII.) Berbamin.

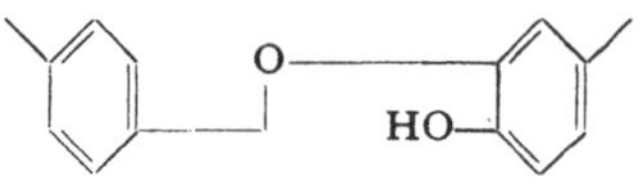

(XLVIII.)
(Sonst wie XLVII.)

Isotetrandrin, $C_{38}H_{42}O_6N_2$, Nadeln oder Würfel, Fp. 182°, $[\alpha]_D^{17} = +146°$ (in Chloroform).

Das Isotetrandrin wurde von KONDO und KEIMATSU (*72, 74*) aus *Stephania cepharantha* HAYATA als eine Nebenbase isoliert und mit O-Methyl-berbamin identifiziert. Nach unveröffentlichten Versuchen von TOMITA und ABE kommt es auch in *Mahonia japonica* DC. vor. Vor kurzem haben TOMITA, FUJITA und MURAI (*191, 193*) mit Hilfe der Natrium-Ammoniak-Methode die Konstitution (XLIX) festgelegt und die Lage der beiden asymmetrischen Zentren ermittelt (—, +).

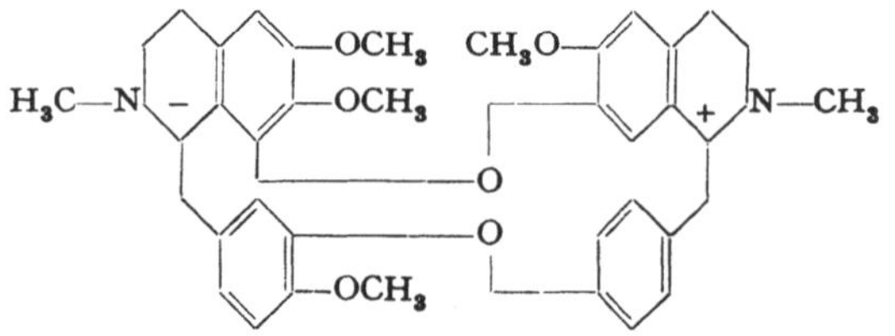

(XLIX.) Isotetrandrin.

Tetrandrin, $C_{38}H_{42}O_6N_2$, Nadeln, Fp. 217°, $[\alpha]_D^{24} = +263,1°$ (in Chloroform).

Diese Base wurde von KONDO und YANO (*116*) aus *Stephania tetrandra* S. MOORE und später von KONDO, NARITA und MURAKAMI (*87*) aus *Menispermum dauricum* DC. gewonnen. KONDO und YANO (*117, 118, 221*) sowie BRUCHHAUSEN, OBEREMBT und FELDHAUS (*11*) haben gezeigt, daß Tetrandrin eine dem Oxyacanthin nahestehende Biscoclaurin-Base darstellt und mit O-Methyl-berbamin eine gemeinsame Methinbase bildet; demzufolge ist Tetrandrin ein optisches Isomer des letzteren. TOMITA, FUJITA und MURAI (*34*) haben durch ihre Spaltungsmethode die Formel als (L) und die asymmetrischen Zentren als (+, +) festgelegt.

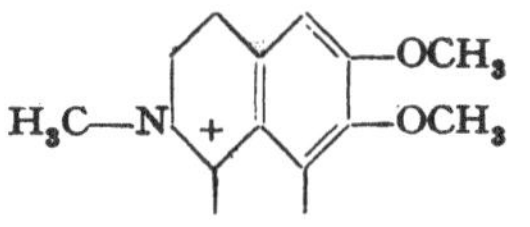

(L.) Tetrandrin.
(Sonst wie XLIX.)

Phaeanthin, $C_{38}H_{42}O_6N_2$, Säule, Fp. 210°, $[\alpha]_D^{30} = -278°$ (in Chloroform).

Das Phaeanthin wurde von SANTOS (*159, 160*) aus *Phaeanthus ebracteoratus* (PRESL) MERRILL isoliert. Eine provisorische Formel wurde von FALTIS, WRANN und KUEHAS (*29*) aufgestellt, doch wurde von KONDO und KEIMATSU (*71, 73, 74*) gezeigt, daß Phaeanthin der optische Antipode des Tetrandrins ist (asymmetrische Zentren: —, —).

Cepharanthin, $C_{37}H_{38}O_6N_2$, amorphes Pulver, $[\alpha]_D^6 = +351°$ (in Chloroform); Benzol-Addukt: Nadeln, Fp. 103°, $[\alpha]_D^6 = +300°$ (in Chloroform).

KONDO, KEIMATSU und YAMASHITA (*76*) haben aus *Stephania cepharantha* HAYATA dieses Alkaloid isoliert, das später von KONDO und TOMITA (*108, 109*) auch in *St. Sasakii* HAYATA aufgefunden wurde. Die letzteren Autoren haben das reine, kristallinische Benzol-Addukt dargestellt. Nach KONDO und KEIMATSU (*71, 74, 75*) muß die Konstitution den Formeln (LI) oder (LII) entsprechen, die durch die Anwesenheit einer Methylendioxy-Gruppe gekennzeichnet sind.

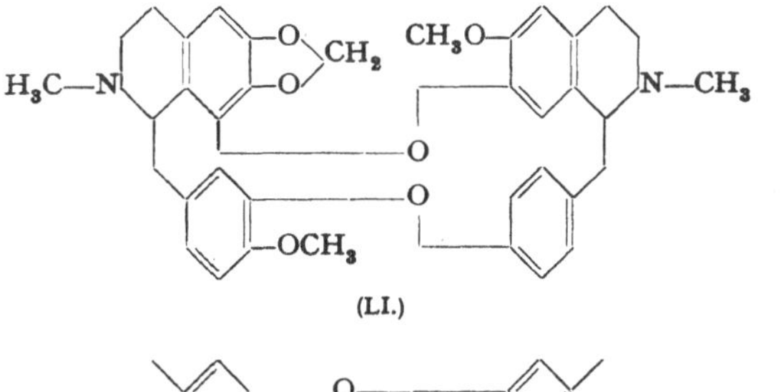

(LI.)

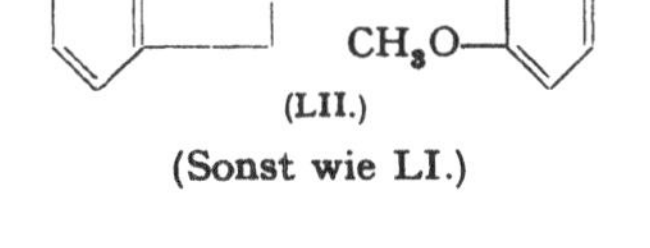

(LII.)
(Sonst wie LI.)

Oxyacanthin, $C_{37}H_{40}O_6N_2$, Nadeln, Fp. 216—217°, $[\alpha]_D^{20} = +279°$ (in Chloroform).

Das Oxyacanthin wurde bereits 1836 von POLEX aus der europäischen Berberitze („Sauerdorn") isoliert und benannt. Es wurde 1886 von

HESSE (*57*) in reinem Zustande gewonnen und von RUEDEL (*156*) sowie von POMMEREHNE (*151*) weiter untersucht. Dieses Studium wurde 1925 von SPÄTH und KOLBE (*177*), PIKL (*181*), GADAMER und BRUCHHAUSEN (*36*) sowie von BRUCHHAUSEN und SCHULTZE (*12*) wieder aufgenommen und schließlich von BRUCHHAUSEN und Mitarbeitern (*10, 11*) zu Ende geführt.

Oxyacanthin liegt, zusammen mit Berbamin, in verschiedenen Berberis-Arten vor; in Japan haben KONDO und TOMITA (*104*) sein Vorkommen auch in *Berberis Thunbergii* bestätigt. Kürzlich wurden die beiden Basen von CHATTERJEE und GUHA (*13, 14, 14a, 15, 15a*) aus indischen Mahonia-Arten isoliert. In Menispermaceae scheint jedoch das Oxyacanthin bisher nicht gefunden worden zu sein; in der älteren Literatur (*220*) finden sich Angaben über sein Vorkommen in *Menispermum canadense* L., während KONDO (*87*) und MANSKE (*125*) dies nicht erwähnen. Hingegen wurden der Oxyacanthin-Reihe zugehörigen Basen, z. B. Epistephanin, Hypoepistephanin und Trilobamin (Daphnolin) aus Menispermaceae isoliert. Inzwischen haben FALTIS und FRAUENDORFER (*23*) in ihren Isochondodendrin-Arbeiten über die Biosynthese der Biscoclaurin-Basen ihre Meinung geäußert. Diese Basen werden wahrscheinlich durch eine enzymatische Dehydrierung von zwei Molen Coclaurin aufgebaut, welche zwei verschiedene Wege befolgen kann und (nach Methylierung) entweder (LIII) oder (LIV) ergibt. So würde einerseits Oxyacanthin, andererseits Berbamin gebildet sein, was von BRUCHHAUSEN und Mitarbeitern (*10, 11*) experimentell bestätigt wurde.

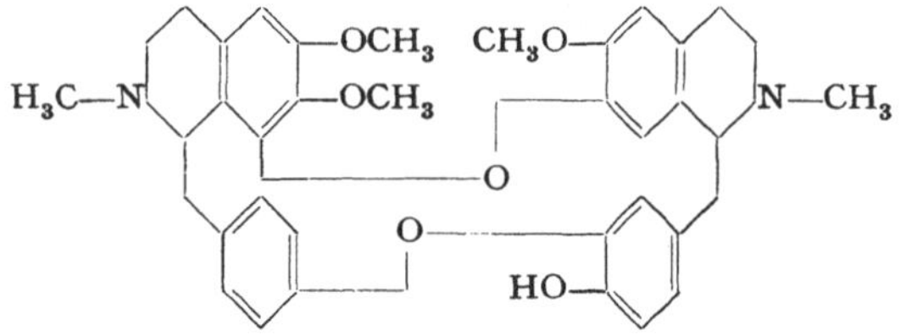

(LIII.)

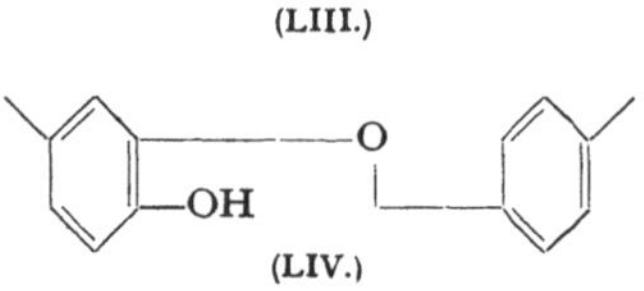

(LIV.)
(Sonst wie LIII.)

Jüngst haben TOMITA und Mitarbeiter (*191*) durch Spaltungsreaktionen am Isotetrandrin indirekt festgelegt, daß dem Oxyacanthin die Formel (LIII) zukommt. Demnach ist die Anordnung der beiden asymmetrischen Zentren im Oxyacanthin-Molekül (+, +). Indessen haben TOMITA und FUJITA (*32, 33*) bei der Ausführung derselben Reaktion mit O-Methyloxyacanthin die folgende merkwürdige Beobachtung gemacht: Obzwar der Abbau zu zwei rechtsdrehenden Phenolbasen (LVI) und (LVII) führt, liefert die letztere unerwarteterweise ein linksdrehendes Methyl-Derivat (LVIII).

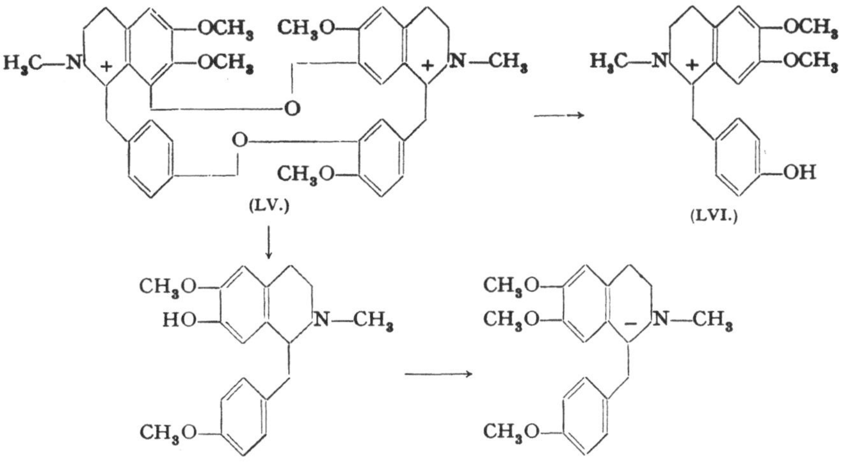

(LV.)

(LVI.)

(LVII.)

(LVIII.)

Diese auffallende Erscheinung wurde von TOMITA so interpretiert, daß das asymmetrische Zentrum des in der Formel rechts stehenden Isochinolin-Kerns bei der Abspaltung des O-Methyl-oxyacanthins (LV, s. oben) seine Konfiguration ändert, was an die WALDENsche Umkehrung von Oxyacanthin nach der weiter unten zitierten Arbeit von BRUCHHAUSEN (9) erinnert.

Repandin, $C_{37}H_{40}O_6N_2$, Kristalle, Fp. 255°, $[\alpha]_D^{15} = -106°$ (in Chloroform).

Das Repandin wurde von BICK und WHALLEY (5) aus *Daphnandra repandula* erhalten und von BICK und TODD (3) als ein optisches Isomere des Oxyacanthins identifiziert, das auch mit der, von BRUCHHAUSEN und SCHULTZE (12) durch Einwirkung von $1/_2$ Mol HCl auf das Oxyacanthin gewonnenen Base [Fp. 260°, $[\alpha]_D^{20} = -94,9°$ (in Methanol)], identisch ist. Nach BRUCHHAUSEN (9) liegt diese optische Form nicht in der Pflanze vor, sondern entsteht beim Aufarbeiten der Rohbase etwa durch WALDENsche Umkehrung.

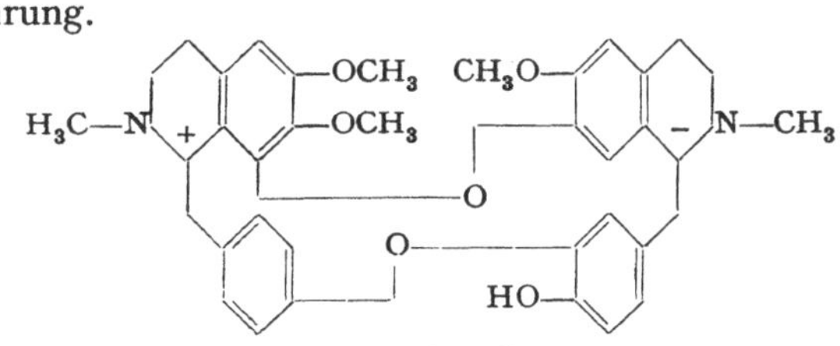

(LIX.) Repandin.

Wie oben angedeutet, kann die Anordnung der beiden asymmetrischen Zentren im Repandin (LIX) als (+, —) bezeichnet werden [vgl. TOMITA und FUJITA (*32, 33*)].

Daphnandrin, $C_{36}H_{38}O_6N_2$, Nadeln, Fp. 280° ($CHCl_3$-Addukt), $[\alpha]_D = +474°$ (in Chloroform).

Das Daphnandrin wurde von PYMAN (*155*) aus *Daphnandra micrantha* zusammen mit Daphnolin und Micranthin isoliert. Nach einer genaueren Untersuchung von BICK, EWEN und TODD (*2*) wurde dieser Base die Struktur (LX) zugeteilt ($R_1 = H$, $R_2 = CH_3$ oder $R_1 = CH_3$, $R_2 = H$).

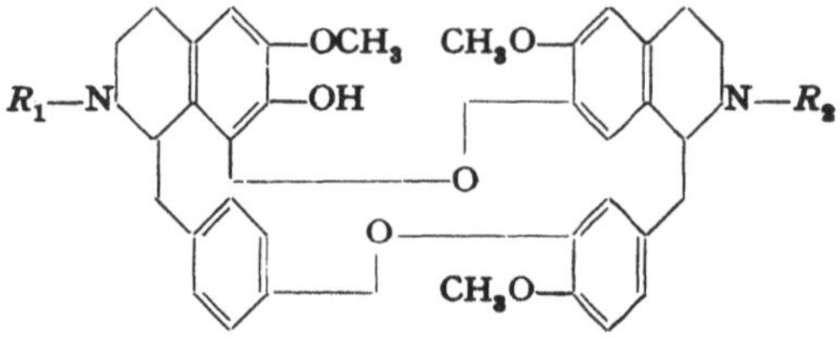

(LX.) Daphnandrin.

Daphnolin (Trilobamin), $C_{35}H_{36}O_6N_2$, Kristalle (Chloroform-Addukt), Fp. 194—196°, $[\alpha]_D = +495°$ (in Chloroform).

Vor kurzem wurde diese Base von BICK und WHALLEY (*7*) sowie von BICK, EWEN und TODD (*2*) eingehend untersucht. Nach ihnen wird eines der beiden Phenol-hydroxyle durch Diazomethan leicht methyliert, was Daphnandrin ergibt. Die Daphnolin-Struktur entspricht dem Symbol (LXI) ($R_1 = H$, $R_2 = CH_3$ oder $R_1 = CH_3$, $R_2 = H$).

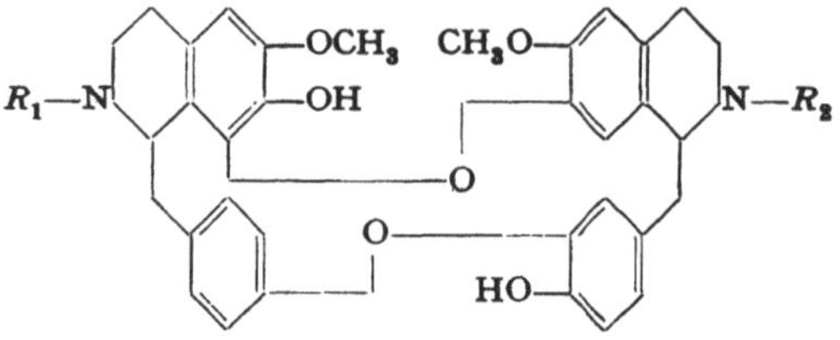

(LXI.) Daphnolin.

KONDO und TOMITA (*105, 187*) isolierten früher aus *Cocculus trilobus* DC. eine Base, Trilobamin genannt; $C_{36}H_{38}O_6N_2$, Fp. 195°, Zp. 215°, $[\alpha]_D^{15} = +356°$ (in Essigsäure). Nach BICK, EWEN und TODD (*2*) ist jedoch diese Verbindung mit Daphnolin identisch.

Aromolin, $C_{36}H_{38}O_6N_2$, Tafeln, Fp. 174—175°, $[\alpha]_D^{17} = +327°$.

Das Aromolin wurde von BICK und WHALLEY (*7*)] aus *Daphnandra aromatica* gewonnen und von BICK, EWEN und TODD (*2*) näher untersucht. Sie fanden diese Base der Oxyacanthin-Gruppe zugehörig und mit

N-Methyl-daphnolin identisch. Die Konstitution entspricht daher (LXI) ($R_1 = R_2 = CH_3$).

Epistephanin, $C_{37}H_{38}O_6N_2$, Säulen, Fp. 200—204°, $[\alpha]_D^{21} = +180,2°$ (in Chloroform).

Dieses Alkaloid wurde von KONDO und SANADA (*94—96*) aus *Stephania japonica* MIERS isoliert. KONDO und TANAKA (*99, 100, 182*) haben bewiesen, daß es der Oxyacanthin-Gruppe angehört und ein Stickstoffatom in der 2-Stellung des 3,4-Dihydro-isochinolinkernes enthält; das

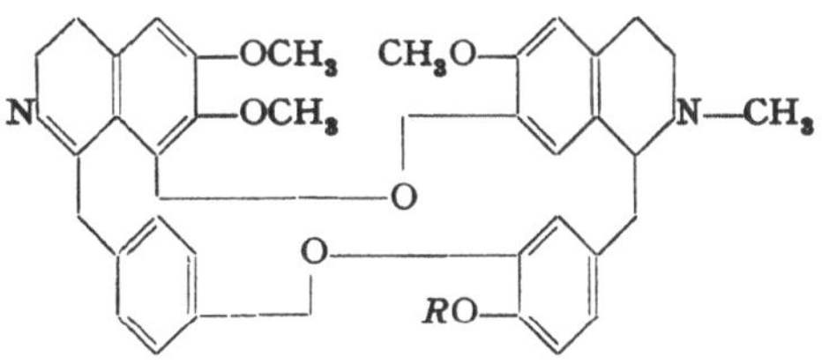

(LXII.) Epistephanin.

N-Methyl-hydroepistephanin-A ist nämlich mit O-Methyl-oxyacanthin identisch. TOMITA und Mitarbeiter (*191, 215*) haben auf Grund einer vergleichenden spektrographischen Untersuchung verschiedener Dihydroisochinoline dem Epistephanin die Formel (LXII; $R = CH_3$) zuerteilt.

Hypoepistephanin, $C_{36}H_{36}O_6N_2$, Kristalle, Fp. 257°, $[\alpha]_D^{30} = +186,6°$ (in Chloroform).

Diese Base wurde von KONDO und SANADA (*94, 95, 157*) aus *Stephania japonica* MIERS (neben Epistephanin) erhalten und „ψ-Epistephanin" genannt. KONDO und NOZOE (*89*) haben sodann auf Grund der Bildung von Epistephanin beim Methylieren sowie der Ermittlung der Stellung des phenolischen Hydroxyls die Formel(LXII; $R = H$) aufgestellt. Gleichzeitig wurde der Name zu Hypoepistephanin abgeändert.

Gruppe IIb.

Isochondodendrin, $C_{36}H_{38}O_6N_2$, Nadeln; Fp. 290° (FALTIS), Fp. 297° (Zers.) (SCHOLTZ), Fp. 316° (Zers.) (KING), $[\alpha]_D^{22} = +47,7°$ (in Pyridin), $[\alpha]_D^{22} = +120°$ (in N/10-HCl).

Diese Base, welche in der Pareira-Wurzel sowie in Curare-Arten enthalten ist, wurde von SCHOLTZ (*165—172*) bereits vor längerer Zeit unter dem Namen „Isochondrodendrin" oder „Isobebeerin" beschrieben. Im Jahre 1912 haben FALTIS und Mitarbeiter (*20, 22—30*) diese Untersuchung wieder aufgenommen und in jener relativ frühen Periode der Alkaloid-Chemie einen wichtigen Fortschritt auf dem Gebiete der Biscoclaurine erzielt. Seither wurde dieses Struktur-Problem von KING

(*62—66*) sowie von DUTCHER (*19*) im Sinne der Formel (LXIII; $R_1 = R_2 = H$) der Lösung zugeführt.

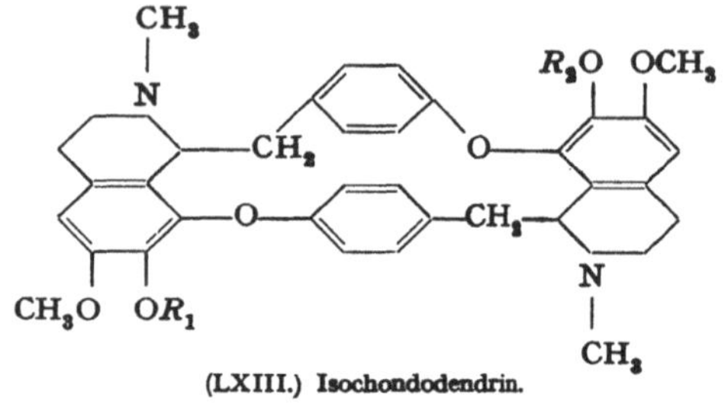

(LXIII.) Isochondodendrin.

Cycleanin (O,O-Dimethyl-isochondodendrin), $C_{38}H_{42}O_6N_2$, Säulen oder Nadeln, Fp. 273—274°, $[\alpha]_D^{20} = -30{,}14°$, $[\alpha]_D^{24} = -15{,}08°$ (in Chloroform).

Das Cycleanin wurde von KONDO, TOMITA und UYEO (*110, 111*) aus *Cyclea insularis* (MAKINO) DIELS und *Stephania cepharantha* HAYATA isoliert und als das O,O-Dimethyl-isochondodendrin (LXIII, $R_1 = R_2 = CH_3$) identifiziert. Später haben TOMITA und SHIRAI (*204*) dieselbe Base in *St. capitata* SPRENG vorgefunden, während von DUTCHER (*19*) deren rechtsdrehende Form aus *Chondrodendron tomentosum* erhalten wurde. Nach TOMITA und FUJITA (*35, 192*) kann das Cycleanin zu 2 Molen des Coclaurin-Typus gespalten werden, ebenso wie die Oxyacanthin-Berbamin-Basen. Das Ergebnis der quantitativen Spaltung ist: 2 Mole *L-1-(4'-Oxybenzyl)-6,7-dimethoxy-N-methyl-1,2,3,4-tetrahydroisochinolin* (LXIV). Demnach soll die Anordnung der beiden asymmetrischen Zentren im Cycleanin der linksdrehenden Form entsprechen.

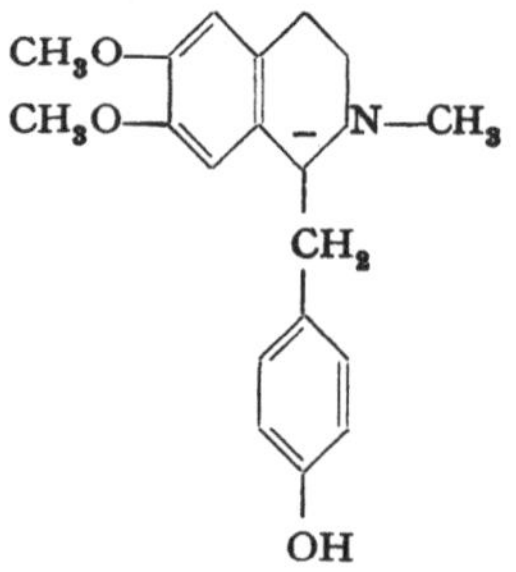

(LXIV.) *L-1-(4'-Oxybenzyl)-6,7-dimethoxy-N-methyl-1,2,3,4,-tetrahydroisochinolin.*

Protocuridin, $C_{36}H_{38}O_6N_2$, Kristalle (mit $^1/_2$ Mol Pyridin), Fp. 295°, $[\alpha]_{5461}^{20} = +7{,}6°$ (HCl-Salz in Wasser).

Das Protocuridin wurde von BOEHM (*8*) in Tubocurare entdeckt. Nach KING (*64, 66*) ist es ein Vertreter des Isochondodendrin-Typus.

Es enthält zwei phenolische Hydroxyle und zwei Methoxyl-Gruppen. Die Stellung eines Hydroxyls entspricht vielleicht derjenigen von OR_2 in (LXV), während die Stellung des anderen unklar ist. Dieses Alkaloid könnte stereoisomer mit Isochondodendrin sein.

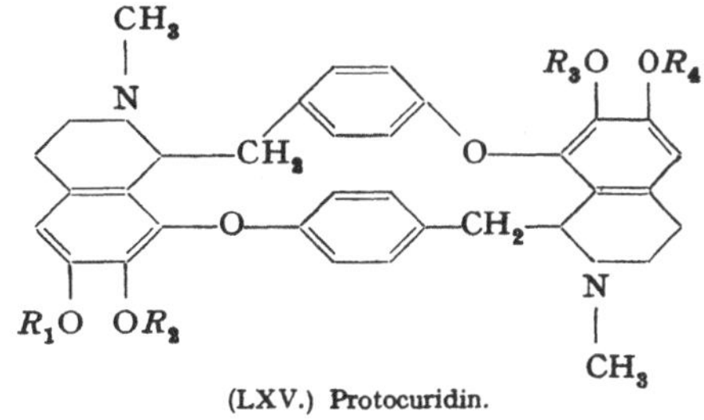

(LXV.) Protocuridin.

Neoprotocuridin, $C_{36}H_{38}O_6N_2$, Tafeln (mit 8 H_2O), Fp. 232° (Zers.); optisch inaktiv.

Diese Base wurde von KING (*64, 66*) neben Protocuridin isoliert und mit der Formel (LXV) belegt ($R_1 = R_4 = H$; $R_2 = R_3 = CH_3$ oder umgekehrt). Neoprotocuridin und Protocuridin sind nicht nur Stellungsisomere (in bezug auf OH und OCH_3), vielmehr müssen sie auch eine verschiedene Anordnung um die asymmetrischen Zentren besitzen.

Bebeerin. $C_{36}H_{38}O_6N_2$, Kristalle, Fp. 161° (Benzol-Addukt), Fp. 213°, Fp. 221° (racem. Form: Fp. 299—300°); $[\alpha]_D^{20} = +332°$ und $-328°$ (in Pyridin).

Die früher als „Bebeeria" (oder Chondrodendrin, Pelosin) bezeichnete Substanz ist ein aus der Pareira-Wurzel gewonnenes Rohbasen-Gemisch. Das *L*-Bebeerin wurde von SCHOLTZ (*166, 167*) aus käuflichem Bebeerinum purum „Merck" in kristallinischer Form isoliert; darauffolgend wurden aus der Pareira-Wurzel die *D*- und die *DL*-Formen gewonnen. Nach SPÄTH, LEITHE und LADECK (*179*) ist *L*-Bebeerin mit dem *L*-Curin (aus Tubocurare) identisch. KING (*66*) hat *D*-Bebeerin aus *Chondrodendron microphyllum* sowie aus *C. candicans* isoliert, während er aus *C. platyphyllum* die *L*-Form gewann. Zur Kenntnis der Bebeerin-Struktur haben

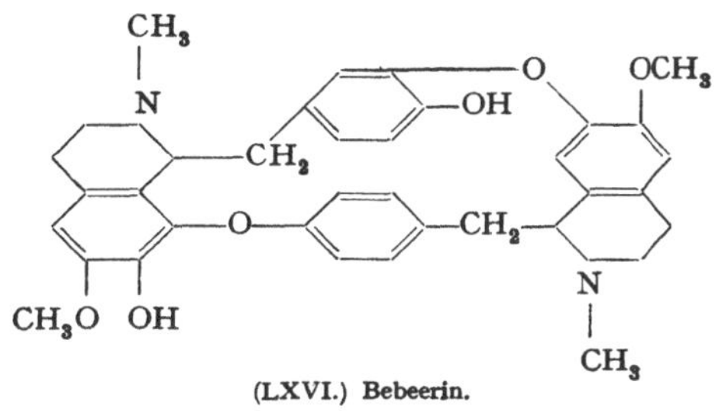

(LXVI.) Bebeerin.

Scholtz (*166*, *167*), dann Späth und Kuffner (*178*), Faltis, Kadiera und Doblhammer (*26*), Faltis, Holzinger, Ita und Schwarz (*25*) sowie King (*63—65,67*) beigetragen, so daß schließlich die Formel (LXVI) bewiesen werden konnte.

Chondrofolin, $C_{35}H_{36}O_6N_2$, Kristalle, Fp. 135° (2 Mole Kristallwasser), $[\alpha]_{5461}^{20} = -280,6°$ (in N/10-HCl).

Das Chondrofolin wurde von King (*66*) aus *Chondrodendron platyphyllum* (aus Bahia) isoliert und dafür eine Formel des Bebeerin-Typus aufgestellt (LXVII, $R_1 = H$; $R_2 = R_3 = R_4 = CH_3$, oder $R_4 = H$; $R_1 = R_2 = R_3 = CH_3$).

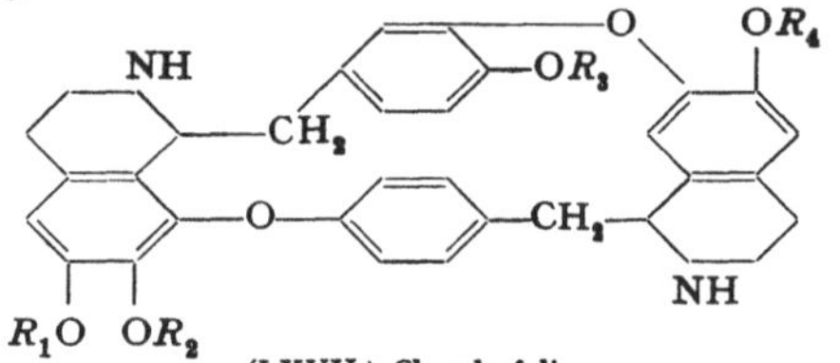

(LXVII.) Chondrofolin.

Tubocurarinchlorid, $C_{38}H_{44}O_6N_2Cl_2$, Tafeln, Fp. 274—275° (Zers.) (5 Mole Kristallwasser), $[\alpha]_{5461}^{20} = +264,8°$, $[\alpha]_D^{22} = +215°$ (in Wasser).

Die Untersuchung des Tubocurarins wurde schon frühzeitig von Boehm begonnen; das kristallinische Chlorid wurde von King isoliert. Dieser Forscher hat das Isomerie-Verhältnis zwischen Tubocurarinchlorid und Curinmethochlorid erkannt (*62—68*) und damit einen bedeutenden Fortschritt erzielt. In den Verbindungen: *D*-Tubocurarinchlorid und *D*-Bebeerin besitzen die Hydroxyl- und die Methoxyl-Gruppen identische Stellungen (LXVIII), doch sind die beiden Basen sterisch verschieden; die asymmetrischen Zentren sind: in der ersteren Verbindung *D* und *L*, dagegen in der letzteren *D* und *D*.

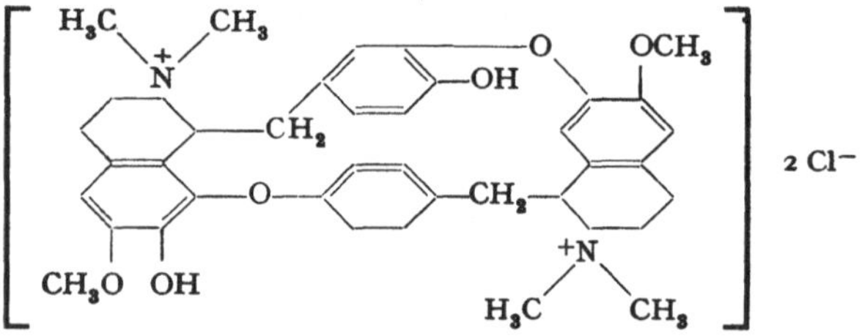

(LXVIII.)

Chondocurin, $C_{36}H_{38}O_6N_2$, Kristalle, Fp. 232—234°, $[\alpha]_D^{24} = +200°$ (in N/10-HCl), oder $+105°$ (in Pyridin).

Das Chondocurin wurde erst von Dutcher (*19*) in kleinen Mengen aus *Chondrodendron tomentosum* R. et P. (Curare aus dem oberen Amazonen-

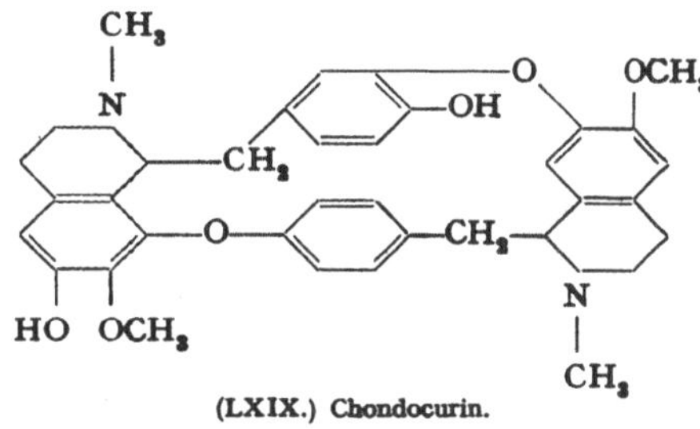

(LXIX.) Chondocurin.

Gebiet) zusammen mit D-Isochondodendrin, D-Isochondodendrin-dimethyläther, D-Tubocurarinchlorid und L-Curin isoliert. Nach DUTCHER sind D-Dimethyl-chondocurin-dijodmethylat und D-Dimethyl-tubocurarinjodid identisch. Der Autor nimmt an, daß dem Chondrocurin die Formel (LXIX) zukommt, wobei aber die Lage der Hydroxyl-Gruppen noch unbewiesen ist.

Gruppe III a.

Trilobin, $C_{36}H_{36}O_5N_2$, Nadeln oder Säulen, Fp. 235—238°, $[\alpha]_D^{20} = = +302,8°$ (in Chloroform).

KONDO und NAKAZATO (*82, 134*) haben aus *Cocculus trilobus* DC. diese Base erhalten, die von KONDO und TOMITA (*101*) auch aus *C. sarmentosus* DIELS isoliert wurde. Nach den letztgenannten Autoren (*102, 106, 184, 185*) stellt sie ein Diphenylendioxyd-Kern enthaltendes, eigenartiges Biscoclaurin-Alkaloid dar. Nach einer Diskussion über die biogenetische Stellung des Trilobins zwischen FALTIS (*21, 25*) einerseits und KONDO und TOMITA (*107*) anderseits wurde die partielle Synthese des Diphenylendioxyd-Kern enthaltenden Molekülteiles von TOMITA und TANI (*208, 209, 183*) durchgeführt und die von FALTIS (*25*) aufgestellte Hypothese bestätigt. Dem Trilobin kommt demnach die Formel (LXX) oder (LXXI) zu. Das Trilobin ist das erste Beispiel eines Naturstoffes, in dem ein Diphenylendioxyd-Kern enthalten ist.

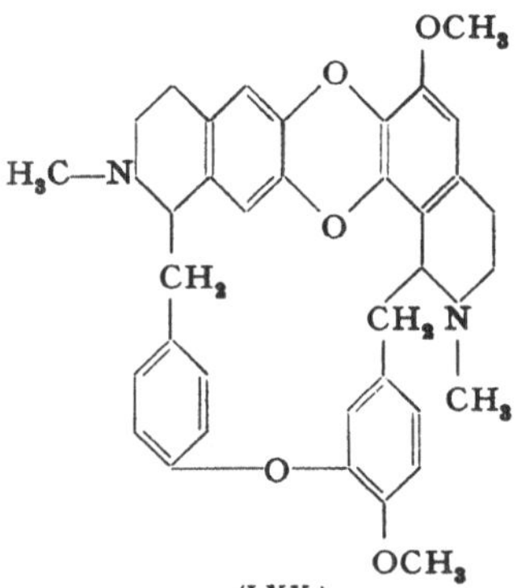

(LXX.) 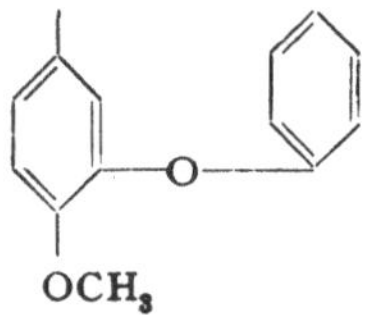   (LXXI.)
         (Sonst wie LXX.)

Isotrilobin, $C_{36}H_{36}O_5N_2$, Nadeln oder Säulen, Fp. 213—215°, $[\alpha]_D^8 = +314{,}8°$ (in Chloroform).

Das Isotrilobin wurde von KONDO und NAKAZATO (*133*) neben Trilobin isoliert und zunächst ,,Homotrilobin" genannt. KONDO und TOMITA haben später die Base als ein Isomeres des Trilobins erkannt und den Namen entsprechend abgeändert. Nach TOMITA und TANI (*208, 209, 183*) steht Isotrilobin zu Trilobin genau in der gleichen Beziehung, wie nach früheren Angaben, Oxyacanthin zu Berbamin; d. h. falls dem einen die Formel (LXX) zukommt, so muß das andere (LXXI) sein.

Menisarin, $C_{36}H_{34}O_6N_2$, Tafeln, Fp. 203° (Zers.), $[\alpha]_D^{12} = +149{,}4°$ (in Chloroform).

KONDO und TOMITA (*103*) haben dieses Alkaloid aus *Cocculus sarmentosus* DIELS in kleinen Mengen dargestellt. Nach ihnen (*186*) enthält das Menisarin-Molekül einen Diphenylendioxyd-Kern und ferner eine Methoxyl-Gruppe mehr als Trilobin oder Isotrilobin; ferner liegt ein 3,4-Dihydroisochinolin-Ring vor. TOMITA und TANI (*208*) haben, unter Berücksichtigung der möglichen Wege der Biosynthese, das Symbol (LXXII; $R = CH_3$) und (LXXIII; $R = CH_3$) für Menisarin zur Diskussion gestellt.

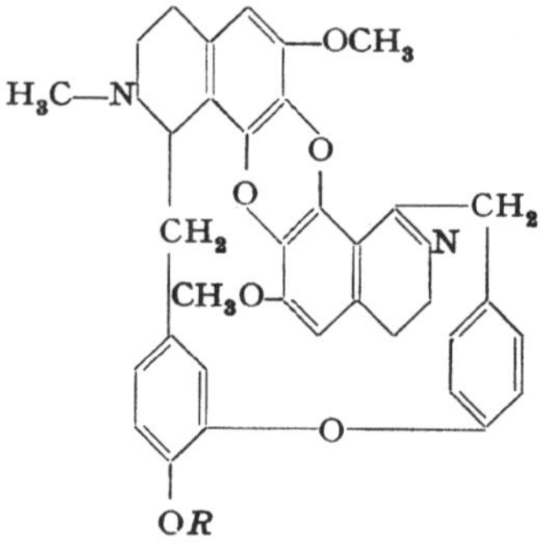

(LXXII.) 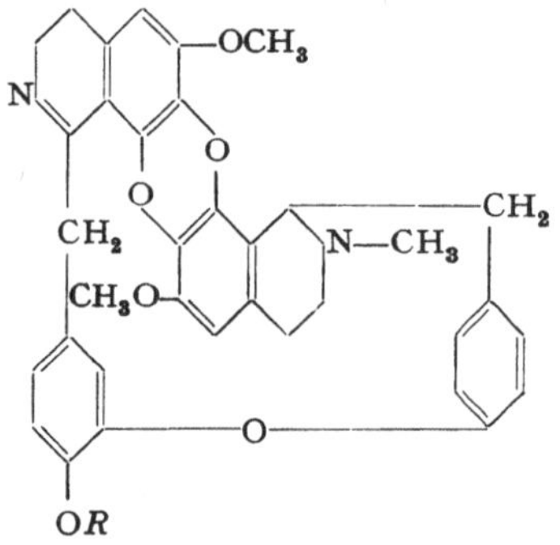 (LXXIII.)

Normenisarin, $C_{35}H_{32}O_6N_2$, körnige Kristalle, Fp. 223°, $[\alpha]_D^{18} = +190{,}3°$ (in Chloroform).

Das Normenisarin wurde von KONDO und TOMITA (*105, 188*) aus *Cocculus trilobus* DC. in kleiner Menge gewonnen. Da diese Base beim Methylieren in Menisarin übergeht, muß ein phenolisches Hydroxyl frei sein, und die Stelle desselben wird auf Grund biogenetischer Erwägungen, wie in den Formeln (LXXII; $R = H$) oder (LXXIII; $R = H$) verzeichnet, angenommen.

Micranthin, $C_{34}H_{32}O_6N_2$, Nadeln, Fp. 194—196°, $[\alpha]_D^{22} = -231°$ (in Chloroform).

Die Base wurde von PYMAN (*155*) aus *Daphnandra micrantha* isoliert und von BICK und TODD (*4*) genauer untersucht; es wurde ihr die dem Menisarin nahestehende Formel (LXXIV) zugeteilt ($R_1 =$ H und $R_2 =$ = CH_3 oder umgekehrt; $R_3 =$ H und $R_4 = CH_3$ oder umgekehrt).

Laut einer Privatmitteilung von Dr. BICK ist O,O-Dimethyl-N-methyl-micranthin-methylmethin mit N-Methyl-dihydro-menisarin-methylmethin identisch.

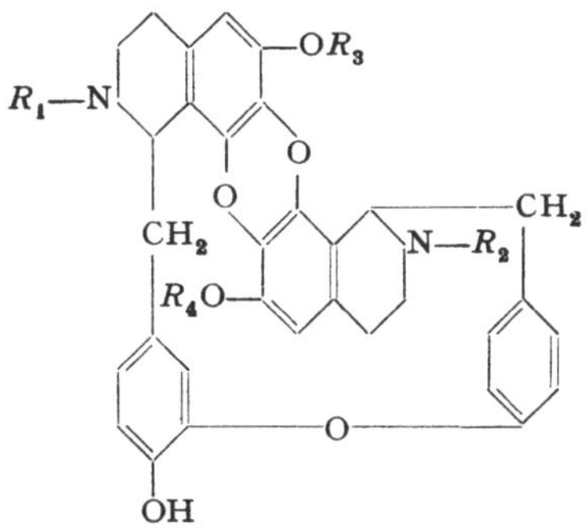

(LXXIV.) Micranthin.

Gruppe IIIb.

Insularin, $C_{37}H_{38}O_6N_2$, amorphes Pulver, $[\alpha]_D^7 = +27{,}95°$ (in Chloroform); Dijodmethylat: Nadeln, Zp. 293°.

Das Insularin wurde von KONDO und YANO (*115*) aus *Cyclea insularis* (MAKINO) DIELS isoliert; seine Konstitution wurde von TOMITA und UYEO (*111, 210—214*) abgeklärt. Es stellt eine Biscoclaurin-Base dar und ist durch das Vorhandensein einer eigentümlichen Äther-Brücke charakterisiert. Die Des-Base des HOFMANNschen Abbaus, wenn mit Ozon und dann mit Permanganat oxydiert, liefert die Insularinsäure $C_{17}H_{12}O_9$ (LXXV) sowie 2,3-Dimethoxy-diphenyläther-5,6,4′-tricarbonsäure (LXXVI). Der Insularinsäure-methylester, in Eisessig gelöst und mit PtO_2 katalytisch hydriert, wurde zu 3,4-Dioxy-5-methoxy-phthalsäure-dimethylester (LXXVII) und Hexahydro-*m*-tolylsäure-methylester (LXXVIII) gespalten. Durch Einwirkung von HBr-Eisessig auf den Insularinsäuremethylester und Verseifung entsteht die Säure (LXXIX), welche durch Pd-Kohle entbromt und mit Diazomethan methyliert werden kann. Verseift man dann den letzteren Methylester, so entsteht schließlich die 2,3-Dimethoxy-2′-methyl-diphenyläther-5,4′-dicarbonsäure (LXXX).

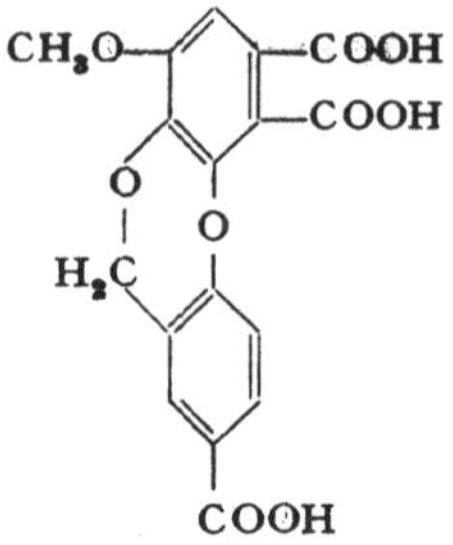

(LXXV.) Insularinsäure.

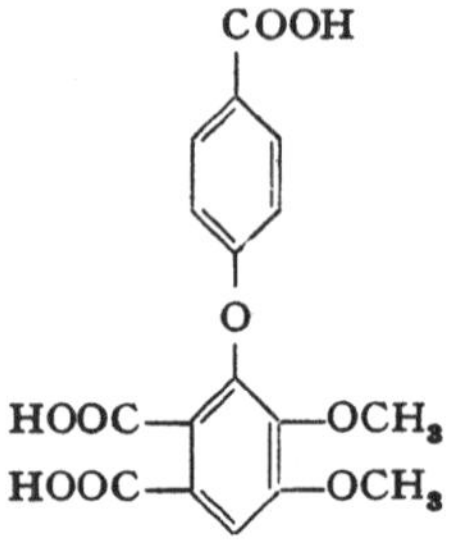

(LXXVI.) 2,3-Dimethoxy-diphenyläther-5,6,4'-tricarbonsäure.

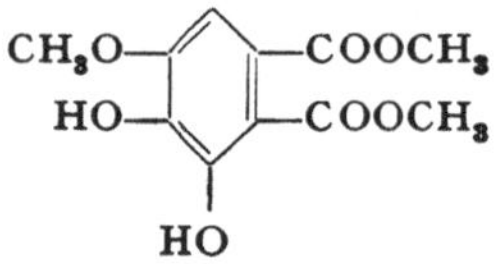

(LXXV)-Methylester →

(LXXVII.)

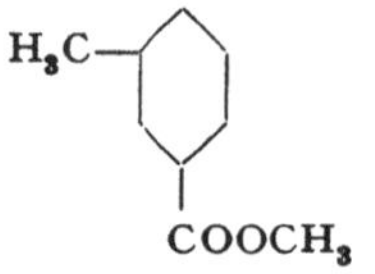

(LXXVIII.)

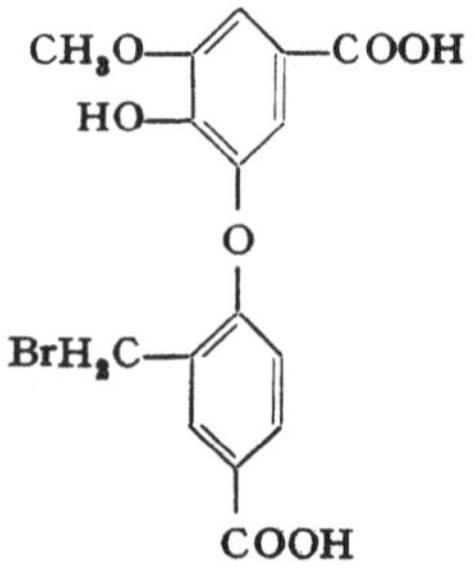

(LXXIX.) 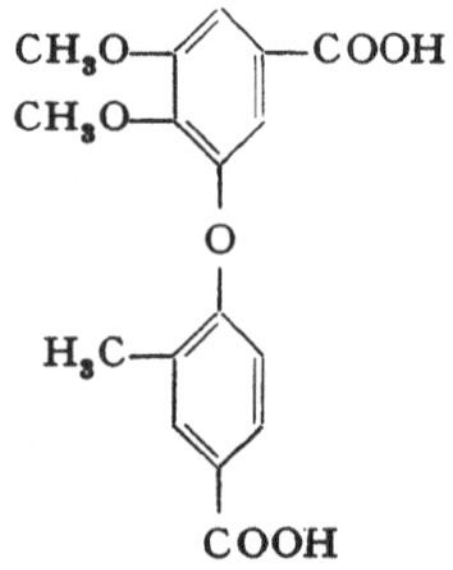 → (LXXX.) 2,3-Dimethoxy-2'-methyl-diphenyläther-5,4'-dicarbonsäure.

Auf Grund dieser Reaktionen wurde der Insularinsäure die Formel (LXXV) und dem Insularin selbst (LXXXI) zuerteilt.

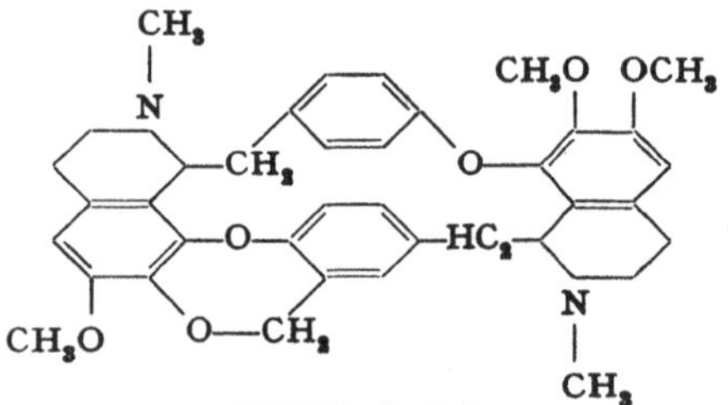

(LXXXI.) Insularin.

6. Strukturell ungeklärte Basen.

Protostephanin (*81, 95, 97, 112—114*), $C_{21}H_{25(27)}O_6N = C_{16}H_{11(13)}(OCH_3)_4(NCH_3)$, Kristalle, Fp. 75° (2 Mole Kristallmethanol), $[\alpha]_D = +3.44°$. — *Metaphanin* (*93*), $C_{18}H_{19}O_3N = C_{15}H_{10}(OH)(OCH_3)_2(NCH_3)$, Kristalle, Fp. 229°; optisch inaktiv. — *Stephanolin* (*94*), $C_{27}H_{29}O_3N_2(OH)(OCH_3)_4$, Nadeln, Fp. 186°, $[\alpha]_D = -255.4°$ (in Chloroform). — *Homostephanolin* (*94*), $C_{32}H_{44}O_7N_2$, Kristalle, Fp. 232°, $[\alpha]^D = -255.6°$ (in Chloroform). — „*VIII-Base*" (*94*), $C_{31}H_{26}O_5N_2$, Kristalle, Fp. 102—103°, $[\alpha]_D^{27} = -83.33°$ (in Chloroform). — *Hasubanonin* (*98*), $C_{17}H_{19}N(OCH_3)_3(OH)(CO)$, Kristalle, Fp. 116°, $[\alpha]_D^{25} = -219.5°$ (in Chloroform). Struktur wahrscheinlich Sinomenin-artig. Die oben erwähnten sechs Basen wurden von KONDO und Mitarbeitern aus *Stephania japonica* MIERS isoliert.

Repandulin (*5, 6*), $C_{35}H_{32}O_4(OCH_3)_2(CH_2O_2)(NCH_3)_2$, gelbe Kristalle, Zp. 160°, $[\alpha]_D^{25} = +443°$ (in Chloroform). Repandulin wurde von BICK aus *Daphnandra repandula* und *D. Dielsii* nebst anderen Basen erhalten. — *Acutumin* (*49*), $C_{15(16)}H_{15}O_4 \cdot (OCH_3)_2(CO)(NCH_3)$, Nadeln, Fp. 240°, $[\alpha]_D = +60.2°$; von GOTO und SUDZUKI als eine Nebenbase aus *Sinomenium acutum* isoliert. — *Diversin* (*80, 92*), $C_{20}H_{27}O_5N$, amorph, Fp. etwa 80—93°, $[\alpha]_D^{17} = +7°$ (in Chloroform). Aus den oben genannten Pflanzen wurde es von KONDO und Mitarbeitern als eine Nebenbase des Sinomenins gewonnen. — *Tiliacorin* (*61*), $C_{30}H_{27}O_2N(OCH_3)_2$, Kristalle, Fp. 260—261°, $[\alpha]_D = = +105.3°$; von VAN ITALLIE aus *Tiliacora racemosa* COLEBR. isoliert.

Ambalin (*161*), $C_{21}H_{26}O_4(OCH_3)_2(CH_2O_2)(CO)(NCH_3)_2$, Kristalle, Fp. 123°, $[\alpha]_D^{26} = +143.2°$ (in Chloroform); wurde von SANTOS und QUIBILAN aus *Pycnarrhena manillensis* VIDAL isoliert, zusammen mit *Ambalinin* (*162*), $C_{15}H_{12}O(OCH_3)_2(NCH_3)$; Kristalle, Fp. 203—204°. — *Fangchinolin* (*17*), $C_{37}H_{40}O_6N_2$, Kristalle, Fp. 237 bis 238°, $[\alpha]_D^{18} = +255.1°$ (in Chloroform); wurde von CHUANG, HSING, KAO und CHANG aus chinesischen Han-fang-chi nebst dem Tetrandrin gewonnen. Es soll ein Demethyl-tetrandrin sein und phenolische OH-Gruppen an den Benzolringen der beiden Isochinolinkerne enthalten. — *D-Tomentocurin* (*69*), Kristalle, Fp. 265° (Zers.), $[\alpha]_D^{17} = +210°$ (in N/10-HCl). Von KING in kleiner Menge aus *Chondrodendron tomentosum* (aus Peru) isoliert. — *Gindarin*, $C_{21}H_{25}O_4N$, Kristalle, Fp. 147°, *Gindaricin*, $C_{18}H_{19}O_3N$, Kristalle, Fp. 193°, und *Gindarinin*, als Nitrat $C_{17}H_9N(OCH_3)_4 \cdot HNO_3$, Kristalle, Fp. 248° (Zers.) wurden kürzlich von CHAUDHRY und SIDDIQUI (*16*) aus *Stephania glabra* (Indien) isoliert.

Cyclein (aus *Cyclea peltata*), *Sangolin* (aus *Tinospora Bakis* MIERS), *Menispermin* und *Paramenispermin* (aus *Anamirta paniculata*) wurden bisher nur mangelhaft charakterisiert.

7. Optische Isomerie der Biscoclaurin-Basen.

Über die optischen Isomerien in der Oxyacanthin-Berbamin-Reihe wurde bereits früher von BRUCHHAUSEN, OBEREMBT und FELDHAUS (*11*) berichtet. Nach diesen Autoren ist das Problem bequem lösbar, wenn man die Drehungsbeiträge der beiden Asymmetrie-Zentren (*A* und *B*) im genannten Molekül $A = \pm 70°$ und $B = \pm 200°$ gleichsetzt.

In der Berbamin-Reihe (LXXXII), z. B. beim Tetrandrin, Isotetrandrin (O-Methyl-berbamin) und Phaeanthin erkennt man sogleich, daß das Tetrandrin ($[\alpha]_D = +263°$) dem Ausdruck ($+A+B$), das Iso-

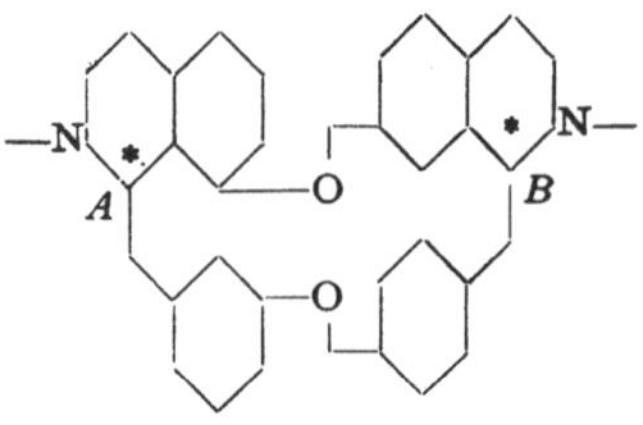

(LXXXII.)

tetrandrin ($[\alpha]_D = +146°$) ($-A + B$) und das Phaeanthin ($[\alpha]_D = -278°$) ($-A - B$) entspricht, und demnach eine vierte Base mit ($+A - B$) noch zu entdecken bleibt.

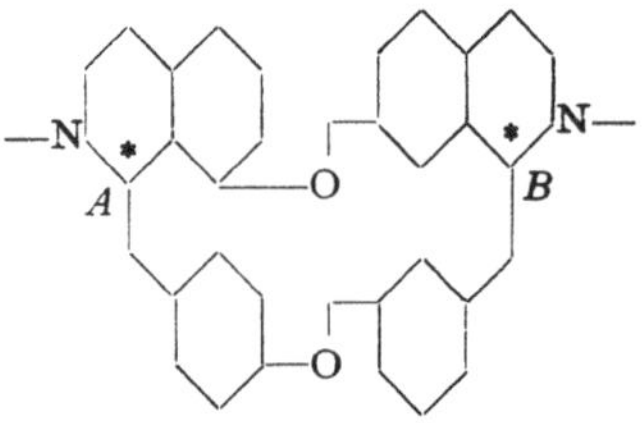

(LXXXIII.)

Betreffend die Oxyacanthin-Reihe (LXXXIII), [strukturisomer mit (LXXXII)] ist bekannt, daß das O-Methyl-oxyacanthin (N-Methyl-dihydro-epistephanin-A) ($[\alpha]_D = +279°$) der Summe ($+A + B$) und das Repandin ($[\alpha]_D = -106°$) dem Ausdruck ($+A - B$) entspricht. Nachdem das durch asymmetrische Hydrierung des Epistephanins gebildete Dihydro-epistephanin-B ($[\alpha]_D = +92{,}0°$, HCl-Salz in Wasser) dem Ausdruck ($-A + B$) entspricht, so bleibt in diesem Falle wohl noch eine derzeit unbekannte Base mit ($-A - B$) aufzufinden.

Diese experimentelle Schlußfolgerung von BRUCHHAUSENs Hypothese wurde durch die erwähnte Spaltungsreaktion von TOMITA und Mitarbeitern (*32—35, 59, 191—193, 196, 197*) weiter bestätigt. Hierbei wurde die Tatsache vermerkt, daß alle auf jenem Wege bereiteten Coclaurine ein zwischen $\pm 80°$ und $100°$ liegendes spez. Drehvermögen zeigen. Demnach ist die spez. Drehung einer Biscoclaurin-Base weit höher als die Summe der durch Spaltung mit Natrium und Ammoniak erhaltenen beiden Benzyl-isochinoline (z. B. Tetrandrin: $+263°$; Spaltbasen: $+84°$ und $+88{,}5°$, in Chloroform). Früher haben TOMITA und UYEO (*214*) beim Abbau des Insularins die auffallende Tatsache bemerkt, daß, trotz Abwesenheit asymmetrischen Kohlenstoffs, ein Abbauprodukt sich als optisch aktiv erwies. Dies wurde als ein durch molekulare Asymmetrie verursachtes Phänomen erklärt. Somit könnte das genannte Exaltations-Problem auf Grund der Annahme von Asymmetrie seine Lösung finden, obzwar dafür noch kein passendes Beispiel bekannt ist.

8. Charakterisierung der Biscoclaurin-Basen.

Die Naturstoffe, welche in ihren Molekülen Äther-Brücken enthalten, sind sehr weit verbreitet. In basischen Substanzen ist die Methoxyl-Gruppe häufig. Wie u. a. in der Klasse des Morphins, verbindet die Äther-Brücke oft Benzol- und Hydrobenzol-Ringe. Diphenyläther-Brücken sind außerhalb der Biscoclaurin-Basen weniger zahlreich. Das Cularin (LXXXIV), kürzlich von MANSKE (126, 127) untersucht, kann als ein Diphenyläther enthaltendes Coclaurin aufgefaßt werden. Als ein besonderer Fall erscheint Thyroxin, das animalischen Quellen entstammt; bekanntlich ist es eine Aminosäure mit jodierter Diphenyläther-Struktur. Als stickstoff-freie Pflanzenstoffe, welche die Diphenyläther-Gruppe enthalten, sind die Flechtenstoffe zu nennen; sie wurden u. a. von ASAHINA und Mitarbeitern (1) weitgehend untersucht.

(LXXXIV.) Cularin.

In den Biscoclaurin-Basen liegt eine besondere Abart der Äther-Brücken vor, nämlich die Diphenylendioxyd-Gruppierung (LXXXV) oder Depsidan-Gruppierung (LXXXVI), so in Trilobin, Menisarin und Insularin. Der Depsidan-Kern im Insularin-Molekül entspricht dem Depsidon-Kern (LXXXVII) in Flechtenstoffen (195). Das Vorkommen von Diphenyläther, Diphenylenoxyd und Depsidon in Flechtenstoffen einerseits und von Diphenyläther, Diphenylendioxyd und Depsidan in Biscoclaurin-Basen anderseits zeigt eine erstaunliche Harmonie in der Architektur dieser Produkte der Phytosynthese.

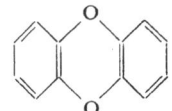 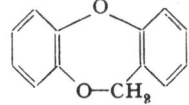 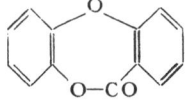

(LXXXV.) Diphenylendioxyd.   (LXXXVI.) Depsidan.   (LXXXVII.) Depsidon.

VI. Biogenetische Betrachtungen über Biscoclaurin-Basen.

Wie schon lange von FALTIS, KING, BRUCHHAUSEN, KONDO, TOMITA angenommen wurde, ist es jetzt klar, daß die Biosynthese der Biscoclaurin-

Schema 1.

Biosynthese des Magnolins, Daphnolins (Trilobamins), Dauricins, Oxyacanthins, Berbamins, Tetrandrins, Isotetrandrins und Phaeanthins:

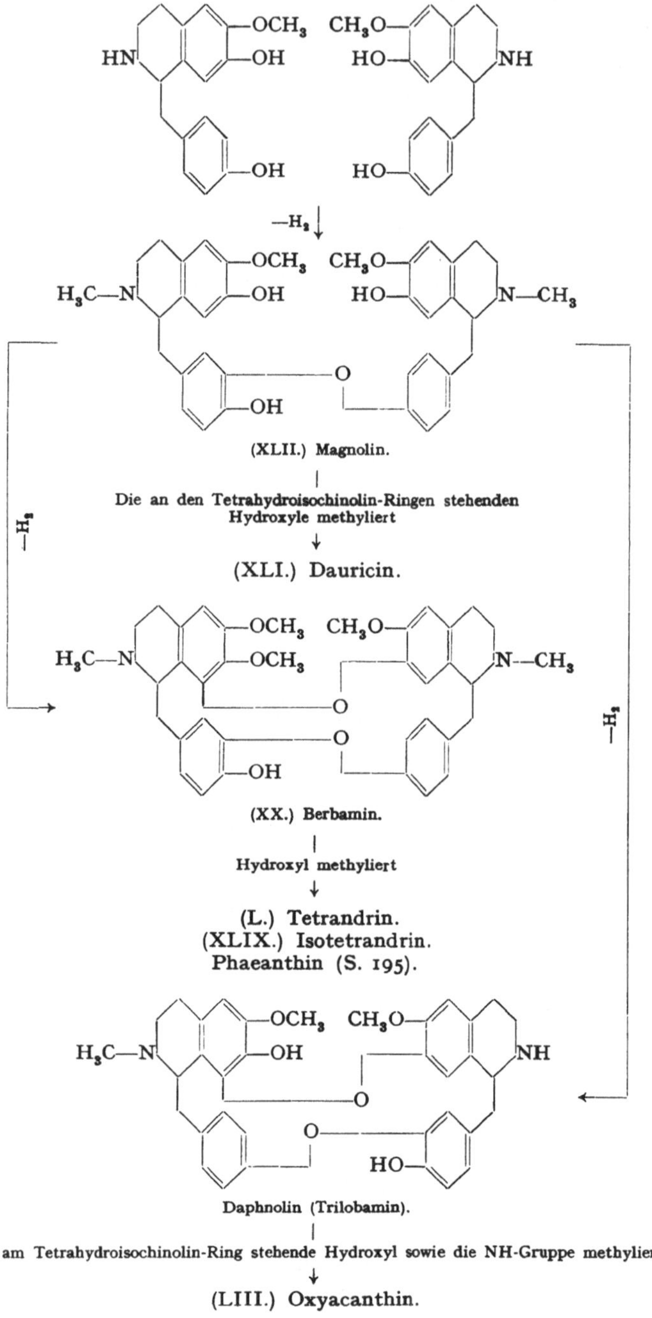

(XLII.) Magnolin.
|
Die an den Tetrahydroisochinolin-Ringen stehenden Hydroxyle methyliert
↓
(XLI.) Dauricin.

(XX.) Berbamin.
|
Hydroxyl methyliert
↓
(L.) Tetrandrin.
(XLIX.) Isotetrandrin.
Phaeanthin (S. 195).

Daphnolin (Trilobamin).
|
Das am Tetrahydroisochinolin-Ring stehende Hydroxyl sowie die NH-Gruppe methyliert
↓
(LIII.) Oxyacanthin.

Schema 2.

Biosynthese des Trilobins, Isotrilobins, Cepharanthins, Menisarins und Micranthins.

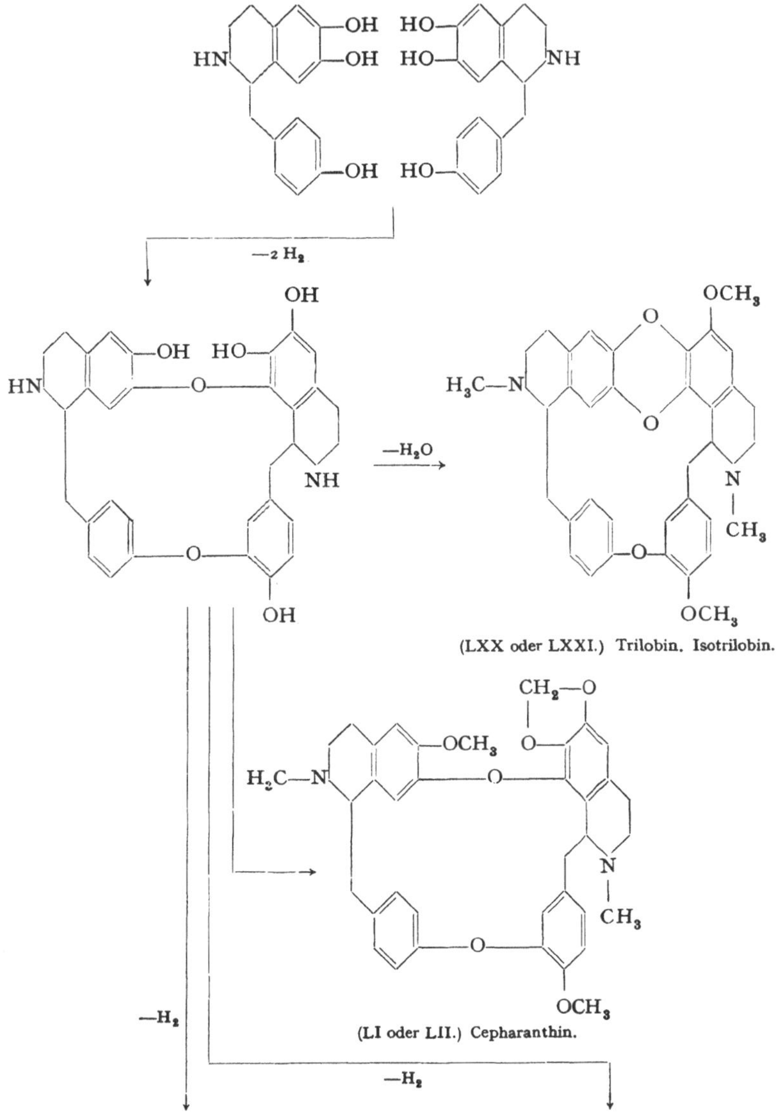

(LXX oder LXXI.) Trilobin. Isotrilobin.

(LI oder LII.) Cepharanthin.

(LXXII oder LXXIII.) Menisarin.
(Formeln auf S. 204.)

(LXXIV.) Micranthin.
(Formel auf S. 205.)

Basen leicht vonstatten gehen könnte, wenn zwei Mole des Coclaurins (oder einer ähnlichen Mutterbase) durch Dehydrierung verknüpft und dann

Schema 3.
Biosynthese des Isochondodendrins, Cycleanins, Bebeerins und Insularins.

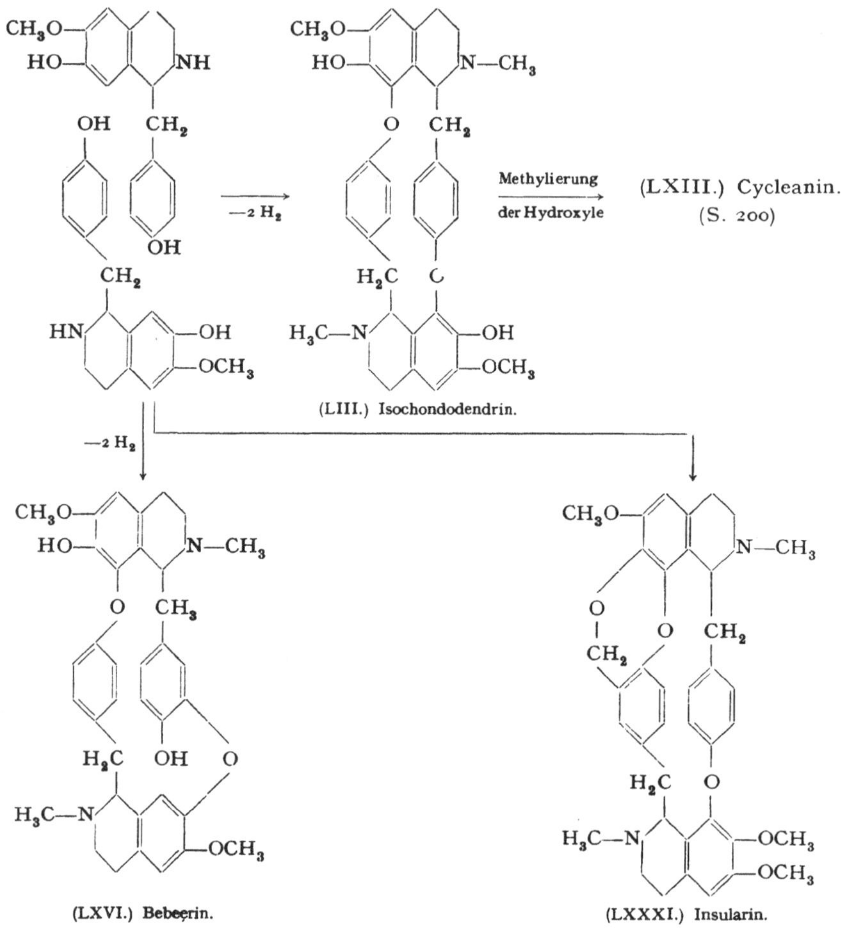

(LIII.) Isochondodendrin.

$\xrightarrow{\text{Methylierung der Hydroxyle}}$ (LXIII.) Cycleanin. (S. 200)

(LXVI.) Bebeerin.

(LXXXI.) Insularin.

methyliert würden (Schemata 1 und 3). Dies wurde durch frühere Teilsynthesen und neuere Abspaltungsreaktionen von TOMITA bewiesen. Für die Vertreter der Gruppe IIIa könnte man natürlich ebenso annehmen, daß z. B. das Trilobin durch Dehydrierung und nachfolgende Dehydratisierung aus zwei Molen Norcoclaurin in der Pflanze aufgebaut wird (Schema 2, S. 211; in den Schemata blieben die jeweiligen Asymmetrie-Zentren unberücksichtigt).

Die Biosynthese von Dihydroisochinolinen, wie z. B. Epistephanin, könnte einen ähnlichen Verlauf nehmen (Schema 4, S. 213), ausgehend von *p*-Oxy-homobenzoyl-dioxy-phenyläthylamin und Nor-coclaurin. Eine

Methylierung würde dann zu Epistephanin, eine Hydrierung und Methylierung zu Oxyacanthin führen. Auch die Biosynthese von Menisarin würde einen ähnlichen Weg befolgen. Außerdem könnten das Sinomenin (Morphin-Typus), ferner das Crebanin sowie Stephanin (Apomorphin-Typus) und auch die meisten Berberine aus Coclaurin als Mutter-Base in ähnlicher Weise aufgebaut werden, obgleich in seltenen Fällen, wie beim Salicifolin, eine Proto-Base dem Coclaurin vorangehen könnte.

Schema 4.
Biosynthese des Oxyacanthins und Epistephanins.

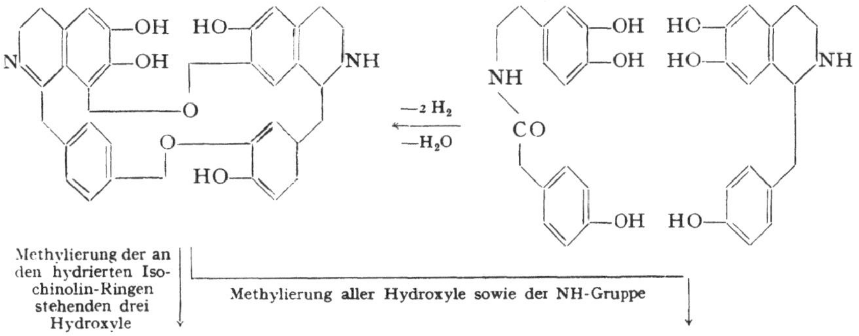

(LIII.) Oxyacanthin.   (LXII.) Epistephanin.

VII. Medizinische Anwendungen.

Bereits vor längerer Zeit haben die Columbo-Wurzel sowie die Pareira-Wurzel in Europa, ferner *Sinomenium acutum* R. et W., *Cocculus trilobus* DC., *Stephania japonica* MIERS, *C. laurifolius* DC. u. a. in Japan, ferner Hanfang-chi und Mu-fang-chi in China als Volksheilmittel mannigfache Anwendungen gefunden. Als erstes chemisches Individuum wurde das Sinomenin (60) gegen Neuralgie und Rheumatismus angewandt.

Einschlägige pharmakologische Untersuchungen wurden von TSURUTA (216, 217), HORIUCHI (58), OKADA (142—149) und MATSUOKA (132) veröffentlicht. Die Biscoclaurin-Basen wurden von HASEGAWA, sowie Mitarbeiter (52, 53) meistenteils in vitro betreffs ihre, auf die Entwicklung der Tuberkel-Bazillen ausgeübte, hemmende Wirkung untersucht. Auch im Verlaufe von in vivo Versuchen hat es sich gezeigt, daß das Cepharanthin eine besonders stark positive Wirkung ausübt. Ferner wurde Cepharanthin als ein Hemmungsmittel im Kampfe gegen Lepra erkannt (55). Kürzlich hat dieselbe Base eine neue Anwendung gefunden, nämlich gegen den Keuchhusten (54), da sie die Vermehrung des Antikörpers begünstigt (173). Das von KING untersuchte Tubocurarinchlorid wurde vor kurzem als ein Mittel für die Curare-Wirkung in die Chirurgie eingeführt.

Die Ammoniumhaloide der Biscoclaurin-Basen besitzen fast alle mehr oder weniger dieselbe Wirkung (*18, 128—130*).

Literaturverzeichnis.

1. ASAHINA, Y.: Flechtenstoffe. Fortschr. Chem. organ. Naturstoffe **2**, 27 (1939); Neuere Entwicklungen auf dem Gebiete der Flechtenstoffe, ebenda **8**, 207 (1951).
2. BICK, I. R. C., E. S. EWEN and A. R. TODD: Alkaloids of *Daphnandra* Species. II. Daphnandrine, Daphnoline and Aromoline. J. chem. Soc. (London) **1949**, 2767.
3. BICK, I. R. C. and A. R. TODD: Alkaloids of *Daphnandra* Species. I. Repandine. J. chem. Soc. (London) **1948**, 2170.
4. — — Alkaloids of *Daphnandra* Species. III. Micranthine. J. chem. Soc. (London) **1950**, 1606.
5. BICK, I. R. C. and T. G. WHALLEY: Alkaloids of Queensland Flora. I. Alkaloids of *Daphnandra repandula*. Univ. Queensland Papers, Dept. Chem. **1**, No. 28, 8. (1946) [Chem. Abstr. **41**, 1390 (1947)].
6. — — Alkaloids of Queensland Flora. II. Alkaloids of *Daphnandra dielsii*. Univ. Queensland Papers, Dept. Chem. **1**, No. 30, 4 (1947) [Chem. Abstr. **42**, 3909 (1948)].
7. — — Alkaloids of Queensland Flora. III. Alkaloids of *Daphnandra aromatica*. Univ. Queensland Papers, Dept. Chem. **1**, No. 33, 7 (1948) [Chem. Abstr. **43**, 6787 (1949)].
8. BOEHM, R.: Über Curare und Curarealkaloide. Arch. Pharmaz. Ber. dtsch. pharmaz. Ges. **235**, 660 (1897).
9. BRUCHHAUSEN, F. v.: Ist Repandin Begleiter des Oxyacanthins in der Berberitze? Arch. Pharmaz. Ber. dtsch. pharm. Ges. **283**, 44 (1950).
10. BRUCHHAUSEN, F. v. und P. H. GERICKE: Die Konstitution des Oxyacanthins. II. Arch. Pharmaz. Ber. dtsch. pharm. Ges. **269**, 115 (1931).
11. BRUCHHAUSEN, F. v., H. OBEREMBT und A. FELDHAUS: Über das Oxyacanthin und Berbamin. III. Liebigs Ann. Chem. **507**, 144 (1933).
12. BRUCHHAUSEN, F. v. und H. SCHULTZE: Zur Konstitution des Oxyacanthins. Arch. Pharmaz. Ber. dtsch. pharm. Ges. **267**, 617 (1929).
13. CHATTERJEE, R. and M. P. GUHA: The Alkaloids of *Mahonia acanthifolia*. Science and Culture **15**, 163 (1949) [Chem. Abstr. **44**, 2706 (1950)].
14. — — Studies on Mahonia genus. I. The Alkaloids of *Mahonia griffithii*. J. Amer. pharmac. Assoc. **39**, 181 (1950).
14a. — — Studies on Mahonia genus. II. *Mahonia acanthifolia* DON. J. Amer. pharmac. Assoc. **39**, 577 (1950).
15. — — Studies on Mahonia genus. III. *Mahonia borealis* and *Mahonia simonsii*. J. Amer. pharmac. Assoc. **40**, 36 (1951).
15a. — — Studies on Mahonia genus. IV. *Mahonia leschenaultii*, *Mahonia manipurensis* and *Mahonia sikkimensis*. J. Amer. pharmac. Assoc. **40**, 229 (1951).
16. CHAUDHRY, G. R. and S. SIDDIQUI: The Alkaloids of Indian Stephanias. I. Isolation of Three Crystalline Alkaloids from Tubers of *Stephania glabra*. J. Sci. Indian Res., Ser. B, **9**, No. 4, 79 (1950) [Chem. Abstr. **45**, 823 (1951)].
17. CHUANG, C. K., C. Y. HSING, Y. S. KAO und K. J. CHANG: Untersuchungen über die Alkaloide von Han-fang-chi. Fangchinolin, ein Demethyl-tetrandrin. Ber. dtsch. chem. Ges. **72**, 519 (1939).
18. CRAIG, L. E.: Curariform Activity and Chemical Structure. Chem. Reviews **42**, 285 (1948).

19. DUTCHER, J. D.: Curare Alkaloids from *Chondodendron tomentosum* RUIZ and PAVON. J. Amer. chem. Soc. **68**, 419 (1946).
20. FALTIS, F.: Alkaloide der Pareira-Wurzel. Mh. Chem. **33**, 873 (1912).
21. — Bemerkung zu der Veröffentlichung von H. KONDO und Mitarbeitern über die Konstitution des Trilobins und Tetrandrins. Liebigs Ann. Chem. **499**, 301 (1932).
22. FALTIS, F. und H. DIETERICH: Über die Konstitution des Isochondodendrins. VI. Ber. dtsch. chem. Ges. **67**, 231 (1934).
23. FALTIS, F. und H. FRAUENDORFER: Über die Konstitution des Isochondodendrins IV. Ber. dtsch. chem. Ges. **63**, 806 (1930).
24. FALTIS, F. und T. HECZKO: Beiträge zur Konstitution des Isochondodendrins und des Thebains. Monatsh. Chem. **43**, 377 (1924).
25. FALTIS, F., L. HOLZINGER, P. ITA und R. SCHWARZ: Über Biscoclaurin-Alkaloide: Die Konstitution des Chondodendrins und des Trilobins. Ber. dtsch. chem. Ges. **74**, 79 (1941).
26. FALTIS, F., K. KADIERA und F. DOBLHAMMER: Über die Konstitution des Chondodendrins. Ber. dtsch. chem. Ges. **69**, 1269 (1936).
27. FALTIS, F. und F. NEUMANN: Alkaloide der Pareira-Wurzel. II. Das Isochondodendrin. Monatsh. Chem. **42**, 311 (1921).
28. FALTIS, F. und A. TROLLER: Über die Konstitution des Isochondodendrins. II. Ber. dtsch. chem. Ges. **61**, 345 (1928).
29. FALTIS, F., S. WRANN und E. KUEHAS: Über die Konstitution des Isochondodendrins. V. Liebigs Ann. Chem. **497**, 69 (1932).
30. FALTIS, F. und K. ZWERINA: Über die Konstitution des Isochondodendrins. III. Ber. dtsch. chem. Ges. **62**, 1034 (1929).
31. FINKELSTEIN, J.: The Synthesis of *dl*-Coclaurine. J. Amer. chem. Soc. **73**, 550 (1951).
32. FUJITA, E.: On the Structure of Biscoclaurine Alkaloids. VII. Cleavage of O-Methyloxyacanthine by Metallic Sodium in Liquid Ammonia. (Alkaloids of Menispermaceae. LXXXIX.) J. pharmac. Soc. Japan **72**, 213 (1952).
33. — On the Structure of Biscoclaurine Alkaloids. VIII. Cleavage of O-Methyloxyacanthine by Metallic Sodium in Liquid Ammonia. (Alkaloids of Menispermaceae. XC.) J. pharmac. Soc. Japan **72**, 217 (1952).
34. FUJITA, E. and F. MURAI: On the Structure of Biscoclaurine Alkaloids. III. Cleavage of Tetrandrine by Metallic Sodium in Liquid Ammonia. (Alkaloids of Menispermaceae. LXXXII.) J. pharmac. Soc. Japan **71**, 1039 (1951).
35. — — On the Structure of Biscoclaurine Alkaloids. IV. Cleavage of Cycleanine by Metallic Sodium in Liquid Ammonia. (Alkaloids of Menispermaceae. LXXXIII.) J. pharmac. Soc. Japan **71**, 1043 (1951).
36. GADAMER, J. und F. v. BRUCHHAUSEN: Zur Kenntnis des Oxyacanthins. Arch. Pharmaz. Ber. dtsch. pharm. Ges. **264**, 193 (1926).
37. GOTO, K.: Sinomenine and Dehydrosinomenine. J. agric. Chem. Soc. Japan **2**, 89 (1926).
38. — A Hofmann Degradation of Sinomenine. Proc. Imp. Acad. (Tokyo) **2**, 167 (1926).
39. — On Sinomenol and Disinomenol. Proc. Imp. Acad. (Tokyo) **2**, 414 (1926).
40. — On Color Reactions of Sinomenine and Sinomenol. Bull. chem. Soc. Japan **4**, 103 (1929).
41. — Tuduranin, ein neues Alkaloid aus *Sinomenium acutum*. Liebigs Ann. Chem. **521**, 175 (1935).
42. — Sinomenine. Kagaku Jikkengaku (Methoden der experimentellen Chemie) Abt. II, Bd. **13**, 455 (1945).

43. Goto, K.: Recent Advance of Sinomenine Researches. J. japan. Chem. 3, Suppl. 2,1 (1949).
44. Goto, K., R. Inaba und H. Nozaki: Synthese von N-Methyl-tuduraninmethyläther. Liebigs Ann. Chem. 530, 142 (1937).
45. Goto, K. and Z. Kitasato: The Constitution of Siactine (1-Tetrahydroepiberberine). J. chem. Soc. (London) 1930, 1234.
46. Goto, K. und H. Shishido: Die Konstitution von Tuduranin. Liebigs Ann. Chem. 539, 262 (1939).
47. Goto, K. and H. Sudzuki: Studies on Bimolecular Alkaloids. I. On Disinomenine and ψ-Disinomenine. Bull. chem. Soc. Japan 4, 107 (1929).
48. — — Studies on Bimolecular Alkaloids. II. Reduction of Disinomenine and ψ-Disinomenine. Bull. chem. Soc. Japan 4, 129 (1929).
49. — — Sinomenine and Disinomenine. IX. On Acutumine and Siactine. Bull. chem. Soc. Japan. 4, 220 (1929).
50. — — On the Position of the Double Linking in Sinomenine. Bull. chem. Soc. Japan 4, 244 (1929).
51. — — On Sinomenine Hydrate. Bull. chem. Soc. Japan 4, 271 (1929).
52. Hasegawa, S.: Studies on the Chemotherapy of Tuberculosis. Biscoclaurine Alkaloids. I. 1st. ed. Ichikawa: The Chemotherapeutic Institute for Medical Research. 1942.
53. — Studies on the Chemotherapy of Tuberculosis. Jap. J. exp. Medicine 20, 60 (1949).
54. Hasegawa, S. and K. Takahashi: The Effect of Cepharanthine on Pertussis. Jap. J. exp. Medicine 20, 229 (1949).
55. Hashimoto, T.: The Cepharanthine Treatment of Leprocy. Jap. J. exp. Medicine 20, 461 (1950).
56. Henry, T. A.: The Plant Alkaloids. 4. Ed., p. 350. Philadelphia and Toronto: The Brakiston Company. 1949.
57. Hesse, O.: Zur Kenntnis der Alkaloide der Berberideen. Ber. dtsch. chem. Ges. 19, 3190 (1886).
58. Horiuchi, K.: Über die pharmakologischen Wirkungen der Alkaloide von japanischen Sinomenium- und Cocculus-Arten: Dauricin, Epistephanin und Insularine. Mitt. med. Ges. Tokyo 45, 740 (1931).
59. Inubushi, Y.: On the Structure of Biscoclaurine Alkaloids. IX. Cleavage of Berbamine by Metallic Sodium in Liquid Ammonia. (Alkaloids of Menispermaceae. XCI.) J. pharmac. Soc. Japan 72, 220 (1952).
60. Ishiwari, N.: An Alkaloid of *Sinomenium diversifolius*. Chugai Iji Shimpo 959, 1 (1920) [Chem. Abstr. 15, 1508 (1921)].
61. Itallie, L. van und A. J. Steenhauer: Untersuchung der Rinde von *Tiliacora acuminata* Miers. Pharmac. Weekbl. 59, 1381 (1922) (Chem. Zbl. 1923, I, 548).
62. King, H.: Curare Alkaloids. I. Tubocurarine. J. chem. Soc. (London) 1935, 1381.
63. — Curare Alkaloids. II. Tubocurarine and Bebeerine. J. chem. Soc. (London) 1936, 1276.
64. — Curare Alkaloide. III. Pot-curare. J. chem. Soc. (London) 1937, 1472.
65. — Curare Alkaloide. IV. Bebeerine and Tubocurarine. Orientation of Phenol Groups. J. chem. Soc. (London) 1939, 1157.
66. — Curare Alkaloids. V. Alkaloids of some Chondrodendron Species and the Origin of *Radix Pareirae Bravae*. J. chem. Soc. (London) 1940, 737.
67. — Curare Alkaloids. VI. Alkaloids from *Chondrodendron tomentosum*. J. chem. Soc. (London) 1947, 936.
68. — Curare Alkaloids. VII. Constitution of Dextro-tubocurarine Chloride. J. chem. Soc. (London) 1948, 265.

69. KING, H.: Curare Alkaloide. VIII. Examination of Commercial Curare. Chondrodendron tomentosum and Anomospermum grandifolium. J. chem. Soc. (London) 1948, 1945.
70. KONDO, H.: Chemistry of Cepharanthine and Analogous Compounds. Hasegawa's Chemotherapy of Tuberculosis 1, 154 (1942).
71. KONDO, H. und I. KEIMATSU: Alkaloide von Stephania cepharantha HAYATA. II. (Alkaloide von Sinomenium- und Cocculus-Arten. XXXVIII.) J. pharmac. Soc. Japan 55, 121 (1935).
72. — — Alkaloide von Stephania cepharantha HAYATA. III. (Alkaloide von Sinomenium- und Cocculus-arten. XL.) J. pharmac. Soc. Japan 55, 234 (1935).
73. — — Bemerkung zur Konstitution des Phaeanthins. Ber. dtsch. chem. Ges. 68, 1503 (1935).
74. — — Alkaloide von Stephania cepharantha HAYATA. IV. Bemerkung zur Konstitution des Phaeanthins. (Alkaloide von Sinomenium- und Cocculus-Arten. XLIII.) J. pharmac. Soc. Japan 55, 894 (1935).
75. — — Alkaloide von Stephania cepharantha HAYATA. V. Über die Konstitution des Cepharanthins. (Alkaloide von Sinomenium- und Cocculus-Arten. XLVIII.) Ber. dtsch. chem. Ges. 71, 2553 (1938); J. pharmac. Soc. Japan 58, 906 (1938).
76. KONDO, H., I. KEIMATSU und Y. YAMASHITA: Alkaloide von Stephania cepharantha HAYATA. I. (Alkaloide von Sinomenium- und Cocculus-Arten. XXXVII.) J. pharmac. Soc. Japan 54, 620 (1934).
77. KONDO, H. und T. KONDO: Alkaloide von Cocculus laurifolius DC. (Alkaloide von Sinomenium- und Cocculus-Arten. VI.) J. pharmac. Soc. Japan 45, 876 (1925).
78. — — Über das Alkaloid, Coclaurin, von Cocculus laurifolius DC. J. prakt. Chem. (2) 126, 24 (1930).
79. KONDO, H. and T. MATSUNO: Alkaloids of Stephania rotunda LOUREIRO. I. (Alkaloids of Menispermaceae. LXXII.) J. pharmac. Soc. Japan 64, A 8, B 113 (1944).
80. KONDO, H. und T. NAKAJIMA: Diversin. II. (Alkaloide von Sinomenium- und Cocculus-Arten. IV.) J. pharmac. Soc. Japan 512, 783 (1924).
81. KONDO, H., T. NAKAMURA, M. FUJII und T. SATO: Über die Alkaloide von Stephania japonoca MIERS. XI. Protostephanin. IV. (Alkaloide von Menispermaceae. LXXVI.) Annu. Rep. Itsuu Lab. 1, 41 (1950).
82. KONDO, H. und T. NAKAZATO: Alkaloide von Cocculus trilobus DC. Trilobin. (Alkaloide von Sinomenium- und Cocculus-Arten. III.) J. pharmac. Soc. Japan 44, 691 (1924).
83. KONDO, H. und Z. NARITA: Alkaloid von Menispermum dauricum DC. I. (Alkaloide von Sinomenium- und Cocculus-Arten. XIV.) J. pharmac. Soc. Japan 47, 279 (1927).
84. — — Über Dauricin. II. (Alkaloide von Sinomenium- und Cocculus-Arten. XXVI.) J. pharmac. Soc. Japan 49, 688 (1929).
85. — — Über Dauricin. III. (Alkaloide von Sinomenium- und Cocculus-Arten. XXIX.) J. pharmac. Soc. Japan 50, 589 (1930).
86. — — Die Konstitution des Dauricins. Ber. dtsch. chem. Ges. 63, 2420 (1930).
87. KONDO, H., Z. NARITA und M. MURAKAMI: Alkaloide von Menispermum canadense L. (Alkaloide von Menispermaceae. LIII.) J. pharmac. Soc. Japan 61, 375, (1941).
88. KONDO, H., Z. NARITA und S. UYEO: Die Konstitution des Dauricins. IV. (Alkaloide von Sinomenium- und Cocculus-Arten. XXXIX.) Ber. dtsch. chem. Ges. 68, 519 (1935); J. pharmac. Soc. Japan. 55, 369 (1935).

89. Kondo, H. und T. Nozoe: Alkaloide von *Stephania japonica* Miers. IX. Struktur von Hypoepistephanin (ψ-Epistephanin). (Alkaloide von Menispermaceae. LXV.) J. pharmac. Soc. Japan **63**, 333 (1943).
90. Kondo, H. und E. Ochiai: Über die Konstitution des Sinomenins. Liebigs Ann. Chem. **470**, 224 (1929).
91. — — Über die Konstitution des Dihydrothebakodins bzw. des Dehydroxytetrahydrokodeins und des β-Tetrahydro-desoxy-kodeins. Ber. dtsch. chem. Ges. **63**, 646 (1930).
92. Kondo, H., E. Ochiai und T. Nakajima: Neue Alkaloide aus *Sinomenium acutum* R. et W. I. Über Sinomenin und Diversin. (Alkaloide von Sinomenium- und Cocculus-Arten. I.) J. pharmac. Soc. Japan **497**, 511 (1923).
93. Kondo, H. und T. Sanada: Über die Alkaloide von *Stephania japonica* Miers. I. (Alkaloide von Sinomenium- und Cocculus-Arten. V.) J. pharmac. Soc. Japan **514**, 1034 (1924).
94. — — Über die Alkaloide von *Stephania japonica* Miers. IV. (Alkaloide von Sinomenium- und Cocculus-Arten. XXII.) J. pharmac. Soc. Japan **48**, 1141 (1928).
95. — — Über die Alkaloide von *Stephania japonica* Miers. II. (Alkaloide von Sinomenium- und Cocculus-Arten. XII.) J. pharmac. Soc. Japan **47**, 177 (1927).
96. — — Über die Alkaloide von *Stephania japonica* Miers. III. (Alkaloide von Sinomenium- und Cocculus-Arten. XVIII.) J. pharmac. Soc. Japan **47**, 930 (1927).
97. Kondo, H., M. Satomi and T. Odera: Alkaloids of *Stephania japonica* Miers. XII. Protostephanine. V. (Alkaloids of Menispermaceae. LXXVII.) Annu. Rep. Itsuu Lab. **1**, 45 (1950).
98. — — — Alkaloids of *Stephania japonica* Miers. XIV. On Hasubanonine. I. (Alkaloids of Menispermaceae. LXXXIV.) Annu. Rep. Itsuu Lab. **2**, 35 (1951).
99. Kondo, H. und K. Tanaka: Alkaloide von *Stephania japonica* Miers. VII. (Alkaloide von Menispermaceae. LXIII.) J. pharmac. Soc. Japan **63**, 267 (1943).
100. — — Alkaloide von *Stephania japonica* Miers. VIII. (Alkaloide von Menispermaceae. LXIV.) J. pharmac. Soc. Japan **63**, 273 (1943).
101. Kondo, H. und M. Tomita: Über die Alkaloide von *Cocculus sarmentosus* Diels. I. (Alkaloide von Sinomenium- und Cocculus-Arten. XIII). J. pharmac. Soc. Japan **47**, 265 (1927).
102. — — Über Trilobin und Homotrilobin. IV. (Alkaloide von Sinomenium- und Cocculus-Arten. XXI.) J. pharmac. Soc. Japan **48**, 659 (1928).
103. — — Über die Alkaloide von *Cocculus sarmentosus* Diels. II. (Alkaloide von Sinomenium- und Cocculus-Arten. XXX.) J. pharmac. Soc. Japan **50**, 633 (1930).
104. — — Über die Alkaloide von *Berberis Thunbergii* DC. I. Arch. Pharmaz. Ber. dtsch. pharm. Ges. **268**, 1 (1930); J. pharmac. Soc. Japan **50**, 309 (1930).
105. — — Über phenolische Alkaloide von *Cocculus trilobus* DC. I. (Alkaloide von Sinomenium- und Cocculus-Arten. XXXII.) Arch. Pharmaz. Ber. dtsch. pharm. Ges. **269**, 433 (1931); J. pharmac. Soc. Japan **51**, 451 (1931).
106. — — Über das Trilobin und Isotrilobin. VI. (Alkaloide von Sinomenium- und Cocculus-Arten. XXXV.) Liebigs Ann. Chem. **497**, 104 (1932); J. pharmac. Soc. Japan **52**, 856 (1932).
107. — — Stereochemische und biogenetische Betrachtungen über die Biscoclaurin-Alkaloide. (Alkaloide von Sinomenium- und Cocculus-Arten. XLV.) Arch. Pharmaz. Ber. dtsch. pharm. Ges. **274**, 65 (1936); J. pharmac. Soc. Japan **55**, 914 (1935).
108. — — Alkaloids of *Stephania Sasakii* Hayata. I. (Alkaloids of Menispermaceae. L.) J. pharmac. Soc. Japan **59**, 542 (1939).

109. KONDO, H., M. TOMITA, M. SATOMI und T. IKEDA: Über die Alkaloide.von *Stephania cepharantha* HAYATA. VI. (Alkaloide von Sinomenium- und Cocculus-Arten. XLIX.) J. pharmac. Soc. Japan **58**, 920 (1938).
110. KONDO, H., M. TOMITA und S. UYEO: Über das Methyl-isochondodendrin. (Alkaloide von Sinomenium- und Cocculus-Arten. XLVI.) Ber. dtsch. chem. Ges. **70**, 1891 (1937).
111. — — Alkaloids of *Cyclea insularis* (MAKINO) DIELS. II. J. pharmac. Soc. Japan **62**, 534 (1942).
112. KONDO, H. und T. WATANABE: Über die Alkaloide von *Stephania japonica* MIERS. VI. (Alkaloide von Sinomenium- und Cocculus-Arten. XLVII.) J. pharmac. Soc. Japan **58**, 268 (1938).
113. — — Über die Alkaloide von *Stephania japonica* MIERS. X. Protostephanin. III. (Alkaloide von Menispermaceae. LXXV.) Annu. Rep. Itsuu Lab. **1**, 37 (1950).
114. — — Alkaloids of *Stephania japonica* MIERS. XIII. Protostephanin. VI. (Alkaloids of Menispermaceae. LXXVIII.) Annu. Rep. Itsuu Lab. **1**, 50 (1950).
115. KONDO, H. und K. YANO: Alkaloide von *Cissampelos insularis* MAKINO. I. Alkaloide von Sinomenium- und Cocculus-Arten. XV.) J. pharmac. Soc. Japan **47**, 815 (1927).
116. — — Alkaloide von *Stephania tetrandra* S. MOORE. I. (Alkaloide von Sinomenium- und Cocculus-Arten. XIX.) J. pharmac. Soc. Japan **48**, 107 (1928).
117. — — Alkaloide von *Stephania tetrandra* S. MOORE. II. (Alkaloide von Sinomenium- und Cocculus-Arten. XXV.) J. pharmac. Soc. Japan **49**, 315 (1929).
118. — — Alkaloide von *Stephania tetrandra* S. MOORE. IV. Konstitution des Tetrandrins. (Alkaloide von Sinomenium- und Cocculus-Arten. XXXIV.) Liebigs Ann. Chem. **497**, 90 (1932); J. pharmac. Soc. Japan **52**, 827 (1932).
119. KONDO, T.: Die Konstitution von Coclaurin. II. (Alkaloide von Sinomenium- und Cocculus-Arten. X.) J. pharmac. Soc. Japan **46**, 1029 (1926).
120. — Die Konstitution von Coclaurin. III. (Alkaloide von Sinomenium- und Cocculus-Arten. XX.) J. pharmac. Soc. Japan **48**, 324 (1928).
121. — Die Konstitution von Coclaurin. IV. (Alkaloide von Sinomenium- und Cocculus-Arten. XXIII.) J. pharmac. Soc. Japan **48**, 1156 (1928).
122. — Ergänzung über die Konstitution des Coclaurins. (Alkaloide von Sinomenium- und Cocculus-Arten. ˙XXVIII.) J. pharmac. Soc. Japan **50**, 427 (1930).
123. KRANZFELDER, A. L., J. J. VERBANC and F. J. SOWA: The Cleavage of Diphenyl Ethers by Sodium in Liquid Ammonia. II. Meta Substituted Diphenyl Ethers. J. Amer. chem. Soc. **59**, 1488 (1937).
124. LEITHE, W.: Die Konfiguration des optisch aktiven α-Phenyl-äthylamins sowie der Basen vom Typus des Laudanosins und des Tetrahydroberberins. Ber. dtsch. chem. Ges. **64**, 2827 (1931).
125. MANSKE, R. H. F.: An Alkaloid from *Menispermum canadense* L. Canad. J. Res., Sect. B **21**, 17 (1943.)
126. — The Alkaloids of Fumariaceous Plants. XLIII. The Structures of Cularine and of Cularimine. J. Amer. chem. Soc. **72**, 55 (1950).
127. MANSKE, R. H. F. and A. E. LEDINGHAM: Synthesis and Reactions of Some Dibenzoxepins. J. Amer. chem. Soc. **72**, 4797 (1950).
128. MARSH, D. F. and D. A. HERRING: Curariform Activity of N-Methylberbamine and N-Methylisotetrandrine. J. Pharmacol. exp. Therapeut. **97**, 19 (1949).
129. — — The Curariform Activity of the Menispermaceous Alkaloids. Experientia **6**, 31 (1950).
130. MARSH, D. F., D. A. HERRING and C. K. SLEETH: The Curariform Activity of N-Methyloxyacanthine. J. Pharmacol. exp. Therapeut. **95**, 100 (1949).

131. MATSUNO, T.: Alkaloids of *Stephania rotunda* LOUREIRO. II. (Alkaloids of Menispermaceae. LXXIII.) J. pharmac. Soc. Japan **64**, A. 52, B. 274 (1944).
132. MATSUOKA, M.: Pharmacological Studies on Cycleanine. Folia pharmacol. japon. **44**, No. 1, 65 (1948).
133. NAKAZATO, T.: Homotrilobin, das zweite Alkaloid aus *Cocculus trilobus* DC. (Alkaloide von Sinomenium- und Cocculus-Arten. VII.) J. pharmac. Soc. Japan **46**, 461 (1926).
134. — Über Trilobin. II. (Alkaloide von Sinomenium- und Cocculus-Arten. VIII.) J. pharmac. Soc. Japan **46**, 465 (1926).
135. OCHIAI, E.: Konstitution von Sinomenin. II. (Alkaloide von Sinomenium- und Cocculus-Arten. II.) J. pharmac. Soc. Japan **503**, 8 (1924).
136. — Konstitution von Sinomenin. III. (Alkaloide von Sinomenium- und Cocculus-Arten. IX.) J. pharmac. Soc. Japan **538**, 1008 (1926).
137. — Konstitution von Sinomenin. IV. (Alkaloide von Sinomenium- und Cocculus-Arten. XI.) J. pharmac. Soc. Japan **539**, 17 (1927).
138. — Konstitution von Sinomenin. V. (Alkaloide von Sinomenim- und Cocculus-Arten. XVI.) J. pharmac. Soc. Japan **549**, 913 (1927).
139. — Konstitution von Sinomenin. VI. (Alkaloide von Sinomenium- und Cocculus-Arten. XVII.) J. pharmac. Soc. Japan **549**, 923 (1927).
140. — Ergänzung über die Konstitution des Sinomenins. I. J. pharmac. Soc. Japan **49**, 556 (1929).
141. — Beiträge zur Konstitution des Sinomenins. II. J. pharmac. Soc. Japan **50**, 360 (1930).
142. OKADA, M.: Pharmacology of Alkaloids of the Biscoclaurine Type. IV. Studies on the Fate of Cepharanthine in the Body. I. Folia pharmacol. japon. **40**, No. 3, 34 (1944).
143. — Pharmacology of Alkaloids of the Biscoclaurine Type. V. Studies on the Fate of Cepharanthine in the Body. II. Folia pharmacol. japon. **47**, No. 2, 95 (1951).
144. OKADA, M., A. ASODA, Z. HAMADA and I. HONDA: Pharmacology of Alkaloids of the Biscoclaurine Type. VIII. Studies on the Curariform Activity of Quaternary Ammonium Derivatives of Alkaloids of the Biscoclaurine Type. Folia pharmacol. japon. **47**, No. 2, 97 (1951).
145. OKADA, M. und S. FUSE: Pharmakologische Studien über die Biscoclaurin-Alkaloide. I. Über die Wirkung des Trilobins und seiner Spaltprodukte. Proc. jap. pharmacol. Soc., 10th Annual Meeting. 1936.
146. — — Pharmakologische Studien über die Biscoclaurin-Alkaloide. II. Über die diuretische Wirkung von dicarbonsauren Salzen des Diphenyläthers und seiner Homologe. Proc. jap. pharmacol. Soc., 11th Annual Meeting. 1937.
147. OKADA, M., S. FUSE und S. WATANABE: Pharmakologische Studien über die Biscoclaurin-Alkaloide. III. Über die Wirkungen von Trilobin, Isotrilobin, Tetrandrin, Isotetrandrin und Cepharanthin. Proc. jap. pharmacol. Soc., 16th Annual Meeting. 1942.
148. OKADA, M., S. HIRAOKA, S. SHIINA and M. YOSHINO: Pharmacology of the Biscoclaurine Type. VI. Studies on Hemolysis by Cepharanthine and Isotetrandrine. Folia pharmacol. japon. **47**, No. 2, 96 (1951).
149. OKADA, M., T. MIMURA, S. MATSUMOTO, S. SHIINA and N. ABE: Pharmacology of Alkaloids of the Biscoclaurine Type. VII. Studies on the Action of Cepharanthine upon Pigment Cells and Reticuloendotherial System. Folia pharmacol. japon. **47**, No. 2, 96 (1951).
150. PALLARES, E. S. and H. M. GARZA: Structure of the Alkaloid of Yoloxochitl. Arch. Biochemistry **16**, 275 (1948) [Chem. Abstr. **42**, 5036 (1948)].

151. POMMEREHNE, H.: Beiträge zur Kenntnis der Alkaloide von *Berberis aquifolium*. Arch. Pharmaz. Ber. dtsch. pharmaz. Ges. **233**, 127 (1895).
152. PROSKOURNINA, N. F.: Alkaloids of *Magnolia fuscata*, III. The Structure of Magnolamine. J. Gen. Chem. (USSR) **16**, 129 (1946) [Chem. Abstr. **41**, 460 (1947)].
153. PROSKOURNINA, N. F. and A. P. ORECHOV: On the Alkaloids of *Magnolia fuscata*. I. J. Gen. Chem. (USSR) **9**, 126 (1936) (Chem. Zbl. **1939**, I, 423).
154. — — Alkaloids of *Magnolia fuscata*. II. The Structure of Magnoline. J. Gen. Chem. (USSR) **10**, 707 (1940) [Chem. Abstr. **35**, 2520 (1941)].
155. PYMAN, F. L.: The Alkaloids of *Daphnandra micrantha*. J. chem. Soc. (London) **1914**, 1679.
156. RUEDEL, C.: Beiträge zur Kenntnis der Alkaloide von *Berberis aquifolium* und *Berberis vulgaris*. Arch. Pharmaz. Ber. dtsch. pharmaz. Ges. **229**, 631 (1891).
157. SANADA, T.: Alkaloide von *Stephanaia japonica* MIERS. V. (Alkaloide von Sinomenium- und Cocculus-Arten. XXXIII.) J. pharmac. Soc. Japan **51**, 509 (1931).
158. SANTOS, A. C.: The Structure of Berbamine. Dissert. Westfälische Wilhelms-Univ., Münster. **1929**, 5 [Chem. Abstr. **24**, 1647 (1930)].
159. — Alkaloide von *Phaeanthus ebracteoratus* (PRESL.) MERRIL. Über Phaeanthin. I. Rev. Filipina. Med. Farmac. **22**, 11 (1931) [Chem. Zbl. **1932**, I, 395].
160. — Zur Kenntnis des Phaeanthins. II. Ber. dtsch. chem. Ges. **65**, 472 (1932).
161. SANTOS, A. C. und G. Q. QUIBILAN: Ambalin, ein neues Alkaloid aus *Pycnarrhena manillensis* VIDAL. Univ. Philipp. nat. appl. Sci. Bull. **3**, 353 (1933) [Chem. Zbl. **1934**, II, 1473].
162. SANTOS, A. C. und VALLANOS: Alkaloide von *Pycnarrhena manillensis* VIDAL. II. Univ. Philipp. nat. appl Sci.. Bull. **4**, 338 (1935).
163. SARTORETTO, P. A. and F. J. SOWA: The Cleavage of Diphenyl Ethers by Sodium in Liquid Ammonia. I. *Ortho* and *para* Substituted Diphenyl Ethers. J. Amer. chem. Soc. **59**, 603 (1937).
164. SASAKI, T.: Curare Like Substance in Magnoliaceous Plants. Fukuoka Acta med. **14**, 391 (1921).
165. SCHOLTZ, M.: Über Bebirin. Ber. dtsch. chem. Ges. **29**, 2054 (1896).
166. — Bebeerin und Buxin. Arch. Pharmaz. Ber. dtsch. pharmaz. Ges. **236**, 530 (1898),
167. — Über Pelosin. Arch. Pharmaz. Ber. dtsch. pharmaz. Ges. **237**, 199 (1899).
168. — Über die Alkaloide der Pareira-Wurzel. Arch. Pharm. Ber. dtsch. pharmaz. Ges. **244**, 555 (1906).
169. — Über die Alkaloide der Pareira-Wurzel. Arch. Pharmaz. Ber. dtsch. pharmaz. Ges. **249** 408 (1911).
170. — Über die Alkaloide der Pareira-Wurzel. Arch. Pharmaz. Ber. dtsch. pharmaz. Ges. **251**, 136 (1913).
171. — Die Entmethylierung des Isobebeerins. Arch. Pharmaz. Ber. dtsch. pharmaz. Ges. **253**, 622 (1915).
172. SCHOLTZ, M. und O. KOCH: Über die Alkaloide der Pareira-Wurzel. Arch. Pharmaz. Ber. dtsch. pharmaz. Ges. **252**, 513 (1914).
173. SHIMURA, J., K. TAKAHASHI, Y. MIYAZAWA, M. ITO, K. KIKUCHI, T. SHINOZUKA and Y. OKAMOTO: Studies on the Mechanism of Antibody Formation. Jap. J. exp. Medicine **20**, 443 (1950).
174. SHIRAI, H.: Alkaloids of *Stephania capitata* SPRENG. III. (Alkaloids of Menispermaceae. LXVII.) J. pharmac. Soc. Japan **63**, 517 (1943).
175. — Alkaloids of *Stephania capitata* SPRENG. II. (Alkaloids of Menispermaceae. LXVI.) J. pharmac. Soc. Japan **63**, 532 (1943).

176. SHIRAI, H.: Über die Alkaloide von *Stephania capitata* SPRENG. IV. (Alkaloide von Menispermaceae. LXX.) J. pharmac Soc. Japan 64, B. 208 (1944).
177. SPÄTH, E. und A. KOLBE: Zur Kenntnis des Oxyacanthins. Ber. dtsch. chem. Ges. 58, 2280 (1925).
178. SPÄTH, E. und F. KUFFNER: Über Curare-Alkaloide. II. Zur Konstitution des Curins (Bebeerins). Ber. dtsch. chem. Ges. 67, 55 (1934).
179. SPÄTH, E., W. LEITHE und F. LADECK: Über Curare-Alkaloide. I. Die Konstitution des Curins. Ber. dtsch. chem. Ges. 61, 1698 (1928).
180. SPÄTH, E. und E. MOSETTIG: Über die Synthese des Siactins. Ber. dtsch. chem. Ges. 64, 2048 (1931).
181. SPÄTH, E. und J. PIKL: Zur Konstitution des Oxyacanthins. Ber. dtsch. chem. Ges. 62, 2251 (1929).
182. TANAKA, K.: Alkaloide von *Stephania japonica* MIERS. IX. Struktur des Epistephanins. (Alkaloide von Menispermaceae. LXVIII.) J. pharmac. Soc. Japan 64, 28 (1944).
183. TANI, T.: Über die Konstitution von Trilobin und Isotrilobin. IX. (Alkaloide von Menispermaceae. LXI.) J. pharmac. Soc. Japan 62, 481 (1942).
184. TOMITA, M.: Über die Konstitution des Trilobins und Homotrilobins. V. (Alkaloide von Sinomenium- und Cocculus-Arten. XXXI.) J. pharmac. Soc. Japan 50, 1035 (1930).
185. — Über das Trilobin und Isotrilobin. VII. (Alkaloide von Sinomenium- und Cocculus-Arten. XXXVI.) J. pharmac. Soc. Japan 52, 889 (1932).
186. — Über die Alkaloide von *Cocculus sarmentosus* DIELS. III. (Alkaloide von Sinomenium- und Cocculus-Arten. XLI.) J. pharmac. Soc. Japan 55, 637 (1935).
187. — Über die phenolischen Alkaloide von *Cocculus trilobus* DC. II. (Alkaloide von Sinomenium- und Cocculus-Arten. XLII.) J. pharmac. Soc. Japan 55, 646 (1935).
188. — Über die phenolischen Alkaloide von *Cocculus trilobus* DC. III. (Alkaloide von Sinomenium- und Cocculus-Arten. XLIV.) J. pharmac. Soc. Japan 55, 911 (1935).
189. TOMITA, M., T. ASADA and Y. WATANABE: Alkaloids and Nitrogen-free Substance of *Parabaena hirsuta* (BECC.) DIELS. (Alkaloids of Menispermaceae. LXXXVI.) J. pharmac. Soc. Japan 72, 203, (1952).
190. TOMITA, M. and E. FUJITA: Structure of Magnolamine. (Alkaloids of Menispermaceae. LXXIX.) J. pharmac. Soc. Japan 70, 411 (1950).
191. TOMITA, M., E. FUJITA and F. MURAI: On the Structure of Biscoclaurine Alkaloids. I. Cleavage of Isotetrandrine by Metallic Sodium in Liquid Ammonia and the Structure of Alkaloids in Oxyacanthine-Berbamine Series. (Alkaloids of Menispermaceae. LXXX.) J. pharmac. Soc. Japan 71, 226 (1951).
192. — — — — Cleavage of Biscoclaurine Alkaloids by Metallic Sodium in Liquid Ammonia. J. pharmac. Soc. Japan 71, 301 (1951).
193. — — — On the Structure of Biscoclaurine Alkaloids. II. Cleavage of Isotetrandrine by Metallic Sodium in Liquid Ammonia. (Alkaloids of Menispermaceae. LXXXI.) J. pharmac. Soc. Japan 71, 1035 (1951).
194. TOMITA, M., E. FUJITA and T. NAKAMURA: Structure of Magnolamine. II. (Alkaloids of Magnoliaceae. II.) J. pharmac. Soc. Japan 71, 1075 (1951).
195. TOMITA, M., Y. INUBUSHI and F. KUSUDA: Synthesis of Depsidone. J. pharmac. Soc. Japan 64, 173 (1943).
196. TOMITA, M., Y. INUBUSHI and H. NIWA: On the Structure of Biscoclaurine Alkaloids. V. Cleavage of Diphenyl Ether and Diphenylene Dioxide Derivatives by Metallic Sodium in Liquid Ammonia. (Alkaloids of Menispermaceae. LXXXVII.) J. pharmac. Soc. Japan 72, 206 (1952).

197. TOMITA, M., Y. INUBUSHI and H. NIWA: On the Structure of Biscoclaurine Alkaloids. VI. Cleavage of Isotetrandrine by Metallic Potassium in Liquid Ammonia. (Alkaloids of menispermaceae. LXXXVIII.) J. pharmac. Soc. Japan **72**, 211 (1952).
198. TOMITA, M., Y. INUBUSHI and M. YAMAGATA: Alkaloids of *Magnolia obovata* THUNB. (Alkaloids of Magnoliaceae. I.) J. pharmac. Soc. Japan **71**, 1069 (1951).
199. TOMITA, M. and H. KITAGISHI: Alkaloids of *Stephania Sasakii* HAYATA. III. (Alkaloids of Menispermaceae. LXIX.) J. pharmac. Soc. Japan **64**, 240 (1944).
200. TOMITA, M. and F. KUSUDA: Alkaloids of *Cocculus laurifolius* DC. (Alkaloids of Menispermaceae. XCII.) J. pharmac. Soc. Japan **72**, 280 (1952).
201. TOMITA, M., K. NAKAGUCHI and S. TAKAGI: Synthesis of *dl*-Coclaurine. (Alkaloids of Menispermaceae. LXXXV.) J. pharmac. Soc. Japan **71**, 1046 (1951).
202. TOMITA, M. and T. NAKANO: Alkaloids of *Magnolia salicifolia* MAXIM. (Alkaloids of Magnoliaceae. III.) J. pharmac. Soc. Japan **72**, 197 (1952).
203. — — Alkaloids of *Magnolia salicifolia* MAXIM. II. Synthesis of O-methyl-salicifoline Iodide. (Alkaloids of Magnoliaceae. IV.) J. pharmac. Soc. Japan **72**, 281 (1952).
204. TOMITA, M. und H. SHIRAI: Über die Alkaloide von *Stephania capitata* SPRENG. I. (Alkaloide von Menispermaceae. LIV.) J. pharmac. Soc. Japan **62**, 381 (1942).
205. — — Alkaloids of *Stephania Sasakii* HAYATA. II. (Alkaloids of Menispermaceae. LXII.) J. pharmac. Soc. Japan **63**, 233 (1943).
206. TOMITA, M. und T. TANI: Über die Alkaloide von *Fibraurea chloroleuca* MIERS. (Alkaloide von Menispermaceae. LI.) J. pharmac. Soc. Japan **61**, 247 (1941).
207. — — Über die Alkaloide von *Coscinium Blumeanum* MIERS. (Alkaloide von Menispermaceae. LII.) J. pharmac. Soc. Japan **61**, 251 (1941).
208. — — Über die Konstitution von Trilobin und Isotrilobin. VIII. (Alkaloide von Menispermaceae. LV.) J. pharmac. Soc. Japan **62**, 468 (1942).
209. — — Eine neue synthetische Methode der asymmetrisch substituierten Diphenylendioxyd-Derivate. (Diphenylendioxyd-Derivate. XVI.) J. pharmac. Soc. Japan **62**, 476 (1942).
210. TOMITA, M. and S. UYEO: Alkaloids of *Cyclea insularis* (MAKINO) DIELS. III. Degradation of Insularine. (Alkaloids of Menispermaceae. LVII.) J. chem. Soc. Japan **64**, 64 (1943).
211. — — Alkaloids of *Cyclea insularis* (MAKINO) DIELS. IV. Constitution of Insularic Acid. I. (Alkaloids of Menispermaceae. LVIII.) J. chem. Soc. Japan **64**, 70 (1943).
212. — — Alkaloids of *Cyclea insularis* (MAKINO) DIELS. V. Constitution of Insularic Acid. II. (Alkaloids of Menispermaceae. LIX.) J. chem. Soc. Japan **64**, 77 (1943).
213. — — Alkaloids of *Cyclea insularis* (MAKINO) DIELS. VI. Constitution of Insularic Acid. III. (Alkaloids of Menispermaceae. LX.) J. chem. Soc. Japan **64**, 142 (1943).
214. — — Alkaloids of *Cyclea insularis* (MAKINO) DIELS. VII. Constitution of Insularine. (Alkaloids of Menispermaceae. LXI.) J. chem. Soc. Japan **64**, 147 (1943).
215. TOMITA, M., S. UYEO, S. SAWA, K. DOI and T. MIWA: Structure of Epistephanine from its Absorption Spectrum. (Alkaloids of Menispermaceae. LXXI.) J. pharmac. Soc. Japan **69**, 22 (1949).
216. TSURUTA, S.: Pharmacological Studies of Menispermaceous Plants. I. Alkaloid of *Cocculus laurifolius*. DC. Folia pharmacol. japon. **2**, 269 (1926).
217. — Pharmacological Studies of Menispermaceous Plants. II. Alkaloid of *Cocculus trilobus* DC. Folia pharmacol. japon. **3**, 280 (1926).

218. WATT, G. W.: Reactions of Organic and Organometallic Compounds with Solutions of Metals in Liquid Ammonia. Chem. Reviews **46**, 317 (1950).
219. WEBER, F. C. and F. J. SOWA: The Cleavage of Diphenyl Ethers by Sodium in Liquid Ammonia. III. 4,4'-Disubstituted Diphenyl Ethers. J. Amer. chem. Soc. **60**, 94 (1938).
220. WEHMER, C.: Die Pflanzenstoffe. II. Aufl., S. 329. Jena: Gustav Fischer. 1929.
221. YANO, K.: Alkaloide von *Stephania tetrandra* S. MOORE. III. (Alkaloide von Sinomenium- und Cocculus-Arten. XXVII.) J. pharmac. Soc. Japan **50**, 224 (1930).
222. YUNUSOV, S.: Alkaloids of *Cocculus laurifolius*. I. J. Gen. Chem. (USSR) **20**, 368 (1950) [Chem. Abstr. **44**, 6582 (1950)].
223. — Alkaloids of *Cocculus laurifolius*. II. J. Gen. Chem. (USSR) **20**, 1514 (1950) [Chem. Abstr. **45**, 2490 (1951)].

(Eingelaufen am 31. Dezember 1951.)

Naturally Occurring Coumarins.

By F. M. Dean, Liverpool*.

Contents.	Page
I. General Structural Features	226
II. The Chemistry of the Coumarin System	229
Conversions and Degradation	229
The Synthesis of Coumarins	235
Theoretical Considerations	237
III. Occurrence, Isolation and Determination	239
IV. Some Biochemical Properties	240
V. Simple Coumarins	242

Coumarin 242. — Dihydro-coumarin 243. — Umbelliferone 243. — Herniarin 243. — Aesculetin 244. — Scopoletin 245. — Fabiatrin 245. — Ayapin 245. — Citropten 245. — Daphnetin 246. — Fraxetin 247. — Fraxidin 248. — Isofraxidin 249. — Fraxinol 249. — Eugenin 249. — 5-Geranoxy-7-methoxycoumarin 249. — Suberosin 250. — Collinin 250. — Brayleyanin 250. — Umbelliprenin 251. — Toddalolactone 251. — Aculeatin 252. — Auraptene 252. — Ostruthin 253. — Osthenol 254. — Osthol 254. — Ammoresinol 256. — Dicoumarol 257.

VI. Furanocoumarins	257

Psoralene 260. — Angelicin 261. — Bergapten 262. — Bergaptol 263. — Isobergapten 264. — Xanthotoxin 264. — Xanthotoxol 265. — Sphondin 265. — Sphondylin 265. — Isopimpinellin 265. — Pimpinellin 266. — Isoimperatorin 266. — Oxypeucedanin 267. — Ostruthol 267. — Imperatorin 268. — Bergamottin 269. — Phellopterin 269. — Byakangelicol 270. — Byakangelicin 270. — Ferulin 271. — Nodakenetin 272. — Marmesin 272. — Peucedanin 273. — Athamantin 275.

VII. Chromeno-α-pyrones	276

Xanthyletin 277. — Seselin 278. — Xanthoxyletin 278. — Luvangetin 280. — Alloxanthoxyletin 281. — Braylin 281.

VIII. 3:4-Benzcoumarins	282

2′:3″-Dihydroxydibenz-α-pyrone 282. — 4:6:4:6′-Dihydroxydiphenic acid dilactone 282. — Ellagic acid 283. — 4 : 4′-Dihydroxy-6:6′-dimethoxydiphenic acid dilactone 284.

References	285

* The Author is greatly indebted to Professor A. Robertson for his interest and advice throughout the writing of this review.

I. General Structural Features.

It is striking that apart from coumarin, hydrocoumarin, and dicoumarol, all naturally occurring coumarins have an oxygen atom at the 7-position i. e. they can be regarded as derivatives of umbelliferone, which is one of the most widely distributed compounds of this class. The simpler members of the group are hydroxylated or alkoxylated coumarins derived from polyhydric phenols containing a resorcinol system,—a fact which accounts for the prominence of the umbelliferone nucleus; thus aesculetin is derived from hydroxyquinol, daphnetin from pyrogallol, citropten from phloroglucinol, and fraxetin from 1:2:3:4-tetrahydroxybenzene. Hydroxyl groups have not been found in position 3 of the coumarin system, and only two authentic cases (dicoumarol and ammoresinol) are known with hydroxyl in position 4. Frequently the oxygen at site 7 is combined in a third ring which may be fused in either of the two possible positions with respect to the α-pyrone system. Only one case, that of *allo*xanthoxyletin, is known where any other oxygen atom is concerned in the formation of the third ring (pp. 281).

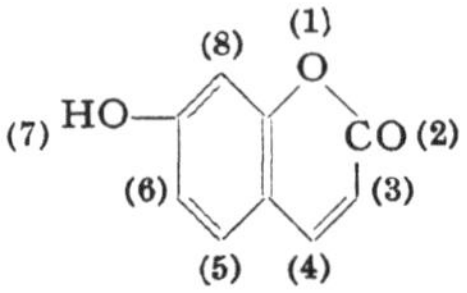

Umbelliferone. 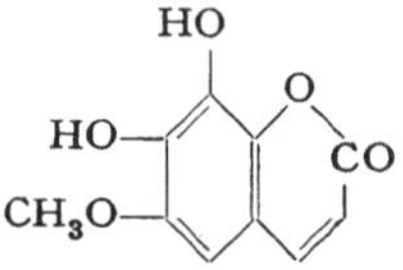   Fraxetin.

Of interest is the frequent occurrence of the isoprene system with these coumarins whilst chains of one, two, or three units have been found to be attached either as alkyl side-chains or to oxygen as ethers, e. g. in *iso*imperatorin, ostruthin, and ammoresinol.

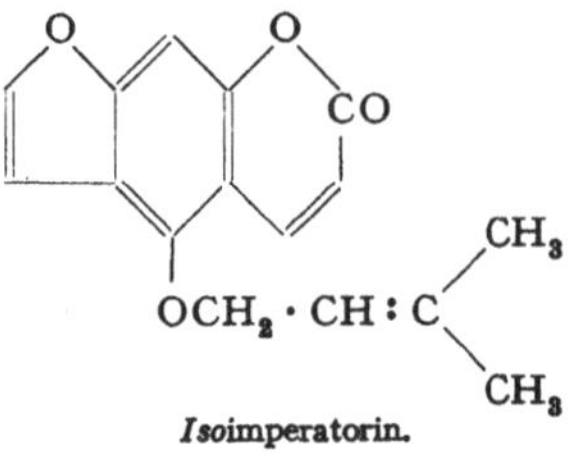

*Iso*imperatorin.

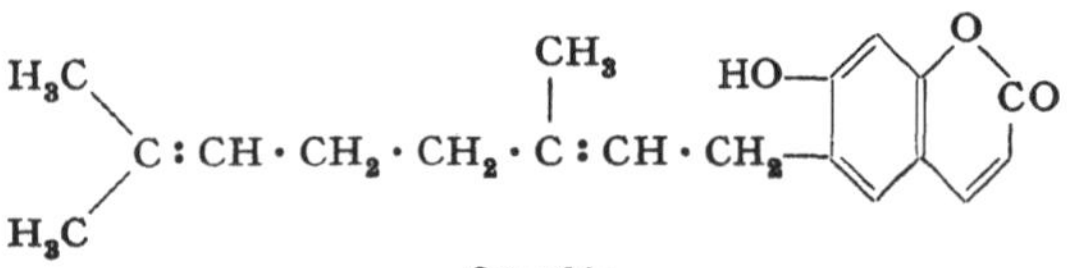

Ostruthin.

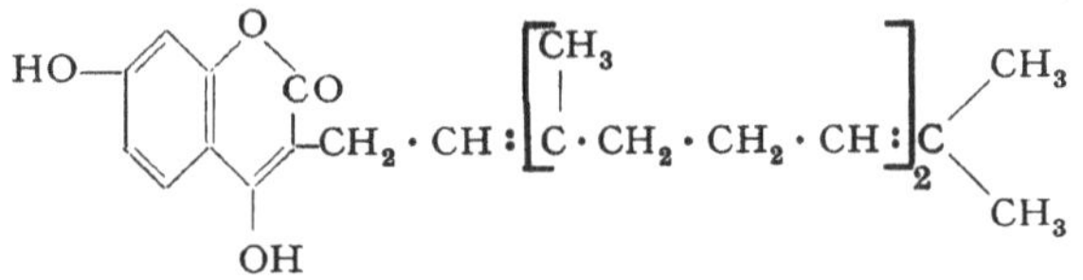

Ammoresinol.

(structure shown)

These isoprenoid units often form the basis of a second heterocyclic ring as in peucedanin and xanthyletin, and in other cases they carry oxygen as epoxides or as 1 : 2-diols, as, e. g. in oxypeucedanin (related to *iso*imperatorin), byakangelicin, and perhaps also in nodakenetin.

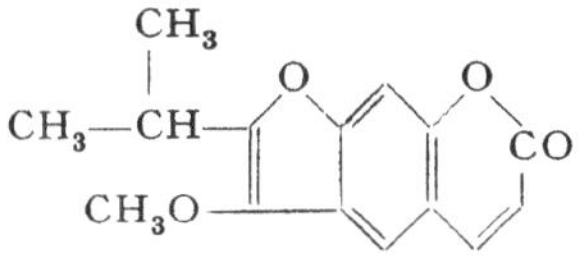

Peucedanin. 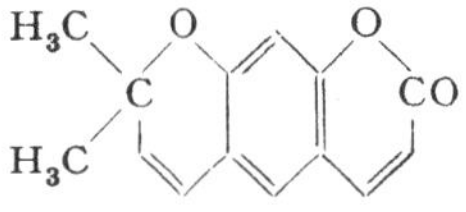   **Xanthyletin.**

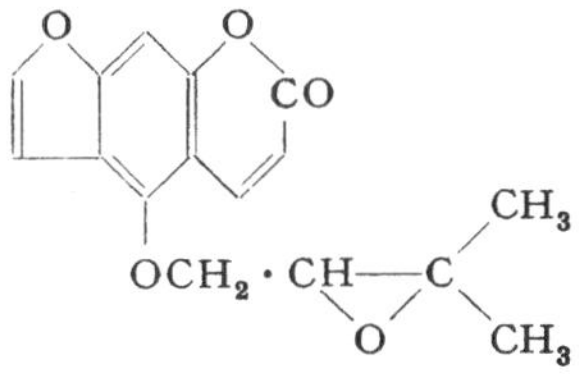

Oxypeucedanin. 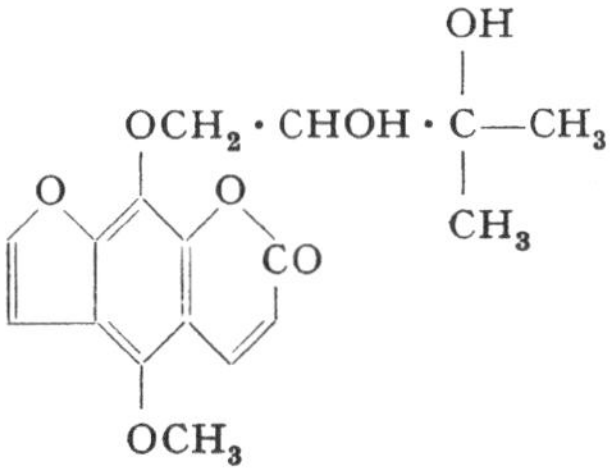   **Byakangelicin.**

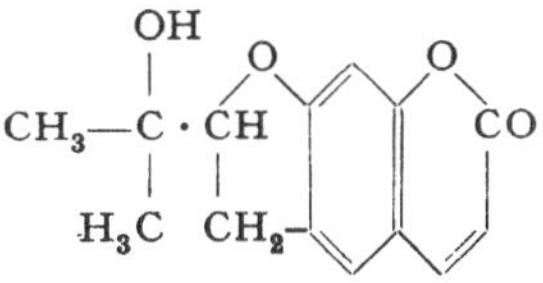

Nodakenetin.

In a few cases Späth and his co-workers have succeeded in reacting umbelliferone with either isoprene itself or with the related α-hydroxy-*iso*propylacetylene to give, for example, seselin. However, the yields were very poor. In several cases ring-closure has been effected with coumarins having adjacent hydroxyl and isoprenoid groups to give pyranocoumarins [a six-membered ring is formed in preference to a five-membered ring with this kind of side-chain (*88*)], while epoxidation of the isoprene system followed by conversion of the epoxide to the glycol has also been carried out successfully.

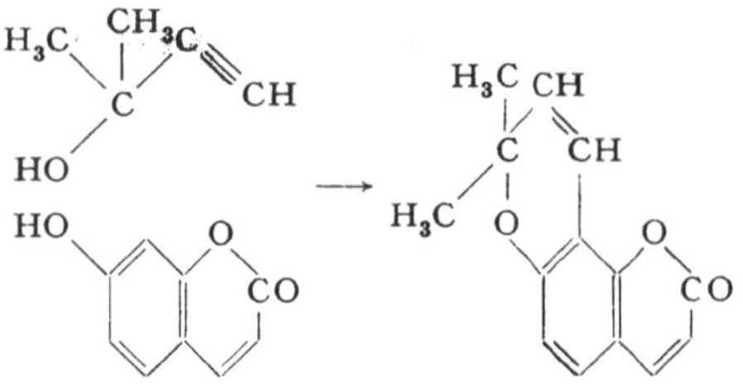

$$\text{Umbelliferone} \longrightarrow \text{Seselin.}$$

ROBINSON's theories of biogenesis postulated that the coumarin system arose by union of C_6 and C_3 units whilst PAVOLINO (1932) suggested pentose sugars as the basis of coumarin biosynthesis, which would account for the great variety of positions occupied by hydroxyl groups. SPÄTH has made the tentative suggestion that the furan system might arise from an isoprene residue, but this concept does not fit easily into a general picture of biosynthetic origins:

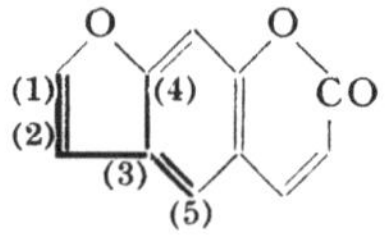

As yet, methyl substituents have not been found in natural coumarins, though the closely related chromones often contain them; nor have coumarins derived from naphthalene or associated with steroid nuclei or nitrogen systems of any kind been discovered in nature, though many of them have been prepared artificially. Except the curiously constructed compound ellagic acid which is very wide-spread, 3 : 4-benzocoumarins are very rare; and LEDERER considers that the benzocoumarins of castoreum are breakdown products of ellagic acid contained in the food of the animal. Ellagic acid itself probably arises from the facile oxidative linking of two gallic acid molecules.

Hydrocoumarin is unique in possessing a reduced pyrone ring; and peucedanin is the only natural coumarin known to carry an enolic methoxyl group.

In addition to their occurence in the free state, most hydroxycoumarins are also found as glycosides, and a few, e. g. coumarin, occur as glucosides of the corresponding coumarinic acid, hydrolysis of which furnishes the coumarin, whilst the stable coumaric acid is obtained if hydrolysis is preceeded by the conversion of the aglucone residue to the *trans* form. In view of the fact that these glucosides are easily hydrolysed by the usual reagents and especially by the enzymes which occur in the plants, it is possible that many more remain to be discovered.

II. The Chemistry of the Coumarin System.
Conversions and Degradation.

Coumarin may be regarded as 5 : 6-benz-α-pyrone, or, more conveniently, as the lactone of cis-o-hydroxycinnamic acid (coumarinic acid) which is not known in the free state because it cyclises to coumarin as soon as it is liberated from solutions of its salts. Yellow sodium coumarinate is obtained by the action of hot dilute sodium hydroxide solution on coumarin, but if the treatment is prolonged isomerisation occurs and the sodium salt of trans-o-hydroxycinnamic acid (coumaric acid) results (Scheme 1). Free coumaric acid is quite stable because the hydroxyl and carboxyl groups are too far apart for interaction to occur. The formation of coumarates from coumarinates is catalysed by yellow mercuric oxide, and coumarates are obtained directly by decomposition of the addition products of coumarins with mercuric salts or sodium bisulfite. Conversely, coumaric acids may be reconverted to coumarinic acids (i. e. coumarins) by irradiation with ultraviolet light, and sometimes merely by heating, although thermal isomerisation is accompanied by decarboxylation to styrenes, a side-reaction which may be avoided by fusion of the methyl coumarate when the coumarin is formed by loss of methanol.

Treatment of an alkaline solution of a coumarin with a methylating agent (methyl sulfate or methyl iodide) results in the formation of an o-methoxycinnamic acid in which cis-trans isomerism may be observed without the interference of lactonisation. Studies of cis-trans isomerism in natural coumarins have been carried out by GRUBER (40) and details of the methylative ring-opening are given by RECHYLER (68) as well as CANTER and ROBERTSON (27). The o-methoxycinnamic acids may be reduced to dihydroderivatives or oxidised to benzoic acids, reactions providing an excellent method of distinguishing coumarins from chromones, which often simulate coumarins in alkaline degradations.

The chemical behaviour of coumarin is explained only partly by the lactonic structure, for the α-pyrone ring is rather difficult to open with alkali while acids do not affect it at all, this tendency of the ring to remain closed being illustrated very well by the fact that even carbonic acid will liberate coumarins from the salts of coumarinic acids. Again, rapid hydrogenation of the 3 : 4-double bond is possible only in the presence of palladium charcoal, under some pressure, whereas hydrogenation of the salts or esters of coumarinic or coumaric acids is facile at ordinary pressures. These facts allow side-chains containing double bonds to be saturated without affecting the coumarin double bond which can be hydrogenated subsequently. Some methods of reduction, e. g. by means of zinc dust and alkali or sodium amalgam, cause the formation of certain amounts of bimolecular reduction products the structures of which have not been clarified completely, while lithium aluminum hydride is said to reduce coumarin to both 3-[2'-hydroxyphenyl]propanol and o-hydroxycinnamyl alcohol (Scheme 1).

Scheme 1.

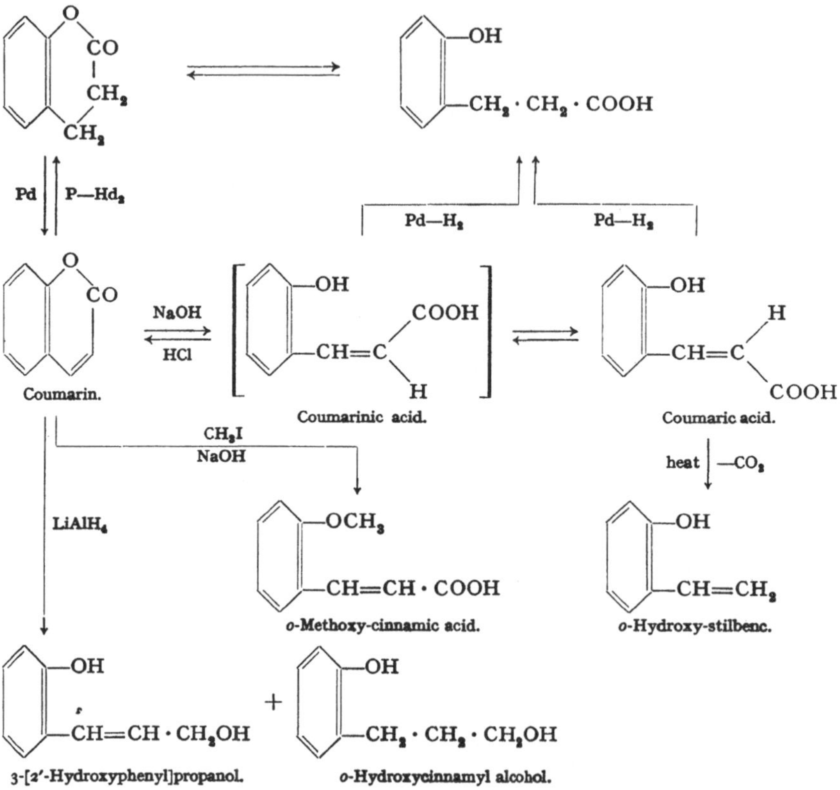

[Reaction scheme showing Coumarin and its derivatives: Coumarinic acid, Coumaric acid, o-Methoxy-cinnamic acid, o-Hydroxy-stilbene, 3-[2'-Hydroxyphenyl]propanol, and o-Hydroxycinnamyl alcohol, with reagents Pd, P—Hd$_2$, NaOH/HCl, Pd—H$_2$, CH$_3$I/NaOH, LiAlH$_4$, heat/—CO$_2$]

The dihydrocinnamic acids formed by *reduction* of coumarinic or coumaric acids are quite stable in the free state, but on distillation they lactonise readily forming dihydrocoumarins. SPÄTH and GALINOVSKY (*89*) have established that dihydrocoumarins can be dehydrogenated to coumarins in good yield by the use of palladium charcoal at 220°, though some decarboxylation to o-ethylphenols occurs at the same time.

The coumarin nucleus is comparatively resistant to *oxidation*. Ozone reacts preferentially with ethylene links in aliphatic side-chains, then with furan double-bonds, and finally with the pyrone system. Chromic acid also attacks side-chains leaving the pyrone ring intact, but permanganates are not very selective in their action and are often employed to break down a complex molecule to derivatives of the basic phenol. Peracids such as perbenzoic acid do not affect the coumarin system, and are used to prepare epoxides from coumarins having allylic side-chains; however, alkaline persulfates have been shown by BARGELLINI and MONTI (8) to effect nuclear oxidation resulting in the formation of 6-hydroxycoumarins, a reaction which has some synthetic use (65).

The degradation of coumarins by the prolonged action of hot concentrated *alkali* or by fusion with alkali is still used to ascertain the nature of the benzenoid nucleus; thus umbelliferone gives rise to resorcinol and bergapten to phloroglucinol, but it must be observed that the reaction has lost some of the importance which was attached to it at one time because the products could arise either from coumarins or from chromones.

If it is desired to retain the pyrone skeleton while other parts of the molecule are being attacked, the coumarin may be converted into the more stable dihydrocoumarin or dihydrocinnamic acid which are able to resist alkaline degradation and the milder oxidising agents. When oxidation is carried far enough, usually by nitric acid or permanganates, the hydrogenated system appears as succinic acid, which, if not obtained from the original compound under the same conditions, is indicative of the coumarin system, provided that care is taken to ensure that the succinic acid does not arise from side-chains.

Although coumarins do not possess any true *carbonyl* activity, they react with some carbonyl reagents. Thus one molecule of hydroxylamine adds to the 3 : 4-double bond and a second causes the lactone ring to open with the formation of a hydroxamic acid; and sodium bisulfite adds to coumarin more easily than to cinnamates giving dihydrocoumarin-4-sulfonic acid:

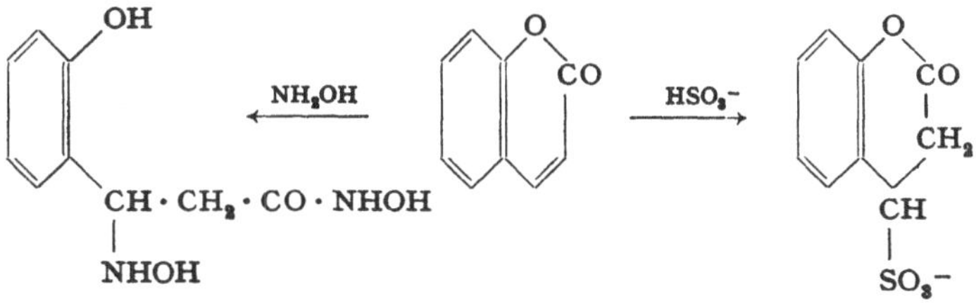

Hydroxamic acid.   Dihydrocoumarin-4-sulfonic acid.

A sequence of reactions which was used to identify the coumarin system in earlier work involves the addition of halogen to the pyrone double bond followed by the removal of one molecule of hydrogen halide by treatment with a base. When heated with alcoholic alkali, the resulting 3-halogeno-coumarins are converted into coumarilic acids:

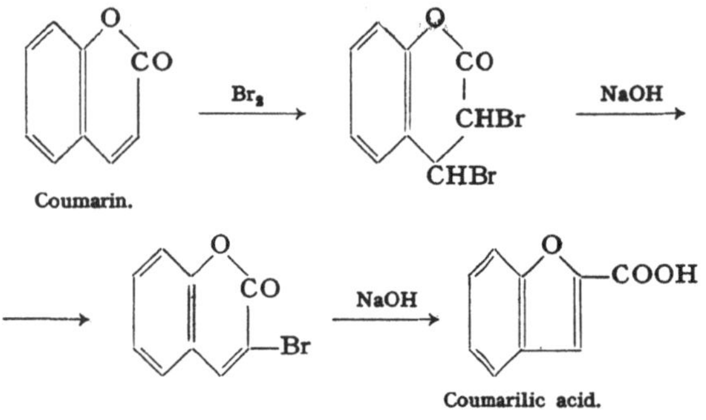

[Scheme 1: Coumarin → (Br₂) → dibromide with CHBr-CHBr → (NaOH) → 3-bromocoumarin → (NaOH) → Coumarilic acid]

Although it is common practice to *methylate* phenolic coumarins by the use of diazomethane, it has been shown that this reagent does attack the 3:4-double bond of coumarin forming a pyrazoline derivative (*131*); but methylation by means of methyl iodide or methyl sulfate in the presence of potassium carbonate is usually satisfactory as this base is too weak to open the pyrone ring.

Nitration of coumarin affects positions 6 and 8, and the products are easily reduced to aminocoumarins without affecting the heterocyclic ring.

Certain reactions of umbelliferone possess particular interest since they provide materials suitable for the elaboration of more complex molecules e. g. the *Duff reaction* (Scheme 2) which has been found to be the best method of preparing aldehydes from hydroxycoumarins, provides 8-formylumbelliferone; and 7-allyl-oxycoumarin undergoes the *Claisen migration* to form 8-allyl-umbelliferone. The *Fries rearrangement* of esters of umbelliferone results in the production of 8-acyl-7-hydroxy-

Scheme 2.

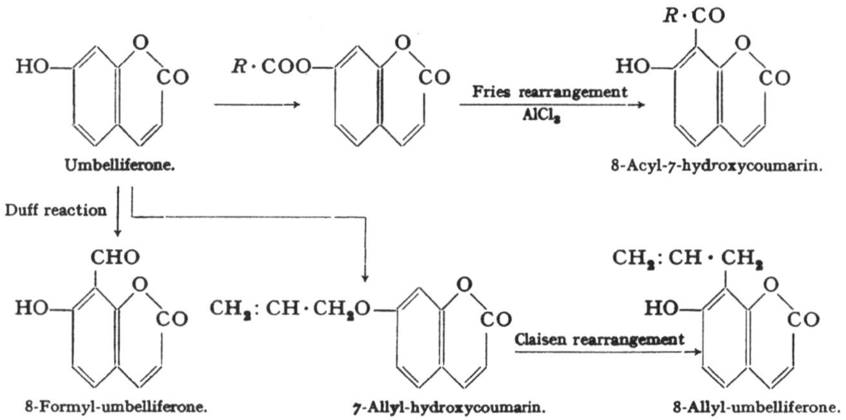

[Umbelliferone → R·COO-coumarin → (Fries rearrangement, AlCl₃) → 8-Acyl-7-hydroxycoumarin; Duff reaction → 8-Formyl-umbelliferone; → 7-Allyl-hydroxycoumarin → (Claisen rearrangement) → 8-Allyl-umbelliferone]

coumarins which suffer loss of the pyrone ring rather than loss of the acyl group when attacked by alkali, so providing a synthesis of 2-acylresorcinols (*Nidhone synthesis*). If position 8 is blocked, then 7-hydroxycoumarins do react at position 6, but only with considerable difficulty.

An interesting feature of coumarin chemistry, and one which is reminiscent of cinnamic acids, is the *dimerisation* of coumarins under the influence of light giving products which are usually formulated in an analogous fashion to the truxillic acids, as, for instance, in the work of WESSELY and PLAICHINGER (*142*). Herniarin dimerises smoothly when the solid is irradiated with light of wavelength 366 mμ, but light of shorter wavelength has little effect and light of longer wavelength causes the formation of a yellowish-red substance. The formulation of the dimer of herniarin as given below (*A*) does not appear to be so well supported as the *cyclo*-butane structure.

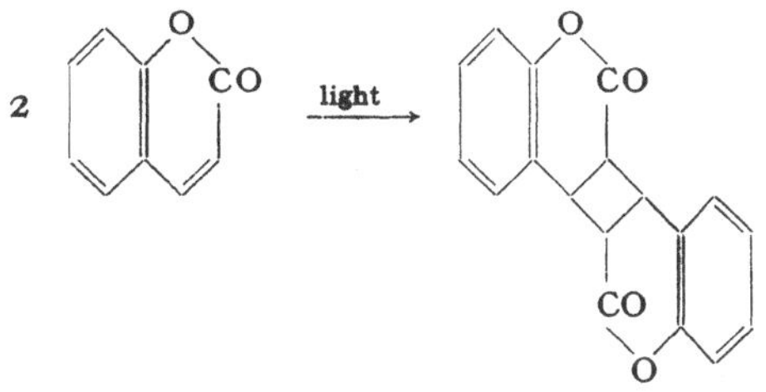

Dimer of coumarin.

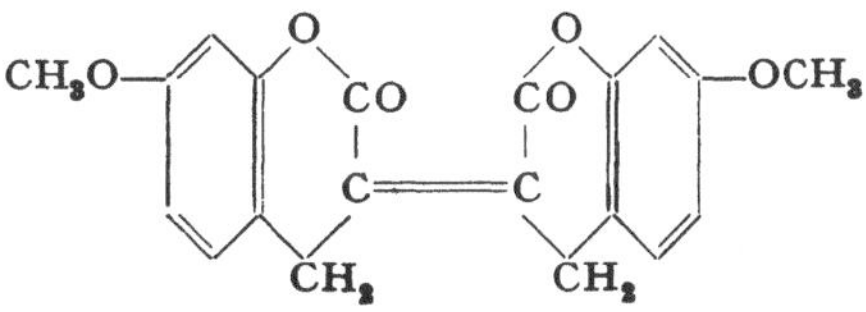

(*A*) Dimer of herniarin (7-methoxy-coumarin)?

Coumarins having a hydroxyl group at position 4 differ considerably from the other hydroxycoumarins because they contain a highly enolised β-ketoester system which enables them to dissolve in solutions of bicarbonates. Their acidic hydroxyl is easily attacked by many reagents, e. g. hot *aniline* furnishes a 4-anilinocoumarin, and phosphorus pentachloride replaces the hydroxyl by chlorine (Scheme 3). Reactions associated with the active methylene of a β-ketoester are also exhibited by 4-hydroxycoumarins which undergo the MICHAEL condensation, couple with diazonium salts, and react with aldehydes (see dicoumarol, p. 257). Hydrolysis of this type of coumarin may involve either ketonic or acidic fission of the β-ketoester residue, so that either the hydroxyacetophenone or the corresponding salicylic acid may be obtained; however, the analogy fails in the case of the ferric reaction because the steric disposition of the enolic system does not favour the formation of complexes, and so 4-hydroxy-coumarins give only a yellow colour with ferric chloride.

The reaction between *diazomethane* and 4-hydroxycoumarins deserves comment. ARNDT and his co-workers (5a) have discovered recently that although the major products are 4-methoxy-coumarins, small quantities of 2-methoxychromones are also produced and appear to be derived from the 2-hydroxychromones which are tautomers of the coumarins (Scheme 3).

Scheme 3.

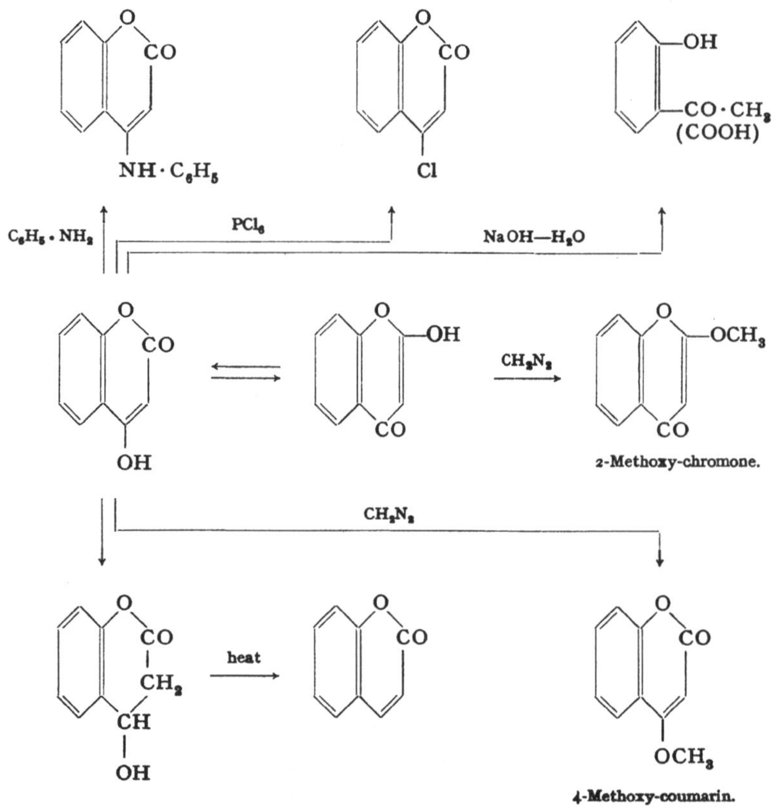

While hydrogenation of a 4-hydroxycoumarin is not easy, it does give rise to a 4-hydroxydihydrocoumarin but because this is a β-hydroxy-ester it tends to lose the elements of water with the production of an unsubstituted pyrone ring.

The Synthesis of Coumarins.

Synthesis of the coumarin system always involves the use of a phenol or a derivative of a phenol, which in the case of natural coumarins is usually a salicylaldehyde. The wellknown method of VON PECHMANN employs the reaction between a phenol, formylacetic acid (in the form of malic acid), and sulfuric acid, and does not seem to give rise to the isomeric chromones, which are sometimes formed when acetoacetic ester and its derivatives are employed instead of malic acid.

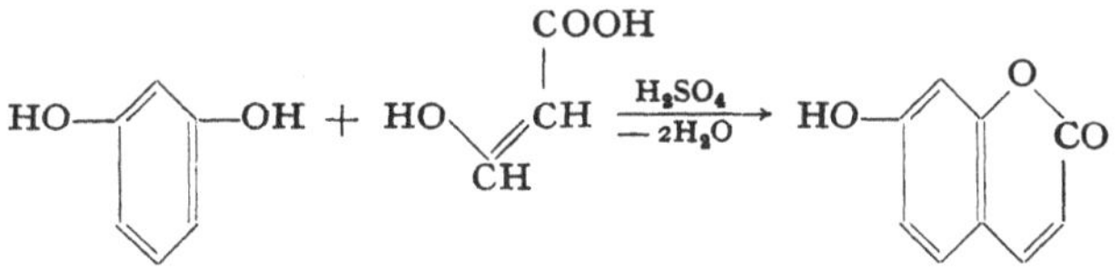

$$HO-C_6H_4-OH + HO-CH=CH-COOH \xrightarrow[-2H_2O]{H_2SO_4} \text{coumarin}$$

This method has been modified occasionally by SPÄTH who heated the acetate of the phenol with the sodium derivative of formylacetic ester, but variants of PERKIN's method are better, since there is no ambiguity concerning the point of ring-closure. This reaction is based on the condensation of a salicylaldehyde and a derivative of acetic acid, the classical example being the synthesis of coumarin itself, although better results are obtained by replacing acetic acid by malonic acid and a tertiary base,

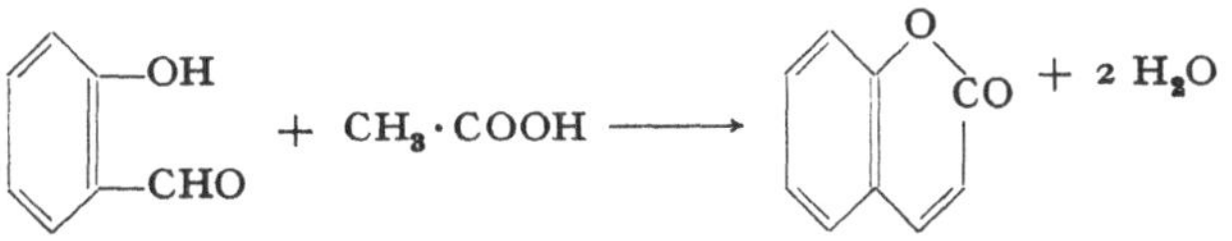

$$\text{salicylaldehyde} + CH_3 \cdot COOH \longrightarrow \text{coumarin} + 2 H_2O$$

when a coumarin-3-carboxylic acid is formed which is easily decarboxylated by heat. Alternatively, the intermediate acid can be obtained by condensing the aldehyde with alkaline solutions of cyanacetic acid followed by hydrolysis of the yellow nitrogenous product (Scheme 4).

Scheme 4.

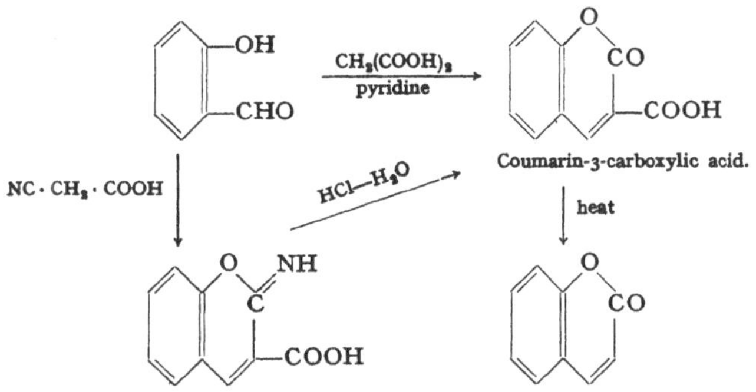

salicylaldehyde $\xrightarrow[\text{pyridine}]{CH_2(COOH)_2}$ Coumarin-3-carboxylic acid $\xrightarrow{\text{heat}}$ coumarin

with $NC \cdot CH_2 \cdot COOH$ giving an iminocoumarin intermediate, hydrolyzed ($HCl-H_2O$) to coumarin-3-carboxylic acid.

The superiority of these methods to the original PERKIN synthesis lies in the fact that resin formation is reduced because the reaction with an active methylene group is rapid, and loss caused by the formation of coumaric acids is impossible. Similar condensations with o-methoxy-aldehydes give derivatives of coumaric acid which must be isomerised before ring-closure can occur.

Scheme 5.

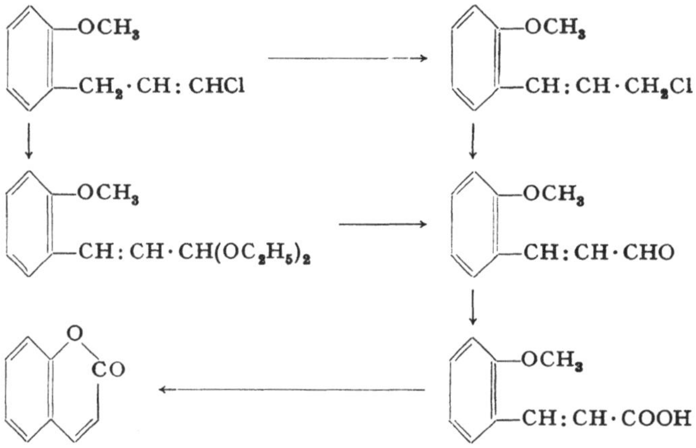

Some of the simplest natural coumarins have been prepared by BERT (*13*) by a new method in which 1 : 3-dichloropropylene is condensed with a phenolic ether by the FRIEDEL-CRAFTS reaction, or with an ether of an o-bromophenol by the GRIGNARD reaction. Scheme 5 indicates the two routes by which the product can be converted into an unsaturated aldehyde that is transformed into the coumarin by oxidation and demethylation.

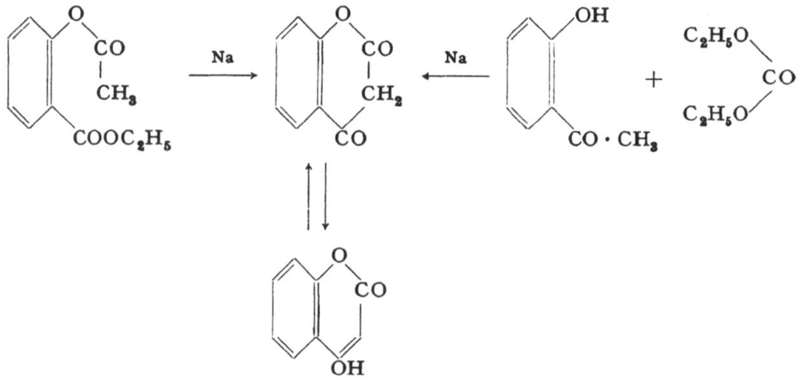

The formation of a 4-hydroxycoumarin involves rather different methods. The best routes make use of the CLAISEN reaction, especially between o-hydroxyacetophenones and ethyl carbonate [BOYD and ROBERTSON (22a)], but if only a phenol is obtainable, the reaction with a malonic ester or cyan-acetic ester can be employed:

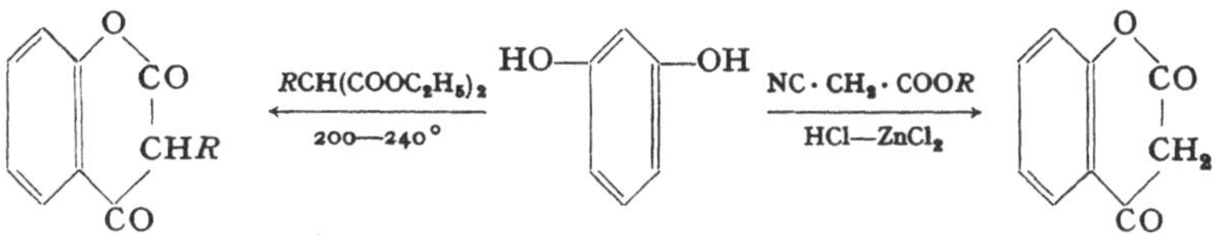

$$\text{[coumarin-CHR(CO)]} \xleftarrow[200-240°]{RCH(COOC_2H_5)_2} \text{HO-C_6H_4-OH} \xrightarrow[HCl-ZnCl_2]{NC \cdot CH_2 \cdot COOR} \text{[coumarin-CH_2(CO)]}$$

Theoretical Considerations.

Some attempts have been made [SESHADRI et al. (7); THAKOR and SHAH (135)] to interpret the behaviour of coumarins by means of *electronic theories*. The dipole moment of coumarin is 4.51×10^{-19} e. s. u. and is compatible with the types of structure (B, C, D, E) which have been supposed to contribute to the normal state of coumarin. The polar contributions would account for the addition complexes which coumarin

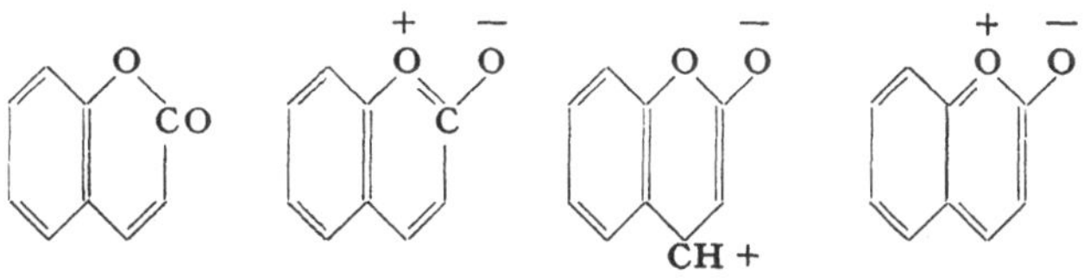

(B)   (C)   (D)   (E)

forms e. g. with hydrocoumarin, and the green complex with mellitic anhydride. Only the first two types of structure can be written for dihydrocoumarin so that it is to be expected that the coumarin ring should be more difficult to open by hydrolysis than the hydrocoumarin ring, and that once open, it should close the more easily of the two.

The unshared electrons of the oxygen hetero-atom can enter into resonance with the π-electrons of either the benzene ring or the carbonyl group, so that its electronic effects on both systems are modified. Thus, although positions 6 and 8 are active towards cationoid reagents, they are less so in coumarin than the corresponding positions in o-hydroxycinnamic acid where the oxygen is able to interact with the benzene ring only. Further, since ester-type resonance is reduced by the benzene ring, the π-electrons of the carbonyl group and the 3 : 4-double bond are able to interact to a greater degree than the corresponding systems in the cinnamates. This results in considerable cationoid activity at position 4 (type *D*) and so groups such as HSO_3^- are added more easily to coumarins than to cinnamates.

New considerations arise with 7-hydroxycoumarin, i. e. umbelliferone (*F*). Further structures, notably (*G*), can in this case contribute to the normal state of the molecule, so that ring-opening or hydrogenation which would destroy much of the resonance energy are now more difficult than with the parent coumarin.

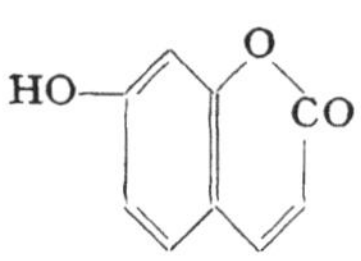

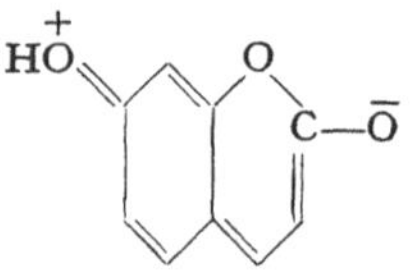

A structure of type (*G*) cannot be written for 6-hydroxy-coumarins, which indeed do not differ greatly in these reactions from coumarin itself, but 5-hydroxycoumarin would be expected to simulate the behaviour of 7-hydroxycoumarin, except that the state (*I*) would be much less important than (*G*), since it is ortho-quinonoid as opposed to the para-quinonoid system of umbelliferone, an effect which is noticeable in fluorescence phenomena. Neither 7- nor 5-hydroxycoumarin fluoresce in the neutral state, but in alkaline media the former exists as the ion (*H*) which possesses a strong blue fluorescence, while the corresponding ion from (*I*) does not. Both ions are yellow as would be expected from their structure. 7-Methoxycoumarin does not fluoresce in alkaline media since it cannot form an ion of type (*H*) but, like other hydroxycoumarins, it is able to fluoresce in concentrated sulfuric acid solution, probably because of the occurrence of ions of type (*J*).

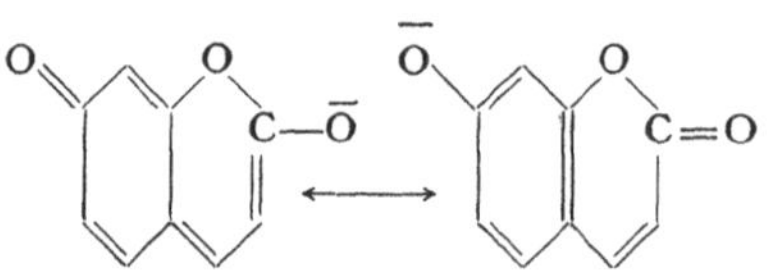

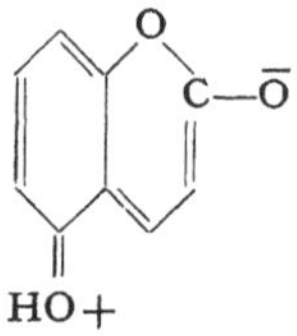

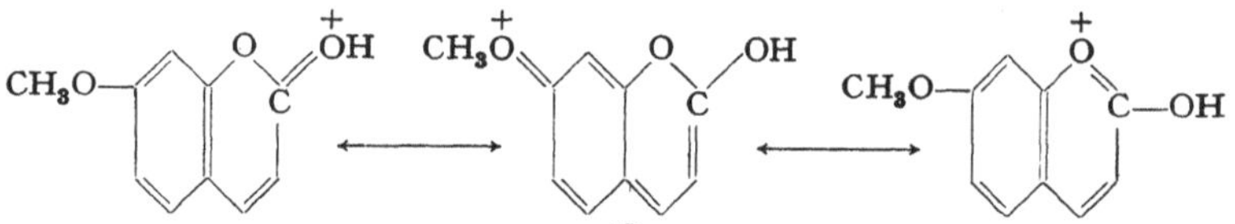

Insertion of hydroxyl into position 5 of umbelliferone greatly reduces the fluorescence, presumably because it can conjugate with the unsaturated system of the ion; insertion of hydroxyl at position 6 does reduce the intensity of fluorescence, but not so markedly, while the appearance of fluorescence in other polyhydroxy-coumarins and their ethers seems to be unpredictable at present. (These theories have been carried further with derivatives of coumarin which do not occur naturally.)

Structures (C) and (E) for coumarin are electronically analogous to naphthalene and in fact it has been shown that the Raman spectrum of coumarin bears similarity to both coumarone and naphthalene. This resemblance appears also in umbelliferone which is easily substituted in position 8 but not in position 6, just as a β-naphthol is attacked almost entirely in position 1; and while 5- and 8-hydroxycoumarins, like α-naphthol, are attacked at either of the positions *ortho* or *para* to the hydroxyl.

None of the above considerations apply to hydrocoumarin derivatives, which are devoid of fluorescence, behave as normal lactones, and show no unusual orientation phenomena in substitution reactions.

III. Occurrence, Isolation, and Determination.

Occurring in various parts of plants, coumarins can usually be isolated from the roots, trunks, leaves or fruits, but frequently several related members are found together making their isolation difficult. They are of common occurrence in legumes, citrus fruits, grasses, and orchids; and the parent compound has been isolated from approximately eighty different species of plants. In 1937 SPÄTH (78) published a list of coumarins and the plants from which they had been obtained, therefore the present article mentions only those sources of coumarins discovered since that time. The few coumarins which have been found outside the plant world are those 3:4-benzocoumarins which LEDERER (53) has shown to be constituents of the scent-glands of the beaver.

Many coumarins can be extracted from plant materials by suitable solvents (e. g. ether, benzene, acetone); coumarin and dihydrocoumarin are easily isolated by steam-distillation.

If the coumarin is present as a glycoside, the aglycone may be isolated after preliminary hydrolysis, alternatively the glycoside may be extracted by means of alcohol or water. In some cases crude preparations of coumariniferous plants are used as drugs, e. g. *Ch'uan sen* and *Semen angelicae* and these form good sources of several coumarins.

Natural coumarins are either neutral or phenolic and appear in the corresponding fractions of the plant extract. In the subsequent isolation it is usual to take advantage of the fact that the coumarin ring is opened by warm dilute alkali with the formation of the sodium coumarinate, thus allowing the removal of neutral material. On decomposition of the sodium salt with acid the coumarinic acid cyclises spontaneously, regenerating the coumarin which is thus easily separated from acids formed by the hydrolysis of the other constituents; however, the method is not ideal because it is hardly possible to avoid some degradation of the coumarins. Chromatographic analysis is useful after the initial separation is completed.

Certain coumarins (e. g. umbelliferone, scopoletin) can be isolated from plant tissues by sublimation, but this process may involve thermal degradation with the formation of artefacts, as in the case of 4:7-dihydroxy-3-methylcoumarin which is formed from a more complex coumarin (ammoresinol) when umbelliferous resin is heated. KOFLER and GEYR (1934), and FISCHER and EHRLICH (1936) have adapted sublimation to the characterisation and estimation of coumarins on a very small scale.

Several coumarins have been characterised by the crystalline derivatives which they form with iodine and mercuric salts.

As a rule, the estimation of coumarins is carried out by various colour reactions either with or without the preliminary opening of the pyrone ring, examples being the ferric reaction given by coumarins after fission by sodium sulfite, the colour obtained with ammonium persulfate, and the method of ROBERTS and LINK (70) in which selective extraction of coumarins from sweet clover tissue is followed by a coupling reaction with diazotised p-nitraniline. Many coumarins can be identified by their characteristic fluorescence phenomena (39).

IV. Some Biochemical Properties.

In common with other unsaturated lactones, the natural coumarins exert varied and pronounced effects on living organisms. These effects have often been modified in synthetic coumarins so as to produce compounds of therapeutic value. A large literature exists on these and related topics to which only brief reference can be made.

The pleasant and persistant odour of coumarin is of great use in perfumery, and hydrocoumarin has a closely similar odour, possibly because it is oxidised to coumarin before it exerts its effect. The simplest methoxy- and hydroxy-coumarins have a faint coumarin-like odour, but this disappears in more complex molecules. The blow-fly *Cynomyia cadaverina* can be trained to react to the odour of coumarin by association with its food, a fact which allowed a study of the olfactory system of the insect to be made.

Coumarin inhibits the growth of lettuce and clover, and is a more powerful inhibitor of the germination of barley seed than thiourea, possibly because the sulfhydryl groups of the enzymes are inactivated by addition to the 3:4-double bond of the coumarin. BEST (14) isolated scopoletin from decapitated tobacco plants infected with the virus of tomato wilt, where it was readily detected by its strong blue fluorescence. Similar observations have been made with virus infected tubers and with oat roots; in the latter case GOODWIN and KAVANAGH (38) noted that those parts of five day old *Avena* roots which contained the most scopoletin showed the slowest growth but, on the other hand, some of the coumarin glycosides are reported to stimulate the growth of cress roots and other plants, perhaps because the 3:4-double bond participates in oxidation-reduction systems.

The strong absorption of ultraviolet light by umbelliferone and many other coumarins has led to their use in sunburn prevention, whilst the concomitant bluish fluorescence makes them useful as "whitening agents" in soap powders and as indicators in titration and chromatographic analysis. ASAI (6) studied the variation of the content of daphnin in the leaves of *Daphne odora* THUNB. and discovered that the newly opening leaf-buds contain as much as 27% of their dry weight of daphnin but that the fully grown leaf has only 6–7% and this falls to a constant value of 1–3% until the leaves drop off. He suggests that daphnin acts as a protection for the young buds from harmful short-wave radiation.

Coumarin is a narcotic for some animals (e. g. rabbits, earthworms) and a sedative and hypnotic for mice. Some larger animals (e. g. dogs, horses) can be killed by coumarin, but moderate quantities have no very marked effect on man except that it has been reported recently that coumarin has a true curare-like activity. A study of the contraction of various muscles by coumarin was made by DE MOURA CAMPOS.

IIDA (50) gave frogs and mice intravenous injections of fraxin which paralysed the central nervous system. The same coumarin had a pressor effect on rabbits, and also induced antipyresis and diuresis, the last observation leading to the discovery that fraxin causes a rise in the excretion of uric acid and is superior to atophan (cincophene) in the treatment of gout.

WASICKY (78) found that pimpinellin, ostruthin and peucedanin have a comparatively slight toxic effect on rats, mice, and guinea-pigs, whereas PRIESS, and ROST, and SIEBURG, and especially SPÄTH and KUFFNER (107) demonstrated that many natural coumarins have powerful effects on fresh-water fish. The simpler coumarins have little action, but the more complex members, particularly the furanocoumarins, are toxic in extremely small doses. At first the fishes are stimulated, then they float on their backs; later they stop moving and finally die. Angelicin induces narcosis, during which the fish do not respond to external stimuli, but they return to normal in fresh water.

Cattle which have eaten spoiled sweet clover (improperly cured hay) often bleed to death from minor injuries because their blood can no longer clot properly. In the course of this "sweet clover disease" haemorrhages may occur in any part of the body, and massive haematomas are frequent. This effect which had been attributed vaguely to the coumarin content of the plant, though it was known that the same condition could not be induced by coumarin itself. In the course of their important studies LINK et al. (26, 132) isolated the active principle from spoiled clover in pure, crystalline form and showed it to be methylene-*bis*-[4-hydroxy-3-coumarinyl], now commonly called "dicoumarol". Injections of this substance cause a delayed coagulability of the blood.

Antagonists to dicoumarol are the antihæmorrhagic naphthoquinones such as vitamin K. A comprehensive survey of this whole field has been offered by LINK (54a).

BAYERLE and MARX (9) suggest that dicoumarol hinders the formation of prothrombin, so that there is a relative increase in the antithrombin and antiprothrombin levels but they consider that the liver is damaged by this treatment. That the liver is in fact the site of dicoumarol activity has been established by the use of dicoumarol with the methylene group containing labelled carbon.

Many coumarins have been synthesised and examined for their antivitamin K effect (cf. 54a) but only those having at least one intact 4-hydroxycoumarin ring possess pronounced activity; the advantage is that more soluble derivatives of this very insoluble substance have become available. The action of dicoumarol has been attributed to its breakdown into formaldehyde, but LINK found that the important substance is salicylic acid which he has shown can be derived from all coumarins having antivitamin K activity and which does have a short-term action similar to that of dicoumarol.

BOSE and SEN (21) have found that ayapin and ayapanin are the active principles of the leaves of *Eupatorium ayapana*, a plant used for its haemostatic properties. A study of the relation between structure and haemostatic activity has been made for coumarins.

In addition to its vitamin P activity, aesculin is useful for the identification of bacteria; e. g. ORLA-JENSEN (64) showed that lactic acid bacteria and most of the streptococci can hydrolyse aesculin in the substrate, an action easily detected by the development of a black colour with ferric ions; and recently the method has been used to differentiate a pneumococcus from a patient suffering from meningitis, from enterococci, which can split aesculin.

V. Simple Coumarins.

Coumarin, $C_9H_6O_2$, m. p. 67–8°, has been isolated from the following sources since 1937.

Melilotus dentata; *Torresea ceariensis* FR.; *Pastinaca sativa*; *Laserpitium latifolium* L.; *Petroselinum sativum*; *Lavandula spica* DESC. (*L. delphinensis* JORD.), and *Verbascum thapsus*, where it is associated with rotenone.

Melilotoside, $C_{15}H_{18}O_8 \cdot H_2O$, m. p. 240–241° (decomp.); $[\alpha]_D$ 64.1°, was extracted from the leaves of *Melilotus officinalis* LAM. (*M. altissima*

THUILL., *M. arvensis* WALLR.) by CHARAUX who noted that hydrolysis of melilotoside by means of emulsin or dilute acids resulted in the formation of *D*-glucose and coumaric acid, whilst SHINODA and IMAIDA (76) also identified coumarin in the hydrolysate, and considered that the glucoside changed to some extent from the *trans* to the *cis* modification at the moment of fission since coumaric acid itself was not isomerised by emulsin. SHINODA and IMAIDA synthesised melilotoside by interaction of helicin (salicylaldehyde *D*-glucoside) and malonic acid in the presence of pyridine and piperidine.

$$\underset{\text{Helicin.}}{\overset{-O \cdot C_6H_{11}O_5}{\underset{-CHO}{\bigcirc}}} \xrightarrow[-H_2O\ -CO_2]{CH_2(COOH)_2} \underset{\text{Melilotoside.}}{\overset{-O \cdot C_6H_{11}O_5}{\underset{-CH:CH \cdot COOH}{\bigcirc}}}$$

$$\downarrow$$

$$\overset{-O \cdot C_6H_{11}O_5}{\underset{-CH_2 \cdot CH_2 \cdot COOH}{\bigcirc}}$$

Dihydro-coumarin (Hydrocoumarin; Melilotol; Melilotic Anhydride), $C_9H_8O_2$, m. p. 25°, b. p. 272°, was originally obtained by steam-distillation of *Melilotus officinalis* LAM. by PHIPSON.

In 1922 NAVEZ reported without giving details that melilotic acid occurred as a glucoside together with coumarin; so LUTZMANN (55) synthesised such a glucoside by catalytic hydrogenation of melilotoside and noted that it sintered at 143–5°, melted at 173° and that in water it had $[\alpha]_D^{20} - 56.1°$.

Umbelliferone (Skimmetin; Dichrin A), $C_9H_6O_3$, m. p. 223–4°, is found by itself and together with other coumarins. It occurs in the umbelliferous resins, galbanum, asafaetida, and ammoniacum and has more recently been found in the bark of *Aegle marmelos* CORREA and in *Dichroa febrifuga* LOUR. From the wood of *Skimmia japonica* THBG. EIJKMAN extracted a substance called skimmin which on hydrolysis gave a sugar and the aglycone skimmetin. MAUTHNER prepared umbelliferone glucoside but considered that his material was not the same as skimmin, but repetition of this work by SPÄTH and NEUFELD (*110a*) showed that skimmin did give umbelliferone and glucose on hydrolysis, and that MAUTHNER's synthetic substance is identical with the natural product.

Herniarin (Ayapanin; 7-Methoxycoumarin), $C_{10}H_8O_3$, m. p. 114°, first found in the leaves of *Herniania hirsuta* L. by BARTH and HERZIG, is easily prepared by methylation of umbelliferone and is also a constituent of lavender oil, the flowers of *Matricaria chamonilla* L. and the leaves of *Eupatorium triplinerve* VAHL (*E. ayapana* VENT.). It has a pleasant coumarin-like odour and is a haemostatic agent.

Aesculetin (Cichorigenin, Crategin), $C_9H_6O_4$, m. p. 270° (decomp.). is also a very common coumarin, especially in the *Fraxinus* species. Originally isolated from *Aesculus hippocastanum* L. and the seeds of *Euphorbia lathyris* L., it has more recently been found, associated with fraxetin, in the bark of *Aesculus turbinata* BLUME. It reduces FEHLING's solution, and gives a green ferric reaction as well as a dimethyl ether.

It is most easily available by the PECHMANN reaction on 1:3:4-triacetoxybenzene, and it has also been obtained by BERT's method.

Aesculetin.

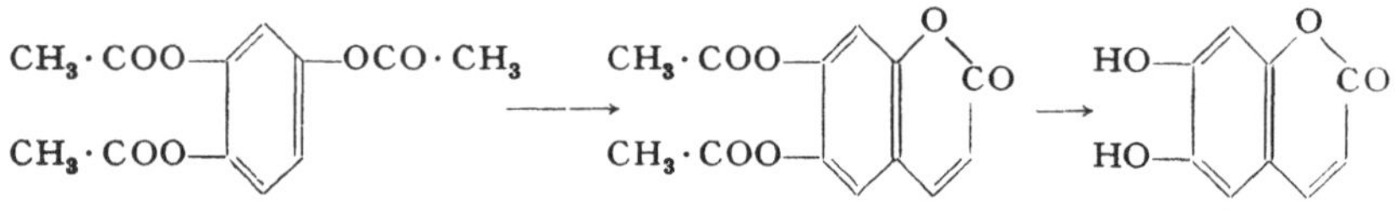

Two aesculetin glucosides are known, viz. *aesculin*, $C_{15}H_{16}O_9 \cdot 2\,H_2O$, m. p. 160° (205°, anhydr.), $[\alpha]_D^{15}$ — 146°, from the horse-chestnut, and the less common *cichoriin*, $C_{15}H_{16}O_9$, m. p. 216°, $[\alpha]_D$ — 104.5°, from chicory. MERZ (57) elucidated the structures of these compounds by hydrolysis which in both cases gave equimolecular proportions of glucose and aesculetin.

Aesculetin. Cichoriin. Scopolin.

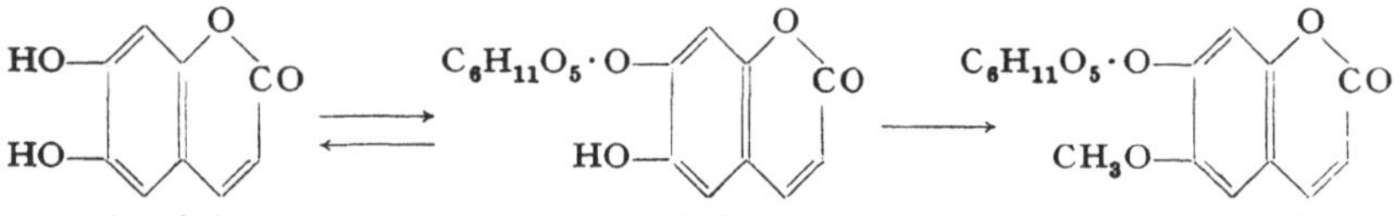

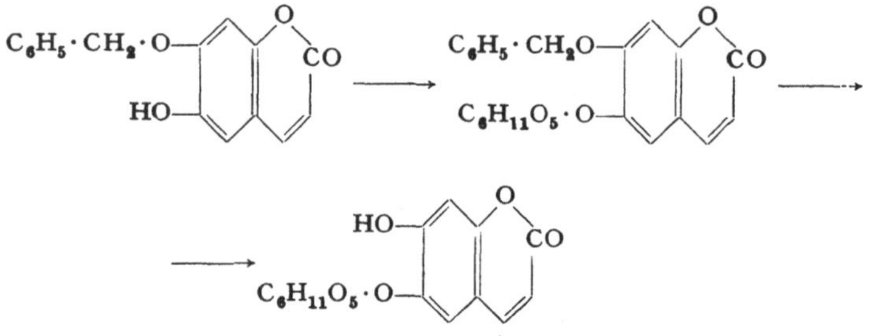

Since diazomethane methylated cichoriin to yield scopolin from which scopoletin (see below) of known structure was obtained by hydrolysis, cichoriin was formulated as the 7-β-glucoside and aesculin as the 6-β-glucoside of aesculetin. HEAD and ROBERTSON (44) proved that aesculetin reacted with O-tetraacetyl-α-glucosidyl bromide in alkaline media to give the tetraacetate not of aesculin (as had been supposed

Aesculin.

originally) but of cichoriin, which was isolated in the free state after deacetylation. MERZ and HAGEMANN (58) obtained synthetic aesculin by attaching the glucose residue to the 7-benzyl ether of aesculetin by standard methods, and then removing the benzyl group by catalytic hydrogenation (p. 244).

Scopoletin (Chrysatropic acid, Gelseminic acid), $C_{10}H_8O_4$, m. p. 203°, is aesculetin 6-methyl ether and occurs naturally as its glucoside scopolin, $C_{16}H_{19}O_9$, m. p. 218°, in the rhizome of *Scopolia japonica* MAXIM., and in the free state in *Murraya exotica* L. On hydrolysis scopolin yields glucose and scopoletin which contains one methoxyl group; it also results from partial methylation of aesculetin by means of alkaline dimethyl sulfate, and can be degraded to 4-methoxyresorcinol, from which it was synthesised by HEAD and ROBERTSON (43) as follows:

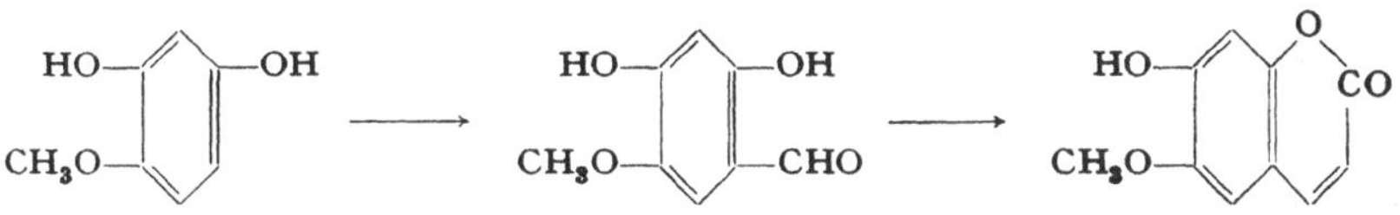

HO—⟨ ⟩—OH   HO—⟨ ⟩—OH   HO—⟨ ⟩=O
                                              \\CO
CH_3O—⟨ ⟩ CH_3O—⟨ ⟩—CHO CH_3O—⟨ ⟩

4-Methoxyresorcinol.   Scopoletin.

Although scopolin was once thought to be a bioside, MERZ (57) showed that it could be synthesised by methylation of cichoriin which had been proved to be a simple glucoside in previous work.

BOSE and MOOKERJEE (20) have recently found scopolin in the petals of *Murraya exotica* L. and it also occurs in the stem-bark of *Diospyros maritima* BLUME, while scopoletin has been shown to be the fluorescent agent in various virus infected plants (2, 14, 38).

Fabiatrin, $C_{21}H_{26}O_{13}$, m. p. 236–8°, from *Fabiana imbricata* RUIZ and PAV., has been proved by synthesis to be the β-primeveroside of scopoletin [CHAUDHURY, HOLLAND, and ROBERTSON (29)]. It was found to be necessary to introduce the disaccharide as a whole instead of attaching the pentose to the preformed glucoside.

Ayapin, $C_{10}H_6O_4$, m. p. 220–1°, is obtainable from the fresh leaves of *Eupatorium triplinerve* VAHL. (*E. ayapana* VENT.) and according to SPÄTH, BOSE and SCHLÄGER (80) it is the methylene ether of aesculetin.

Aesculetin Dimethylether, $C_{11}H_{10}O_4$, m. p. 144°, is found in the unripe fruits of *Artemisia capillaris* (63a).

The Formosan (but not the Japanese) variety of this plant also contains a phenol $C_{10}H_8O_4$, m. p. 205°, which is closely similar to, but not identical with, scopoletin.

Citropten (Limettin), $C_{11}H_{10}O_4$, m. p. 147°, is a constituent of the deposits of lime and lemon oils etc. TILDEN and BURROWS (137) isolated phloroglucinol and acetic acid from alkali fusions and later SCHMIDT (75) synthesised citropten by methylation of 5:7-dihydroxycoumarin prepared from phloroglucinaldehyde [cf. also HEYES and ROBERTSON (*46*)].

Phloroglucinaldehyde. Citropten.

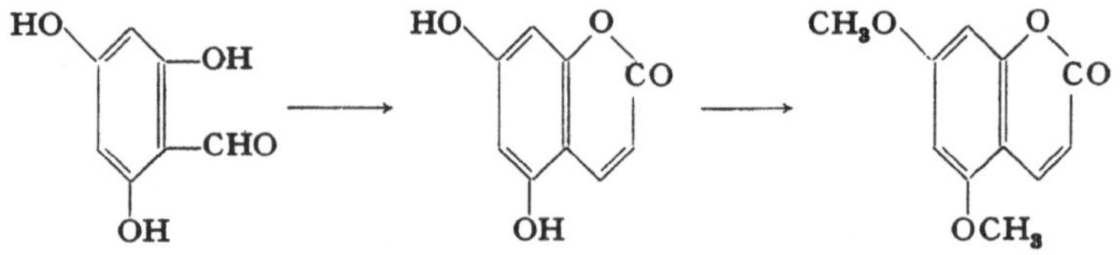

Daphnetin, $C_9H_6O_4$, m. p. 256°, has long been known as one product of hydrolysis of daphnin, $C_{15}H_{16}O_9$, m. p. 215° (223–4°), $[\alpha]_D^{22}$—114.7° (in methanol), a glycoside isolated from various species of ash. Recently, daphnetin was found in the seeds of *Euphorbia lathyris* L.

This coumarin, which possesses two hydroxyl groups, reduces FEHLING'S solution and gives a green colouration with ferric chloride, has been synthesised by the condensation of malic acid with pyrogallol.

The reaction between daphnetin and O-tetraacetyl-α-glucosidyl bromide supplied LEONE with a glucoside similar to daphnin. WESSELY and STURM (*143*) proved that the synthetic material was 8-glucosido-daphnetin, by methylation, removal of the sugar residue, and ethylation of the aglucone, a sequence which resulted in 8-ethoxy-7-methoxycoumarin. The latter was clarified by ethylative ring-opening and oxidation of the product to 2:3-diethoxy-4-methoxy-benzoic acid.

Daphnetin.

2:3-Diethoxy-4-methoxy-benzoic acid. 8-Ethoxy-7-methoxy-coumarin.

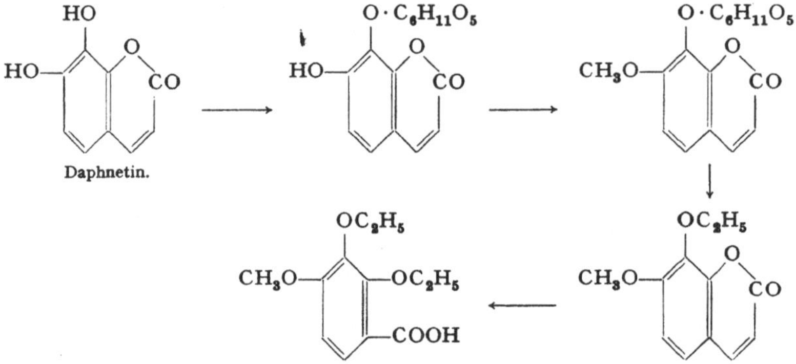

Further work by the latter authors and by HATTORI (*42*) proved that LEONE'S synthetic glucoside was not identical with daphnin, since the natural glucoside gave 8-ethoxy-7-methoxycoumarin when ethylated first, then hydrolysed, and the freed hydroxyl methylated. Therefore daphnin is 7-glucosidodaphnetin, the synthesis of which was carried out by GANDINI (*37*) who observed that partial acetylation of daphnetin gave the 8-acetyl derivative, the sodium salt of which he treated with O-tetraacetyl-α-glucosidyl bromide and obtained a mixture of 7- and 8-glucosidodaphnetin; from this material pure daphnin could be isolated.

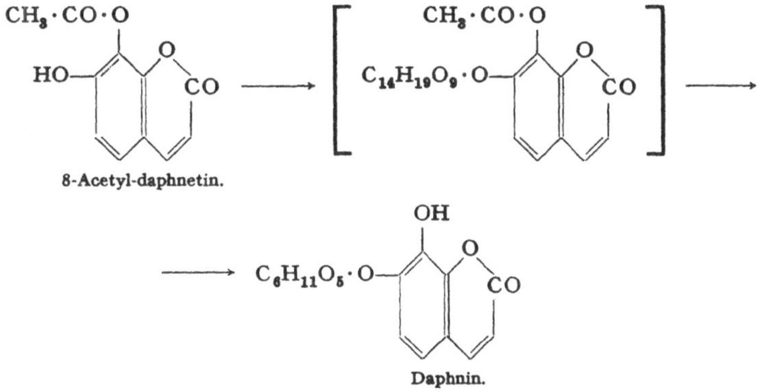
$CH_3 \cdot CO \cdot O$
HO
O
CO
$CH_3 \cdot CO \cdot O$
$C_{14}H_{19}O_9 \cdot O$
O
CO
OH
$C_6H_{11}O_5 \cdot O$
O
CO

8-Acetyl-daphnetin.

Daphnin.

Instead of the usual hydrolysis procedures, daphnin is easily split by heating in a high vacuum, according to VON CHRISTIANI and HOFFMANN (*30*).

According to RINDL (*69*), ***Anthrosolen polycephalus*** C. A. MEY (*Gnidia polycephala* GILG.; "January Bossie") contains a glycoside, $C_{15}H_{16}O_9 \cdot H_2O$, m. p. 197–8° (decomp.), $[\alpha]_D^{20}$ − 65.6°; $[\alpha]_D^{23}$ − 61.9° (in water) which is not identical with daphnin, although it furnishes daphnetin when hydrolysed by acids or emulsin.

Fraxetin, $C_{10}H_8O_5$, m. p. 228°, can be extracted from the bark of *Aesculus turbinata* BLUME. It was obtained by SALM-HORSTMAR in 1857 by hydrolysis of fraxin, $C_{16}H_{18}O_{10}$, m. p. 205°, and suspected to be a monomethyl ether of a trihydroxycoumarin by KORNER and BIGINELLI. WESSELY and DEMMER (*138*) oxidised the tetramethyl ether of the cinnamic acid moiety ex fraxetin, and obtained 2:3:4:5-tetramethoxy-

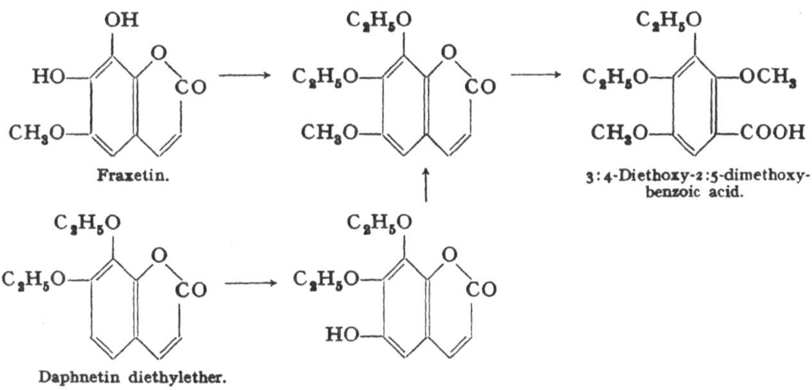
OH
HO
O
CO
CH_3O
C_2H_5O
C_2H_5O
O
CO
CH_3O
C_2H_5O
C_2H_5O
OCH_3
CH_3O
COOH
C_2H_5O
C_2H_5O
O
CO
C_2H_5O
C_2H_5O
O
CO
HO

Fraxetin.

3:4-Diethoxy-2:5-dimethoxy-benzoic acid.

Daphnetin diethylether.

benzoic acid, whereas the diethyl ether of fraxetin when subjected to ring-opening, methylation, and oxidation gave 3:4-diethoxy-2:5-dimethoxybenzoic acid. These reactions established that fraxetin was 6:7:8-trihydroxycoumarin 6-methylether, a view substantiated by the preparation of fraxetin diethylether by BARGELLINI persulfate oxidation of daphnetin diethylether, followed by methylation (*139*).

The total synthesis of fraxetin proved to be very difficult. By interaction of hydrogen peroxide and 2:3-dihydroxy-4-methoxybenzaldehyde, SPÄTH and DOBROVOLNY (*86*) prepared apionol-1-methyl ether, which condensed with the sodium derivative of formyl acetic ester to give a small yield of fraxetin (see also sphondin, p. 265).

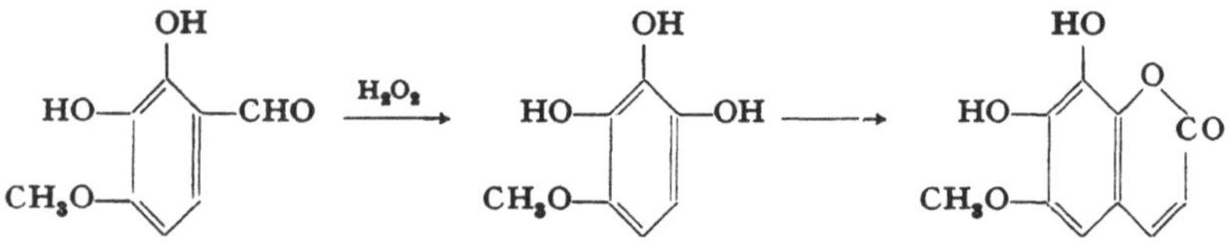

2:3-Dihydroxy-4-methoxy-benzaldehyde. Apionol-1-methylether. Fraxetin.

Methylation of fraxin, followed by hydrolytic removal of the glucose residue and ethylation of the resulting phenol, furnished WESSELY and DEMMER (*139*) with 6:7-dimethoxy-8-ethoxycoumarin which was synthesised by methylation of the product of BARGELLINI oxidation of 8-ethoxy-7-methoxycoumarins. Thus the identity of fraxin with 8-glucosidofraxetin was proved.

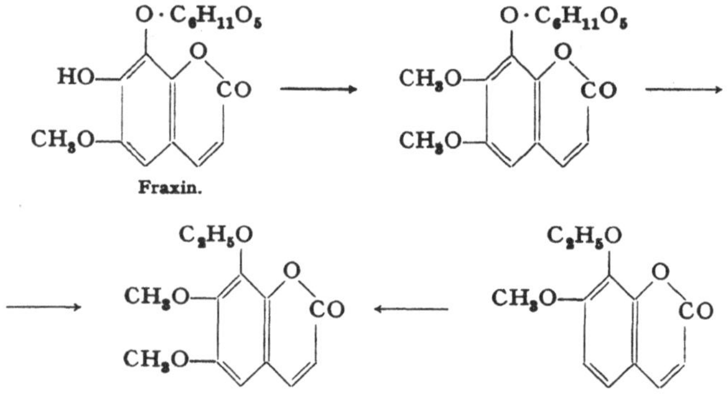

Fraxin.

6:7-Dimethoxy-8-ethoxycoumarin.

Fraxidin, $C_{11}H_{10}O_5$, m. p. 196—7°, obtained from the mother liquors of fraxinol, was shown by SPÄTH and JERZMANOWSKA-SIENKIEWICZOWA (*95*) to be identical with 6:7-dimethoxy-8-hydroxycoumarin which had been prepared by WESSELY and DEMMER (*139*) from fraxin.

Isofraxidin, $C_{11}H_{10}O_5$, m. p. 148–9°, was found with fraxidin and gave 6:7:8-trimethoxycoumarin when methylated. It contained two methoxyl groups; its non-identity with fraxidin and 6-hydroxy-7:8-dimethoxycoumarin indicated that it was 7-hydroxy-6:8-dimethoxycoumarin. Partial methylation of fraxetin with diazomethane gave a mixture of fraxidin and *iso*fraxidin (96).

Fraxinol, $C_{11}H_{10}O_5$, m. p. 172–3°, was isolated by SPÄTH and JERZMANOWSKA-SIENKIEWICZOWA (97) from the hydrolysate of the ether extract of the fresh bark of the ash and shown to contain two methoxy groups and to give 5:6:7-trimethoxycoumarin when methylated. It was synthesised from 2:5-dihydroxy-4:6-dimethoxybenzaldehyde by standard methods.

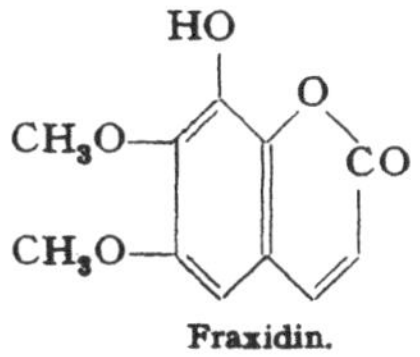

Fraxidin.

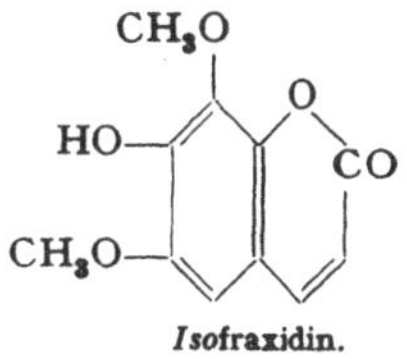

Isofraxidin.

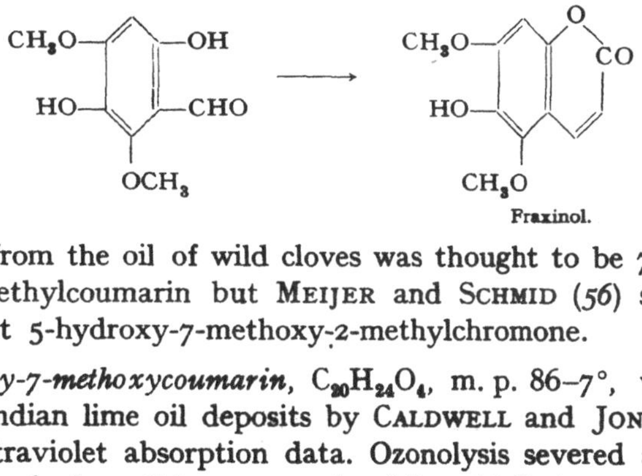

Fraxinol.

Eugenin from the oil of wild cloves was thought to be 7-hydroxy-5-methoxy-4-methylcoumarin but MEIJER and SCHMID (56) showed that it was in fact 5-hydroxy-7-methoxy-2-methylchromone.

5-Geranoxy-7-methoxycoumarin, $C_{20}H_{24}O_4$, m. p. 86–7°, was isolated from West Indian lime oil deposits by CALDWELL and JONES (25) who record its ultraviolet absorption data. Ozonolysis severed acetone and laevulic aldehyde from this compound, while acid fission resulted in the production of 5-hydroxy-7-methoxycoumarin which was identified by its methylation to citropten, its non-identity with the isomeric 7-hydroxy-5-methoxycoumarin, its greenish ferric reaction, and its positive indophenol test.

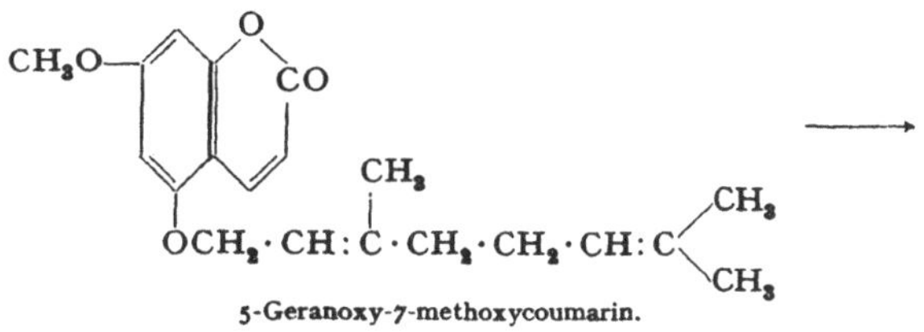

5-Geranoxy-7-methoxycoumarin.

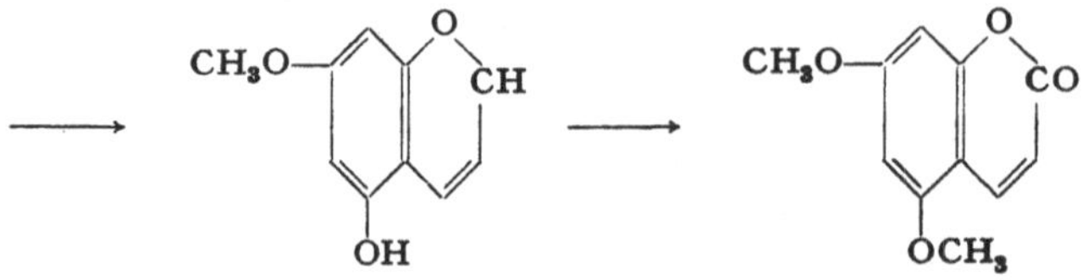

5-Hydroxy-7-methoxycoumarin.   Citropten.

Suberosin, $C_{15}H_{16}O_3$, m. p. 87.5°, obtained from the bark of *Zanthoxyl suberosum* by EWING, HUGHES and RITCHIE *(34)*, was shown to be 6-(γ:γ-dimethylallyl)-7-methoxycoumarin mainly by its non-identity with osthol (p. 254) and its oxidation to acetone and 7-methoxycoumarin-6-carboxylic acid (or the corresponding aldehyde). This oxidation is unlike that of osthol but is strongly reminiscent of the oxidation of the methyl ether of ostruthin.

Collinin, $C_{20}H_{24}O_4$, m. p. 67–8°, occurs in the bark of *Flindersia collina*. ANET, BLANKS, and HUGHES *(3)* treated the substance with hydrochloric acid and isolated daphnetin-8-methyl ether (which they synthesised) and geranyl chloride, compounds which were shown to reform collinin under appropriate conditions.

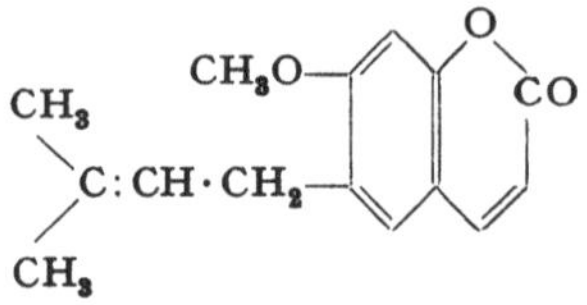

Suberosin.

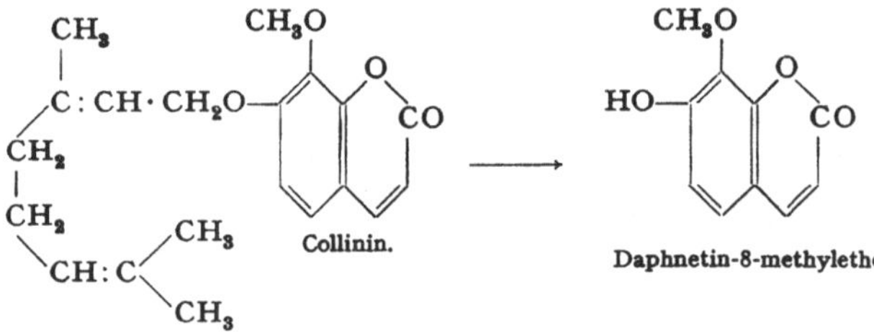

Collinin.   Daphnetin-8-methylether.

Brayleyanin, $C_{20}H_{24}O_4$, m. p. 95°, occurs with braylin. ANET, HUGHES, and RITCHIE *(4)* hydrogenated this substance using RANEY nickel and isolated a phenol, $C_{15}H_{16}O_4$, which was synthesised from the sodium salt of scopoletin and γ:γ-dimethylallyl bromide by a reaction involving a CLAISEN migration.

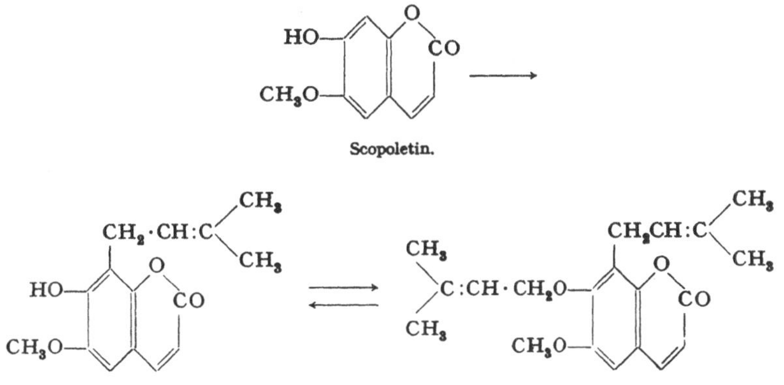

Scopoletin. [structure]

Brayleyanin. [structure]

Repetition of the allylation then gave an ether which was identical with natural brayleyanin.

Umbelliprenin, $C_{24}H_{30}O_3$, m. p. 61–3°, from extracts of angelica seed was split by sulfuric acid in acetic acid solution into umbelliferone and a non-homogeneous oil which suggested to SPÄTH and VIERHAPPER (*126*) that umbelliprenin was an ether. The ether side-chain contained fifteen carbon atoms, and included three C-methyl groups (estimated according to KUHN and ROTH), and three double bonds since catalytic hydrogenation resulted in the absorption of four molecules of hydrogen one of which was taken up by the pyrone double-bond. In the absence of further evidence, umbelliprenin is regarded as a farnesyl ether.

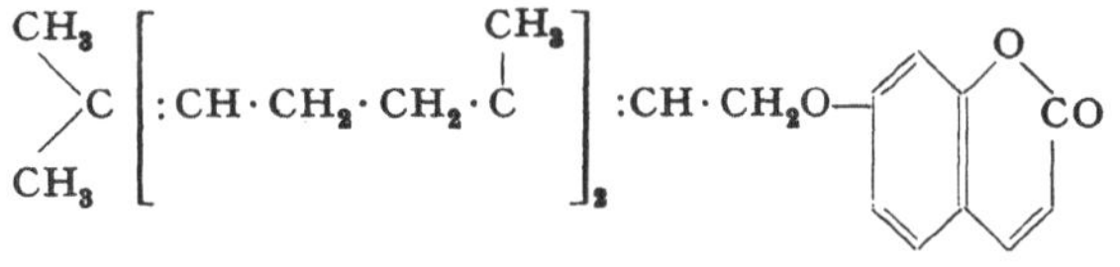

Umbelliprenin. [structure]

Toddalolactone (Aculeatin hydrate), $C_{16}H_{20}O_6$, m. p. 132.5°, from *Toddalia aculeata* PERS. was found by DEY and PILLAY (*31*) to behave as a coumarin and also to possess two alcoholic hydroxyl groups, one of which was rather unreactive and therefore probably tertiary. SPÄTH and his co-workers (*84, 85*) proved the substance to be a 1:2-glycol by periodic acid oxidation which gave an aldehydo-coumarin. By a sequence of ring-opening, ethylation and vigorous oxidation, toddalolactone was broken down to 2-ethoxy-4:6-dimethoxybenzene-1:3-dicarboxylic acid, which was synthesised by an unambiguous route and served to orientate the original coumarin.

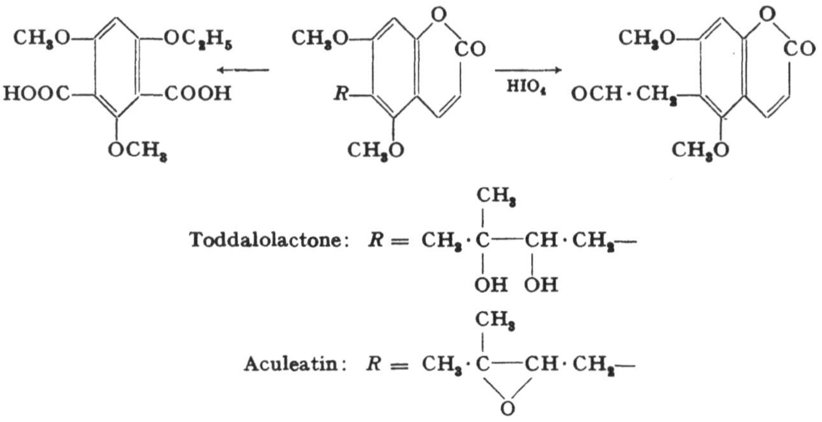

$$\text{CH}_3\text{O}-\underset{\underset{\text{OCH}_3}{|}}{\overset{\overset{-\text{OC}_2\text{H}_5}{|}}{\bigcirc}}-\text{COOH} \quad \xleftarrow{} \quad \text{CH}_3\text{O}-\underset{\underset{\text{CH}_3\text{O}}{|}}{\overset{}{\bigcirc}}\overset{\text{O}}{\underset{}{\bigcirc}}_{\text{CO}} \quad \xrightarrow{\text{HIO}_4} \quad \text{OCH·CH}_2-\underset{\underset{\text{CH}_3\text{O}}{|}}{\overset{\overset{\text{CH}_3\text{O}}{|}}{\bigcirc}}\overset{\text{O}}{\underset{}{\bigcirc}}_{\text{CO}}$$

$$\text{Toddalolactone:} \quad R = \text{CH}_3\cdot\underset{\underset{\text{OH}}{|}}{\overset{\overset{\text{CH}_3}{|}}{\text{C}}}-\underset{\underset{\text{OH}}{|}}{\text{CH}}\cdot\text{CH}_2-$$

$$\text{Aculeatin:} \quad R = \text{CH}_3\cdot\overset{\overset{\text{CH}_3}{|}}{\underset{\underset{\text{O}}{\diagdown\diagup}}{\text{C}}}-\text{CH}\cdot\text{CH}_2-$$

Aculeatin, $C_{16}H_{18}O_5$, m. p. 113°, occours with toddalolactone, and adds one molecule of water to give aculeatin hydrate which according to DUTTA (*33*) is closely similar to, and probably identical with, toddalolactone. Aculeatin is therefore considered to be the epoxide related to toddalolactone.

Auraptene, $C_{15}H_{16}O_4$, m. p. 91°, is a constituent of the deposits from orange-peel oil. BÖHME *et al.* (*15–17*) noted that dihydroauraptene, obtained by hydrogenation, behaved as a lactone but that dihydroauraptenic acid could not be recyclised, nor did the cinnamic acid obtained by methylative ring-opening contain a new methoxyl group, although it did exhibit *cis-trans* isomerism. An ethylene oxide system was detected in auraptene by its hydration to a glycol by means of 1% oxyalic acid, and by the isomerisation of aurapten to *iso*auraptene which exhibited ketonic properties. Oxidation of auraptene with chromic acid or lead tetraacetate split auraptene into acetone, and either an aldehyde or the corresponding acid, ostholic acid (7-methoxy-8-coumarinylacetic acid) which gave rise to 7-methoxy-8-methylcoumarin on decarboxylation.

The curious behaviour of auraptene shown after the pyrone ring was opened could now be attributed to interaction of the epoxide system with the freed phenolic hydroxyl group, but further work has not settled this matter finally, although perphthalic acid oxidation of osthol gave an epoxide (*not* auraptene) which was isomerised to a ketone identical with *iso*auraptene.

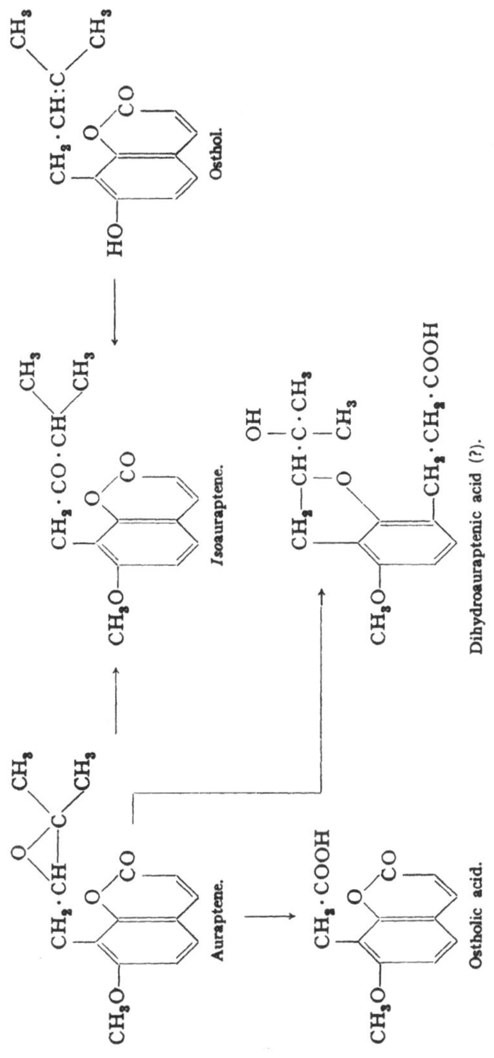

Ostruthin, $C_{19}H_{22}O_3$, m. p. 119°, originally isolated by HERZOG and KROHN from the *Imperatoria* rhizome was eventually assigned the structure given below by SPÄTH and KLAGER (*104*). The substance behaved as a hydroxycoumarin, and the cinnamic acid obtained by the action of alkali and dimethyl sulfate could be oxidised to 2:4-dimethoxybenzene-1:3-dicarboxylic acid. Hydrogenation of ostruthin resulted in the uptake of one molecule of hydrogen by the pyrone ring and of two by the side-chain, after which oxidation gave rise to 4:8-dimethylnonan-1-oic acid, while under appropriate conditions, ostruthin could be oxidised to acetone or to methylheptenone, which established the positions of the double bonds in the side-chain. That ostruthin does indeed contain a coumarin system was proved by oxidation of its methyl ether to an aldehyde, the oxime of which was converted into 6-cyano-7-methoxycoumarin.

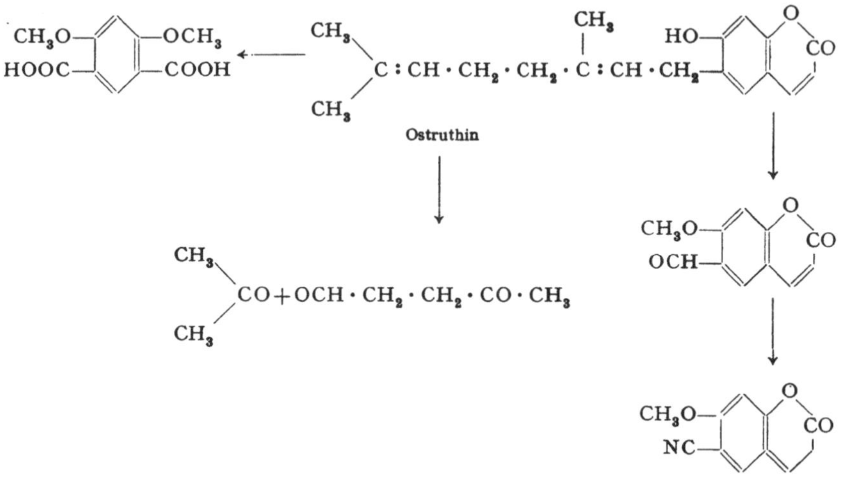

Osthenol, $C_{14}H_{14}O_3$, m. p. 124–5°, is 7-hydroxy-8-(γ:γ-dimethylallyl)-coumarin since it reacts with diazomethane to give osthol (*82*). BOTTOMLEY and WHITE (*22*) extracted vellein, $C_{20}H_{24}O_8 + 0.5\ H_2O$, m. p. 189°, from *Velleia discophora* F. MUELL and found that it gave dihydroseselin on acid hydrolysis, but that emulsin hydrolysis yielded glucose and osthenol, whereupon they concluded that vellein was the β-glucoside of osthenol.

Osthol, $C_{15}H_{16}O_3$, m. p. 83–4°, is one constituent of the *Imperatoria* rhizome; the elucidation of its structure is due to BUTENANDT and MARTEN (*24*), as well as to SPÄTH and PESTA (*115*). Vigorous oxidation of osthol gave rise to 2-hydroxy-4-methoxybenzoic acid, while chromic acid oxidation generated acetone, the other product being an acid, ostholic acid, which on decarboxylation yielded 7-methoxy-8-methyl-coumarin. Oxidation of tetrahydro-osthol from catalytic hydrogenation gave succinic and *iso*caproic acids (p. 255).

Dihydro-osthol, obtained by partial hydrogenation of osthol, was synthesised by the condensation of malic acid with the decarboxylation product of tetrahydrotubaic acid-4-methyl ether (*41, 124, 134*) (p. 255).

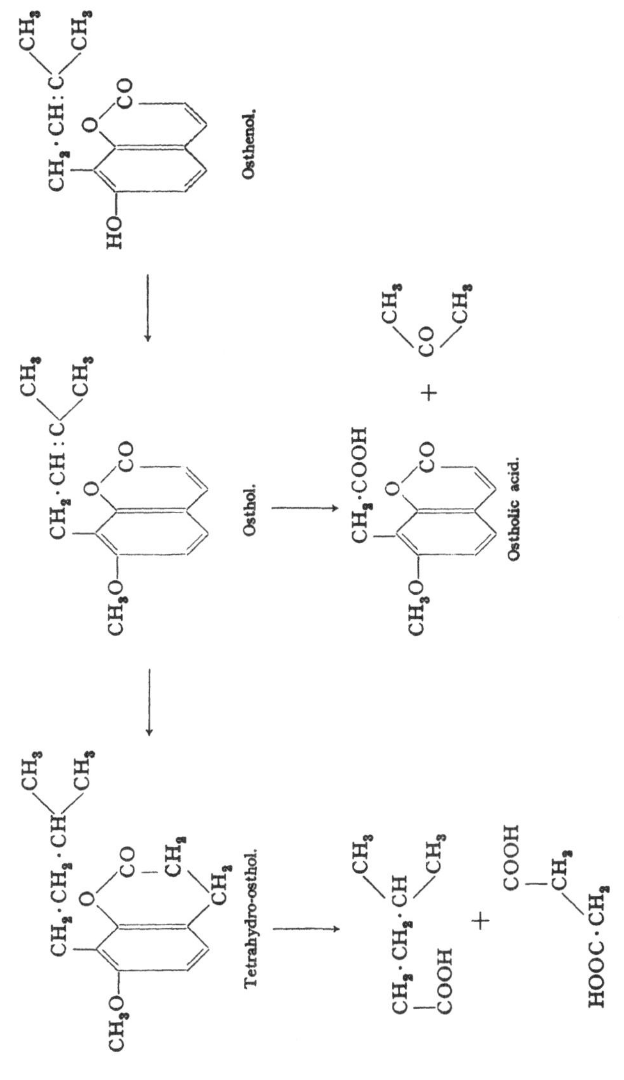

Osthenol.

Osthol. + Ostholic acid.

Tetrahydro-osthol.

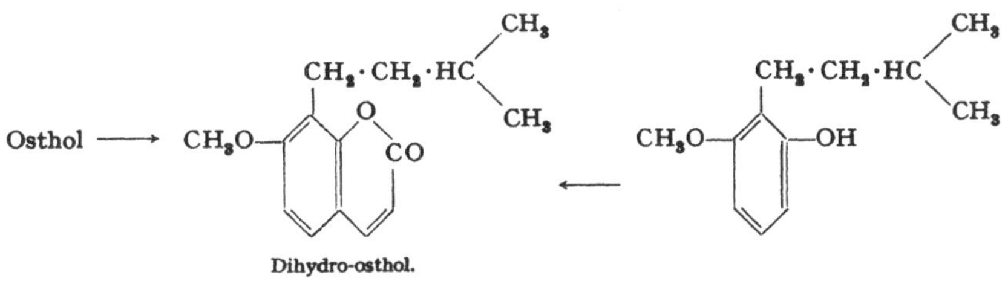

Osthol → Dihydro-osthol.

The synthesis of osthol was carried out by SPÄTH and HOLZEN (*92*) from 2-hydroxy-4-methoxybenzaldehyde by the following sequence,

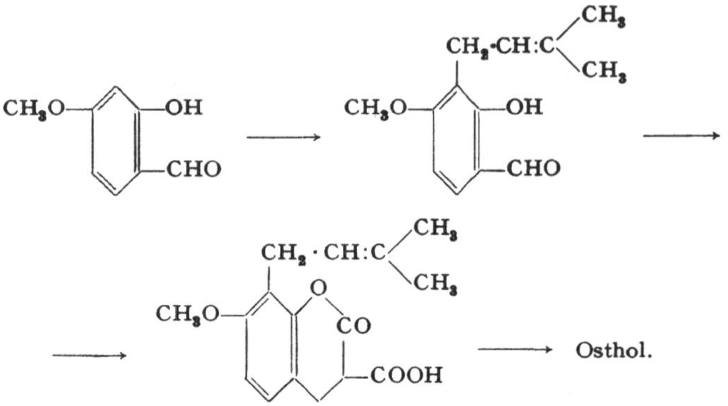

and in confirmation of the position allocated to the side-chain, YAMASHITA (*144*) observed that synthetic 6-*iso*amyl-7-methoxycoumarin was not identical with dihydro-osthol. Although osthol was demethylated by aluminium chloride in benzene, osthenol was not produced because under the influence of the catalyst the side-chain added benzene (*101*).

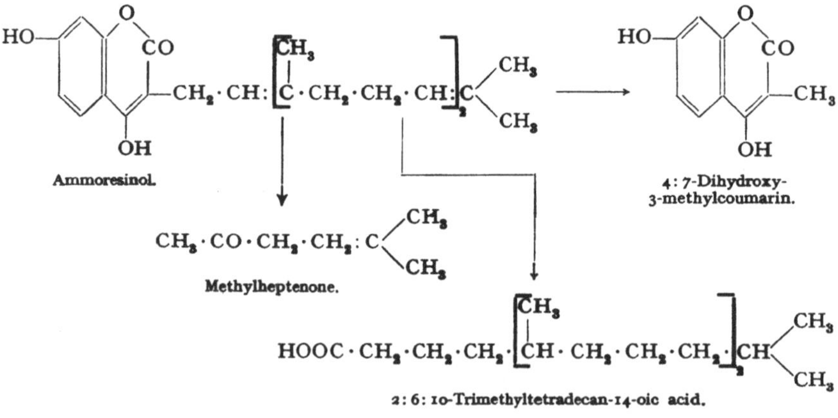

Ammoresinol, $C_{24}H_{30}O_4$, m. p. 107–8°, was isolated from gum ammoniacum by CASPARIS and MICHEL, and its constitution was clarified by SPÄTH, SIMON, and LINTNER (*122*). Ammoresinol diacetate could be oxidised to β-resorcylic acid, while dry distillation of ammoresinol itself supplied 4:7-dihydroxy-3-methylcoumarin. The 4-hydroxycoumarin system is difficult to hydrogenate, so that the six atoms of hydrogen absorbed by ammoresinol were presumably taken up by a side-chain, which appeared as 2:6:10-trimethyltetradecan-14-oic acid when the hexahydride was oxidised by permanganates. Furthermore, ozonolysis of ammoresinol diacetate gave methyl heptenone supporting the formula given below, which was confirmed by the concurrent work of KUNZ and HOOPS (52) and RAUDNITZ et al. (67), though the latter has disputed the chain-length of the C_{17}-acid.

Dicoumarol [3:3'-methylene-*bis*-(4-hydroxycoumarin)], $C_{19}H_{12}O_6$, m. p. 288–9°, was discovered by CAMPBELL and LINK (26) to be the haemorrhagic principle of spoiled sweet clover (cf. p. 241). Investigations by LINK and his co-workers (48, 132) established that two highly acidic, enolic hydroxyl groups were present and that caustic potash melts gave a theoretical yield of salicylic acid, while milder treatment with alkali furnished 1:3-*bis*(2-hydroxybenzoyl)-ethane which was synthesised from *o*-methoxybenzoylacetic ester. When heated with aniline, dicoumarol yielded 4-anilino-coumarin, which led to the formulation of the compound as 3:3'-methylene-*bis*-(4-hydroxycoumarin), identical with the substance prepared many years before by ANSCHÜTZ who obtained it by condensation of two molecules of 4-hydroxycoumarin with one of formaldehyde.

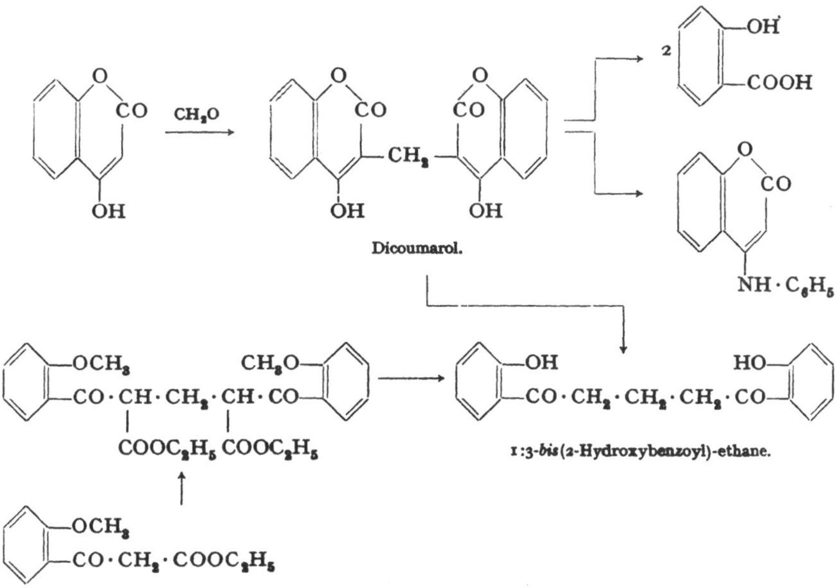

VI. Furanocoumarins.

Furanocoumarins exhibit the chemical properties of both coumarins and coumarones. The furan ring is unaffected by either acid or alkaline hydrolysis, though it is removed, together with the α-pyrone ring, by fusion with alkalis. The non-benzenoid double bond of the furan ring is difficult to hydrogenate, so that selective hydrogenation of a furanocoumarin furnishes a 3':4'-dihydro-α-pyronocoumarone, and further hydrogenation a 3':4'-dihydro-α-pyronocoumaran. On the other hand the furan ring is rapidly attacked by ozone to give an aldehydocoumarin, although further ozonolysis attacks the pyrone ring also and phenolic dialdehydes are produced. Permanganate oxidation is not usually specific and only phenolic carboxylic acids are formed, but chromic acid oxidation is milder and often attacks only side-chains or converts the benzene ring to a quinonoid system.

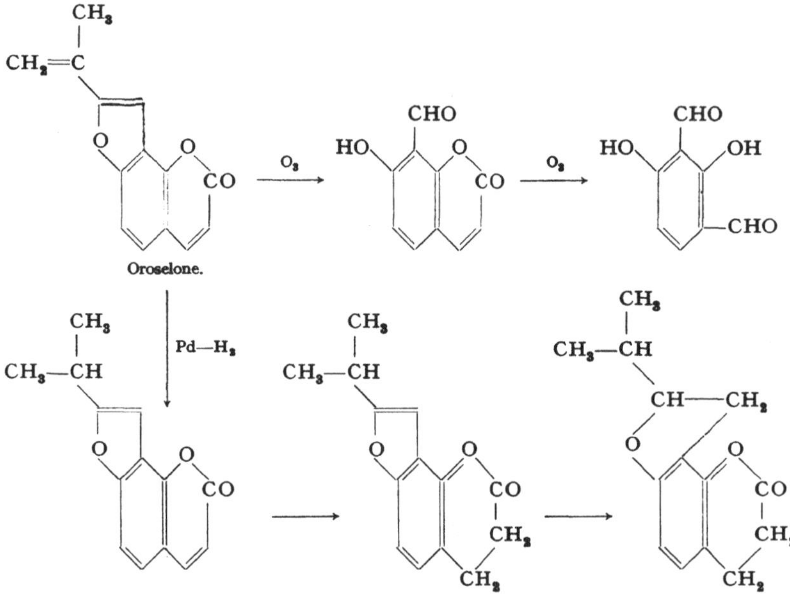

It is of major importance in this field that alkaline hydrogen peroxide destroys both the α-pyrone ring and the benzene ring of a furanocoumarin, and leaves furan-2:3-dicarboxylic acid as the main fragment.

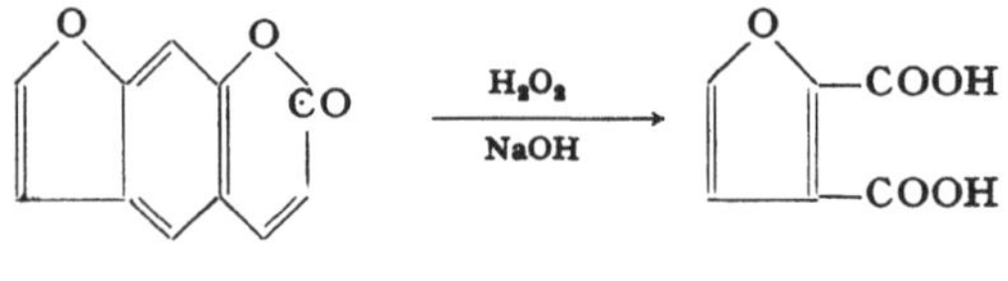

This reaction provides direct evidence for the existence of the furan ring in many of these compounds, but unfortunately there is no record of a substituted acid being produced from any furanocoumarin substituted in the furan ring. SPÄTH and KLAGER (102) have been careful to prove that furan-2:3-dicarboxylic acid does not arise by ring contraction of the umbelliferone system.

Use of aluminium chloride for demethylating methoxyfuranocoumarins is limited by the reaction of the furan system with the reagent; for example; in the presence of benzene and aluminium chloride, psoralene forms 7-hydroxy-6-(1 : 2-diphenylethyl)-coumarin (*51*).

Syntheses of furanocoumarins have been carried out beginning with either the coumarin or the coumarone nucleus and adding the third ring. Since umbelliferone derivatives are substituted so easily in position 8, they are good starting materials for the preparation of angular furano coumarins. Thus 8-formylumbelliferone may be condensed with iodoacetic ester and the product readily hydrolysed to give the corresponding phenoxyacetic acid which, when heated with acetic anhydride, cyclises by an internal PERKIN condensation and is decarboxylated spontaneously yielding angelicin:

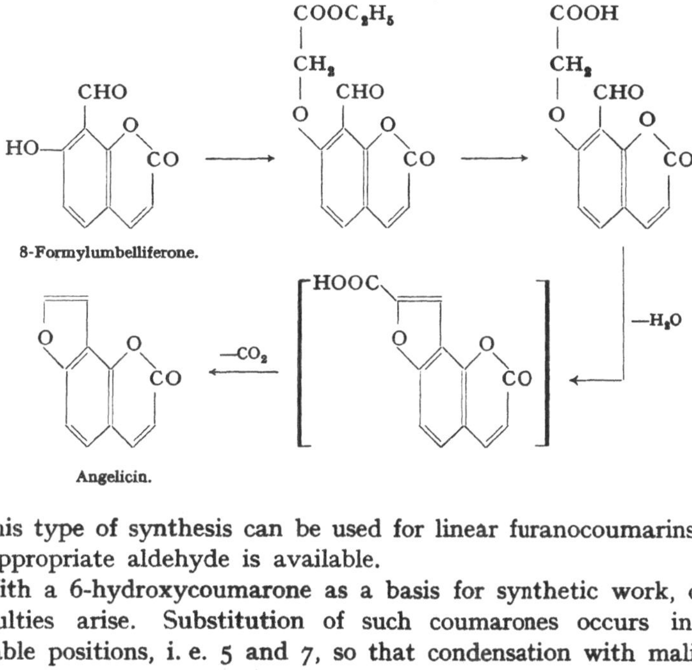

This type of synthesis can be used for linear furanocoumarins when the appropriate aldehyde is available.

With a 6-hydroxycoumarone as a basis for synthetic work, certain difficulties arise. Substitution of such coumarones occurs in both available positions, i. e. 5 and 7, so that condensation with malic acid is ambiguous. Again, the GATTERMANN aldehyde synthesis affects position 2 of the furan ring and leaves the phenolic ring untouched. It is usual to overcome these difficulties by use of a 6-hydroxycoumaran, in which only position 5 is attacked in substitution reactions, so that an α-pyrone ring can be added by any of the general methods; however, it is necessary to effect dehydrogenation of the resulting compound to obtain the furanocoumarin, and although this may be carried out with palladium catalysts, the reaction gives very poor yields. Nevertheless, this method allows synthesis of linear furanocoumarins which would be very difficult to prepare in any other way. The synthesis of psoralene is

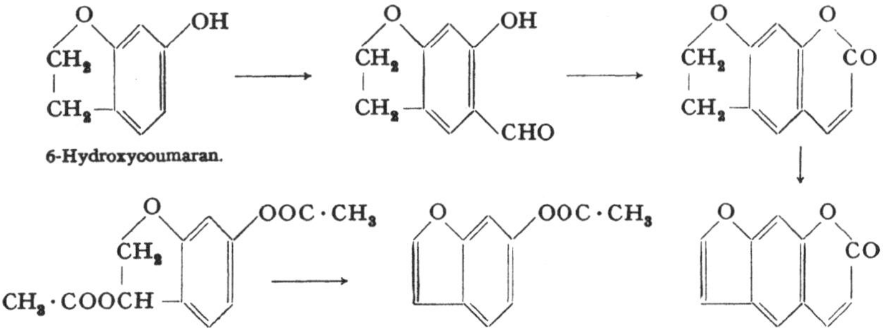

[6-Hydroxycoumaran → intermediate CHO → Psoralene; and 3:6-Diacetoxycoumaran → OOC·CH₃ intermediate → Psoralene]

an example. To avoid the dehydrogenation step, SPÄTH has employed 3-acetoxycoumarans because the furan double bond is formed by a facile loss of the elements of acetic acid under very mild conditions; but unfortunately the preparation of 3-acetoxycoumarans is not very easy.

Peucedanin is unique in this field, and although it is formally a furanocoumarin, its chemistry is dominated by that of its coumaran-3-one nucleus.

Psoralene (Ficusin), $C_{11}H_6O_3$, m. p. 171°, from the seeds of *Psoralea corylifolia* L., was shown to yield a resodicarboxylic acid by permanganate oxidation [SPÄTH, MANJUNATH, PAILER, and JOIS (*108*)] and to behave like a coumarin towards alkali. OKAHARA obtained ficusin from *Ficus carica* L., and found that alkaline hydrogen peroxide oxidised it to furan 2:3-dicarboxylic acid; ficusin and psoralene were shown to be identical by SPÄTH, OKAHARA, and KUFFNER (*111*).

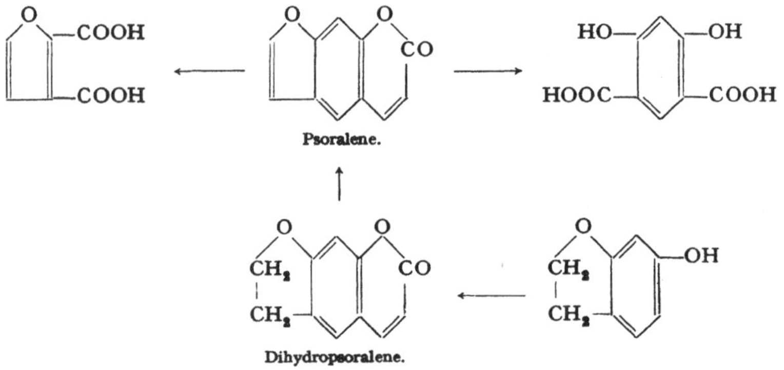

[Psoralene and Dihydropsoralene reaction scheme]

The synthesis of psoralene involved the condensation of malic acid with 6-hydroxycoumaran, followed by dehydrogenation with a palladium black catalyst, a process improved somewhat by FOSTER, ROBERTSON, and BUSHRA (35) and by HORNING and REISNER (48), who obtained dihydropsoralene from 5-formyl-6-hydroxycoumaran by modified PERKIN condensations.

During a study of cardiac glycosides, STOLL, PEREIRA, and RENZ (133) isolated from the seeds of *Coronilla glauca* L. a glycoside which melted at 125°, had $[\alpha]_D^{20}$—61.3° (in water), and was hydrolysed by emulsin or dilute acids to psoralene and glucose. Boiling water induced a change from the *cis* to the *trans* form, after which hydrolysis gave only the stable furanocoumaric acid.

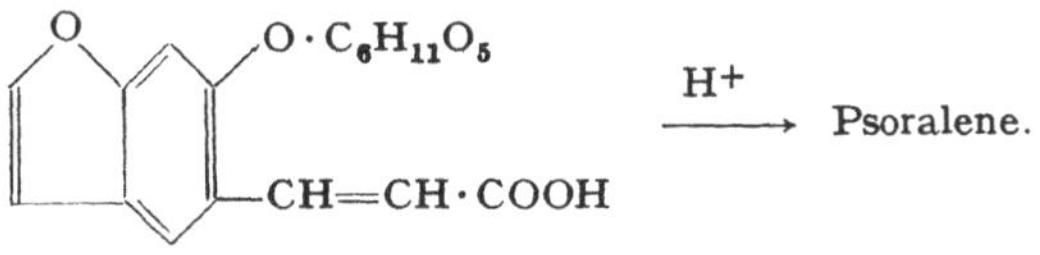

Glycoside (*cis* form). $\xrightarrow{H^+}$ Psoralene.

Angelicin (*Iso*psoralene), $C_{11}H_6O_3$, m. p. 139°, was isolated from the root of *Archangelica officinalis* HOFFM. (*Angelica archangelica* L.) by SPÄTH and PESTA (116). (The name "angelicin" was once used to designate sitosterol.) Angelicin could be oxidised and the resulting acid decarboxylated to give umbelliferone, whilst hydrogen peroxide yielded furan-2:3-dicarboxylic acid, and another oxidation sequence furnished 2:4-dimethoxy*iso*phthalic acid, thus proving that angelicin possessed the structure of an angular furanocoumarin.

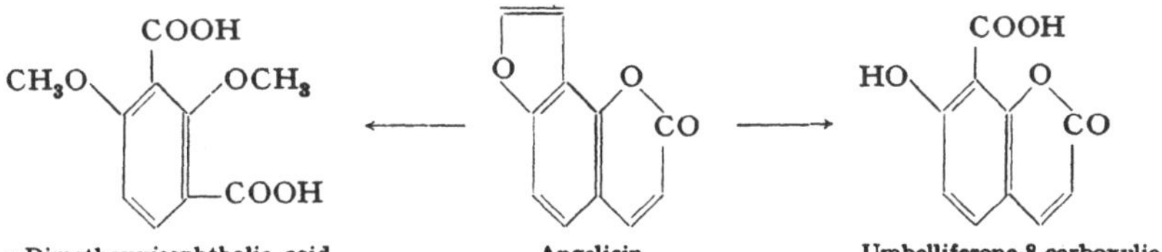

2:4-Dimethoxy*iso*phthalic acid. ← Angelicin. → Umbelliferone-8-carboxylic acid.

The sodium salt of 8-formylumbelliferone was condensed with iodoacetic ester, and the product was hydrolysed to the phenoxyacetic acid which was cyclised and decarboxylated to give material identical with natural angelicin (113, 114).

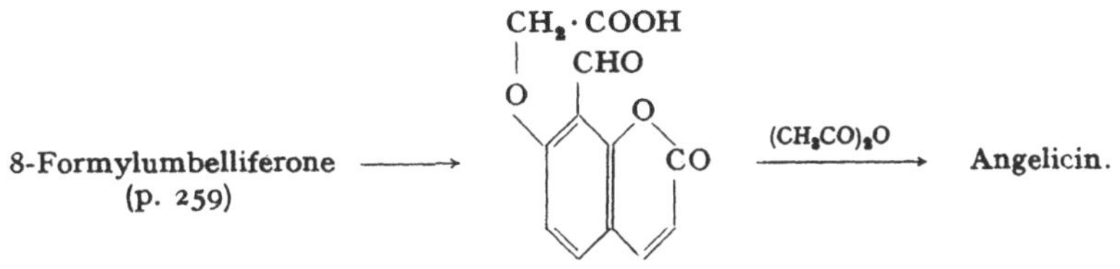

8-Formylumbelliferone (p. 259) → [structure] $\xrightarrow{(CH_3CO)_2O}$ Angelicin.

Bergapten (Bergamot camphor, Heraclin, Majudin), $C_{12}H_8O_4$, m. p. 192° (sublimes), was originally studied by POMERANZ, whose work indicated that it was a furanocoumarin derived from phloroglucinol. THOMS and BAETCKE (*136*) reduced nitrobergapten to aminobergapten which was oxidised by chromic acid to a quinone, with loss of its methoxyl group- reactions best explained by a linear furanocoumarin structure.

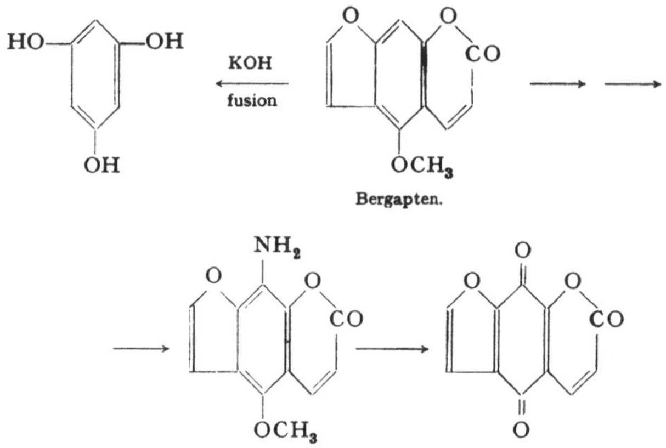

Bergapten.

Bergapten-quinone.

A wholly satisfactory synthesis of bergapten has not been achieved. HOWELL and ROBERTSON (*49*) having obtained *apo*xanthoxyletin (6-formyl-7-hydroxy-5-methoxycoumarin) by degradation of xanthoxyletin built up the furan ring as in the case of angelicin, whereas SPÄTH, WESSELY, and KUBICZEK (*130*) started from 4:6-dihydroxycoumaran

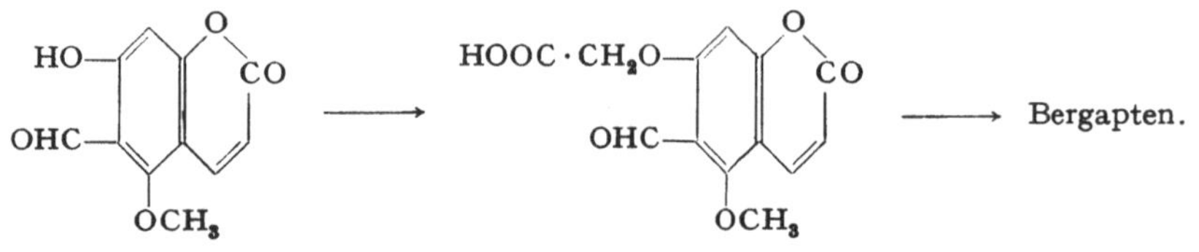

*Apo*xanthoxyletin.

and by means of a carefully controlled acetylation converted it into 3:4:6-triacetoxycoumarone, from which 3:4:6-triacetoxycoumaran resulted by a rather difficult hydrogenation. Condensation of the triacetoxycoumaran with sodioformylacetic ester in a sealed tube gave a mixture of products which, after methylation, was separated into a small quantity of bergapten and a large amount of *allo*bergapten which possesses a type of structure not yet found in nature.

Bergapten. *Allo*bergapten.

Bergaptol.

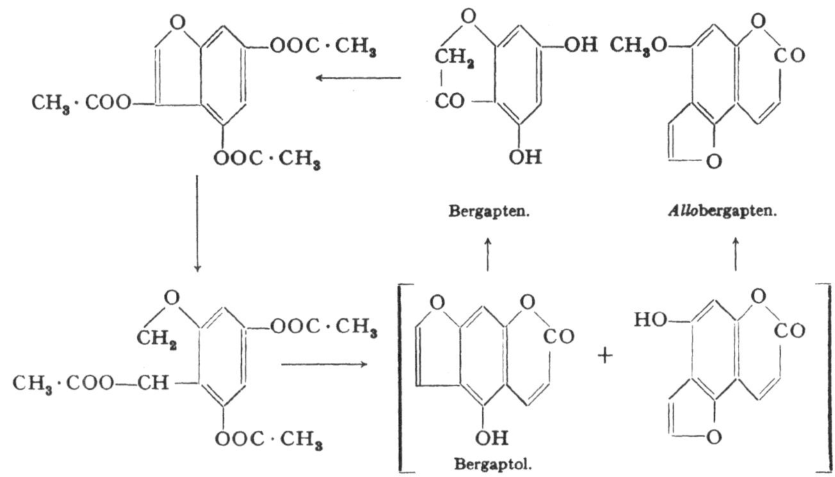

"Heraclin" from the seeds of *Heracleum sphondylium* L., and "Majudin" from the fruits of *Ammi majus* have been identified with bergapten, which also occurs in *Heracleum nepalanese* D. DON, *Citrus Medica* L. (*Citrus acida*) with other furanocoumarins, in *Ligusticum acutilobum* SIEB. et ZUCC., and in *Semen angelicae*.

Bergaptol, $C_{11}H_6O_4$, m. p. 278°, was found in Calabrian bergamot oil after a special search had been made for it by SPÄTH and SOCIAS (*123*). With diazomethane, bergapten was obtained, but alkaline methyl sulfate gave *iso*bergapten because of ring-opening followed by cyclisation in a new position. Synthesis (*106*) of bergaptol was achieved by the method used for bergapten, except of course for the omission of the final methylation.

Bergaptol.

*Iso*bergapten.

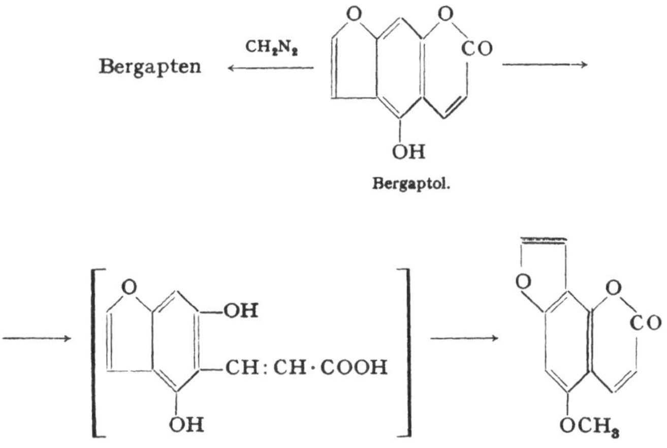

Isobergapten, $C_{12}H_8O_4$, m. p. 224°, occurs in *Pimpinella saxifraga* L. and in *Heracleum sphondylium* L. Wessely *et al.* (*141*) showed that sodium amalgam reduction of bergapten and *iso*bergapten gave dihydrocinnamic acid derivatives which furnished the same dimethoxy esters after methylation. This allows only one structural possibility for *iso*bergapten, which has been synthesised *via* bergaptol (see above).

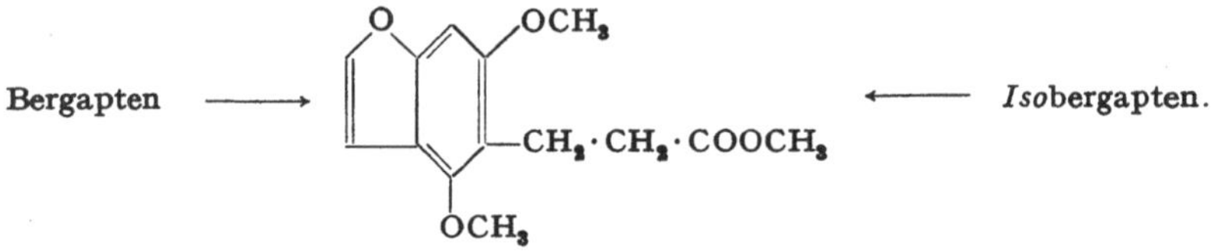

Bergapten
Isobergapten.
O
OCH₃
CH₂·CH₂·COOCH₃
OCH₃

Xanthotoxin (Ammoidin), $C_{12}H_8O_4$, m. p. 145–6°, was obtained from *Xanthoxylum senegalense* (*Fagara zanthoxyloides* Lam.) by Thoms and also by Priess. The former author showed that xanthotoxin was a methoxylactone which could be broken down by alkali fusion to 2:3:4-trihydroxybenzoic acid, whereupon he suggested that a furanocoumarin system was present. In further work Thoms obtained bergaptene-quinone by a nitration, reduction, and oxidation sequence on xanthotoxin, thus establishing the constitution of the substance.

OH
HO
OH
COOH
OCH₃
O
O
CO
O
O
O
CO
O

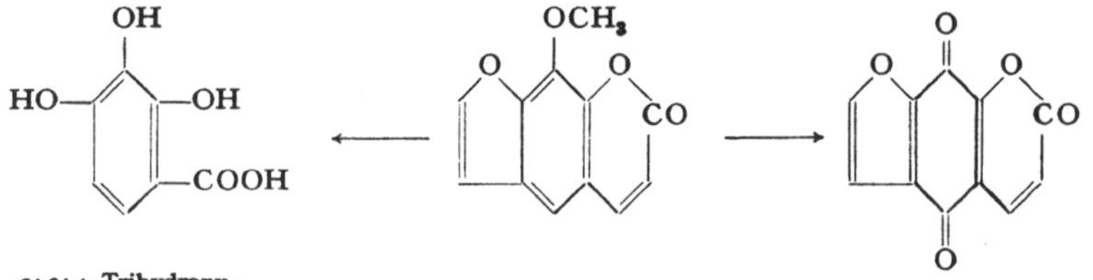

2:3:4-Trihydroxybenzoic acid. Xanthotoxin. Bergaptene-quinone.

Späth and Pailer (*112*) synthesised xanthotoxin by the condensation of malic acid with 6:7-dihydroxycoumaran, followed by methylation and dehydrogenation using palladium. The yield was poor.

OH
O
CH₂
CH₂
OH
OH
O
O
CH₂
CH₂
CO
OCH₃
O
O
CH₂
CH₂
CO
OH
O
O
CO
OCH₃
O
O
CO

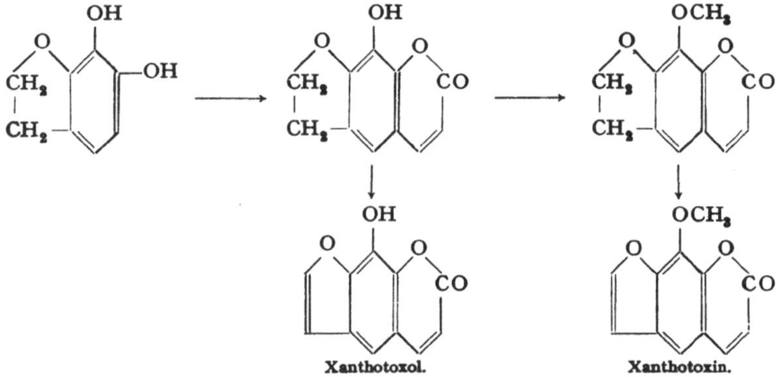

Xanthotoxol. Xanthotoxin.

Xanthotoxin is found in the mature berries of *Luvanga scandens* HAM., is identical with ammoidin from the fruit of *Ammi majus* L., and is a constituent of *Semen angelicae*.

Xanthotoxol, $C_{11}H_6O_4$, m. p. 252–3°, was discovered amongst the coumarins obtained from the seeds of *Archangelica officinalis* HOFFM. (*Angelica archangelica* L.) by SPÄTH and VIERHAPPER (*127*), who made a very careful search for it. On methylation, xanthotoxin is obtained, while demethylation of xanthotoxin, using magnesium iodide at 160–170°, furnishes xanthotoxol which was synthesised by dehydrogenation of dihydroxanthotoxol, an intermediate in the synthesis of xanthotoxin.

Sphondin, $C_{12}H_8O_4$, m. p. 189–191°, was isolated in small quantities from the rhizome of *Heracleum sphondylium* L. by SPÄTH and SCHMID (*118*). Hydrogen peroxide gave rise to furan-2:3-dicarboxylic acid, and ozonolysis to an aldehyde which was synthesised by DUFF's reaction on scopoletin and which gave fraxetin by DAKIN's reaction.

Sphondin. Scopoletin. Furan-2: 3-dicarboxylic acid. Fraxetin.

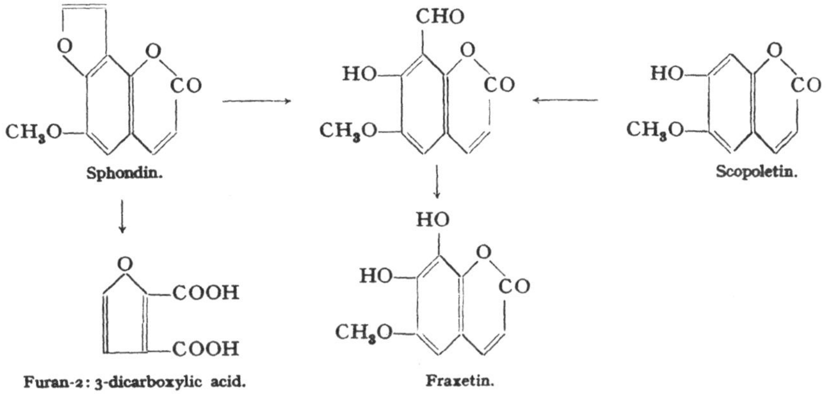

Sphondylin, $C_{12}H_8O_4$, m. p. 161–3°, accompanies sphondin, has the properties of a coumarin, and can be oxidised to furan-2:3-dicarboxylic acid.

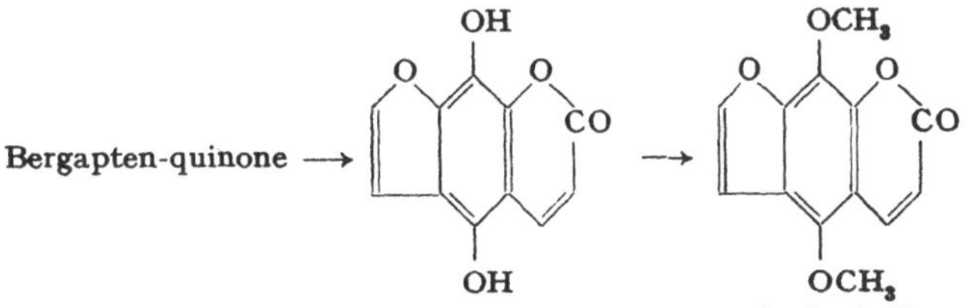

Isopimpinellin, $C_{13}H_{10}O_5$, m. p. 151°, occurs in the roots of *Pimpinella saxifraga* L., in *Luvanga scandens* HAM., in *Heracleum sphondylium* L.,

Isopimpinellin.

and in lime oil. Investigated by WESSELY and KALLAB (*141*), it behaved as a dimethoxy-furanocoumarin, and was synthesised by sulfur dioxide reduction of bergapten-quinone to the corresponding quinol, followed by methylation.

A *Phenol*, $C_{12}H_8O_5$, occurs in *Semen angelicae* and on methylation gives *iso*pimpinellin (*128*). It contains one methoxyl group, but its exact constitution is not yet certain, though one possible isomer has been synthesised (see phellopterin).

Pimpinellin $C_{13}H_{10}O_5$, m. p. 119°, is found in association with *iso*pimpinellin. WESSELY and KALLAB (*141*) used standard methods to show that it was a dimethoxy-furanocoumarin isomeric with *iso*pimpinellin, and gave it the accompanying formula, but a rigid proof is lacking. Two isomeric dimers can be formed from pimpinellin by the action of light, and these regenerate pimpinellin when heated.

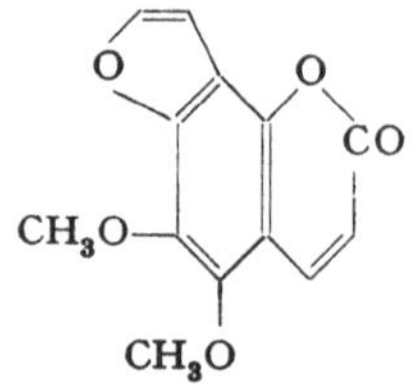

Pimpinellin.

Isoimperatorin $C_{16}H_{14}O_4$, m. p. 109°, is a component of *Peucedanum ostruthium* KOCH (*Imperatoria ostruthium* L.). SPÄTH and KAHOVEC (*98*) established the presence of furan and coumarin rings and showed that acid caused fission of a C_5 residue (which appeared as amyl alcohol, after hydrogenation), and left bergaptol. As the substance took up three molecules of hydrogen in catalytic hydrogenation, the ether side-chain must have contained one double bond, located by chromic acid oxidation which split off acetone. Therefore *iso*imperatorin was the $\gamma:\gamma$-dimethylallyl ether of bergaptol; nevertheless, synthesis by interaction of bergaptol sodium and $\gamma:\gamma$-dimethylallyl bromide, though successful, was rendered difficult by the hindered nature of the hydroxyl group [SPÄTH and DOBROVOLNY (*87*)].

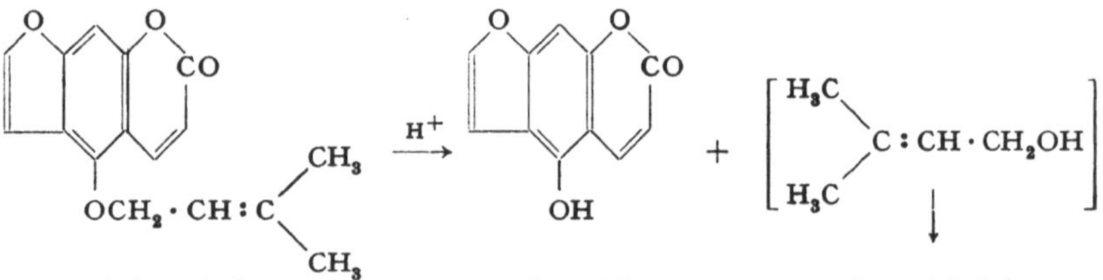

*Iso*imperatorin.   Bergaptol.   *Iso*amyl alcohol.

Oxypeucedanin (Hydroxypeucedanin), $C_{16}H_{14}O_5$, m. p. 142–3°, is found with *iso*imperatorin, to which it is closely related. It is a constituent of *Prangos patularia* LINDL. It is slightly laevorotatory.

Early work was carried out by HERZOG and KROHN, as well as by BUTENANDT and MARTEN, and was completed by SPÄTH and KLAGER (*102*).

The isolation of phloroglucinol and furan-2:3-dicarboxylic acid by appropriate means, and the lactonic properties of the substance together with the isolation of succinic acid by nitric acid oxidation of hydrogenated oxypeucedanin established the main outlines of the molecule. Acid fission of oxypeucedanin furnished bergaptol, but there remained one oxygen atom which was thought to be present in an ether system as it was not responsive to the usual tests. An ethylene oxide structure was supported by the ready addition of the elements of water and hydrogen halides to oxypeucedanin, and by the fact that isomerisation to the ketone *iso*-oxypeucedanin could be effected.

The position of the oxide link was elucidated by chromic acid oxidation of oxypeucedanin which gave acetone and oxypeucedaninic acid, and hydrogen peroxide oxidation of *iso*oxypeucedanin, which yielded *iso*-butyric acid. Finally (*94*), perbenzoic acid oxidation of *iso*imperatorin resulted in synthetic oxypeucedanin.

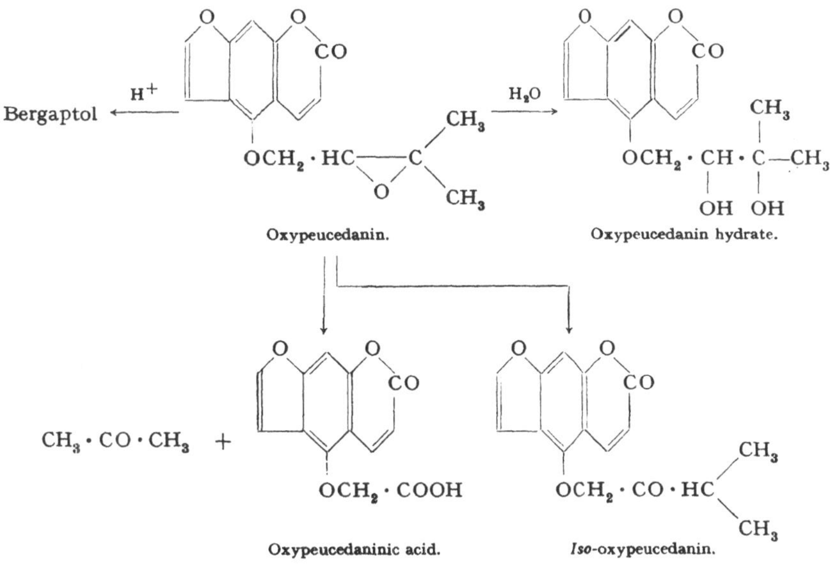

Ostruthol, $C_{21}H_{22}O_7$, m. p. 136–7°, $[\alpha]_D^{22} + 8.36°$ (in acetone), $[\alpha]_D^{15} -18.3°$ (in pyridine), first noted by HERZOG and KROHN in 1909, was shown to have the characteristics of a furanocoumarin by SPÄTH and von Christiani (83). Key reactions in further work were, acid fission of the substance to bergaptol and alkaline hydrolysis which gave a mixture of angelic acid and (±)-oxypeucedanin hydrate, a substance apparently identical with racemic oxypeucedanin hydrate apart from the difference in optical rotation. Thus ostruthol seemed to be an angelic ester of oxypeucedanin hydrate, a view supported by the formation of a monoacetate, a hexahydro-derivative, and of *iso*oxypeucedanin on heating with phosphorus pentoxide; but which of the two available hydroxyl groups is esterified is not clear. Späth and von Christiani favour the formulation of ostruthol as a secondary ester, because the substance may be distilled without change under conditions in which a tertiary ester would yield an olefin.

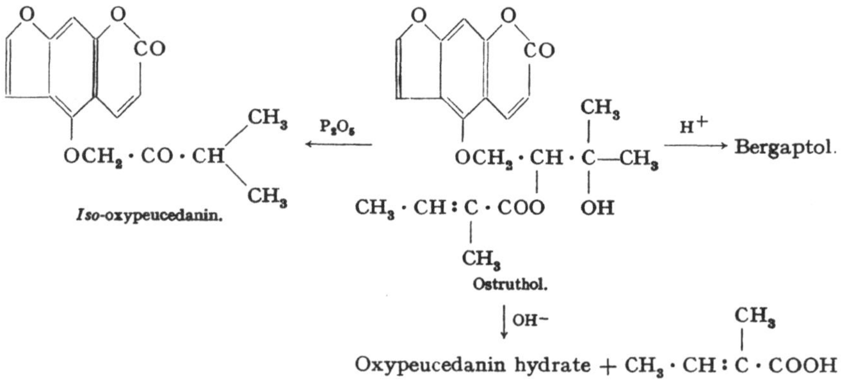

$$\text{Iso-oxypeucedanin} \xleftarrow{P_2O_5} \text{Ostruthol} \xrightarrow{H^+} \text{Bergaptol}$$

$$\downarrow OH^-$$

Oxypeucedanin hydrate $+ CH_3 \cdot CH \colon C(CH_3) \cdot COOH$

Imperatorin (Marmelosin, Ammidin), $C_{16}H_{14}O_4$, m. p. 102–3°. This name was given by Osan in 1831 to a mixture of substances extracted from the rhizome of masterwort, and was later applied to the fifth pure substance isolated from this mixture. Späth and Holzen (93) were unable to detect hydroxyl, methoxyl, or carbonyl groups; however, the compound had the lactonic properties of a coumarin and it could be oxidised to furan-2:3-dicarboxylic acid. As in the case of *iso*imperatorin, acid cleavage gave a phenol (in this case xanthotoxol) and a C_5 residue which appeared as *iso*amyl alcohol, after hydrogenation. Catalytic hydrogenation of imperatorin furnished a hexahydro derivative, which yielded succinic and α-hydroxy-*iso*butyric acid on oxidation. Chromic acid oxidation of imperatorin gave acetone; therefore the substance was formulated as the γ:γ-dimethylallyl ether of xanthotoxol. In agreement with this formula, hydrogenation caused some fission of the allylic side-chain, and a phenol resulted. The characteristic melting-point phenomena of imperatorin (re-solidification and remelting as the temperature rises), are due to a Claisen *p*-migration of the allylic system forming *allo-*

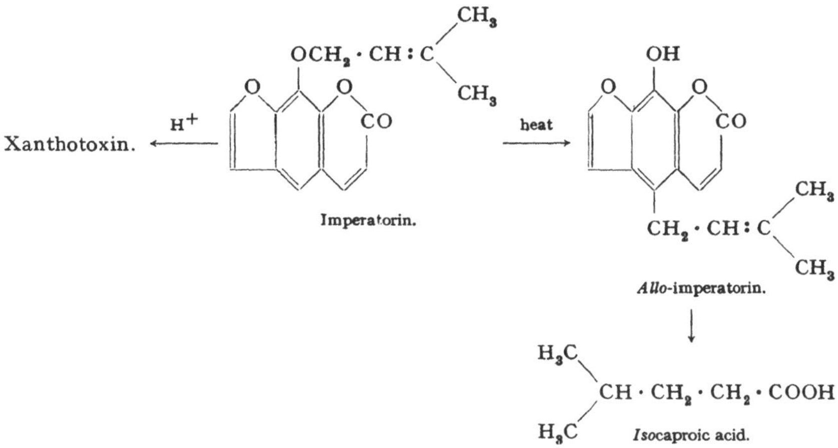

Xanthotoxin. ←H^+— Imperatorin —heat→ Allo-imperatorin ↓ Isocaproic acid: $(CH_3)_2CH\cdot CH_2\cdot CH_2\cdot COOH$

imperatorin, a phenol which could be oxidised to furan-2:3-dicarboxylic acid, or, after hydrogenation, to *iso*caproic acid.

SPÄTH and HOLZEN (*94*) also succeeded in forming synthetic imperatorin from the sodium salt of xanthotoxol and a large excess of isoprene hydrobromide, and showed that perbenzoic acid oxidation of imperatorin gave oxyimperatorin, a substance which has not yet been found in nature.

Bergamottin (Bergaptin), $C_{21}H_{22}O_4$, m. p. 59–61°, is found in bergamot oil. SPÄTH and KAINRATH (*100*) noted that bergamottin decomposed at 180–190° to give bergaptol, which was also obtained, together with geraniol, by acid fission. Bergamottin contains no active hydrogen, a fact consistent with its formulation as bergaptol geranyl ether, but it has not yet been synthesised.

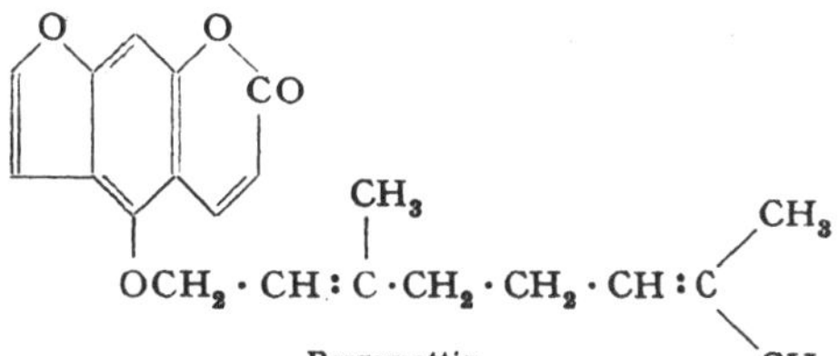

Bergamottin.

Phellopterin, $C_{17}H_{16}O_5$, m. p. 102°, was first isolated in 1940 by NOGUCHI and KAWANAMI from the fruit of *Phellopterus littoralis* BENTH., and later from the root and fruit of *Angelica glabra* MAKINO (*59, 60, 62, 63*), and shown to be related to bergapten by chromic acid oxidation to bergapten-quinone. Towards acids or catalytic hydrogenation phellopterin behaved as a $\gamma:\gamma$-dimethylallyl ether and gave 8-hydroxy-5-methoxypsoralene, which had been synthesised already from aminobergapten by the diazo reaction. The Hydrogenation of phellopterin gave a dihydro derivative which was synthesised from 8-hydroxy-5-methoxypsoralene and an *iso*amyl halide. In 1939 the same authors described the formation of the $\gamma:\gamma$-dimethylallyl ether of the above phenol as part of a synthesis of byakangelicol (see below), but no direct comparison of the natural and synthetic substances seems to have been published.

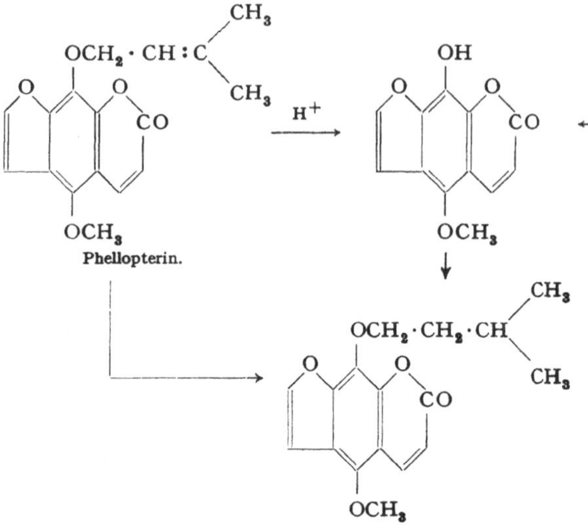

Byakangelicol, $C_{17}H_{16}O_6$, m. p. 106°, $[\alpha]_D$ 34.77° (pyridine) was obtained in 1938 from *Angelica glabra* MAKINO by NOGUCHI and KAWANAMI (*61*). The seeds of *Heracleum nepalanese* D. DON also contain this substance.

The exact constitution is in doubt (see ferulin), but the substance seems to be the epoxide of phellopterin. Thus, appropriate oxidations gave bergapten-quinone, acetone, and α-hydroxy*iso*butyric acid, while acid fission resulted in 8-hydroxy-5-methoxypsoralene identical with material synthesised from aminobergapten.

The substance behaved like an ethylene oxide in that it added the elements of water under the influence of 1% oxalic acid, and gave the diacetate of the resulting glycol with acetic anhydride. Perbenzoic acid oxidation of the $\gamma:\gamma$-dimethylallyl ether of 8-hydroxy-5-methoxypsoralene yielded optically inactive material which was alleged to be identical with natural (optically active) byakangelicin. Both natural and synthetic materials gave the same "*iso*byakangelicolic acid" with hot alkali.

Byakangelicin, $C_{17}H_{18}O_7 \cdot H_2O$, m. p. 117–8° (125–6°, anhydrous), $[\alpha]_D^{25}$ 24.62° (in pyridine), is faintly yellow and was shown by NOGUCHI and KAWANAMI to be closely similar to byakangelicol (see refs. to byakangelicol and phellopterin) and, like this substance, to be an ether of 8-hydroxy-5-methoxypsoralene. Acetone and α-hydroxy*iso*butyric acids were amongst the products of oxidation. In addition byakangelicin formed a diacetate and when heated with phosphorus pentoxide in toluene it lost the elements of water and gave rise to a ketone called anhydro-byakangelicin which could be oxidised to *iso*butyric acid. Therefore byakangelicin was considered to be the diol related to byakangelicol, a view substantiated by the hydration of natural or synthetic byakangelicol to byakangelicin.

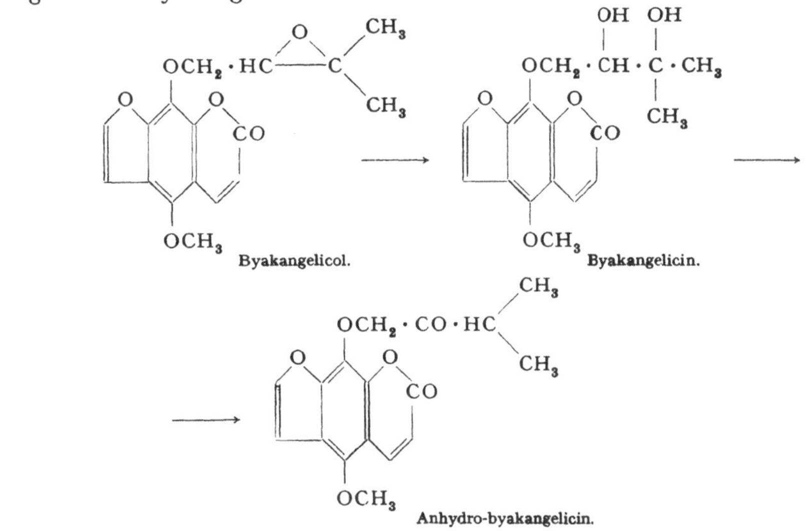

Byakangelicol.

Byakangelicin.

Anhydro-byakangelicin.

Ferulin, $C_{17}H_{16}O_6$, m. p. 87°, $[\alpha]_D^{23} + 27.31°$, has been isolated from the fruit of *Ferula alliacea* BOISS by BOSE and CHAUDHURI (*18*). When

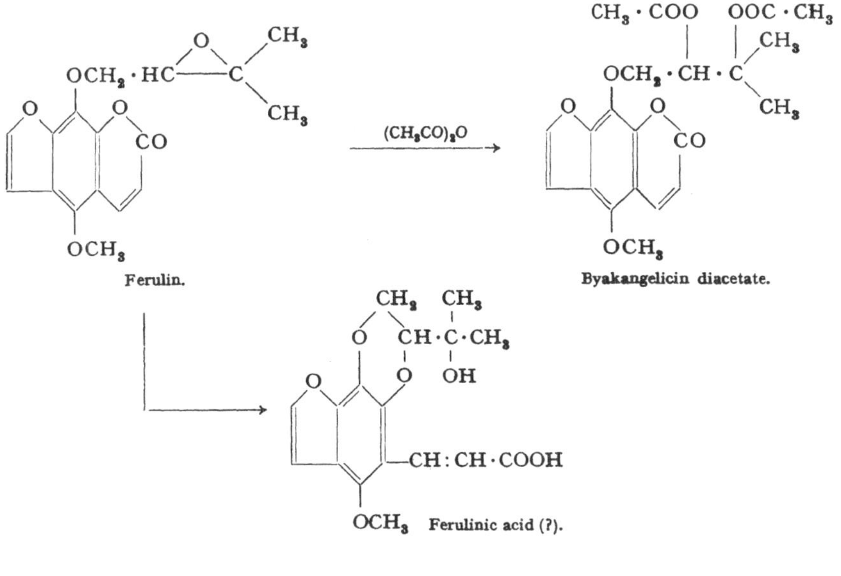

Ferulin.

$(CH_3CO)_2O$

Byakangelicin diacetate.

Ferulinic acid (?).

treated with acetic anhydride and sodium acetate, it gave rise to the diacetate of byakangelicin, while the action of 10% alkali resulted in ferulinic acid $C_{17}H_{18}O_7$, m. p. 194–5°, isomeric with *iso*byakangelicolic acid. The Indian authors claim that the formula previously advanced for byakangelicol belongs by right to ferulin, but the matter needs further investigation.

Nodakenetin, $C_{14}H_{14}O_4$, m. p. 192°, $[\alpha]_D^{24}$–25.4°, occurs as the glucoside nodakenin $C_{20}H_{24}O_9$, m. p. 215°, in *Peucedanum decursivum* Maxim. Arima (*5*) noted that nodakenetin showed no acidity or carbonyl activity, but that it gave a mono-acetate indicating the presence of one alcoholic hydroxyl group, and that it had the lactonic properties of a coumarin. Resorcinol resulted from alkali fusions, and umbelliferone-6-carboxylic acid from chromic acid oxidation.

Späth and Kainrath (*99*), observing that nodakenetin resembled oreoselone (see peucedanin), dehydrated it with phosphorus pentoxide at 120–150° and obtained optically inactive anhydro-nodakenetin which absorbed four atoms of hydrogen in catalytic reductions giving desoxy-dihydro-oreoselone. The hydroxyl group is not yet located exactly, but its position in the formula below is consistent with the oxidation of nodakenetin to acetone, and the comparative resistance of nodakenetin to dehydration which is very facile with 3-hydroxycoumarans. When heated with β-pentaacetyl-*D*-glucose and a trace of *p*-toluene sulfonic acid, nodakenetin formed a little nodakenin tetraacetate, which yielded a compound identical with natural nodakenin when treated with ammonia [Späth and Tyray (*125*)].

Nodakenin. Nodakenetin.

Desoxydihydro-oreoselone. Anhydro-nodakenetin.

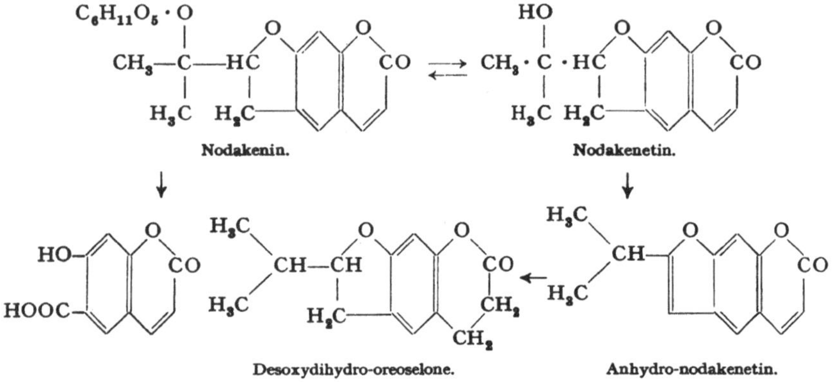

Marmesin, $C_{14}H_{14}O_4$, m. p. 189.5°, $[\alpha]_D^{34}$ 26.8°, was isolated from the bark of *Aegle marmelos* Correâ by Chatterjee and Mitra (*28*) who showed that it was the optical antipode of nodakenetin (formula above).

Peucedanin, $C_{15}H_{14}O_4$, m. p. 109°, was found in *Peucedanum officinale* by SCHLATTER in 1833. Early work had shown that the substance contained a resorcinol nucleus, that it could be hydrolysed by alkali or acids to give oreoselone, $C_{14}H_{12}O_4$, and that it was poisonous to fish and accelerated the alcoholic fermentation of yeast. These facts interested SPÄTH and caused him to begin his investigations of natural coumarins.

SPÄTH et al. (*103, 105*) noted that both peucedanin and oreoselone were optically inactive and that only the latter gave an acetate, while the former contained a methoxyl group which was lost during hydrolysis. They regarded peucedanin as the methyl ether of an enolic form of the ketone oreoselone.

In alkaline media, oreoselone absorbed two atoms of hydrogen giving an acid which on heating furnished the monolactone dihydrooreoselone, oxidation of which yielded succinic acid. Oxidation of oreoselone itself gave α-hydroxy*iso*butyric acid or *iso*butyric acid, according to the conditions.

Tetrahydro-peucedanin resulted from hydrogenation of peucedanin, and on distillation it lost the elements of methanol, yielding a product which could then absorb two more atoms of hydrogen forming desoxy-dihydro-oreoselone.

Oreoselone was oxidised by permanganates to α-resodicarboxylic acid that suggested a linear arrangement for the three rings which SPÄTH, KLAGER and SCHLÖSSER (*105*) thought were present. The formulae put forward by these authors are given below.

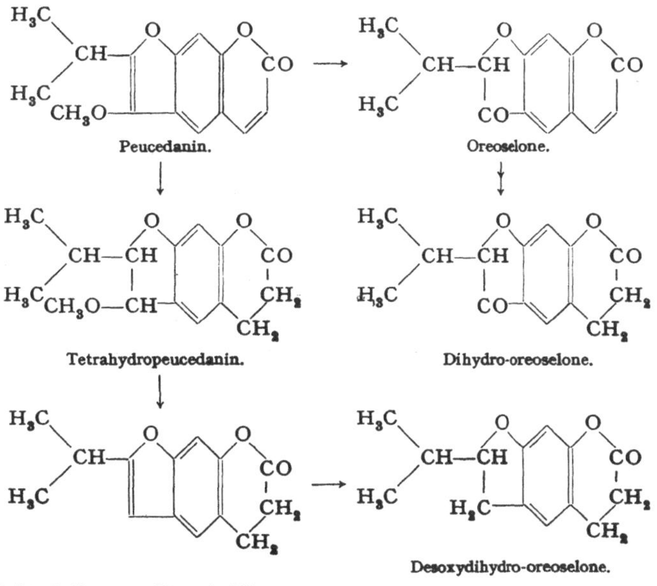

Further, alkali fusion of dihydro-oreoselone resulted in dihydro-umbelliferone, while the α-ketol structure of oreoselone accounted for its facile reduction of FEHLING's solution.

After much difficulty VON BRUCHHAUSEN and HOFFMANN (*23*) achieved a satisfactory synthesis of oreoselone in which umbelliferone-6-carboxylic acid was treated with a mixture of propionyl chloride and thionyl chloride, and the product was converted to a diazomethyl ketone which cyclised to a coumaranone. The latter condensed with acetone and the product when subjected to selective hydrogenation gave rise to synthetic oreoselone.

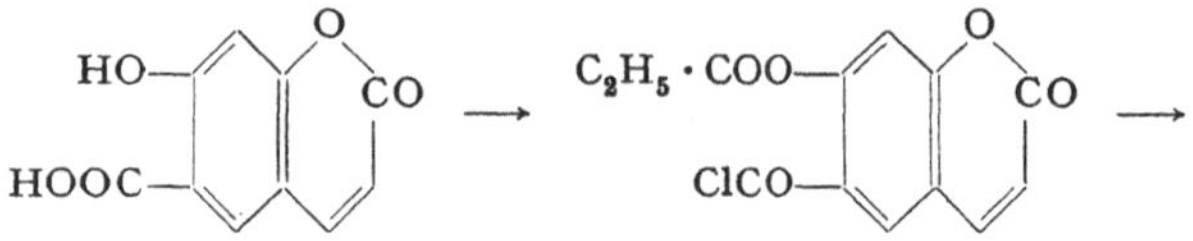

Umbelliferone-6-carboxylic acid.

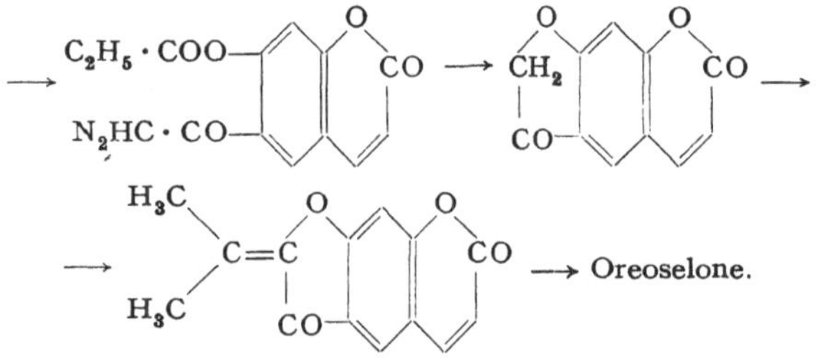

A simpler synthesis has recently been published by GAIND, GUPTA, RAY, and SAREEN (*36*), and is illustrated below.

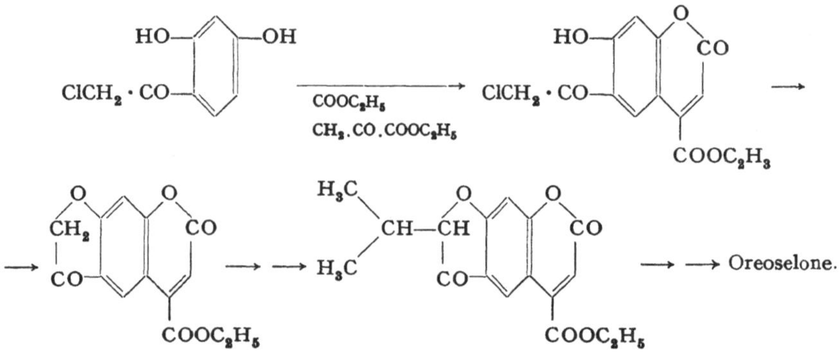

SCHMID and EBNÖTHER (*74*) have succeeded in converting oreoselone into peucedanin. The more common methods of obtaining enol ethers from ketones did not succeed, nor did attempts to remove the elements of methanol from the dimethylketal, but peucedanin was formed smoothly when oreoselone was treated with a 37% solution of aluminium chloride in anhydrous methanol.

Athamantin, $C_{24}H_{30}O_7$, m. p. 58–60°, though isolated from *Peucedanum oreoselinum* MÖNCH (*Athamanta oreoselinum* L.) in 1844, was not adequately investigated until 1940, when SPÄTH and his co-workers (*119, 117*) clarified its structure.

It was known already that athamantin when treated with acids lost "valeric acid" giving oroselone, a substance similar to, but not identical with, oreoselone. Oroselone, $C_{14}H_{10}O_3$, m. p. 188–9°, behaved as a coumarin, possessed one inert oxygen atom, and was not optically active. Under appropriate conditions, oroselone was attacked by ozone to give either 2:4-dihydroxy*iso*-phthalaldehyde or 8-formyl-umbelliferone, so that it could be formulated provisionally as a substituted angelicin.

Hydrogenation of oroselone quickly saturated one double bond giving dihydro-oroselone which could be oxidised to *iso*butyric acid; tetrahydro-oroselone formed more slowly, and furnished *iso*butyric, succinic, and oxalic acids, on oxidation. Prolonged hydrogenation yielded hexahydro-oroselone. These relationships are illustrated below.

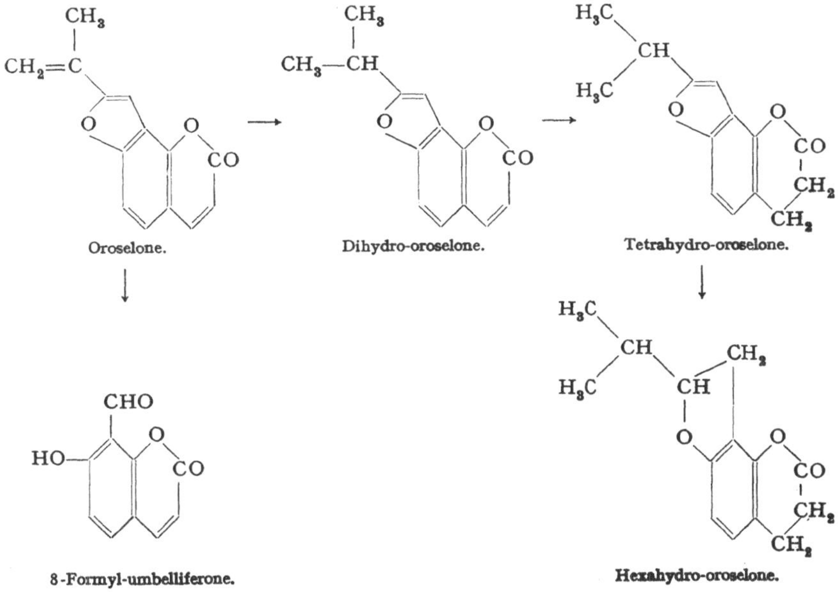

Re-investigation of athamantin showed that two molecules of *iso*valeric acid and one of oroselone were formed on hydrolysis, thus accounting for all the atoms of the original molecule. It seemed necessary to find a way in which two *iso*valeric acid molecules could be added to two double bonds of oroselone. As athamantin exhibited the lactonic properties of a coumarin, SPÄTH and SCHMID (*119*) adopted the accompanying formula, which also accounted for the facile loss of the *iso*valeric acid molecules and for the production of a little acetone when the crude diol (obtained by very mild alkaline hydrolysis) was oxidised. Nevertheless, other possibilities are not excluded.

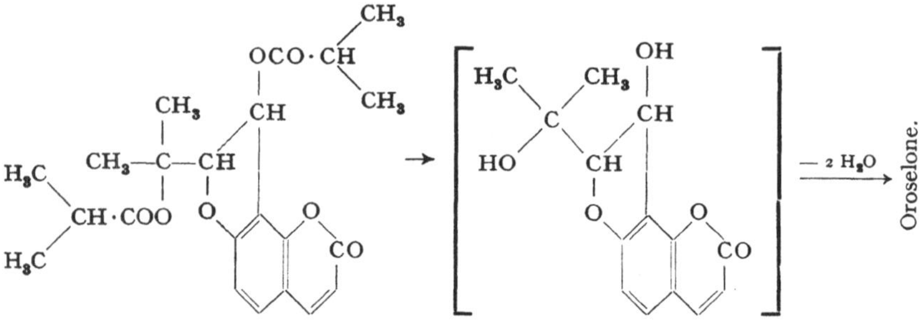

$$\text{Athamantin} \rightarrow [\text{intermediate}] \xrightarrow{-2H_2O} \text{Oroselone.}$$

VII. Chromeno-α-pyrones.

The coumarins of this group can be regarded as derivatives of 2:2-dimethyl-Δ^3-chromene; and much of the difficulty involved in establishing their structures arises from the fact that this system is not easily distinguishable from that of an α-*iso*propyl-furan or an α-*iso*propylidine-coumaran. The double bond of the chromene ring is reactive, being readily hydrogenated and readily severed by ozonolysis, a conversion which is abnormal since it results in fission of an ether link and gives rise to an o-hydroxyaldehyde. The α-pyrone ring is not attacked so easily and survives these reactions. It is known now that fission with the production of acetone by prolonged alkaline hydrolysis is a characteristic of 2:2-dimethylchromenes and is a reaction which lends support to the existence of a six-membered as opposed to a five-membered heterocyclic ring. Oxidation of these substance by permanganates results in α-hydroxy-*iso*butyric acid, but as this acid is obtained by the action of permanganates on *iso*butyric acid, its formation can only be taken as evidence for the structure of the carbon chain and not as evidence for the size of the ring.

In the case of seselin (p. 278) it was found that oxidation produced a substituted phenoxydimethylacetic acid which could have arisen only from a chromene ring, but in other cases the presence of this type of ring has been established by hydrogenation of the chromene double bond, followed by removal of the α-pyrone ring from the comparatively stable 2:2-dimethylchroman and synthesis of the product.

Synthesis of chromenes is not easy and so far only the reaction between α-hydroxy*iso*propylacetylene and umbelliferone has given a natural chromenopyrone. GRIGNARD solutions react with coumarins giving a chromene as one product, but the reaction has not been adapted to work in this field. On the other hand, the ring-closure of o-(3:3-dimethylallyl)-phenols provides a good route to the 2:2-dimethylchroman ring system, so that synthesis of the dihydro-derivatives of chromeno-α-pyrones has met with some success; and in addition, α-*iso*propylcoumarans isomeric with the chromans may be prepared by unambiguous methods and shown to be different from the natural products.

Any specific illustrations of these reactions would seem to be unnecessary as this group of coumarins is very limited.

Xanthyletin, $C_{14}H_{12}O_3$, m. p. 128.5°, was isolated with some difficulty from the bark of *Zanthoxylum americanum* MILL., and also from *Luvanga scandens* HAM. and *Citrus Medica* L. (*C. acida*). It was found to resemble xanthoxyletin by BELL and ROBERTSON (*11*). Hydroxyl, methoxyl, and carbonyl groups were absent, but the substance behaved like a coumarin and very vigorous alkaline hydrolysis produced acetone and resorcinol. The ready formation of dihydroxanthyletin by catalytic hydrogenation showed that one reactive double bond was present.

Ozonolysis of xanthyletin furnished an *ortho* hydroxyaldehyde which was not 8-formylumbelliferone, and which was reduced by catalytic hydrogenation to a compound identical with synthetic 6-methylumbelliferone. On the other hand, dihydroxanthyletin was attacked by ozone (*10*) to give an o-hydroxyaldehyde which also resulted from 7-hydroxy-2:2-dimethylchroman by GATTERMANN's synthesis, and regenerated dihydroxanthyletin in the cyanacetic acid synthesis.

By condensing umbelliferone with isoprene or α-hydroxy*iso*propylacetylene, SPÄTH and his co-workers (*90, 109*) obtained small amounts of dihydroxanthyletin and xanthyletin, respectively, but although these reactions are of great interest, they have little value in structural determinations.

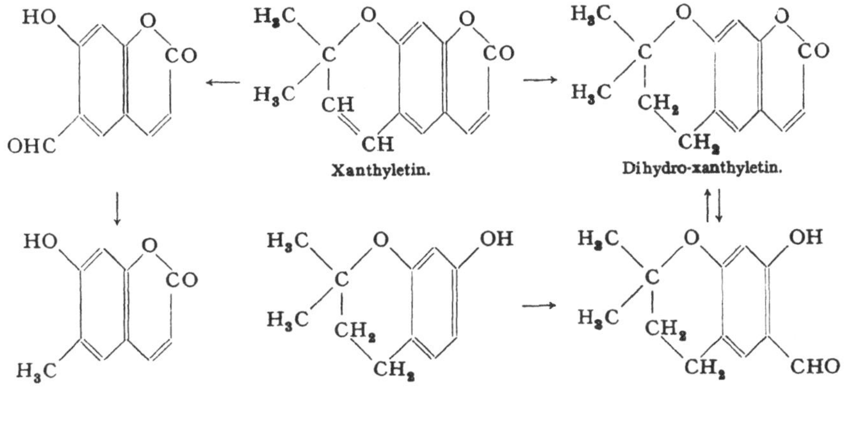

Seselin, $C_{14}H_{12}O_3$, m. p. 120°, obtained from *Seseli indicum* WIGHT and ARN. by BOSE and GUHA (*19*), was shown to be a neutral, unsaturated lactone free from methoxyl groups. SPÄTH and NEUFELD (*110*) discovered seselin in *Skimmia Japonica* THUNB. and SPÄTH, BOSE, MATZKE, and GUHA (*79*) undertook an examination of it.

When hydrogenated, seselin formed dihydroseselin and subsequently tetrahydroseselin. Treatment of seselin with sulfuric acid in acetic acid gave umbelliferone, ozonolysis generated acetone and 2:4-dihydroxy-*iso*phthalaldehyde, and finally, permanganate oxidation furnished α-hydroxy-*iso*butyric acid.

Oxidation of the methyl ether of the coumarinic acid corresponding to seselin followed by esterification of the tricarboxylic acid produced gave rise to an ester which was synthesised by a route involving the interaction of the sodium salt of 8-formylumbelliferone and methyl α-brom*iso*butyrate, thus establishing the existence of a 2:2-dimethyl-chromene ring. The reaction between umbelliferone and α-hydroxy-*iso*propylacetylene (see xanthyletin, p. 277) gave mainly seselin (*90*, *191*) while dihydroseselin is formed by ring closure of osthenol (*22*, *88*).

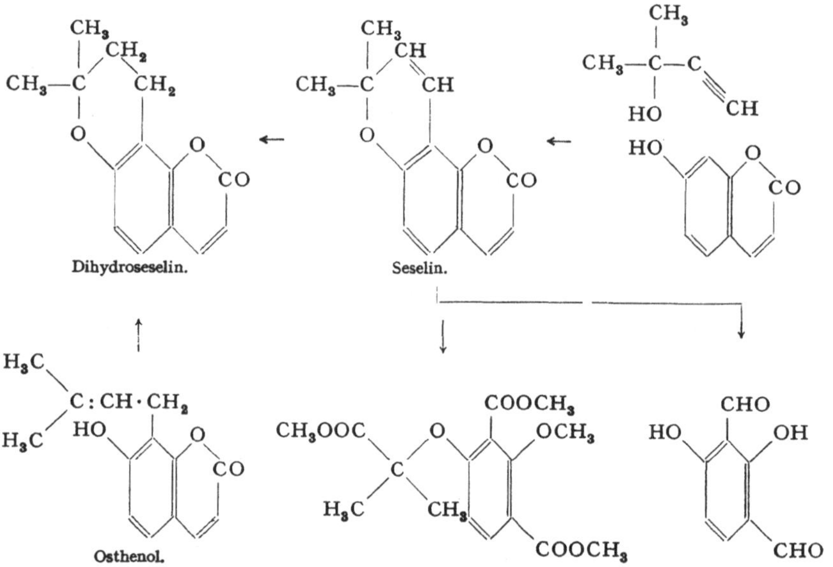

Xanthoxyletin (Xanthoxylin N), $C_{15}H_{14}O_4$, m. p. 132°, was noted in "prickly ash bark" in 1906, and later in *Zanthoxylum americanum* MILL. by BELL, ROBERTSON, and SUBRAMANIAM (*12*) who isolated acetone and phloroglucinol monomethyl ether by alkaline degradation of xanthoxyletin, and obtained α-hydroxy*iso*butyric acid from permanganate oxidations. Both di- and tetrahydro-xanthoxyletin were prepared by catalytic hydrogenation.

The same authors found that ozonolysis of xanthoxyletin led to a compound $C_{10}H_5O_4(OCH_3)$ which was called *apo*xanthoxyletin and which had the properties of an *ortho*-hydroxyaldehyde. Reduction of the aldehyde group to methyl led to desoxy-*apo*xanthoxyletin, the methyl ether of which was not identical with synthetic 5:7-dimethoxy-8-methylcoumarin, though both substances furnished the same 2:4:6-trimethoxy-3-methyl-cinnamic acid on ring-opening and methylation. *Apo*xanthoxyletin was given the formula below.

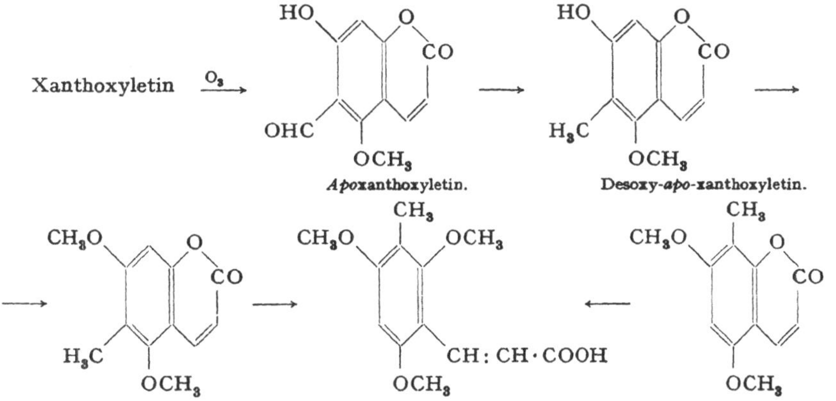

Desoxy-*apo*xanthoxyletin was synthesised much later (73), but, in the meantime, its ethyl ether was shown by ROBERTSON and SUBRAMANIAM (71) to give the same 2:6-dimethoxy-4-ethoxy-3-methylcinnamic acid by methylative ring-opening as that obtained from synthetic 5-methoxy-7-ethoxy-8-methylcoumarin. It followed that the free hydroxyl of desoxy-*apo*xanthoxyletin was in position 7 of the coumarin system, and that xanthoxyletin must have a linear configuration.

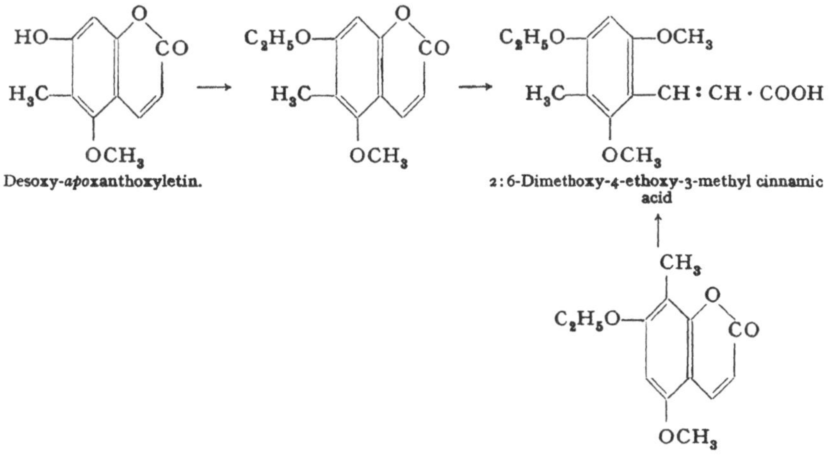

The work of DIETERLE and KRUTA (*32*) confirmed the "linear" structure for xanthoxyletin, since it established that bergapten and xanthoxyletin both gave the same dialdehyde with excess ozone.

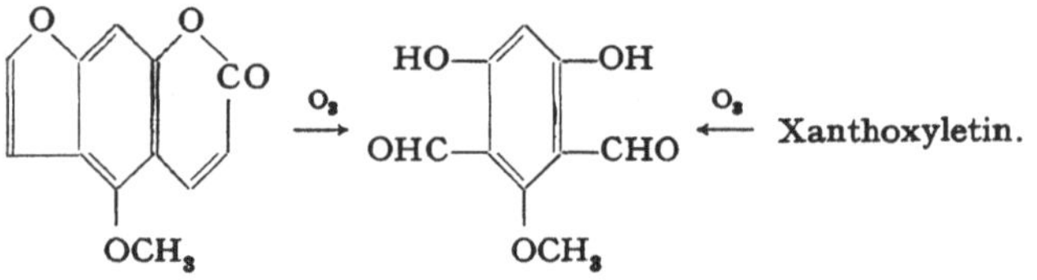

The existence of a chromene ring in xanthoxyletin was demonstrated by ozonolysis of dihydro-xanthoxyletin which gave rise to an *o*-hydroxy-aldehyde that regenerated dihydro-xanthoxyletin by the cyanacetic acid condensation, followed by hydrolysis and decarboxylation of the product. The methyl ether of this aldehyde could be oxidised to the corresponding acid which, on decarboxylation gave 5:7-dimethoxy-2:2-dimethyl-chroman, identical with synthetic material. As this chroman was an oil, the identification was achieved by means of carbonyl derivatives of the aldehyde formed in the GATTERMANN reaction.

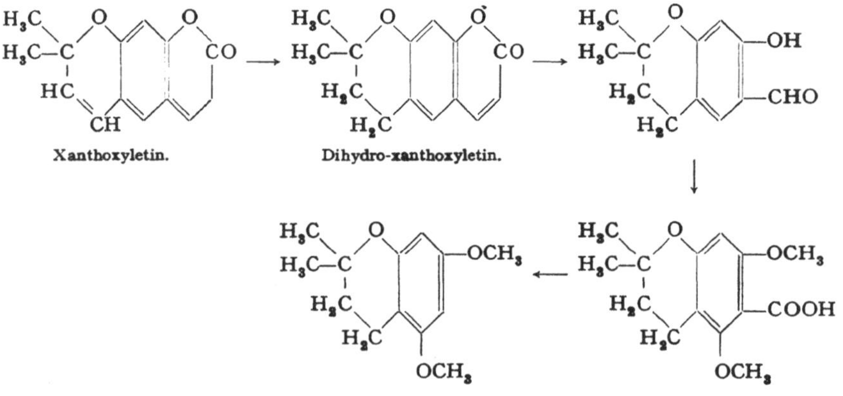

Luvangetin, $C_{15}H_{14}O_4$, m. p. 108–9°, isolated from *Luvanga scandens* HAM., was shown by SPÄTH, BOSE, SCHMID, DOBROVOLNY, and MOOKERJEE (*81*) to react with hydrogen in the presence of a catalyst in two stages giving dihydro- and tetrahydro-luvangetin, of which the first could be oxidised to α-hydroxy*iso*butyric acid and the second to succinic acid. Ozonolysis of luvangetin gave a compound formulated as 6-formyl-8-methoxyumbelliferone; and, after opening of the heterocyclic rings by hydrogen bromide and phosphorus, methylation and oxidation produced 2:3:4-trimethoxybenzoic acid. SPÄTH and SCHMID (*120*) obtained

synthetic luvangetin by condensation of 8-O-methyldaphnetin with
α-hydroxy*iso*propylacetylene.

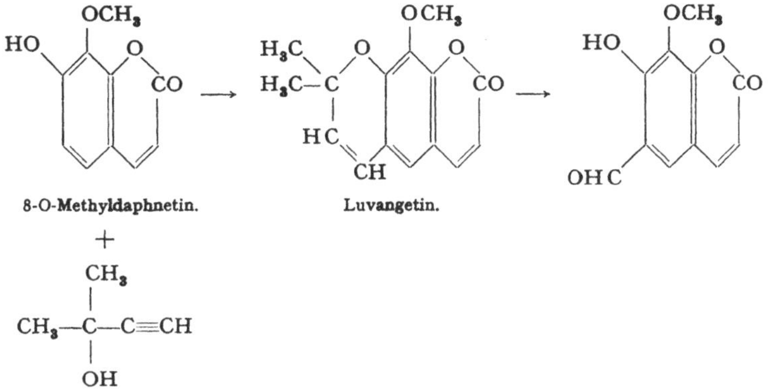

8-O-Methyldaphnetin. → Luvangetin. →

+

$$CH_3-\underset{\underset{OH}{|}}{\overset{\overset{CH_3}{|}}{C}}-C\equiv CH$$

Alloxanthoxyletin, $C_{15}H_{14}O_4$, m. p. 115.5°, obtained from *Zanthoxylum americanum* MILL by ROBERTSON and SUBRAMANIAM (72) was found to be closely similar to xanthoxyletin. The main difference was that ozonolysis of dihydro*allo*xanthoxyletin gave 8-formyl-7-hydroxy-5-methoxy-2:2-dimethyl-chroman, which was identified by synthesis, and regenerated dihydro*allo*xanthoxyletin by the cyanacetic acid condensation.

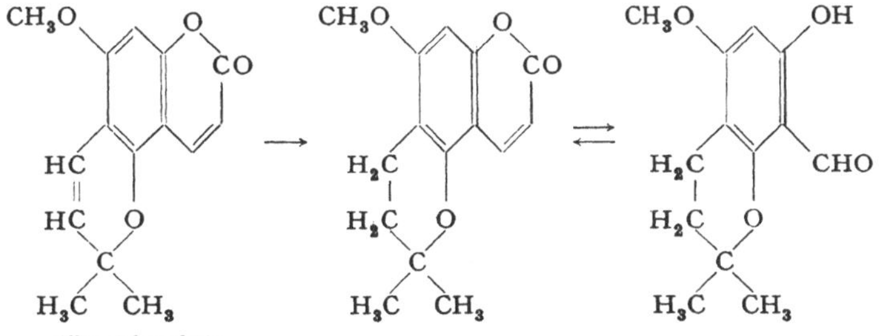

*Allo*xanthoxyletin.

Braylin, $C_{15}H_{14}O_4$, m. p. 150°, from the bark of *Flindersia brayleyana* F. MUELL gave a dihydro- and then a tetrahydro-derivative (*4*). On the basis of its relation to brayleyanin and its ultraviolet absorption characteristics it was given the accompanying formula. Oxidation furnished α-hydroxy*iso*butyric acid, and alkaline hydrolysis finally resulted in the formation of acetone and 4-methoxyresorcinol. O-Methyltetrahydrobraylinic acid could be cyclised to a ketone, which would be impossible with any isomer.

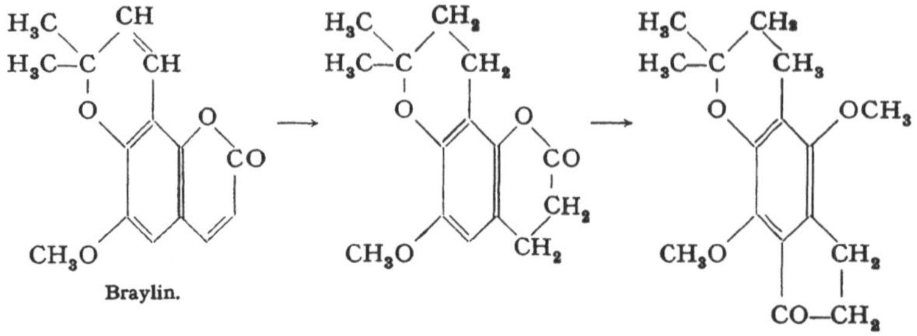

VIII. 3:4-Benzocoumarins.

2′:3″-Dihydroxydibenz-α-pyrone, $C_{13}H_{18}O_4$, m. p. > 360° (decomp.), is one of the two yellow pigments of castoreum (the secretion of the scent gland of the beaver). LEDERER (53) purified the substance by means of its colourless diacetate and found that two phenolic hydroxyl groups and one lactonic system were present. Zinc dust distillation resulted in the formation of fluorene, $C_{13}H_{10}$, while fusion with alkali furnished *m*-hydroxybenzoic acid. These facts are embodied in the structure given below, which was further supported by the positive GRIESSMAYER reaction and confirmed by LEDERER and POLONSKY by a synthesis which utilised the condensation of 2-bromo-5-hydroxybenzoic acid with resorcinol in the presence of copper salts, a reaction investigated by ADAMS et al. (*1*).

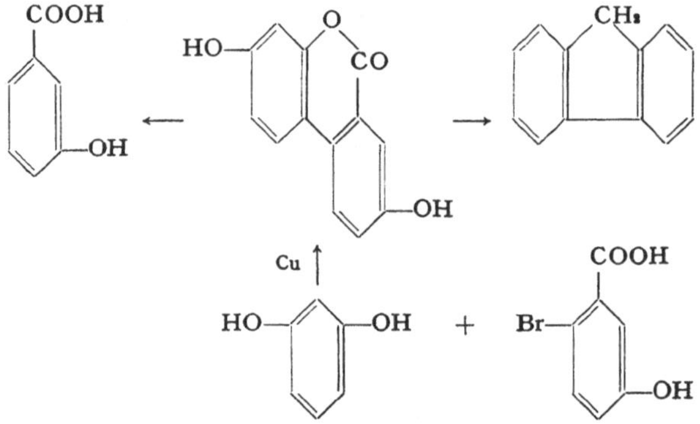

4:6:4′:6′-Dihydroxydiphenic acid dilactone, $C_{14}H_6O_6$, m. p. > 360°, is the second yellow pigment of castoreum, and was found by LEDERER (53) to contain two phenolic groups and two lactonic systems. Fluorene results from zinc dust distillation and β-resorcylic acid from alkali fusion. That a 4:4′-dihydroxydiphenyl nucleus was present was strongly supported by a positive GRIESSMAYER reaction. (When warmed with nitric acid containing a little nitrous acid, derivatives of 4:4'-dihydroxydiphenyl develop a red colour which becomes apparent on dilution.)

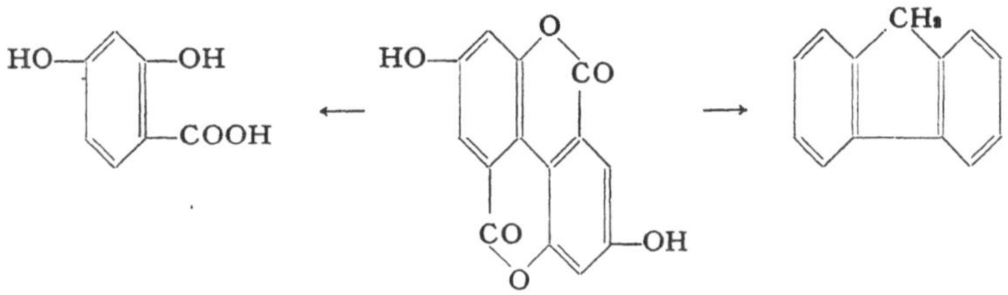

4: 6, 4': 6'-Dihydroxydiphenic acid dilactone.

Ellagic acid, $C_{14}H_{16}O_8$, m. p. $> 360°$, has been known since CHEVREUL isolated it in 1818 from oak galls. It is one of the most widely distributed natural yellow colouring matters since it can be produced by hydrolysis of ellagitannins, and is usually associated with gallotannins. Some animals which feed on plant material collect ellagic acid into lumps ("bezoar stones") in their stomachs.

The easy oxidation of gallic acid to ellagic acid (by air, arsenic acid, persulfuric acid, hydrogen peroxide etc.) ensures a ready supply of material, and throws some light on its constitution, but although many workers had made contributions to the problem, it was not until PERKIN and NIERENSTEIN (66) favoured a structure originally put forward by SCHIFF that any formula was accepted generally.

Ellagic acid was known to form a tetraacetate and a tetramethyl ether, and to furnish hexahydroxy-diphenyl when fused with alkali. Zinc dust distillation resulted in the formation of fluorene, a reaction typical of 3:4-benzocoumarins.

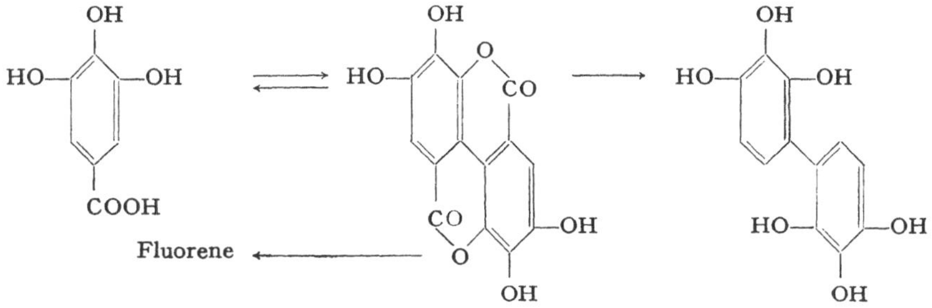

Ellagic acid.

Mild treatment with alkali permitted the isolation of a compound $C_{13}H_8O_7$ which contained five hydroxyl groups and still gave rise to fluorene when heated with zinc dust.

Ellagic acid ⟶ [structure] ⟶ Fluorene.

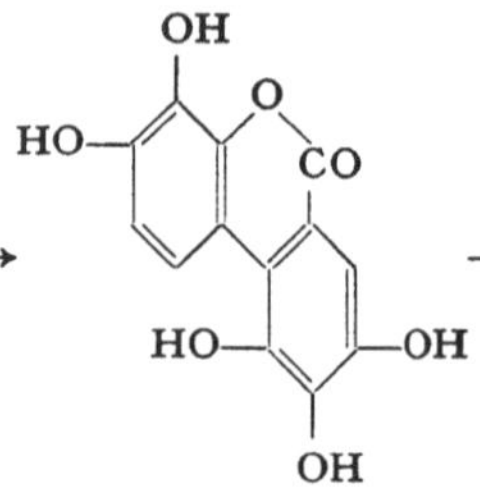

HERZIG and POLLAK (45) provided evidence for the two lactone rings by showing that alcoholic potash and methyl iodide opened either one or two rings in O-tetramethylellagic acid to give compounds which on hydrolysis yielded the corresponding acids.

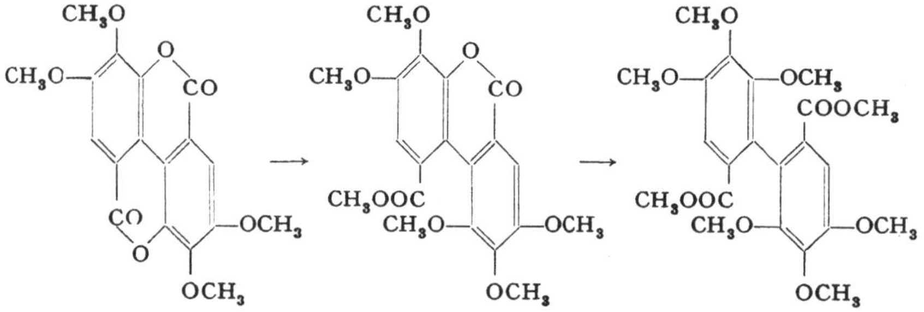

Ellagic acid gives a green ferric reaction, a yellow solution in alkali, and a positive GRIESSMAYER test.

4:4'-Dihydroxy-6:6'-dimethoxydiphenic acid dilactone, $C_{16}H_{10}O_8$, m. p. 337–8°, was extracted from the roots of *Euphorbia formosanum* HAY by SHINODA and KUN (77). It reacted with hydrogen iodide to yield ellagic acid, and as it gave no green colour with ferric chloride, the two methoxyl groups were thought to be in different rings. Since it gave a positive GRIESSMAYER test, LEDERER (53) considered the free hydroxyl groups to be in the 4:4'-positions of the diphenyl system, so that (A) would be the constitution of this compound, and (B) the structure of the isomeric dimethyl ether which is obtained by methylation of ellagic acid itself and gives no GRIESSMAYER test.

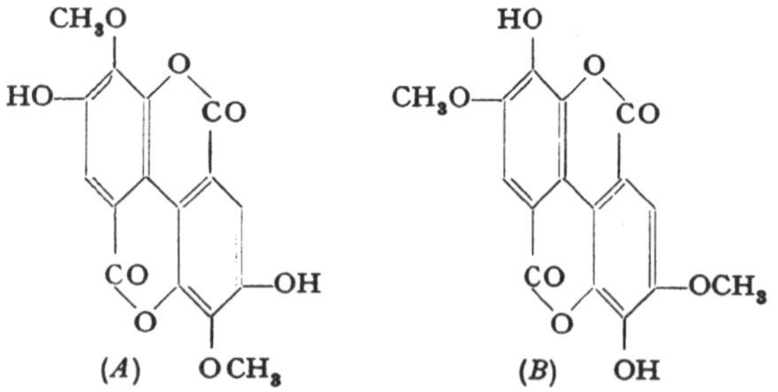

References.

1. ADAMS, R., D. C. PEASE, J. H. CLARK and B. R. BAKER: Structure of Cannabinol. I. Preparation of an Isomer, 3-Hydroxy-1-*n*-amyl-6,6,9-trimethyl-6-dibenzopyran. J. Amer. chem. Soc. **62**, 2197 (1940).
2. ANDREAE, W. A.: The Isolation of a Blue Fluorescing Compound Scopoletin from Green Mountain Potato Tubers Infected with Leaf Roll Virus. Canad. J. Res., Sect. C **26**, 31 (1948).
3. ANET, L., F. R. BLANKS and G. K. HUGHES: The Chemical Constituents of Australian *Flindersia* Species. I. Collinin, 7-Geranyloxy-8-methoxycoumarin. Austral. J. Sci., Ser. A **2**, 127 (1949).
4. ANET, L., G. K. HUGHES and E. RITCHIE: The Chemical Constituents of Australian *Flindersia* Species. II. Braylin and Brayleyanin. Austral. J. Sci., Ser. A **2**, 608 (1949).
5. ARIMA, J.: Über die Konstitution des Nodakenins, eines neuen Glucosids von *Peucedanum decursivum* MAXIM. II. Bull. chem. Soc. Japan **4**, 113 (1929).
5a. ARNDT, F., L. LOEWE, R. UN und E. AYÇA: Ber. dtsch. chem. Ges. **84**, 319 (1951).
6. ASAI, T.: Über das Vorkommen und die physiologische Bedeutung des Daphnins bei *Daphne odora*. Acta phytochim. (Tokyo) **5**, 9 (1930).
7. BALAIAH, V., T. R. SESHADRI and V. VENKATESWARLU: Visible Fluorescence and Chemical Constitution of Compounds of the Benzopyrone Group. III. Further Study of Structural Influences in Coumarins. Proc. Indian Acad. Sci., Ser. A **16**, 68 (1942).
8. BARGELLINI, G. e L. MONTI: Richerche sulle cumarine. Gazz. chim. ital. **45** (i), 90 (1915).
9. BAYERLE, H. und R. MARX: Über Cumarinderivate als Antithrombotika. Biochem. Z. **319**, 18 (1948).
10. BELL. J. C., W. BRIDGE and A. ROBERTSON: Constituents of the Bark of *Zanthoxylum americanum* MILL. IV. The Constitution of Xanthyletin. J. chem. Soc. (London) **1937**, 1542.
11. BELL, J. C. and A. ROBERTSON: Constituents of the Bark of *Zanthoxylum americanum* MILL. II. Xanthyletin. J. chem. Soc. (London) **1936**, 1828.
12. BELL, J. C., A. ROBERTSON and T. S. Subramaniam: The Constituents of the Bark of *Zanthoxylum americanum* MILL. I. Xanthoxyletin. J. chem. Soc. (London) **1936**, 627.
13. BERT, L.: Sur une nouvelle méthode générale de préparation synthétique des coumarins. C. R. hebd. Séances Acad. Sci. **214**, 230 (1942).
14. BEST, R. J.: A Fluorescent Substance Present in Plants. II. Isolation of the Substance in a Pure State and its Identification as 6-Methoxy-7-hydroxy-1:2-benzpyrone. Austral. J. exp. Biol. med. Sci. **22**, 251 (1944).
15. BÖHME, H. und G. PIETSCH: Zur Kenntnis des Auraptens. Ber. dtsch. chem. Ges. **72**, 773 (1939).
16. BÖHME, H. und E. SCHNEIDER: Oxydativer Abbau und Konstitution des Auraptens. Ber. dtsch. chem. Ges. **72**, 780 (1939).
17. — — Zur Konstitution der Auraptensäuren. IV. Mitt. über Stearopten des Pomeranzenschalenöls. Arch. Pharmaz. Ber. dtsch. pharmaz. Ges. **279**, 213 (1941).
18. BOSE, P. K. and J. C. CHAUDHURY: Constitution of Coumarins Isolated from *Ferula alliacea*. Ann. biochem. exp. Med. **6**, 1 (1946).
19. BOSE, P. K. and N. C. GUHA: The Crystalline Constituents of *Seseli indicum*. Sci. and Cult. **2**, 326 (1936).

20. BOSE, P. K. and A. MOOKERJEE: Natural Glucosides. I. The Constitution of the Glucoside Present in *Murraya exotica*. J. Indian chem. Soc. **14**, 489 (1937).
21. BOSE, P. K. and P. B. SEN: New Haemostatic Agents. I. Experiments with Ayapanin and Ayapin. Ann. biochem. exp. Med. **1** (4), 311 (1941).
22. BOTTOMLEY, W. and D. E. WHITE: The Chemistry of Western Australian Plants. V. Vellein from *Velleia discophora*. Austral. J. Sci., Ser. A **4**, 107 (1951).
22a. BOYD, J. and A. ROBERTSON: A New Synthesis of 4-Hydroxycoumarins. J. Chem. Soc. (London) **1948**, 174.
23. BRUCHHAUSEN, F. v. und H. HOFFMANN: Synthese des Oreoselons. Ber. dtsch. chem. Ges. **74**, 1584 (1941).
24. BUTENANDT, A. und A. MARTEN: Untersuchungen über pflanzliche Fisch- und Insektengifte. V. Über die Inhaltsstoffe der Meisterwurz (*Imperatoria Ostruthium*). Liebigs Ann. Chem. **495**, 187 (1932).
25. CALDWELL, A. G. and E. R. H. JONES: Constituents of Expressed West Indian Lime Oil. J. chem. Soc. (London) **1945**, 540.
26. CAMPBELL, H. A. and K. P. LINK: Studies on the Hemorrhagic Sweet Clover Disease. IV. The Isolation and Crystallization of the Hemorrhagic Agent. J. biol. Chemistry **138**, 21 (1941).
27. CANTER, F. W. and A. ROBERTSON: Conversion of 7-Hydroxy-3:4-dimethylcoumarin into 2:4-Dimethoxy-α,β-dimethyl-cinnamic Acid. J. chem. Soc. (London) **1931**, 1875.
28. CHATTERJEE, A. and S. S. MITRA: On the Constitution of the Active Principles Isolated from the Matured Bark of *Aegle marmelos*, Correâ. J. Amer. chem. Soc. **71**, 606 (1949).
29. CHAUDHURY, D. N., R. A. HOLLAND and A. ROBERTSON: The Synthesis of Glycosides. XII. Fabiatrin. J. chem. Soc. (London) **1948**, 1671.
30. CHRISTIANI, A. v. und C. HOFFMANN: Über die Spaltung von Glycosiden im Hochvacuum. Mikrochim. Acta (Wien) **2**, 93 (1937).
31. DEY, B. B. und P. P. PILLAY: Chemische Untersuchungen von *Toddalia aculeata* PERS. II. Über Toddalolactone. Arch. Pharmaz. Ber. dtsch. pharmaz. Ges. **273**, 223 (1935).
32. DIETERLE, H. und E. KRUTA: Über einen Inhaltsstoff von *Xanthoxylum fraxineum* WILD. Arch. Pharm. Ber. dtsch. pharmaz. Ges. **275**, 45 (1937).
33. DUTTA, P.: On the Constitution of Natural Coumarins of *Toddalia aculeata*. J. Indian chem. Soc. **19**, 425 (1942).
34. EWING, J., G. K. HUGHES and E. RITCHIE: The Chemical Constituents of Australian *Zanthoxylum* Species. I. Suberosin, 6-(3:3-Dimethylallyl)-7-methoxycoumarin. Austral. J. Sci., Ser. A **3**, 342 (1950).
35. FOSTER, R. T., A. ROBERTSON and A. BUSHRA: Furano-Compounds. VII. A Synthesis of 2:3-Dihydropsoralene. J. chem. Soc. (London) **1948**, 2254.
36. GAIND, K. N., I. S. GUPTA, J. N. RAY and K. N. SAREEN: Experiments on the Synthesis of Oreoselone. II. J. Indian chem. Soc., ind. News Edit. **23**, 370 (1946).
37. GANDINI, A.: Sintesi della daphnina. Gazz. chim. ital. **70**, 611 (1940).
38. GOODWIN, R. H. and F. KAVANAGH: The Isolation of Scopoletin, a Blue-Fluorescing Compound from Oat-Roots. Bull. Torrey bot. Club **76**, 255 (1949).
39. — — Fluorescence of Coumarin Derivatives as a Function of p_H. Arch. Biochemistry **27**, 152 (1950).
40. GRUBER, W.: Über substituierte Cumarsäuren. Monatsh. Chem. **75**, 14 (1944).
41. HALLER, H. L. and F. ACREE, Jr.: Synthesis of 8-Isoamyl-7-methoxycoumarin (Dihydro-osthol). J. Amer. chem. Soc. **56**, 1389 (1934).
42. HATTORI, S.: Constitution of Daphnetin. J. pharmac. Soc. Japan **50**, 539 (1930) [Chem. Abstr. **24**, 4787 (1930)].

43. HEAD, F. S. H. and A. ROBERTSON: Hydroxy-carbonyl Compounds. I. A Synthesis of Scopoletin. J. chem. Soc. (London) 1931, 1241.
44. — — Syntheses of Glycosides. XI. Cichoriin. J. chem. Soc. (London) 1939, 1266.
45. HERZIG, J. und J. POLLAK: Zur Konstitution der Ellagsäure. III. Mitt. über Laktonfarbstoffe. Monatsh. Chem. 29, 263 (1908).
46. HEYES, R. G. and A. ROBERTSON: Hydroxy-carbonyl Compounds. XII. 5:7-Dihydroxycoumarin. J. chem. Soc. (London) 1936, 1831.
47. HUEBNER, C. F. and K. P. LINK: Studies in the Hemorrhagic Sweet Clover Disease. VI. The Synthesis of the δ-Diketone Derived from the Hemorrhagic Agent through Alkaline Degradation. J. biol. Chemistry 138, 529 (1941).
48. HORNING, E. C. and D. B. REISNER: Furocoumarin Studies. Synthesis of Psoralene and Related Furocoumarins. J. Amer. chem. Soc. 72, 1514 (1950).
49. HOWELL, W. N. and A. ROBERTSON: Furano-Compounds. I. A Synthesis of Bergapten. J. chem. Soc. (London) 1937, 293.
50. IIDA, G.: Chemische und pharmakologische Untersuchungen über das Fraxin, einen Bestandteil von *Fraxinus borealis* NAKAI. Tôhoku J. exp. Med. 25, 454 (1935).
51. KRISHNASWAMY, B. and T. R. SESHADRI: Synthetic Experiments in the Benzopyrone Series. VI. Action of Aluminium Chloride on Angelicin, Psoralene, and Related Compounds. Proc. Indian Acad. Sci., Ser. A 16, 151 (1942).
52. KUNZ, K. und L. HOOPS: Über die Harzbestandteile des Ammoniacums. II. Die Konstitution des Ammoresinols. Ber. dtsch. chem. Ges. 69, 2174 (1936).
53. LEDERER, E.: Chemistry and Biochemistry of Some Mammalian Secretions and Excretions. J. chem. Soc. (London) 1949, 2115.
54. LIMAYE, D. B.: Syntheses in the Furocoumarin Group. II. Karanjelin Way of Synthesising Furocoumarins as Illustrated on 7:8-Furocoumarin. Rasayanam (J. Progr. chem. Sci.) 1, 1 (1936).
54a. LINK, K. P.: The Anticoagulant from Spoiled Sweet Clover Hay. The Harvey Lectures 39, 162 (1943/44).
55. LUTZMANN, H.: *Cis-trans* Umlagerung der o-Oxyzimtsäureglucoside, über das Glucosid der o-Hydrocumarsäure und das Vorkommen des Cumarins in der Tonkabohne. Ber. dtsch. chem. Ges. 73, 632 (1940).
56. MEIJER, TH. M. und H. SCHMID: Über die Konstitution des Eugenins. Helv. chim. Acta 31, 1603 (1948).
57. MERZ, K. W.: Über das Cichoriin und die Konstitution des Aesculins und des Skopolins. Arch. Pharmaz. Ber. dtsch. pharmaz. Ges. 270, 476 (1932).
58. MERZ, K. W. und W. HAGEMANN: Synthese des Aesculins. Naturwiss. 29, 650 (1941)·
59. NOGUCHI, T. und M. KAWANAMI: Über die Bestandteile der Wurzel von *Angelica glabra* MAKINO. I. Ber. dtsch. chem. Ges. 71, 344 (1938).
60. — — Über die Bestandteile der Wurzel von *Angelica glabra* MAKINO. II. Ber. deutsch. chem. Ges. 71, 1428 (1938).
61. — — Constituents of *Ligusticum acutilobum*. VIII. Constitution of Byakangelicol. J. pharmac. Soc. Japan 59, 755 (1939) [Chem. Abstr. 34, 2346, (1940)].
62. — — Constituents of *Phellopterus littoralis* BENTH. J. pharmac. Soc. Japan 60, 57 (1940) [Chem. Abstr. 34, 3717 (1940)].
63. — — Constituents of Umbelliferae. XI. Constituents of *Angelica glabra* MAKINO. J. pharmac. Soc. Japan 61, 77 (1941) [Chem. Abstr. 36, 464 (1942)].
63a. OHTA, T.: Components of the Fruit of *Artemisia capillaris*. J. pharmac. Soc. Japan 66, 11 (1946).
64. ORLA-JENSEN, A. D.: Application of Aesculin for the Identification of Bacteria. Acta pathol. microbiol. scand. 11, 312 (1934).

65. Parikh, R. J. and S. Sethna: Elb's Persulphate Oxidation. I. Oxidation of Some coumarin Derivatives. J. Indian chem. Soc. **27**, 369 (1950).
66. Perkin, A. G. and M. Nierenstein: Some Oxidation Products of the Hydroxybenzoic Acids and the Constitution of Ellagic Acid. I. J. chem. Soc. (London) Trans. **87**, 1412 (1905).
67. Raudnitz, H., F. Petrů, E. Diamant, K. Neurad und K. Lanner: Über das Ammoresinol. Ber. dtsch. chem. Ges. **69**, 1956 (1936).
68. Reychler, A.: Some Derivatives of Coumarin. Bull. Soc. chim. France, Mém. (4), **3**, 551 (1908). [Chem. Abstr. **2**, 3232 (1908)].
69. Rindl, M.: The Crystalline Glycoside from *Gnidia polycephala* (January Bossie). II. South African J. Sci. **30**, 455 (1933).
70. Roberts, W. L. and K. P. Link: Determination of Coumarin and Melilotic Acid. A Rapid Micromethod for Determination in Melilotus Seed and Green Tissue. Ind. Engng. Chem., Analyt. Edit. **9**, 438 (1937).
71. Robertson, A. and T. S. Subramaniam: Constituents of *Zanthoxylum americanum* Mill. III. The Constitution of Xanthoxyletin. J. chem. Soc. (London) **1937**, 286.
72. — — Constituents of the Bark of *Zanthoxylum americanum* Mill. V. The Structure of *allo*Xanthoxyletin. J. chem. Soc. (London) **1937**, 1545.
73. Robertson, A. and W. B. Whalley: Synthesis of Deoxy*apo*xanthoxyletin. J. chem. Soc. (London) **1951**, 1935.
74. Schmid, H. und A. Ebnöther: Synthese des Peucedanins. Helv. chim. Acta **34**, 1983 (1951).
75. Schmidt, E.: Über das Citropten (Citronenölstearopten, Citronkamfer, Citropten, Limettin). Arch. Pharmaz. Ber. dtsch. pharmaz. Ges. **242**, 288 (1904).
76. Shinoda, J. und M. Imaida: Über das Cumaringlycosid und eine Synthese des Melilotosides. J. pharmac. Soc. Japan. **54**, 107 (1934).
77. Shinoda, J. und C. P. Kun: Ein Bestandteil von *Euphorbia formanosa* Hay. I. J. pharmac. Soc. Japan. **51**, 50 (1931).
78. Späth, E.: Die natürlichen Cumarine. Ber. dtsch. chem. Ges. **70 A**, 83 (1937).
79. Späth, E., P. K. Bose, J. Matzke und N. C. Guha: Die Cumarine von *Seseli indicum* und die Konstitution des Seselins. Ber. dtsch. chem. Ges. **72**, 821 (1939).
80. Späth, E., P. K. Bose und J. Schläger: Konstitution und Synthese von Ayapin. XXVI. Mitteil. über natürliche Cumarine. Ber. dtsch. chem. Ges. **70**, 702 (1937).
81. Späth, E., P. K. Bose, H. Schmid, E. Dobrovolny und A. Mookerjee: Über die Konstitution des Luvangetins. LIV. Mitteil. über natürliche Cumarine. Ber. dtsch. chem. Ges. **73**, 1361 (1940).
82. Späth, E. und J. Bruck: Die Konstitution des Osthenols. XXIX. Mitteil. über natürliche Cumarine. Ber. dtsch. chem. Ges. **70**, 1023 (1937).
83. Späth, E. und A. von Christiani: Über pflanzliche Fischgifte. VII. Mitteil. Die Konstitution des Ostruthols (aus *Imperatoria Ostruthium*). Ber. dtsch. chem. Ges. **66**, 1150 (1933).
84. Späth, E., B. B. Dey und E. Tyray: Die Konstitution des Toddalo-lactons. XLI. Mitteil. über natürliche Cumarine. Ber. dtsch. chem. Ges. **71**, 1825 (1938).
85. — — — Die Strukturformel des Toddalolactons. XLIV. Mitteil. über natürliche Cumarine. Ber. dtsch. chem. Ges. **72**, 53 (1939).
86. Späth, E. und E. Dobrovolny: Synthese des Fraxetins, des Fraxidins und des Iso-fraxidins. XLII. Mitteil. über natürliche Cumarine. Ber. dtsch. chem. Ges. **71**, 1831 (1938).

87. SPÄTH, E. und E. DOBROVOLNY: Synthese des Isoimperatorins und des Oxypeucedanins. XLIII. Mitteil. über natürliche Cumarine. Ber. dtsch. chem. Ges 72, 52 (1939).
88. SPÄTH, E., K. EITER und T. MEINHARD: Chroman- und Cumaron-Ringschlüsse bei einigen natürlichen Cumarinen. LIX. Mitteil. über natürliche Cumarine. Ber. dtsch. chem. Ges. 75, 1623 (1942).
89. SPÄTH, E. und F. GALINOVSKY: Über die Dehydrierung von Dihydro-cumarinen. X. Mitteil. über katalytische Dehydrierungs-Vorgänge. Ber. dtsch. chem. Ges. 70, 235 (1937).
90. SPÄTH, E. und R. HILLEL: Die Synthese des Xanthyletins. LI. Mitteil. über natürliche Cumarine. Ber. dtsch. chem. Ges. 72, 2093 (1939).
91. — — Die Synthese des Seselins. XLVI. Mitteil. über natürliche Cumarine. Ber. dtsch. chem. Ges. 72, 963 (1939).
92. SPÄTH, E. und H. HOLZEN: Synthese des Osthols. X. Mitteil. über natürliche Cumarine. Ber. dtsch. chem. Ges. 67, 264 (1934).
93. — — Über pflanzliche Fischgifte. V. Mitteil. Die Konstitution des Imperatorins (aus *Imperatoria Ostruthium*). Ber. dtsch. chem. Ges. 66, 1137 (1933).
94. — — Teilsynthese des Imperatorins und Darstellung des Oxy-imperatorins. XV. Mitteil. über natürliche Cumarine. Ber. dtsch. chem. Ges. 68, 1123 (1935).
95. SPÄTH, E. und Z. JERZMANOWSKA-SIENKIEWICZOWA: Über Fraxidin und Isofraxidin. XXVII. Mitteil. über natürliche Cumarine. Ber. dtsch. chem. Ges. 70, 1019 (1937).
96. — — Partialsynthese von Fraxidin und Isofraxidin sowie Synthese eines anderen Abkömmlings des 6,7,8-Trioxy-cumarins. XXXII. Mitteil. über natürliche Cumarine. Ber. dtsch. chem. Ges. 70, 1672 (1937).
97. — — Über Fraxinol, einen neuen Inhaltsstoff der Eschenrinde. XXV. Mitteil. über natürliche Cumarine. Ber. dtsch. chem. Ges. 70, 698 (1937).
98. SPÄTH, E. und L. KAHOVEC: Über pflanzliche Fischgifte. VI. Mitteil. Die Konstitution des Isoimperatorins (aus *Imperatoria Ostruthium*). Ber. dtsch. chem. Ges. 66, 1146 (1933).
99. SPÄTH, E. und P. KAINRATH: Die Konstitution des Nodakenins aus *Peucedanum decursivum* MAXIM. XX. Mitteil. über natürliche Cumarine. Ber. dtsch. chem. Ges. 69, 2062 (1936).
100. — — Über Bergamottin und über die Auffindung von Limettin in Bergamottöl. XXIV. Mitteil. über natürliche Cumarine. Ber. dtsch. chem. Ges. 70, 2272 (1937).
101. — — Über die Einwirkung von Aluminiumbromid und Benzol auf das Osthol. XXXVIII. Mitteil. über natürliche Cumarine. Ber. dtsch. chem. Ges. 71, 1662 (1938).
102. SPÄTH, E. und K. KLAGER: Über pflanzliche Fischgifte. IV. Mitteil. Die Konstitution des Oxy-peucedanins (aus *Imperatoria Ostruthium*). Ber. dtsch. chem. Ges. 66, 914 (1933).
103. — — Über pflanzliche Fischgifte. II. Mitteil. Zur Konstitution von Peucedanin und Oreoselon (aus *Peucedanum officinale*). Ber. dtsch. chem. Ges. 66, 749 (1933).
104. — — Über natürliche Cumarine. XII. Mitteil. Die Konstitution des Ostruthins (aus *Imperatoria Ostruthium*). Ber. dtsch. chem. Ges. 67, 859 (1934).
105. SPÄTH, E., K. KLAGER und C. SCHLÖSSER: Über die Konstitution von Peucedanin und Oreoselon. Ber. dtsch. chem. Ges. 64, 2203 (1931).
106. SPÄTH, E. und G. KUBICZEK: Synthese des Bergaptols und des Iso-bergaptens. XXX. Mitteil. über natürliche Cumarine. Ber. dtsch. chem. Ges. 70, 1253 (1937).
107. SPÄTH, E. (und F. KUFFNER): Die natürlichen Cumarine und ihre Wirkung auf Fische. Monatsh. Chem. 69, 75 (1936).

108. SPÄTH, E., B. L. MANJUNATH, M. PAILER und H. S. JOIS: Synthese und Konstitution des Psoralens. Ber. dtsch. chem. Ges. **69**, 1087 (1936).
109. SPÄTH, E. und W. MOČNIK: Synthesen des Dihydroxanthyletins. XXXV. Mitteil. über natürliche Cumarine. Ber. dtsch. chem. Ges. **70**, 2276 (1937).
110. SPÄTH, E. und O. NEUFELD: Über das Vorkommen von Seselin in japanischen Skimmia-Arten. XXXVI. Mitteil. über natürliche Cumarine. Ber. dtsch. chem. Ges. **71**, 353 (1938).
110 a. — — Über das Skimmin. XXXVII. Mitteil. über natürliche Coumarine. Rec. trav. chim. Pays-Bas **57**, 535 (1938).
111 SPÄTH, E., K. OKAHARA und F. KUFFNER: Die Identität von Ficusin mit Psoralen. Ber. dtsch. chem. Ges. **70**, 73 (1937).
112. SPÄTH, E. und M. PAILER: Synthese des Xanthotoxins. XVII. Mitteil. über natürliche Cumarine. Ber. dtsch. chem. Ges. **69**, 767 (1936).
113. — — Über natürliche Cumarine. XIII. Mitteil. Synthese des Angelicins (aus *Angelica Archangelica*). Ber. dtsch. chem. Ges. **67**, 1212 (1934).
114. — — Über eine neue Synthese des Angelicins (aus *Angelica Archangelica* L.). XIV. Mitteil. über natürliche Cumarine. Ber. dtsch. chem. Ges. **68**, 940 (1935).
115. SPÄTH, E. und O. PESTA: Über pflanzliche Fischgifte. III. Mitteil. Konstitution des Osthols (aus *Imperatoria Ostruthium*). Ber. dtsch. chem. Ges. **66**, 754 (1933).
116. — — Die Konstitution des Angelicins (aus *Angelica Archangelica* L.). XI. Mitteilung über natürliche Cumarine. Ber. dtsch. chem. Ges. **67**, 853 (1934).
117. SPÄTH, E., N. PLATZER und H. SCHMID: Über die Konstitution des Oroselons. LII. Mitteil. über natürliche Cumarine. Ber. dtsch. chem. Ges. **73**, 709 (1940).
118. SPÄTH, E. und H. SCHMID: Die Konstitution des Sphondins. LVI. Mitteil. über natürliche Cumarine. Ber. dtsch. chem. Ges. **74**, 595 (1941).
119. — — Über die Konstitution des Athamantins. LIII. Mitteil. über natürliche Cumarine. Ber. dtsch. chem. Ges. **73**, 1309 (1940).
120. — — Synthese des Luvangetins. LV. Mitteil. über natürliche Cumarine. Ber. dtsch. chem. Ges. **74**, 193 (1941).
121. SPÄTH, E. und A. F. J. SIMON: Über die Cumarine der Wurzel von *Heracleum Sphondylium* L. XVI. Mitteil. über natürliche Cumarine. Monatsh. Chem. **67**, 344 (1936).
122. SPÄTH, E., A. F. J. SIMON und J. LINTNER: Die Konstitution des Ammoresinols. XIX. Mitteil. über natürliche Cumarine. Ber. dtsch. chem. Ges. **69**, 1656 (1936).
123. SPÄTH, E. und L. SOCIAS: Über Bergaptol, einen neuen Inhaltsstoff des calabrischen Bergamottöls. VIII. Mitteil. über natürliche Cumarine. Ber. dtsch. chem. Ges. **67**, 59 (1934).
124. SPÄTH, E., S. TAKEI und S. MIYAJIMA: Synthese von Dihydro-osthol aus einem Abbauprodukt des Rotenons. Ber. dtsch. chem. Ges. **67**, 262 (1934).
125. SPÄTH, E. und E. TYRAY: Über die Konstitution des Nodakenins aus *Peucedanum decursivum* MAXIM. L. Mitteil. über natürliche Cumarine. Ber. dtsch. chem. Ges. **72**, 2089 (1939).
126. SPÄTH, E. und F. VIERHAPPER: Über die Konstitution des Umbelliprenins. XXXIX. Mitteil. über natürliche Cumarine. Ber. dtsch. chem. Ges. **71**, 1667 (1938).
127. — — Xanthotoxol, ein neuer Naturstoff aus *Semen Angelicae*, und über die Totalsynthese von Xanthotoxol und Imperatorin. XXIII. Mitteil. über natürliche Cumarine. Ber. dtsch. chem. Ges. **70**, 248 (1937).
128. — — Über Cumarine der Droge *Semen Angelicae*. XL. Mitteil. über natürliche Cumarine. Monatsh. Chem. **72**, 179 (1938).

129. SPÄTH, E., F. WESSELY und G. KUBICZEK: Synthese des Allo-bergaptens. XXII. Mitteil. über natürliche Cumarine. Ber. dtsch. chem. Ges. **70**, 243 (1937).
130. — — — Synthese des Bergaptens. XXIV. Mitteil. über natürliche Cumarine. Ber. dtsch. chem. Ges. **70**, 478 (1937).
131. SPENCER, E. Y. and G. F. WRIGHT: The Action of Diazomethane on Lactones and on Lignins. J. Amer. chem. Soc. **63**, 2017 (1941).
132. STAHMANN, M. A., C. F. HUEBNER and K. P. LINK: Studies on the Hemorrhagic Sweet Clover Disease. V. Identification and Synthesis of the Hemorrhagic Agent. J. biol. Chemistry **138**, 513 (1941).
133. STOLL, A., A. PEREIRA und J. RENZ: Das Furocumarin und die β-D-Glucosidofurocumarinsäure aus den Samen von Coronilla-Arten. Helv. chim. Acta. **33**, 1637 (1950).
134. TAKEI, S., S. MIYAJIMA and M. ONO: Rotenone, the Active Constituent of Derris Root. XVI. Synthesis of Dihydro-osthol from Tetrahydrotubaic Acid. Sci. Pap. Inst. physic. chem. Res. **24**, 1 (1934). [Chem. Abstr. **28**, 4730 (1934)].
135. THAKOR, V. M. and N. M. SHAH: Reactivity in the Coumarin Ring System. J. Univ. Bombay, Ser. A **16**, 38 (1947) [Chem. Abstr. **42**, 4171 (1948)].
136. THOMS, H. und E. BAETCKE: Die Konstitution des Bergaptens. Ber. dtsch. chem. Ges. **45**, 3705 (1912).
137. TILDEN, W. A. and H. BURROWS: The Constitution of Limettin. J. chem. Soc. (London) **81**, 508 (1902).
138. WESSELY, F. und E. DEMMER: Die Konstitution des Fraxetins. Ber. dtsch. chem. Ges. **61**, 1279 (1928).
139. — — Konstitution und Eigenschaften des Fraxins. Ber. dtsch. chem. Ges. **62**, 120 (1929).
140. WESSELY, F. und K. DINJAŠKI: Über die Lichteinwirkung auf Stoffe vom Typus der Furo-cumarine. Monatsh. Chem. **64**, 131 (1934).
141. WESSELY, F. und F. KALLAB: Über die Inhaltsstoffe der Wurzel von *Pimpinella saxifraga*. I. Monatsh. Chem. **59**, 161 (1932).
142. WESSELY, F. und I. PLAICHINGER: Über die Konstitution der Photodimerisate der Cumarine und Furo-coumarine. Ber. dtsch. chem. Ges. **75**, 971 (1942).
143. WESSELY, F. und K. STURM: Zur Konstitution des Daphnins. Ber. dtsch. chem. Ges. **62**, 115 (1929); **63**, 1299 (1930).
144. YAMASHITA, M.: Synthesis of Dihydro-*iso*-osthol. Note on the Constitution of Osthol. Bull. chem. Soc. Japan **8**, 276 (1933).

(Received, April 15, 1952.)

The Biosynthesis of Proteins and Peptides, including Isotopic Tracer Studies.
By H. BORSOOK, Pasadena, California.

Contents.

	Page
I. Introduction	293
1. The Theory of Endogenous and Exogenous Protein Metabolism	294
2. The Theory of Protein Metabolism as a Dynamic Steady State	294
a) Indirect Evidence	294
b) Direct Evidence	297
c) Lability of Enzyme Proteins	298
II. The Measurement of Protein Turnover	299
III. Incorporation of Labeled Amino Acids *in vivo*	300
1. N^{15}-labeled Amino Acids as Tracers	300
2. C^{14}- and S^{35}-labeled Amino Acids as Tracers	303
a) In Normal Tissues	303
b) In Tumors	305
c) Influence of Hormones	305
d) Incorporation of Foreign Amino Acids	308
IV. Incorporation of Labeled Amino Acids *in vitro*	309
1. Incorporation of Carbon Dioxide into Amino Acids	310
2. Net Synthesis of Protein *in vitro*	312
3. Comparison of Incorporation of Amino Acids *in vivo* and *in vitro*	313
4. Amino Acid Incorporation in Different Cell Fractions	313
5. The Nucleus, Amino Acid Incorporation, and the Maintenance of the Amino Acid Pattern in Proteins	314
6. Nucleic Acids, Protein Synthesis and Amino Acid Incorporation into Proteins	315
7. Normal, Foetal and Tumor Tissue	315
8. Effect of Concentration of Labeled Amino Acid on its Rate of Incorporation	315
9. Does Incorporation of One Amino Acid Require the Presence of Others?	316
a) Feeding Experiments	316
b) *In vivo* Experiments with Single Labeled Amino Acids	316
c) *In vitro* Experiments with Labeled Amino Acids	317
V. The Biological Significance of the High Lability of the Proteins in the Cell	319
VI. Mechanism of Peptide Bond Synthesis	319
1. Heats and Free Energies of Formation of Some Amino Acids and Peptides (Solids)	319

	Page
2. Free Energies of Formation of Some Peptides in Aqueous Solution	320
3. The Effect of pH on the Free Energy Change in Peptide Formation	321
4. Peptide Synthesis by Proteases and Peptidases	322
a) Classification of Enzymatic Peptide Syntheses According to the Sign and Magnitude of the Free Energy Change ($-\Delta F$)	322
b) Peptide Syntheses where $-\Delta F$ is Positive and Large	323
c) Peptide Syntheses where $-\Delta F$ is Small and the Peptide is Relatively Insoluble	325
d) Plastein Formation	327
e) Peptide Synthesis in an Exchange Reaction during Hydrolysis (Transamidation and Transpeptidation)	328
f) Peptide Synthesis from Amino Acid Esters	333
5. Glutamo- and Asparto-Transferases	334
6. Syntheses where $-\Delta F$ is Negative and Large, Coupled with High Energy Phosphate	336
a) Synthesis of Glutamine	336
b) Synthesis of Hippuric Acid	337
c) Synthesis of p-Aminohippuric Acid	338
d) Synthesis of Ornithuric Acids	339
e) Synthesis of Glutathione	341
VII. Mechanism of Amino Acid Incorporation into Proteins	343
1. Effect of Inhibitors	343
2. Amino Acid Incorporation and Phosphorylation	346
3. Heat-Stable Co-factors for Amino Acid Incorporation	346
4. Is Amino Acid Incorporation Synthesis of Protein *de novo* or an Exchange?	347
5. The Possibility of Peptides as Intermediates in Protein Synthesis	348
6. The Linkage of Incorporated Amino Acids	349
References	352

I. Introduction.

It has been the hope of the biochemist for more than half a century to study *in vitro* the biological, i. e. enzymatic synthesis of protein. Of course, the hope has not yet been realized; but the outlook is brighter, or at least, less dim now. Certain misconceptions have been recognized, and with the removal of these obstructions approaches to the core of the problem appear to have been opened. This advance was made possible by the new experimental tool of isotope-labeled tracers. In most *in vitro* chemical reactions the products are formed in large enough quantities to be measured. This is not the case in protein biosynthesis, first because the reaction is very slow compared with ordinary chemical reactions, and second because a relatively large amount of protein is needed as enzyme system to make the reaction go at all; and any small increase in protein that may occur has to be seen against the background of the large amount of protein present initially. The latter difficulty is now being circumvented, to some extent, by the use of antibodies to precipitate small amounts of specific proteins formed, [PETERS and ANFINSEN (*129, 130*), KESTON and DREYFUS (*100*)]. But even in these experiments it is possible that there has not been an increase in the mass of total protein, but only a transformation of one tissue protein into another.

1. The Theory of Endogenous and Exogenous Protein Metabolism.

Until about 1939 experiments on the mechanism of the biological synthesis of protein were few and intermittent. All of these were attempts at the *in vitro* enzymatic synthesis of protein, and the only positive results were syntheses of mixtures of insoluble and unnatural proteins from enzymatic protein hydrolysates containing large polypeptides.

There was an intellectual obstruction. It was the theory of protein metabolism in animals epitomized in the terms "endogenous" and "exogenous" metabolism; this theory is associated with the name of FOLIN (*57, 58*). According to this theory, the structural parts of the animal body are subject to a continual but small wear and tear and repair. The terms "structural" and "protein" were used almost as if they were synonymous, and the sum total of changes involved in this wear and tear and repair was designated "endogenous" protein metabolism. Because the organism was viewed as an engine, and because the structural substance of an engine wears away only very slowly compared with the rate at which it consumes fuel, it followed that only a small fraction of the protein in the food is needed for ,,endogenous" metabolism. In other words there was very little protein synthesis in an adult animal in nitrogen equilibrium. The remainder of the protein in the food, the bulk of it, is quickly hydrolysed and burned and the fragments excreted. This moiety was designated "exogenous" protein metabolism. FOLIN's theory was useful; it was based on data which he was the first to obtain, by methods which he had devised or greatly improved. But the underlying concept of the organism as a machine was not a biological concept at all. It is now proved that FOLIN misinterpreted his data; because he, and with him the next two generations, forced the data into the strait-jacket concept of the organism as a combustion engine.

2. The Theory of Protein Metabolism as a Dynamic Steady State.

a) Indirect Evidence.

In the intervening years it was recognized that the theory needed to be modified somewhat, but the modifications proposed were patchwork. In 1935 (*30*) the fundamentals of the theory were challenged by BORSOOK and KEIGHLEY, and an alternative proposed. The new theory envisaged about half the nitrogen excreted in a day by an animal in nitrogen balance as coming from the catabolism of tissue protein. This tissue loss is made good by an equal amount of amino acids coming from the food and synthesized into protein. Since an adult animal attains nitrogen balance over a wide range of nitrogen intake, it followed that the rate of tissue protein breakdown is proportional to the nitrogen intake. In short, tissue protein was envisaged, not as inert structural substance, but as labile. In an animal in nitrogen balance there is a dynamic steady state of equal breakdown and synthesis; and, if one writes the steady state as protein $\rightleftharpoons$ amino acids $\rightleftharpoons$ catabolic products, it is maintained far over from thermodynamic equilibrium toward the side of protein. Energy is required in the active system of the call to maintain the steady state so far from thermodynamic equilibrium, and this energy is continually poured in by the unceasing oxidations.

The evidence adduced in support of the proposed new theory was mainly indirect; some came from crude labeling experiments, and some was inferred from data on the free energy of formation of the peptide bond (see below). The most cogent indirect evidence was the near constancy of the concentration of free amino acids in the blood and tissues regardless of the nutritional state of the animal.

The argument ran somewhat as follows: in passing from a high to a low level of nitrogen balance or *vice versa*, there is about a ten-day lag in the attainment of the new balance. The diminution in excess urinary nitrogen in passing from the high to the low dietary level corresponds to that of a first order reaction. Since the animal was in nitrogen balance at both levels it was to be expected, if the excess nitrogen was all "exogenous", (a) that the attainment of the new level would have come much sooner, and (b) until balance was attained at the lower level the excess, eventually excreted, would be found in the blood and tissues in the non-protein nitrogen. Neither expectation from this theory was realized in the experiment. The excess nitrogen in the animal is protein; and the rate of its breakdown is proportional to the level of dietary protein. Protein in the tissues is, therefore, in a more labile state than envisaged in the theory of "endogenous" and "exogenous" metabolism of protein. There were data in the literature that when muscle protein was breaking down rapidly and the products resynthesized to gonadal protein, the concentrations of amino acid and total non-protein nitrogen in the blood and tissues hardly changed (Table 2). This is the case also in growing seedlings when seed protein is rapidly transformed to stem and leaf protein (Table 3) [BORSOOK and DUBNOFF (*28*)]. It was inferred, then, in the proposed new theory, that extensive breakdown of protein is continually going on; and, in nitrogen balance, an equal amount of synthesis occurs; and that the rates of breakdown and synthesis are governed by the external supply of protein (as amino acids). The last inference is today an experimental fact called "turnover".

Just how narrow is the range of variation of amino acid concentration in the tissues is not yet widely appreciated. Accordingly, some representative data are presented in the following tables (Tables 1, 2 and 3).

Table 1. Concentration of Free Amino N in Tissues of Fasting Dog (*186*).

Duration of Fast	Free Amino Nitrogen			
	Muscle		Blood	
	mg./100 gm. tissue	Neg. Log. of Molar Conc.	mg. per cent	Neg. Log. of Molar Conc.
5 hours	66	1.33	—	—
24 ,,	54	1.41	6	2.37
48 ,,	51	1.44	6	2.37
96 ,,	56	1.40	7	2.30
144 ,,	66	1.33	5	2.44
288 ,,	61	1.36	5	2.44

Table 2. Concentration of Free Amino N in Salmon En Route to Spawning Grounds (*77*).

Distances from Sea in Miles	Per Cent Fat-free Muscle		
	Protein N	Amino Acid N	Neg. Log. Conc.* Amino Acid N
0	20.3	0.061	1.46
130	19.9	0.076	1.36
210	19.7	0.078	1.35
710	12.3	0.074	1.38
Spawning Grounds	13.8	0.084	1.34

Table 3. Concentration of Free Amino N in Growing Pea Seedlings (*27*).

Days	Free Amino N/100 g. Water					
	Dark			Light		
	Weight gm.	mg. %	Neg. Log. Molar Conc. of amino N	Weight gm.	mg. %	Neg. Log. Molar Conc. of amino N
1	0.5064	140	1.00	0.4786	147	0.98
2	0.5021	132	1.03	0.4812	138	1.01
3	0.4791	150	1.03	0.5113	133	1.02
germination	—	—	—	—	—	—
4	0.488	212	0.82	—	—	—
5	—	—	—	0.4753	265	0.72
6	0.4734	228	0.79	—	—	—
7	—	—	—	0.4834	280	0.70
9	0.4306	—	—	—	316	0.65
12	—	255	0.75	0.4463	—	—

* Assuming muscle contained 80% water.

The average concentration of free amino acids in the plasma is 4.8 mg.%, and the normal range is 4—5.65 (*15*). It is customary to express only the concentration of hydrogen ions in logarithmic form. When the concentration of free amino acids is expressed as a negative logarithm of the molar concentration, the normal range in the blood of animals varies very little from 2.40. In the tissues of the salmon *en route* to the spawning grounds, during which time the protein concentration decreased from 20.3 to 13.8% of fat-free muscle, the negative logarithm of the amino acid concentration changed only from 1.46 to 1.34. It is clear that the amino acid concentration in the tissues does not vary significantly more than does the pH. The variation of the blood sugar is as small; thus, the normal blood sugar is from 80–120 mg.% and the renal threshold is 140 mg.%. Expressed as a negative logarithm of the molar concentration, these values are respectively: 2.35, 2.21, and 2.11.

The changes in the growing pea seedling (Table 3) are particularly interesting because the form of the organism changes markedly. A small change occured as soon as germination began and leaves appeared. In seedlings grown in the dark there was very little change in the free amino acid concentration after germination, although the plant had grown to a height of nine inches. Even the intervention of photosynthesis in plants grown in the light brought about only a small change.

There is no question of the great lability of the proteins in the growing plant, and in the fasting salmon *en route* to the spawning grounds who breaks down his muscle and from the breakdown products synthesizes the proteins of the gonads. The processes of breakdown and synthesis in plants and animals are so regulated as to maintain the concentration of free amino acids nearly constant. Of course, the free amino acid concentration is not absolutely constant. After a protein meal the amino acid concentration in the blood rises quickly and remains above "normal" for some hours. It has been found in the case of at least one amino acid, tryptophane, that its steady level in the blood varies with the amount of this amino acid in the diet [SCHWEIGERT *et al.* (*156*)]. But these changes are relatively small.

b) Direct Evidence.

The direct proof that amino acids in the proteins of the adult animal in nitrogen balance are rapidly exchanging with amino acids from the diet came from the studies of SCHOENHEIMER and his colleagues on the effect of feeding N^{15}-labeled amino acids and ammonia (*149*). They found that amino acids are continually and rapidly being deaminized and reaminated; free ammonia is extensively used for reamination (*141, 173*); and that the peptide bonds of the protein chains are continually being broken and reconstituted. Synthesis of amino acids and protein occurs both when these are abundantly supplied in the diet and when the animal is fasting [TARVER and SCHMIDT (*178*)]. Proteins with specific functions, specific antibody proteins, for example, undergo the same rapid synthesis and disintegration as the average, immunologically inert, serum proteins (*149*).

The rate of amino acid exchange is different in different tissues, and the relative difference is characteristic of the tissue. Thus SCHOENHEIMER's group (*151*) after feeding an adult rat *L*-leucine for three days found the labeled leucine in the proteins of different organs in the following relative amounts: serum 100, intestinal wall 89, kidney 82, spleen 65, liver 56, heart 53, testes 46, muscle 18, hemoglobin 17, and skin 11. These relative rates are somewhat different when the tissues are analysed sooner after the administration of the labeled amino acid (see below).

After an animal is fed a labeled amino acid so long that its proteins contain a relatively high concentration of the labeled amino acid, then, when the labeled amino acid is withdrawn from the diet there is a steady decline in the concentration of the labeled amino acid in the tissues. The tissues that incorporate the amino acid fastest show the most rapid decline and *vice versa*. Eventually the proteins of all tissues have the same concentration of the labeled amino acid (*1, 160*).

The foregoing experiments countered the argument of adherents of the older theory that only the so-called "reserve" proteins undergo rapid disintegration and re-synthesis. Later studies have strengthened the new theory. Thus, it was found in four different fractions of rabbit liver cells, nuclei, mitochondria, microsomes and the particle-free supernatant solution, that the most rapid rate of labeled amino acid incorporation was in the microsome fraction [BORSOOK et al. (*22*), KELLER (*98*)]. As the different fractions of the liver cell have different rates of anabolism, it was to be expected that they would have different rates of catabolism, i. e. loss of nitrogen during protein starvation. This was found to be so (*127, 158, 187*). Relatively the greatest losses occurred in the mitochondrial and microsome fractions, and little or no loss in the nuclear fraction. Similar changes occurred in hepatomatous as compared with normal liver and in rats fed carcinogenic compounds (*50, 134–137, 155*). Mitochondria and microsomes are among the most active components of liver, enzymatically. The greatest losses in protein occur in them.

c) *Lability of Enzyme Proteins.*

That the loss is not restricted to "storage protein" but to active enzymes is indicated in the findings that maintenance of rats on low or non-protein diets is associated with loss in liver of catalase, alkaline phosphatase, xanthine dehydrogenase, cathepsin, and arginase [MILLER (*126*)].

Refeeding a high protein diet led to prompt restoration of enzyme activity and total liver protein. When bovine pancreas slices were incubated in Ringer's solution containing $C^{14}O_2$, the subsequently isolated ribonuclease was radioactive and approximately twice as radioactive as the average of the slice proteins [ANFINSEN (28)]. The $C^{14}O_2$ had been incorporated largely into the carboxyl groups of aspartic and glutamic acids, and these amino acids in the proteins are labeled [ANFINSEN et al. (29)]. Evidently, enzymes as proteins participate actively in the anabolism and catabolism of proteins in general.

In accord with these findings, xanthine oxidase activity of liver has been found to be a sensitive index of the availability of dietary proteins. During inanition the relative loss of xanthine oxidase activity exceeds that of total liver proteins [MILLER (125)]; increasing the levels of protein in the diet increased the liver xanthine oxidase activity [WESTERFIELD and RICHERT (201)]. Using this enzyme activity as a criterion it was found that methionine in dietary casein is not readily available to the rat, whereas acid-hydrolyzed casein or a mixture of purified amino acids corresponding to the composition of casein are utilized more completely [LITWACK et al. (116), WILLIAMS and ELVEHJEM (202)].

II. The Measurement of Protein Turnover.

Protein synthesis is a slow reaction whether it is measured as net gain of protein in a growing animal or as incorporation of a labeled amino acid where the change in the amount of protein is usually too small to be measured. It is slow, that is, when compared with rates of ordinary chemical reactions; from the biological point of view, its speed is something to wonder at. Because the reaction is, in absolute terms, so slow it was impossible to study the process in any detail *in vivo*, let alone *in vitro*, until isotopes became so readily available that isotope-labeled amino acids could be synthesized and used relatively freely. It was with this new tool that SCHOENHEIMER and his colleagues demonstrated that conditions were now favorable for a direct attack on the problem of the biological synthesis of protein. These conditions were, as discussed above, the realization that even in adult animals proteins are being rapidly synthesized, the demonstration that there is a rapid incorporation of labeled amino acids into tissue proteins, and that this process can be observed easily with isotope-labeled amino acids.

As matters stand now, it appears that *in vivo* studies can provide information on what may be called the physiology of the process, *in vitro* studies on the mechanisms. The following considerations indicate the experimental problem.

A weanling, 50-gm. rat on a good diet can grow 25 gm. in a week (*170*). A 50-gm. rat contains 7.8 gm. of protein (*188*) and it gains in a week 3.9 gm. of protein. On the basis of the initial weight the rate of synthesis of protein is 3 mg. per gram of protein per hour; or, taking an average equivalent weight of amino acids, as 110, and that a protein contains 20 different amino acids, the rate of synthesis is an average of 1.36 μ-Equiv. of each amino acids per gram of original protein in the animal per hour. With N^{15}-labeled amino acids some milligrams of nitrogen, i. e. of the order of 1 mEquiv., are needed to measure amino acid incorporation. Of this 1 mEquiv. 0.05% or 0.5 μEquiv. are N^{15}. This concentration of the isotope was attained in *in vivo* experiments in the adult rats in about three days. A relatively large quantity of protein is needed, because the labeled amino acid must be isolated. For example, 0.5 mEquiv. of a labeled amino acid is contained in the hydrolysate of about 1.0 gm. of protein or 12 gm. of fresh tissue. If the tissue is liver, two or three rats are needed for a single experimental datum. Where milligrams of an N^{15}-labeled amino acid is needed for a measurement, micrograms labeled with C^{14} suffice. One tenth of a microgram of a labeled amino acid with a specific activity of 100000 counts per minute per milligram (c. p. m./mg.) can be measured. We have been able to use amino acids with specific activities as low as 3000 c. p. m./mg. With C^{14}-labeled amino acids it was possible to carry out experiments *in vitro* of one hour duration with 100 mg. of fresh tissue, whereas with N^{15} the shortest experiments were for three days and in a whole animal.

The amino nitrogen of amino acids exchanges freely in metabolism with that of others. Hydrogen in certain positions exchanges with the hydrogen of water spontaneously, and in other positions it exchanges freely under the influence of enzymes. The carbon of an amino acid exchanges far less freely with that of other amino acids; its very long half life, and ease of measurement make C^{14} preferable to either N^{15} or deuterium in most protein synthesis studies.

III. Incorporation of Labeled Amino Acids in Vivo.

1. N^{15}-labeled Amino Acids as Tracers.

In one of their first studies SCHOENHEIMER *et al.* (*150*) fed N^{15}-tyrosine (I) to rats in nitrogen balance and found that only 50—60% of the marked nitrogen was excreted in ten days; 25% of the labeled nitrogen in the body remained attached to the carbon chain of tyrosine, the remainder to the carbon chains of other amino acids. The α-amino nitrogen of most amino acids is transferred qualitatively in a similar manner, although in some cases, e. g. lysine (II) (*200*) and threonine (III) [ELLIOTT and NEUBERGER (*54*)] the α-amino group is not reconstituted once deamination has occurred. The nitrogen of ammonia is extensively incorporated into α-amino groups of amino acids (*141, 173*).* By using *L*-leucine (IV) labeled

(p) HO · C$_6$H$_4$ · CH$_2$ · CH · COOH
          |
         *NH$_2$
(I.) Tyrosine (* = N^{15}).

CH$_2$ · CH$_2$ · CH$_2$ · CH$_2$ · CH · COOH
| |
NH$_2$ *NH$_2$
(II.) Lysine (* = N^{15}).

CH$_3$ · CHOH · CH · COOH
                |
               *NH$_2$
(III.) Threonine (* = N^{15}).

with both N^{15} and deuterium in stable positions (*151*) it was determined that half the leucine of the liver proteins was replaced by dietary leucine in about seven days. On the assumption that the rates of replacement of all the other amino acids are the same, seven days is the "half-life" of rat liver protein. By following the rate of disappearance of N^{15}-glycine (V)

CH$_3$†
      \\
       CH · CH$_2$ · CH · COOH
      / |
CH$_3$†   *NH$_2$
(IV.) Leucine (* = N^{15} and † = deuterium).

H$_2$Ṅ · CH$_2$ · COOH
(V.) Glycine (* = N^{15}).

SHEMIN and RITTENBERG (*160*) deduced a figure for the "half-life" for rat liver protein of six days. Later SPRINSON and RITTENBERG (*174*) developed a mathematical expression for estimating the continual rate of amino acid incorporation into the total body proteins of a living animal after feeding a single dose of a labeled amino acid, N^{15} glycine, where the label, in this case N^{15}, was excreted practically entirely in the urine. Although the equation developed by these authors is based on certain unproved assumptions, these assumptions appear reasonable, and the estimates obtained agree well with other, including direct, determinations.

As an example, a human subject on a normal diet was given 10 mg. of N^{15}-glycine per kilo. 30% of the N^{15} appeared in the urine in the first 24 hours; and it was inferred that 70% of the dietary glycine was used during that day in the formation of tissue protein. In the first 48 hours 40% of the N^{15} appeared in the urine; after this time the excretion of N^{15} was very low.

The curve of N^{15} excretion plotted against time fitted the equation,

$$\lambda_E = \lambda_0 \frac{E}{E+S}\left(1 + e^{-\left(\frac{E+S}{P}\right)t}\right)$$

where λ_E = mEquiv. N^{15} excreted; λ_0 = mEquiv. N^{15} fed at zero time; S = gm. nitrogen synthesized into protein per day; P = gm. nitrogen

---

* SCHOENHEIMER (*149*) gives the picture of amino nitrogen transfer *in vivo* in some detail. Discussion of the mechanisms involved is beyond the scope of this review.

in the metabolic pool (assumed to be constant); and $t =$ time in days. λ_E, λ_0, E and t are experimental data; S, and P are then estimated by means of the equation. Table 4 contains data obtained by SPRINSON and RITTENBERG and estimates they made by means of the above equation of the rate of protein synthesis and of the size of the metabolic pool of nitrogen in men.

Table 4. Rate of Protein Synthesis and Size of Nitrogen Pool in Human Subjects (*174*).

Expt. No.	Weight of Subject	E^*	$\dfrac{100\,\lambda_E}{\lambda_0}$		A	B	S	P
			0—24 hrs.	0—48 hrs.				
	kg.	gm.	per cent	per cent			gm.	gm.
1	75.0	17.2	34.7	42.0	0.450	1.21	0.280	0.421
2	64.4	14.0	33.0	42.5	0.488	1.08	0.228	0.412
3	57.6	31.1	62.0	82.9	0.926	1.18	0.043	0.494
4	63.5	10.4	22.4	32.9	0.457	0.72	0.195	0.500

$E^* =$ Total nitrogen excreted in the first 24 hours.

$$A = \frac{E}{E+S};\ B\ \frac{E+S}{P}$$

$S\ =$ gm. of N synthesized into protein in 24 hours per kilo of body weight.
$P\ =$ gm. of N in the metabolic pool per kilo of body weight.

Table 5. Amino Acid Turnover in Proteins *In Vivo* (*174*).

Animal	Tissue	Amino Acid	Total amino acids incorporated into protein. μM/gm. protein per hour	Estimated average rate of incorporation of a single amino acid. μM/gm. protein per hour
Man	Whole animal	Glycine-N^{15}	4.15	0.45
	Viscera*	,,	26.3	2.6
Rat	Whole animal	,,	19.3	1.9
	Viscera*	,,	56.3	5.6

Averages of such estimates from a number of subjects for the human and the rat, recalculated in rate units to make the results comparable to where the incorporation of a single amino acid was measured, are shown in Table 5. The directly determined values are only those for the whole animal. The values for viscera come from earlier work (*152, 160*). The values for total amino acid incorporation are for total N^{15}, i. e. as glycine in which form the N^{15} was administered and as the other amino acids

* *Viscera* = Liver, plasma, spleen, heart, kidneys, small intestine, pancreas, endocrines.

to which the N^{15} of the glycine had been transferred. When C^{14}-labeled amino acids were used, the C^{14} in most cases does not transfer and one measures the incorporation of a single amino acid. In order to compare the N^{15} with the C^{14} values it is necessary to convert total N^{15} incorporation into the proteins into an average value for a single amino acid. This was done in the last column of Table 5. The total N^{15} incorporation was divided by 10. If the N^{15} introduced as glycine had been transferred equally among all the free amino acids of the nitrogen metabolic pool, division by 20 would have been more nearly correct; but this is not the case. The N^{15} concentration in the nitrogen of the glycine in the liver proteins was about 9% of that fed, of the other amino acids it ranged from about 0.5 to 1% (*149*). Accordingly, the factor used to convert the total N^{15} incorporation into an average value per individual amino acid was takens as 0.1. It is a very rough estimate but it gives the order of magnitude.

The average half-life of the total protein, computed from the foregoing values, is in the human 80 days, and in the rat 17 days. SPRINSON and RITTENBERG (*174*) estimated that of the total turnover in the human 41% occurs in the liver and plasma, and in the rat 25%.

The order of magnitude of the energy requirement of these rates of turnover can be computed as follows. There are approximately 156 gm. of protein per kg. of body weight. A turnover rate of 4.1×10^{-3} mEquiv. of amino acid per gm. of protein per hour corresponds to $156 \times 4.1 \times 10^{-3} \times 10^{-3} = 0.650$ mEquiv. per kg. per hour. The free energy of a peptide bond is approximately 3500 calories or 3.5 Calories (see below). On the basis of one peptide bond resynthesized per equivalent of amino acid turned over the free energy requirement is $3.5 \times 0.65 \times 10^{-3} = 2.3 \times 10^{-3}$ Calories. If we take $\Delta H = 1.5 \Delta F$ (it may be 0.5 ΔF or less), then the heat requirement of the above turnover rate is $1.5 \times 2.3 \times 10^{-3} = 3.5 \times 10^{-3}$ Calories. The basal metabolism per kg. is approximately 1 Calorie, or 300 times the requirement of the protein turnover on the basis of one peptide bond resynthesized per equivalent of amino acid turned over. Even on the basis of the resynthesis of 2 peptide bonds per equivalent of amino acid turned over, the basal metabolism is still greatly in excess of the energy (ΔH) required.

2. C^{14}- and S^{35}-labeled Amino Acids as Tracers.

a) In Normal Tissues.

Table 6 summarizes results of some direct measurements of incorporation of C^{14}-labeled amino acids in mice [BORSOOK et al. (*22*)]. The values at 1 hour in the viscera are practically the same as the average rates per individual amino acid in Table 5 estimated from excretion of previously incorporated N^{15}. In all cases the incorporation into the plasma proteins began later than in the viscera, but eventually attained a higher value than in the liver or most of the other organs. The injection experiments indicate more clearly than those with continuous feeding the relatively great speed with which the viscera incorporate circulating amino acids into their proteins and conversely the speed with which they liberate amino acids. The immediate impact of an injected amino acid is largely dissipated in about an hour; from then on redistribution of amino acids released from visceral proteins dominates the picture.

Table 6. Amounts of Labeled Amino Acids Found in Visceral and Plasma Proteins of Mice at Successive Time Intervals after Intravenous Injection. Results Expressed as μ-Mols/gm. Protein (22).

Time after injection (minutes)	Glycine		Histidine		Leucine		Lysine	
	Viscera	Plasma	Viscera	Plasma	Viscera	Plasma	Viscera	Plasma
10	0.53	0.07	0.66	0.11	1.4	0.36	1.1	0
20	1.4	0.30	1.3	0.65	3.0	0.15	1.7	0.04
30	2.6	1.23	2.9	1.52	3.6	1.33	1.7	0.25
60	2.5	1.31	3.1	3.10	3.6	4.23	1.9	1.11
120	2.2	1.69	2.8	4.37	4.0	4.21	2.0	1.25
240	2.4	2.18	3.2	5.00	4.0	5.47	1.6	0.94

Table 7. Incorporation *In Vivo* of Injected Labeled Amino Acids into the Proteins of Rat Tissues. Results Expressed as μ-Mols./Gm. Protein.

Animal	Rat	Rat	Rat	Rat	Rat	Rabbit
Labeled amino acid injected	Glycine		L-Serine	L-Tryptophane	DL-Tyrosine	L-Leucine
Injected/100 gm. body weight	13.9	13.9	3.0	0.10	3.0	5.0
Hours after injection	0.25	6	8	6	8	6	0.5
Bone marrow	0.6	6.45	—	—	—	—	—
Brain	0	0.3	—	—	0.48	0.14	—
Heart	—	—	0.86	—	0.49	—	1.7
Intestine	—	—	3.94	—	3.02	—	4.5
Intestinal mucosa	0.9	11.1	—	0.053	—	1.93	—
Kidney	0.45	5.85	2.36	—	1.63	1.46	2.6
Liver	0.75	6.15	2.80	0.34	1.93	0.63	3.8
Lung	0.45	3.75	—	—	—	—	—
Muscle, skeletal	0	0.45	0.47	0.0046	0.22	0.06	1.7
Plasma	0.15	5.4	—	0.053	2.42	1.10	0.7
Red blood cells	0	0.45	—	—	0.23	0	0.2
Spleen	0.15	4.35	2.28	0.0034	1.06	—	3.7
Stomach	—	—	—	—	1.41	—	—
Testes	0.15	1.5	—	—	0.91	0.36	—
Reference	(76)	(76)	(145)	(111)	(145)	(206)	(22)

Tables 7 and 9 give a somewhat more detailed picture of the differences in rates of incorporation and subsequent release of amino acids in different tissues. Shortly after administration of an amino acid the most rapid incorporation is in the intestinal mucosa and bone marrow cells with the liver, spleen and kidney next. Later the plasma proteins contain more than the liver, and still later the content in skeletal muscle begins to rise. Eventually, after six to twelve days, the values in all the tissues tend toward a common value, the average for the body proteins as a whole (*160*).

It has been found that the metabolic activity of different components of a cell is qualitatively and quantitatively different (*48, 56, 87, 88, 101, 133*). Analogous differences have been found in the rate at which different cell components incorporate labeled amino acids (Table 8). The highest rate was in the microsomes. These aggregates contain the highest concentration of ribosenucleic acid (*98, 148*). A number of workers (*31, 37–40, 90, 172*) have suggested that a close relation exists between ribosenucleic acid and protein synthesis.

Table 8. Incorporation In Vivo of Labeled Amino Acids into Intracellular Fractions of Guinea Pig Liver (*21*), 30 Minutes after Intravenous Injection of 1.6 mg./100 gm. Body Weight. Results Expressed as μ-Mols./gm. Protein.

Fraction	Labeled Amino Acid			
	Glycine	L-Histidine	L-Leucine	L-Lysine
Nuclear fraction	0.56	1.5	2.3	1.3
Mitochondria	0.60	1.2	1.1	1.6
Microsomes	1.2	3.1	4.3	2.9
Supernatant	0.69	1.2	1.8	1.6

b) In Tumors.

Tumors *in vivo* incorporate labeled amino acids about as fast as such active tissues as liver or kidney (*63, 78, 89, 109, 110, 139, 160, 185, 206, 212–214*). SHEMIN and RITTENBERG found that transplantable sarcoma R-39 releases labeled amino acid (glycine-N^{15}) from its proteins at about one quarter the rate of liver (*160*). In all other cases the release of the labeled amino acid from the tumor was at about the same rate as from liver and kidney.

c) Influence of Hormones.

That hormones must influence protein synthesis is a commonplace; this applies not only to the growth hormone *per se* (*173*) but probably to many of the others; loss of weight is a salient feature of diabetes and of hyperthyroidism; the balance between synthesis and breakdown of

Table 9. Relative Rates of Incorporation *In Vivo* of Labeled Amino Acids

Animal	Rat	Rat	Rat	Rat	Rat	Dog
Amino Acid	C^{14}-glycine	C^{14}-glycine	C^{14}-glycine	N^{15}-glycine	S^{35}- Glutathione	DL-S^{35}- Methionine
Route of administration	Injected	Injected	Injected	In food	Injected	Injected
Dose: mg./100 gm. body weight	13.9	13.9	3.0	Tracer	60—120	0.67
Time after administration	0.25 hrs.	6 hrs.	8 hrs.	Daily for 3 days	18 hrs.	2 hrs.
Bone marrow	80	104	—	—	—	—
Brain	0	4	—	—	—	—
Diaphragm	—	—	—	—	—	—
Heart	—	—	30	—	—	—
Intestine, wall	—	—	—	70	—	—
,, whole	—	—	140	—	—	—
,, small, muscle	—	—	—	—	—	—
,, large, ,,	—	—	—	—	—	—
Kidney	60	95	84	—	114	92
Liver	100	100	100	100	100	100
Lung	60	60	—	—	—	—
Lymph nodes	—	—	—	—	—	59
Muscle, skeletal	0	7	16	21	—	—
Mucosa, intestine	120	180	—	—	413	169
,, stomach	—	—	—	—	—	—
Pancreas	—	—	—	—	—	69
Plasma (or serum)	20	87	—	127	—	71
,, albumin	—	—	—	—	—	69
,, fibrin	—	—	—	—	—	135
,, globulin	—	—	—	—	—	76
Red blood cells (or hemoglobin)	0	7	—	32	—	—
Skin	—	—	—	—	—	—
Spleen	20	70	81	—	65	49
Stomach, muscle	—	—	—	—	—	—
Submaxillary gland	—	—	—	—	—	—
Testes	20	24	—	—	—	—
Thymus	—	—	—	—	—	—
Thyroid	—	—	—	—	—	—
Reference	(76)	(76)	(145)	(149)	(209)	(177)

protein is affected by the pituitary and adrenal sex hormones (*102–107, 108, 140*). Observations are now beginning to be reported on hormonal effects on incorporation of labeled amino acids. The difficulty of interpretation here, as in the case of tumors, is to determine whether the final net effect, acceleration or depression, is the resultant mainly of a change in the rate of incorporation or in the rate of breakdown and consequent liberation of the labeled amino acid. FRIEDBERG and GREENBERG (*65*) found that in both normal and hypophysectomized rats injection of the anterior pituitary growth hormone caused a 70% greater incorporation of into Animal Tissues. Results Expressed as Per Cent of Value in Liver.

Dog	Dog	Rat	Rat	Dog	Rat	Rat	Rat	Rat
DL-S³⁵-Methionine	DL-S³⁵-Methionine	DL-S³⁵-Methionine	DL-S³⁵-Methionine	DL-S³⁵-Methionine	L,S³⁵-Leucine	L-C¹⁴-Serine	L-C¹⁴-Tryptophane	DL-C¹⁴-Tyrosine
Injected	Injected	Injected	Injected	Per os	Per os	Injected	Injected	Injected
0.67	Trace	3.0	4.0	0.37	Trace	0.10	3.0	5.0
5 hrs.	6 hrs.	6 hrs.	12 hrs.	4 days	Daily for 3 days	6 hrs.	8 hrs.	6 hrs.
89	—	—	—	—	—	—	—	—
—	—	—	40	19	—	—	24	22
—	—	—	—	12	—	—	—	—
—	—	—	43	17	—	—	25	—
—	—	—	—	—	94	—	—	—
—	—	—	—	—	—	—	156	—
—	—	—	—	37	—	—	—	—
—	—	—	—	62	—	—	—	—
106	160	127	187	—	146	151	84	231
100	100	100	100	100	100	100	100	100
—	104	—	—	37	—	—	—	—
49	—	—	—	—	—	—	—	—
—	17	18	37	6	32	14	11	10
165	347	175	250	106	—	—	—	531
—	—	—	—	197	—	—	—	—
65	169	—	—	81	—	—	—	—
143	—	158	143	—	177	155	125	174
97	—	—	—	100	—	—	—	—
226	—	—	—	93	—	—	—	—
152	—	—	—	87	—	—	—	—
—	—	—	—	2	30	—	12	—
—	—	—	—	—	19	—	—	—
77	165	—	—	25	117	100	54	—
—	—	—	—	18	—	—	—	—
—	—	—	—	60	—	—	—	—
—	—	—	50	—	81	—	47	57
—	82	—	—	—	—	—	—	—
—	—	—	—	36	—	—	—	—
(177)	(67)	(67)	(67)	(178)	(151)	(111)	(145)	(206)

S³⁵-methionine into the proteins of skeletal muscle, but not of liver. The injection of dehydrocorticosterone into adrenalectomized rats reduced slightly the rate of incorporation of S³⁵-labeled methionine into the proteins of all the tissues examined (liver, mucosa, plasma, kidney, muscle) (*64*). This result is in accord with the interpretation of LOTSPEICH (*118*) on the effects of insulin, alloxan, and pituitary hormones on the amino acid concentration in the blood. Taking a decrease in amino acids in the blood as evidence of protein synthesis, and an increase as evidence of breakdown, LOTSPEICH found that carbohydrate promotes synthesis in the normal animal and breakdown in the diabetic; adrenocorticotrophic hormone (ACTH) was, in this sense, diabetogenic; insulin in the diabetic animal, like carbohydrate in the normal, promoted synthesis. HOBERMAN (86) adduced evidence that in the alloxan diabetic rat there is decreased synthesis rather than increased catabolism.

This conclusion is supported by the observations of FORKER et al. (60) on the rates of incorporation of S^{35}-methionine (VI) into the skeletal muscle proteins of normal and diabetic (depancreatized) dogs; five hours

$$CH_3 \cdot \overset{*}{S} \cdot CH_2 \cdot CH_2 \cdot \underset{\underset{NH_2}{|}}{CH} \cdot COOH$$

(VI.) Methionine (* = S^{35}).

after the injection of the amino acid more than twice as much was found in the skeletal muscle proteins of the normal animals. And injections of insulin increased the incorporation in the diabetic animal three to four times. In so short a period as five hours after injection of the labeled amino acid, the loss of labeled amino acid previously incorporated into the proteins is negligible and the rate of incorporation may be taken as a measure solely of the ability of a tissue to incorporate.

We may conclude, therefore, that insulin facilitates incorporation of amino acids into proteins, and that in its absence this process is slower. What the effect of insulin may be on the rate of breakdown of tissue proteins has not yet been studied.

d) Incorporation of Foreign Amino Acids.

Two experiments have been reported in which a labeled amino acid, not a normal constituent of animal proteins, was administered to animals and its metabolic course followed. α-Aminoadipic acid (VII) injected into mice was catabolized at about the same rate as L-lysine, but none was incorporated into any of the proteins [BORSOOK et al. (22)]. This result might have been taken to indicate that an animal can incorporate into its proteins only its normal amino acid constituents. However, this conclusion is no longer admissible.

$$HO\overset{*}{O}C \cdot CH_2 \cdot CH_2 \cdot CH_2 \cdot \underset{\underset{NH_2}{|}}{CH} \cdot COOH \qquad CH_3 \cdot \overset{*}{S} \cdot CH_2 \cdot \underset{\underset{NH_2}{|}}{CH} \cdot COOH$$

(VII.) α-Aminoadipic acid (* = C^{14}).   (VIII.) Ethionine (* = S^{35}).

LEVINE and TARVER (112) have demonstrated that ethionine (VIII) is incorporated into rat tissue proteins. The degree of incorporation was one tenth or less that of amino acids normally constituent in proteins, and the "half-life" of the ethionine was also much less than that of normal amino acids that had been incorporated into the corresponding proteins. But the lower degree of incorporation and the shorter "half-life" do not take away from the importance of the finding that an "unnatural" amino acid can be incorporated into animal proteins. It is now easier to understand the synthesis in animals of such "foreign" proteins as antibodies, which are different from any pre-existing protein in the animal. The finding that ethionine is incorporated adds to the cogency of the suggestion by WORK and WORK (210) that amino acid analogues might be useful chemotherapeutic agents. They said:

"Proteins are ... species specific and therefore some species-specific synthetic processes must go on during growth to build up their amino acid patterns. A logical point of attack would then seem to be the mechanisms concerned with protein synthesis ... To upset protein synthesis, the organism might be presented with amino acid analogues capable of being built up in peptide linkage but incapable of proper function in the completed protein. Alternatively peptides might be synthesized capable of inhibiting by their unnatural configuration the build-up or breakdown essential to vital function. The recognition of gramicidin as a peptide and of penicillin and streptomycin as closely related to amino acids is suggestive of further profitable exploration in this field.

Provided we know enough about the building blocks used by micro-organisms, we may be able to imitate these sufficiently closely for the analogue to be caught up in a synthetic process for which it is unsuited. Before a metabolite analogue can play such a part, it must, however, be sufficiently stable to resist breakdown during metabolism and it must also resemble a natural metabolite so closely that an enzyme, a highly selective and specific catalyst, must be unable to reject it in preference to its natural substrate."

The incorporation of unnatural amino acids into proteins is another example of the versatility of enzymes. One of the most important general ideas established by the work of M. BERGMANN is that even the, hitherto, sacrosanct proteases and peptidases can hydrolyse "unnatural" substrates.

IV. Incorporation of Labeled Amino Acids in Vitro.

Table 10 summarizes measurements of the incorporation of labeled amino acids into tissue proteins *in vitro*. The tissues were obtained from the chick, guinea pig, mouse, rabbit and rat; and whole tissues or cells, slices, homogenates and homogenate fractions were used. In one case, the incorporation of lysine, the system was a crude enzyme preparation. The values given are in many cases, because insufficient details are available, only approximations or estimates. They are also, in most instances, too low for two reasons: one is that in an experiment of several hours duration the rate steadily declined with time, so that in the first hour, for example, the rate was faster than the average rate for the whole period; the other reason is that these values were obtained before optimal conditions had been found. Nevertheless, the values in Table 10 give the order of magnitude.

There are a number of observations where it was not possible to deduce from the published data an estimate even of the order of magnitude of the incorporation rates; yet these experiments contribute to the body of data.

Thus it was found (2) that rabbit bone marrow cells converted labeled acetate to aspartic and glutamic acids and these amino acids were then incorporated into the proteins. Phenylalanine was incorporated into the proteins of rat liver homogenate (*131*), glycine into the proteins of intestinal tissue (*205*), methionine into the proteins of rat liver homogenates (*131*), serine, after its formation from glycine, into the proteins of rat liver homogenates (*208*) and of rabbit reticulocytes (*25*). Glycine and methionine were incorporated faster by foetal than adult rat liver homogenates (*66*) and faster by regenerating than normal adult liver (*75*). Liver homogenates of folic acid deficient rats incorporated glycine more slowly than rats on a diet supplemented with folic acid (*184*).

1. Incorporation of Carbon Dioxide into Amino Acids.

Certain experiments in Table 10 are especially interesting. $C^{14}O_2$, as bicarbonate, was evidently quickly incorporated into the carboxyl groups of aspartic (IX) and glutamic acids (X) and the amino acids were then incorporated into the proteins. The combined processes of amino acid formation and incorporation were so rapid that the final value of amino acid incorporated was as high as the most rapid incorporation of a labeled

$$\text{HOO}\overset{*}{\text{C}} \cdot \text{CH}_2 \cdot \text{CH} \cdot \text{COOH} \qquad \text{HOO}\overset{*}{\text{C}} \cdot \text{CH}_2 \cdot \text{CH}_2 \cdot \text{CH} \cdot \overset{*}{\text{C}}\text{OOH}$$
$$\phantom{\text{HOOC} \cdot \text{CH}_2 \cdot} | \phantom{\text{CH} \cdot \text{COOH}} \qquad \phantom{\text{HOOC} \cdot \text{CH}_2 \cdot \text{CH}_2 \cdot} | $$
$$\phantom{\text{HOOC} \cdot \text{CH}_2 \cdot} \text{NH}_2 \phantom{\text{CH} \cdot \text{COOH}} \qquad \phantom{\text{HOOC} \cdot \text{CH}_2 \cdot \text{CH}_2 \cdot} \text{NH}_2$$

(IX.) Aspartic acid (* = C^{14}).   (X.) Glutamic acid (* = C^{14}).

amino acid directly presented to the tissue [ANFINSEN (*3*, *5*), PETERS and ANFINSEN (*129*, *130*)]. It would be interesting to measure which of the two processes, amino acid formation from CO_2 or incorporation of aspartic acid and glutamic acid, was the faster. One of these two processes must be extraordinarily rapid.

Table 10. Incorporation of Labeled Amino Acids into Tissue Proteins *In Vitro*.

Tissue	Labeled Compounds	Incubation time (hours)	µM/gm. protein per hour	References
Liver slices, chicken	$C^{14}O_2$	3.4	1.15–3.3 in slice proteins;	(*129*, *130*)
			40–256 in serum albumin	(*129*, *130*)

Tissue	Labeled Compounds	Incubation time (hours)	μM/gm. protein per hour	References
Liver slices, rabbit	$C^{14}O_2$	3	3.2	(51)
,, ,, rat	Alanine	3.5	0.6	(214)
,, ,, ,, ...	,,	2	0.085	(62)
Liver hepatoma slices, rat	,,	3.5	5.1	(5)
,, slices, foetal rat	,,	3.5	3.6	(5)
,, homogenate, mouse	,,	4	0.06	(203)
,, ,, rat	,,	4	0.04	(203)
,, ,, foetal rat	,,	4	0.50	(203)
Tumor homogenate, mouse	,,	4	0.3	(203)
Bone marrow cells, rabbit	Glycine	1	0.5	(23)
Diaphragm, rat	,,	1	0.1	(24)
Reticulocytes, rabbit, in saline	,,	1	0.78	(25)
,, ,, in plasma	,,	1	1.29	(25)
Liver slices, rat	,,	3.5	3.0	(214)
,, hepatoma slices, rat	,,	3.5	8.9	(214)
,, homogenate, chicken	,,	1–1.5	0.11–0.25	(184)
,, ,, rat	,,	4	0.08	(203)
,, ,, ,, ...	,,	2	0.18 as glycine 0.12 as serine	(208)
,, ,, mouse	,,	4	0.08	(203)
,, ,, foetal rat	,,	4	0.55	(203)
Tumor homogenate, mouse	,,	4	0.08	(203)
Liver, "nuclei", guinea pig..........	,,	1	0.125	(21)
,, mitochondria, guinea pig	,,	1	0.10	(21)
,, microsomes, guinea pig	,,	1	0.019	(21)
Blood, duck	Histidine	12	0.83	(159)
Reticulocytes, rabbit in saline	,,	1	1.08	(25)
,, ,, in plasma......	,,	1	1.75	(25)
Bone marrow cells, rabbit	Leucine	1	2.9	(23)
Diaphragm, rat	,,	1	0.1	(24)
Reticulocytes, rabbit in saline	,,	1	0.7	(25)
,, ,, in plasma......	,,	1	2.0	(25)
Liver, guinea pig, "nuclei"...........	,,	1	0.15	(21)
Bone marrow cells, rabbit	Lysine	1	1.8	(23)
Diaphragm, rat	,,	1	0.1	(24)
Reticulocytes, rabbit, in saline	,,	1	0.99	(25)
,, ,, in plasma	,,	1	2.51	(25)
Liver, "nuclei", guinea pig..........	,,	1	2.1	(21)
,, mitochondria, guinea pig	,,	1	1.6	(21)
,, microsomes, guinea pig	,,	1	0.45	(21)
,, particle free supernatant, guinea pig	,,	1	1.6	(21)
E. coli, resting	Methionine	6	0.004–0.11	(121, 122)
Liver slices, rat	,,	2	0.13–0.27 as cystine; 0.19–0.40 as methionine	(123, 208)

2. Net Synthesis of Protein in Vitro.

The other experiments that call for special mention also used $C^{14}O_2$ as the label, aspartic acid and glutamic acid labeled in their carboxyl groups were formed, and the labeled amino acids were incorporated into the proteins. Chicken liver slices were employed. In these experiments a net synthesis of protein could be demonstrated [PETERS and ANFINSEN (*129, 130*)]. The synthesized protein had the physical, chemical and immunological properties of serum albumin. An extra 0.42 mg. of this protein was formed per gram of liver (wet weight) in four hours. An extraordinarily high rate of 256 μ-Mols. of labeled CO_2 per gram of serum albumin per hour was observed in the best experiments. Taking a molecular weight of serum albumin as 35000, it can be calculated that 0.012 μ-Mols. of this albumin was synthesized per gram of liver. In the total serum albumin at the end of the experiment 37.5% of the dicarboxylic amino acid carboxyl groups were labeled. This value is within a factor of approximately 3 of the value that would prove synthesis of protein *de novo* from amino acids. There are not sufficient data from the experiments to decide whether this factor of 3 is sufficiently close or too low to render a verdict in favor or against synthesis *de novo* from amino acids. For example, one might assume that the average specific activity of the free aspartic acid and glutamic acid was less than that of the $C^{14}O_2$ because the slices contained quantities of these amino acids in the free state, initially. Accordingly the specific activity of these amino acids incorporated into the protein would be less than that of the $C^{14}O_2$, even on the basis of *de novo* synthesis. The specific activity of the labeled CO_2 was certainly lowered by the CO_2 arising from respiration of the slices; this, also, would tend to lower specific activity of the aspartic and glutamic acids formed and subsequently incorporated.

The synthesized protein discussed above was that found in the saline solution at the end of the experiment, i. e. it was excreted or secreted by the slices. PETERS and ANFINSEN (*129, 130*) presented evidence that the incorporation of labeled amino acids (or the protein synthesis) must have occurred at the cell membrane. The ratio of the specific activity of protein in the medium to that of the protein in the slices was 16.5/1, whereas the ratio of the specific activity of the serum albumin within the slices to that of the total protein in the slices was only 0.94/1. The question arises, then, whether the protein that was found in the medium ever appeared in the interior of the cell. The evidence indicates that it may not have. Many questions regarding both cellular and biochemical mechanisms then arise. It would be unprofitable to discuss these questions until more data become available.

3. Comparison of Incorporation of Amino Acids in Vivo and in Vitro.

From the *in vitro* studies the following general conclusions may be drawn.

Every amino acid (in the form in which it occurs in proteins) that has been presented has been found to be incorporated into the proteins; amino acids formed from CO_2 and from acetate were incorporated; in the one case of an "unnatural" isomer, D-lysine, the evidence indicates that it was not incorporated [BORSOOK *et al.* (*18*)]. The rate of incorporation in intact cells was found to be of the same order of magnitude as in the corresponding cells *in vivo*; and the relative speeds in different tissues and of different amino acids were also similar *in vitro* to those *in vivo*. Cell damage, i. e. slices compared with intact cells, or homogenates with slices, reduced the rate of incorporation. In two cases, rabbit bone marrow cells (*23*) and rabbit reticulocytes (*25*), after lysis there was no incorporation. However, the important point is that some incorporation of certain labeled amino acids can occur in homogenized tissues. It is to be expected that, as in so many other instances, when these systems are studied further, means will be found, through addition of appropriate co-factors and by better methods of preparing the homogenate, to augment greatly the rate of incorporation. There are, already, examples available in this field (see below).

4. Amino Acid Incorporation in Different Cell Fractions.

An homogenate is neither an enzyme, an enzyme system, nor a mixture of soluble and insoluble proteins. A large fraction of the suspension consists of organized particles such as nuclei, mitochondria and microsomes. When working with homogenates we are not free of organized structures, and the level of organisation of the particles is much higher than that of proteins. It may be for the synthesis of molecules so large, so complex, and so specific in pattern as protein, that the enzyme system itself must be equally large, complex and organized. Nevertheless, information can be obtained from homogenates and homogenate fractions which could not be obtained, at least as directly, from intact cells. One such example is in Table 10 (p. 310). All four fractions of liver homogenate, three particulate fractions and the particle free supernatant solution differed in the rate at which they incorporated *in vitro* different amino acids; and in the case of lysine, the three particulate fractions on the one hand, and the supernatant solution on the other, had different

$$CH_3 \cdot CH \cdot \overset{*}{C}OOH$$
$$|$$
$$NH_2$$

(XI.) Alanine ($* = C^{14}$).

optimal conditions [BORSOOK et al. (*18, 21*)]. Another example was reported by SIEKEVITZ and ZAMECNIK (*161*): The incorporation of labeled alanine (XI) by a mixture of mitochondria and microsomes of rat liver was accelerated two- to three-fold by aerobic oxidation of α-ketoglutarate or succinate, or by anaerobic oxidation by ferricyanide of α-ketoglutarate or malate; adenosinetriphosphate (ATP) accelerated incorporation whether aerobically or anaerobically; and mitochondria acted synergistically on the incorporation by microsomes.

It may be concluded from such observations as the foregoing that different components of the liver cell have different rates of protein anabolism. If the theory of the dynamic steady state of proteins discussed above is valid, it follows that the components of the liver cell that incorporate amino acids into their proteins most rapidly, would also lose them most rapidly, i. e. their rate of catabolism would be greatest. This has not been tested in animals in nitrogen balance; however, experimental results are available from animals in protein starvation. It was found that relatively the greatest losses of protein occurred in the mitochondrial and microsome fractions, and there was little or no loss in the nuclear fraction (*127, 158, 187*). Similar changes occured in hepatomatous as compared with normal liver and in rats fed carcinogenic compounds (*50, 134–137, 156*). These experimental findings are in accord with, indeed they provide strong support for the theory.

5. The Nucleus, Amino Acid Incorporation, and the Maintenance of the Amino Acid Pattern in Proteins.

Of more biological general interest is the conclusion which must be drawn from the findings in Table 10, p. 310, on individual liver cell fractions that the incorporation of amino acids into proteins, does not, in the adult cell, necessarily depend on direct participation of the nucleus. *In vivo* a short time after the injection of a labeled amino acid, the greatest incorporation was found in the microsomes (Table 8, p. 305). SIEKEVITZ and ZAMECNIK (*161*) found the same *in vitro* with *DL*-alanine and rat liver cell fractions; the relative degrees of incorporation they found were: microsomes, 15; mitochondria, 6; homogenate, 3; and supernatant, 1. The relative speeds of incorporation among the cell fractions may be different with different amino acids it is different with lysine. But in any event the *in vitro* findings with separated fractions excluded as an explanation of the *in vivo* findings that the incorporation occurred first in the nucleus, and the labeled protein (or peptide) was then transferred to extranuclear particles and the particle-free cytoplasm.

The amino acid pattern of proteins is an inherited characteristic and it is a reasonable assumption that this pattern is preserved after a labeled acid is incorporated. Since the extra-nuclear enzyme systems can incorporate labeled amino acids into the proteins independently of the nucleus, the question arises: at what stage in development does the nucleus impose its hereditary influence on the extra-nuclear enzyme systems? The ability of extra-nuclear enzyme systems to incorporate amino acids independently of the nucleus also makes it both easier to apprehend the synthesis of proteins foreign to the cell, such as antibodies, or of proteins containing an amino acid which is not a normal constituent, as has been found in the case of ethionine.

6. Nucleic Acids, Protein Synthesis, and Amino Acid Incorporation into Proteins.

There is a considerable body of circumstantial biological evidence relating nucleic acid to protein synthesis. A number of authors (*31, 32, 38, 45, 51, 90, 91, 148, 172, 183*) associate ribosenucleic acid with cytoplasmic protein synthesis; desoxyribosenucleic acid, which is confined to the nucleus, is not, however, excluded from this process. It is interesting in this connection that nuclei, mitochondria, microsomes, and the particle-free solution of adult liver homogenates, all incorporate labeled amino acids into their proteins *in vivo* and *in vitro* (*21*), and that, under favorable conditions the most rapid incorporation occurs *in vitro*, as it does *in vivo*, in the microsomes (*22, 98*), which have the highest concentration of ribosenucleic acid (*148*). Similarly in the course of the development of reticulocytosis in rabbits there is a 30–40-fold increase in ribosenucleic acid and a parallel increase in the ability to incorporate labeled amino acids; the increase in desoxyribosenucleic acid is only 2–3-fold (*89*). What is lacking at present is direct evidence of participation of ribosenucleic acid in protein synthesis, or in amino acid acid incorporation into proteins. So far the evidence is only circumstantial.

7. Normal, Foetal and Tumor Tissue.

The collected observations in Table 10 are in accord that foetal tissues, whether as slices or homogenates, incorporate labeled amino acids faster than adult tissues. It may be said also that, in general, the incorporation into foetal or tumor slices or homogenates is rapid, more so than in the corresponding normal tissue. This is the case *in vitro*; *in vivo*, however, the incorporation may be the same in the tumor as in the normal tissue (*213*).

8. Effect of Concentration of Labeled Amino Acid on its Rate of Incorporation.

In all cases but one so far studied, the rate of incorporation *in vitro* of labeled amino acids was a logarithmic function of the initial concentration of the labeled amino acid up to a certain optimum concentration

(0.003 to 0.001 M). Higher concentrations were inhibitory. In the case of lysine in liver homogenate (but not in other tissues) (*18*), the relation of rate of incorporation to initial concentration was linear. But this incorporation of lysine is exceptional in a number of other respects (see below). The general conclusion is that the concentrations of labeled amino acids from which tissues can incorporate them *in vitro* into their proteins are of the order of those in the blood. And the dependence of the rate of incorporation on concentration of labeled amino acid is greatest in the physiological range of concentration (*18, 20, 23, 24*). The fact that the relation between rate of incorporation and initial concentration is logarithmic (in most cases at least) may be part of the explanation of a finding, such as that a 30-fold increase in the amount of glycine injected only doubled the rate of its incorporation into the liver (*76*); it may explain the logarithmic decrease in nitrogen excretion in passing from a high to a low level of nitrogen balance (*119*).

9. Does Incorporation of One Amino Acid Require the Presence of Others?

a) Feeding Experiments.

It has been found in feeding experiments that an indispensable amino acid is ineffective for growth, for recovery from protein depletion, or for maintenance, unless it is fed or injected within a few hours of other necessary amino acids (*36, 69–74, 81, 83, 147, 211*). The interpretation placed on this finding has been that all the amino acids must be present at concentrations greater than the fasting levels for protein synthesis to occur.

b) *In Vivo Experiments with Single Labeled Amino Acids.*

In support of this interpretation the observations have been cited of SANADI and GREENBERG (*145*) discussed above, that young rats on a tryptophane-deficient diet incorporated intraperitoneally-injected C^{14}-labeled tryptophane more slowly than litter mates receiving tryptophane in their diet.

Against this interpretation stand the other observations of SANADI and GREENBERG that animals on a diet deficient in phenylalanine showed only slight reduction in the rates of incorporation of C^{14}-labeled tryptophane and glycine. As pointed out above, all the observations of SANADI and GREENBERG are in accord with a quite different interpretation, that animals on deficient diets eat less, their total protein turnover is slower, and accordingly the directly measured incorporation of labeled amino acids is slower than on a normal adequate diet when the animals eat more and their total protein turnover is accordingly higher.

More cogent evidence against the interpretation placed on the feeding experiments are those from direct observations on the fate of either injected or fed labeled amino acids. After the injection or feeding of a single labeled amino acid, whether dispensable or indispensable, it is found extensively incorporated into the proteins of the animal in a few minutes. Extensive incorporation occurs whether the animal is normally fed or fasting [TARVER and SCHMIDT (*178*)].

c) *In Vitro Experiments with Labeled Amino Acids.*

Still more cogent evidence against a simple interpretation of the results of the feeding experiments appeared from *in vitro* experiments in which the effect was measured of incubating a labeled amino acid with and without other amino acids. In the first experiments of this kind labeled glycine, leucine and lysine were incubated separately and together. The results summarized in Table 11 show that the counts in

Table 11. *In Vitro* Incorporation of Labeled Amino Acids Added Separately and Together.

(Results Expressed as Counts Per Minute Per Milligram of Protein.)

	Guinea Pig Liver Mitochondria	Rat Diaphragm	Rabbit Bone Marrow Cells
Glycine....................	6.22	0.37	1.75
L-Leucine..................	2.76	0.37	2.14
L-Lysine...................	4.03	1.12	4.27
Glycine + Leucine + Lysine..	13.05 Calcd. 13.01	1.73 Calcd. 1.86	8.57 Calcd. 8.16
Reference..................	(*20*)	(*24*)	(*23*)

the protein when all three amino acids were added together was the sum of when the three were added separately. In other experiments the addition of two non-radioactive amino acids together with one which was radioactive did not affect the uptake of the latter. Nor did the addition of a mixture of amino acids approximating the composition of casein or of hemoglobin affect the uptake of labeled glycine, leucine or lysine. The conclusion drawn from these experiments was that each amino acid is incorporated independently of the others, a conclusion in accord with the results of administration of single labeled amino acids *in vivo*.

More recently experiments such as those in Table 11 were repeated with rabbit reticulocytes using four labeled amino acids—glycine, histidine, leucine, and lysine [BORSOOK *et al.* (*25*)]. The experiments were carried out in saline and in plasma (because it had been found that plasma contains an unidentified factor, not an amino acid, that accelerates the incorporation of labeled amino acids into reticulocyte proteins). Typical results summarized in Table 12 show that, as in the previous experiments, the counts in the protein when the cells were incubated with labeled glycine, leucine and lysine was the arithmetical sum of when these three amino acids were added separately. This was the case both in the saline and plasma reaction mixtures. On the other hand, whenever histidine was included in the combination of amino acids, the counts in the protein were a little greater than the sum of the amino acids separately.

Table 12. *In Vitro* Incorporation by Rabbit Reticulocytes of Labeled Glycine, *L*-Histidine, *L*-Leucine and *L*-Lysine when Incubated Separately and Together.

(Results Expressed as Counts Per Minute Per Mg. Protein) (*25*).

Labeled Amino Acid	Incubated in Saline		Incubated in Plasma	
	Obs.	$\frac{\text{Obs.}}{\text{Calcd.}}$ %	Obs.	$\frac{\text{Obs.}}{\text{Calcd.}}$ %
Glycine	4.4		10.0	
Histidine	8.0		15.1	
Leucine	4.65		10.5	
Lysine	4.2		10.6	
Gly. + His. + Leu.	21.6	127	41.5	131
Gly. + Leu. + Lys.	13.35	101	31.4	101
Gly. + His + Lys.	20.3	122	38.5	108
His. + Leu. + Lys.	23.4	139	39.1	108
Gly. + His. + Leu. + Lys.	28.2	133	53.0	115

The following additional findings need to be taken into account in the interpretation of the results in Table 12. The incorporation of labeled amino acids into reticulocytes is accelerated in saline by histidine, phenylalanine and valine, and in plasma also by leucine. None of the other amino acids occuring in proteins, whether singly or together, had any effect. Of the above four amino acids only histidine had any accelerating effect when added alone; leucine, phenylalanine and valine require histidine to exert their accelerating effects. The synergistic action of histidine shown in Table 12, and of leucine, phenylalanine and valine are associated with a specific mechanism, which has not yet been elucidated. With this qualification it appears, in the light of evidence available so far, that each amino acid can be incorporated independently of others. No clear evidence has been found in either *in vivo* or *in vitro* studies with labeled amino acids indicating that a mixture of all the amino acids is required for the incorporation of any single amino acid into proteins. As matters stand now there is a discrepancy between, on the one hand the feeding experiments, and on the other hand experiments on the incorporation of labeled amino acids. All the experimental facts have been abundantly confirmed; the discrepancy is in the interpretation.

V. The Biological Significance of the High Lability of the Proteins in the Cell.

Before proceeding to review the information available on the detailed mechanism of protein turnover and of amino acid incorporation into proteins, it is worthwhile to consider the following general question.

The proteins in the cell are labile, they are continually losing and re-incorporating amino acids; of what use can it be to the organism that its proteins are not fixed structures, but exist in a dynamic steady state? Any answer at the present time can be only speculative, and, necessarily, it will be teleological. A cell is exposed to such vicissitudes as growth, reproduction, changing food supply, infection and the production of antibodies and of new adaptive enzymes, hypertrophy in response to increased work, to name but a few varying conditions. It is easier to envisage its ability to adapt itself to such changing circumstances, given that its proteins are labile; it would be much more difficult to envisage this adaptability if the proteins were rigid and immutable structures. Adaptability implies flexibility and mutability. As pointed out above the basal metabolism of a cell supplies sufficient energy for the most rapid rates of amino acid incorporation that have been observed. And it may be that the continual loss and re-incorporation of amino acids into proteins is essential to maintenance of the organization as a dynamic system, and to the speed of biochemical processes in the cell.

VI. Mechanisms of Peptide Bond Synthesis.

1. Heats and Free Energies of Formation of Some Amino Acids and Peptides (Solids).

The formation of a simple peptide from two amino acids represents a gain in free energy. The order of magnitude of this free energy change is so large that at equilibrium the dipeptide is nearly completely hydrolyzed; in other words peptide formation from free amino acids does not proceed spontaneously to any significant extent.

Table 13 contains the free energies of formation of amino acids and peptides, from which the free energy change in the synthesis of a peptide can be calculated (*17*). The free energy change is related to the equilibrium constant by the following equation:

$$-\widetilde{\Delta F} = RT\, 2.303 \log K.$$

The following approximate calculation shows the relation of the equilibrium constant to the degree of synthesis at equilibrium of the dipeptide from its constituent amino acids. Where the initial concentration of each of the amino acids is represented by a, the equilibrium concentration

Table 13. The Heats and Free Energies of Formation of Some Amino Acids and Peptides (17).

Compound	$\Delta F°_{310.6}$	$\widetilde{\Delta H}_{298.1}$	$\widetilde{\Delta F}_{298.1}$	$\widetilde{\Delta F}_{310.6}$
DL-alanine	−86830	−132370	−89110	−87300
DL-alanylglycine	−113190	−186410	−117570	−114680
Benzoic acid	−57660			
Benzoate ion			−51175	−49710
Glycine	−86920	−122500	−89140	−87710
Glycylglycine	−115200	−175960	−118060	−115630
Hippuric acid	−86160			
Hippurate ion			−81000	−78590
Hippurylglycine	−114710	−191700	−115580	−112380
DL-Leucine	−80350	−150900	−81760	−78870
DL-Leucylglycine	−108225	−203620	−110900	−107065
Water	−56200	−68320	−56690	−56200

ΔF^0 refers to the pure substance; $\widetilde{\Delta H}$, $\widetilde{\Delta F}$ refer to the substance in the standard state in aqueous solution, which is 1 M activity except in the case of water where it is mol-fraction = 1.

of the dipeptide by $a\alpha$, and the equilibrium concentration of each of the amino acids by $a(1-\alpha)$, the equilibrium expression is:

$$\frac{a\alpha}{a^2[1-\alpha]^2} = K.$$

Since α is very small with respect to 1, this expression is simplified to $\alpha = Ka$. In dilute solutions, for example where each of the amino acids is initially $M/10$, α would be equal to $K \times 10^{-1}$. In the synthesis of DL-alanylglycine under these conditions, α would be of the order of magnitude of 0.001×10^{-1}, or 10^{-4}.

2. Free Energies of Formation of Some Peptides in Aqueous Solution.

The data (26) in Table 14 show that the free energy of formation of a peptide in solution is greater than that calculated from the solids. In the tissues, where the concentrations of the amino acids are of the order of magnitude of 0.001 M or less, the degree of synthesis by mass action is infinitesimal.

The data also show that the free energy change in the condensation of two amino acids is significantly greater than in the formation of hippuric acid from benzoic acid and glycine.

There is a suggestion that more energy is needed to condense two free amino acids, to form a dipeptide, than when only one of the reactants is a zwitterion in which the positive and negative charges are in α-position to each other. In other words a charge near to the bond to be synthesized increases the energy required for the synthesis. This is a reasonable assumption.

Table 14. **Free Energies and Equilibrium Constants of Formation of Some Peptide Bonds** (26).

Reaction	Pure Substance $-\varDelta F°_{310.6}$	Aqueous Solution $-\widetilde{\varDelta F°}_{310.6}$	$K_{310.6}$
DL-Alanine + glycine → DL-alanylglycine + water	−3560	−4230	0.00106
2 Glycine → glycylglycine + water	−2440	−3590	0.00299
Benzoate ion + glycine → hippurate ion + water		−2630	0.01415
Benzoic acid + glycine → hippuric acid + water	−2220		
Hippuric acid + glycine → hippurylglycine + water	−2170		
DL-Leucine + glycine → DL-leucylglycine + water	−2850	−3416	0.00252

The heats of reaction, i. e. $-\widetilde{\varDelta H}$, in the formation of alanylglycine, glycylglycine, and leucylglycine are respectively, −140, −720, and 1460 calories. These are all larger taking the sign into account, or numerically smaller ignoring the sign, than the correspondieg values of $-\widetilde{\varDelta F}$. From the values of $-\widetilde{\varDelta H}$ alone one might have concluded that the synthesis of alanylglycine might go by mass action to a measurable extent; the value of $-\widetilde{\varDelta F}$ shows that this is not the case. In the early stages of peptic digestion where the products are nearly all peptides (varying in size) the heat evolved is on the average, in the case of β-lactoglobulin, 1300 cal. per mole of peptide bond split [HAUGAARD and ROBERTS (82)] which is approximately the same as that in the hydrolysis of leucylglycine.

The general conclusion would be, then, that the largest free energy hump in the biosynthesis of protein is in the formation of small peptides from free amino acids. From then on the synthesis to protein would be, as far as the energy required is concerned, much easier. This may well be the explanation of the energetics in some of the peptide transfers studied by M. BERGMANN and others after him (see below).

3. The Effect of pH on the Free Energy Change in Peptide Formation.

It is interesting to examine the possible effect of the pH on the equilibrium degree of synthesis of a dipeptide. At the pH of the tissues the preponderant ionic species for most amino acids and dipeptides is the zwitterion. The equilibrium expression for the reaction of amino acids A and B to form the dipeptide AB is:

$$\frac{[AB \pm]}{[A \pm][B \pm]} = K.$$

Designating the ionization constant in acid as k_1 and in alkali as k_2, the ionization constants for A, B, and AB respectively as k_1^A, k_2^A, k_1^B, k_2^B, k_1^{AB} and k_2^{AB}, the equilibrium constant for the reaction is:

$$K = \frac{\Sigma[AB]}{\Sigma[A]\,\Sigma[B]} \cdot$$

$$\cdot \frac{(k_1^{AB}[H^+])\,([H^+]^2 + k_1^A[H^+] + k_1^A k_2^A)\,([H^+]^2 + k_1^B[H^+] + k_1^B k_2^B)}{([H^+]^2 + k_1^{AB}[H^+] + k_1^{AB} k_2^{AB})\,(k_1^A[H^+])\,(k_1^B[H^+])} \cdot$$

For the synthesis of alanylglycine, k_1^A is 4.6×10^{-3}, k_2^A is 1.35×10^{-10}, k_1^B is 4.5×10^{-3}, k_2^B is 1.7×10^{-10}, k_1^{AB} is 6.8×10^{-4}, and k_2^{AB} is 3.8×10^{-9} (85). From these data the values of the ionization term from pH 3 to 9 can be computed and are given in Table 15. It is seen that, given the low equilibrium constant for the synthesis, the ionization term does not significantly increase the degree of synthesis at any pH over this range; it is always very small.

Table 15. *Variation of Ionization Term with pH in the Equilibrium Expression in the Synthesis of Alanylglycine from Alanine and Glycine.*

pH	Value of ionization term
3	0.60
4	0.91
5	0.99
6	1.00
7	1.00
8	0.99
9	0.28

4. Peptide Synthesis by Proteases and Peptidases.

a) Classification of Enzymatic Peptide Syntheses According to the Sign and Magnitude of the Free Energy Change ($-\Delta F$).

Although the fact that the free energy of formation of peptide bonds (in small peptides) from amino acids corresponds to an equilibrium position in dilute solution beyond 99% hydrolysis, and aware of the fact, a number of authors have been inclined to the view that peptide and protein synthesis may be catalyzed by proteases and peptidases. The free energy data indicate that, whether or not this is the case, the synthesis of small peptides from amino acids cannot, under physiological conditions, be a simple mass action reversal of hydrolysis. On the other hand the condensation of large peptides may be promoted by proteases alone.

In recent times M. BERGMANN has been the most notable proponent of the view that proteases and peptidases participate in protein and peptide synthesis *in vivo*. He was led to this view by a series of highly original and brilliantly conducted experiments. The differences among the reactions he and his group studied arose, he considered, from differences in enzyme-substrate specificity. Except insofar as the formation of a relatively insoluble product tended to drive the reaction in the direction of the latter, the authors took no account of the free energy change of the reaction, i. e. of the equilibrium constant. It is the free energy change, its sign and magnitude, i. e. the magnitude of the equilibrium constant, which is the fundamental distinction. Distinguishing characteristics of the categories of reactions studied are:

(a) where $-\Delta F$ for the condensation is positive in sign and so large that the reaction proceeds spontaneously nearly to completion; (b) where $-\Delta F$ for the condensation is positive in sign but the order of magnitude is small, and as a result the direction of the reaction is affected by the concentrations of reactants and products; and (c) where $-\Delta F$ is negative in sign for the condensation, i. e. the reaction is an hydrolysis, but during the enzyme-catalysed hydrolysis there is an exchange at the locus of hydrolysis on the enzyme with another reactant in the reaction mixture. Finally a reaction may consist of a combination of the above three types.

b) Peptide Syntheses where $-\Delta F$ is Positive and Large.

An example of the first category is as follows. A solution containing 4.2% of carbobenzoxy-glycine (XII) and 3.7% of aniline (XIII) was incubated with activated papain at 40° and pH 4.6. Carbobenzoxy-glycine anilide (XIV) was formed in 80% yield. The optimum pH and the necessity for activation by cysteine, glutathione or HCN were the same as for the

$$C_6H_5 \cdot CH_2 \cdot O \cdot CO \cdot NH \cdot CH_2 \cdot COOH$$
(XII.) Carbobenzoxy-glycine.

↓ aniline (XIII.)

$$C_6H_5 \cdot CH_2 \cdot O \cdot CO \cdot NH \cdot CH_2 \cdot CO \cdot NH \cdot C_6H_5$$
(XIV.) Carbobenzoxy-glycine anilide.

$$C_6H_5 \cdot CO \cdot NH \cdot CH_2 \cdot COOH \xrightarrow{\text{aniline}} C_6H_5 \cdot CO \cdot NH \cdot CH_2 \cdot CO \cdot NH \cdot C_6H_5$$
(XV.) Hippuric acid. (XVI.) Hippuric acid anilide.

hydrolytic action of the enzyme. Analogously, hippuric acid (XV) and aniline gave hippuric acid anilide (XVI). Acetyl, benzoyl and carbobenzoxy derivatives of alanine, leucine and phenylalanine yielded with aniline or phenylhydrazine the corresponding anilides or phenylhydrazides (XVII); acetyl-*L*-phenylalanyl-*L*-glutamic acid (XVIIa) and *p*-toluenesulfonyl-glycine gave with aniline the corresponding anilides (XVIII) [BERGMANN and FRAENKEL-CONRAT (*10, 11*)]. Similar reactions were catalysed by bromelin and cathepsin, the proteolytic enzymes respectively of pineapple and pig liver. Under the conditions that promoted the above syntheses hippurylamide was completely hydrolysed; there was no synthesis of the amide from hippuric acid and ammonia.

WALDSCHMIDT-LEITZ and KÜHN (*198*) have recently studied in detail the synthesis of hippurylanilide (XVI), the type of synthesis first found by BERGMANN *et al.* (*10, 11*). Using nearly equivalent amounts of hippuric acid and aniline, with papain as enzyme, there was 94% synthesis of the anilide. The equilibrium was approached from both sides. They could not obtain synthesis of hippurylamide from hippuric acid and ammonia,

$$CH_3 \cdot CO— \qquad\qquad —NH \cdot \underset{\underset{CH_3}{|}}{CH} \cdot CO—$$

acetyl   alanyl

$$C_6H_5 \cdot CO— \qquad\qquad —NH \cdot \underset{\underset{CH_2 \cdot CH(CH_3)_2}{|}}{CH} \cdot CO—$$

benzoyl   leucyl

$$C_6H_5 \cdot CH_2 \cdot O \cdot CO— \qquad\qquad —NH \cdot \underset{\underset{CH_5}{|}}{CH} \cdot CO—$$

carbobenzoxyl   phenylalanyl

(XVII.) Acyl and aminoacyl groups connected with each other and with —NH·CH or —NH·NH·C₆H₅.

$$CH_3 \cdot CO—NH \cdot \underset{\underset{C_6H_5}{|}}{CH} \cdot CO—NH \cdot \underset{\underset{CH_2 \cdot CH_2 \cdot COOH}{|}}{CH} \cdot CO \cdot NH \cdot C_6H_5$$

(XVII a.) Acetyl-*L*-phenylalanyl-*L*-glutamylanilide.

$$(p)\,CH_3 \cdot C_6H_4 \cdot SO_2—NH \cdot CH_2 \cdot CO—NH \cdot C_6H_5$$

(XVIII.) *p*-Toluenesulfonyl-glycine-anilide.

the amide was completely and rapidly hydrolysed. The condensation with hippuric acid occurred with aniline, *o*-, *m*-, and *p*-toluidine, *o*- and *p*-aminophenol, *o*-anisidin, *p*-aminobenzoic acid, sulfanilamide, and *o*- and *p*-phenylenediamine. The following compounds were inactive: N-methylaniline, *o*-aminobenzoic acid, sulfanilic acid, α-aminopyridine, adenine, benzylamine, cyclohexamine, and ammonia.

From the equilibrium data it is possible to calculate the free energy of formation of hippurylanilide under the conditions applied. The value is approximately 5000 calories at 37°, whereas the free energy of formation of small peptides and of amides is of the order of magnitude of —3500 calories. In other words, the formation of the anilide proceeds spontaneously, as the authors found but the formation of peptides and amides does not. This is the explanation of the failure to observe condensation of hippuric acid and ammonia.

The foregoing "peptide" syntheses can, at best, be only analogues of peptide synthesis *in vivo* from amino acids. Even granting that they are analogues of the *in vivo* process, it would still be necessary to find the physiological analogues of the acetyl, benzoyl and carbobenzoxy derivatives of the amino acids, and of aniline and phenylhydrazine. This leaves the problem where it was before: we need to find the reactive intermediates and to learn the mechanisms by which they are formed. Almost certainly these mechanisms involve coupling with energy-yielding reactions in such a manner that reactive intermediates with higher chemical potential than free amino acids are formed. And it may very well then turn out that the enzyme promoting the coupling is not a protease or peptidase as, for example, in glycogen or sucrose synthesis, where glucose-1-phosphate is synthesized into glycogen or sucrose not by an amylase or saccharase but by specific phosphorylases.

c) Peptide Syntheses where —ΔF *is Small and the Peptide is Relatively Insoluble.*

Examples of the second category of condensations catalysed by proteolytic enzymes are the following: by papain, benzoyl-L-leucine + + L-leucine anilide (XIX) to benzoyl-L-leucyl-L-leucine anilide [BERGMANN and FRAENKEL-CONRAT (*11*)]; benzoylphenylalanine + leucine anilide to benzoyl-phenylalanyl-leucine anilide (XX); carbobenzoxyphenylalanyl-glycine + tyrosine amide to carbobenzoxy-phenylalanyl-glycyl-tyrosine amide (XXI) [BERGMANN and FRUTON (*13*)]; by chymotrypsin benzoyl-L-tyrosine + glycine anilide to benzoyl-L-tyrosyl-glycine anilide (XXII) (*12*);

$$C_6H_5 \cdot CO-NH \cdot CH \cdot CO-NH \cdot CH \cdot CO-NH \cdot C_6H_5$$
$$\underset{CH_2 \cdot CH(CH_3)_2}{\overset{CH_2 \cdot CH(CH_3)_2}{|}}$$

(XIX.) Benzoyl-leucyl-leucine anilide.

$$C_6H_5 \cdot CO-NH \cdot CH \cdot CO-NH \cdot CH \cdot CO-NH \cdot C_6H_5$$
$$\underset{CH_2 \cdot C_6H_5}{|} \quad \underset{CH_2 \cdot CH(CH_3)_2}{|}$$

(XX.) Benzoyl-phenylalanyl leucine anilide.

$$C_6H_5 \cdot CH_2 \cdot O \cdot CO-NH \cdot CH \cdot CO-NH \cdot CH_2 \cdot CO-NH \cdot CH \cdot CO \cdot NH_2$$
$$\underset{CH_2 \cdot C_6H_5}{|} \qquad \underset{CH_2 \cdot C_6H_4 \cdot OH(p)}{|}$$

(XXI.) Carbobenzoxy-phenylalanyl-glycyl-tyrosine amide.

$$C_6H_5 \cdot CO-NH \cdot CH \cdot CO-NH \cdot CH_2 \cdot CO \cdot NH-C_6H_5$$
$$\underset{CH_2 \cdot C_6H_4 \cdot OH(p)}{|}$$

(XXII.) Benzoyl-tyrosyl-glycine anilide.

$$C_6H_5 \cdot CO-NH \cdot CH \cdot CO-\!\!-\!\!-NH \cdot CH \cdot CO-NH \cdot C_6H_5$$
$$\underset{CH_2 \cdot C_6H_4 \cdot OH(p)}{|} \quad \underset{CH_2 \cdot CH(CH_3)_2}{|}$$

(XXIII.) Benzoyl-tyrosyl-leucine anilide.

benzoyl-L-tyrosine + leucine anilide to benzoyl-tyrosyl-leucine anilide (XXIII) (*13*). The amide radical of benzoyl-L-tyrosyl-glycine amide (XXI) was hydrolysed off by chymotrypsin. In these reactions the enzyme was specific for the L-form of the amino acids whose derivatives participated in the reactions; even in the reaction of acetyl-DL-phenylalanyl-glycine with aniline, only acetyl-L-phenyl-alanyl-glycine anilide was formed, although glycine, with which the aniline combined, has no asymmetric carbon [BERGMANN and BEHRENS (9)].

In the above reactions two amino acids were combined in peptide linkage. There was no coupling with an energy yielding reaction, nor was there an hydrolytic step in which part of the energy that might have been released was used for peptide formation. It is certain therefore, in view of the data in Table 14 (p. 321) that benzoyl, carbobenzoxy or phenylalanyl substitution on one of the amino acids and aniline or amide on the other so increased the free energy contents (i. e. the chemical potential) of the substituted amino acids that they could be condensed in peptide linkage by the enzyme without coupling with another free energy yielding reaction. There are no free energy data available on amino acids substituted as above, and to this extent the interpretation given is uncertain. But there is little room for doubt.

To some extent the formation in measurable amounts of peptides, such as the above, is promoted by their relative insolubility. They appear to be intermediates in hydrolysis. The following is an example of such an intermediate (13). Glycyl-leucine (XXV) is not hydrolysed by papain-HCN but it is hydrolysed when acetylphenylalanyl-glycine is

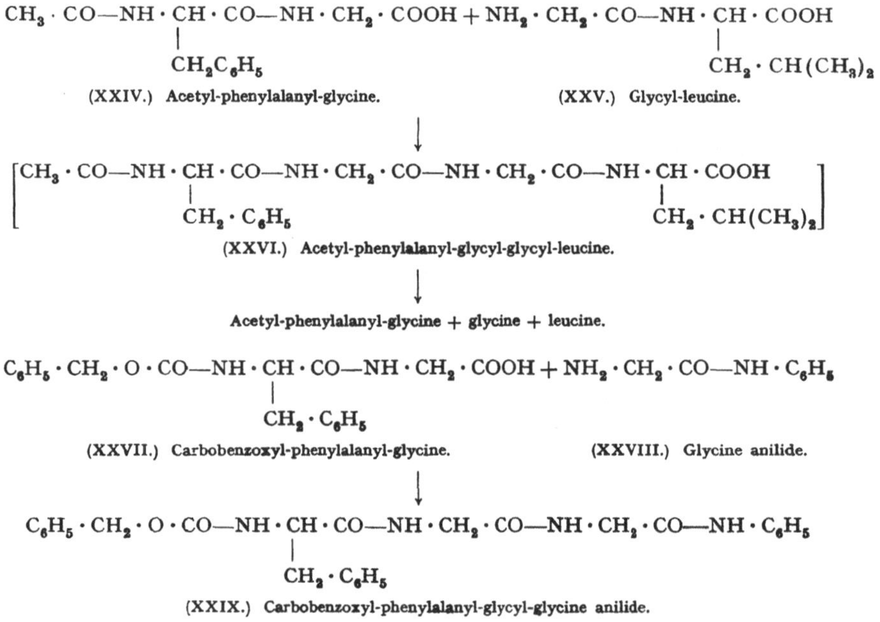

$$CH_3 \cdot CO-NH \cdot \underset{\underset{CH_2C_6H_5}{|}}{CH} \cdot CO-NH \cdot CH_2 \cdot COOH + NH_2 \cdot CH_2 \cdot CO-NH \cdot \underset{\underset{CH_2 \cdot CH(CH_3)_2}{|}}{CH} \cdot COOH$$

(XXIV.) Acetyl-phenylalanyl-glycine.   (XXV.) Glycyl-leucine.

↓

$$\left[CH_3 \cdot CO-NH \cdot \underset{\underset{CH_2 \cdot C_6H_5}{|}}{CH} \cdot CO-NH \cdot CH_2 \cdot CO-NH \cdot CH_2 \cdot CO-NH \cdot \underset{\underset{CH_2 \cdot CH(CH_3)_2}{|}}{CH} \cdot COOH \right]$$

(XXVI.) Acetyl-phenylalanyl-glycyl-glycyl-leucine.

↓

Acetyl-phenylalanyl-glycine + glycine + leucine.

$$C_6H_5 \cdot CH_2 \cdot O \cdot CO-NH \cdot \underset{\underset{CH_2 \cdot C_6H_5}{|}}{CH} \cdot CO-NH \cdot CH_2 \cdot COOH + NH_2 \cdot CH_2 \cdot CO-NH \cdot C_6H_5$$

(XXVII.) Carbobenzoxyl-phenylalanyl-glycine.   (XXVIII.) Glycine anilide.

↓

$$C_6H_5 \cdot CH_2 \cdot O \cdot CO-NH \cdot \underset{\underset{CH_2 \cdot C_6H_5}{|}}{CH} \cdot CO-NH \cdot CH_2 \cdot CO-NH \cdot CH_2 \cdot CO-NH \cdot C_6H_5$$

(XXIX.) Carbobenzoxyl-phenylalanyl-glycyl-glycine anilide.

added to the reaction mixture; and the final products are glycine, leucine and acetylphenylalanyl-glycine (XXIV). Acetylphenylalanyl-glycyl-glycyl-leucine (XXVI) was postulated as an intermediate. As evidence for such an intermediate the following was cited. Carbobenzoxy-*L*-phenylalanyl-glycine (XXVII) + glycine anilide (XXVIII) give carbobenzoxy-*L*-phenylalanyl-glycyl-glycine anilide (XXIX), which is so insoluble that it precipitates out soon after it is formed and so is not hydrolysed further. It was argued that replacing the carbobenzoxy residue by acetyl rendered the intermediate peptide too soluble, and hence the reaction proceeded to hydrolysis. The difference between glycyl-leucine and glycine anilide was not taken into account in this argument, and was, presumably, considered as unimportant here. The difference may very well be significant, not only with respect to enzyme specificity and solubility of the intermediate but also to the free energy change in the subsequent formation of the peptide bond.

d) Plastein Formation.

These examples of peptide linkage may be considered as analogous to plastein formation. Plastein formation is a phenomenon that has been known for a long time. It consists in the enzymatic synthesis of an insoluble, high-molecular peptide, or mixture of such, from a concentrated partial (peptic) hydrolysate of protein. The enzyme usually employed for this reversal of hydrolysis has been pepsin [BORSOOK (*16*), WASTENEYS and BORSOOK (*199*)]. FOLLEY (*59*) questioned whether peptide bonds were reconstituted in plastein formation [see also ECKER (*53*)]. SALTER and PEARSON (*144*) and COLLIER (*46, 47*) confirmed the original findings; COLLIER employed papain as well as crystalline pepsin. NORTHROP (*128*) was inclined to the view that peptide bonds were synthesized although the protein formed was, of course, different from the original source of the partial hydrolysate from which the plastein was formed.

VIRTANEN and co-workers (*189–192*) have recently taken up the study of plastein formation. They found that in plasteins the free amino nitrogen was two to three per cent of the total nitrogen, and estimated that, on the average, the plastein contained 40 amino acid residues. Cryoscopic and viscosity measurements gave molecular weights that varied from 2500 to 10000. This and the low free amino nitrogen VIRTANEN *et al.* interpreted as indicating that plastein consists of cyclic peptides. No plastein was obtained when the initial reaction mixture consisted of either free amino acids or a mixture of di- and tri-peptides. Beginning with peptides of greater complexity, plastein formation occurred.

TAUBER (*179–182*) observed plastein formation in peptic digests of several proteins with chymotrypsin as enzyme (trypsin is inactive in the synthesis). Hitherto, plastein formation had always been carried out in the neighborhood of pH 4; with chymotrypsin the reaction proceeds rapidly at pH 7.3. "The average molecular weights of the synthetic products as determined in the analytical centrifuge are estimated to be in the range 250000–400000." As in the case of plasteins formed by pepsin, those formed by chymotrypsin are hydrolysed in dilute suspension by pepsin, chymotrypsin or trypsin.

In plastein synthesis the free energy of formation of the peptide bonds is small, as attested to by the reversal of hydrolysis merely by concentrating certain enzymatic hydrolytic products; the synthetic product is insoluble which tends to drive the reaction towards synthesis. In these two respects plastein formation resembles reactions of the type, benzoyl-L-leucine + L-leucine anilide → benzoyl-L-leucyl-leucine anilide (XIX).

Whatever the biological significance of plastein formation may be (and on this point there is at present no evidence *pro* or *con*), the phenomenon is interesting in that it indicates that the condensation of certain large peptides entails only a small free energy change. No doubt enzyme—substrate specificity as well as energy relations are involved. As stated above some of the reactions studied by BERGMANN and his collaborators with synthetic substrates appear to be analogous to plastein formation. Synthetic substrates were first introduced by BERGMANN in the study of protease and peptidase action in general, and it is their use that has elucidated the action mechanism of these enzymes. It may well be that the use of synthetic substrates may, similarly, clarify not only plastein formation, but also the configurational requirements for condensation by proteases of certain, not necessarily very large, peptides, *in vivo* and *in vitro*.

e) Peptide Synthesis in an Exchange Reaction during Hydrolysis (Transamidation and Transpeptidation).

The third category of reactions discovered by BERGMANN involving proteases is an exchange during hydrolysis. The first example found was when hippurylamide (XXX) was hydrolysed by papain in the presence

$$C_6H_5 \cdot CO \cdot NH \cdot CH_2 \cdot CONH_2$$
(XXX.) Hippurylamide.

of aniline (*10*). Under these conditions hippuric acid anilide (XVI, p. 323) was formed faster than from hippuric acid and aniline, indicating some direct replacement of the amide group by aniline. Another example was the formation of benzoyl-L-leucine anilide (XVII) from benzoyl-L-leucine and glycine anilide; here the glycine residue in glycine anilide was replaced directly by benzoyl-L-leucyl (*11*).

This class of reactions has been investigated further and extended in a new direction by FRUTON and his collaborators. FRUTON (*68*) digested benzoyl-glycine amide with cysteine-activated papain in the presence of $N^{15}H_3$. At all stages before complete hydrolysis, the amide radical in the unhydrolysed benzoyl-glycine amide (XXXI) was found to be

$$C_6H_5 \cdot CO-NH \cdot CH_2 \cdot CONH_2 \xrightarrow{\overset{*}{N}H_3} C_6H_5 \cdot CO-NH \cdot CH_2 \cdot CO\overset{*}{N}H_2$$

(XXXI.) Benzoyl-glycine amide.   (* = N^{15})

partially replaced by the $N^{15}H_3$ from the medium. The substrate was practically completely hydrolysed by the enzyme, yet the amount of replacement was greater than the calculated equilibrium amount. But the degree of replacement was always much less than the degree of hydrolysis. The highest degree of replacement reported [JOHNSTON *et al.* (*93*)] was 3.5% (theoretical maximum 50%) when 41% hydrolysis had occurred. When there was no hydrolysis there was no replacement.

It was found that the optimum pH for replacement was several units higher than that for hydrolysis. At pH 7.9 they found a rough proportionality between hydrolysis and replacement, the degree of replacement always much less than that of hydrolysis. The data in Table 15 (p. 322) that a high pH tends to favor peptide synthesis, where the ionization constant of the amino group of the peptide is higher than that of the constituent amino acids. This may account for the observation that interchange in the course of hydrolysis was greater at higher pH values. As far as the actual amount of synthesis is concerned, both the free energy data and the observations of FRUTON *et al.* indicate that this effect is small.

An analogous reaction was found to be catalysed by activated papain between acylamino acid amides and hydroxylamine to form hydroxamic acids: $R \cdot CONH_2 + NH_2OH \rightleftharpoons R \cdot CONHOH + NH_3$. Substrates with hydroxylamine were: benzoyl-*L*-arginine amide (XXXII), carbobenzoxy-

$$C_6H_5 \cdot CO-NH \cdot \underset{\underset{CONH_2}{|}}{CH} \cdot CH_2 \cdot CH_2 \cdot NH \cdot C{=}NH$$
$$\phantom{C_6H_5 \cdot CO-NH \cdot CH \cdot CH_2 \cdot CH_2 \cdot NH \cdot C{=}} \diagdown NH_2$$

(XXXII.) Benzoyl-arginine amide.

L-isoglutamine (XXXIII), carbobenzoxy-*L*-isoasparagine (XXXIV), benzoyl-glycine amide (XXXI), carbobenzoxy-*L*-serine amide (XXXV), carbobenzoxy-*L*-methionine amide (XXXVI), and carbobenzoxy-*D*-methionine amide. The *D*-compound was neither hydrolysed nor did it yield the hydroxamic acid. In the case of all the others there was a proportionality between the degree of hydrolysis and hydroxamic acid formation. A lower pH favored hydrolysis; a higher pH, hydroxamic acid formation, except

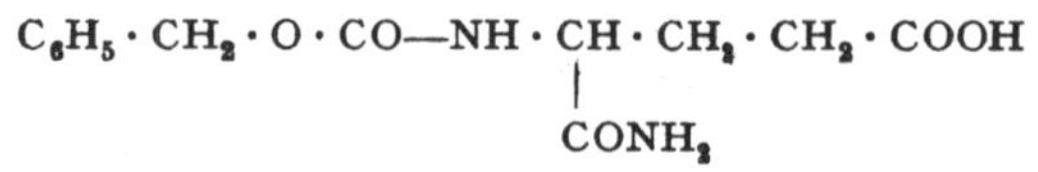

$C_6H_5 \cdot CH_2 \cdot O \cdot CO-NH \cdot CH \cdot CH_2 \cdot CH_2 \cdot COOH$
|
$CONH_2$

(XXXIII.) Carbobenzoxy-isoglutamine.

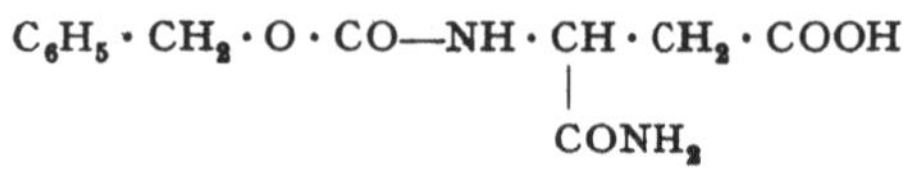

$C_6H_5 \cdot CH_2 \cdot O \cdot CO-NH \cdot CH \cdot CH_2 \cdot COOH$
|
$CONH_2$

(XXXIV.) Carbobenzoxy-isoasparagine.

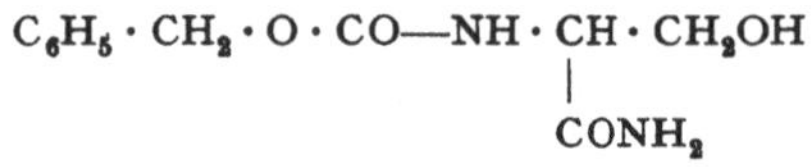

$C_6H_5 \cdot CH_2 \cdot O \cdot CO-NH \cdot CH \cdot CH_2OH$
|
$CONH_2$

(XXXV.) Carbobenzoxy-serine amide.

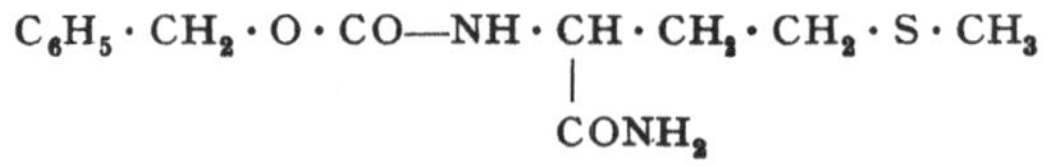

$C_6H_5 \cdot CH_2 \cdot O \cdot CO-NH \cdot CH \cdot CH_2 \cdot CH_2 \cdot S \cdot CH_3$
|
$CONH_2$

(XXXVI.) Carbobenzoxy methionine amide.

in the case of benzoyl-L-arginine amide. The authors report that they have found similar results with cathepsin as enzyme. The hydroxamic acid was always a small fraction of the ammonia liberated from the amide.

FRUTON proposed that the mechanism of hydrolysis of a peptide or amide bond involves "activation" of the C=O bond in which step it is converted to —C—OH, and that free ammonia, hydroxylamine, a free amino group of an amino acid peptide or amide, or water compete for the unsubstituted bond on the "activated" carbonyl group. When the addition is with water, there is hydrolysis; when it is with any of the other radicals or compounds, there is replacement. Evidence in favor of this mechanism was found (94) in the following reaction with chymotrypsin. Benzoyl-1-tyrosyl-glycine amide ("BTGA") was incubated with glycine amide containing N^{15} in the glycine nitrogen. After

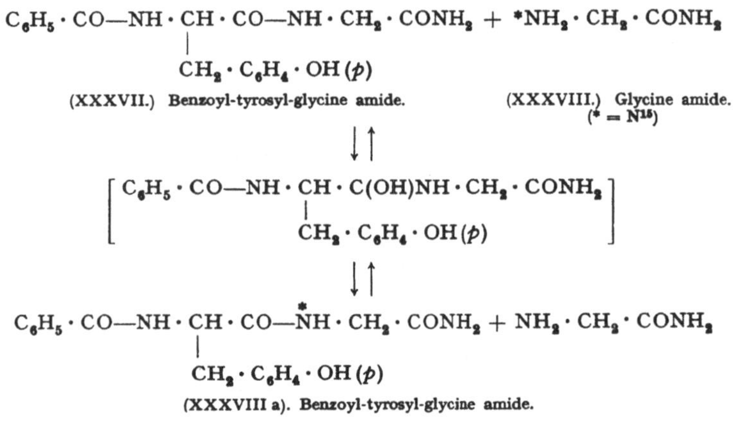

$C_6H_5 \cdot CO-NH \cdot CH \cdot CO-NH \cdot CH_2 \cdot CONH_2 + {}^*NH_2 \cdot CH_2 \cdot CONH_2$
|
$CH_2 \cdot C_6H_4 \cdot OH\,(p)$

(XXXVII.) Benzoyl-tyrosyl-glycine amide.   (XXXVIII.) Glycine amide.
(* = N^{15})

↓↑

$\left[\begin{array}{l} C_6H_5 \cdot CO-NH \cdot CH \cdot C(OH)NH \cdot CH_2 \cdot CONH_2 \\ | \\ CH_2 \cdot C_6H_4 \cdot OH\,(p) \end{array} \right]$

↓↑

$C_6H_5 \cdot CO-NH \cdot CH \cdot CO-\overset{*}{N}H \cdot CH_2 \cdot CONH_2 + NH_2 \cdot CH_2 \cdot CONH_2$
|
$CH_2 \cdot C_6H_4 \cdot OH\,(p)$

(XXXVIII a). Benzoyl-tyrosyl-glycine amide.

some time, when 42% hydrolysis of BTGA had occurred, the remaining BTGA was isolated and N^{15} was found in the tyrosyl-glycine peptide bond. The amount of N^{15} indicated 17% replacement of this nitrogen. The reaction is formulated in the scheme (p. 333), the postulated intermediate being enclosed in brackets (XXXVII, XXXVIII).

This conversion was designated as a transpeptidation reaction since the exchange occurred at the peptide bond*.

FRUTON's hypothesis regarding the mechanism of this and similar transfer reactions can be formulated in another way. Instead of figuring the existence of an activated carbonyl group as a free radical, as it were, the reaction may be considered as, first, enolization of the peptide (or amide) bond and cleavage on the enzyme surface with the unsatisfied valence bond of the carbonyl group attached to the enzyme in such manner that all, or nearly all, of the free energy of the original peptide (or amide) bond now resides in the carbonyl-enzyme bond. This bond can then be broken by water, resulting in hydrolysis, by free ammonia or hydroxylamine as in transamidation, or by the NH_2 group of a substituted amino acid as in transpeptidation. Envisaged in this manner the mechanism is analogous to disaccharide and glycogen synthesis in which the OH of a phosphate radical participates in the condensation, with the liberation of inorganic phosphate instead of water as in the hydrolysis of amides and peptides.

Further experiments with chymotrypsin indicated that there is not only a substrate specificity for hydrolysis but also for transfer (replacement) reactions. Under the conditions in which the above transpeptidation occurred, benzoyl-L-tyrosine amide incubated with $N^{15}H_4NO_3$ underwent a negligible replacement of the amide radical by $N^{15}H_2$ from the medium, although 28% hydrolysis of the amide had occurred.

Still another type of transfer reaction during hydrolysis has been reported by HANES et al. (80). Glutathione (XXXIX) was submitted to the action of a kidney extract containing 0.001 M Mg^{++} and in addition either leucine, phenylalanine (XVII) or valine (XL). The initial concentration of glutathione and of the amino acids was 0.05 M. After some hydrolysis of the glutathione filter paper chromatography of the hydrolysate indicated that a small amount of γ-glutamylpeptides of the

---

* In a paper wich became available after the manuscript was submitted for publication DOBRY, FRUTON and STURTEVANT [J. biol. Chem. *195*, 149 (1952)] present data that $-\widetilde{\Delta F}$ for the reaction, N-benzoyltyrosine + glycine amide → → N-benzoyl-tyrosyl-glycine amide is — 361 calories at 37.5°, corresponding to an equilibrium constant of 0.558. At the concentrations of reactants used about 5 per cent synthesis occurs by mass action. These data take away from the necessity of the intermediate proposed by FRUTON, since the conventional reversibility of mass action could account for the interchange observed.

added amino acid had been formed. As far as the published evidence permits estimation, it appears that the degree of hydrolysis of the glutathione was greater than the formation of the glutamylpeptide. The authors

$$\text{HOOC} \cdot \text{CH} \cdot \text{CH}_2 \cdot \text{CH}_2 \cdot \text{CO—NH} \cdot \text{CH} \cdot \text{CO—NH} \cdot \text{CH}_2 \cdot \text{COOH}$$
$$| \qquad\qquad\qquad\qquad |$$
$$\text{NH}_2 \qquad\qquad\qquad\qquad \text{CH}_2\text{SH}$$

(XXXIX.) Glutathione (γ-glutamyl-cysteinyl-glycine).

$$(\text{CH}_3)_2\text{CH} \cdot \text{CH} \cdot \text{COOH}$$
$$|$$
$$\text{NH}_2$$

(XL.) Valine.

proposed that in the cleavage of the cysteinyl-glycine residue of the glutathione the γ-carboxyl of the glutamic acid condenses with the free amino acid that may be in solution. It appears that here, as in the reactions described by FRUTON et al. (68, 93, 94), there is a carbonyl activation in the course of hydrolysis, and amino acids can, to some extent, compete with water for the activated carbonyl group.

HANES et al. (80) obtained active enzyme extracts from pig kidney, the acetone powder of pig kidney, and fresh ox pancreas. Extracts of liver were inactive. It is pertinent to recall in this connection that the amino acid turnover of glutathione in liver is very active [WAELSCH and RITTENBERG (197)]. In their studies on the enzymatic synthesis of glutathione, JOHNSTON and BLOCH (92) found that the synthesizing enzyme was present in the liver of a number of animals but was absent from rabbit kidney and rabbit intestine. The enzyme system used by HANES et al. was present in kidney and absent from liver. Furthermore JOHNSTON and BLOCH found that, in their preparations, the enzyme activity which was responsible for the hydrolysis of glutathione was not involved in, and could be separated, from the enzyme system taking part in the synthesis.

In accord with the foregoing observations ZAMECNIK and FRANTZ (213), and FRANTZ and LOFTFIELD (61) reported that, when glycyl-glycine (XLI)

$$\text{NH}_2 \cdot \text{CH}_2 \cdot \text{CO—NH} \cdot \text{CH}_2 \cdot \text{COOH}$$

(XLI.) Glycyl-glycine.

is hydrolysed by a dipeptidase in the presence of C^{14}-labeled glycine, at equilibrium the theoretical amount (as indicated by the free energy data) of the labeled glycine is found in the dipeptide; but, when the reaction is stopped short of equilibrium, the amount of labeled glycine in the dipeptide is greater than the calculated equilibrium amount. Similar findings were obtained with a carboxypeptidase, but, as in the cases studied by FRUTON et al. (68, 93), the replacement in the unhydrolysed dipeptide was low.

The foregoing observations have revived interest in BERGMANN's hypothesis that peptidases and proteases may *in vivo* promote peptide synthesis, and support it to the extent that they have shown that peptidases may give rise, in the course of hydrolysis, to new and different peptides. The findings of FRUTON *et al.* have shown that replacement may occur at a normal peptide linkage, of HANES *et al.* that a free amino acid may be incorporated into a peptide bond. It should be noted that in the latter case the peptide involved contains a γ-carboxyl, and not a carboxyl α to an amino group. That peptidases are not involved in the incorporation of labeled amino acids into tissue proteins is indicated by the finding that metal ions, such as cobalt, copper, iron, manganese, and nickel, on which the activity of a number of peptidases has been shown to depend, [SMITH (*164*)], were found to be either inhibitory or without any effect, [BORSOOK *et al.* (*19, 20*)]. As far as peptidases and proteases promoting protein synthesis by tranfer is concerned it must be noted that in the above experiments the amount of new peptide formed was always much less than the initial substrate hydrolysed; the new peptide was not more complex than that hydrolysed; and where a free amino acid was incorporated into a new peptide, it was always terminal. On the other hand when glycine, histidine, leucine or lysine were incorporated (*in vitro*) into the proteins (mainly hemoglobin) of rabbit reticulocytes (*25*), none of the incorporated amino acid was found in an end position (at least the NH_2 group was not free); and further all four amino acids could be incorporated simultaneously without competing with each other. The fact that every amino acid which has been tested has been found to be incorporated into tissue proteins argues strongly against most of the incorporated amino acids being terminal.

f) Peptide Synthesis from Amino Acid Esters.

A new type of enzymatic synthesis of peptides has been found by BRENNER *et al.* (*33*). They incubated *DL*-methionine or *DL*-threonine esters with chymotrypsin and found that, in addition to the hydrolysis of the ester, after one or two days incubation, a large amount of a mixture of peptides (XLII) of varying molecular weight was formed. The usual

$$NH_2 \cdot \underset{R}{CH} \cdot CO-[NH \cdot \underset{R}{CH} \cdot CO]_x-NH \cdot \underset{R}{CH} \cdot COOR'$$

(XLII.) Methionine peptides ($R = CH_2 \cdot CH_2 \cdot S \cdot CH_3$; $R' = H$ or alkyl; $x = 3$ to 10^5).

reaction mixture consisted of 100 parts of ester and five parts of water. The reaction stops when the reaction product reaches so high a molecular weight that it becomes insoluble. No peptide formation occurred with only free methionine in the reaction mixture. Beginning with *DL*-methionine-isopropyl ester they isolated from the reaction mixture L-methionyl-L-methionine. Higher peptides were left unidentified (*34*). Similarly from DL-threonine-isopropylester they obtained, after partial hydrolysis of a mixture of more complex peptides, L-threonyl-L-threonine (*35*). Evidently chymotrypsin acts only on the L-form of the ester and amino acids.

The course of the reaction was followed by filter paper chromatography, and, from the evidence thus obtained, the authors concluded that the lower-molecular peptides were formed first, and these then reacted with the ester lengthening the chain. Accordingly, their view of the mechanism of the reaction is that some of the ester is hydrolysed first; the free amino acid then condenses with unhydrolysed ester, with the eliminiation of the alcohol and the formation of the peptide. The high-molecular weight peptide is slowly hydrolysed by chymotrypsin. In no case did the authors observe peptide synthesis without simultaneous appearance of free amino acids. In the case of methionine ester, one part (by weight) of methionine was liberated as the free amino acid. With threonine esters relatively more peptide was formed. With phenylalanine, tyrosine, and tryptophane esters, there was more free amino acids. Most of the work was done with methionine esterified with a wide variety of aliphatic and aromatic alcohols.

This type of peptide synthesis by a protease [it could also be called an esterase, chymotrypsin is an esterase (*95–97, 157, 165*)] is, then, to be added to those found by BERGMANN and by FRUTON. All its features are in accord with the transpeptidation reaction mechanism proposed by FRUTON (XXXVII—XXXVIII, p. 330). There is no information, at present, on the biological significance of this type of synthesis of peptides *via* amino acid esters.

5. Glutamo- and Asparto-Transferases.

The transamidation reactions involving either ammonia or hydroxylamine discussed so far may be characterized as follows. The amino acid amide involved can be glycine or other amino acids, the acceptor is not restricted to glutamic or aspartic acid; the transfer depends on, and its degree is less than, that of the concomitant hydrolysis; it does not depend on adenosine triphosphate (ATP) or adenosine diphosphate (ADP); it is not coupled with an energy or phosphate transfer system. Another type of transamidase reaction was discovered by WAELSCH *et al.* and shortly thereafter the field was broadened by STUMPF *et al.* The characteristics which distinguish it from the transamidation reactions studied by FRUTON *et al.* appear to be as follows. The amide (or hydroxylamine) transfer is restricted to the β-carboxyl of aspartic acid or the γ-carboxyl of glutamic acid; the transfer is independent of hydrolysis of the amide grouping; it does depend on ATP or ADP, arsenate or phosphate, and Mn^{++}. It shares with the other type of transamidation its independence of energy or phosphate transfer, notwithstanding its dependence on ATP or ADP, arsenate or phosphate, and Mn^{++}.

WAELSCH et al. (153, 195) first found the enzyme in the washed cells and cell extracts of a number of micro-organisms. The reactions catalysed may be represented schematically as follows (* designates N^{15}):

(a) glutamo $-NH_2 + \overset{*}{N}H_4OH \rightarrow$ glutamo $-\overset{*}{N}H_2 + NH_4OH$;

(b) glutamo $-NH_2 + NH_2OH \rightarrow$ glutamo $-NHOH + NH_3$.

Ammonia and a number of (but not all) amino acids in high concentration are inhibitory. The transfer of ammonia was proved by using $N^{15}H_3$ in the medium and finding the labeled nitrogen, after incubation with the enzyme, in the amide radical (196). The enzyme extract used effected the transfer between asparagine and glutamine and ammonia and hydroxylamine but with no other amides. The reaction was not inhibited by cyanide, fluoride or iodoacetate. Acetone powder prepared from extracts of mammalian tissues also were effective; they required activation by Mn^{++}, and their activity was enhanced further by phosphate; they had no glutaminase activity (193). GROSSOWICZ et al. (79) found that addition of ATP or ADP was without effect; but this was probably because the enzyme system had not been sufficiently purified to show this dependence (see below).

STUMPF et al. (52, 175, 176) found a similar enzyme in pumpkin seedlings. After purifying the enzyme the following activating agents were found to be necessary: Mn^{++}, ATP or ADP, and arsenate or phosphate. Insufficiently purified enzyme preparations did not show the ATP or ADP dependence, resembling in this respect the enzyme preparations used by WAELSCH. The latter's enzyme preparation resembles that of STUMPF so closely in all other respects that it seems highly probable that it too, when sufficiently purified, will show dependence on ATP or ADP. In the purified enzyme from pumpkin seedlings arsenate was two to three times as effective as phosphate; and addition of phosphate to a reaction medium containing arsenate reduced the activity, indicating a competition between arsenate and phosphate for the enzyme locus. Experiments designed to detect phosphate transfer between the medium and ATP or ADP could not find that any had occurred, nor was any evidence found of a glutamylphosphate intermediate. Only glutamine was reactive with the pumpkin seedling enzyme; asparagine, a number of other amides including nicotinamide and coenzyme I and also glutathione were inactive. As with the enzyme studied by WAELSCH et al., iodoacetate and cyanide were not inhibitory, nor were azide, diisopropyl-fluorophosphate, dinitrophenol or malonate. Ammonia in concentrations 10^{-2} M, and 5×10^{-2} M, and certain amino acids, inhibited 50 and 100%, respectively, resembling also in this respect the transferases of WAELSCH *et al.*

WAELSCH (*194*) in a recent review summarized his views on the possible biological significance of these transferases as follows:

> "It is probable that neither hydroxylamine nor ammonia is the biological substrate of the transfer reaction. Possible substrates appear to be those amino acids whose natural isomers have been found to inhibit the formation of glutamohydroxamic acid by glutamotransferase. Participation of amino acids in the exchange reaction would lead to the formation of γ-glutamyl or β-aspartyl peptides. The wide distribution of the glutamo — and asparto — transferase potencies gives support to the hypothesis that the naturally occuring amides, glutamine and asparagine, are implicated in the mechanism of peptide formation."

Of course, it is too early to say, but it would appear that the reservations regarding the function of the transfer enzyme of HANES *et al.*, set out above, apply also to the asparto- and glutamo-amide transferases.

6. Syntheses where $-\Delta F$ is Negative and Large, Coupled with High Energy Phosphate.

a) *Synthesis of Glutamine.*

There is a third category of enzymatic synthesis of glutamine in which glutamine is built up from glutamic acid and ammonia. The free energy of formation of the amide bond is of the same order of magnitude as that of simple peptides from amino acids. This category is distinguished from the previous two as follows: It is not a transfer reaction, but a synthesis; nor is it accompanied by any hydrolysis; the γ-carboxyl of glutamic acid is the only acceptor of the free base (ammonia, hydroxyl-

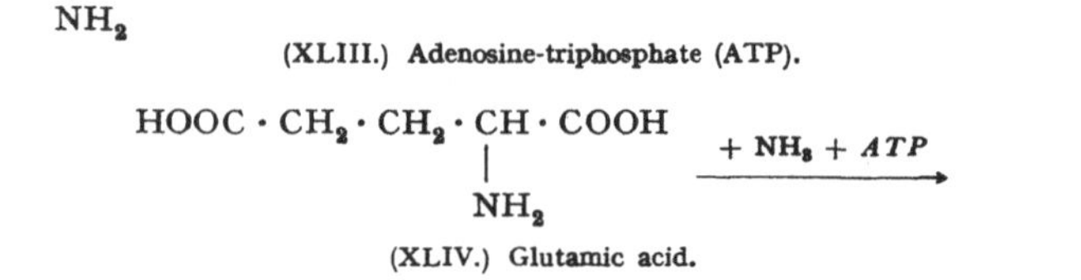

(XLIII.) Adenosine-triphosphate (ATP).

$$HOOC \cdot CH_2 \cdot CH_2 \cdot CH(NH_2) \cdot COOH \xrightarrow{+ NH_3 + ATP}$$

(XLIV.) Glutamic acid.

$$\longrightarrow NH_2 \cdot CO \cdot CH_2 \cdot CH_2 \cdot CH(NH_2) \cdot COOH + ADP + H_3PO_4$$

(XLIV a). Glutamine.

amine, hydrazine or methylamine); it depends on adenosinetriphosphate (ATP) (XLIII), but this is a coupled reaction in which inorganic phosphate is liberated from ATP in an amount equivalent to the amide synthesized; it is inhibited by fluoride in low concentration. Like the asparto- and glutamo-transferases it requires either Mg^{++} (optimum concentration 0.01 M) or Mn^{++} (optimum concentration 0.002 M); but the maximum activation with Mn^{++} is only half that with Mg^{++}.

This synthesis of glutamine was elucidated, and the enzyme somewhat purified, simultaneously and independently by SPECK (*167–169*) and ELLIOTT (*55–56*). The reaction mechanism is given in (XLIV, XLIVa). The enzyme was found in the brain, liver and kidney of a number of animals and in *Staphylococcus aureus*. Active cell-free extracts and acetone powders of these were prepared. Adenosine monophosphate (AMP) and adenosine diphosphate (ADP) could not replace ATP. Intermediates, such as glutamyl-phosphate were looked for, but none was detected. For maximum activity a reducing agent, e. g. cysteine, is required, presumably the activity of the enzyme involves sulfhydryl groups. In addition to fluoride two other inhibitors were found, methionine-sulfoxide and crystal violet. The methionine-sulfoxide competes with glutamate for the locus of activity on the enzyme.

b) Synthesis of Hippuric Acid.

Peptides may be viewed as acylamides; there are a number of such amides of great biological importance; in addition to asparagine and glutamine there may be cited acetylcholine, acetylsulfonamide, and citrulline. The mechanisms by which they are synthesized may be, and have been considered as being possibly simple models of the more complicated processes of amino acid incorporation into proteins and of protein biosynthesis. Among themselves these mechanisms have certain similarities, but also differences. In the hope of obtaining possibly a better vantage point for considering amino acid incorporation into proteins and protein biosynthesis, we shall consider in this review the *in vitro* synthesis of some quasi- and actual peptides. The enzymatic mechanisms by which the immediate precursors of the latter compounds are condensed are of course interesting, indeed, important in themselves.

It has long been known that animals form hippuric acid from benzoic acid and glycine. Hippuric acid is only one of a number of such quasi-peptide syntheses known to occur. The free energy of formation of the —CO·NH— bond of hippuric acid (XVI) is of the same order of magnitude as that of simple peptides (Table 14, p. 321). The synthesis of hippuric acid from benzoic acid and glycine was first attained *in vitro* with kidney and liver slices (dog, guinea pig, rabbit, and rat), [BORSOOK and DUBNOFF (*26*)]. In the course of some hours 60–75% of added benzoic acid and glycine were condensed to hippuric acid, although the thermodynamic equilibrium point was at less than 1% synthesis. Obviously, so high a yield of hippuric acid could not have occurred by simple mass action, there must have been coupling with an energy yielding reaction. Accordingly, as was to be expected, inhibition of respiration by 0,001 M potassium cyanide also completely inhibited the synthesis, as did also anaerobiosis. It was found possible later to carry out the reaction in a guinea pig liver homogenate; and indirect evidence was obtained that ATP was involved, i. e. that the free energy driving the reaction came from the high energy phosphate bond of ATP, which in turn was formed by oxidative phosphorylation (27). In the course of these studies N-phosphorylated glycine and benzoyl-phosphate were tested as possible intermediates, but were found to be no more active or less active than glycine and benzoic acid (28).

c) Synthesis of p-Aminohippuric Acid.

A more searching study was begun at about this time and independently by COHEN and MCGILVERY (42) of an analogous reaction, the synthesis of p-aminohippuric acid (PAH) (XLV) from p-aminobenzoic acid (PAB) and glycine. PAB is a constituent of naturally occuring peptides (7, 138). COHEN and MCGILVERY first used slices of rat liver and kidney; their findings were, in general, similar to those in the synthesis of hippuric acid. Contributing toward insight into the coupled energy yielding reaction they found the following inhibitions, expressed as per cent: arsenite (0.01 M) 95; azide (0.001 M) 44; fluoride (0.01 M) 90; HCN (0.001 M) 99; iodoacetate (0.01 M) 97; and malonate (0.001 M) 35.

(p) $H_2N \cdot C_6H_4 \cdot CO-NH \cdot CH_2 \cdot COOH$ $HOOC \cdot CH_2 \cdot NH-CO \cdot CH_3$
  (XLV.) p-Aminohippuric acid.   (XLVI.) N-Acetyl-glycine.

The authors then succeeded in carrying out the reaction in rat liver homogenates (43), and in this enzyme system were able to add the following information regarding the mechanism: The optimum pH is 7.5, replacement of Na^+ by K^+ was stimulating, Ca^{++} was inhibitory, and Mg^{++} was stimulating. Addition of cytochrome C was necessary for maximum synthesis. Addition of ATP ($3 \times 10^{-3} M$) permitted some synthesis anaerobically, but much less than under oxygen without added ATP. Addition of a number of metabolites whose oxidation is known to lead to ATP formation increased the yield of PAH somewhat or permitted the reaction to proceed for a longer time. Hexose-diphosphate, Coenzyme I, and co-carboxylase were without effect, nicotinamide was inhibitory, N-acetylglycine (XLVI) was inactive. Glyoxalate and ammonia could not substitute for glycine. One of the general mechanisms suggested by RITTENBERG and SHEMIN (142), for protein synthesis is the formation of N-acylated, specifically N-acetylated, amino acids as intermediates; there is sufficient chemical potential as a consequence of the acylation to condense the nitrogen with the carboxyl of another amino acid. The inactivity of N-acetylated glycine (XLVI) definitely excludes this mechanism in the synthesis of p-aminohippuric acid; and tends, also, to exclude condensation of a ketoacid derivative of an amino acid with the nitrogen of another amino acid as an intermediate in peptide synthesis [HERBST and SHEMIN (84)].

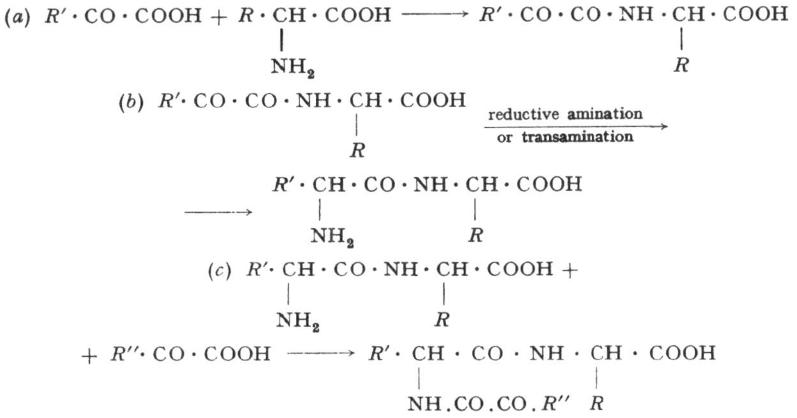

$$
\begin{aligned}
&(a)\ R' \cdot CO \cdot COOH + R \cdot \underset{\underset{NH_2}{|}}{CH} \cdot COOH \longrightarrow R' \cdot CO \cdot CO \cdot NH \cdot \underset{\underset{R}{|}}{CH} \cdot COOH \\
&(b)\ R' \cdot CO \cdot CO \cdot NH \cdot \underset{\underset{R}{|}}{CH} \cdot COOH \xrightarrow[\text{or transamination}]{\text{reductive amination}} \\
&\longrightarrow R' \cdot \underset{\underset{NH_2}{|}}{CH} \cdot CO \cdot NH \cdot \underset{\underset{R}{|}}{CH} \cdot COOH \\
&(c)\ R' \cdot \underset{\underset{NH_2}{|}}{CH} \cdot CO \cdot NH \cdot \underset{\underset{R}{|}}{CH} \cdot COOH + \\
&+ R'' \cdot CO \cdot COOH \longrightarrow R' \cdot \underset{\underset{NH.CO.CO.R''}{|}}{CH} \cdot CO \cdot NH \cdot \underset{\underset{R}{|}}{CH} \cdot COOH
\end{aligned}
$$

(d) Repeated reamination, prolonging the chain.

(XLVII.) Polypeptide formation by condensation of ketoacid derivative with amino group of amino acid or peptide [proposed by HERBST and SHEMIN (84)].

In a still later paper COHEN and MCGILVERY (44) purified the enzyme further and then found that ATP in the absence of oxidizable metabolite promoted p-aminohippuric acid synthesis anaerobically, that adenosinemonophosphate was active even aerobically only under conditions in which it was phosphorylated to ATP, that N-phosphoglycine was inactive, as were Coenzymes I and II, cocarboxylase and pyridoxalphosphate. Recently, CHANTRENNE (41) has demonstrated that Coenzyme A as well as ATP are needed for hippuric acid synthesis. He found also that benzoylphosphate was unable to replace benzoic acid plus ATP, thus excluding the formation of free benzoylphosphate as an intermediate. The mechanism which has been suggested [BARKER (8)] is that the reactive intermediate is benzoyl-Coenzyme A, which on reaction with glycine forms hippuric acid and free Coenzyme A. The ATP which also is necessary, may participate in the formation of benzoyl-Coenzyme A, or in the activation of glycine, or in the transfer reaction.

d) Synthesis of Ornithuric Acids.

MCGILVERY and COHEN (120) later extended their studies to the formation of ornithuric acids from α-, and δ-benzoyl-L-ornithine and

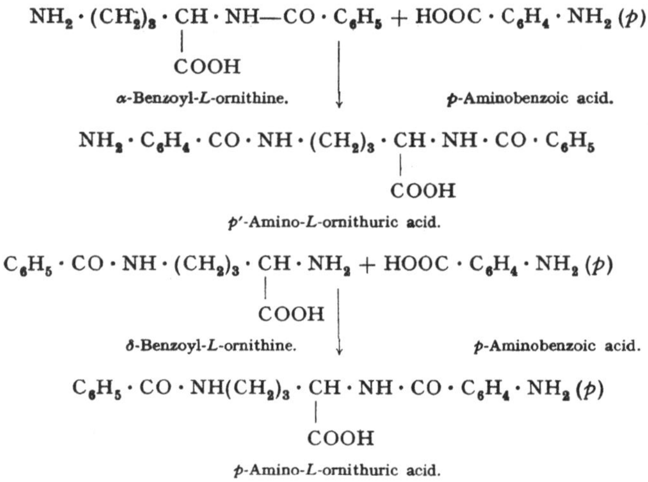

$$NH_2 \cdot (CH_2)_3 \cdot \underset{COOH}{CH} \cdot NH - CO \cdot C_6H_5 + HOOC \cdot C_6H_4 \cdot NH_2 \, (p)$$

α-Benzoyl-L-ornithine. ↓ p-Aminobenzoic acid.

$$NH_2 \cdot C_6H_4 \cdot CO \cdot NH \cdot (CH_2)_3 \cdot \underset{COOH}{CH} \cdot NH \cdot CO \cdot C_6H_5$$

p′-Amino-L-ornithuric acid.

$$C_6H_5 \cdot CO \cdot NH \cdot (CH_2)_3 \cdot \underset{COOH}{CH} \cdot NH_2 + HOOC \cdot C_6H_4 \cdot NH_2 \, (p)$$

δ-Benzoyl-L-ornithine. ↓ p-Aminobenzoic acid.

$$C_6H_5 \cdot CO \cdot NH(CH_2)_3 \cdot \underset{COOH}{CH} \cdot NH \cdot CO \cdot C_6H_4 \cdot NH_2 \, (p)$$

p-Amino-L-ornithuric acid.
(XLVIII.) Synthesis of ornithuric acids.

p-aminobenzoic acid [see scheme (XLVIII)]. They succeeded in obtaining enzymatic activity in an acetone powder of the sedimentable macro-particles of chicken liver homogenate, and found that the conditions for the synthesis of the ornithuric acids, and the indications of the mechanisms involved, were the same as in the synthesis of p-aminohippuric acid.

KIELLEY and SCHNEIDER (101) localized the enzyme for p-aminohippuric acid synthesis in their mouse liver homogenates as almost entirely in the mitochondria; and they adduced further evidence implicating oxidative phosphorylation leading to ATP in the synthesis of p-aminohippuric acid. The data show a relation between increase in PAH synthesis on the one hand, and an increase in high energy phosphate (ATP) on the other, but this relation, from the data reported, is hardly the "direct" parallel the authors saw in their results.

SARKAR et al. (146) confirmed the localization of the hippuric acid (XV) synthesizing enzyme in the mitochondria of rat liver homogenate, found that the synthesis was abolished by dinitrophenol (see below for the significance of this observation), and that benzoic acid can be replaced by some other aromatic and heterocylic carboxylic acids (phenylacetic and cholic acids were inactive), and that glycine could not be replaced by either β-alanine or taurine.

As mentioned above N-acetylated (in general, N-acylated) amino acids have been suggested as possible, even probable intermediates in peptide and protein biosynthesis. The positive evidence cited in support is that animals do acetylate unnatural amino acids and amines (142). This possibility, as a general mechanism, was excluded in the synthesis of p-aminohippuric acid. It was also excluded in protein synthesis in *E. coli*. SIMMONDS *et al.* (*162*) employed two mutants, one requiring L-phenylalanine, the other L-tyrosine for growth. They found that acetyl-L-phenylalanine (XLIX), and acetyl-L-tyrosine (L), even in high

$$C_6H_5 \cdot CH_2 \cdot \underset{\underset{\text{COOH}}{|}}{CH} \cdot NH \cdot CO \cdot CH_3 \qquad (p)\ HO \cdot C_6H_4 \cdot CH_2 \cdot \underset{\underset{\text{COOH}}{|}}{CH} \cdot NH \cdot CO \cdot CH_3$$

(XLIX.) Acetyl-phenylalanine.   (L.) Acetyl-tyrosine.

concentrations were unable to support growth of the respective mutants. There were similar negative or unfavorable results with dehydropeptides of phenylalanine, where the corresponding (normal) peptides were favorable, e. g. glycyl-dehydrophenylalanine did not support growth, glycyl-phenylalanine did. Thus dehydropeptides were excluded in this organism as intermediates in protein synthesis.

The foregoing studies on the synthesis of hippuric acid and its analogues have definitely implicated the high energy phosphate bond of ATP in their synthesis. But the mechanism of ATP's participation is still unknown. In the synthesis of glutamine it was shown that one mole of inorganic phosphate was formed from ATP per mole of amide synthesized. This has not been shown in the hippuric acid syntheses. In the synthesis of glutamine, once ATP was provided, it was immaterial whether the reaction was aerobic or anaerobic, or whether oxidizable metabolites and the enzyme systems for their oxidation were present or not. In the hippuric acid syntheses ATP was much more effective under aerobic conditions in the presence of oxidizable metabolites and the appropriate enzyme systems that produced high energy phosphate. The explanation offered, and it appears probable, is that the enzyme preparations used contained also an enzyme that hydrolysed ATP rapidly and hence continual and rapid re-synthesis of ATP was necessary to keep the hippuric acid synthesis going. Yet the results with the intermediates that might be expected to be formed by the action of ATP, N-phosphoglycine or benzoylphosphate, were conclusively negative.

e) Synthesis of Glutathione.

We are, at present, faced with a somewhat similar situation in the case of the enzymatic synthesis *in vitro* of glutathione (XXXIX, p. 332). This reaction has been studied by JOHNSTON and BLOCH (*92*). The method these authors employed was the incorporation of labeled glycine and glutamic acid into glutathione. The reaction proceeds in cell-free pigeon liver homogenates and in acetone powder preparations of the homogenate. Evidence was presented that, under certain conditions, a net synthesis of glutathione occurred. In other words, the extent of incorporation of labeled amino acids was several times greater than could be accounted for by amino acid exchange in the glutathione originally present in the enzyme preparation. In the most active and purest preparations, the condensation did not proceed unless either ATP or ADP were added to the reaction mixture, and data are given which indicate that one pyrophosphate bond was utilized per peptide bond synthesized. The enzyme system requires phosphate, Mg^{++}, and ATP.

SNOKE and ROTHMAN (166), in BLOCH'S laboratory, were able to inactivate the enzyme synthesizing glutathione from the three amino acids and retain the enzyme condensing γ-L-glutamyl-L-cysteine with glycine. ATP was necessary for both syntheses.

Some of the data reported by JOHNSTON and BLOCH cannot, at present, be reconciled with their interpretation that a high energy phosphate bond is coupled stoichiometrically in the synthesis of each peptide bond of glutathione. In some instances, under anaerobic conditions adenosinemonophosphate (AMP) was much more active than ATP. With their purest enzyme preparation AMP did not become active until after a lag period of 40 minutes, whereas ATP was active from the outset. The autors interpreted their findings with AMP and the positive, but somewhat lower activity of inosinic and guanylic acids as indicating that such nucleotides are converted to the corresponding pyrophosphate, e. g. AMP to ATP, by concomitant glycolysis, which their purification of the glutathione synthesizing enzymes had not eliminated sufficiently. This does not explain their finding that, in some instances, AMP was more effective than ATP.

The role of ATP is not yet clear. We have the monition of the glutamotransferases, discussed above, that require ATP or ADP, Mg^{++} or Mn^{++}, and phosphate (or arsenate), and yet there is no net utilization of high energy phosphate bond energy. What are some of the alternative possibilities if the function of ATP or ADP is not to supply the energy necessary from their pyrophosphate bonds to form reactive substrate intermediates? The glutamo-transferases suggest that either AMP, ADP, ATP, plus Mg^{++} or Mn^{++}, may be required for the proper configuration of the enzyme locus or for bringing substrates and enzyme within the reactive sphere. Another possibility, especially when one is dealing with organized particles or cells, is that AMP, ADP and ATP may have a stabilizing function on the enzyme system. And if ATP and ADP are continually being hydrolysed, continuous oxidative phosphorylation would be needed to maintain an adequate concentration of the nucleotides, which in turn would maintain the stability of the enzyme system. But neither stabilization of the enzyme nor bringing the substrates together at the reaction locus on the enzyme will suffice where free energy is needed for the synthesis. The remaining alternative is that the primary contribution of the free energy by ATP is to the enzyme rather than to the substrates.

The less purified pigeon liver enzyme preparations contained, as well as the glutathione synthesizing enzymes, another that hydrolysed glutathione. JOHNSTON and BLOCH eventually obtained a preparation that actively synthesized glutathione, but from which the hydrolysing enzyme was absent. This, then, eliminates from the mechanism of the synthesis the concomitant hydrolysis that appears to be a necessary condition in the amide and peptide syntheses studied by FRUTON et al., and HANES et al., discussed above.

VII. Mechanism of Amino Acid Incorporation into Proteins.

1. Effect of Inhibitors.

We shall turn now to the chemical mechanisms of amino acid incorporation into proteins and of protein synthesis. Nothing positive and specific is known. The best that can be done at present is to compare the factors known, on the one hand, to influence amino acid incorporation and protein synthesis, *in vitro*, and on the other, the simpler transfer reactions and peptide syntheses discussed above.

Before doing so two questions need to be considered. One is, "Is the mechanism of incorporation in any one tissue the same for all amino acids?"; and the other "Is the mechanism of incorporation the same in every tissue?"

On the first question the evidence indicates that the mechanism of incorporation of most, if not all amino acids, in any one tissue is the same.

Thus in rabbit bone marrow cells (*23*) the incorporation of glycine, leucine or lysine was inhibited completely by anaerobiosis, 0.001 M arsenite and 0.001 M 2,4-dinitrophenol. Arsenate (0.001 M) inhibited the incorporation of glycine, leucine and lysine, 96, 77, and 20% respectively, and azide (0.001 M) 84, 77, and 68% respectively. In rat diaphragm (*24*) anaerobiosis, arsenite (0.001 M) and 2,4-dinitrophenol (0.001 M) inhibited completely the incorporation of the same three amino acids, arsenate (0.001 M) inhibited their incorporation 67, 58, and 63% respectively, and azide (0.001 M) 85, 78, and 85% respectively.

The similarity of the degree of incomplete inhibition of the three amino acids in each tissue is, possibly, a stronger indication that the incorporation mechanism of the three amino acids is the same than the similarity of the complete inhibition.

A more extensive test was carried out with rabbit reticulocytes (*25*). There were four amino acids and more inhibitors. The following is a summary of the results, expressed as per cent inhibition of the incorporation of glycine, histidine, leucine and lysine respectively.

Anaerobiosis; it was never complete and very variable, the average inhibitions of many experiments are 55, 23, 63, and 25%. Arsenate (0.001 M): 29, 18, 20, 46. Arsenite (0.001 M): 90, 95, 96, 97. Azide (0.001 M): 30, 19, 13, 30. 2,4-Dinitrophenol (0.001 M): 93, 83, 81, 86. Diethyldithiocarbamate (0.001 M): 19, 19, 37, 53. $\alpha\alpha'$-Dipyridyl (0.001 M): 64, 46, 62, 69. Fluoride (0.02 M): 99, 97, 99, 100. Fluoride (0.001 M): 0, 0, 1, 0. Hydroxylamine (0.02 M): 89, 89, 93, 95. Hydroxylamine (0.001 M): 0, 0, 0, 0. Iodoacetate (0.001 M): 88, 54, 51, 79. Ammonium molybdate (0.001 M): 42, 55, 61, 33.

The degree of inhibition of each of the four amino acids with each inhibitor is sufficiently similar to conclude that the mechanism of the incorporation of each of the four amino acids is very similar, and also similar in the three tissues, rabbit bone marrow cells, rat diaphragm, and rabbit reticulocytes.

This conclusion cannot be generalized. In guinea pig liver homogenate lysine is incorporated into the proteins by two different enzyme mechanisms (*18*). The enzyme in one is confined to the particle-free solution (*21*), its optimum pH is in the neighborhood of 6.5, and added Ca^{++} (optimum conc., 0.004 M) is obligatory; the other enzyme is largely in the mitochondria, its optimum pH is near to 7.5 and added Ca^{++} accelerates somewhat but is not obligatory. The following summary shows the striking differences in the effects of inhibitors; the numbers give the degree of inhibition respectively of the first (acid-calcium) and the second (alkaline) reactions (%).

Anaerobiosis: 0, 24. Arsenate (0.001 M): 27, 2. Arsenite (0.001 M): 67, 0. Azide (0.001 M): 43, 8. 2,4-Dinitrophenol (0,001 M): 28, 0. Fluoride (0.02 M): 95, 14.

The following findings in *E. coli* (*17*) indicate that the mechanism of amino acid incorporation into their proteins differs from that in the foregoing animal tissues. It was found that the amino acids (glycine, histidine, leucine, and lysine) were rapidly incorporated not only into the growing organisms but also in cultures in which rapid lysis was in progress, and also in the culture medium remaining after the cells (living and dead) and debris has been centrifuged down. Arsenite (0.001 M), which inhibited growth 50% did not inhibit amino acid incorporation. Azide (0.001 M) inhibited growth completely and amino acid incorporation 30–50%. 2,4-Dinitrophenol (0.001 M), $\alpha\alpha'$-dipyridyl (0.001 M) and iodoacetate (0.002 M) inhibited both growth and amino acid incorporation nearly completely. Penicillin G (100–1000 units per ml.) lysed the bacteria but did not (under certain conditions) inhibit amino acid incorporation; and extensive amino acid incorporation occurred in the cell-free supernatant after centrifugation of the culture. The experiments with *E. coli* showed that growth (i. e. ability to form colonies) is experimentally separable from amino acid incorporation. In the early stages of lysis by penicillin there is a considerable increase in bacterial protein, it may be doubled or more; this synthesis of protein stops within two hours. On the other hand, ability to incorporate amino acids into the proteins continues for another eight to ten hours in both the remaining rapidly diminishing number of cells and in the proteins in the external culture medium.

There are consistent (with one exception) differences in the effects of inhibitors between the proved transfer reactions and the proved syntheses.

The glutamo-transferases are not inhibited by cyanide (0.01–0.005 M), 2,4-dinitrophenol (0.005 M), iodoacetate (0,005 M) (*79*, *176*). Fluoride (0.01 M) inhibits the enzyme from pumpkin seadlings completely (*176*), but 0.1 M fluoride does not inhibit the enzyme from *Proteus vulgaris* (*79*). On the other hand cyanide (0.002 M) completely inhibits the synthesis of serum albumin (*130*), and 0.001 M the synthesis of hippuric acid (*26*) and p-aminohippuric acid (*42*). 2,4-Dinitrophenol (0,005 M) inhibited the synthesis of serum albumin 98%. Iodoacetate (0,01 M) inhibited the synthesis of p-aminohippuric acid 97%, and 0.001 M the synthesis of serum albumin 90%. Fluoride (0.02 M) inhibited p-aminohippuric acid synthesis of serum albumin 56%.

The differences between the transfer and synthetic reactions in the effects of inhibitors are striking, because both classes of reactions are activated by ATP, Mg^{++} or Mn^{++} and phosphate. The transfer reactions are activated by arsenate, the synthesis of serum albumin is inhibited.

Classified according to the effects of inhibitors, the amino acid incorporation reactions belong with the synthetic reactions. They are inhibited, in varying degree, by arsenate, arsenite, azide, cyanide, dinitrophenol and fluoride. The one exception is the incorporation of lysine by the alkaline reaction (see above) of guinea pig liver homogenate; it is not affected by the above inhibitors and, in these respects, resembles the transfer reactions.

Since respiration can participate only indirectly, even in the synthetic reactions, by providing a reactive intermediate either through ATP or some other means, it is not surprising that the effects of anaerobiosis are inconsistent and throw no light on the immediate mechanisms of either peptide synthesis, protein synthesis, or amino acid incorporation. The following is a summary of the reported inhibitory effects of anaerobiosis, expressed as per cent inhibition.

Synthesis of hippuric acid, 100 (*26*); synthesis of serum albumin 77 (*130*); incorporation of glycine, leucine and lysine into rabbit bone marrow cells (*23*) and rat diaphragm 100 (*24*); incorporation of glycine, histidine, leucine and lysine into rabbit reticulocytes (*25*) variable and incomplete; incorporation of *DL*-alanine into rat liver slices 90 (*62*); glycine into rat liver mitochondria 90 (*207*); lysine by the acid-calcium reaction in guinea pig liver homogenate, 0, and by the alkaline reaction 24 (*18*).

The following is a summary of the effects of oxidation and phosphorylation inhibitors in addition to those which have been cited above.

In "resting" *E. coli* (*122*) the incorporation of S^{35}-methionine was inhibited by azide (0.005 M) 100%, 2,4-dinitrophenol (0.001 M) 96%, and fluoride (0.02 M) 100%. Malonate, at concentrations of 0.02, 0.01 and 0.005 M respectively, inhibited the incorporation of glycine in foetal rat liver homogenate 75, 65, and 40%; and the incorporation of alanine, 74, 60, and 40% (*204*).

The latter observations indicate that glycine and alanine are incorporated by the same mechanism.

2. Amino Acid Incorporation and Phosphorylation.

The most consistent and most powerful inhibitor of amino acid incorporation in peptide and protein synthesis is 2,4-dinitrophenol (see above). This is the strongest evidence in favor of the participation of high-energy phosphate bonds, because 2,4-dinitrophenol uncouples many (but not all) coupled oxidation and phosphorylation reactions (*49, 117*). The coupling of phosphorylation in peptide synthesis was first suggested by LIPMANN (*114, 115*) before the action of dinitrophenol was known. As phosphorylation is normally coupled with respiration, protein and peptide synthesis will ultimately depend, if phosphorylation is a necessary intermediate step, on respiration. But since anaerobic respiration does permit phosphorylation, amino acid incorporation might occur under anaerobic conditions, and this, as pointed out above, does occur in some cases. SPIEGELMAN and KAMEN (*171*) had suggested that nucleic acids may act as phosphate donors; their later work casts doubt on this suggestion, a doubt which they, themselves, expressed (*172*). As we have seen the question is still open on what is phosphorylated, no phosphorylated intermediates have yet been found, and except in the case of the synthesis of glutamine, and probably of glutathione, no certain stoichiometric correspondence has been established between synthesis and liberation of inorganic phosphate from ATP or ADP, a necessary consequence if the energy of the high-energy phosphate bond is used in the synthesis. And it is possible that in some cases ATP is necessary, not for the direct donation of energy, but for the activation or stabilization of the enzyme system.

3. Heat-Stable Co-factors for Amino Acid Incorporation.

Evidence of a different kind has recently been found that the mechanisms of incorporation in one tissue of at least a number of amino acids have several important features in common. The incorporation of glycine, histidine, leucine and lysine into the proteins of rabbit reticulocytes (*25*) is several times faster in plasma than in Ringer's solution. This acceleration is caused by two groups of factors. One consists of certain specific amino acids, histidine, leucine, phenylalanine and valine. The other factor has not yet been identified; it is not a protein and all the common amino acids, familiar co-factors, vitamins, metabolites, metals and inorganic salts have been excluded. It is present in the plasma of every mammal so far tested, in their normal erythrocytes, in rabbit reticulocytes, liver, spleen and yeast. It is heat stable. During pre-incubation this factor disappears from the reticulocytes, and the latter then lose activity; on the addition of the factor their activity is partially restored. The factor is, therefore, labile in the cells, there is probably an enzyme in the cells that destroys its activity, since in the absence of enzymes the factor can be boiled without loss of activity. In low concentrations the degree of acceleration of incorporation is proportional to the concentration of the stimulating factor.

The wide distribution of the factor suggests that its function is more extensive than merely the incorporation of labeled amino acids into reticulocytes. That the incorporation of labeled glycine, histidine, leucine and lysine is accelerated in a similar manner by a mixture of histidine, leucine, phenylalanine and valine, and that the accelerating effect of this amino acid mixture on the incorporation of each of the four amino acids is considerably augmented, and to the same degree, by the unidentified factor, these findings are strong evidence that the mechanism of incorporation of the four labeled amino acids used is the same or very similar. The isolation and identification of the factor is now under way.

4. Is Amino Acid Incorporation Synthesis of Protein *de novo* or an Exchange?

In one of their first papers on amino acid turnover in proteins *in vivo* SCHOENHEIMER *et al.* (*151*) recognized that labeled amino acid incorporation may occur by two different mechanisms:

"There are two general reactions possible which might lead to amino acid replacement: (1) complete breakdown of the proteins into its units followed by resynthesis or (2) only partial replacement of units. Metabolic studies with isotopes indicate only end-results but not intermediate steps of a reaction. We have no indication as to what had happened to the protein molec le in the animals. Both reactions are conceivable.

We still have no clear indication. The question raised by SCHOENHEIMER *et al.* may be stated as two extreme alternatives: (*a*) Amino acid incorporation into proteins occurs by synthesis of protein *de novo* from amino acids; or (*b*) it occurs by an enzyme-mediated exchange of amino acids bound in the protein with the corresponding free amino acids in the medium, without hydrolysis of the moieties on either side of the two peptide bonds split in the exchange. It is possible that amino acids are incorporated by both methods.

A way to a direct experimental attack on this problem has not yet been found. The interpretation of what indirect evidence there is, is a matter merely of opinion. Thus SIMPSON *et al.* (*163*) observed that ethionine (VIII) inhibited, *in vivo*, the incorporation of both methionine (its naturally occurring homologue containing one CH_2 more in the chain) and also glycine, both to about the same extent. Administration of methionine relieved the ethionine inhibition. The authors interpreted these findings as favoring, "superficially at least", incorporation of amino acids by *de novo* synthesis of protein from the free amino acids. But they are aware of many other possible interpretations and do not stress the point. Arguing against *de novo* synthesis is the fact that amino acid incorporation is not dependent on the presence of all the other amino acids, or on the nutritional state of the animal. Under certain experimental conditions (see above) the evidence indicates that amino acids are incorporated independently of each other, both in fast and slow reacting tissues.

If incorporation represented synthesis *de novo* from amino acids, one would expect the presence of other amino acids to have a great effect. But against this it can be argued that the tissues, as used, contained sufficient amounts of all the other amino acids. In one case there is clear evidence of direct incorporation in the absence of other amino acids. SCHWEET (*154*) has prepared and somewhat purified by precipitation a soluble enzyme from guinea pig liver homogenate that incorporates lysine. No other free amino acids or metabolites need to be added to the reaction mixture. But this incorporation of lysine appears to be a special case (see above).

5. The Possibility of Peptides as Intermediates in Protein Synthesis.

Where there is a net synthesis of protein, as in the production of serum albumin by chicken liver slices (*130*), the protein must have been synthesized either from free amino acids or peptides, or from both. That peptides may be intermediates in the process is suggested by the finding [BORSOOK *et al.* (*19*)] that when guinea pig liver homogenate was incubated with labeled amino acids a peptide was formed containing a much higher concentration of the labeled amino acid than in the proteins; with labeled leucine the latter was incorporated into the peptide and none into the proteins. A similar peptide fraction is one of the major products of the peptic hydrolysis of a number of animal proteins, and it is present in relatively large amounts in the liver and other animal organs. These findings suggest that peptides may be not only structural, but also metabolic intermediates in protein anabolism and catabolism.

ANFINSEN and STEINBERG (*6*) concluded that peptides are intermediates in the synthesis of ovalbumin by the hen's oviduct. They incubated minced oviduct with $C^{14}O_2$, crystallized the ovalbumin present after the incubation, and then digested it with a bacterial enzyme that hydrolyses the protein to plakalbumin and three peptides: (*a*) a hexapeptide (3 alanines, 1 aspartic acid, 1 glycine and 1 valine); (b) a tetrapeptide (1 alanine, 1 aspartic acid, 1 glycine and 1 valine); and (c) a dipeptide (alanyl-alanine). The aspartic acid was separated from the hexapeptide and from the plakalbumin and the specific activity of the aspartic acid from each source compared. The specific activity of the aspartic acid in the hexapeptide was always greater than that from the plakalbumin, the ratio varied from 1.3 to 3.5. Similarly, labeled alanine was found to be incorporated unequally into different parts of the protein (4).

If the synthesis (or incorporation) had occurred directly from a pool of free amino acids, the specific activities would have been the same throughout the protein molecule. ANFINSEN and STEINBERG concluded from the finding that they were different that synthesis from a pool of free amino acids was excluded, and suggested that peptides, into which the labeled amino acid had been incorporated, were intermediates in the synthesis. But the latter explanation does not explain their findings. Labeled aspartic acid (or alanine) was found both in the plakalbumin and hexapeptide moieties. If the labeled amino acid was incorporated into both moieties *via* the same labeled peptide, then their specific activities would have been the same. Even if the peptides were not the same, but if they had been synthesized from free amino acids containing aspartic acid (or alanine) the specific activities of the amino acids in the protein eventually synthesized would have been the same at every locus in the protein. There would appear to be in general three explanations for the difference in specific activity of the same amino acid situated in different parts of the protein molecule. One is that a preexisting protein molecule is hydrolysed, and that it is reconstituted by using one or more parts containing the amino acid in question unchanged, and other parts either synthesized *de novo* from the amino acid pool or from peptides into which the labeled amino acid had been incorporated. A second explanation would be that the protein was synthesized *de novo* from peptides of different origin, one group synthesized from the amino acid pool and one group not. The third possible explanation is that there was direct interchange of the free (labeled) aspartic acid or alanine with aspartic acid or alanine in the protein, without disruption of the remainder of the molecule; and the interchange occurred at different rates at different parts of the protein molecule, the rates being affected by the amino acids with which the aspartic acid or alanine were linked.

6. The Linkage of Incorporated Amino Acids.

Until now we have not considered the question of the mode of linkage of amino acids incorporated into proteins. A major assumption underlying nearly all the work is that radioactivity in the protein after incubating a tissue with a radioactive amino acid, or finding N^{15} in an amino acid after hydrolysis of the protein following an experiment with an N^{15}-labeled amino acid, means that the labeled amino acid had been incorporated into it by peptide bonds. The whole structure of interpretation rests on this assumption. It has been rigorously proved in only one case, and the proof is almost complete in only one other.

It may be said, at once, that since every amino acid that has been tested, indispensable or dispensable, *in vivo* or *in vitro*, has been found to be incorporated, it argues strongly for incorporation by peptide linkage. There are simply not enough side chain loci that might conceivably combine with so many and so different amino acids. But to a limited extent such combination can and does occur. MELCHIOR and TARVER (*123*) found that after S^{35}-methionine had been incubated with liver slices, a large fraction of the radioactivity "bound" to the proteins was released by treating them with thioglycolic acid; S^{35}-methionine (and cystine formed from it) were bound to the proteins by other than peptide linkage. Similarly, when S^{35}-ethionine was incubated with plasma, some of the ethionine was liberated from the proteins by monoethyleneglycol (*112*). When glycine was incubated with rat liver homogenate (*207*), a large fraction of the counts in the protein were released after solution of the protein in dilute alkali followed by dialysis, and by heating at 90° with 5% trichloroacetic acid (*203, 208*).

In many experiments the labeled (C^{14}) amino acid was recovered as such after acid hydrolysis of the protein. This showed that the radioactivity was carried into the protein by the labeled amino acid in a form that the amino acid could be recovered as such after complete acid hydrolysis. One commonly used test is treatment of the protein with the ninhydrin reagent which liberates CO_2 from the α-carboxyl group of an amino acid when both the carboxyl and amino groups are free; in the case of aspartic acid, but not glutamic acid, both carboxyl groups are split off as CO_2. When an amino acid labeled in its carboxyl group is merely adsorbed on an amino acid, its carboxyl group is decomposed and labeled CO_2 is released. However, even in the case of a protein that has ostensibly incorporated a carboxyl-C^{14}-labeled amino acid and does not yield labeled CO_2 on treatment with the ninhydrin reagent, and the labeled amino acid is recovered as such after complete hydrolysis, this evidence is not proof that the labeled amino acid was incorporated by peptide bonds.

It would be proof if a peptide could be isolated and identified among the partial hydrolysis products of a radioactive protein. This was done in the experiment cited above (*6*) where C^{14}-labeled egg albumin was formed by minced hen's oviduct and radioactivity was found in the hexapeptide of the partial enzymatic hydrolysate. Proof, almost as rigorous, was furnished by WINNICK et al. (*208*) in the case of carboxyl-

C^{14}-glycine incorporated into rat liver homoengate proteins. On treatment with ninhydrin the following results were obtained; no $C^{14}O_2$ was released from the unhydrolysed protein; all the C^{14} was released from the protein after complete acid hydrolysis; no $C^{14}O_2$ was released after peptic or tryptic digestion; 75% of the C^{14} was released as $C^{14}O_2$ following digestion with a mixture of trypsin, chymotrypsin, carboxypeptidase and erepsin.

Table 16. Results of Treatment of Proteins Obtained After Incubation with C^{14}-Labeled Glycine, L-Histidine, L-Leucine or L-Lysine (25).

Treatment	Amino acid incorporated			
	* Glycine	* Histidine	** Leucine	* Lysine
None. c. p. m./mg. protein............	4.30	7.10	14.3	6.87
Boiled with 5 per cent trichloroacetic acid. Specific activity as per cent of original protein....................	101	95	99	95
Dissolved in alkali and dialysed. Specific activity as per cent of original protein	101	98	99	100
Radioactivity released by ninhydrin. Per cent of original protein	0	1	1	0
Oxidized with performic acid. Specific activity as per cent of original protein	108	106	109	101
Hydrolysed. Radioactivity in isolated amino acid corresponding to that with which it was incubated as per cent in original protein....................	79 glycine; 14 serine	101	103	99
Per cent radioactivity in amino acid corresponding to that with which it was incubated and with NH_2 group free in the protein	<1	—	<1	<1

Table 16 contains an example of results obtained when the proteins of rabbit reticulocytes, after the incorporation of labeled amino acids, were submitted to a variety of treatments. All the results are in accord with the interpretation that the labeled amino acids were incorporated by peptide bonds. Practically none of the incorporated amino acids were in an end (NH_2) group position (histidine was not determined because of methodical difficulties). This is in accord with the finding of PORTER and SANGER (*132*) who determined the end (NH_2) groups in the hemoglobin of a number of animals. In the donkey, horse, human (adult and fetal) valine was the only end group amino acid; in the cow, goat and sheep the only end group amino acids were methionine and valine. Evidently

---

* Mixed proteins of reticulocytes, mainly hemoglobin.
** Purified hemoglobin from the reticulocytes.

the labeled amino acids were incorporated in the reticulocyte proteins into the interior of the molecule, and, hence probably by peptide bonds.

The evidence at hand as far as it goes is in accord with the interpretation that the labeled amino acids taken up by tissue proteins, *in vivo* or *in vitro*, are incorporated by peptide bonds. There are no discrepancies and the evidence is quite varied.

The preparation of this article and the experiments by the author and his collaborators were supported by a contract between the Office of Naval Research, Department of the Navy, and the California Institute of Technology, Division of Biology (NR 164304). They were also supported (in part) by a research grant from the National Institutes of Health, Public Health Service, and by a grant from the American Cancer Society.

References.

1. ABDOU, I. A. and H. TARVER: Plasma Protein. II. Relationship between Circulating and Tissue Protein. J. biol. Chemistry **190**, 781 (1951).
2. ABRAMS, R., J. M. GOLDINGER and E. S. G. BARRON: Synthesis of Protein and Other Cell Substances from Acetic Acid in Isolated Bone Marrow. Biochim. Biophys. Acta **5**, 74 (1950).
3. ANFINSEN, C. B.: Radioactive Crystalline Ribonuclease. J. biol. Chemistry **185**, 827 (1950).
4. — The Nature of Intermediates in Protein Synthesis. Science (New York) **114**, 683 (1951).
5. ANFINSEN, C. B., A. BELOFF, A. B. HASTINGS and A. K. SOLOMON: The *In Vitro* Turnover of Dicarboxylic Amino Acids in Liver Slice Proteins. J. biol. Chemistry **168**, 771 (1947).
6. ANFINSEN, C. B. and D. STEINBERG: Studies on the Biosynthesis of Ovalbumin. J. biol. Chemistry **189**, 739 (1951).
7. ANGIER, R. B., J. H. BOOTHE, B. L. HUTCHINGS, J. H. MOWAT. J. SEMB, E. L. R. STOKSTAD, Y. SUBBAROW, C. W. WALLER, D. B. CONSULICH, M. J. FAHRENBACH, M. E. HULTQUIST, E. KUH, E. H. NORTHEY, D. R. SEEGER, J. P. SICKELS and J. M. SMITH, Jr.: The Structure and Synthesis of the Liver L. Casei Factor. Science (New York) **103**, 667 (1946).
8. BARKER, H. A.: Recent Investigations on the Formation and Utilization of Active Acetate. In: W. D. McELROY and B. GLASS, A Symposium on Phosphorus Metabolism, pp. 240–241. Baltimore: John Hopkins Press. 1951.
9. BERGMANN, M. and O. K. BEHRENS: On the Assymmetric Course of the Enzymatic Synthesis of Peptide Bonds. J. biol. Chemistry **124**, 7 (1938).
10. BERGMANN, M. and H. FRAENKEL-CONRAT: The Rôle of Specificity in the Enzymatic Synthesis of Proteins. Syntheses with Intracellular Enzymes. J. biol. Chemistry **119**, 707 (1937).
11. — — — — The Enzymatic Synthesis of Peptide Bonds. J. biol. Chemistry **124**, 1 (1938).
12. BERGMANN, M. and J. S. FRUTON: Some Synthetic and Hydrolytic Experiments with Chymotrypsin. J. biol. Chemistry **124**, 321 (1938).
13. — — The Significance of Coupled Reactions for the Enzymatic Hydrolysis and Synthesis of Proteins. Ann. Yew York Acad. Sci. **45**, 409 (1944).
14. BLOCH, K.: A Heat Stable Co-factor for Glutathione Synthesis. Federat. Proc. (Amer. Soc. exp. Biol.) **10**, 163 (1951).
15. BODANSKY, O.: Introduction to Physiological Chemistry, p. 280. New York: J. Wiley and Sons. 1938.

16. BORSOOK, H.: Protein Turnover and Incorporation of Labeled Amino Acids into Tissue Proteins *In Vivo* and *In Vitro*. Physiologic. Rev. **30**, 206 (1950).
17. — (unpublished).
18. BORSOOK, H., C. L. DEASY, A. J HAAGEN-SMIT, G. KEIGHLEY and P. H. LOWY: The Incorporation of Labeled Lysine into the Proteins of Guinea Pig Liver Homogenate. J. biol. Chemistry **179**, 689 (1949).
19. — — — — — A Peptide Fraction in Liver. J. biol. Chemistry **179**, 705 (1949).
20. — — — — — Uptake of Labeled Amino Acids by Tissue Proteins *In Vitro*. Federat. Proc. (Amer. Soc. exp. Biol.) **8**, 589 (1949).
21. — — — — — The Uptake *In Vitro* of C^{14}-Labeled Glycine, L-Leucine, and L-Lysine by Different Components of Guinea Pig Liver Homogenate. J. biol. Chemistry **184**, 529 (1950).
22. — — — — — Metabolism of C^{14}-Labeled Glycine, L-Histidine, L-Leucine and L-Lysine. J. biol. Chemistry **187**, 839 (1950).
23. — — — — — Incorporation *In Vitro* of Labeled Amino Acids into Bone Marrow Cell Proteins. J. biol. Chemistry **186**, 297 (1950).
24. — — — — — Incorporation *In Vitro* of Labeled Amino Acids into Rat Diaphragm Proteins. J. biol. Chemistry **186**, 309 (1950).
25. — — — — — Incorporation *In Vitro* of Labeled Amino Acids into Proteins of Rabbit Reticulocytes. J. biol. Chemistry **196**, 669 (1952).
26. BORSOOK, H. and J. W. DUBNOFF: The Biological Synthesis of Hippuric Acid *In Vitro*. J. biol. Chemistry **132**, 307 (1940).
27. — — Synthesis of Hippuric Acid in Liver Homogenate. J. biol. Chemistry **168**, 397 (1947).
28. — — (unpublished).
29. BORSOOK, H. and H. M. HUFFMAN: Some Thermodynamical Considerations of Amino Acids, Peptides, and Related Substances. In: C. L. A. SCHMIDT, Chemistry of the Amino Acids and Proteins, p. 822. Springfield, Ill. — Baltimore. 1938.
30. BORSOOK, H. and G. L. KEIGHLEY: The Continuing Metabolism of Nitrogen in Animals. Proc. Roy. Soc. (London), Ser. B **118**, 488 (1935).
31. BRACHET, J.: Recherches sur la synthèse de l'acide thymonucléique pendant le développement de l'œuf d'oursin. Arch. Biol. (Paris) **44**, 519 (1933).
32. — The Metabolism of Nucleic Acid during Embryonic Development. Cold Spring Harbor Symp. Quant. Biology **12**, 18 (1947).
33. BRENNER, M., H. R. MÜLLER und R. W. PFISTER: Eine neue enzymatische Peptidsynthese. Helv. chim. Acta **33**, 568 (1950).
34. BRENNER, M. und R. W. PFISTER: Enzymatische Peptidsynthese. Isolierung von enzymatisch gebildetem L-Methionyl-L-methionin und L-Methionyl-L-methionyl-L-methionin. Helv. chim. Acta **34**, 2085 (1951).
35. BRENNER, M., E. SAILER und K. RÜFENACHT: Enzymatische Peptidsynthese. Peptidbildung aus DL-Threonin-isopropyl-ester. Helv. chim. Acta **34**, 2096 (1951).
36. CANNON, P. R., C. H. STEFFEE, L. J. FRAZIER, D. A. ROWLEY and P. C. STEPTO: The Influence of Time of Ingestion of Essential Amino Acids upon Utilization in Tissue-Synthesis. Federat. Proc. (Amer. Soc. exp. Biol.) **6**, 390 (1947).
37. CASPERRSON, T. und K. BRANDT: Nucleotidumsatz und Wachstum bei Preßhefe. Protoplasma **35**, 507 (1940/41).
38. CASPERRSON, T., H. LANDSTRÖM-HYDÉN und L. AQUILONIUS: Cytoplasmanukleotide in eiweißproduzierenden Drüsenzellen. Chromosoma **2**, 111 (1941–1944).

39. CASPERRSON, T. and J. SCHULTZ: Nucleic Acid Metabolism of the Chromosomes in Relation to Gene Reproduction. Nature (London) 142, 294 (1938).
40. CASPERRSON, T. und B. THORELL: Der endozellulare Eiweiß- und Nukleinsäure-Stoffwechsel im embryonalen Gewebe. Chromosoma 2, 132 (1941—1944).
41. CHANTRENNE, H.: The Requirement for Coenzyme A in the Enzymatic Synthesis of Hippuric Acid. J. biol. Chemistry 189, 227 (1951).
42. COHEN, P. P. and R. W. McGILVERY: The Formation of p-Aminohippuric Acid by Rat Liver Slices. J. biol. Chemistry 166, 261 (1946).
43. — — Peptide Bond Synthesis. II. The Formation of p-Amino-hippuric Acid by Liver Homogenates. J. biol. Chemistry 169, 119 (1947).
44. — — Peptide Bond Synthesis. III. On the Mechanism of p-Aminohippuric Acid Synthesis. J. biol. Chemistry 171, 121 (1947).
45. COHEN, S. S.: The Synthesis of Bacterial Viruses in Infected Cells. Cold Spring Harbor Sympos. quantitat. Biol. 12, 35 (1947).
46. COLLIER, H. B.: The Problem of Plastein Formation. I. The Formation of Plastein by Papain. Canad. J. Res. Sect. B 18, 255 (1940).
47. — The Chemical Changes Involved in Plastein Formation by Papain and by Pepsin. Canad. J. Res. Sect. B 18, 272 (1940).
48. COTZIAS, G. C. and V. P. DOLE: Metabolism of Amines. II. Mitochondrial Localization of Monoamine Oxidase. Proc. Soc. exp. Biol. Med. 78, 157 (1951).
49. CROSS, R. J., J. V. TAGGART, G. A. COVO and D. E. GREEN: Studies on the Cyclophorase System. VI. The Coupling of Oxidation and Phosphorylation. J. biol. Chemistry 177, 655 (1949).
50. CUNNINGHAM, L., A. C. GRIFFIN and J. M. LUCK: Effect of a Carcinogenic Azo Dye on Liver Cell Structure. Isolation of Nuclei and Cytoplasmic Granules. Cancer Res. 10, 194 (1950).
51. DAVIDSON, J. N.: Some Factors Influencing the Nucleic Acid Content of Cells and Tissues. Cold Spring Harbor Sympos. quantitat. Biol. 12, 50 (1947).
52. DELWICHE, C. C., W. D. LOOMIS and P. K. STUMPF: Amide Metabolism in Higher Plants. II. The Exchange of Isotopic Ammonia by Glutamyl Transferase. Arch. Biochemistry 33, 333 (1951).
53. ECKER, P. G. E.: The Ultracentrifuge Study of Plastein. J. gen. Physiol. 30, 399 (1946/47).
54. ELLIOTT, D. F. and A. NEUBERGER: Irreversibility of the Deamination of Threonine in the Rabbit and Rat. Biochemic. J. 46, 207 (1950).
55. ELLIOTT, W. H.: Adenosinetriphosphate in Glutamine Synthesis. Nature (London) 161, 128 (1948).
56. — Adenosinetriphosphate in Glutamine Synthesis. Biochemic. J. 42, V (1948).
57. FOLIN, O.: A Theory of Protein Metabolism. Amer. J. Physiol. 13, 117 (1905).
58. FOLIN, O. and W. DENIS: Protein Metabolism from the Standpoint of Blood and Tissue Analyses. J. biol. Chemistry 11, 87 (1912).
59. FOLLEY, S. J.: Note on the Preparation and Fractionation of the α-Naphthylisocyanate Compound of Plastein. Biochemic. J. 27, 151 (1933).
60. FORKER, L. L., I. L. CHARKOFF, C. ENTENMAN and H. TARVER: Formation of Muscle Protein in Diabetic Dogs, Studied with S^{35}-Methionine. J. biol. Chemistry 188, 37 (1951).
61. FRANTZ, I. D., Jr. and R. B. LOFTFIELD: Equilibrium and Exchange Reactions Involving Peptides, Amino Acids, and Proteolytic Enzymes. Federat. Proc. (Amer. Soc. exp. Biol.) 9, 172 (1950).
62. FRANTZ, I. D., Jr., R. B. LOFTFIELD and W. W. MILLER: Incorporation of C^{14} from Carboxyl-Labeled DL-Alanine into the Proteins of Liver Slices. Science (New York) 106, 544 (1947).

63. FRANTZ, I. D., Jr., P. C. ZAMECNIK; J. W. REESE and M. L. STEPHENSON: The Effect of Dinitrophenol on the Incorporation of Alanine Labeled with Radioactive Carbon into the Proteins of Slices of Normal and Malignant Rat Liver. J. biol. Chemistry **174**, 773 (1948).
64. FRIEDBERG, F.: The Action of Dehydrocorticosterone in the Regulation of Protein Turnover Studied with S^{35} Labeled Methionine. Euclides **109**, 116 (1950).
65. FRIEDBERG, F. and D. M. GREENBERG: The Effect of Growth Hormone on the Incorporation of S^{35} of Methionine into Skeletal Muscle Protein of Normal and Hypophysectomized Animals. Arch. Biochemistry **17**, 193 (1948).
66. FRIEDBERG, F., M. P. SCHULMAN and D. M. GREENBERG: The Effect of Growth on the Incorporation of Glycine Labeled with Radioactive Carbon into the Protein of Liver Homogenates. J. biol. Chemistry **173**, 437 (1948).
67. FRIEDBERG, F., H. TARVER and D. M. GREENBERG: The Distribution Pattern of Sulfur-Labeled Methionine in the Protein and the Free Amino Acid Fraction of Tissues after Intravenous Administration. J. biol. Chemistry **173**, 355 (1948).
68. FRUTON, J. S.: Rôle of Proteolytic Enzymes in Biosynthesis of Peptide Bonds. Yale J. Biol. Med. **22**, 263 (1950).
69. GEIGER, E.: Experiments with Delayed Supplementation of Incomplete Amino Acid Mixtures. J. Nutrit. **34**, 97 (1947).
70. — The Rôle of the Time Factor in Feeding Supplementary Proteins. J. Nutrit. **36**, 813 (1948).
71. — — The Importance of the Time Element in Feeding of Growing Rats: Experiments with Delayed Supplementation of Protein. Science (New York) **108**, 42 (1948).
72. — — The Rôle of the Time Factor in Protein Synthesis. Science (New York) **111**, 594 (1950).
73. — Extra Caloric Function of Dietary Components in Relation to Protein Utilization. Federat. Proc. (Amer. Soc. exp. Biol.) **10**, 670 (1951).
74. GEIGER, E., E. B. HAGERTY and H. D. GATCHELL: Transformation of Tryptophan to Nicotinic Acid Investigated with Delayed Supplementation of Tryptophan Arch. Biochemistry **23**, 315 (1949).
75. GREENBERG, D. M., F. FRIEDBERG, M. P. SCHULMAN and T. WINNICK: Studies on the Mechanism of Protein Synthesis with Radioactive Carbon-Labeled Compounds. Cold Spring Harbor Sympos. quantitat. Biol. **13**, 113 (1948).
76. GREENBERG, D. M. and T. WINNICK: Studies in Protein Metabolism with Compounds Labeled with Radioactive Carbon. II. The Metabolism of Glycine in the Rat. J. biol. Chemistry **173**, 199 (1948).
77. GREENE, C. H.: Changes in Nitrogenous Extractives in the Muscular Tissue of the King Salmon During the Fast of Spawning Migration. J. biol. Chemistry **39**, 457 (1919).
78. GRIFFIN, A. C., S. BLOOM, L. CUNNINGHAM, J. D. TERESI and J. M. LUCK: The Uptake of Labeled Glycine by Normal and Cancerous Tissues in the Rat. Cancer **3**, 316 (1950).
79. GROSSOWICZ, N., E. WAINFAN, E. BOREK and H. WAELSCH: The Enzymatic Formation of Hydroxamic Acids from Glutamine and Asparagine. J. biol. Chemistry **187**, 111 (1950).
80. HANES, C. S., F. J. R. HIRD and F. A. ISHERWOOD: Synthesis of Peptides in Enzymatic Reactions Involving Glutathione. Nature (London) **166**, 288 (1950).
81. HARTE, R. A., J. J. TRAVERS and P. SARICH: The Effect on Rat Growth of Alternated Protein Intakes. J. Nutrit. **35**, 287 (1948).
82. HAUGAARD, G. and R. M. ROBERTS: Heats of Organic Reactions. XIV. The Digestion of β-Lactoglobulin by Pepsin. J. Amer. chem. Soc. **64**, 2664 (1942).

83. HENDERSON, R. and R. S. HARRIS: Concurrent Feeding of Amino Acids. Federat. Proc. (Amer. Soc. exp. Biol.) **8**, 385 (1949).
84. HERBST, R. M. and D. SHEMIN: The Synthesis of Peptides by Transamination. J. biol. Chemistry **147**, 541 (1943).
85. HITCHCOCK, D.: Amphoteric Properties of Amino Acids and Proteins. In: C. L. A. SCHMIDT, Chemistry of the Amino Acids and Proteins, p. 596. Springfield, Ill. — Baltimore. 1938.
86. HOBERMAN, H. D.: Measurement of Rates of Protein Degradation and Protein Loss in Fasting Animals. J. biol. Chemistry **188**, 797 (1951).
87. HOGEBOOM, G. H.: Separation and Properties of Cell Components. Federat. Proc. (Amer. Soc. exp. Biol.) **10**, 640 (1951).
88. HOGEBOOM, G. H. and W. C. SCHNEIDER: Cytochemical Studies of Mammalian Tissues. III. Isocitric Dehydrogenase and Triphosphopyridine Nucleotide-Cytochrome C Reductase of Mouse Liver. J. biol. Chemistry **186**, 417 (1950).
89. HOLLOWAY, B. J. and S. H. RIPLEY: Nucleic Acid Content of Reticulocytes and its Relation to Uptake of Radioactive Leucine *In Vitro*. J. biol. Chemistry **196**, 695, (1952).
90. HYDÉN, H.: Protein Metabolism in the Nerve Cell and Reproduction. Acta physiol. Scand. **6**, Suppl. 17, 1 (1943).
91. — The Nucleoproteins in Virus Reproduction. Cold Spring Harbor Sympos. quantitat. Biol. **12**, 104 (1947).
92. JOHNSTON, R. B. and K. BLOCH: Enzymatic Synthesis of Glutathione. J biol. Chemistry **188**, 221 (1951).
93. JOHNSTON, R. B., M. J. MYCEK and J. S. FRUTON: Catalysis of Transamidation Reactions by Proteolytic Enzymes. J. biol. Chemistry **185**, 629 (1950).
94. — — — Catalysis of Transamidation Reactions by Chymotrypsin. J. biol. Chemistry **182**, 205 (1950).
95. KAUFMAN, S. and H. NEURATH: Inhibition of Chymotrypsin by Structural Analogs of Specific Substrates. Arch. Biochemistry **21**, 245 (1949).
96. — — Structural Requirements of Specific Substrates for Chymotrypsin. II. An Analysis of the Contribution of the Structural Components to Enzymatic Hydrolysis. Arch. Biochemistry **21**, 437 (1949).
97. KAUFMAN, S., H. NEURATH and G. W. SCHWERT: The Specific Peptidase and Esterase Activities of Chymotrypsin. J. biol. Chemistry **177**, 793 (1949).
98. KELLER, E. B.: Turnover of Proteins of Cell Fractions of Adult Rat Liver *In Vivo*. Federat. Proc. (Amer. Soc. exp. Biol.) **10**, 206 (1951).
99. KEMEN, A. J., S. W. HUNTER, G. E. MOORE and C. R. HITCHCOCK: Distribution of Tracer Doses of Methionine Tagged with Radiosulfur in Normal and Neoplastic Tissue. Cancer Res. **9**, 174 (1949).
100. KESTON, A. and J.-C. DREYFUS: Tracer Studies in Protein Synthesis: Antibody Formation by Spleen Slices. Federat. Proc. (Amer. Soc. exp. Biol.) **10**, 206 (1951).
101. KIELLEY, R. K. and W. C. SCHNEIDER: Synthesis of *p*-Aminohippuric Acid by Mitochondria of Mouse Liver Homogenates. J. biol. Chemistry **185**, 869 (1950).
102. KOCHAKIAN, C. D.: The Protein Anabolic Effects of Steroid Hormones. Vitamins and Hormones **4**, 255 (1946).
103. — The Effect of Dose and Nutritive State on the Renotrophic and Androgenic Activities of Various Steroids. Amer. J. Physiol. **145**, 549 (1946).
104. — Comparison of Protein Anabolic Property of Various Androgens in the Castrated Rat. Amer. J. Physiol. **160**, 53 (1950).
105. — Comparison of Protein Anabolic Properties of Testosterone Propionate and Growth Hormone in the Rat. Amer. J. Physiol. **160**, 66 (1950).

106. KOCHAKIAN, C. D. and B. BEALL: Comparison of the Protein Anabolic Property of Testosterone Propionate in the Male and Female Rat. Amer. J. Physiol. 160, 62 (1950).
107. KOCHAKIAN, C. D., J. H. HAMM and M. N. BARTLETT: Effect of Steroids on the Body Weight, Temporal Muscle and Organs of the Guinea Pig. Amer. J. Physiol. 155, 242 (1948).
108. KOCHAKIAN, C. D., J. G. MOE and J. DOLPHIN: Protein Anabolic Property of Testosterone Propionate in Adrenalectomized and Normal Rats. Amer. J. Physiol. 162, 581 (1950).
109. LE PAGE, G. A. and C. HEIDELBERGER: Incorporation of Glycine-2-C^{14} into Proteins and Nucleic Acids of Normal and Neoplastic Rat Tissues. Federat. Proc. (Amer. Soc. exp. Biol.) 9, 195 (1950).
110. — — Incorporation of Glycine-2-C^{14} into the Proteins and Nucleic Acids of the Rat. J. biol. Chemistry 188, 593 (1951).
111. LEVINE, M. and H. TARVER: On the Synthesis and some Applications of Serine-β-C^{14}. J. biol. Chemistry 184, 427 (1950).
112. — — Studies on Ethionine. III. Incorporation of Ethionine into Rat Proteins. J. biol. Chemistry 192, 835 (1951).
113. LI, C. H. and H. M. EVANS: The Properties of the Growth and Adrenocorticotrophic Hormones. Vitamins and Hormones 5, 197 (1947).
114. LIPMANN, F.: Metabolic Generation and Utilization of Phosphate Bond Energy. Adv. Enzymology 1, 99 (1941).
115. — Mechanism of Peptide Bond Formation. Federat. Proc. (Amer. Soc. exp. Biol.) 8, 597 (1949).
116. LITWACK, G., J. N. WILLIAMS, Jr., F. FEIGELSON and C. A. ELVEHJEM: Xanthine Oxidase and Liver Nitrogen Variation with Dietary Protein. J. biol. Chemistry 187, 605 (1950).
117. LOOMIS, W. F. and F. LIPMANN: Reversible Inhibition of the Coupling between Phosphorylation and Oxidation. J. biol. Chemistry 173, 807 (1948).
118. LOTSPEICH, W. D.: Relations between Insulin and Pituitary Hormones in Amino Acid Metabolism. J. biol. Chemistry 185, 221 (1950).
119. MARTIN, C. J. and R. ROBISON: The Minimum Nitrogen Expenditure of Man and the Biological Value of Various Proteins for Human Nutrition. Biochemic. J. 16, 407 (1922).
120. MCGILVERY, R. W. and P. P. COHEN: Enzymatic Synthesis of Ornithuric Acids. J. biol. Chemistry 183, 179 (1950).
121. MELCHIOR, J. B., O. KLIOZE and I. M. KLOTZ: Further Studies of the Synthesis of Protein by *Escherichia Coli*. J. biol. Chemistry 189, 411 (1951).
122. MELCHIOR, J. B., M. MELLODY and l. M. KLOTZ: The Synthesis of Protein by Non-proliferating *Escherichia Coli*. J. biol. Chemistry 174, 81 (1948).
123. MELCHIOR, J. B. and H. TARVER: Studies in Protein Synthesis *In Vitro*. I. On the Synthesis of Labeled Cystine (S^{35}) and its Attempted Use as a Tool in the Study of Protein Synthesis. Arch. Biochemistry 12, 301 (1947).
124. — — Studies on Protein Synthesis *In Vitro*. II. On the Uptake of Labeled Sulfur by the Proteins of Liver Slices Incubated with Labeled Methionine (S^{35}). Arch. Biochemistry 12, 309 (1947).
125. MILLER, L. L.: Changes in Rat Liver Enzyme Activity with Acute Inanition. Relation of Loss of Enzyme Activity to Liver Protein Loss. J. biol. Chemistry 172, 113 (1948).
126. — The Loss and Regeneration of Rat Liver Enzymes Related to Diet Protein. J. biol. Chemistry 186, 253 (1950).

127. MUNTWYLER, E., S. SEIFTER and D. M. HARKNESS: Some Effects of Restriction of Dietary Protein on the Intracellular Components of Liver. J. biol. Chemistry **184**, 181 (1950).
128. NORTHROP, J.: Plastein Formation from Pepsin,and Trypsin. J. gen. Physiol. **30**, 377 (1946/47).
129. PETERS, T., Jr. and C. B. ANFINSEN: The Production of Radioactive Serum Albumin by Liver Slices. J. biol. Chemistry **182**, 171 (1950).
130. — — Net Production of Serum Albumin by Liver Slices. J. biol. Chemistry **186**, 805 (1950).
131. PETERSON, E. A., D. M. GREENBERG and T. WINNICK: Characteristics of the Amino Acid Incorporation System of Liver Homogenates. Federat. Proc. (Amer. Soc. exp. Biol.) **9**, 214 (1950).
132. PORTER, R. R. and F. SANGER: The Free Amino Groups of Haemoglobin. Biochemic. J. **42**, 287 (1948).
133. POTTER, V. R., R. O. RECKNAGEL and R. B. HURLBERT: Intracellular Enzyme Distribution; Interpretations and Significance. Federat. Proc. (Amer. Soc. exp. Biol.) **10**, 646 (1951).
134. PRICE, J. M., E. C. MILLER and J. A. MILLER: The Intracellular Distribution of Protein, Nucleic Acids, Riboflavin and Protein-Bound Aminoazo Dye in the Livers of Rats Fed p-Dimethyl-aminoazobenzene. J. biol. Chemistry **173**, 345 (1948).
135. PRICE, J. M., E. C. MILLER, J. A. MILLER and G. M. WEBER: Studies on the Intracellular Composition of Livers from Rats Fed Various Aminoazo Dyes. I. 4-Aminoazobenzene, 4-Dimethylaminoazobenzene, 4'-Methyl and 3'-Methyl-4-Dimethylaminoazobenzene. Cancer Res. **9**, 398 (1949).
136. — — — — Studies on the Intracellular Composition of Livers from Rats, Fed Various Aminoazo Dyes. II. 3'-Methyl-2'-Methyl-, and 2-Methyl-4-Dimethylaminoazobenzene, 3-Methyl-4-Monomethylaminoazobenzene, and 4'-Fluoro-4-Dimethylaminoazobenzene. Cancer Res. **10**, 18 (1950).
137. PRICE, J. M., J. A. MILLER, E. C. MILLER and G. M. WEBER: Studies on the Intracellular Composition of Liver and Liver Tumor from Rats Fed 4-Dimethylaminoazo-benzene. Cancer Res. **9**, 96 (1949).
138. RATNER, S., M. BLANCHARD, A. F. COBURN and D. E. GREEN: Isolation of a Peptide of p-Aminobenzoic Acid from Yeast. J. biol. Chemistry **155**, 689 (1944).
139. REID, J. C. and H. B. JONES: Radioactivity Distribution in the Tissues of Mice Bearing Melanosarcoma after Administration of DL-Tyrosine Labeled with Radioactive Carbon. J. biol. Chemistry **174**, 427 (1948).
140. REIFENSTEIN, E. C., F. ALLBRIGHT, Jr. and S. L. WELLS: Accumulation, Interpretation, and Presentation of Data Pertaining to Metabolic Balances, Notably Those of Calcium, Phosphorus, and Nitrogen. J. clin. Endocrin. **5**, 367 (1945); Correction, ibid. **6**, 232 (1946).
141. RITTENBERG, D., R. SCHOENHEIMER and A. S. KESTON: Studies in Protein Metabolism. IX. The Utilization of Ammonia by Normal Rats on a Stock Diet. J. biol. Chemistry **128**, 603 (1939).
142. RITTENBERG, D. and D. SHEMIN: The Metabolism of Proteins and Amino Acids. Annu. Rev. Biochem. **15**, 247 (1946).
143. RUTMAN, R., E. DEMPSTER and H. TARVER: Genetic Differences in Methionine Uptake by Surviving Tissues. J. biol. Chemistry **177**, 491 (1949).
144. SALTER, W. T. and O. H. PEARSON: The Enzymatic Synthesis from Thyroid Diiodotyrosine Peptone of an Artificial Protein which Relieves Myxedema. J. biol. Chemistry **112**, 579 (1935/36).

145. SANADI, D. R. and D. M. GREENBERG: Effect of Amino Acid Deficiencies on Incorporation of Radioactive-Carbon Labeled Amino Acids into Animal Proteins. Proc. Soc. exp. Biol. Med. 69, 162 (1948).
146. SARKAR, N., M. FULD and D. E. GREEN: Studies on the Synthesis of Hippuric Acid. Federat. Proc. (Amer. Soc. exp. Biol.) 10, 242 (1951).
147. SCHAEFFER, A. J. and E. GEIGER: Cataract Development in Animals with Delayed Supplementation of Tryptophane. Proc. Soc. exp. Biol. Med. 66, 309 (1947).
148. SCHNEIDER, W. C.: Nucleic Acids in Normal and Neoplastic Tissues. Cold Spring Harbor Sympos. quantitat. Biol. 12, 169 (1947).
149. SCHOENHEIMER, R.: The Dynamic State of Body Constituents. Cambridge, Mass.: Harvard Univ. Press. 1942.
150. SCHOENHEIMER, R., S. RATNER and D. RITTENBERG: Studies in Protein Metabolism. VII. The Metabolism of Tyrosine. J. biol. Chemistry 127, 333 (1939).
151. — — — Studies on Protein Metabolism. X. The Metabolic Activity of Body Proteins Investigated with L(—) Leucine Containing Two Isotopes. J. biol. Chemistry 130, 703 (1939).
152. SCHOENHEIMER, R., S. RATNER, D. RITTENBERG and M. HEIDELBERGER: The Interaction of Blood Proteins of the Rat with Dietary Nitrogen. J. biol. Chemistry 144, 541 (1942).
153. SCHOU, M., N. GROSSOWICZ, A. LAJTHA and H. WAELSCH: Enzymatic Formation of Glutamo-Hydroxamic Acid from Glutamine in Mammalian Tissue, Nature (London) 167, 818 (1951).
154. SCHWEET, R. S.: (unpublished).
155. SCHWEIGERT, B. S., B. T. GUTHNECK, J. M. PRICE, J. A. MILLER and E. C. MILLER: Amino Acid Composition of Morphological Fractions of Rat Livers and Induced Liver Tumors. Proc. Soc. exp. Biol. Med. 72, 495 (1949).
156. SCHWEIGERT, B. S., H. E. SAUBERLICH, C. A. ELVEHJEM and C. A. BAUMANN: Free Tryptophane in Blood and Urine. J. biol. Chemistry 164, 213 (1946).
157. SCHWERT, G. W., H. NEURATH, S. KAUFMAN and J. E. SNOKE: The Specific Esterase Activity of Trypsin. J. biol. Chemistry 172, 221 (1948).
158. SEIFTER, S., E. MUNTWYLER and D. M. HARKNESS: Some Effects of Continued Protein Deprivation, with and without Methionine Supplementation, on Intracellular Liver Components. Proc. Soc. exp. Biol. Med. 75, 46 (1950).
159. SHEMIN, D., J. M. LONDON and D. RITTENBERG: The Synthesis of Protoporphyrin *In Vitro* by Red Blood Cells of the Duck. J. biol. Chemistry 183, 757 (1950).
160. SHEMIN, D. and D. RITTENBERG: Some Interrelationships in General Nitrogen Metabolism. J. biol. Chemistry 153, 401 (1944).
161. SIEKEVITZ, P. and P. C. ZAMECNIK: *In Vitro* Incorporation of 1-C^{14}-DL-Alanine into Protein of Rat-Liver Granular Fractions Federat. Proc. (Amer Soc. exp. Biol.) 10, 246 (1951).
162. SIMMONDS, S., E. L. TATUM and J. S. FRUTON: The Utilization of Phenylalanine and Tyrosine Derivatives by Mutant Strains of *Escherichia Coli*. J. biol. Chemistry 169, 91 (1947).
163. SIMPSON, M. V., E. FARBER and H. TARVER: Studies on Ethionine: Inhibition of Protein Synthesis in Intact Animals. J. biol. Chemistry 182, 81 (1950).
164. SMITH, E. L.: Catalytic Action of Metal Peptidases. Federat. Proc. (Amer. Soc. exp. Biol.) 8, 581 (1949).
165. SNOKE, J. E. and H. NEURATH: Structural Requirements of Specific Substrates for Chymotrypsin. I. The Contribution of the Secondary Peptide Group. Arch. Biochemistry 21, 351 (1949).

166. SNOKE, J. E. and F. ROTHMAN: Glutathione Synthesis from Glutamyl Cysteine and Glycine. Federat. Proc. (Amer. Soc. exp. Biol.) **10**, 249 (1951).
167. SPECK, J. F.: The Enzymatic Synthesis of Glutamine. J. biol. Chemistry **168**, 403 (1947).
168. — The Synthesis of Glutamine in Pigeon Liver Dispersions. J. biol. Chemistry **179**, 1387 (1949).
169. — The Enzymatic Synthesis of Glutamine; A Reaction Using Adenosine Triphosphate. J. biol. Chemistry **179**, 1405 (1949).
170. SPECTOR, H. and H. H. MITCHELL: Paired Feeding in the Study of the Counteraction by Nicotinic Acid and Tryptophane of the Growth-Depressing Effect of Corn in Rats. J. biol. Chemistry **165**, 37 (1946).
171. SPIEGELMAN, S. and M. D. KAMEN: Genes and Nucleoproteins in the Synthesis of Enzymes. Science (New York) **104**, 581 (1946).
172. — — Some Basic Problems in the Relation of Nucleic Acid Turnover in Protein Synthesis. Cold Spring Harbor Sympos. quantitat. Biol. **12**, 211 (1947).
173. SPRINSON, D. B. and D. RITTENBERG: The Rate of Utilization of Ammonia for Protein Synthesis. J. biol. Chemistry **180**, 707 (1949).
174. — — The Rate of Interaction of the Amino Acids of the Diet with the Tissue Proteins. J. biol. Chemistry **180**, 715 (1949).
175. STUMPF, P. K. and W. D. LOOMIS: Observations on a Plant Amide Enzyme System Requiring Manganese and Phosphate. Arch. Biochemistry **25**, 451 (1950).
176. STUMPF, P. K., W. D. LOOMIS and C. MICHELSON: Amide Metabolism in Higher Plants. I. Preparation and Properties of Glutamyl Transferase from Pumpkin Seedling. Arch. Biochemistry **30**, 126 (1951).
177. TARVER, H. and W. O. REINHARDT: Methionine Labeled with Radioactive Sulfur as an Indicator of Protein Formation in the Hepatectomized Dog. J. biol. Chemistry **167**, 395 (1947).
178. TARVER, H. and C. L. A. SCHMIDT: Radioactive Sulfur Studies. I. Synthesis of Methionine. II. Conversion of Methionine Sulfur to Taurine Sulfur in Dogs and Rats. III. Distribution of Sulfur in the Proteins of Animals Fed Sulfur or Methionine. IV. Experiments *In Vitro* with Sulfur and Hydrogen Sulfide. J. biol. Chemistry **146**, 69 (1942).
179. TAUBER, H. T.: Protein Synthesis by Chymotrypsin. J. Amer. chem. Soc **71**, 2952 (1949).
180. — Synthesis of High Molecular-Weight Protein-Like Substances by Chymotrypsin. Federat. Proc. (Amer. Soc. exp. Biol.) **9**, 237 (1950).
181. — Synthesis of Protein-Like Substances by Chymotrypsin. J. Amer. chem. Soc. **73**, 1288 (1951).
182. — Synthesis of Protein-Like Substances by Chymotrypsin from Dilute Peptic Digests and their Electrophoretic Patterns. J. Amer. chem. Soc. **73**, 4965 (1951).
183. THORELL, B.: The Relation of Nucleic Acids to the Formation and Differentiation of Cellular Proteins. Cold Spring Harbor Sympos. quantitat. Biol. **12**, 247 (1947).
184. TOTTER, J. R., B. KELLEY, P. L. DAY and R. R. EDWARDS: The Metabolism of Glycine by Folic Acid-Deficient Chick Liver Homogenates. J. biol. Chemistry **186**, 145 (1950).
185. TYNER, E. P., C. HEIDELBERGER and G. A. LE PAGE: Rates and Synthesis and Turnover of Proteins and Nucleic Acid Purines in the Rat. Federat. Proc. (Amer. Soc. exp. Biol.) **10**, 262 (1951).
186. VAN SLYKE, D. D. and G. M. MEYER: The Effects of Feeding and Fasting on the Amino Acid Content of the Tissues J. biol. Chemistry **16**, 231 (1913).

187. VENDRELY, C. et R. VENDRELY: L'acide ribonucléique des mitochondries et des microsomes du foie et ses variations au cours du jeûne protéique. C. R. hebd. Séances Acad. Sci. 230, 333 (1950).
188. VIERORDT, H.: Anatomische, physiologische und physikalische Daten und Tabellen. Jena. 1906.
189. VIRTANEN, A. I. and H. K. KERKKONEN: On the Chemical Nature of Plasteins. Acta chem. Scand. 1, 140 (1947).
190. — — Structure of Plasteins. Nature (London) 161, 888 (1948).
191. VIRTANEN, A. I., H. K. KERKKONEN, M. HAKALA und T. LAAKSONEN: Die Synthese von Polypeptiden durch die Wirkung von Pepsin. Naturwiss. 37, 139 (1950).
192. VIRTANEN, A. I., H. K. KERKKONEN, T. LAAKSONEN and M. HAKALA: Plastein, a Mixture of Higher-Molecular Polypeptides Synthesized by Proteolytic Enzymes. Acta chem. Scand. 3, 520 (1949).
193. WAELSCH, H.: Glutamotransferase Activity in Mammalian Tissue Extracts. Federat. Proc. (Amer. Soc. exp. Biol.) 10, 266 (1951).
194. — Glutamic Acid and Cerebral Function. Adv. Protein Chem. 6, 299 (1951).
195. WAELSCH, H., E. BOREK, N. GROSSOWICZ and M. SCHOU: Glutamo- and Asparto-Transferases. Federat. Proc. (Amer. Soc. exp. Biol.) 9, 242 (1950).
196. WAELSCH, H., P. OWADES, E. BOREK, N. GROSSOWICZ and M. SCHOU: The Enzyme-Catalyzed Exchange of Ammonia with the Amide Group of Glutamine and Asparagine. Arch. Biochemistry 27, 237 (1950).
197. WAELSCH, H. and D. RITTENBERG: Glutathione. II. The Metabolism of Glutathione Studied with Isotopic Ammonia and Glutamic Acid. J. biol. Chemistry 144, 53 (1942).
198. WALDSCHMIDT-LEITZ, E. u. K. KÜHN: Über die enzymatische Synthese von Peptidbindungen. Hoppe-Seyler's Z. physiol. Chem. 285, 22 (1950).
199. WASTENEYS, H. and H. BORSOOK: The Enzymatic Synthesis of Protein. Physiologic. Rev. 10, 110 (1930).
200. WEISSMAN, N. and R. SCHOENHEIMER: The Relative Stability of L(+) Lysine in Rats Studied with Deuterium and Heavy Nitrogen. J. biol. Chemistry 140, 779 (1941).
201. WESTERFIELD, W. W. and D. A. RICHERT: Dietary Effects on Liver Xanthine Oxidase. Federat. Proc. (Amer. Soc. exp. Biol.) 8, 265 (1949).
202. WILLIAMS, J. N., Jr. and C. A. ELVEHJEM: The Relation of Amino Acid Availability in Dietary Protein to Liver Enzyme Activity. J. biol. Chemistry 181, 559 (1949).
203. WINNICK, T.: Studies on the Mechanism of Protein Synthesis in Embryonic and Tumor Tissues. I. Evidence Relating to the Incorporation of Labeled Amino Acids into Protein Structure in Homogenates. Arch. Biochemistry 27, 65 (1950).
204. — Studies on the Mechanism of Protein Synthesis in Embryonic and Tumor Tissues. II. Inactivation of Fetal Rat Liver Homogenates by Dialysis, and Reactivation by the Adenylic Acid System. Arch. Biochemistry 28, 338 (1950).
205. WINNICK, T., F. FRIEDBERG and D. M. GREENBERG: Incorporation of C^{14}-Labeled Glycine into Intestinal Tissue and its Inhibition by Azide. Arch. Biochemistry 15, 160 (1947).
206. — — — Studies in Protein Metabolism with Compounds Labeled with Radioactive Carbon. I. Metabolism of DL-Tyrosine in the Normal and Tumor-Bearing Rat. J. biol. Chemistry 173, 189 (1948).
207. — — — The Utilization of Labeled Glycine in the Process of Amino Acid Incorporation by the Protein of Liver Homogenate. J. biol. Chemistry 175, 117 (1948).

208. WINNICK, T., E. A. PETERSON and D. M. GREENBERG: Incorporation of C^{14}-Glycine into Protein and Lipide Fractions of Homogenates. Arch. Biochemistry **21**, 235 (1949).
209. WOODWARD, G. E.: S^{35}-Glutathione Preparation from Yeast and Tracer Studies in Cancerous and Non-Cancerous Rats. J. Franklin Inst. **251**, 557 (1951).
210. WORK, T. S. and E. WORK: The Basis of Chemotherapy, p. 227. New York: Interscience Publ. 1948.
211. YESHODA, K. M. and M. DAMODARAN: Amino-Acids and Proteins in Haemoglobin Formation. Biochemic. J. **41**, 382 (1947).
212. ZAMECNIK, P. C.: The Use of Labeled Amino Acids in the Study of the Protein Metabolism of Normal and Malignant Tissues: A Review. Cancer Res. **10**, 659 (1950).
213. ZAMECNIK, P. C. and I. D. FRANTZ, Jr.: Peptide Bond Synthesis in Normal and Malignant Tissue. Cold Spring Harbor Sympos. quantitat. Biol. **14**, 199 (1949).
214. ZAMECNIK, P. C., l. D. FRANTZ, Jr., R. B. LOFTFIELD and M. L. STEPHENSON: Incorporation *In Vitro* of Radioactive Carbon from Carboxyl-Labeled DL-Alanine and Glycine into Proteins of Normal and Malignant Rat Livers. J. biol. Chemistry **175**, 299 (1948).

(Received, January 14, 1952.)

The Enzymes of Nucleoside Metabolism.
By HERMAN M. KALCKAR, Copenhagen.

With 1 Figure.

Contents.
                                        Page

Introduction 363
I. The Preparation of Nucleosides 364
II. The Enzymes of Nucleoside Metabolism 365
   1. Purine Nucleoside Phosphorylase 367
   2. Pyrimidine Nucleoside Phosphorylase 372
   3. Trans-N-Glycosidase 372
   4. Ribosidase 374
   5. Phosphoribomutase 375
   6. Degradation and Synthesis of Ribose-Phosphoric Esters 376
   7. Nucleoside Deaminases 378
III. Phospho-Ribosides 381
   1. Preparation and Properties of Ribose-1-phosphate 381
   2. Enzymatic Synthesis of Ribosides 382
   3. Preparation and Properties of Deoxyribose-1-phosphate 385
   4. Enzymatic Synthesis of Hypoxanthine Deoxyriboside 386
IV. Trans-N-Glycosidic Reactions 387
   1. Non-participation of Ribose-1-Phosphate and Deoxyribose-1-Phosphate in Trans-N-glycosidic Reactions 387
   2. Trans-N-Glycosidic Reactions in the Deoxyribose Nucleoside Series . 387
   3. Enzymatic Formation of New Deoxyribosides 388
V. Phosphorylation of Nucleosides 390
VI. Incorporation of Purines and Pyrimidines into Nucleic Acids 391
   In vivo Studies with Labelled Purines 391. — *In vivo* Studies with Labelled Pyrimidines 392. — *In vitro* Studies with Labelled Purines 393. — Studies on the Amphibian and Echinoderm Egg 394. — Studies on Micro-organisms 394.

References 395

Introduction.

Purines and pyrimidines constitute some of the most essential building stones of the living organism. They are found mainly as complexes bound to pentoses, partly as low- and partly as high-molecular compounds. Free purines occur predominantly as excretion products or as deposits, usually under or on the integuments of vertebrates and invertebrates.

The low-molecular purines or pyrimidine compounds are more or less abundant in the form of nucleotides in various tissues (adenosine polyphosphates, fermentation and oxidation coenzymes, uracil-dinucleotide, vitamin B_{12} etc.). The bulk of purines and pyrimidines are found in the high-molecular nucleic acids which are of two main types: (a) pentose nucleic acid, a substance predominant in the cytoplasm of the cell, and also a constituent of the nucleus, and (b) deoxyribonucleic acid* which, with a few exceptions, occurs exclusively in cellular nuclei.

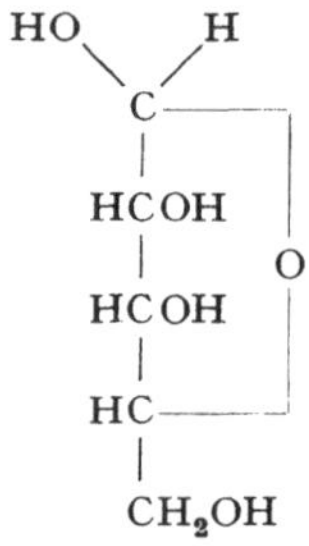

β-D-Ribose. β-D-2-Deoxyribose.

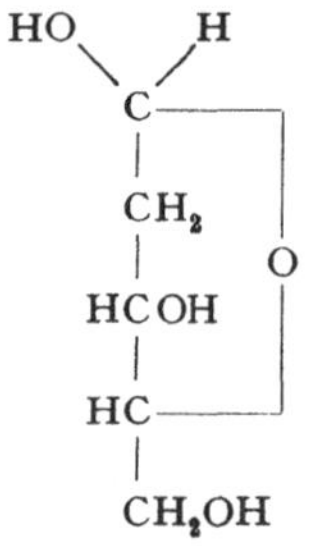

The present review is going to discuss our most recent knowledge about the biological function of ribosides and deoxyribosides. This survey will encompass studies conducted by *in vitro* techniques, mainly concerning the enzymes of nucleoside *metabolism* as well as some investigations performed on intact organisms during which growth responses are recorded and metabolic turnover processes are followed by means of isotope-labelled metabolites.

For a more detailed biochemical treatment of nucleosides and nucleotides as growth substances for microorganisms cf. a survey written by McNutt (76a) (p. 401). The chemistry of the nucleotides was recently discussed in our series by Kenner (61a) and that of the sugar phosphates by Leloir (68a).

Some structural formulas as well as the numbering of purine and pyrimidine systems appear on pp. 403 and 404.

I. The Preparation of Nucleosides.

Since the present survey deals mainly with the metabolism of nucleosides, the chemistry of nucleosides as well as the preparation of this group of substances will be mentioned only very briefly.

The purine and pyrimidine nucleosides can be prepared by enzymatic depolymerization and dephosphorylation of nucleic acids. The preparation of ribosides and deoxyribosides from either nucleic acid by means of

---

* The pretix *Desoxy* is also widely used.

intestinal enzymes was described about twenty years ago by the Freiburg School [BIELSCHOWSKY and KLEIN (*10*); KLEIN (*64*); KLEIN and THANN-HAUSER (*65*)]. A modification of the preparation of adenine deoxyriboside has been proposed by BRADY (*11*). A more dependable way for obtaining adenine deoxyribosides has been developed by they use of lead hydroxide hydrolysis (*28 a*) and of counter current techniques (*108*).

The *in vitro* synthesis of natural adenosine was carried out by the Cambridge School [DAVOLL, LYTHGOE and TODD (*26*); LYTHGOE and TODD (*73*)]. It has also been shown that the ribosides of yeast ribose nucleic acid are β-glycosides (*73*). The isolation of uric acid riboside from beef blood [DAVIS et al. (*25*)] and beef liver [FALCONER and GULLAND (*33*)] should also be mentioned here. Formation of nucleosides can also be obtained from a number of 5-phospho nucleotides by incubation with 5-nucleotidase [HEPPEL and HILMOE (*47*)].

II. The Enzymes of Nucleoside Metabolism.

It has been recognized for some time that mammalian tissues contain certain enzymes *(nucleosidases)* which bring about a cleavage of purine nucleosides. KLEIN (*63*) who followed these reactions by means of reduction titrations of the liberated sugar component, observed that the rate of cleavage of purine deoxyribosides was as high as that of the corresponding ribosides. He found, furthermore, that addition of phosphate or arsenate brought about a marked increase in the enzymatic cleavage. The pyrimidine ribosides and deoxyribosides are split by the action of a group of enzymes 372 different from those attacking the purine nucleosides. DISCHE (*29*) who studied the degradation of adenosine in human erythrocyte hemolysates, found that the cleavage process was accompanied by an uptake of inorganic phosphate and that the ribose moiety (as determined by the orcinol reaction) gradually disappeared.

The steadily increasing interest in the function of nucleosides, especially as growth factors, made it desirable to develop specific micro methods which could enable investigators to determine in a mixture free and bound purines and pyrimidines, respectively. The method to be described combines two techniques, viz. ultraviolet spectrophotometry and enzymatic assays. Since purine compounds show an intense light absorption in the region between 2500 and 2800 Å, ultraviolet spectrophotometry offers great sensitivity in an analysis of these compounds. Yet, the lack of specificity of this method would greatly limit its use if it were not possible to combine a differential analysis with the ultraviolet technique. The most specific differential technique would be the use of highly specific enzymes which would induce such changes in the purine skeleton as to cause marked alterations in the spectrum. It is fortunate that a number of enzymes which catalyze the interconversion of purine compounds quite often show a high degree of specificity [SCHMIDT (94, 95)]. This constitutes the basis for several sensitive and selective optical micro methods for the determination of free and combined purines [KALCKAR (52, 53, 56)]. Following are some examples in which use has been made of this technique.

Hypoxanthine riboside (inosine) and the corresponding deoxyriboside can be determined by the combined addition of two enzymes viz. milk xanthine oxidase [BALL (8)] and liver nucleosidase, in the presence of inorganic phosphate [KALCKAR (52)]. Liver nucleosidase brings about the liberation of hypoxanthine from the nucleosidic linkage. Xanthine oxidase removes hypoxanthine by catalytic oxidation to uric acid, a conversion which is accompanied by a very substantial increase in optical density around 2900 Å (λ_{max}, 2930 Å at p_H 7.5). Evidently, the marked alterations in density at proper wave lengths may serve as a basis for a specific assay technique of a substrate, e.g., hypoxanthine riboside, as well as for a study of an enzyme, e.g., nucleosidase. In the course of a study of this seemingly hydrolytic enzyme, it was found that phosphate is an essential component in the enzymatic cleavage of hypoxanthine riboside, and that this splitting was accompanied by an uptake of inorganic phosphate [KALCKAR (55)]. These observations which will be discussed below (p. 367), led to the conclusion that liver nucleosidase is not a hydrolytic but a phosphorolytic enzyme; and its name was, therefore, changed to "liver nucleoside phosphorylase".

It is possible, by using xanthine oxidase and nucleoside phosphorylase, to determine a mixture of hypoxanthine and hypoxanthine riboside in amounts down to 0.5 μg. of purine per ml., with an accuracy of about 10 per cent. The application of micro cuvettes permits the estimation of less than 0.1 μg. of purine.

Guanosine and adenosine can be determined by combining the enzymes mentioned with fractionated preparations of the corresponding deaminases. Adenosine can also be estimated more specifically by the sole addition of intestinal deaminase. The deamination of adenine nucleosides is accompanied by a marked decrease in optical density at 2650 Å and an increase at 2400 Å [KALCKAR (53)]. In the presence of large amounts of protein, it is, however, advantageous to operate at 2930 Å at which wave length the protein absorbs much less light than in the range between 2600 and 2800 Å.

In general, the advantage of the "differential enzymatic spectrophotometry" over microbiological methods lies in the circumstance that the latter do not differentiate between various purines or nucleosides whereas the former method does, especially if the course of the enzymatic reactions is followed at two or three different wave lengths. Moreover, the existence of a bacterial ribosidase (59) makes it possible to differentiate between ribosides and deoxyribosides.

1. Purine Nucleoside Phosphorylase.

It was mentioned in the previous section that this enzyme which by earlier authors had been called "nucleosidase" and considered to be a hydrolytic enzyme, requires inorganic phosphate as well as hypoxanthine riboside (or guanine) in order to bring about the fission of the ribosidic linkage. Since it had been known from CORI's work (23) on the enzymatic splitting of polysaccharides as well as from DOUDOROFF's investigations (30) on the cleavage of sucrose by a bacterial enzyme that phosphate is a participant in these reactions and that 1-phospho compounds could be isolated, the observation made in the study of nucleoside phosphorylase justified a search for a suspected 1-phosphoribose compound.

Preliminary experiments had indicated that the incubation of liver nucleoside phosphorylase with inosine and inorganic phosphate gave rise to the formation of a new phosphoric ester which was highly acid-sensitive even at room temperature [KALCKAR (55)]. It was possible to detect an uptake of inorganic phosphate, whose estimation revealed that 1 mol. of phosphate was incorporated into an acid-labile ester for each mol. of liberated purine. The quantitative determination of this ester was greatly facilitated by employing the LOWRY-LOPEZ (71) phosphate estimation method which operates at p_H 4. The chemical properties of this new 1-phospho-ribose compound will be described later.

The following Section summarizes some of our recent knowledge concerning the properties of purine nucleoside phosphorylase.

Mammalian Purine Nucleoside Phosphorylase.

This enzyme occurs in most animal tissues. It is particularly abundant in liver and spleen whereas skeleton muscle contains only small amounts. Its localization within the liver cells has been studied by PRICE (84) using the HOGEBOOM technique. It was persistently found that more than 90 per cent of the enzyme was confined to the liquid phase which is obtained after centrifugation of the microsomes at 80000 g. It was observed recently that cell nuclei well isolated in non-aqueous media contain the main portion of nucleoside phosphorylase and adenosine deaminase (99 a).

Fractionation of the enzymes was obtained from rat and calf liver by means of centrifugation and ammonium sulfate precipitation [KALCKAR (55); FRIEDKIN and KALCKAR (38)].

The rat livers were usually perfused with chilled saline in order to remove all blood before mincing. The liver was minced and homogenized in distilled water with a Waring Blendor. After centrifugation at 3000 rpm. the supernatant was subjected to ammonium sulfate fractionation by the addition of 100 mg. of solid ammonium sulfate for each ml. of liquid plus 0.43 ml. of saturated ammoniacal ammonium sulfate (p_H 8.6). After some hours' standing at room temperature this mixture was spun at 10000 rpm. for 20 min. The enzyme was then precipitated from the clear supernatant by further addition of 200 mg. of ammonium sulfate per ml. After standing overnight the active fraction was centrifuged at 10000 rpm. The precipitate was dissolved in a small volume of twice distilled water and the solution dialyzed in cellophane against re-distilled water. The clear solution was frozen and stored at $-20°$ which resulted only in a slight loss of activity over a period of months.

Rat liver nucleoside phosphorylase was precipitated between 0.4 and 0.6 of ammonium sulfate saturation at $-23°$ C. Its activity was expressed as micro moles of hypoxanthine liberated per 10 min. upon addition of enzyme to a standard assay solution of inosine (100 μg. per ml. of 0.1 M phosphate at p_H 8) and an excess of xanthine oxidase. In the presence of the latter, the hypoxanthine released is immediately converted to uric acid which in turn can be quantitatively estimated by the increase of light absorption at 2930 Å as outlined above (p. 366).

The amount of protein was estimated by the ultraviolet absorption at 2800 and 2600 Å [KALCKAR (54)]. A freshly prepared and fractionated nucleoside phosphorylase is able to bring about the phosphorolysis of 1.8 μmoles of inosine per 10 min. per mg. of protein.

Recently, two other procedures were described which yield highly active preparations. In the method of ROWEN and KORNBERG (89), the starting material was beef liver-acetone powder.

This powder is extracted with 0.1 M Na_2HPO_4. The solution is fractionated with ammonium sulfate at p_H 5.0. The active fraction is dissolved in water, adsorbed on calcium phosphate gel and eluted with dilute phosphate buffer, p_H 7.5. Then follows adsorption on C_γ-alumina gel and elution with $M/50$ phosphate buffer, p_H 7.5. This final fraction represents a 60-fold purification over the crude extract, with a yield of 17 per cent.

PRICE (84) has isolated the most active nucleoside phosphorylase preparation described so far by a combination of alcohol fractionation and ionic effects as follows.

The clear red supernatant (from 2 hr. centrifugation at 0°, 10000 rpm.) is cooled to $-5°$ and 54 per cent ethanol is added until the mixture is adjusted to 18 per cent ethanol. The precipitate contains the activity leaving the hemoglobin in the supernatant. The main part of the guanase is also left in the latter. The precipitate is extracted twice, at $-5°$ with ethanol acetate mixtures (p_H 5.3), first with 14 per cent and then with 10 per cent ethanol. The activity remains in the precipitate which is suspended in 4 vol. of cold $M/50$ acetate buffer (p_H 5.2), centrifuged at 0° and dialyzed against cold $M/50$ acetate buffer, p_H 6.3. The active solution is treated with increasing amounts of tricalcium phosphate gel. The active fraction which is adsorbed after removal of other constituents is eluted first with acetate buffer and then with phosphate buffer.

This method allows a 100- to 200-fold purification, and the purine nucleoside phosphorylase thus obtained is free of phosphoribomutase, and adenosine deaminase, and is very low in guanase.

The specificity of liver nucleoside phosphorylase depends on the nature of the nitrogenous base as well as on that of the carbohydrate moiety. Hypoxanthine- and guanine ribosides undergo rapid enzymatic phosphorolysis; xanthine riboside, a slow phosphorolysis [FRIEDKIN (37)], while adenine riboside is not attacked at all. If adenine ribosides suffer phosphorolysis, this conversion is solely attributed to the presence of more or less potent adenosine deaminases. WANG and LAMPEN (105) report that an enzyme from *Lactobacillus pentosus* directly catalyzes the phosphorolysis or arsenolysis of adenosine. It is difficult indeed to predict which nucleosides will be attacked by nucleoside phosphorylase. For instance, according to FRIEDKIN (36), azaguanine, surprisingly enough, is capable of reacting with ribose-1-phosphate or deoxyribose-1-phosphate in the presence of horse liver purine nucleoside phosphorylase. He has separated by paper chromatography the ribosides and desoxyribosides of these purines and, furthermore, isolated the crystalline azaguanine deoxyriboside. The observation made by ROWEN and KORN-

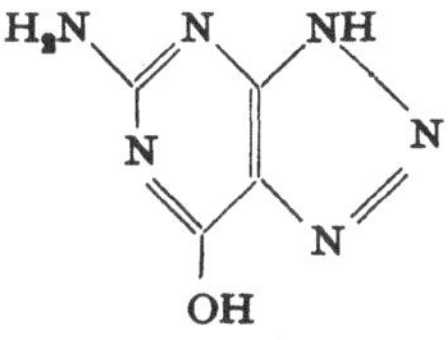

Azaguanine. 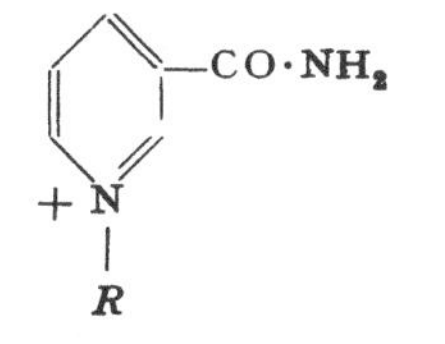   Nicotinamide riboside (R = ribose group). β-D-Ribose-1-phosphate.

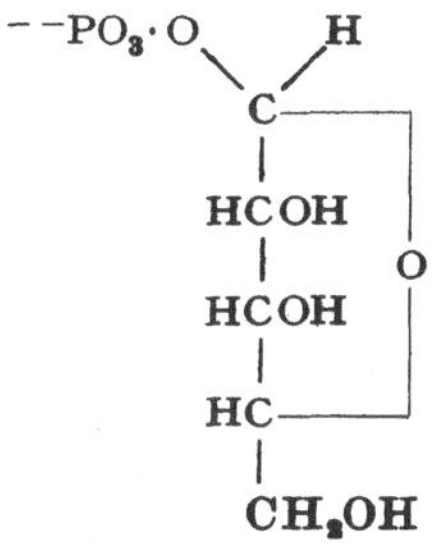

BERG (89) that nicotinamide riboside undergoes enzymatic phosphorolysis is also rather startling. These authors found that the enzyme isolated from beef liver catalyzes the following reaction:

Nicotinamide riboside + orthophosphate $\rightleftarrows$ nicotinamide + ribose-1-phosphate + H^+.

The liberation of hydrogen ion is due to the fact that a strong quarternary base, e.g., nicotinamide riboside, is converted into a weaker base, viz. nicotinamide. Evidently, the equilibrium must be strictly dependent on the p_H (89).

Synthesis of nicotinamide riboside as determined by fluorimetric methods could be demonstrated (89) at p_H 6. The equilibrium constant,

$$K_{eq} = \frac{\text{[nicotinamide riboside]} \times \text{[phosphate]}}{\text{[nicotinamide]} \times \text{[ribose-1-phosphate]} \times \text{[H+]}}$$

is approximately 10 which is very close to the value found in the inosine system [KALCKAR et al. (49)]. The nicotinamide riboside phosphorylase system is strongly inhibited by the addition of inosine (89).

Uric acid riboside [DAVIS, NEWTON and BENEDICT (25)] is not attacked. Pyrimidine ribosides undergo catalytic phosphorolysis for which, however, a different enzyme is responsible (cf. p. 372).

It should also be mentioned that ribose-1-phosphate is one of the specific substrates of nucleoside phosphorylase since, as will be explained in the following Sections (cf. p. 372), the ester just mentioned can be converted to inosine and phosphate, in the presence of hypoxanthine and enzyme. It is also possible to demonstrate an exchange between ribose-1-phosphate and inorganic phosphate if the latter is labelled with P^{32} [KALCKAR (59)]. The rate of exchange is, however, increased two-to threefold upon the addition of inosine (59). The seemingly direct exchange between ribose-1-phosphate and inorganic phosphate is analogous to that described for glucose-1-phosphate by DOUDOROFF et al. (31).

With regard to the effect of the carbohydrate part of the substrate, the following observations have been made. Glucose-1-phosphate is inactive. Hypoxanthine- and guanine-deoxyribosides undergo enzymatic phosphorolysis as rapidly as the corresponding ribosides. This feature does not apply to enzymes extracted from microorganisms (p. 394). Curiously enough a seemingly minor alteration in the riboside structure such as replacing furanoriboside-1-phosphate (the naturally occurring structure) by synthetic pyranoriboside-1-phosphate* made the compound inert towards the enzyme.

Neither isoguanosine nor adenine thiomethyl-riboside is attacked by mammalian nucleoside phosphorylase [SCHAEDEL et al. (91)].

The Relation Between Purine Riboside Phosphorylase and the Corresponding Deoxyriboside Phosphorylase.

As mentioned, the liver enzyme catalyzes the phosphorolytic cleavage of hypoxanthine ribosides and hypoxanthine deoxyribosides with about the same rate. It would be of interest to ascertain whether this preparation consists of a single enzyme or of two enzymes. If only a single enzyme is involved, then in the case of two substrates which are as similar in structure as hypoxanthine riboside and hypoxanthine deoxyriboside, one would expect the MICHAELIS-MENTEN constants to be essentially equal. Kinetic studies carried out with the spectrophotometric technique (38)

---

* A sample of this compound was kindly given us by Professor A. R. TODD (Cambridge).

showed that the respective constants for the two substrates were identical. Yet it is still conceivable that in spite of this identity (within experimental errors), separate enzymes were operating; in this case if hypoxanthine riboside was present at a concentration which yields maximum reaction rates, then the addition of hypoxanthine deoxyriboside should still further increase the velocity of the process as measured by the rate of uric acid formation. However, the same velocity was observed when the two substrates were present, either separately or jointly, in the incubation mixture at concentrations corresponding to the maximum rate for each substrate.

These observations are in accordance with the assumption that only a single enzyme is present, that it is completely saturated with respect to one substrate, and hence operating at maximum velocity. The addition of the second substrate, having the same affinity for the enzyme as the first one, can therefore not bring about any significant increase in the catalytic phosphorolysis. The absolute proof that only a single enzyme is involved in this process in the liver fractions described must, evidently, await the isolation of the enzyme in pure form.

It was found that a number of microorganisms contain purine riboside phosphorylase [KALCKAR (59)]. A study of dialyzed extracts obtained from *Escherichia coli* and *Lactobacillus casei* revealed that the purine ribosides are undergoing phosphorolytic fission at a rate which is five to ten times higher than that of the corresponding deoxyribosides [KALCKAR (59); HOFFMANN and MANSON (50)]. It should also be mentioned at this point that a group of lactic acid bacteria which more or less specifically require deoxyribosides as growth factors seems to be devoid of purine nucleoside phosphorylase. The significance of this observation will be discussed later.

Interfering Enzymes in Relation to the Assay of Purine Nucleoside Phosphorylase.

The purine nucleoside phosphorylase can be assayed partly by ultraviolet spectrophotometry and partly by the formation of acid-labile phosphate esters. For a useful application of the first-mentioned technique it is necessary that xanthine oxidase be added in excess and that uricase be absent. The latter would cause a degradation of uric acid and hence conceal the liberation of purine to be observed by the optical test. For a judgment of the specificity of substrates it is also important to remove such enzymes which could alter the structure of the nucleosides to be assayed. The most common interfering enzymes of this type are adenosine deaminase, guanase, and phosphatase. In certain bacterial extracts (e.g., *Lactobacillus helveticus*, *L. delbrückii*), the existence of a potent hydrolytic enzyme specific for the riboside series (ribosidase;

see below, p. 374) would completely conceal the existence of riboside phosphorolysis. Under such circumstances it is necessary, to separate the two enzymes or to take advantage of the fact that the ribosidase does not catalyze the hydrolysis of ribose-1-phosphate.

In order to follow the formation of ribose-1-phosphate, not only ordinary phosphatases should be removed but also a specific transmutase which effects the migration of the phosphate group in ribose-1-phosphate from the 1- to the 5-position. This enzyme, termed phosphoribomutase, will be described later (p. 375).

If one is satisfied with following the phosphorolysis by means of the ordinary tests for the uptake of inorganic phosphate, the presence of phosphoribomutase is not disturbing, provided that enzymes which bring about more or less complete cleavage of ribose phosphate have been removed.

2. Pyrimidine Nucleoside Phosphorylase.

DEUTSCH and LASER (27) found that bone marrow and kidney contain an enzyme which catalyzes the cleavage of pyrimidine nucleosides, ribosides as well as deoxyribosides. According to KLEIN (63), the enzyme which splits off purines from nucleosides (purine nucleoside phosphorylase) is not identical with pyrimidine nucleoside phosphorylase.

MANSON and LAMPEN (78) have recently drawn our attention towards the metabolism of pyrimidine nucleosides. They made the following observations. Incubation of thymidine with kidney or bone marrow extracts gives rise to liberation of thymine provided that the digest contains phosphate or arsenate. The deoxyribose part is preserved in the presence of arsenate as the free sugar (79) but decomposed in the presence of phosphate. In thymus extract MANSON found deoxyribose-5-phosphate as an endproduct of thymidine cleavage (77).

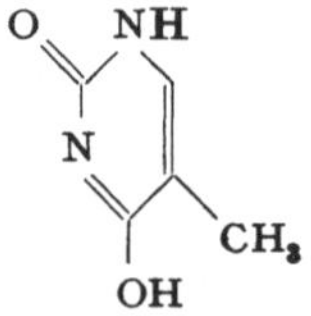

Thymine.

FRIEDKIN and ROBERTS (39) have recently shown that deoxyribose-1-phosphate and thymine yielded thymidine, in the presence of a fractionated enzyme preparation obtained from calf kidney.

3. Trans-N-Glycosidase.

MCNUTT (75, 76) observed that a crude, dialyzed extract from *Lactobacillus helveticus* catalyzes an exchange between purines and pyrimidines linked to deoxyriboside. This exchange is not phosphorolytic but seemingly direct, according to the scheme:

$$N^1\text{-deoxyriboside} + N^2 \rightleftharpoons N^2\text{-deoxyriboside} + N^1,$$

where N^1 and N^2 signify the nitrogen in a variety of purines or pyrimidines.

These types of exchange reactions were demonstrated by a combination of microbiological assays with acid hydrolysis. It is well-known that purine deoxyribosides are highly labile, whereas pyrimidine deoxyribosides are acid-stable. When, for example, the enzymatic digest at the onset of the experiment contained a free pyrimidine and a purine deoxyriboside, upon incubation, a significant amount of pyrimidine deoxyriboside had been formed, the conversion would manifest itself by the appearance of an acid-stable growth factor originating from an acid-labile one (76).

The trans-N-glycosidase can also be demonstrated in the ultraviolet spectrophotometer as follows: A dialyzed extract of *Lactobacillus helveticus* is incubated with hypoxanthine deoxyriboside and xanthine oxidase, in the absence of phosphate. If thymine is added, a marked increase in the rate of liberation of hypoxanthine is observed, made manifest by catalytic oxidation to uric acid and spectral reading at 2930 Å. The effect of thymine can be replaced by the addition of other pyrimidines or purines (see chapter IV, p. 387). Addition of phosphate has no effect; correspondingly, deoxyribose-1-phosphate is not attacked by this system.

Since the trans-N-glycosidase has not as yet been subjected to purification, only scanty data are available as to its properties. The catalytic activity of a crude, dialyzed *Lactobacillus helveticus* extract is variable.

A rather typical example of the rate of conversion was furnished by the following experiment: Hypoxanthine deoxyriboside in an amount corresponding to 0.1 μmol. was incubated with 1 μmol. of thymine in the presence of enzyme (corresponding to 0.42 mg. dry bacterial substance) at 37°; p_H was 8.5 and the volume 400 μl. Within six hours equilibrium was attained and about 80 per cent of the hypoxanthine deoxyriboside had been converted into the corresponding thymine compound.

Extracts prepared from *Lactobacillus delbrückii* show about the same trans-N-glycosidic activity as those of *Thermobacterium acidophilus R 26*. Living cells of *Leuconostoc citrovorum* were inactive in this system. These differences may have something to do with the fact that the two first-mentioned strains especially *Th. acidophilus* are able to respond to a variety of deoxyribosides whereas *L. citrovorum* responds exclusively to thymidine.

Does the presence or absence of a particular enzyme in bacterial extracts entitle us to establish correlations with nutritional requirements? This is probably still open to criticism. An investigation of artificially induced nutritional mutations in bacterial strains such as *Lactobacillus casei* or *Escherichia coli* with respect to some of these enzyme systems would seem to be justified in order to clarify this problem.

It will be mentioned later that ribosidase is present in extracts of a broad variety of bacterial strains requiring deoxyribosides as growth factors. It is interesting to compare the content of trans-N-glycosidase and nucleoside phosphorylase in two lactic acid bacterial strains, one of which *(L. delbrückii)* requires deoxyribosides and the other *(L. casei)* responds to free purines and pyrimidines (Table 1).

Table 1. Incubation of Hypoxanthine Deoxyriboside (80 μg.) with Bacterial Extracts (0.5 mg. protein) for 2 Hours at 35° and pH 7.0; Analysis of Protein-free Filtrates for Uric Acid [KALCKAR (59)].

Enzyme	Purine liberated (μg.)		
	no addition	PO_4^{---} added	thymine added
None	0	0	0
Lactobacillus casei	3.4	32.0	3.4
L. delbrückii	4.8	4.0	15.8

4. Ribosidase.

Ribosidase is an enzyme which catalyzes the cleavage of various purine ribosides, presumably by hydrolysis, since neither inorganic phosphate nor the presence of a base is required for its activity. It was found invariably in extracts from a group of lactic acid bacteria which under more or less specific conditions require deoxyribosides as growth factors e.g., extracts from *Lactobacillus delbrückii*, *L. helveticus*, *L. Leichmanni*, *Thermobacterium acidophilus*, and *Th. lactis 1* [KALCKAR (59)]. Extracts from *L. casei*, a lactic acid bacterium which requires purines but is able to synthesize deoxyribosides, contain nucleoside phosphorylase but neither ribosidase nor trans-N-glycosidase. The latter two enzymes seem also to be absent from *Escherichia coli* and from mammalian tissues.

Ribosidase brings about rapid liberation of hypoxanthine from inosine but not from hypoxanthine deoxyriboside. Unlike the purine nucleoside phosphorylases, ribosidase is able to attack adenosine directly, i.e., incubation of the latter with ribosidase gives rise to a liberation of free adenine. The fate of the ribose molecule has not yet been explored. It is important to emphasize that 1-phospho-furano-riboside (formula, p. 369) does not undergo hydrolysis if incubated with ribosidase. This enzyme may be of value in conducting assays for ribo-nucleoside compounds since it does not split deoxyribosides (59).

CARTER (*18*) has recently purified and described a non-phosphorolytic uridine nucleosidase occurring in yeast plasmolysates. This enzyme does not attack adenosine, guanosine, inosine, cytidine or thymidine. However, according to HEPPEL (*46*) yeast contains, (besides inosine phosphorylase), a ribosidase acting on the purine nucleosides.

5. Phosphoribomutase.

SCHLENK and WALDVOGEL (*93*) observed that incubation of adenosine with extracts obtained from various tissues results in the disappearance of pentose from the sample [DISCHE (*29*)], accompanied by an esterification of phosphate into an acid-stable linkage. This observation indicated an enzymatically catalyzed migration of the phosphate group of ribose-1-phosphate into another position, either 2, 3 or 5. The following observations support the assumption that ribose-5-phosphate is one of the primary rearrangement products of ribose-1-phosphate.

(a) Ribose-5-phosphate when added to various tissue extracts undergoes degradation during which the pentose disappears and, among other products, hexose-6-phosphate is formed. This is the same type of process which can be observed when adenosine or inosine is used as a substrate. Ribose-3-phosphate is inert in this system.

(b) COHEN and MCNAIR SCOTT (*22, 74*) have obtained evidence for the formation of ribose-5-phosphate by enzymatic oxidation of 6-phosphogluconic acid which is an oxidation product of hexose-6-phosphate. This has been confirmed by HORECKER and SMYRNIOTIS (*51*).

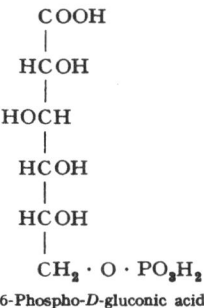

6-Phospho-*D*-gluconic acid.

(c) Muscle extracts which contain phosphoribomutase catalyze the conversion of ribose-5-phosphate to inosine, i.e., when this 5-ester is incubated with dialyzed muscle extract plus liver nucleoside phosphorylase plus hypoxanthine, about five per cent of the ribose-5-phosphate is converted to inosine [SABLE and KALCKAR (*90*); KLENOW (*66*)]:

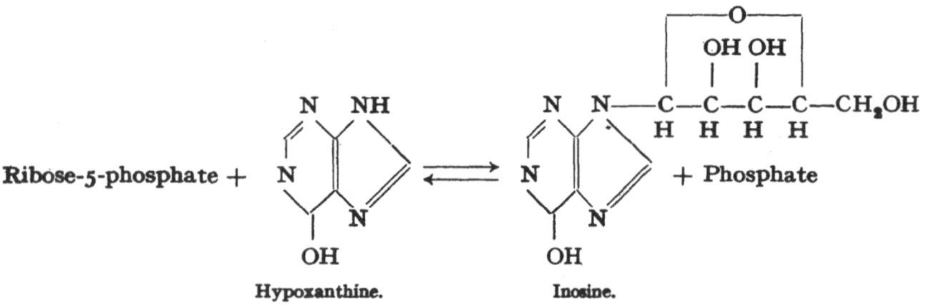

Ribose-5-phosphate + Hypoxanthine ⇌ Inosine + Phosphate

(Structures: Hypoxanthine and Inosine shown; inosine ribose chain: —C(OH)H—C(OH)H—CH—CH—CH₂OH with ring O)

In a study of phosphoribomutase cognizance must be taken of the fact that the enzyme is readily destroyed by air bubbling [ABRAMS and KLENOW (*1*)] and that the presence of even very moderate amounts of salts inhibits this enzyme as well as phosphoglucomutase [KLENOW (*66*); CORI et al. (*24*)]. It is likely that ribo- and the glucomutase are identical. According to KLENOW (*66*), synthetic α-glucose-1,6-diphosphate is able to increase the activity of phosphoribomutase. It was subsequently found (*66*) that ribose-1-phosphate plus α-glucose-1,6-diphosphate yield in the presence of phosphoribomutase a new diphosphopentose which is also able to activate purified phosphoribomutase. It is most likely that this is ribose-1,5-diphosphate and that this new di-ester acts like the corresponding glucose compound [CARDINI et al. (*17*); SUTHERLAND et al. (*101*)]. Deoxyribose-1-phosphate is also undergoing conversion in the presence of this enzyme, presumably to the 5-ester (*66*).

MANSON and LAMPEN (*77*) have collected chemical evidence proving that 5-ester is accumulating in thymus extracts incubated in the presence of certain purine or pyrimidine deoxyribosides.

6. Degradation and Synthesis of Ribose Phosphoric Esters.

As well known, most living organisms do not ferment or oxidize free ribose, nevertheless, the fermentation or oxidation of ribosides was found to proceed with great rapidity in most biological systems. Thus, LUTWAK-MANN (*72*) as well as STEPHENSON and TRIM (*99*) found that *Escherichia coli* ferments adenosine and inosine rapidly; no ribose could be discovered in the medium. According to DISCHE (*29*) human erythrocytes catalyzed the degradation of purine ribosides; the ribose disappeared and some phosphoric esters were formed which were dephosphorylated in the presence of alkali. The author suggests formation of phospho-triose catalyzed by a sort of aldolase. RACKER (*85*) reports the existence of a bacterial enzyme which catalyzes the oxidation of phospho-gluconic acid. In the same preparations he detected an enzyme which brings about the formation of triose-phosphate from ribose-5-

phosphate. RACKER found that crystalline muscle aldolase catalyzes the condensation of glycolaldehyde with triose-phosphate, yielding a pentose-5-phosphate; this pentose was not identical with ribose. Likewise it was observed (86) that hexose-diphosphate plus muscle aldolase (yielding a triose-phosphate), when incubated with acetaldehyde and an enzyme *ex E. coli* gave rise to the formation of a deoxypentose phosphate. This ester shows reducing power and, on the basis of its stability in acids and of the R_F value in ethanol-acetic acid may be identical with deoxyribose-5-phosphate.

There are two main possibilities with respect to the pathway of riboside formation from hexoses. Ribose phosphate could be formed, primarily as a 5-ester, by an enzymatic condensation between glycol aldehyde (or another C_2 fragment) and glyceraldehyde phosphate as suggested by DISCHE and by RACKER; however, it could also be the product of an oxidative decarboxylation of 2-keto-6-phospho-gluconate.

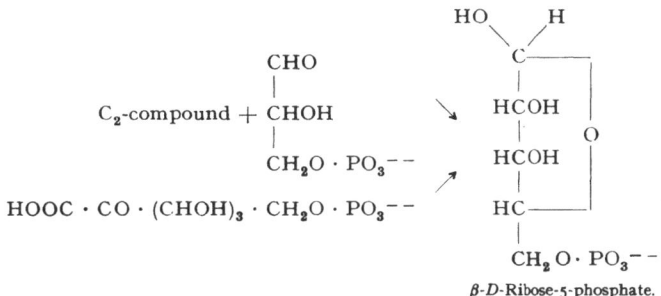

β-D-Ribose-5-phosphate.

This is supported by some observations made on a yeast enzyme system as described by DICKENS (28), LIPMANN (70), and WARBURG and CHRISTIAN (107). The yeast enzyme catalyzes the oxidation of glucose-6-phosphate, through 6-phospho-gluconic acid, to the 2-keto compound which is then rapidly decomposed.

Recently, COHEN and McNAIR SCOTT (22, 74) were able to present a clear-cut demonstration of ribose phosphate formation by enzymatic oxidation of 6-phospho-gluconic acid. The ribose ester was characterized by the strong suppression of the distribution coefficient, organic solvent/water (R_F) by borate. These authors also detected among the phosphate-free components a substance which might be an 1,2-enediol-pentose.

Their observation and another recent contribution to this problem by HORECKER and SMYRNIOTIS (51) are of decisive importance because these findings may explain how a *trans-cis*-hydroxy configuration in glucose can be converted into the *cis*-hydroxy arrangement of ribose. According to HORECKER and SMYRNIOTIS, 6-phospho-gluconate when incubated with a yeast enzyme was converted into a mixture of sugars which gave positive orcinol reaction and hence, were characterized as pentoses. The pentose phosphates were separated on a Dowex-1 column charged with formate ions and eluted with a formic acid buffer p_H 3. The first component, which was the more abundant the earlier the enzymatic oxidation had been interrupted, was a laevorotatory substance that was not oxidizable with bromine and gave positive resorcinol test. It was therefore characterized as a ketopentose phosphoric ester (ribulose phosphate). The dextrorotatory component which usually constituted 75 per cent of the pentose esters was found to be an aldopentose, yielding formate with periodate and forming cis-hydroxy complex with borate and copper ions; hence, it was identified with ribose-5-phosphate. The demonstration of the ketopentose ester as an intermediary stage to ribose-5-phosphate is a noteworthy discovery. HORECKER and SMYRNIOTIS (51) suggest 2-keto-6-phospho-gluconate as precursor of the ketopentose. They also emphasize that the formation of a ketopentose which presumably implies the existence of an enolic compound erases the last trans-hydroxy feature of the glucose series.

It will be recalled that according to SCHLENK and WALDVOGEL (93), ribose-5-phosphate can be enzymatically converted into hexose-6-phosphate. This transformation might either pass through the intermediary compounds described by HORECKER and SMYRNIOTIS or it could be an aldolase reaction in which the C_2 fragment was replaced by a C_3 fragment.

The degradation and synthesis of deoxyribose phosphoric esters may follow analogous trends. MANSON and LAMPEN (79) observed that washed *Escherichia coli* cells decomposed hypoxanthine deoxyriboside (HDR) whereby the hypoxanthine accumulated, but the deoxyribose disappeared. If arsenate was added to HDR plus *E. coli*, free deoxyribose accumulated; presumably deoxyribose-1-arsenate (the proposed primary arsenolysis product) undergoes non-enzymatic hydrolysis instantaneously. The same was observed with ribosides (79). Free deoxyribose and ribose, cannot be metabolized by the great majority of living cells. RACKER's recent observations (86) concerning the formation of a deoxypentose from acetaldehyde and phosphotriose in the presence of an enzyme from *E. coli* has been mentioned.

7. Nucleoside Deaminases.

Adenosine Deaminase. This group of enzymes catalyzes the hydrolysis of the 6-amino group of the adenine molecule which is bound to pentose.

Pharmacological Action of Adenosine Compounds. DRURY and SZENT-GYÖRGYI (32) showed more than twenty years ago that adenosine compounds when injected intravenously into mammals exerted a rapid depression of the arterial blood pressure.

This effect is due mainly to a peripheral vasodilation. There is also a depressing effect on the transmission of heart impulses. Adenosine, 5-adenylic acid and adenosine polyphosphate show about the same effect. Adenine does not possess any pharmacological action; neither do the deaminated products of adenosine compounds such as inosine and inosinic acid. Since the progressive depression in arterial blood pressure caused by extensive muscle lesions is not stopped or reversed by intravenous injections of adenosine or adenylic acid deaminases, it is concluded that the appearance of adenosine compounds in the blood stream cannot be a major factor in the etiology of this condition [KALCKAR and LOWRY (60)]. Intravenous injection of the same type of enzymes in an animal with depressed arterial blood pressure as a result of an injection of adenosine compound brings about a more or less complete restoration of normal blood pressure (60). It was found recently by CLARK et al. (20) that 2-chloro-adenosine exerts a much stronger depressor effect on the arterial blood pressure than does adenosine or 5-adenylic acid. The great increase in potency may be ascribed in part to the circumstance that 2-chloro-adenosine cannot be deaminated enzymatically.

a. Intestinal Adenosine Deaminase.

Active enzyme preparations can be obtained by the method of BRADY (*12*). It is also possible to obtain highly active phosphatase preparations which are very low in protein by a modification of the SCHMIDT-THANNHAUSER technique (*96*) [KALCKAR (*53*)], viz. at the stage in which aluminum gel is added, the adenosine deaminase is present in the precipitate from which it is eluted by means of phosphate. The activity of this enzyme can be studied by a micro method which takes advantage of the marked decrease in the optical density at 2650 Å that is typical for the deamination of adenine compounds to hypoxanthine compounds (*53*).

The intestinal enzyme from calf catalyzes the deamination of adenosine as well as of adenine deoxyriboside. Tissue extracts obtained from rabbit deaminate adenosine at about twice as high a rate as that of adenine deoxyriboside [BRADY (*12*)]. 2-Chloro-adenosine is not deaminated but exerts an inhibitory effect on adenosine [CLARK et al. (20)]. Isoguanosine (2-hydroxy-6-aminopurine-*D*-riboside) and adenine-thiomethyl-pentoside are not deaminated in the presence of the intestinal enzyme (SCHAEDEL et al. (*91*)]. Interestingly enough, 2-amino-adenosine is deaminated to

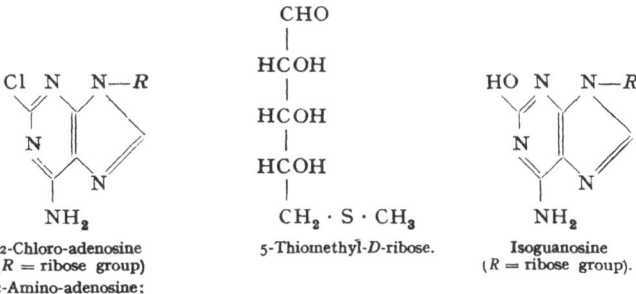

2-Chloro-adenosine
(*R* = ribose group)
(2-Amino-adenosine: NH₂ instead of Cl).

5-Thiomethyl-*D*-ribose.

Isoguanosine
(*R* = ribose group).

guanosine upon incubation with intestinal deaminase [KORNBERG and PRICER (67)].

b. Bacterial Adenosine Deaminases.

(1) Unspecific. *Escherichia coli* and *Lactobacillus casei* contain adenosine deaminases which catalyze the hydrolysis of the 6-amino group in adenosine as well as that in adenine deoxyriboside [KALCKAR, McNUTT and HOFF-JÖRGENSEN (61)].

(2) *L. helveticus* contains an enzyme which is active towards adenosine but inactive towards adenine deoxyriboside (61). A variety of bacteria contain cytosine and cytidine deaminases [SCHLENK (92)]; such processes can also be studied by means of ultraviolet spectrophotometry [CHARGAFF and KREAM (19); WANG et al. (106)].

Cytidine Deaminase [WANG, SABLE and LAMPEN (106)]. The decrease of the extinction at 2820 Å was selected for the spectrophotometric test of deamination of cytidine (or of the deoxyribose analogue) to uridine.

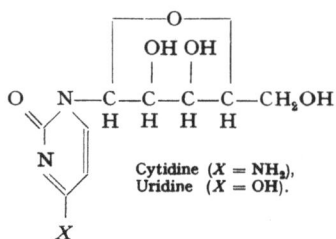

Cytidine ($X = NH_2$),
Uridine ($X = OH$).

Table 2. Action of *Escherichia coli* Extract on Purine and Pyrimidine Derivatives.

[WANG, SABLE and LAMPEN (106).]

(Each cuvette contained 0.1 ml. of extract equivalent to 5 to 10 mg. of wet cells.)

Compound	Concentration ($M \times 10^5$)	Wavelength (mμ.)	Time (min.)	ΔE observed
Adenine	5.71	265	10	0
Adenosine	4.67	265	13	0
Cytosine	8.36	282	14	0
Isocytosine	26.4	282	8	0
Cytidine	10.6	282	21	− 0.354*
Cytosine deoxyriboside	12.1	282	14	− 0.371
Cytidylic acid	4.30	282	17	0
Cytidylic acid + phosphatase	4.30	282	16	− 0.126**
Guanine	5.16	248	10	0
Guanosine	3.35	255	5	0
Uracil	9.01	282	14	0
Uridine	7.72	282	15	0

* Theoretical $\Delta E_{282} = -0.355$.
** Theoretical $\Delta E_{282} = -0.144$.

An *E. coli* strain (No. 15 or 9723 of the American type culture collection) was grown on beef extract (Difco), peptone, yeast extract, and glucose. After 24 hours incubation at 37° the cells were collected by centrifugation. Upon washing they were ground with alumina powder in a chilled mortar for just 5 minutes. The paste was then mixed with 10 to 20 vol. of cold 0.1 M phosphate buffer (p_H 7.6) and, after 30 minutes incubation at 2°, centrifuged at 20000 × g for 10 minutes. The supernatant was used as the enzyme source.

Table 2 illustrates the high degree of specificity of crude *E. coli* extracts, prepared by the above method, with respect to purine and pyrimidine deaminases. It is also noteworthy that the rate of deamination of cytosine deoxyriboside is three to four times higher than that of cytidine. The relatively slower deamination of cytidine does not seem to be attributed to a contamination with heavy metals during the cytidine preparation.

Adenosine deaminase in fungi has been studied by MITCHELL and MC ELROY (*81*).

III. Phospho-Ribosides.

1. Preparation and Properties of Ribose-1-phosphate.

Preparation of Ribose-1-phosphate (cf. formula on p. 369). In order to obtain this compound in a reasonable yield it is necessary to have the mammalian purine nucleoside phosphorylase freed from a number of enzymes which would induce the cleavage of the 1-ester. The phosphorylase solution should contain neither phosphatase nor ribosidase which would cause liberation of phosphate and purine, respectively. Moreover, the phosphoribomutase which catalyzes the transmutation of the 1-ester into the stable 5-ester should be removed as thoroughly as possible. Since the latter enzyme occurs in large amounts in liver and spleen that are also good sources for nucleoside phosphorylase, it is necessary either to purify the phosphorylase until it is freed from mutase or to attempt a specific inhibition of the latter.

As mentioned, it is possible, to purify the nucleoside phosphorylase to such an extent that hardly any mutase activity remains in the preparation. It has also been mentioned that various procedures which do not affect the phosphorylase cause a far-reaching inhibition of the phosphoribomutase. Provided that the phosphorylase had been freed from any other interfering enzyme, the following directions should be followed in order to obtain larger amounts of ribose-1-phosphate from inosine (ribose-1-hypoxanthine):

(a) Equivalent amount of inorganic orthophosphate (no arsenate!). (b) Excess of xanthine oxidase in order to remove, by oxidation, the hypoxanthine liberated by phosphorolysis. (c) Equilibration with oxygen and addition of catalase in order to ascertain that the enzymatic oxidation will proceed with maximum possible velocity.

It is also possible to prepare ribose-1-phosphate from inosine and phosphate using unfractionated liver extracts. However, since such an extract contains considerable amounts of phosphoribomutase which will convert ribose-1-phosphate to an acid-stable ester, it is necessary either to destroy or to inhibit this enzyme without affecting the nucleoside phosphorylase. The destruction of the mutase can be achieved by surface denaturation such as foam formation by means of bubbling [ABRAMS and KLENOW (1)]. Another, more easily reproducible technique by which the phosphoribomutase can be selectively inhibited, is the addition of a variety of salts [CORI et al. (24)]. Thus, using a dialyzed liver extract to which ammonium sulfate was added (until a concentration of 5 per cent was reached), ribose-1-phosphate was obtained in a yield of more than 40 per cent [KLENOW (66)]. The product behaved in the paper chromatogram as a homogeneous phosphate [solvent, propanol-ammonia; HANES and ISHERWOOD (45)]. By using a method analogous to that of FRIEDKIN (35), the crystalline cyclohexylamine salt of ribose-1-phosphate was obtained (66).

KLENOW (66) observed that most ribose-1-phosphate preparations contain traces of a pentose diphosphate which is an activator for phosphoribomutase.

A complete separation of the di-ester from ribose-1-phosphate was accomplished by chromatography on a formate-charged Dowex 2 column by ROWEN and KORNBERG (89).

Properties of Ribose-1-phosphate. This compound is usually isolated in form of its barium salt which is readily soluble in water; for enzymatic experiments the sodium or ammonium salt is used. The ester linkage is fairly stable at neutral and alkaline reactions.

Even at p_H 4 the 1-ester linkage is stable at room temperature for more than 24 hours; at p_H 1 it is hydrolyzed rapidly even at room temperature; at 25° and in 0.5 N sulfuric acid 50 per cent of the ester appears as inorganic phosphate after 2.5 min. incubation [LOWRY and LOPEZ (71)].

Ribose-1-phosphate can withstand boiling at p_H 9. It is non-reducing as determined by the aldose estimation; the appearance of an aldose group parallels the hydrolysis of the phospho ester linkage. These observations as well as the enzymatic formation of inosine strongly suggest the structure of a 1-aldo ester.

For a discussion of the β-furanoside structure cf. LELOIR (68a).

2. Enzymatic Synthesis of Ribosides.

It is possible to achieve an enzymatic synthesis of hypoxanthine riboside from hypoxanthine and ribose-1-phosphate by the following procedure.

The enzymes and the solution to be used should be free of inorganic phosphate. The ammonium salt of ribose-1-phosphate is incubated with hypoxanthine (formula, p. 376) and purine nucleoside phosphorylase. If equimolar amounts of the two substrates are applied, about 80 per cent of the 1-ester and hypoxanthine will be converted into hypoxanthine-1-riboside and inorganic phosphate. Correspondingly, equimolar amounts of inosine and inorganic phosphate in the presence of the enzyme are converted into the 1-ester and free hypoxanthine; and this reaction

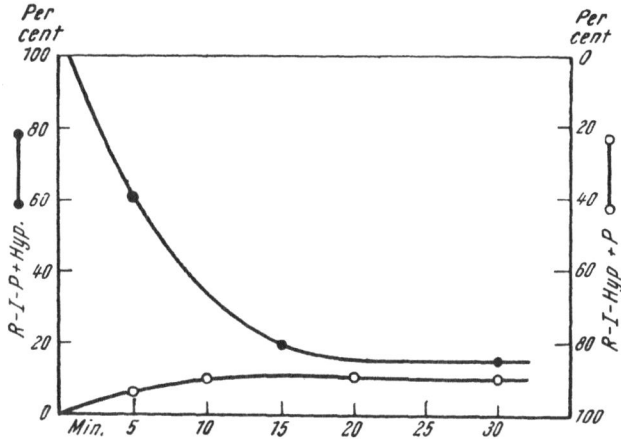

Fig. 1. Graphic illustration of the equilibrium, ribose-1-phosphate (R-1-P) + hypoxanthine (Hyp.) ⇌ ⇌ ribose-1-hypoxanthine (R-1-Hyp.) + phosphate (P). Abscissa, incubation time in minutes; ordinate, concentration of substrate mixture in percentage of initial concentration. The ordinate at the left applies to the mixture R-1-P + Hyp. (●); the ordinate at the right to the mixture R-1-Hyp. + P (○). [From: J. biol. Chemistry 167, 484 (1947).]

proceeds until 10 to 20 per cent of the initial products have been converted. The equilibrium is illustrated in Fig. 1. The equilibrium constant:

$$K = \frac{[\text{inosine}] \times [\text{phosphate}]}{[\text{hypoxanthine}] \times [\text{ribose-1-phosphate}]}$$

was found to be of the order of 10. This equilibrium is dependent on the hydrogen-ion concentration to the extent to which the second dissociation constant of ribose-1-phosphate differs from that of orthophosphate. In analogy with glucose-1-phosphate the pK_2 of ribose-1-phosphate is lower than that of orthophosphate.

FRIEDKIN (37) reports that in the slow enzymatic reaction taking place between equimolar amounts of ribose-1-phosphate and xanthine the equilibrium is not reached until more than 90 per cent of the ribose and xanthine are bound in ribosidic linkage.

The following simple thermodynamic considerations seem pertinent to stress this point. If purine nucleoside phosphorylase is capable of catalyzing an exchange between ribose-1-phosphate and adenine or 2,6-diamino-purine (2-amino-adenine), the detection of such a reaction would be difficult if the equilibrium were much less favorable as compared with hypoxanthine or guanine.

As mentioned previously, ROWEN and KORNBERG (*89*) have found the equilibrium constant,

$$K_{eq} = \frac{[\text{nicotinamide riboside}] \times [\text{phosphate}]}{[\text{1-phosphoriboside}] \times [\text{nicotinamide}] \times [\text{H}^+]}$$

to be approximately 10.

It should be mentioned that, although it has not been possible to observe a phosphorolysis of the glycosidic linkage of nucleotides, some observations indicate the formation of inosinic acid from ribose phosphate and hypoxanthine in the presence of phosphorylase. WAJZER (*102*) found that the ribose phosphate formed from inosine in the presence of crude calf liver enzyme, upon removal of inorganic phosphate (but preserving the enzyme), brings about a substantial disappearance of hypoxanthine. The formation of inosine did not account for more than 60 to 70 per cent; the remainder was accounted for by addition of intestinal phosphatase which indicated the formation of a nucleotide. Fractionation by chemical methods (nucleotide precipitants) pointed in the same direction (*103*).

It is not excluded that these findings may be connected with more recent observations made by WAJZER and BARON (*104*). These authors reported that calf liver extracts fractionated with ammonium sulfate (precipitate obtained by ammonium sulfate concentration between 0.5 to 0.7 of saturation) are able to catalyze the formation of a purine nucleotide from a ribose phosphate (obtained from yeast adenylic acid) and a purine such as hypoxanthine, adenine or guanine (but not xanthine). The yield was very satisfactory, i.e. 50 to 100 per cent of the calculated "ribose-3-phosphate" was converted into a nucleotide.

In the reviewer's opinion ribose-1-phosphate might play a role in nucleotide synthesis through reactions analogous to those discovered recently by LELOIR (*68*). If ribose-1-phosphate is able to form a pyrophosphate linkage with a nucleotide (through reactions with pyrophosphorolysis of adenosine- or uridine-5'-triphosphate, for example), then possibilities for the formation of a cyclic phosphate (1,2 or 1,3) do exist [cf. also FORREST and TODD (*34*)]. A subsequent reaction of such an intramolecular di-ester with a "completed" or "incomplete" purine molecule might have a very good chance of furnishing a high yield of nucleotide.

The interesting studies by GREENBERG (*43*) on pigeon liver, using ^{14}C formate, strongly indicate that inosinic acid is formed primarily, and hypoxanthine and inosine secondarily.

3. Preparation and Properties of Deoxyribose-1-phosphate.

When hypoxanthine deoxyriboside (or the corresponding guanine derivative) is incubated with liver nucleoside phosphorylase, cleavage takes place only in the presence of inorganic phosphate. Yet, no consumption of phosphate could be demonstrated even if the LOWRY and LOPEZ method was used (71). The existence of a phosphorolysis was, however, revealed when inorganic phosphate was determined in the form of ammonium magnesium phosphate [FRIEDKIN and KALCKAR (38)].

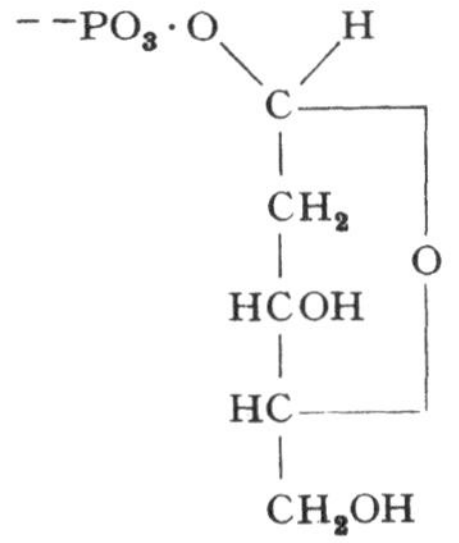

$$\begin{array}{c} --PO_3 \cdot O \diagdown \diagup H \\ C \\ | \\ CH_2 \\ | \quad O \\ HCOH \\ | \\ HC \\ | \\ CH_2OH \end{array}$$

2-Deoxy-D-ribose-1-phosphate.

The deoxyribose-1-phosphate was isolated by the following procedure.

Crystalline guanine deoxyriboside (from thymus nucleic acid) was incubated with a potent calf liver nucleoside phosphorylase to which a 10-fold excess (on a molar basis) of Na_2HPO_4 was added. Tris-(hydroxymethyl)-aminomethane-HCl buffer secured a constant p_H 8.5. After about 24 hours incubation at 20–25° the mixture was cooled to 0°, the liberated xanthine precipitate centrifuged off, and the supernatant deproteinized by three extractions with 3 vols. of *n*-butanol saturated with water. The denatured protein accumulated at the intersurface. The bulk of the remaining purine compounds was found in the butanol phase, whereby the deoxyribose phosphate and the inorganic phosphate remained in the aqueous phase. The inorganic phosphate was removed with ammoniacal barium acetate and, after the elimination of a small precipitate, the barium deoxyribose-phosphate was collected by precipitation with 4 vols. of ammoniacal ethanol. The crude barium salt was redissolved in water and reprecipitated by addition of 15 vols. of dry butanol. From this salt FRIEDKIN (35) prepared the cyclohexylamine salt of deoxyribose-1-phosphate which he obtained in pure, crystalline form (long needles).

The new ester proved to be more acid-labile than any other phosphoric ester hitherto described. It is rapidly hydrolyzed at room temperature even at p_H 4. Thus, incubation of the barium or the crystalline cyclohexylamine salt in an 0.05 M acetate buffer p_H 4 at 20° C. caused a release of inorganic phosphate corresponding to a 50 per cent hydrolysis of the deoxyribose-1-phosphate within 15 min. It was also shown that liberation of free aldehyde accompanies the dephosphorylation of the deoxyribose phosphate. Iodometric oxidation (adapted for microtitration) demonstrated that liberation of aldose sugar took place simultaneously with the appearance of inorganic phosphate. This was accepted as another indication of a 1-phospho linkage in the deoxyribose ester.

4. Enzymatic Synthesis of Hypoxanthine Deoxyriboside.

When deoxyribose-1-phosphate and hypoxanthine (formula; p. 376) are incubated with liver nucleoside phosphorylase in a phosphate-free medium, rapid synthesis of hypoxanthine deoxyriboside takes place [FRIEDKIN and KALCKAR (38)]:

1-phospho-deoxyriboside + hypoxanthine $\rightleftharpoons$ hypoxanthine deoxyriboside + phosphate.

This reaction can be performed directly, using the water-soluble barium or cyclohexylamine salt of the 1-ester. The linkage between deoxyribose and hypoxanthine is probably the same as that found in the deoxyribosides of deoxyribonucleic acid, since the enzymatically synthesized deoxynucleoside shows high growth promoting effect towards *Thermobacterium R 26* [HOFF-JÖRGENSEN, FRIEDKIN and KALCKAR (49)]. Table 3 summarizes an experiment in which the enzymatically formed deoxynucleoside was tested both by the enzymatic-optical and the microbiological techniques [HOFF-JÖRGENSEN (48)].

Table 3. Synthesis of Hypoxanthine Deoxyriboside with Crystalline Cyclohexylamine Deoxyribose-1-phosphate.

[HOFF-JÖRGENSEN, FRIEDKIN and KALCKAR (49).]

25 μl. of calf liver nucleoside phosphorylase (0.4 mg. of protein) were added to a mixture consisting of 2.5 μM of cyclohexylamine deoxyribose-1-phosphate (1.03 mg.) and 2.5 μM of hypoxanthine, in 400 μl. of 0.1 M tris-(hydroxymethyl)-amino-methane-HCl buffer, p$_H$ 7.4; temperature, 23.5°. Aliquots of 50 μl. each were taken at intervals as indicated, and the enzymatic reaction was stopped by heating for 2 min. at 100°. Other conditions are mentioned in the Table.

Experiment No.	Description	Deoxyriboside formed	
		Found in microbiological assays μM	Found by spectrophotometric data μM
1	0 time (before addition of enzyme)	0	
2	2 min. incubation		0.2
3	61 ,, ,,	1.9	2.1
4	75 ,, ,,	1.9	
5	75 ,, ,, enzyme alone	0	

IV. Trans-N-Glycosidic Reactions.

1. Non-participation of Ribose-1-Phosphate and Deoxyribose-1-Phosphate in Trans-N-glycosidic Reactions.

It should be mentioned that attempts to demonstrate deoxyribose-1-phosphate as a growth factor for *Thermobacterium R 26, Lactobacillus delbrückii* or *L. helveticus* have so far indicated that the 1-phospho deoxyribosyl compound cannot replace the 1-purine or 1-pyrimidine ribosyl derivative. It is also unlikely that the mechanism of the deoxyribosyl group transfer from one purine or pyrimidine to another should involve deoxyribose-1-phosphate as an intermediate. The following pertinent observations were made by McNutt (76): (a) None of the purines or pyrimidines which enter into the exchange reaction does react with deoxyribose-1-phosphate in the presence of the *L. helveticus* enzyme. (b) Addition of larger amounts of phosphate does not cause even a slight inhibition of the trans-N-glycosidase reaction. (c) Deoxyribose-1-phosphate, in addition to its inactivity in the exchange mechanism, does not seem to be metabolized by the *L. helveticus* enzyme, i.e., the presence of the latter enzyme does not seem to accelerate the slow non-enzymatic disappearance of deoxyribose-1-phosphate (cf. also McNutt's contribution, on p. 401 of the present volume).

2. Trans-N-Glycosidic Reactions in the Deoxyribose Nucleoside Series.

It has been mentioned previously that a number of bacteria which more or less require deoxyribosides as growth factors contain the enzyme trans-N-glycosidase which, without intervention by inorganic phosphate, brings about an exchange of purines and pyrimidines in deoxyribosides.

For the identification of the compounds formed by the trans-N-glycosidic reaction, among others, paper chromatography was applied. The most useful solvent mixtures for this purpose were found to be: *n*-butanol-water-ethylacetate-morpholine; *n*-butanol-water-ammonia, and *n*-butanol-water-morpholine-methylglycol.

When, for example, hypoxanthine deoxyriboside is incubated in the presence of dialyzed *Lactobacillus helveticus* extract together with thymine or uracil, pyrimidine deoxyribosides are formed and hypoxanthine is liberated. The same type of reaction takes place if guanine deoxyriboside is incubated with thymine or uracil. The reaction is reversible. When, for instance, adenine is incubated with uracil deoxyriboside, adenine deoxyriboside is formed and uracil liberated. Moreover, the enzyme will catalyze an exchange of one pyrimidine for another. Thymine and uracil, for example, were shown to exchange enzymatically with the cytosine of cytosine deoxyriboside. Likewise, the enzyme will catalyze the exchange of one purine for another. It can be shown directly that guanine deoxyribosides plus adenine undergo partial conversion to

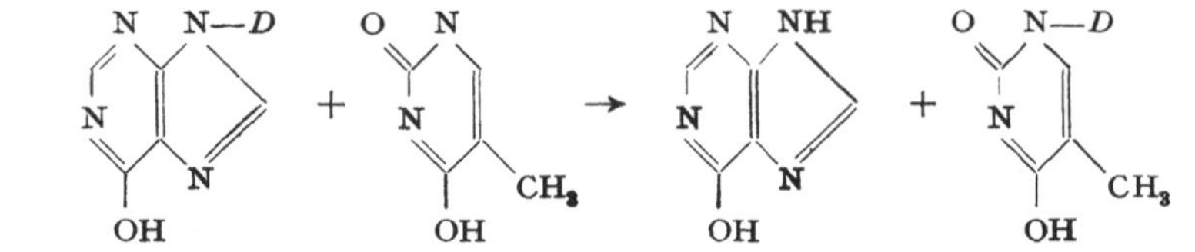

Hypoxanthine-deoxyriboside (D = deoxyriboside group). + Thymine. → Hypoxanthine. + Thymine-deoxyriboside (thymidine).

adenine deoxyribosides and free guanine. In the case of the corresponding interaction between adenine and hypoxanthine deoxyriboside, the demonstration of the formation of adenine deoxyriboside and hypoxanthine would be ambiguous inasmuch as a corresponding result could be obtained by a transamination reaction [STEPHENSON and TRIM (99)].

In order to obtain unambiguous proof of a corresponding reaction, hypoxanthine deoxyriboside plus the enzyme were incubated together with ^{14}C labelled adenine [CLARK and KALCKAR (21)]. The principle of this experiment was as follows: If the bacterial enzyme catalyzes exclusively the transaminase reaction, radioactivity will be confined to the adenine added and the hypoxanthine formed. Should, however, a reaction of the trans-N-glycosidase type be solely catalyzed, the radioactivity will be confined to the added adenine and the adenine deoxyriboside formed. Furthermore, the radioactivity would be distributed in all four components of the equilibrium system in case the bacterial extract should catalyze both types of reaction. The result showed clearly that the bacterial extract exclusively catalyzes the reaction of the trans-N-glycosidase type, since the adenine deoxyriboside isolated from a paper chromatogram spot showed as high specific radioactivity as the ^{14}C compound administered; in contrast, the hypoxanthine was devoid of radioactivity [KALCKAR et al. (61)]. About 75 per cent of the hypoxanthine deoxyriboside (HDR) was converted to adenine deoxyriboside (ADR), when a five- to sixfold excess of free adenine was added.

$$K_{eq} = \frac{[\text{adenine}] \times [\text{HDR}]}{[\text{ADR}] \times [\text{hypoxanthine}]}$$

had the order of magnitude of 5.

3. Enzymatic Formation of New Deoxyribosides.

The bacterial trans-N-glycosidase is able to catalyze the formation of deoxyribosides which have hitherto not been detected in nature. This is especially interesting because this enzyme is endowed with some specificity. A number of purines such as uric acid or 2,6-diamino-purine or some pyrimidine derivatives (methyl thiouracil, 4,5,6-triamino-pyrimidine) were found to be inactive in this test. Yet the enzyme can give rise to the synthesis of deoxyribosides which are supposed to occur exclusively as degradation products of naturally occurring deoxy-nucleosides. There is strong indication, for instance, that xanthine deoxyriboside or uracil deoxyribosides are formed by addition of the two respective bases to an incubation mixture of the enzyme and a deoxyriboside compound such as thymidine [McNUTT (76)]. A nitrogenous base like 5-methyl-cytosine has been described as a constituent of certain microorganisms, e.g. tubercle bacilli. It is not without interest that incubation of 5-methyl-cytosine with guanine deoxyriboside brings about utilization of the latter with the concomitant formation of a new deoxyriboside. McNUTT (76) could show that the unknown deoxyriboside, upon acid hydrolysis, gave rise to an ultraviolet absorbing base which had an R_F value corresponding to that of 5-methyl-cytosine. Moreover, upon deamination of the new deoxyriboside by means of nitrous acid, a deoxyriboside was formed which had growth promoting activity for *Leuconostoc citrovorum* and showed an R_F value corresponding to thymidine (76).

It is of special importance that 4-amino-imidazole-5-carboxamide which is supposed to be a precursor of purines [STETTEN and FOX (*100*); SHIVE et al. (*98*); SHAW and WOOLLEY (*97*)], seems to react with thymidine

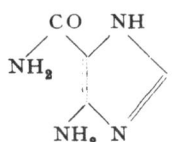

4-Amino-imidazole-5-carboxamide.

in the presence of the enzyme from *Lactobacillus helveticus* (76). The product formed when this "incomplete purine" is incubated with thymidine or deoxyribosides is different from all the other known deoxyribosides. The new deoxyriboside is, interestingly enough, active as a growth factor for *Thermobacterium acidophilus* R 26. The identity of the new compound with the deoxyriboside of 4-amino-imidazole-5-carboxamide seems to be secured by the following observations (76): Upon acid hydrolysis a base was liberated which (a) showed the characteristic R_F value of the free carboxamide, (b) had an absorption maximum at 2640 Å; and (c) gave a positive BRATTON-MARSHALL test. The respective quantitative results gained by the two latter methods were in good agreement.

The demonstration of an enzymatic deoxyriboside synthesis by catalyzing an "incomplete" purine structure such as 4-amino-imidazole-5-carboxamide is noteworthy and may have wider bearings. It is not excluded that the biosynthesis of ribosides and deoxyribosides in mammalian tissues as well as in microorganisms primarily passes through such "incomplete" purine structures.

Table 4 summarizes some experiments on trans-N-glycosidase from a variety of nitrogenous bases [McNUTT (76)].

Table 4. **Ability of Various Purines and Related Compounds to Enter into the Exchange Reaction.** 0.368 μmoles of thymine deoxyriboside, 500 μg. of the given compound in 500 μl. of alanyl-glycine buffer (p_H 8.5), and 500 μl. (1.7 mg. dry weight) of *Lactobacillus helveticus* enzyme (dialyzed 4 days). Incubation, 10 hours at 37° [McNUTT (76)].

Purine or related compound	Millimoles of purine deoxyribose formed
Adenine	0.260
Guanine	0.155*
Hypoxanthine	0.295
Xanthine	0.300
4-Amino-5-imidazole	0.295
Carboxamide HCl	0.303
Formamido malonamamidine HCl	0.020
Uric acid	− 0.020
2,6-Diaminopurine, $1/_2$ H_2SO_4, $1/_2$ H_2O	0.030
5-Methylcytosine**	0.200**
Formyl-4,5,6-triaminopyrimidine	− 0.011
Dihydrothymine***	0.020

V. Phosphorylation of Nucleosides.

As well known, the two main types of nucleotides are those which carry the phosphate group in the 5-position of the pentose (typical representative, muscle adenylic acid) and those which are phosphorylated at the 3rd carbon of the pentose (typical representatives, cleavage products of ribonucleic acids).

An enzymatic phosphorylation of a nucleoside to give a nucleotide was first demonstrated by OSTERN and co-workers (*82*). They found that adenosine is phosphorylated by yeast in the presence of glucose to adenosine-5-phosphate and adenyl-pyrophosphate. CAPUTTO (*16*) reported that yeast juice as well as extracts from liver and kidney contain an enzyme which catalyzes the reaction,

---

* Since guanine is poorly soluble, its concentration could not be brought to that of the other compounds tested.

** In this experiment hypoxanthine deoxyriboside was the donor.

*** 0.440 millimoles of cytosine deoxyriboside (a gift from Dr. H. K. MITCHELL), incubated with 500 μg. of dihydrothymine.

ATP + adenosine → ADP + 5-adenylic acid.

This enzyme was called adenosinekinase. Its p_H optimum lies around 7, whereas the enzyme originating from liver or kidney, surprisingly enough, showed a p_H 5 optimum.

KORNBERG and PRICER (67) have recently purified the yeast enzyme which catalyzes the reaction,

ATP + Adenosine → ADP + adenosine-5-phosphate

and demonstrated that, in conjunction with phosphopyruvate and with pyruvate, phosphokinase and myokinase (from muscle), this enzyme brings about the following reaction:

Adenosine + 3-phosphopyruvate → ATP + 3-pyruvate.

The same reactions take place with 2-amino-adenosine (2,6-diamino-9-β-D-ribofuranosyl-purine). The authors have isolated 2-amino-adenosine triphosphate and shown it to act as a phosphate donor in various enzyme systems. These findings, in conjunction with the observation made by ROWEN and KORNBERG (89) that 2-amino-adenosine, in the presence of adenosine deaminase, can be deaminated to give guanosine, add to the significance of the observation by BROWN and co-workers (14) that administration of labelled 2,6-diamino-purine results in a high concentration of the isotope in the guanine group of the pentose nucleic acid.

VI. Incorporation of Purines and Pyrimidines into Nucleic Acids.

In vivo Studies with Labelled Purines.

An extensive discussion and bibliography of this field will be found elsewhere [BROWN (13); BUCHANAN (15)].

The formation of nucleic acid purines seems to follow two pathways. The first and most physiological one can be studied by means of labelled glycine, formate and ammonia. The non-growing liver incorporates these constituents into the purines ribose nucleic acids (RNA) and deoxyribose nucleic acid (DNA). The rate of incorporation is greatly increased in the regenerating liver. There is evidence that ribosides or ribotides with "incomplete" purine rings are intermediates in this pathway [GREENBERG (43)].

The other mechanism by which the purine of a nucleic acid is renewed can be studied by means of labelled purines. BROWN and his colleagues (13, 14) found that labelled adenine and 2,6-diaminopurine are the only two purines which bring about a significant amount of labelling of the adenine and guanine of the ribonucleic acid of the adult rat liver. The adenine ring seems to be incorporated intact as judged from experiments performed with triple-labelled adenine [MARRIAN et al. (80)]. Guanine, labelled in the 8-position with ^{14}C is incorporated at a rate of not more than 1–2 per cent of that of adenine [BALIS (6)]. Hypoxanthine is not incorporated in this system. The DNA purines of the rat liver take up adenine exclusively during regeneration [FURST et al. (40)], in contrast with the observations made by HAMMARSTEN and his group (44) who used labelled glycine; it was shown that the DNA purines of the non-growing liver are not static as judged on the basis of this method. This has been confirmed by FURST and BROWN (40) as well as by LE PAGE and HEIDELBERGER (69). The rate of renewal of both nucleic acid purines increases during growth and regeneration. Administration of labelled purine nucleosides did not yield any incorporation of labelled nucleic acid purines in the rat liver [KERR et al. (62)]. ROLL and WELIKY (88) have found recently that labelled purine nucleotides can be used in the renewal of RNA purines. It was observed that 3 per cent of the adenine and 1 per cent of the guanine were incorporated into RNA upon the administration of labelled adenylic acid. Guanylic acid was incorporated exclusively in the RNA guanine (2 per cent of the material given).

It has already been pointed out that rat liver nucleoside phosphorylase does not catalyze the incorporation of adenine or diamino purine, the two purines which *in vivo* are incorporated into the rat liver on a large scale. The discrepancy will remain unexplained until labelled pentose compounds will become available. If the ribose moiety of inosine is incorporated into RNA, it may be worth while to consider the possibility of an exchange of purines (e.g., adenine *versus* hypoxanthine) on the nucleotide level. Such an exchange might be of the trans-N-glycosidic type [KALCKAR (57)].

In vivo Studies with Labelled Pyrimidines.

In this section only that part of our limited knowledge about pyrimidine incorporation will be listed which seems most pertinent in a discussion of nucleoside metabolism.

The HAMMARSTEN school has shown that orotic acid* is incorporated into RNA and DNA pyrimidines of the rat (9). The other free pyrimidines

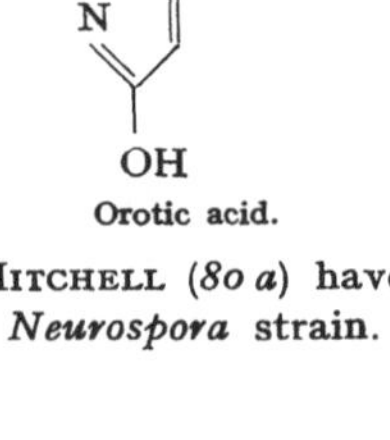

```
      O   NH
       \ / \
        Y   \—COOH
        |   /
        N  /
         \/
         |
         OH
```
Orotic acid.

---

* MICHELSON, DRELL and MITCHELL (80 a) have recently isolated orotic acid riboside from a pyrimidine-less *Neurospora* strain.

including thymine are not incorporated [PLENTL and SCHOENHEIMER (83)]. In striking contrast with the observations made in the purine series, pyrimidine nucleosides are incorporated into nucleic acid. HAMMARSTEN, REICHARD and SALUSTE (44) found that pyrimidine ribosides can be incorporated into RNA and to a small extent, even into the DNA of the rat liver. There occurred less dilution of the isotope with cytidine as a source than with uridine. The fact that there was an unmistakable incorporation of pyrimidine ribosides into DNA poses the problem of a conversion of ribosides into deoxyribosides.

Recently, REICHARD and ESTBORN (87) reported that ^{15}N-labelled pyrimidine deoxyribosides (from *Escherichia coli*, grown in ^{15}NH$_4^+$ containing medium) are utilized by the non-growing rat liver for replacement of DNA pyrimidines. Thymidines were particularly active. Deoxyribo-cytidine contributed both to the cytosine and to the thymine of the DNA while hypoxanthine deoxyriboside did not contribute to the DNA purines. The renewal of the DNA pyrimidines in the non-growing rat liver by pyrimidine deoxynucleosides once more contrasts with the observations made on the administration of adenine. The rate of incorporation of the pyrimidines of DNA was increased at least 3-fold during regeneration. There was no incorporation into PNA. The inertness of free thymine in the intact rat makes it difficult to understand the function of pyrimidine nucleoside phosphorylase which catalyzes the incorporation of thymine but not that of orotic acid.

In general, the possibility of formation of nucleosides with "incomplete" purines and pyrimidines is of interest for an understanding of nucleic acid biosynthesis. This aspect may also be of importance in an attempt to interpret the function of nucleoside phosphorylases.

In vitro Studies with Labelled Purines.

ABRAMS and GOLDINGER (4) found that both adenine and guanine (labelled in 8-position with ^{14}C) are incorporated into the PNA of hyperplastic bone marrow. Both compounds were also incorporated to a small extent into DNA. In the formation of PNA about 10 per cent of the adenine formed guanine, and about the same amount of guanine formed adenine. Incorporation of carboxyl-labelled glycine into nucleic acids was very much smaller than that of purines. The authors also found that labelled adenine gave rise to both labelled hypoxanthine and labelled guanine in the incubation medium, even under anerobiosis when no incorporation into nucleic acids took place. The labelled hypoxanthine had a very high isotope concentration, indicating the formation of this compound through direct deamination of adenine. It might, however, also have been formed through a small metabolic cycle, for instance,

Adenine ⇌ adenosine ⇌ inosine ⇌ hypoxanthine.

KERR et al. (62) reported the incorporation of adenine into PNA by surviving rat liver slices. GOLDWASSER (42) found that rat spleen slices are appreciably more active in the incorporation of adenine into PNA than are liver slices. The same author also observed a formation of acid-insoluble and especially acid-soluble nucleotides from labelled adenine in pigeon liver homogenates.

Studies on the Amphibian and Echinoderm Egg.

According to ABRAMS (2), developing embryos of *Arbacia* incorporate ^{14}C-labelled glycine into the purines of PNA and DNA. The isotope concentration in the latter is about 10-fold higher than in the former. It is therefore rather unlikely that riboside compounds could serve as precursors of DNA in this organism. The isotope concentrations in carbon No. 4 of the DNA purines amounted to about 15—30 per cent of that of the added glycine. Since the glycine in some of the experiments was added at a time when there was only one nucleus, ABRAMS raised the question why the carbon 14 of the DNA purines of embryos, having several hundred nuclei, do not show as high an isotope concentration as the administered glycine. This question is noteworthy since there was no sign of a significantly larger glycine pool. Thus, the possibility of the existence of a purine deoxyriboside pool (endogenous source) would arise.

The existence of a large cytoplasmic pool of deoxyribosides in frog eggs was demonstrated recently by ZEUTHEN and HOFF-JÖRGENSEN (*109*). They found both high- and low-molecular deoxyriboside compounds. It might be of interest to investigate whether or not a part of the deoxyriboside store shows a preponderance of adenine compounds, since according to ABRAMS (*2*) the isotope concentration of DNA adenine was less than half of that of guanine. This ratio could be taken as an indication of the presence of an adenine deoxyriboside store which is about twice as large as the corresponding guanine store.

Studies on Micro-organisms.

The first demonstration that glycine is incorporated into nucleic acid purines was made on yeast by ABRAMS, HAMMARSTEN and SHEMIN (5). A study of the incorporation of labelled adenine into microorganisms offers also a number of interesting aspects. For instance, it should be possible to clarify whether bacterial strains (or artificially produced and selected mutants) which require purines as growth factors are unable to perform a synthesis of nucleic acid purines or whether the added purines are primarily required for other purposes. ABRAMS (*3*) has described a yeast mutant (ultraviolet mutant from *Saccharomyces cerevisiae*) which did not grow unless supplied with adenine or hypoxanthine; and yet, it seemed to synthesize nucleic acid purines from glycine as readily as did the wild type. KALCKAR and McNUTT (*58*) have shown that *Lactobacillus casei* grown in the absence of folic acid incorporates (without isotope dilution) adenine labelled with ^{14}C in the 8-position [CLARK and KALCKAR (*21*)] into both the adenine and the guanine moieties of the nucleic acids.

In the presence of a minute amount of folic acid, the uptake of adenine from the medium is greatly decreased and, correspondingly, the isotope concentrations of the two nucleic acid purines are decreased to the same extent, indicating synthesis of purines from other constituents. BALIS et al. (*7*) report that ^{14}C-labelled guanine added to *L. casei* transmits isotope to both nucleic acid purines. In other words, guanine can also be converted into adenine. Such an interconversion does not exist in yeast in which ^{14}C adenine is incorporated into both of the purines of RNA, whereas labelled guanine was found exclusively in the RNA guanine. Nucleosides and nucleotides were only used to a very slight extent.

The role of B_{12} (cyano-cobaltamine) in the synthesis of deoxyribosides in microorganisms is still obscure. There is no direct proof that it plays a part in reduction of ribosides to deoxyribosides, if such a reaction exists.

References.

1. ABRAMS, A. and H. KLENOW: Phosphoribomutase and Enzymatic Preparation of Ribose-1-Phosphate. Federat. Proc. (Amer. Soc. exp. Biol.) **10**, 153 (1951).
2. ABRAMS, R.: Synthesis of Nucleic Acid Purines in the Sea Urchin Embryo. Exp. Cell Research **2**, 235 (1951).
3. — Purine Synthesis in a Purine-requiring Yeast Mutant. J. Amer. chem. Soc. **73**, 1888 (1951).
4. ABRAMS, R. and J. M. GOLDINGER: Utilization of Purines for Nucleic Acid Synthesis in Bone Marrow Slices. Arch. Biochemistry **30**, 261 (1951).
5. ABRAMS, R., E. HAMMARSTEN and D. SHEMIN: Glycine as a Precursor of Purines in Yeast. J. biol. Chemistry **173**, 429 (1948).
6. BALIS, M. E.: Utilization of Guanine by the Rat. (Abstract of Paper presented at Amer. chem. Soc. Meeting, April, 1951.)
7. BALIS, M. E., G. B. BROWN, G. B. ELION, G. H. HITCHINGS and H. VANDER WERFF: On the Interconversion of Purines by *Lactobacillus casei*. J. biol. Chemistry **188**, 217 (1951).
8. BALL, E. G.: Xanthine Oxidase: Purification and Properties. J. biol. Chemistry **128**, 51 (1939).
9. BERGSTRÖM, S., H. ARVIDSON, E. HAMMARSTEN, N. A. ELIASSON, P. REICHARD and H. VON UBISCH: Orotic Acid, a Precursor of Pyrimidines in the Rat. J. biol. Chemistry **177**, 495 (1949).
10. BIELSCHOWSKY, F. u. W. KLEIN: Experimentelle Studien über den Nucleinstoffwechsel. XXVII. Mitt. Über die fermentative Aufspaltung der Thymusnucleinsäure mit Nucleotidase aus Darmschleimhaut. Die Isolierung der Nucleoside der Thymusnucleinsäure. Hoppe-Seyler's Z. physiol. Chem. **207**, 202 (1932).

11. BRADY, T. G.: Isolation of Adenine-Desoxyriboside from Thymusnucleic Acid. Biochemic. J. **35**, 855 (1941).
12. — Adenosine Deaminase. Biochemic. J. **36**, 478 (1942).
13. BROWN, G. B.: Biosynthesis of Nucleic Acids in the Mammal. Federat. Proc. (Amer. Soc. exp. Biol.) **9**, 517 (1950).
14. BROWN, G. B., P. M. ROLL, A. A. PLENTL and L. F. CAVALIERI: The Utilization of Adenine for Nucleic Acid Synthesis and as a Precursor of Guanine. J. biol. Chemistry **172**, 469 (1948).
15. BUCHANAN, J. M.: Biosynthesis of Purines. J. cellular comparat. Physiol. **38**, Suppl. 1, **143** (1951).
16. CAPUTTO, R.: The Enzymatic Synthesis of Adenylic Acid; Adenosinekinase. J. biol. Chemistry **189**, 801 (1951).
17. CARDINI, C. E., A. C. PALADINI, R. CAPUTTO, L. F. LELOIR and R. E. TRUCCO: The Isolation of the Coenzyme of Phosphoglucomutase. Arch. Biochemistry **22**, 87 (1949).
18. CARTER, C. E.: Partial Purification of a Non-phosphorylytic Uridine Nucleosidase from Yeast. J. Amer. chem. Soc. **73**, 1508 (1951).
19. CHARGAFF, E. and J. KREAM: Procedure for the Study of Certain Enzymes in Minute Amounts and its Application to the Investigation of Cytosine Deaminase. J. biol. Chemistry **175**, 993 (1948).
20. CLARK, D. A., J. DAVOLL, F. S. PHILIPS and G. B. BROWN: Vasodepressor Activity of Adenosine, 2-Chloro-Adenosine and Related Nucleosides. Federat. Proc. (Amer. Soc. exp. Biol.) **10**, 286 (1951).
21. CLARK, V. M. and H. M. KALCKAR: A Synthesis of Adenine Labelled with ^{14}C. J. chem. Soc. (London) **1950**, 1029.
22. COHEN, S. S. and D. B. McNAIR SCOTT: Formation of Pentose Phosphate from 6-Phosphogluconate. Science (New York) **111**, 543 (1950).
23. CORI, C. F., G. T. CORI and A. A. GREEN: Crystalline Muscle Phosphorylase. III. Kinetics. J. biol. Chemistry **151**, 39 (1943).
24. CORI, G. T., S. P. COLOWICK and C. F. CORI: The Enzymatic Conversion of Glucose-1-Phosphoric Ester to 6-Ester in Tissue Extracts. J. biol. Chemistry **124**, 543 (1938).
25. DAVIS, A. R., E. B. NEWTON and S. R. BENEDICT: The Combined Uric Acid in Beef Blood. J. biol. Chemistry **54**, 595 (1922).
26. DAVOLL, J., B. LYTHGOE and A. R. TODD: Experiments on the Synthesis of Purine Nucleosides. XIX. A Synthesis of Adenosine. J. chem. Soc. (London) **1948**, 967.
27. DEUTSCH, W. u. R. LASER: Experimentelle Studien über den Nucleinstoffwechsel. XIX. Mitt. Zur Kenntnis der Nucleosidase. Verhalten einer Nucleosidase aus Rinderknochenmark zu einem Spaltprodukt der Thymusnucleinsäure. Hoppe-Seyler's Z. physiol. Chem. **186**. 1 (1929).
28. DICKENS, F.: Oxidation of Phosphohexonate and Pentose Phosphoric Acids by Yeast Enzymes. I. Oxidation of Phosphohexonate. II. Oxidation of Pentose Phosphoric Acids. Biochemic. J. **32**, 1626 (1938).
28 a. DIMROTH, K., L. JAENICKE und D. HEINZEL: Die Spaltung der Pentosenucleinsäure der Hefe mit Bleihydroxyd. (I. Mitt. über Nucleinsäuren.) Liebigs Ann. Chem. **566**, 206 (1950).
29. DISCHE, Z.: Phosphorylierung der in Adenosin enthaltenen d-Ribose und nachfolgender Zerfall des Esters unter Triosephosphatbildung im Blute. Naturwiss. **26**, 252 (1938).
30. DOUDOROFF, M.: Studies on the Phosphorolysis of Sucrose. J. biol. Chemistry **151**, 351 (1943).

31. DOUDOROFF, M., H. A. BARKER and W. Z. HASSID: Studies with Bacterial Sucrose Phosphorylase. I. The Mechanism of Action of Sucrose Phosphorylase as a Glucose-Transferring Enzyme (Transglucosidase). J. biol. Chemistry 168, 725 (1947).
32. DRURY, A. N. and A. SZENT-GYÖRGYI: The Physiological Activity of Adenine Compounds with Especial Reference to their Action Upon the Mammalian Heart. J. Physiology 68, 213 (1929).
33. FALCONER, R. and J. M. GULLAND: The Constitution of the Purine Nucleosides. VIII. Uric Acid Riboside. J. chem. Soc. (London) 1939, 1369.
34. FORREST, H. S. and A. R. TODD: Nucleotides. V. Riboflavin-5'-Phosphate. J. chem. Soc. (London) 1950, 3295.
35. FRIEDKIN, M.: Desoxyribose-1-Phosphate. II. The Isolation of Crystalline Desoxyribose-1-Phosphate. J. biol. Chemistry 184, 449 (1950).
36. — Azaguanine, Ribose-1-phosphate and Desoxyribose-1-Phosphate. Reported Reaction at Symposium on Phosphorus Metabolism I., McCollum-Pratt Inst., Johns Hopkins Univ., June, 1951.
37. — Enzymatic Synthesis of Desoxyxanthinosine by the Action of Xanthinosine Phosphorylase in Mammalian Tissue. J. Amer. chem. Soc. 74, 112 (1952).
38. FRIEDKIN, M. and H. M. KALCKAR: Desoxyribose-1-phosphate. I. The Phosphorolysis and Resynthesis of Purine Desoxyribose Nucleoside. J. biol. Chemistry 184, 437 (1950).
39. FRIEDKIN, M. and D. ROBERTS: Desoxyribose-1-Phosphate in Nucleic Acid Synthesis. Formation of Thymidine. Federat. Proc. (Amer. Soc. exp. Biol.) 10, 184 (1951).
40. FURST, S. S. and G. B. BROWN: On the Rôle of Glycine and Adenine as Precursors of Nucleic Acid Purines. J. biol. Chemistry 191, 239 (1951).
41. FURST, S. S., P. M. ROLL and G. B. BROWN: On the Renewal of the Purines of the Desoxypentose and Pentose Nucleic Acids. J. biol. Chemistry 183, 251 (1950).
42. GOLDWASSER, E. (unpublished).
43. GREENBERG, G. R.: De Novo Synthesis of Hypoxanthine via Inosine-5-phosphate. J. biol. Chemistry 190, 611 (1951).
44. HAMMARSTEN, E., P. REICHARD and E. SALUSTE: Pyrimidine Nucleosides as Precursors of Pyrimidines in Polynucleotides. J. biol. Chemistry 183, 105 (1950).
45. HANES, C. S. and F. A. ISHERWOOD: Separation of the Phosphoric Esters on the Filter Paper Chromatogram. Nature (London) 164, 1107 (1949).
46. HEPPEL, L. A.: (unpublished).
47. HEPPEL, L. A. and R. J. HILMOE: Purification and Properties of 5-Nucleotidase. J. biol. Chemistry 188, 665 (1951).
48. HOFF-JÖRGENSEN, E.: A Microbiological Assay of Deoxyribonucleosides and Deoxyribonucleic Acids. Biochemic. J. 50, 400 (1952).
49. HOFF-JÖRGENSEN, E., M. FRIEDKIN and H. M. KALCKAR: Desoxyribose-1-Phosphate. III. Comparison of Microbiological and Spectrophotometric Estimations of Enzymatically Produced Purine Desoxyribose Nucleoside. J. biol. Chemistry 184, 461 (1950).
50. HOFFMANN, C. E. and L. A. MANSON: Products of Desoxyribose Nucleosides Degradation by *Escherichia coli*. Federat. Proc. (Amer. Soc. exp. Biol.) 10, 198 (1951).
51. HORECKER, B. L. and P. Z. SMYRNIOTIS: The Enzymatic Production of Ribose-5-Phosphate from 6-Phosphogluconate. Arch. Biochemistry 29, 232 (1950).

52. KALCKAR, H. M.: Differential Spectrophotometry of Purine Compounds by Means of Specific Enzymes. I. Determination of Hydroxypurine Compounds. J. biol. Chemistry **167**, 429 (1947).
53. — Differential Spectrophotometry of Purine Compounds by Means of Specific Enzymes. II. Determination of Adenine Compounds. J. biol. Chemistry **167**, 445 (1947).
54. — Differential Spectrophotometry of Purine Compounds by Means of Specific Enzymes. III. Studies of the Enzymes of Purine Metabolism. J. biol. Chemistry **167**, 461 (1947).
55. — The Enzymatic Synthesis of Purine Ribosides. J. biol. Chemistry **167**, 477 (1947).
56. — The Biological Synthesis of Purine Compounds. Symposia of the Society for Experimental Biology, Number 1, p. 38, Nucleic Acid. 1947.
57. — The Biological Incorporation of Purines and Pyrimidines into Nucleosides and Nucleic Acid. Biochim. et Biophys. Acta **4**, 232 (1950).
58. — Enzymatic Reactions in the Synthesis of Purine Compounds. In: The Harvey Lectures. Springfield, Ill.: Charles C. Thomas. 1949—1950.
59. — Certain Aspects of Biosynthesis of Nucleosides and Nucleic Acids. Publ. Staz. Zool. Napoli **23**, Suppl. (in press).
60. KALCKAR, H. M. and O. H. LOWRY: The Relationship Between Traumatic Shock and the Release of Adenylic Acid Compounds. Amer. J. Physiol. **149**, 240 (1947).
61. KALCKAR, H. M., W. S. MCNUTT and E. HOFF-JÖRGENSEN: Trans-N-Glycosidase as Studied with Carbon 14 Adenine. Biochemic. J. (in press).
61a. KENNER, G. W.: The Chemistry of Nucleotides. Fortschr. Chem. organ. Naturstoffe **8**, 96 (1951).
62. KERR, S. E., K. SERAIDARIAN and G. B. BROWN: On the Utilization of Purines and Their Ribose Derivatives by Yeast. J. biol. Chemistry **188**, 207 (1951).
63. KLEIN, W.: Experimentelle Studien über den Nucleinstoffwechsel. XXXVII. Über Nucleosidase. Hoppe-Seyler's Z. physiol. Chem. **231**, 125 (1935).
64. — Experimentelle Studien über den Nucleinstoffwechsel. XXXIII. Über Adenin-desoxyribosid. Hoppe-Seyler's Z. physiol. Chem. **224**, 244 (1934).
65. KLEIN, W. u. S. J. THANNHAUSER: Experimentelle Studien über den Nucleinstoffwechsel. XXXV. Die Pyrimidinnucleotide aus Thymusnucleinsäure. Hoppe-Seyler's Z. physiol. Chem. **231**, 96 (1935).
66. KLENOW, H. (unpublished).
67. KORNBERG, A. and W. E. PRICER, Jr.: Enzymatic Phosphorylation of Adenosine and 2,6-Diaminopurine Riboside. J. biol. Chemistry (1951) (in press).
68. LELOIR, L. F.: The Enzymatic Transformation of Uridine Diphosphate Glucose Into a Galactose Derivative. Arch. Biochem. and Biophys. **33**, 186 (1951).
68a. — Sugar Phosphates. Fortschr. Chem. organ. Naturstoffe **8**, 47 (1951).
69. LE PAGE, G. A. and C. HEIDELBERGER: Incorporation of Glycine-2-C^{14} Into Proteins and Nucleic Acids of the Rat. J. biol. Chemistry **188**, 593 (1951).
70. LIPMANN, F.: Fermentation of Phosphogluconic Acid. Nature (London) **138**, 588 (1936).
71. LOWRY, O. H. and J. A. LOPEZ: The Determination of Inorganic Phosphate in the Presence of Labile Phosphate Esters. J. biol. Chemistry **162**, 421 (1946).
72. LUTWAK-MANN, C.: The Decomposition of Adenine Compounds by Bacteria. Biochemic. J. **30**, 1405 (1936).
73. LYTHGOE, B. and A. R. TODD: Structure and Synthesis of Nucleotides. Symposia Soc. exp. Biology **1**, 15 (1947).

74. McNAIR SCOTT, D. B. and S. S. COHEN: Enzymatic Formation of Pentose Phosphate from 6-Phosphogluconate. J. biol. Chemistry 188, 509 (1951).
75. McNUTT, W. S.: The Exchange Between Free Purines and Pyrimidines and the Aglucones of Deoxyribosyl Purines and Deoxyribosyl Pyrimidines. Nature (London) 166, 444 (1950).
76. — Trans-N-Glycosidase. Biochemic. J. (in press).
76a. — Nucleosides and Nucleotides as Growth Substances for Microorganisms. Fortschr. Chem. organ. Naturstoffe 9, 401 (1952).
77. MANSON, L. A. and J. O. LAMPEN: The Metabolism of Hypoxanthine Desoxyriboside in Animal Tissues. J. biol. Chemistry 191, 95 (1951).
78. — — Enzymatic Degradation of Thymidine. Federat. Proc. (Amer. Soc. exp. Biol.) 8, 224 (1949).
79. — — Metabolism of Desoxyribosides in *Escherichia coli*. Federat. Proc. (Amer. Soc. exp. Biol.) 9, 397 (1950).
80. MARRIAN, D. H., V. L. SPICER, M. E. BALIS and G. B. BROWN: Purine Incorporation Into Pentose Nucleotides of the Rat. J. biol. Chemistry 189, 533 (1951).
80a. MICHELSON, A. M., W. DRELL and H. K. MITCHELL: A New Ribose Nucleoside from Neurospora: "Orotidine". Proc. Nat. Acad. Sc. (U. S. A.) 37, 396 (1951).
81. MITCHELL, H. K. and W. D. McELROY: Adenosine Deaminase from *Aspergillus Oryzae*. Arch. Biochemistry 10, 351 (1946).
82. OSTERN, P., T. BARANOVSKI u. J. TERSZAKOWEĆ: Über die Phosphorylierung des Adenosins durch Hefe und die Bedeutung dieses Vorganges für die alkoholische Gärung. Hoppe-Seyler's Z. physiol. Chem. 251, 258 (1938).
83. PLENTL, A. A. and R. SCHOENHEIMER: Studies in the Metabolism of Purines and Pyrimidines by Means of Isotopic Nitrogen. J. biol. Chemistry 153, 203 (1944).
84. PRICE, V. (unpublished).
85. RACKER, E.: Enzymatic Formation and Breakdown of Pentose Phosphate. Federat. Proc. (Amer. Soc. exp. Biol.) 7, 180 (1948).
86. — Enzymatic Synthesis of Deoxypentose Phosphate. Nature (London) 167, 408 (1951).
87. REICHARD, P. and B. ESTBORN: Utilization of Desoxyribosides in the Synthesis of Polynucleotides. J. biol. Chemistry 188, 839 (1951).
88. ROLL, P. M. and I. WELIKY: .Utilization of Purine Nucleotides for Nucleic Acid Synthesis by the Rat. Federat. Proc. (Amer. Soc. exp. Biol.) 10, 238 (1951).
89. ROWEN, J. W. and A. KORNBERG: Phosphorolysis of Nicotinamide Riboside. Federat. Proc. (Amer. Soc. exp. Biol.) 10, 240 (1951).
90. SABLE, H. Z. and H. M. KALCKAR (unpublished).
91. SCHAEDEL, M. L., M. J. WALDVOGEL and F. SCHLENK: The Specificity of Adenosine Deaminase and Purine Nucleosidase. J. biol. Chemistry 171, 135 (1947).
92. SCHLENK, F.: Chemistry and Enzymology of Nucleic Acids. Adv. Enzymology 9, 455 (1949).
93. SCHLENK, F. and M. J. WALDVOGEL: Studies on the Metabolism of Some Ribose Derivatives. Arch. Biochemistry 12, 181 (1947).
94. SCHMIDT, G.: Über fermentative Desaminierung in Muskel. Hoppe-Seyler's Z. physiol. Chem. 179, 243 (1928).
95. — Über den fermentativen Abbau der Guanylsäure in der Kaninchenleber. Hoppe-Seyler's Z. physiol. Chem. 208, 185 (1932).
96. SCHMIDT, G. and S. J. THANNHAUSER: Intestinal Phosphatase. J. biol. Chemistry 149, 369 (1943).

97. Shaw, E. and D. W. Woolley: A New and Convenient Synthesis of 4-amino-5-imidazolecarboxamide. J. biol. Chemistry 181, 89 (1949).
98. Shive, W., W W. Ackermann, M. Gordon, M. E. Getzendaner and R. E. Eakin: 5(4)-Amino-4-(5)imidazolecarboxamide, a Precursor of Purines. J. Amer. chem. Soc. 69, 725 (1947).
99. Stephenson, M. and A. R. Trim: The Metabolism of Adenine Compounds by *Bacteria coli* with a Micro-Method for the Estimation of Ribose. Biochemic. J. 32, 1740 (1938).
99a. Stern, H., V. Allfrey, A. E. Mirsky and H. Saetren: Some Enzymes of Isolated Nuclei. J. gen. Physiol. 35, 559 (1952).
100. Stetten, M. R. and C. L. Fox, Jr.: An Amine Formed by Bacteria During Sulfonamide Bacteriostasis. J. biol. Chemistry 161, 333 (1945).
101. Sutherland, E. W., M. Cohn, T. Posternak and C. F. Cori: The Mechanism of the Phosphoglucomutase Reaction. J. biol. Chemistry 180, 1285 (1949).
102. Wajzer, J.: Synthèse enzymatique de nucléotides puriques. I. Arch. Sci. physiol. 1, 485 (1947) [Chem. Abstr. 43, 5064 (1949)].
103. — Synthèse enzymatique de nucléotides puriques. II. Arch. Sci. physiol. 1, 493 (1947) [Chem. Abstr. 43, 5064 (1949)].
104. Wajzer, J. et F. Baron: Réactions enzymatiques entre les esters phosphoriques du ribose et les bases puriques. Bull. Soc. Chim. biol. (Paris) 31, 750 (1949).
105. Wang, T. P. and J. O. Lampen: The Cleavage of Adenosine Cytidine and Xanthosine by *Lactobacillus pentosus*. J. biol. Chemistry 192, 339 (1951).
106. Wang, T. P., H. Z. Sable and J. O. Lampen: Enzymatic Deamination of Cytosine Nucleosides. J. biol. Chemistry 184, 17 (1950).
107. Warburg, O. u. W. Christian: Abbau von Robisonester durch Triphospho-Pyridin-Nucleotid. Biochem. Z. 292, 287 (1937).
108. Weygand, F., A. Wacker und H. Dellweg: Spaltung von Desoxyribonucleinsäure mit Bleihydroxyd und Isolierung der Desoxyriboside durch kontinuierliche Gegenstromverteilung. Z. f. Naturforsch. 6 b, 140 (1951).
109. Zeuthen, E.: Segmentation, Nuclear Growth and Cytoplasmic Storage in Eggs of Echinoderms and Amphibia. Publ. Staz. Zool. Napoli 23, Suppl. (in press).

(Received, November 30, 1951.)

Nucleosides and Nucleotides as Growth Substances for Microorganisms.

By W. S. MC NUTT, Nashville, Tennessee.

Contents.

	Page
Introduction	402
I. Nucleosides and Nucleotides of Ribose	405
1. Coenzyme I, "Desamino-codehydrogenase I," Coenzyme II and Nicotinamide Riboside	405
2. Purine-Nucleosides and Nucleotides	405
a) Growth-promoting Activity	405
b) Growth-inhibiting Activity and the Ability to Reverse Growth-inhibition	409
3. Nucleotides in the Nutrition of *Lactobacillus gayonii*	410
4. Pyrimidine-Nucleosides and Nucleotides	411
a) Growth-promoting Activity	411
b) Growth-inhibiting Activity	412
5. The Biosynthesis of Ribosides and Ribonucleotides	413
A Comparison between Microorganisms and Higher Animals with Regard to Purine Precursors in Nucleic Acid Biosynthesis	413
6. Vitamin B_{12}	417
Microbiological Functions of Vitamin B_{12}	418
Different Forms of Vitamin B_{12}	419
II. Nucleosides and Nucleotides of Desoxyribose	420
1. The Biosynthesis of Desoxyribosides	421
Considerations of the Mode of Formation of the Desoxyribosidic Linkage	422
2. The Growth-promoting Activity of Desoxyribosides and Desoxyribonucleotides	424
a) The Specificity of Certain Desoxyribosides in Eliciting the Growth-response of Bacteria	424
b) The Non-specificity of the Natural Desoxyribosides in Promoting the Growth of Certain Bacteria	424

	Page
3. The Relationship of the Desoxyribosides, Vitamin B_{12}, Reducing Agents, and the "Citrovorum-Factor" in Supporting the Growth of Various Microorganisms	426
Relationship between Certain Reducing Agents and Vitamin B_{12} Requirement	427
The "Citrovorum Factor"	431
References	433

Introduction.

The importance of the naturally occuring nucleosides and nucleotides in the nutrition of a number of microorganisms is now well established. Such organisms as lack these synthetic abilities, by their very defects, betray the additional biosynthetic powers of organisms which do not suffer these metabolic insufficiencies. Thus, although most of the information in this field is as yet of a descriptive nature only, there is reason to hope that, through studies of the additional nutritional demands and the intermediary metabolism of these more fastidious microorganisms, much may eventually be surmised of the general metabolic sequence through which nucleosides, nucleotides, and nucleic acids arise normally in plants and animals.

The discovery that the pyrimidines of pyrimidine ribosides (56) and desoxyribosides* (117), but not the free pyrimidines, are incorporated into the nucleic acids of the rat parallels the earlier work upon experimentally induced mutant strains of *Neurospora* which require uridine or cytidine for growth and for which the pyrimidines, themselves, are either far less active or completely inactive (88).

Certain of the vitamins are known to be involved with steps in the synthesis of the desoxyribosides, and while considerable attention is being directed to this end, the relationships are at present neither defined fully nor understood concisely.

The studies upon the nutritional requirements of certain microorganisms for purines, pyrimidines, nucleosides or nucleotides have supplied results which are of a wider application than one might at first suppose from the narrowness of the field, for which reasons a brief review of this subject may prove of interest.

A number of stuctural formulas of compounds mentioned in the present discussion appear in Table 1.

The chemistry of the nucleotides has been recently reviewed by KENNER (72a) and the enzymes of nucleoside metabolism by KALCKAR (70a) (p. 363 of the present volume).

---
* The prefix, D e o x y is also widely used.

Table 1. Numbering systems and structural formulas.

The *numbering* the pyrimidine and purine nucleoside ring systems will be apparent from the following diagrams:

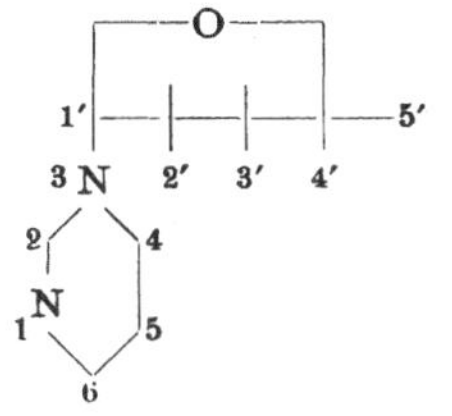

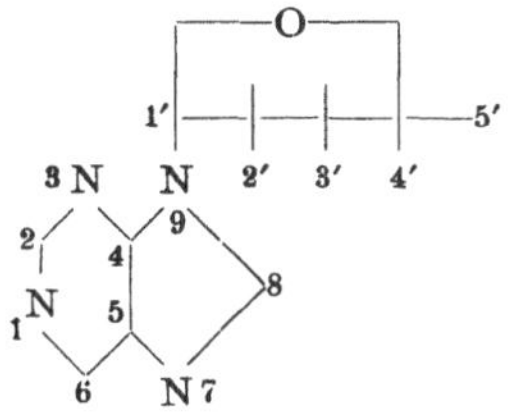

Customary *abbreviations* for the names of the more complex nucleotides are:

ADP for adenosine-5'-diphosphate.

ATP for adenosine-5'-triphosphate.

DPN for diphosphopyridine nucleotide (coenzyme I).

TPN for triphosphopyridine nucleotide (coenzyme II).

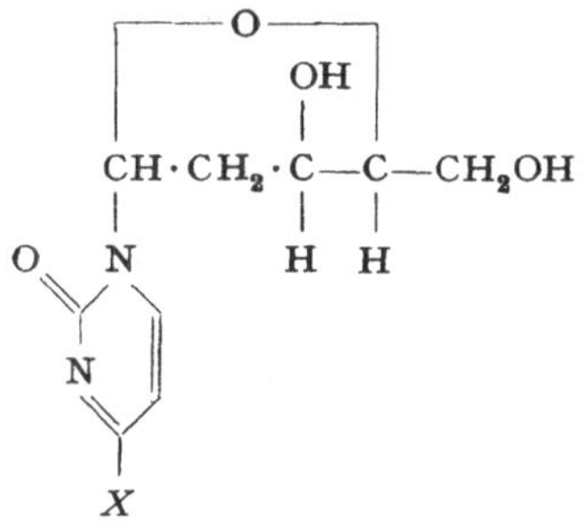

[X = OH] Uracil desoxyriboside,
[X = NH$_2$] Cytosine desoxyriboside.

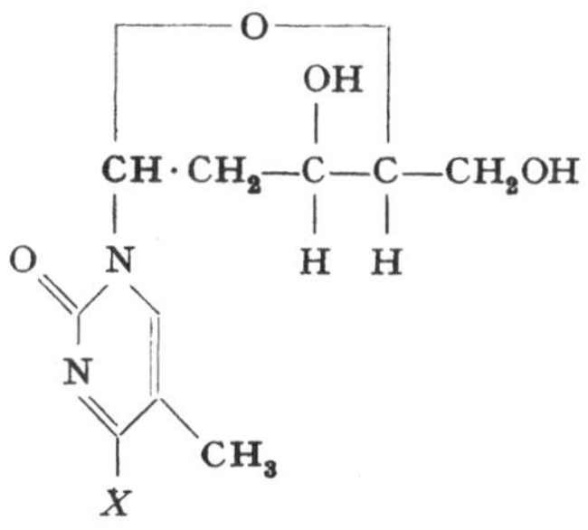

[X = OH] Thymidine,
[X = NH$_2$] 5-Methylcytosine desoxyriboside.

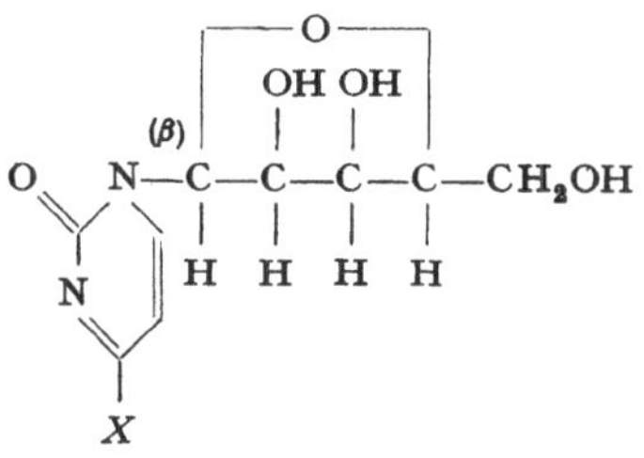

[X = OH] Uridine. [X = NH$_2$] Cytidine.
(The 3'-phosphates are, uridylic acid and citidylic acid.)

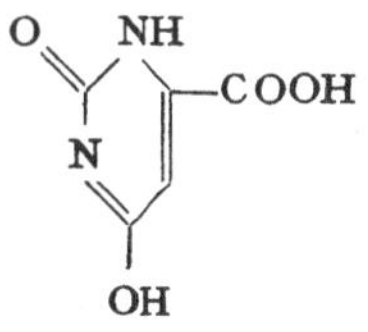

Orotic acid (4-Carboxy-uracil).

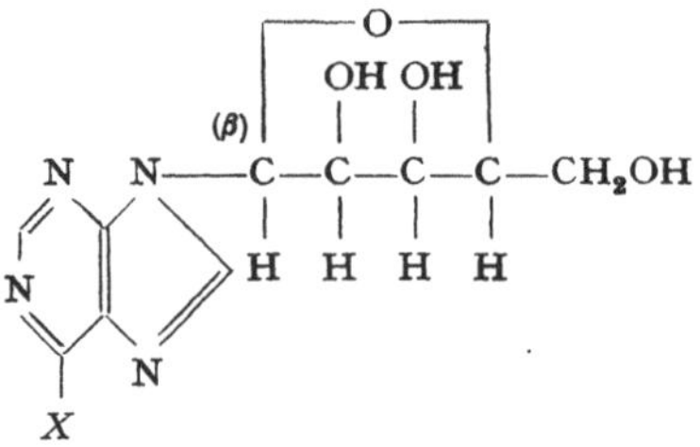

[X = OH] Inosine.
[X = NH₂] Adenosine.
(The 5'-phosphate of inosine is inosinic acid; the monophosphates of adenosine are the adenylic acids.)

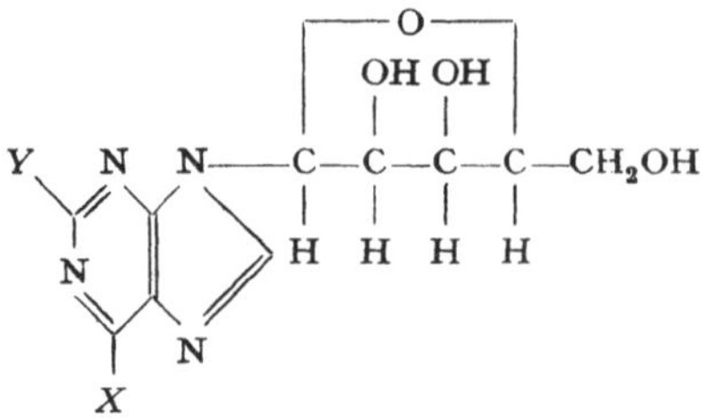

[X = OH, Y = OH₂] Xanthosine.
[X = OH, Y = NH₂] Guanosine.
(The 3'-phosphate of guanosine is guanylic acid.)

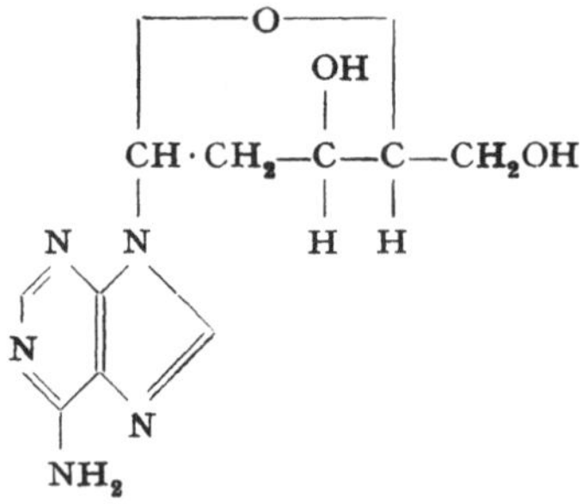

Adenine desoxyriboside.

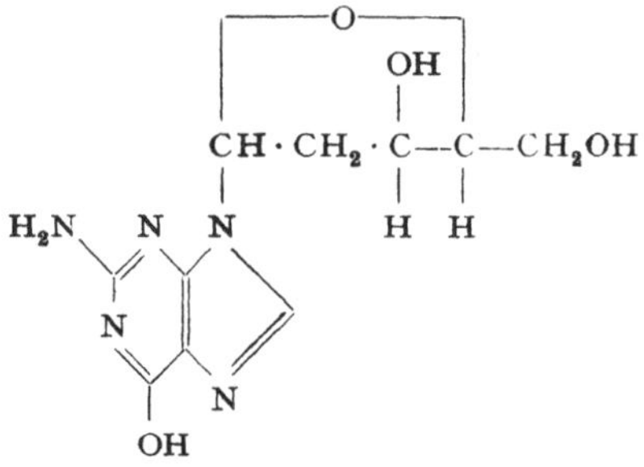

Guanine desoxyriboside.

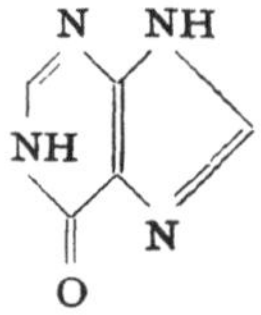

Hypoxanthine.

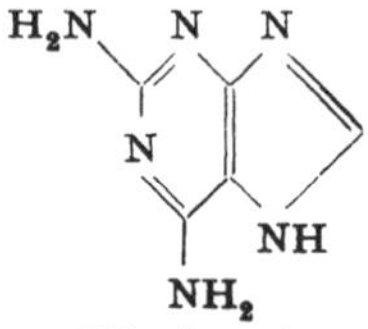

2,6-Diaminopurine.

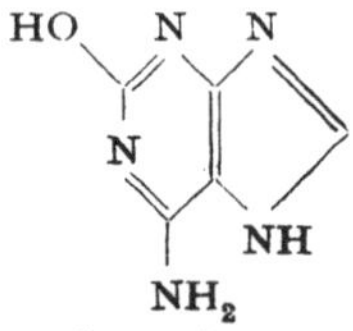

Isoguanine.
(2-Hydroxy-6-aminopurine.)
(Isoguanosine is 9-β-D-ribofuranosido-isoguanine.)

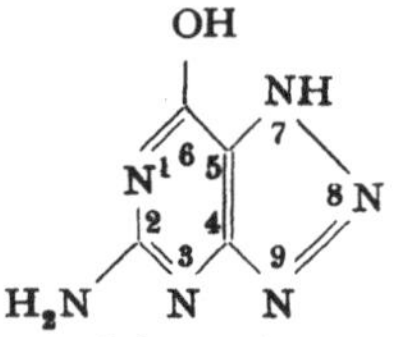

8-Azaguanine.

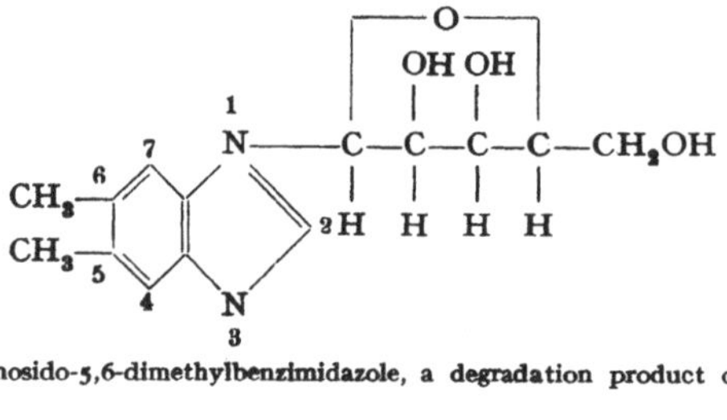

1-α-D-Ribofuranosido-5,6-dimethylbenzimidazole, a degradation product of vitamin B_{12}.

I. Nucleosides and Nucleotides of Ribose.

1. Coenzyme I, "Desamino-codehydrogenase I," Coenzyme II, and Nicotinamide Riboside.

The first indication that nucleotides are of importance in promoting the growth of microorganisms was the finding of LWOFF and LWOFF (89) that an essential growth-requirement of *Hemophilus parainfluenzae* (growth factor "V") was satisfied by coenzyme I or coenzyme II. Adenosine-5'-phosphate, nicotinamide, and nicotinic acid were inactive.

Studies upon the specificity of this requirement were carried further by GINGRICH and SCHLENK (51) with two strains of *Hemophilus parainfluenzae* and five strains of *H. influenzae*. Coenzyme I and reduced coenzyme I were the most active substances examined, minimal concentrations of $1.3 \cdot 10^{-3} \mu g$. per ml. being required to support the growth of these bacteria. "Desamino-codehydrogenase I," prepared by replacing the amino group of the adenine-portion of coenzyme I by a hydroxyl group, was about 60 per cent as active as the natural coenzyme. Nicotinamide riboside was less active than coenzyme I, even in the presence of adenosine, but was more active than coenzyme II. Nicotinamide riboside may, thus, be considered as the limiting structure of the coenzyme which meets the demands of this requirement for these organisms.

2. Purine-Nucleosides and Nucleotides.

a) Growth-promoting activity.

Cells of *Saccharomyces cerevisiae* when subjected to sublethal doses of ultraviolet radiation undergo an increased permeability to certain substances, such that their concentrations are disturbed inside the injured cells (86). The substances which diffused from the injured cells had a stimulating effect on the growth of normal cells when cultured aerobically. Upon the basis of the ultraviolet spectra of the substances which diffused from the injured cells, the presence of adenine nucleotides was suspected. Direct testing showed that adenosine and nucleotides of adenine were active as growth-promoting substances (85, 86). The order of decreasing activity was: Adenosine triphosphate, adenosine-5'-phosphate, adenosine, and adenosine-3'-phosphate (86). The effect of adenosine-3'-phosphate was negligible, and adenine, apparently, was inactive, although this point was not stated explicitly. Guanosine, as well, has been reported to stimulate the rate of growth of this yeast (27). Adenine diphosphate had the effect of elevating the uptake of oxygen by growing or resting cells of *Saccharomyces cerevisiae*, and here, too, adenosine-3'-phosphate was inactive (85).

HILLS (*59*) has demonstrated that an essential requirement for the germination of *Bacillus anthracis* is satisfied by adenosine or adenosine-3'-phosphate. Preparations of ribonucleic acid and desoxyribonucleic acid showed some activity, as did guanylic acid, but adenine and guanine were without effect. Both adenosine and a preparation of its 3'-phosphate were highly active under the conditions of the test. That the activity was due to adenosine and not a contaminant, was established by careful purification and quantitative comparison in a sensitive biological assay which detects as little as 0.2 μg. of adenosine per ml.

This demonstration of the adenosine requirement for the germination of *Bacillus anthracis* is due, undoubtedly, to refinements in the technique as practiced by HILLS, for this organism was grown from the spore-stage by GLADSTONE (*52*) in a synthetic medium containing neither added purines nor adenosine, and by BREWER *et al.* (*13*) in a synthetic medium containing no added adenosine. As pointed out by HILLS (*59*), if only a few spores in the inoculum contained sufficient adenosine to germinate, growth of *Bacillus anthracis* could well ensue in a synthetic medium free of adenosine, since adenosine is not required for the *growth* of this organism.

Adenosine triphosphate acts as a growth-factor for *Hemophilus piscium* (*54*), but its activity very probably involves the well-known ability of this nucleotide to transport phosphoryl groups. The actual growth-requirement for this organism, diphosphothiamine, was some thirty times more active than the nucleotide. Coenzyme I was found to be inactive. Since it has been demonstrated by WEIL-MALHERBE (*152*) that enzyme preparations from yeast catalyze the formation of diphosphothiamine from thiamine and adenosine triphosphate, this common effect of the two substances upon the growth of the bacterium is not surprising. *Neisseria gonorrhoeae* also has a growth-requirement for the phosphorylated forms of thiamine, but the effect of adenosine triphosphate in supporting the growth of this organism was not reported (*83*).

A survey by WILSON (*158*) upon the nutrition of certain group A hemolytic streptococci showed that one-third of the members examined did not respond in growth to adenine but did so to adenosine. Various other purines, guanine desoxyriboside, and ribonucleic acid were active, while the pyrimidines and desoxyribonucleic acid were inactive. With another strain of group A hemolytic streptococci PAPPENHEIMER and HOTTLE (*103*) have shown that the requirement for adenylic acid by this organism is essential only when the CO_2-tension of the medium is low, for at CO_2-tensions of the order of 20 mm. mercury only a slight effect of adenylic acid upon the growth of this strain was observed. It is probable, however, that the effect is not specific for the nucleotide (as opposed to the purine), although this point was not stated categorically.

Nucleic acid adenosine, and adenylic acid produced a stimulation in the growth of *Clostridium sporogenes* when added to a medium containing free purines and uracil, but the effect was not impressive. Guanosine and guanylic acid were inactive (*134*).

The Comparative Activity of Purines, Purine-nucleosides and Nucleotides in Promoting the Growth of Purine-less Microorganisms. Various purine-less microorganisms show marked differences in their response to purine-nucleosides or nucleotides when substituted for the required purine. For some of them the purine, the nucleoside, and the nucleotide are equally active on a molar basis.

Neurospora mutant No. 28610 is an example of this type. This organism has an essential growth-requirement for adenine which may be met as well by an equimolar amount of adenosine. Guanine, though inactive in supporting growth, has a sparing action on adenine, and this sparing action, over a prescribed range of concentration, is served just as well by equimolar amounts of guanine, guanosine or guanylic acid (*40*). *Neurospora* mutant No. 3254 is very similar in that regard. The purine-requirement of this organism is met by hypoxanthine, isoguanine, or adenine. Adenosine, adenosine-3'-phosphate, coenzyme I, or ribonucleic acid will each replace adenine, and the activity of each is dependent upon the amount of adenine the compound contains (*107*). A purine-less mutant strain of *Photobacterium fischeri* responds to guanosine, guanylic acid, and ribonucleic acid as it does to purines (*105*). The absence of any comparison by the author of the relative activities of these substances suggests that the nucleoside and nucleotides were no more active than the purines. Nucleosides and nucleotides of the purines (adenosine, adenylic acid, guanosine, guanylic acid, and inosine triphosphate) were no more active on a molar basis than the free purines in the nutrition of *Lactobacillus casei* (*36*).

In other microorganisms the nucleoside or nucleotide is of greater growth-promoting activity than the purine. Frequently, the differences are not very outstanding. *Tetrahymena geleii*, for example, has a purine requirement for which guanine is the most active of the purines. But guanosine and guanylic acid were twice as active as guanine on a molar basis (*74*). An equimolar amount of guanylic acid was slightly more active than guanosine, and equal weights of adenine and adenylic acid had about the same sparing action on guanine. Also, in satisfying the purine requirement of *Lactobacillus lactis* DORNER, guanylic acid was more active than guanine on a weight basis (*132*). FRIES (*47, 48*) has made a study of certain artificially induced, purine-less, mutant strains of *Ophisotoma multiannulatum*. Growth of the guanine-less mutant strains in response to guanosine was initiated slower than was the growth in response to equimolar amounts of guanine, but a greater extent of growth was achieved. Guanylic acid was also active. A like situation holds for the adenine-less mutants, with equimolar amounts of adenosine or adenosine-3'-phosphate producing more growth than adenine, although the growth was initiated slower. In none of these examples is the greater activity of the purine nucleoside or nucleotide so spectacular as that encountered among the pyrimidine-less mutant strains of *Neurospora* (*88*) or *Ophiostoma* (*49*), in which case the nucleoside or nucleotide is some 10–60 times more active than the pyrimidine.

Certain strains of group A hemolytic streptococci mentioned, differ from the above examples, inasmuch as the organisms responded in growth to adenosine, whereas adenine was without effect.

Finally, there are those purine-less microorganisms for which the nucleosides and nucleotides are less active or completely inactive. ROBBINS and KAVANAGH (*119*) have described a species of *Phycomyces* which required for growth either guanine or hypoxanthine; adenine, isoguanine, and guanosine were inactive. *Leuconostoc citrovorum*, another example of this type, responded less than half as well to guanosine and less than $1/_{10}$ as well to guanylic acid as it did to an equal weight of guanine (*123*); and *Torulopsis utilis* readily incorporated guanine and adenine into its ribonucleic acid, but guanosine, guanylic acid, adenosine, and adenosine-3'-phosphate were utilized to only a very small extent (*73*).

These variations in the ability of purine-less microorganisms to utilize nucleosides and nucleotides possibly arise both from differences in cellular permeability and the activity of specific enzymes.

The specific enzymic effects have been considered by DI CARLO *et al.* (*32*) in studies upon the ability of purine nucleosides and nucleotides to satisfy the nitrogen-requirement of *Torula utilis*. For this yeast purines serve as sources of nitrogen. The nucleosides (adenosine, guanosine, and xanthosine) served as nitrogen-sources but less efficiently than the purines. Adenosine-3'-phosphate was utilized, but adenosine-5'-phosphate, adenosine diphosphate, and triphosphate were ineffective. On the basis of the selectivity of the substrate which was shown by the organism the presence or absence of various enzymes (nucleosidases, purine deaminases, adenosine-3'-phosphate deaminase, etc.) was postulated, but actual isolation and characterization of the suspected enzymes was not carried out.

Torula utilis also used adenine thiomethyl-pentoside as a source of nitrogen (*32*).

The biological activity of this compound in promoting lactation in the rat has been described (*99*) but no important role seems to have been found for it in the nutrition of microorganisms.

Coenzyme A. Although coenzyme A (*99b, 25a, 76b*) and certain products of its enzymic degradation show specific activities in promoting the growth of certain microorganisms (*Acetobacter suboxidans*, *Lactobacillus bulgaricus* etc.) (*76a, 99a, 76b, 137a*), the adenine nucleotide-portion of the coenzyme is not essential for the expression of these specific effects. The organisms in question show differing abilities to utilize free- or "bound-pantothenic acid", but none of them has an actual requirement for the complete coenzyme. *Lactobacillus arabinosus* responds to free pantothenic acid but is able to utilize "bound-pantothenic acid" only after treatment with intestinal phosphatase (*76a*). *Acetobacter suboxidans* responds well to "bound-pantothenic acid" (*76a*), coenzyme A (*99a*), or the enzymically degraded product of coenzyme A which lacks the adenylic acid-portion of the molecule (*25a*). Whether the first of these forms is identical with the last is not yet known. Both are more active than coenzyme A in supporting the growth of *A. suboxidans* while coenzyme A is much more effective than an equimolar amount of pantothenic acid.

For *Lactobacillus bulgaricus*, pantothenic acid at concentrations customarily employed in bacteriological media is without any growth effect whatever. This organism responds to a β-mercaptoethylamino derivative of pantothenic acid (*137a*) which is itself part of the coenzyme A molecule. Disagreement exists in regard to the points of attachment of the phosphoryl groups in coenzyme A. NOVELLI et al. (*99b*) have submitted evidence which suggests that adenosine-5'-phosphoric acid is linked to the rest of the molecule through a pyrophosphoryl group; KING and STRONG (*76b*), however, concluded from their studies that a pyrophosphoryl group did not exist in their coenzyme A preparations.

b) Growth-inhibiting Activity and the Ability to Reverse Growth-inhibition.

Several pyrimidine-less and purine-less microorganisms are inhibited in growth by one or another of the purines, purine nucleosides or nucleotides (*58, 48, 108, 35*). These effects are quite involved and can be somewhat of a nuisance when a quantitative assay of an individual purine is desired (*48*) or the estimation of uridine and cytidine in natural materials (*87, 49*). A detailed investigation of these relationships has been conducted with a pyrimidine-less mutant strain of *Neurospora* by PIERCE and LORING (*108*). When this organism was grown in response to uridine, cytidine, uridylic acid, or cytidylic acid, an inhibition upon growth was exerted by guanosine, guanylic acid, adenosine, or adenosine-3'-phosphate. Adenosine and adenosine-3'-phosphate were about twice as inhibitory as were guanosine and guanylic acid; and adenosine was twice as active as adenosine-3'-phosphate in the inhibition of growth in response to cytidine. Adenine and guanine were without appreciable inhibitory effect.

The guanine-less mutant strains of *Ophiostoma* (*48*) can not be used satisfactorily for the direct assay of guanine in natural materials due to the inhibitory action of some other purines (adenine or hypoxanthine) which are usually present also. Hence, separation of the purines must first be carried out.

This inhibitory activity of adenine was less if the organisms were grown in response to guanosine instead of guanine. When hypoxanthine was the inhibitor, just the reverse was true.

The utilization of desoxyribonucleic acid by *Lactobacillus bifidus* (*136*) was inhibited competitively by ribonucleic acid or purine mononucleotides (adenosine-3'-or-5'-phosphate, or guanylic acid). The order of decreasing inhibition was: Adenylic acid b, adenosine-5'-phosphate, and adenylic acid a, the last being essentially inactive (*135*). Nucleosides of ribose, uridylic acid, cytidylic acid, purines, and pyrimidines had no effect. If vitamin B_{12} (*136*) or thymidine (the desoxyriboside of thymine) (*135*), instead of desoxyribonucleic acid, was used to promote the growth of the organism, the purine-nucleotides exerted no inhibitory action.

2,6-Diamino-purine, at relatively low concentrations, has been shown to be an effective growth-inhibitor for two adenine-less mutant strains of *Neurospora*. It acted in the presence of adenine or hypoxanthine but not in the presence of adenosine or inosine. Guanine inhibited the growth of the organism mentioned in response to hypoxanthine, and guanosine antagonized inosine as well. Neither isoguanine nor isoguanosine was significantly active in promoting or inhibiting growth (*35*).

Adenine nucleotides and nucleosides reversed the sulfonamide-inhibition in the growth of *Eremothecium asbyii* (*127*). Ribonucleic acid, adenosine--3'-or-5'-phosphate, "adenine-cytosine dinucleotide", and adenosine were all active, the effectiveness of each being dependent upon the quantity of adenosine contained in the compound. Adenosine was as efficient as *p*-aminobenzoic acid in reversing the inhibition. When adenine was tested for such effects (*126*) it was found to be $^1/_{10}$ as active as *p*-aminobenzoic acid, but "in certain cases" it was of equal activity. It would appear that the effectiveness of the nucleotides and nucleosides was due principally to the purines (especially adenine) the compounds contained.

Under special conditions of sulfanilamide-inhibition adenine was usually ineffective in meeting the purine-requirement of *Escherichia coli*, while adenosine as well as inosine, xanthine, and guanine were usually completely effective (*128*).

3. Nucleotides in the Nutrition of Lactobacillus gayonii.

The unusual nutritional behavior of a particular strain of *Lactobacillus gayonii* requires a special discussion, for this organism is stimulated in growth specifically by nucleotides while the nucleosides are inactive. Furthermore, either purine- or pyrimidine-nucleotides are active as well as nucleotides of either ribose or desoxyribose. HUTCHINGS *et al.* (*65*) first showed that the growth-requirement was satisfied by any one of the following compounds: guanylic acid, adenosine-3'-phosphate, uridylic or cytidylic acid. The loss of phosphorus from the nucleotides, by the action of a phosphatase, was associated with a loss of microbiological activity; and guanosine, when tested directly, proved to be inactive.

Later, NYGAARD and CHELDELIN (*100*) found that desoxyribonucleotides were also active for this strain. Adenosine-3'-or-5'-phosphate were found to be equally effective; however, ribosides, desoxyribosides, ribose-1- or-5-phosphate and 1-(α,β)-methyl-D-ribopyranoside-3-phosphate were inactive.

Under special conditions involving the use of a very dilute inoculum and a lowered temperature of incubation it could be shown that another strain, less exacting in its nutritional demands, was also stimulated by ribonucleotides. But under these conditions vitamin B_{12}, thymidine, or, to a lesser extent, ascorbic acid could substitute for the ribonucleotide (*100*). This is the only case of its kind yet reported.

Since any one of the nucleotides was able to support the growth of this organism, the metabolic inadequacy most probably lies at the initial phosphorylation step, which difficulty once surmounted, the phosphoryl group might be redistributed to the various nucleosides required for the biosynthesis of the bacterial nucleic acid. Could a trans-phosphorylase of such type be demonstrated in this organism, the finding would parallel the situation with regard to *Lactobacillus helveticus* S, which can use any one of the natural desoxyribosides for growth and which contains an enzyme (a trans-N-glycosidase) that catalyzes the transfer of the desoxyribosyl group from one purine or pyrimidine to another (*92, 90, 91*).

The resolution of each of the four mononucleotides of ribonucleic acid into two forms (*25*) and the suggestion of CARTER and COHN (*24*) that the adenylic acid in hydrolysates of yeast nucleic acid contains adenosine-2'-and-3'-phosphate, have provoked speculation as to the biological significance of these hitherto unknown substances. It would be interesting to know whether *Lactobacillus gayonii* distinguishes between the two forms of the mononucleotides or whether enzymes may be prepared from this organism which catalyze the conversion of one form into the other.

4. Pyrimidine-Nucleosides and Nucleotides.
a) Growth-promoting Activity.

Studies by KIDDER and DEWEY (*74*) have shown that *Tetrahymena geleii* requires a pyrimidine for growth. Either uracil or cytidylic acid satisfied this requirement. Although cytidylic acid was actually slightly more active than uracil (on a molar basis) cytosine, itself, was completely inactive and had no sparing action on uracil.

The purine and pyrimidine requirements for the growth of *Hemophilus parainfluenzae* in a chemically defined medium were investigated by HERBST and SNELL (*58*). Uracil permitted growth of the organism in the absence of purines. Neither cytosine nor thymine could substitute for uracil, but cytidine and cytidylic acid were slightly more active than equimolar amounts of uracil. Cytosine desoxyriboside was approximately half as potent as cytidine, and thymidine was about $1/25$ as active as cytidine. Both uridine and uridylic acid were slightly more effective than uracil on a molar basis.

The pyrimidine-less mutant strains of *Neurospora* (88, 87) and *Ophiostoma* (47, 49) have already been mentioned. The similarity of the pyrimidine requirement of these mutant strains from the two genera is very remarkable: (a) Both responded in growth to uracil. Orotic acid (4-carboxy-uracil), also, is known to be active for the *Neurospora* strains. (b) Neither responded to cytosine, and thymine was either inactive or only slightly active. (c) For each, uridine, cytidine and cytidylic acid were much more active than uracil. In the case of the *Neurospora* strains this greater effectiveness of the pyrimidine nucleosides or nucleotides as compared with uracil or orotic acid was 10–60 fold. (d) Both responded equally to equal amounts of uridine and cytidine.

Each investigator devised a microbiological assay for the quantity of cytidine plus uridine and measured the amount of these substances in hydrolysates of ribonucleic acid. The values agreed and were low for the tetranucleotide theory.

The microbiological assay of the pyrimidine ribosides with the *Neurospora* strain mentioned is complicated by the fact that the purine-nucleosides and nucleotides inhibit the growth of the organism. After chemical separation from interfering compounds and from each other, the uridine and cytidine from hydrolysates of ribonucleic acid were estimated individually by microbiological assay (87).

In contrast to the above microorganisms which show a preference for the pyrimidine ribosides are those microorganisms which show no such preference or those which are actually unable to utilize the pyrimidine ribosides as a source of pyrimidines. *Lactobacillus pentosus*, for example, grew equally well in response to either uridine or uracil (149); and a strain of *Lactobacillus bulgaricus* has been shown to have a specific requirement for orotic acid (160). Other pyrimidines, and pyrimidine nucleosides or nucleotides were inactive.

b) Growth-inhibiting Activity.

5-Chloro-uridine competitively inhibited the growth of a pyrimidine-less strain of *Neurospora* but had no appreciable effect on two pyrimidine-sufficient organisms, *viz.* a wild strain of *Neurospora* and *Mycobacterium tuberculosis* (120). Both 5-bromo- and 5-chloro-uridine reversibly inhibited the growth of the pyrimidine-less strain of *Neurospora* in response to either uridine or cytidine but were without effect when the pyrimidine-requirement was met by uracil (50). Uridine was more effective than cytidine in reversing this inhibition.

Synthetic nucleosides of uracil, thymine, and cytosine which differ from the natural nucleosides in possessing the pyranose ring system of the sugar (as well as sugars other than ribose) were tested for their ability to support or inhibit the growth of certain uracil-less and thymine-less microorganisms. The compounds showed neither effect to any significant extent (*33*, *44*). Some of the nucleosides of thymine possibly had a slight inhibitory effect upon the growth of *Leuconostoc citrovorum* when cultured in the presence of limiting amounts of thymidine (*124*).

Cytidine is reported to repress slightly the growth of a guanine-less *Ophiostoma* strain, in response to guanosine (*48*).

5. The Biosynthesis of Ribosides and Ribonucleotides.

It is reasonable to suppose that investigations upon the growth requirements of microorganisms for purines, pyrimidines, and their derivatives, as well as the effect of special antagonists upon their growth, should aid materially the eventual formulation of a generalized scheme including the synthesis of nucleosides, nucleotides, and nucleic acids in living organisms. Several of these findings have already been applied with success to investigations of nucleic acid-metabolism in higher animals, but the greater part of the data constitute unrelated observations, some of which had better remained unexplored until an expanded knowledge made possible their more graceful reception.

A Comparison between Microorganisms and Higher Animals with Regard to Purine Precursors in Nucleic Acid Biosynthesis. Little is known of the microbiological origin of the purine-ribosides or ribonucleotides, but certain comparisons can be made between microorganisms and higher animals with regard to which of the purines serve as precursors in nucleic acid synthesis.

Much variation exists among microorganisms in this regard. The microbiologist is not alone, however, in his efforts to rationalize this unruly individuality of different species, although he gives the appearance of greater boldness by engaging superior numbers. The metabolism of purines in mammals has been investigated with few species, and even these show variations in regard to which of the purines, when injected, are appreciably incorporated into the nucleic acid of the animal (*21*).

Ingested 2,6-diaminopurine is a precursor of guanine in the ribo- and desoxyribonucleic acids of the rat (*10*) but it is inactive as a source of purines in the nutrition of *Lactobacillus casei* (*60*), a bacterium which responds in growth to guanine and other purines. 2,6-Diaminopurine is also inactive in supporting the growth of purine-less strains of *Escherichia coli* which respond to adenine, guanine, hypoxanthine, xanthine, and, to a lesser extent, isoguanine (*10*). The demonstration of KALCKAR (*70*) that ^{14}C-adenine gives rise to undiluted adenine and guanine of high isotopic concentration in the nucleic acids of *Lactobacillus casei*, if the organism is grown in a pteroylglutamic acid-free medium, resembles the finding of BROWN (20) that ^{15}N-adenine, when ingested, gives rise to adenine and guanine in the ribonucleic acid of the rat. In the nutrition of *Escherichia coli*, also, ^{14}C-adenine gave rise to adenine and guanine of high isotopic concentration in the cellular material of the organism (6). *Tetrahymena geleii*, however, could not convert adenine into guanine. When this organism was grown in the presence of radioactive adenine and suboptimal guanine, the adenine of the nucleic acid possessed 84 per cent of the specific activity of the administered adenine, while the guanine of the nucleic acid had no significant activity (42).

In the Sherman strain of rat adenine (20) and 2,6-diaminopurine (10) are the only purines which are appreciably incorporated into the nucleic acid of the animal, but in the C 57 strain of mouse guanine is also utilized in nucleic acid synthesis to a small but significant extent (21). Guanine serves as a precursor of both adenine and guanine in the cellular material of *Escherichia coli* (42) and in the nucleic acids of *Lactobacillus casei* (3) and *Tetrahymena geleii* (41) but not in *Torulopsis utilis*. This organism utilizes adenine for the synthesis of both the adenine and guanine of its ribonucleic acid but utilizes guanine only for the synthesis of ribonucleic acid-guanine (73). The nucleosides and nucleotides of adenine and guanine were not utilized to any considerable extent by this yeast, an observation in agreement with the findings of HAMMARSTEN and REICHARD (55) that guanosine is not a precursor for the synthesis of nucleic acids in the rat and of KERR et al. (73) that the nucleosides and nucleotides of adenine and guanine do not serve as precursors of purines in the ribonucleic acid of rat-liver slices. According to ELION et al. (36) the nucleosides and nucleotides of the purines are probably degraded by *Lactobacillus casei* before being used as a source of purines.

The inability of these organisms to make use of the purine-nucleosides in place of the purines is just the reverse of the experience in the metabolism of pyrimidines; for in the latter case it is the nucleoside and not the pyrimidine which is the more immediately utilizable. Mention has already been made of the fact that purine-ribosides are more active than purines in supporting the growth of certain microorganisms, but no explanation of these findings has yet been offered.

MITCHELL and HOULAHAN (96) collected genetic evidence from the study of experimentally induced mutant strains of *Neurospora* which indicated the existence of seven reactions concerned with the biosynthesis of adenosine in this species. The last reaction was thought to be the conversion of inosine into adenosine.

Triazolo analogs of the purines* have been employed by KIDDER et al. (75, 76) as growth-inhibitors for the guanine-less protozoan, *Tetrahymena geleii*. From these studies and those upon the growth-promoting properties of the suspected intermediates the conclusion was reached that guanine lies in the direct pathway of nucleic acid synthesis in this organism. Guanine, itself, was thought to be converted into guanosine which might give rise, in part, to adenosine. Alternatively, guanine could be converted into adenine which might subsequently be converted to adenosine. But FRIES (48), from his examination of the nutritional requirements of guanine-less strains of *Ophiostoma*, thought the evidence was insufficient to conclude whether, in this organism, guanine itself lies in the direct pathway of nucleic acid synthesis.

Whether purine-nucleosides arise by ring-closure of the riboside of 4-amino-5-imidazole carboxamide is yet to be investigated (9, 53).

There are also conflicting opinions in regard to what constitutes the direct pathway of pyrimidine-nucleoside synthesis. LORING and PIERCE (88) suggested that cytosine is not a normal intermediate in the biosynthesis of pyrimidine-ribosides in *Neurospora*. MITCHELL and HOULAHAN (97) doubt that any of the pyrimidines (cytosine, uracil or orotic acid) are directly involved in the biosynthesis of the pyrimidine-ribosides in this organism. In their opinion uridine arises through the intermediate formation of an aliphatic N-riboside which undergoes subsequent ring closure. Oxalacetic acid, $HOOC \cdot CH_2 \cdot CO \cdot COOH$, was suggested as a precursor of the aliphatic portion of the intermediate. Uracil and orotic acid, both of which possess slight growth-promoting activity for the organism, were envisaged as opening, by a side reaction, to yield aliphatic compounds which could serve as precursors of the postulated intermediate. Cytidine was thought to arise from uridine (98). FRIES (47) is also of the opinion that uridine, and not uracil, is in the direct pathway of nucleic acid synthesis in *Ophiostoma* and that cytidine arises from uridine.

---

* It has been pointed out by KIDDER and DEWEY (75) that 8-azaguanine (formula, p. 404) exhibits an unusually prolonged antagonism against guanine in the nutrition of *Tetrahymena geleii*. These authors suggest that 8-azaguanine might be converted into nucleosides or nucleotides which are themselves fatal to the organism at some later stage of development. An inhibitor of such type would differ from that usually encountered in which, as is supposed, the antagonist functions only through its effect in blocking the biosynthesis of the essential metabolite. Following this suggestion MITCHELL et al. (98a) were able to demonstrate that 8-azaguanine, as such, was incorporated in small amount into the ribonucleic acid fraction of mouse viscera and served also as a precursor for the guanine and adenine in this fraction. The nucleoside of 8-azaguanine has recently been synthesized enzymatically from desoxyribose-1-phosphate and 8-azaguanine (45a). Whether in fact 8-azaguanine inhibits growth by virtue of its formation of "abnormal" nucleosides, nucleotides, or nucleic acids is not as yet established.

KIDDER et al. (76) believe that *Tetrahymena geleii* has the faculty of linking uracil with ribose to form uridine, and uridine and cytidine were considered as interconvertible. Since both thymine and thymidine had a sparing action upon the requirement for pteroylglutamic acid by this organism, and since thymidine had twice the sparing action of thymine, they concluded that pteroylglutamic acid is involved in at least two reactions having to do with the biosynthesis of thymidine, *viz.* one, with the formation of thymine and, the other, with the formation of the desoxyribosidic linkage of thymidine.

The finding of PAEGE and SCHLENK (*102*) that *Escherichia coli* contains a specific nucleoside phosphorylase which catalyzes the synthesis of uridine from uracil and ribose-1-phosphate (in which system cytosine is inactive) seems to be related to the activity of uracil and the inactivity of cytosine in satisfying the pyrimidine-requirement of certain microorganisms. However, the microbiological activity of orotic acid would not agree with this idea, since this acid was inactive in the above system. *Escherichia coli* did not contain orotic acid decarboxylase.

Orotic acid is known to act as a precursor of uridine and cytidine in the biosynthesis of ribonucleic acid (*1*, *115*) and of thymine and cytosine in the synthesis of desoxyribonucleic acid by the rat (*115*). It also gives rise to the pyrimidines of ribonucleic and desoxyribonucleic acid in rat liver slices (*116*, *151*). ARVIDSON et al. (*1*) have suggested that the incorporation of orotic acid may occur through the intermediate formation of its riboside which could form uridine by decarboxylation. Cytidine could arise from uridine or, alternatively, by amination of orotic acid riboside followed by decarboxylation. No conclusion was reached as to whether orotic acid should be considered as a normal intermediate in the biosynthesis of nucleic acids in this animal. Orotic acid riboside has been isolated from a pyrimidine-less strain of *Neurospora* grown in response to a minimal amount of cytidine sulfate (*95*) but a comparison of the microbiological activity of this riboside with orotic acid and other pyrimidine ribosides has not been reported as yet. REICHARD (*115*) showed that orotic acid contributed toward a more rapid synthesis of cytosine than uracil in the synthesis of ribonucleic acid by the rat. He concluded that the biosynthesis of cytosine was not preceded by that of uracil. In the synthesis of desoxyribonucleic acid, also, cytosine arose from orotic acid at a higher rate than did uracil. HAMMARSTEN et al. (*56*), finding that the administration of uridine resulted in much less incorporation of pyrimidines into the nucleic acid of the rat than that of cytidine, support the views that orotic acid gives rise to cytidine, without the intermediate formation of uridine and that uridine arises from cytidine, just the reverse of the conclusions reached from the studies on the nutrition of *Neurospora* and *Ophiostoma*. The complexity of this question is further compounded by the existence of a *Lactobacillus bulgaricus* strain which has an essential growth-requirement for orotic acid that cannot be met by other pyrimidines, uridine, cytidine, or pyrimidine nucleotides (*160*). WRIGHT et al. (*161*) found that orotic acid was a precursor of pyrimidines in the ribonucleic acid of this bacterium, as others had shown for the rat, and reported further that *DL*-ureido-succinic acid served as an acyclic precursor of the pyrimidine ring.

Studies of the antagonism of purine-ribosides upon the utilization of pyrimidine-ribosides by *Neurospora* led PIERCE and LORING (*108*) to conclude that at least two reactions were involved in this inhibition. One was thought to be the deamination of cytidine to yield uridine, and the other, the utilization of uridine itself for the synthesis of ribonucleic acid. However, FUKUHARA (*50*) has concluded from his studies of the inhibition of 5-chloro-uridine upon the utilization of the pyrimidine-ribosides by this same strain of *Neurospora* that both uridine and cytidine are used as such for the synthesis of nucleic acid, and that the conversion of one riboside to the other must occur at some later stage in the synthesis of nucleic acid.

Upon considering the above findings, what seems particularly striking (apart from the obvious difficulty of achieving a simple and complete interpretation) is the remarkable absence of any established relationship between one or more of the vitamins and the biosynthesis of the ribosidic linkage, whereas pteroylglutamic acid, the "citrovorum factor" (a derivative of pteroylglutamic acid), and vitamin B_{12} are known to bear a relationship to the biosynthesis of desoxyribosides in various microorganisms.

6. Vitamin B_{12}.

No detailed account of the eventful history of vitamin B_{12} need to be inserted here. Several reviews have given a treatment of its discovery, occurrence, biological effects, and assay (*137, 163, 139, 156, 38*).

Products obtained by degradation of the vitamin show it to be a nucleotide of *D*-ribose. BRINK et al. (*14*) isolated 1-α-*D*-ribofuranosido-5,6-dimethyl-benzimidazole; and BUCHANAN et al. (*22*) isolated the nucleotide and showed that the phosphoryl group was attached to either the 3'- or the 2'-position. It will be noted that the benzimidazole-nucleotide resembles the nucleotides of ribonucleic acid as to the point of attachment of the phosphoryl group but differs in having the α- instead of the β-glycosidic configuration. The α-benzimidazole riboside had about $1/_{400}$ the activity of vitamin B_{12} in the rat (*39*) but neither the nucleoside nor the nucleotide has as yet been reported to show vitamin B_{12}-like activity for microorganisms. 5,6-Dimethyl-benzimidazole, itself, was inactive in the *Lactobacillus lactis* DORNER-assay for vitamin B_{12} but possessed some activity in the assay with the rat (*37*).

Microbiological Functions of Vitamin B_{12}. Vitamin B_{12} is known to be involved with several biochemical processes in the nutrition of microorganisms. One of its functions has to do with the biosynthesis of desoxyribosides while most of its other functions may be considered formally as biochemical reactions which involve one-carbon fragments. These include its role in the biosynthesis of methionine, serine, purines, and thymine (or pteroylglutamic acid). That all of these latter functions belong to one category is not, in fact, proved; it is merely a reasonable guess which allows a convenient classification.

Still other effects of vitamin B_{12} in the metabolism of microorganisms are not known with certainty to be related to any of the above functions.

Since thymidine had been found to replace the requirement for vitamin B_{12} by lactobacilli, WRIGHT et al. (*162*) assumed that vitamin B_{12} functioned specifically in the biosynthesis of thymidine. When, however, desoxyribosides other than thymidine were shown to replace the requirement for vitamin B_{12} in the nutrition of a number of lactic acid bacteria, KITAY et al. (*77*) suggested that vitamin B_{12} might be involved with the biosynthesis of desoxyribose-1-phosphate.

This suggestion was based upon analogy with the known action of mammalian nucleoside phosphorylase in catalyzing the synthesis and phosphorolysis of certain purine-desoxyribosides. Desoxyribose-1-phosphate was envisaged as a precursor common to the bacterial synthesis of all the desoxyribosides.

There is no evidence that it is, and the subsequent demonstration that desoxyribose-1-phosphoric acid was inactive in the same enzyme system, from certain bacteria, which catalyzed the interconversion of desoxyribosides (*90, 91*) seems to leave little excuse for believing that vitamin B_{12} functions in this way*. Exactly how vitamin B_{12} and the desoxyribosides are functionally related remains obscure.

A description of the relationships of vitamin B_{12} and the desoxyribosides in the nutrition of certain bacteria follows in Chapter II (p. 420).

When HUTNER et al. (*64*) discovered that vitamin B_{12} was an essential growth-factor for *Euglena gracilis* var. *bacillaris* and that thymidine was inactive, a second function of vitamin B_{12} was indicated. DAVIS and MINGIOLI (*31*) have obtained experimentally induced mutant strains of *Escherichia coli* which show a similar nutritional behavior. Methionine was found to replace the requirement for vitamin B_{12}, while homocysteine

---

* This reasoning involves the assumption that desoxyribose-1-phosphate, prepared by the action of mammalian nucleoside phosphorylase on guanine desoxyriboside, possesses the biologically active configuration for the bacterial nucleosidase. It is not known whether enzymically produced desoxyribose-1-phosphate has the α- or the β-configuration. Since, however, no effect of phosphate ion on the interconversion of desoxyribosides was apparent, there are no positive indications that desoxyribose-1-phosphate is involved as an intermediate.

was inactive, from which DAVIS and MINGIOLI concluded that vitamin B_{12} was concerned with the methylation of homocysteine.

The role of vitamin B_{12} in this and other functions in the nutrition of animals has been reviewed by EMERSON and FOLKERS (38).

Vitamin B_{12} has been implicated as one of the factors limiting the multiplication of bacteriophage T_4 r in *Escherichia coli* (121). The effect of vitamin B_{12} in reducing the burst-time and increasing the average burst-size is due, in all probability, to its action on processes in the host rather than to a direct action on the parasite. The supposition of ROBERTS and SANDS (121) that the ultimate function of vitamin B_{12} in this relationship involves the biosynthesis of thymidine seems unlikely, in view of the findings by other authors that the biosynthesis of thymidine is not among the limiting functions of vitamin B_{12} which have been demonstrated in the nutrition of *Escherichia coli*. SHIVE (126) found that vitamin B_{12} ("erythrotin") reversed the inhibitory effect of sulfanilamide upon the growth of *Escherichia coli*. Under suitable conditions methionine, purines, serine, and thymine (or pteroylglutamic acid), each partially replaced vitamin B_{12} in this reversal of inhibition, while thymidine and other desoxyribosides were ineffective. These results were interpreted to mean that, in this organism, vitamin B_{12} is involved in the biosynthesis of methionine, purines, serine, thymine (or pteroylglutamic acid) and at least one additional substance of unknown character. The unknown metabolite was inferred, since the known metabolites replaced vitamin B_{12} incompletely. The limiting functions of vitamin B_{12} in this organism involved, therefore, processes other than the biosynthesis of desoxyribosides.

Vitamin B_{12} has the effect of increasing the rate at which stored cells of a particular strain of *Escherichia coli* are able to oxidize such substrates as fatty acids, hydroxyacids, and aminoacids, but the nature of this action is unknown (101).

Different Forms of Vitamin B_{12}. Several vitamin B_{12}'s are known which differ merely by the grouping which is coordinated with the cobalt. In the nomenclature employed by KACZKA et al. (69) the grouping coordinated with the cobalt appears as the prefix to the parent name "*cobalamin.*" Vitamin B_{12}, isolated from natural sources includes the coordination complex of cobalt with the CN group (15, 69, 23); it is called *cyano-cobalamin*. Its microbiological activity is not destroyed by autoclaving it with the medium (*Lactobacillus leichmannii* assay) (57, 19). Vitamin B_{12a}, isolated from media in which *Streptomyces aureofaciens* had grown (110, 109) and prepared from vitamin B_{12} by catalytic hydrogenation (68, 143) differs from vitamin B_{12} in having (H_2O) coordinated with the cobalt (69, 23); hence, its name *hydroxo-cobalamin*. It undergoes inactivation when autoclaved with the medium (*Lactobacillus leichmannii* assay) but this destruction may be prevented by the addition of reducing agents (57, 19). Cyano-cobalamin is converted into hydroxo-cobalamin upon exposure to light (146) and undergoes a partial conversion to hydroxo-cobalamin when subjected to chromatography on paper (159), but this effect is also due to the action of light upon cyano-cobalamin during the chromatographic process (28). Cyano-cobalamin may be regenerated from hydroxo-cobalamin by treatment with cyanide ion (*69, 155, 7*), and cyano-cobalamin itself forms complexes having one or two moles of cyanide in addition to the cyano-group already present in the vitamin (*26, 7*). The conversion of hydroxo-cobalamin to cyano-cobalamin by cyanide ion is the most probable explanation for the observations that the addition of potassium cyanide to bacterial media prevented the destruction of hydroxo-cobalamin during autoclaving (*Lactobacillus leichmannii* assay) (*140*) and raised the activity of hydroxo-cobalamin in supporting the growth of *Lactobacillus lactis* DORNER up to the value given by cyano-cobalamin (*29*).

"Vitamin B_{12b}" (*143, 16, 110, 109, 19*) is almost certainly another name for hydroxo-cobalamin (*67, 155, 57*) and has been assumed to be such in the foregoing discussion. Cyano-cobalamin, when autoclaved with thiomalic acid is converted into a new form of the vitamin which is more active microbiologically than the untreated substance (*Lactobacillus leichmannii* assay) (*19*). Several other modifications of the vitamin have been prepared by chemical means and tested for microbiological activity. These include chloro-cobalamin, sulfato-cobalamin, a crystalline product prepared by treating cyano-cobalamin with H_2S (*69*), a thiocyanate-analog (*23*), amino-cobalamin (*28*), nitroso-cobalamin (*19*) and a complex with histidine (*28*). All are microbiologically active. It is probable that each of them may be reconverted to cyano-cobalamin by reaction with cyanide ion (*69, 28*).

Vitamin B_{12} is known to exist also in microbiologically inactive, non-dialyzable complexes from which the vitamin may be released by heating (*144, 122, 106, 11*). Oxidation of the vitamin with hydrogen peroxide in hydrochloric acid solution converts it into a colorless substance which competitively antagonizes vitamin B_{12} in the nutrition of *Lactobacillus leichmannii* 4797 (*8*).

ROBBINS *et al.* (*118*) have made an investigation of the source of vitamin B_{12} which sustains *Euglena* in its natural habitat. Many actinomycetes and soil bacteria were able to synthesize vitamin B_{12}. They suggest that the synthetic activity of microorganisms, especially actinomycetes and bacteria, and not the higher plants, is the original source of vitamin B_{12} in nature. Some of the soil bacteria also required vitamin B_{12}, but neither thymidine nor methionine was tested for its ability to replace this requirement (*84*).

II. Nucleosides and Nucleotides of Desoxyribose.

The nucleosides of desoxyribose were first implicated in the nutrition of microorganisms by the demonstration of SHIVE *et al.* (*130*) that thymidine was several times more active than pteroylglutamic acid in reversing the inhibitory action of methyl-pteroylglutamic acid upon the growth of *Leuconostoc mesenteroides*. Then followed in very short order reports that thymidine acted as an essential growth factor for various lactic acid bacteria (*138*), that thymidine could replace the vitamin B_{12} requirement in the nutrition of lactic acid bacteria (*162, 131*), and that several desoxyribosides, other than thymidine, also acted as growth-factors for certain of these organisms (*77, 80, 61*). Thymidine was shown, as well, to bear a relationship to the nutritional requirement of *Leuconostoc citrovorum* for the "citrovorum factor" (*125, 124*), a derivative of pteroylglutamic acid (*4, 66, 129, 17, 43, 111*).

1. The Biosynthesis of Desoxyribosides.

From the investigations upon the growth-promoting activity of the natural desoxyribosides in the nutrition of certain microorganisms, little has been deduced, as yet, which contributes to our understanding of their biosynthesis. KIDDER et al. (76) believe that pteroylglutamic acid is concerned with the biosynthesis of the desoxyribosidic linkage of thymidine in the metabolism of *Tetrahymena geleii*. WRIGHT et al. (162) assumed from their studies upon the growth-requirements of lactobacilli that thymidine arises from thymine through one or more reactions catalyzed by vitamin B_{12}. Thymine, under appropriate conditions replaces the requirement for pteroylglutamic acid by many bacteria, but comparisons of the relative effectiveness of thymine and thymidine in this relationship have not usually been made. In the nutrition of *Streptococcus faecalis* STOKES (141) has shown that thymine and thymidine behave similarly in the replacement of pteroylglutamic acid. The effectiveness of thymidine was due merely to the amount of thymine the molecule contained.

Although thymine effectively replaces the requirement for pteroylglutamic acid by many bacteria, if cultured in otherwise complete media, it is not effective for all. *Leuconostoc mesenteroides* is an example of the latter type. In this organism thymine will not (but thymidine will) reverse the growth inhibitory effect of methyl-pteroylglutamic acid. WILLIAMS et al. (157) believe that thymidine does not arise, in this organism, through the intermediate formation of thymine. Pteroylglutamic acid would thus be assumed to play a role in the synthesis of thymidine as contrasted with thymine, but to what extent selective permeability might be expected to confuse conclusions of this kind is not always clear. For instance, it could be argued that the bacterium is permeable to thymidine and impermeable to thymine. Pteroylglutamic acid, acting catalytically, would need to be absorbed in only small amounts to produce sufficient thymine inside the cell to support its growth. Thymidine might then arise from thymine without the involvement of pteroylglutamic acid in any additional function concerning the biosynthesis of thymidine.

Pteroylglutamic acid and vitamin B_{12} are known to be involved in the nutrition of *Lactobacillus casei*, specifically with the synthesis of desoxyribonucleic acid. Culturing the organism in suitable media, rendered partially deficient in pteroylglutamic acid, resulted in a considerably lowered content of desoxyribonucleic acid, while the content of ribonucleic acid was essentially unchanged (112). Also, both pteroylglutamic acid and vitamin B_{12} were shown to enhance and to have an additive effect on the synthesis of desoxyribonucleic acid by this same organism but were without any pronounced effect on the synthesis of ribonucleic acid (114).

Many organisms are known which require thymidine or another desoxyriboside for their growth and for which pteroylglutamic acid and the "citrovorum factor" are ineffective. For most, but not all, of these organisms vitamin B_{12} meets the requirement.

Considerations of the Mode of Formation of the Desoxyribosidic Linkage. The formation of the desoxyribosidic linkage might conceivably arise through the agency of bacterial enzymes (93), comparable to nucleoside phosphorylase from mammals (46), or through a series of reactions which are less well characterized and which involve the formation of nucleotides by interaction of ribose-3-phosphoric acid and purines (148). What relationship vitamin B_{12} could bear to these reactions is very obscure. Another possibility is that desoxyribosides are not formed biochemically through a joining of the sugar and the aglycone but arise from preformed ribosides, in which case thymidine and 5-methylcytosine desoxyriboside might be considered as arising by subsequent methylation of the pyrimidine-portion of the molecule. However, here again, vitamin B_{12} would have to be related functionally to the desoxygenation of one or more ribosides at position 2 of the ribose.

HAMMARSTEN *et al.* (56) have found that the cytosine of administered cytidine gives rise to cytosine and thymine in the desoxyribonucleic acid of the rat. Since the pyrimidines are not, themselves, incorporated into the nucleic acids of the rat, the conclusion was reached that the desoxyribosides arose from the riboside without the loss of the aglycone. The evidence was given as proof that in both the ribosides and the desoxyribosides the glycosyl group is linked to the pyrimidine at the same position*.

---

* Although it seems very likely that the point of attachment of the glycosyl group to the pyrimidine is identical in both the ribosides and the desoxyribosides, this evidence fails to actually establish it. This proof is weakened by the very possibility of trans-N-glycosidation in the nucleic acid-metabolism of the rat such as that which is catalyzed by certain bacterial enzymes (90, 91). Once this possibility is admitted, there is no reason *a priori* to suppose that the glycosyl group could not migrate to the other nitrogen atom. RABATÉ (113) has shown, for example, that enzyme preparations from *Salix purpurea* L. which catalyze the transfer of the glucosyl group of O-glycósides, catalyzed the exchange of the β-glucosyl group of salicin from the phenolic to the primary alcoholic group of the radical without prior hydrolysis. A trans-glucosidase from the potatoe (Q-enzyme) has been shown by PEAT and his associates (5) to catalyze molecular rearrangements of amylose and starch without hydrolysis or the intervention of phosphate. A number of 1,4-linkages of the amylose-chain were transformed into 1,6-linkages, thus forming branched molecules of the amylopectin-type. PAZUR and FRENCH (104) demonstrated that a trans-glucosidase from *Aspergillus oryzae* catalyzed the conversion of maltose (1,4-linkage) into isomaltose (1,6-linkage) but failed to catalyze the synthesis of disaccharide from glucose and glucose-1-phosphate.

BADDILEY's (2) suggestion that the evidence of HAMMARSTEN *et al.* (56) shows that the ribosides and desoxyribosides each have β-configurations is not open to

REICHARD and ESTBORN (*117*) have shown that the conversion of cytidine into cytosine desoxyriboside is irreversible in the rat, for cytosine desoxyriboside did not give rise to cytidine in the ribonucleic acid of that animal.

Methods are now available which permit an investigation of the possible conversion of ribosides into desoxyribosides in the metabolism of microorganisms. Ribosides, labeled with ^{14}C in both the ribosyl group and the aglycones might be prepared (*6, 45*). Sensitive and quantitative microbiological tests for desoxyribosides were described (*63*), and suitable microorganisms are available for examining not only the possible conversion of ribosides into desoxyribosides but the relationship of vitamin B_{12} or other vitamins to such a process.

Mention has already been made of a strain of *Hemophilus parainfluenzae* which required uracil for growth (*58*). Cytosine did not have the ability to substitute for uracil, but cytidine and cytosine desoxyriboside were active, although the activity of cytosine desoxyriboside was only one half that of cytidine. How is the activity of cytosine desoxyriboside to be explained? Are we to assume that the organism mentioned converts cytosine desoxyriboside into cytidine? This conclusion may be avoided if several assumptions are permitted: The organism must not contain a cytosine deaminase (cytosine is inactive) and must have a cytosine desoxyriboside deaminase which converts the compound into uracil desoxyriboside. This compound might then be split by a nucleosidase to yield uracil, the pyrimidine derivative which supports the growth of the organism. Since cytosine desoxyriboside was actually less active than uracil (on a molar basis) the above sequence is very possible. Or, again, it could just as well be assumed that the organism is permeable to the nucleosides of cytosine but not to free cytosine. Cytosine might, then, be considered as arising inside the cell through the action of a nucleosidase.

In the biosynthesis of certain bacterial viruses there are bare suggestions that the transfer of purines and pyrimidines from host to parasite takes place through nucleosides or nucleotides. KOCH (*79*) states that the transfer of purines from *Escherichia coli* to bacteriophage T_6 probably does not proceed through the free purines; and the evidence of WEED and COHEN (*150*) suggests that in the parasitism of *Escherichia coli* strain B by bacterial viruses $T_6 I^+$ and $T_6 r$, units of bacterial desoxyribonucleic acid, larger than nucleosides or nucleotides, are not involved in the formation of desoxyribonucleic acid by the parasite.

---

this criticism, perhaps, since trans-glycosidases apparently show great specificity as to the configuration at the point of enzymic attack.

2. The Growth-promoting Activity of Desoxyribosides and Desoxyribonucleotides.

a) The Specificity of Certain Desoxyribosides in Eliciting the Growth-response of Bacteria.

Some twenty-five microorganisms, most of them lactic acid bacteria, are now known which have an essential growth-requirement that is satisfied by a natural desoxyriboside. The majority of them responded equally in growth to equimolar amounts of the five naturally occuring desoxyribosides examined (*77, 61, 80, 92, 78*). Certain of the organisms, however, exhibited a preference—in several cases a marked preference— for a particular desoxyriboside or desoxyribonucleotide. *Lactobacillus delbrückii* ATCC 9649, for example, responded appreciably to only thymidine over short periods of incubation. With longer incubation periods the organism responded to the other desoxyribosides, but the response was not as strong as that to thymidine (*78*). Only thymidine supports the growth of *Leuconostoc citrovorum* (*77, 66*); uracil desoxyriboside and even 5-methylcytosine desoxyriboside were inactive for this organism (*91*). For several strains of *Lactobacillus acidophilus* cytosine desoxyriboside permitted more rapid and heavier growth than that obtained with any of the other desoxyribosides (*78*). Cytosine desoxyribonucleotide was also found to have a greater stimulatory effect upon the growth of *Lactobacillus leichmannii* and *L. lactis* than had the other desoxyribonucleotides examined, while delays in growth occurred in response to adenine desoxyribonucleotide and thymidylic acid (*133*). Thymidylic acid was reported, also, to be less active than thymidine in an undefined assay with *L. arabinosus*, and replaced the requirement for pteroylglutamic acid (*147*).

To what extent these specific effects are due to the selective permeability which the different bacteria exhibit is not known, but with the possibility of preparing ^{14}C-purines, -pyrimidines (*6, 45*), and -desoxyribosides (*72*) an investigation may now be made of the permeability of these bacteria to the substances in question.

b) The Non-specificity of the Natural Desoxyribosides in Promoting the Growth of Certain Bacteria.

As previously mentioned, the majority of bacteria which showed a growth-response to desoxyribosides responded equally to equimolecular amounts of the five natural desoxyribosides examined. HOFF-JÖRGENSEN (*62, 63*) has taken advantage of this fact to devise a sensitive and quantitative microbiological assay of the desoxyribosides. The organism employed, *Thermobacterium acidophilus* R 26, is easily grown, is unaffected during a twenty-four-hour period of incubation by the presence of vitamin B_{12}, reducing agents, or ribosides, is only very slightly affected by desoxyribonucleic acid, and responds quantitatively to equimolar amounts of the desoxyribosides of adenine, guanine, hypoxanthine, cytosine, uracil, and thymine (*63, 90, 91*). 5-Methylcytosine desoxyriboside is also active for this organism (*91*). The individual desoxyribosides can be separated from one another by paper chromatography and estimated quantitatively (*90, 91*).

Through the use of appropriate enzymes which release desoxyribonucleic acid from tissues and which degrade it to nucleosides or nucleotides (desoxyribonucleotides are active for this organism) HOFF-JÖRGENSEN (*63*) has estimated the quantity of desoxyribonucleic acid in bacterial cells. Since quantitative analysis may be carried out with as little as a few micrograms of desoxyribonucleic acid, this method should prove to be of considerable value.

The curious fact, that just any one of the desoxyribosides is sufficient to promote the growth of these organisms, suggests that the organisms have the ability to convert one desoxyriboside into all the other desoxyribosides which are included in the bacterial desoxyribonucleic acid. And since each desoxyriboside is equally effective (on a molar basis), this interconversion must be quite efficient.

An investigation of this question has been carried out by McNUTT (*90, 91*). Enzyme-preparations from *Lactobacillus helveticus* S catalyzed the transfer of the desoxyribosyl group from a number of desoxyribosides to many different purines and pyrimidines (*90, 91, 72*). A number of known desoxyribosides not previously synthesized biochemically, as well as certain desoxyribosides not previously described, were produced enzymically by reacting a given desoxyriboside with the desired purine or pyrimidine. Since the enzyme preparation which brought about these interconversions failed to catalyze the synthesis of desoxyribosides from desoxyribose-1-phosphate plus the purine or pyrimidine (or from desoxyribose plus the purine or pyrimidine) the enzyme was considered to be a trans-N-glycosidase, analogous to the trans-O-glycosidases obtained from *Pseudomonas saccharophila* (*34*) and *Aspergillus oryzae* (*104*). The analogy with the *Aspergillus* enzyme is more complete, since no synthesis of disaccharide resulted when glucose-1-phosphate plus glucose was incubated with the enzyme (*104*); whereas the enzyme preparation from *Pseudomonas saccharophila* catalyzed the synthesis of sucrose from glucose-1-phosphate plus fructose (*34*).

The enzyme which catalyzed the transfer of the desoxyribosyl group from one purine or pyrimidine to another was shown to be present, as well, in two other desoxyriboside-less bacteria and may be of wide distribution among similar organisms. If the operation of a trans-desoxyribosidase of this type in living bacteria is assumed, the lack of specificity of the desoxyriboside-requirement in these organisms becomes less puzzling.

Whether such enzymes have a more general importance in the synthesis of nucleic acids in plants and animals is unknown. KRITSKIĬ (82) reports that purified mammalian nucleoside phosphorylase, under suitable conditions, catalyzed the transfer of the ribosyl group from guanine to hypoxanthine, although the inosine, which he assumed to be formed, was not actually characterized. The findings that purines and pyrimidines, in general, are not incorporated into the nucleic acids of the rat might seem, at first glance, to labor against the view that enzymes of this type play a role in the synthesis of nucleic acids in this animal. But what is more to the point is the following experiment: Let a pyrimidine nucleoside (or purine nucleotide), labeled with ^{14}C in the glycosyl group, be administered to the rat (these compounds are known to act as precursors of pyrimidines or purines in the nucleic acid of this animal). Following its administration, let the nucleic acids be separated, the nucleosides obtained, and the specific activity of the glycosyl group of each determined. Could it be shown that the glycosyl group had been distributed among the various purines and pyrimidines present in the nucleic acid, the findings would render more probable the possibility that trans-N-glycosidation occurs in this animal. This question cannot be answered at the present time, since the investigators, who have studied the incorporation of nucleosides into the nucleic acids of the rat, have invariably employed nucleosides which were labeled only in the aglycone.

3. The Relationship of the Desoxyribosides, Vitamin B_{12}, Reducing Agents, and the "Citrovorum-Factor" in Supporting the Growth of Various Microorganisms.

The interrelationships of desoxyribosides, vitamin B_{12}, reducing agents, and the "citrovorum factor" in supporting the growth of microorganisms are very complex and have not yet been adequately interpreted. Representative examples of microorganisms have been selected (Table 2) to illustrate the differences exhibited by various microorganisms in their growth-response to one or more of these substances. In general, most of the organisms which responded to thymidine grew in its absence if vitamin B_{12} or desoxyribonucleic acid were added to the medium (78). Less commonly, ascorbic acid or other reducing agent was effective in substituting for thymidine, but thymine, high levels of folic acid (except in *Leuconostoc citrovorum*), cobalt salts, ribosides, and 2-desoxy-D-ribose were ineffective.

For some organisms, of which *Lactobacillus acidophilus* S and *L. helveticus* S are examples, thymidine, vitamin B_{12}, and reducing agents had very similar effects upon growth. The requirement for thymidine by *L. helveticus* S could even be eliminated by merely autoclaving the medium for a longer period of time. For many others, typified by *L. leichmannii* ATCC 7830, thymidine and vitamin B_{12} produced very similar effects upon growth, but ascorbic acid was less effective on both the rate and the amount of growth.

In a few cases, of which *Lactobacillus acidophilus* ATCC 332 is an example, ascorbic acid promoted more rapid and frequently heavier growth than that resulting with thymidine, while vitamin B_{12} was inactive under the conditions employed. *L. bifidus*, *Thermobacterium acidophilus* R 26, and *L. delbrückii* ATCC 9649 responded to neither vitamin B_{12} nor reducing agents. While the concentration of thymidine required to produce a given growth-response was roughly the same for all of the organisms examined, marked differences were shown by various individuals in their quantitative response to a given concentration of vitamin B_{12}.

For example, five times as much vitamin B_{12} was required by *Lactobacillus acidophilus* ATCC 4355 to produce a given response in growth as that required by *L. leichmannii* ATCC 7830.

KODITSCHEK et al. (*81*) have shown that *L. lactis* DORNER, when cultured aerobically, produces hydrogen peroxide and that the quantity of vitamin B_{12} required for the growth of the organism is related to the amount of hydrogen peroxide present in the culture. It is not known, however, whether this demand for greater quantities of vitamin B_{12} by certain of these bacteria is due to their greater production of hydrogen peroxide.

Relationship between Certain Reducing Agents and Vitamin B_{12} Requirement. How reducing agents replace the requirement for vitamin B_{12} in the nutrition of certain microorganisms is imperfectly understood. WELCH and WILSON (*153*) showed that *Lactobacillus leichmannii*, an organism which responds to vitamin B_{12} or reducing agents, when grown in a medium containing charcoal treated trypsinized casein, failed to respond to reducing agents when cultured in an aminoacid medium. They suggested that the activity of the reducing agent was due to its reduction of a microbiologically inactive, oxidized form of vitamin B_{12} (present in the trypsinized casein) to the active substance. This idea was further supported by the fact that the trypsinized casein, when treated separately with reducing agents, supported growth of the organism in the aminoacid medium. Although this explanation fits the results obtained with many microorganisms, there are certain exceptions. As was noted above, microorganisms do exist for which reducing agents are effective in supporting growth and for which vitamin B_{12}, under the same conditions, is inactive. Furthermore, reducing agents are effective in supporting the growth of certain bacteria in media which contain no enzymically digested casein.

Table 2. Differences in Growth Exhibited by Various Bacteria in Response to the Desoxyribosides, Vitamin B_{12}, and Certain Reducing Agents.

Organism	Growth Response to:				Remarks	References
	Thymidine	Desoxy-ribosides other than thymidine	Vitamin B_{12}	Reducing agents		
Lactobacillus acidophilus S	+	+	+	+	Ascorbic acid or other reducing agent was effective in supporting growth in media containing no enzymically digested casein. Cytosine desoxyriboside supported better growth than did any other desoxyriboside.	(92, 78)
Lactobacillus helveticus S	+	+	+	+	The requirement for thymidine could be eliminated by merely autoclaving the medium for a longer period of time.	(92, 78)
Lactobacillus leichmannii ATCC 7830	+	+	+	±	Ascorbic acid did not permit as rapid or as heavy growth as that obtained in response to thymidine or vitamin B_{12}. The reducing agent was effective only in the presence of enzymically digested casein.	(77, 78)

Organism					Comments	References
Lactobacillus acidophilus ATCC 332	+	⊥	—	+	Ascorbic acid permitted growth which was frequently heavier and always more rapid than that induced by thymidine. Desoxyribosides other than thymidine were less effective.	(78)
Lactobacillus delbrückii ATCC 9649	+	⊥	—			(77, 92, 78)
Lactobacillus bifidus	+	+	—	—		(145)
Lactobacillus acidophilus ATCC 4355	+	+	⊥	(not stated)	Cytosine desoxyriboside was more effective than the other desoxyribosides. This organism required five times as much vitamin B_{12} as Lactobacillus leichmannii ATCC 7830 did in order to produce a given amount of growth.	(78)
Leuconostoc citrovorum ATCC 8081	+	—	—	—	Thymidine could be replaced by small amounts of 5-formyl-5,6,7,8-tetrahydropteroyl-glutamic acid.	(124, 77, 78, 111)
Mutant strains of Escherichia coli	—	(not stated)	+	(not stated)	Vitamin B_{12} was replaced by methionine.	(31)
Escherichia coli	—	—	+	—	Vitamin B_{12} could be replaced in part by methionine, thymine (or pteroyl-glutamic acid), serine, or purines but not by desoxyribosides or desoxyribo nucleotides.	(128, 133, 147)

To explain the behavior of these exceptional cases it has been suggested that below a certain oxidation-reduction potential of the medium the organism might have the ability to synthesize vitamin B_{12} or its equivalent (92). CUTHBERTSON and LLOYD (30) believe that *Lactobacillus lactis* DORNER, when cultured in the presence of reducing agents, synthesizes a substance capable of replacing vitamin B_{12}. This supposed ability would resemble superficially the observation of STOKES et al. (142) that the pyridoxine-less mutant strain of *Neurospora sitophila* is able to synthesize its own pyridoxine at p_H values of 5.8 or higher.

Even this explanation fails to take into account those bacteria which respond to reducing agents and which do not respond to vitamin B_{12} itself, for example, *Lactobacillus acidophilus* ATCC 332. In this case, it has been suggested, the organism might not be able to convert vitamin B_{12} into the metabolically active form of the vitamin, or the organism may be impermeable to vitamin B_{12} but may synthesize its own supply in the presence of a lowered oxidation-reduction potential in the medium (78).

In regard to the desoxyriboside-less organisms which do not respond to either vitamin B_{12} or reducing agents (for example, *Lactobacillus delbrückii* ATCC 9649), it might be assumed that the bacterium lacks the apoenzyme responsible for the synthesis of the desoxyriboside (78). Such easy assumptions are of little scientific usefulness, however, unless they initiate a concrete experiment.

The lack of any established relationship between vitamin B_{12} and the desoxyribosides in the nutrition of *Escherichia coli* introduces additional complexity into these questions. If vitamin B_{12} performs some vital function in the biosynthesis of desoxyribosides, how does it happen that vitamin B_{12}-less mutant strains of *Escherichia coli* (Table 2, p. 428) grow and reproduce in the absence of added desoxyribosides, provided that methionine is present?

Although it is true that the varying amounts of the purines and pyrimidines in the cells of different strains of *Escherichia coli* leave much doubt as to whether the composition of the nucleic acids in the organism is absolutely constant (94), surely the biosynthesis of desoxyribonucleic acid must be essential for the growth and reproduction of these organisms. It seems reasonable, therefore, to assume that desoxyribosides arise in the metabolism of *E. coli* by an alternative route to that in the desoxyriboside-less bacteria. Perhaps there is some significance to the fact that a representative of the latter group, *Lactobacillus helveticus* S, is apparently devoid of nucleoside phosphorylase (91, 71), while *Escherichia coli* is known to contain this enzyme which acts on both hypoxanthine desoxyriboside and thymidine (93). In the group of bacteria of which *L. helveticus* S is characteristic, the original source of the desoxyribosidic linkage may involve a process requiring vitamin B_{12} (desoxygenation of one of the ribosides at carbon 2', for example), whereas in *E. coli* vitamin B_{12} may function in nucleic acid synthesis only insofar as it catalyzes the formation of purines and pyrimidines, one or more of which react in the nucleoside phosphorylase-system to yield the parent source of the desoxyribosides.

The "Citrovorum Factor". *Leuconostoc citrovorum* ATCC 8081 is a special case among the thymidine-less bacteria. The thymidine-requirement for this organism, in contrast to that of the other bacteria (*78*), can be replaced by minute amounts of the "citrovorum factor".

SAUBERLICH and BAUMANN (*125*) first showed that this organism has an essential growth requirement which is satisfied by substances present in liver preparations. The unknown factor was proved to be different from folic acid, vitamin B_{12} or thymidine. Of these, folic acid was active only at very high levels, and thymidine, though active, was less so than preparations of the unknown substance which they designated the "citrovorum factor." The latter also reversed the inhibitory action of 4-amino-pteroylglutamic acid upon the growth of this organism. Thymidine, in much greater amount, also reversed this inhibition, but vitamin B_{12} was ineffective (*124*). BOND et al. (*12*) have concentrated substances from natural materials which reverse the inhibitory effect of methyl-pteroylglutamic acid upon the growth of *Lactobacillus casei*.

Their most active preparations ("folinic acids") were highly effective in supporting the growth of *Leuconostoc citrovorum*, 0.001 μg. per ml. supporting half-maximal growth (*4*). Upon mild acid hydrolysis of active preparations of the "citrovorum factor" a substance was obtained which corresponded in microbiological activity and chromatographic R_F value to pteroylglutamic acid (*4, 66*). Preparations which were highly active for *L. citrovorum* could be obtained by submitting pteroylglutamic acid to a series of chemical treatments. These consisted of (a) formylation, (b) catalytic hydrogenation of the reaction mixture in the presence of a reducing agent, and (c) autoclaving the resulting product in the presence of a reducing agent (*129*).

A microbiologically active crystalline compound isolated from this reaction mixture was assigned the structure of 5-formyl-5,6,7,8-tetrahydropteroylglutamic acid by POHLAND et al. (*111*). A crystalline compound showing high activity for *Leuconostoc citrovorum* was also obtained by BROCKMAN et al. (*17*) from reaction mixtures prepared in the same way. Tetrahydropteroylglutamic acid had 5000 times the activity of pteroylglutamic acid and 2.5 per cent of the activity of the "citrovorum factor" for this organism (*18*).

WEYGAND et al. (*154*) reported that 6-aminopteroylglutamic acid, when subjected to catalytic hydrogenation or formylation followed by catalytic hydrogenation, gave products showing "citrovorum factor" activity. Their potencies, however, were only $1/_{200}$ or less of that of preparations from pteroylglutamic acid similarly formylated and reduced. Further work showed that an impurity in the sample employed, and not 6-aminopteroylglutamic acid, was responsible for the biological effect which resulted from the chemical treatment (*153 a*).

The relationship of thymidine and the "citrovorum factor" in the nutrition of *L. citrovorum* appears to be similar to that of thymidine and pteroylglutamic acid in the nutrition of *L. mesenteroides* (p. 420

and 421). *L. citrovorum*, perhaps, merely lacks the ability to convert pteroylglutamic acid into its metabolically active form. Since thymidine is the only desoxyriboside which supports the growth of this organism in place of the "citrovorum factor" (*138*, *66*, *91*), it has been assumed that this substance is involved specifically with the biosynthesis of thymidine (*4*, *66*), but no precise scheme has as yet been formulated to explain this relationship.

JUKES *et al.* (*66*) have presented the following diagram (Scheme 1) in an attempt to interpret their results.

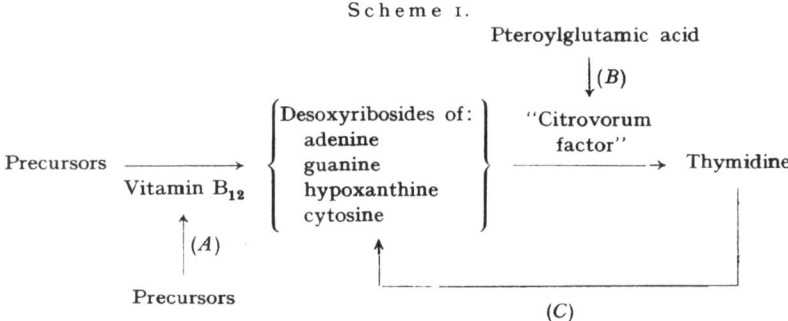

Scheme 1.

The nutritional behavior of two bacterial species was examined. *Lactobacillus leichmannii* responded to vitamin B_{12}, thymidine, or any of four other natural desoxyribosides. Administration of thymidine dispensed with the need for both vitamin B_{12} and pteroylglutamic acid. *L. leichmannii* was assumed to be able to carry out reaction (*B*) but not (*A*).

Leuconostoc citrovorum responded to the "citrovorum factor" or thymidine but not to vitamin B_{12} or any of the other four natural desoxyribosides. This organism was assumed to be able to carry out reaction (*A*) but not (*B*) (Scheme 1).

JUKES *et al.* (*66*) suggest that vitamin B_{12} is involved in the biosynthesis of the desoxyribosides of adenine, guanine, hypoxanthine, and cytosine. *Lactobacillus leichmannii*, being vitamin B_{12}-less, cannot synthesize these desoxyribosides, but *Leuconostoc citrovorum* can. These desoxyribosides act as precursors of thymidine when the "citrovorum factor" is present. Due to the inability of *Leuconostoc citrovorum* to convert pteroylglutamic acid into the "citrovorum factor", these desoxyribosides cannot give rise to thymidine in this organism. *Lactobacillus leichmannii*, on the other hand, can bring about the synthesis of thymidine from the other desoxyribosides if they are administered to the organism, since it is able to convert the supplied pteroylglutamic acid into the "citrovorum factor." Each organism has the ability to carry out reaction (*C*), a conversion in which neither vitamin B_{12} nor the "citrovorum factor" seems to be involved.

Scheme is not very satisfying due to its lack of exactness.

For example, does each of the desoxyribosides which is represented to act as a precursor of thymidine arise independently through a vitamin B_{12}-catalyzed reaction, or does vitamin B_{12} function more specifically in the biosynthesis of a single parent source of the desoxyribosides, from which original source the other desoxyribosides arise by trans-desoxyribosidation? Furthermore, if reaction (C) is carried out by a trans-desoxyribosidase such as that present in extracts from *Lactobacillus helveticus* S one would expect thymine to enter into the exchange reaction just as well as the pyrimidines and purines of the other desoxyribosides which are represented in the scheme. It might be supposed, therefore, that culturing *Leuconostoc citrovorum* in the presence of any one of the microbiologically inactive natural desoxyribosides plus thymine would result in the growth of the organism. This supposition, when tested, proved to be wrong, however, for the incubation of *L. citrovorum* in the presence of each of the four natural desoxyribosides plus thymine in no case resulted in any detectable growth (91).

The possibility that this organism is impermeable to thymine (and permeable to thymidine) has not been excluded. With this view in mind the "citrovorum factor" might be assumed to catalyze reactions which produce thymine inside the cell. Thymidine might then conceivably arise from the other desoxyribosides by trans-desoxyribosidation. The same argument has already been mentioned in connection with *L. mesenteroides*, an organism in which the limiting function of pteroylglutamic acid is apparently involved with the biosynthesis of thymidine in contradistinction to thymine (p. 420 and 421). An investigation is required to establish whether cell-free enzyme-preparations from *L. citrovorum* have the ability to catalyze the formation of thymidine from thymine plus one of the natural desoxyribosides. Such enzyme-preparations obtained from *Lactobacillus delbrückii* ATCC 9649, an organism which also shows a specific requirement for thymidine, had this ability (91).

It will be recalled that under special experimental conditions ribonucleotides, vitamin B_{12}, thymidine, and (to a lesser extent) ascorbic acid had like effects in stimulating the growth of a strain of *Lactobacillus gayonii*. This is the only microorganism yet reported in which a nutritional relationship has been demonstrated between ribonucleotides and thymidine. It is not known how direct this relationship is. Among the bacteria which have an essential growth-requirement for thymidine or its microbiological equivalent (under the traditional conditions employed in carrying out a microbiological assay), none of them which has been tested for its response to ribonucleotides has shown significant growth.

References.

1. ARVIDSON, H., N. A. ELIASSON, E. HAMMARSTEN, P. REICHARD, H. v. UBISCH and S. BERGSTRÖM: Orotic Acid as a Precursor of Pyrimidines in the Rat. J. biol. Chemistry **179**, 169 (1949).

2. BADDILEY, J.: Nucleic Acids, Purines, and Pyrimidines. Annu. Rev. Biochem. **20**, 149 (1951).
3. BALIS, M. E., G. B. BROWN, G. B. ELION, G. H. HITCHINGS and H. VANDER WERFF: On the Interconversion of Purines by *Lactobacillus casei*. J. biol. Chemistry **188**, 217 (1951).
4. BARDOS, T. J., T. J. BOND, J. HUMPHREYS and W. SHIVE: Relationship of the Folinic Acid Group and the *Leuconostoc citrovorum* Factors. J. Amer. chem. Soc. **71**, 3852 (1949).
5. BARKER, T. J., E. J. BOURNE, I. A. WILKINSON and S. PEAT: The Enzymic Synthesis and Degradation of Starch. VII. The Mechanism of Q-Enzyme Action. J. chem. Soc. (London) **1950**, 93.
6. BARRY, J. M., M. G. BANKS and A. L. KOCH: Transfer of Purines and Pyrimidines from Bacterial Host to Bacteriophage Progeny. Federat. Proc. (Amer. Soc. exp. Biol.) **9**, 148 (1950).
7. BEAVEN, G. H., E. R. HOLIDAY, E. A. JOHNSON, B. ELLIS and V. PETROW: The Chemistry of Anti-pernicious Anaemia Factors. VI. The Mode of Combination of Component α in Vitamin B_{12}. J. Pharm. and Pharmacol. **2**, 944 (1950).
8. BEILER, J. M., J. N. MOSS and G. J. MARTIN: Formation of a Competitive Antagonist of Vitamin B_{12} by Oxidation. Science (New York) **114**, 122 (1951).
9. BEN-ISHAI, R., B. VOLCANI and E. D. BERGMAN: 4-Amino-imidazole-5-carboxamide a Precursor of Purines in *Escherichia coli*. Arch. Biochemistry **32**, 229 (1950).
10. BENDICH, A., S. S. FURST and G. B. BROWN: On the Rôle of 2,6-Diaminopurine in the Biosynthesis of Nucleic Acid Guanine. J. biol. Chemistry **185**, 423 (1950).
11. BIRD, O. D. and B. HOEVEL: The Vitamin B_{12}-Binding Power of Proteins. J. biol. Chemistry **190**, 181 (1951).
12. BOND, T. J., T. J. BARDOS, M. E. SIBLEY and W. SHIVE: The Folinic Acid Group, a Series of New Vitamins Related to Folic Acid. J. Amer. chem. Soc. **71**, 3852 (1949).
13. BREWER, C. R., W. G. MCCULLOUGH, R. C. MILLS, W. G. ROESSLER, E. J. HERBST and A. F. HOWE: Studies on the Nutritional Requirements of *Bacillus anthracis*. Arch. Biochemistry **10**, 65 (1946).
14. BRINK, N. G., F. W. HOLLY, C. H. SHUNK, E. W. PEEL, J. J. CAHILL and K. FOLKERS: Vitamin B_{12}. IX. 1-α-D-Ribofuranosido-5,6-dimethylbenzimidazole, a Degradation Product of Vitamin B_{12}. J. Amer. chem. Soc. **72**, 1866 (1950).
15. BRINK, N. G., F. A. KUEHL, Jr. and K. FOLKERS: Vitamin B_{12}: The Identification of Vitamin B_{12} as a Cyano-Cobalt Coordination Complex. Science (New York) **112**, 354 (1950).
16. BROCKMAN, J. A., Jr., J. V. PIERCE, E. L. R. STOKSTAD, H. P. BROQUIST and T. H. JUKES: Some Characteristics of a Crystalline Compound Derived from Vitamin B_{12}. J. Amer. chem. Soc. **72**, 1042 (1950).
17. BROCKMAN, J. A., Jr., B. ROTH, H. P. BROQUIST, M. E. HULTQUIST, J. M. SMITH, Jr., M. J. FAHRENBACH, D. B. COSULICH, R. P. PARKER, E. L. R. STOKSTAD and T. H. JUKES: Synthesis and Isolation of a Crystalline Substance with the Properties of a New B Vitamin. J. Amer. chem. Soc. **72**, 4325 (1950).
18. BROQUIST, H. P., M. J. FAHRENBACH, J. A. BROCKMAN, Jr., E. L. R. STOKSTAD and T. H. JUKES: "Citrovorum Factor" Activity of Tetrahydropteroylglutamic Acid. J. Amer. chem. Soc. **73**, 3535 (1951).

19. BROQUIST, H. P., E. L. R. STOKSTAD and T. H. JUKES: Further Observations on the Microbiological Assay of Vitamin B_{12} Using *Lactobacillus leichmannii*. Proc. Soc. exp. Biol. Med. 76, 806 (1951).
20. BROWN, G. B.: Studies of Purine Metabolism. Cold Springs Harbor Symposia on Quant. Biol. 13, 43 (1948).
21. BROWN, G. B., A. BENDICH, P. M. ROLL and K. SUGIURA: Utilization of Guanine by C 57 Black Mouse Bearing Adenocarcinoma E_0 771. Proc. Soc. exp. Biol. Med. 72, 501 (1949).
22. BUCHANAN, J. G., A. W. JOHNSON, J. A. MILLS and A. R. TODD: The Isolation of a Phosphorus-containing Degradation Product from Vitamins B_{12} and B_{12c}. Chem. and Ind. 1950, 426.
23. BUHS, R. P., E. G. NEWSTEAD and N. R. TRENNER: An Analog of Vitamin B_{12}. Science (New York) 113, 625 (1951).
24. CARTER, C. E. and W. E. COHN: Separation of Three Naturally Occurring Adenine Ribonucleotides by Paper Chromatography and Ion-exchange. Federat. Proc. (Amer. Soc. exp. Biol.) 8, 190 (1949).
25. COHN, W. E.: Heterogeneity in Pyrimidine Nucleotides. J. Amer. chem. Soc. 72, 2811 (1950).
25a. COLOWICK, S. P. and N. O. KAPLAN: Carbohydrate Metabolism. Annu. Rev. Biochem. 20, 513 (1951).
26. CONN, J. B., S. L. NORMAN and T. G. WARTMAN: The Equilibrium Between Vitamin B_{12} (Cyanocobalamin) and Cyanide Ion. Science (New York) 113, 658 (1951).
27. COOK, E. S., S. A. G. CRONIN, C. W. KREKE and S. T. M. WALSH: Relation of Nucleotides and Nucleosides to Proliferation-promoting Factors Produced by Ultra-violet Irradiated Yeast Cells. Nature (London) 152, 474 (1943).
28. COOLEY, G., B. ELLIS, V PETROW, G. H. BEAVEN, E. R. HOLIDAY and E. A. JOHNSON: The Chemistry of Anti-pernicious Anaemia Factors. VII. Some Transformations of Vitamin $B_{12}b$. J. Pharm. and Pharmacol. 3, 271 (1951).
29. COOPERMAN, J. M., R. DRUCKER and B. TABENKIN: Microbiological Assays for Vitamin B_{12}: A Cyanide Enhancement Effect. J. biol. Chemistry 191, 135 (1951).
30. CUTHBERTSON, W. F. J. and J. T. LLOYD: The Assay of Vitamin B_{12}. I. Factors Affecting the Response of *Lactobacillus lactis* DORNER ATCC 8000 to Vitamin B_{12}. J. gen. Microbiol. 5, 416 (1951).
31. DAVIS, B. D. and E. S. MINGIOLI: Mutants of *Escherichia coli* Requiring Methionine or Vitamin B_{12}. J. Bacteriol. 60, 17 (1950).
32. DI CARLO, F. J., A. S. SCHULTZ and D. K. MCMANUS: The Assimilation of Nucleic Acid Derivatives and Related Compounds by Yeast. J. biol. Chemistry 189, 151 (1951).
33. DITTMER, K., I. GOODMAN, D. VISSER and H. P. MCNULTY: Some Microbiological Properties of Synthetic Nucleosides. Proc. Soc. exp. Biol. Med. 69, 40 (1948).
34. DOUDOROFF, M., H. A. BARKER and W. Z. HASSID: Studies With Bacterial Sucrose Phosphorylase. I. The Mechanism of Action of Sucrose Phosphorylase as a Glucose-transferring Enzyme (Transglucosidase). J. biol. Chemistry 168, 725 (1947).
35. DRELL, W.: Effect of 2,6-Diaminopurine on Purine Metabolism in *Neurospora*. Federat. Proc. (Amer. Soc. exp. Biol.) 10, 177 (1951).
36. ELION, G. B. and G. H. HITCHINGS: Antagonists of Nucleic Acid Derivatives. III. The Specificity of the Purine Requirement of *Lactobacillus casei*. J. biol. Chemistry 185, 651 (1950).

37. EMERSON, G., N. G. BRINK, F. W. HOLLY, F. KONIUSZY, D. HEYL and K. FOLKERS: Vitamin B_{12}. VIII. Vitamin B_{12}-Like Activity of 5,6-Dimethylbenzimidazole and Tests on Related Compounds. J. Amer. chem. Soc. **72**, 3084 (1950).
38. EMERSON, G. and K. FOLKERS: Water-soluble Vitamins. Annu. Rev. Biochem. **20**, 559 (1951).
39. EMERSON, G., F. W. HOLLY, C. H. SHUNK, N. G. BRINK and K. FOLKERS: Vitamin B_{12}. XII. Vitamin B_{12}-Like Activity of α- and β-Ribazole. J. Amer. chem. Soc. **73**, 1069 (1951). ·
40. FAIRLEY, J. L., Jr. and H. S. LORING: Growth-promoting Activities of Guanine, Guanosine, Guanylic Acid, and Xanthine for a Purine-deficient Strain of *Neurospora*. J. biol. Chemistry **177**, 451 (1949).
41. FLAVIN, M. and S. GRAFF: Utilization of Guanine for Nucleic Acid Biosynthesis by *Tetrahymena geleii*. J. biol. Chemistry **191**, 55 (1951).
42. — — The Utilization of Adenine for Nucleic Acid Biosynthesis by *Tetrahymena geleii*. J. biol. Chemistry **192**, 485 (1951).
43. FLYNN, E. H., T. J. BOND, T. J. BARDOS and W. SHIVE: A Synthetic Compound With Folinic Acid Activity. J. Amer. chem. Soc. **73**, 1979 (1951).
44. FOX, J. J. and I. GOODMAN: The Synthesis of Nucleosides ot Cytosine and 5-Methylcytosine. J. Amer. chem. Soc. **73**, 3256 (1951).
45. FRESCO, J. R. and A. MARSHAK: Isolation of C^{14}-Labelled Purines and Pyrimidines from Biological Materials. Federat. Proc. (Amer. Soc. exp. Biol.) **10**, 45 (1951).
45a. FRIEDKIN, M.: Azaguanine, Ribose-1-phosphate, and Desoxyribose-1-phosphate. Symposium on Phosphorus Metabolism I., McCollum-Pratt Institute, Johns Hopkins Univ., June 1951.
46. FRIEDKIN, M. and H. M. KALCKAR: Desoxyribose-1-phosphate. I. The Phosphorolysis and Resynthesis of Purine Desoxyribose Nucleosides. J. biol. Chemistry **184**, 437 (1950).
47. FRIES, N.: Mutant Strains of *Ophiostoma multiannulatum* Requiring Components of Different Nucleotides. Ark. Botanik **33** A, 1 (1947).
48. — The Effects of Different Purine-compounds on the Growth of Guanine-deficient *Ophiostoma*. Physiol. Plantarum **2**, 78 (1949).
49. — *Ophiostoma multiannulatum* (HEDGC. & DAVIDS) as a Test-object for the Determination of Pyridoxine and Various Nucleotide Constituents. Ark. Botanik **1**, 271 (1950).
50. FUKUHARA, T. K. and D. W. VISSER: Pyrimidine Nucleoside Antagonists. J. biol. Chemistry **190**, 95 (1951).
51. GINGRICH, W. and F. SCHLENK: Codehydrogenose I and Other Pyrimidinium Compounds as V-Factor for *Hemophilus influenzae* and *H. parainfluenzae*. J. Bacteriol. **47**, 535 (1944).
52. GLADSTONE, G. P.: Inter-relationships Between Aminoacids in the Nutrition of *Bacillus anthracis*. Brit. J. exp. Pathol. **20**, 189 (1939).
53. GREENBERG, G. R.: *De novo* Synthesis of Hypoxanthine *via* Inosine-5-phosphate and Inosine. J. biol. Chemistry **190**, 611 (1951).
54. GRIFFIN, P. J.: Cocarboxylase and Adenosine Triphosphate as Growth Factors for *Hemophilus piscium*. Arch. Biochemistry **30**, 100 (1951).
55. HAMMARSTEN, E. and P. REICHARD: Inability of Guanosine to Act as a Precursor of Polynucleotides. Acta chem. Scand. **4**, 711 (1950).
56. HAMMARSTEN, E., P. REICHARD and E. SALUSTE: Pyrimidine Nucleosides as Precursors of Pyrimidines in Polynucleotides. J. biol. Chemistry **183**, 105 (1950).

57. HENDLIN, D. and M. H. SOARS: Comparative Microbiological Studies With Vitamins B_{12} and B_{12a}. J. biol. Chemistry 188, 603 (1951).
58. HERBST, E. J. and E. E. SNELL: The Nutritional Requirements of *Hemophilus parainfluenzae* 7901. J. Bacteriol. 58, 379 (1949).
59. HILLS, G. M.: Chemical Factors in the Germination of Spore-bearing Aerobes. The Effect of Yeast Extract on the Germination of *Bacillus anthracis* and its Replacement by Adenosine. Biochemic. J. 45, 353 (1949).
60. HITCHINGS, G. H., G. B. ELION, E. A. FALCO, P. B. RUSSELL, M. B. SHERWOOD and H. VANDER WERFF: Antagonists of Nucleic Acid Derivatives. I. The *Lactobacillus casei* Model. J. biol. Chemistry 183, 1 (1950).
61. HOFF-JÖRGENSEN, E.: Difference in the Growth-promoting Effect of Desoxyribosides and Vitamin B_{12} on Three Strains of Lactic Acid Bacteria. J. biol. Chemistry 178, 525 (1949).
62. — A Microbiological Method for the Determination of Desoxyribosides. Abstr. 1st Internat. Congr. Biochem., p. 292, Cambridge (1949).
63. — A Microbiological Assay of Deoxyribonucleosides and Deoxyribonucleic Acid. Biochemic. J. 30, 400 (1952).
64. HUTNER, S. H., L. PROVASOLI, E. L. R. STOKSTAD, C. E. HOFFMANN, M. BELT, A. L. FRANKLIN and T. H. JUKES: The Assay of the Anti-Pernicious Anemia Factor with *Euglena*. Proc. Soc. exp. Biol. Med. 70, 118 (1949).
65. HUTCHINGS, B. L., N. H. SLOANE and E. BOGGIANO: The Nutrition of *Lactobacillus gayonii*. J. biol. Chemistry 162, 737 (1946).
66. JUKES, T. H., H. P. BROQUIST and E. L. R. STOKSTAD: Vitamin B_{12} and "Citrovorum Factor" in the Nutrition of *Lactobacillus leichmannii* and *Leuconostoc citrovorum*. Arch. Biochemistry 26, 157 (1950).
67. KACZKA, E. A., R. G. DENKEWALTER, A. HOLLAND and K. FOLKERS: Vitamin B_{12}. XIII. Additional Data on Vitamin B_{12}. J. Amer. chem. Soc. 73, 335 (1951).
68. KACZKA, E. A., D. E. WOLF and K. FOLKERS: Vitamin B_{12}. V. Identification of Crystalline Vitamin B_{12a}. J. Amer. chem. Soc. 71, 1514 (1949).
69. KACZKA, E. A., D. E. WOLF, F. A. KUEHL, Jr. and K. FOLKERS: Vitamin B_{12}: Reactions of Cyano-Cobalamin and Related Compounds. Science (New York) 112, 354 (1950).
70. KALCKAR, H. M.: Enzymatic Reactions in the Synthesis of Purine Compounds. In: The Harvey Lectures 1950/51. Springfield, Ill.: Charles C. Thomas (in press).
70a. — The Enzymes of Nucleoside Metabolism. Fortschr. Chem. organ. Naturstoffe 9, 363 (1952).
71. KALCKAR, H. M. and W. S. MCNUTT (unpublished).
72. KALCKAR, H. M., W. S. MCNUTT and E. HOFF-JÖRGENSEN: *Trans*-N-glycosidase as Studied with Carbon-14 Adenine. Biochemic. J. 30, 397 (1952).
72a. KENNER, G. W.: The Chemistry of Nucleotides. Fortschr. Chem. organ. Naturstoffe 8, 96 (1951).
73. KERR, S. E., K. SERAIDARIAN and G. B. BROWN: On the Utilization of Purines and Their Riboside Derivatives by Yeast. J. biol. Chemistry 188, 207 (1951).
74. KIDDER, G. W. and V. C. DEWEY: Studies in the Biochemistry of *Tetrahymena*. XIV. The Activity of Natural Purines and Pyrimidines. Proc. nat. Acad. Sci. (USA) 34, 566 (1948).
75. — — The Biological Activity of Substituted Purines. J. biol. Chemistry 179, 181 (1949).
76. KIDDER, G. W., V. C. DEWEY, R. E. PARKS, Jr. and M. R. HEINRICH: Further Studies on the Purine and Pyrimidine Metabolism of *Tetrahymena*. Proc. nat. Acad. Sci. (USA) 36, 431 (1950).

76a. KING, T. E., L. M. LOCHER and V. H. CHELDELIN: Pantothenic Acid Studies. III. A Pantothenic Acid Conjugate Active for *Acetobacter suboxidans*. Arch. Biochemistry **17**, 483 (1948).
76b. KING, T. E. and F. M. STRONG: Some Properties of Coenzyme A. J. biol. Chemistry **189**, 325 (1951).
77. KITAY, E., W. S. MCNUTT and E. E. SNELL: The Non-specificity of Thymidine as a Growth Factor for Lactic Acid Bacteria. J. biol. Chemistry **177**, 993 (1949).
78. — — — Desoxyribosides and Vitamin B_{12} as Growth Factors for Lactic Acid Bacteria. J. Bacteriol. **59**, 727 (1950).
79. KOCH, A. L.: Purine Metabolism During Bacteriophage Growth. Federat. Proc. (Amer. Soc. exp. Biol.) **10**, 210 (1951).
80. KOCHER, V. u. O. SCHINDLER: Desoxyribonucleoside als Wachstumsfaktoren für *Lactobacillus lactis* in B_{12}-freier Nährlösung. Internat. Z. Vitaminforsch. **20**, 441 (1949).
81. KODITSCHEK, L. K., D. HENDLIN and H. B. WOODRUFF: Investigations on the Nutrition of *Lactobacillus lactis* DORNER. J. biol. Chemistry **179**, 1093 (1949).
82. KRITSKIĬ, G. A.: On the Mechanism of the Phosphorolysis Reaction. C. R. (Doklady) Acad. Sci. URSS **70**, 667 (1950) [Chem. Abstr. **44**, 7367 (1950)].
83. LANKFORD, C. E. and P. K. SKAGGS: Cocarboxylase as a Growth Factor for Certain Strains of *Neisseria gonorrhoea*. Arch. Biochemistry **9**, 265 (1946).
84. LOCHHEAD, A. G. and R. H. THEXTON: Vitamin B_{12} as a Growth Factor for Soil Bacteria. Nature (London) **167**, 1034 (1951).
85. LOOFBOUROW, J. R.: Rôle of Adenine Nucleotide and Growth Factors in Increased Proliferation Following Damage to Cells. Nature (London) **150**, 349 (1942).
86. — Intercellular Hormones. V. Evidence that the Proliferation Promoting Effect of Damaged-Cell Products is Attributable to Adenine Nucleotides and Known Growth Factors. Biochemic. J. **36**, 737 (1942).
87. LORING, H. S., G. L. ORDWAY and J. G. PIERCE: A Method of Assay for Cytidine and Uridine by Means of a Pyrimidine-deficient Strain of *Neurospora*. J. biol. Chemistry **176**, 1123 (1948).
88. LORING, H. S. and J. G. PIERCE: Pyrimidine Nucleosides and Nucleotides as Growth Factors for Mutant Strains of *Neurospora*. J. biol. Chemistry **153**, 61 (1944).
89. LWOFF, A. and M. LWOFF: Studies on Codehydrogenase I. Nature of Growth Factor "V". Proc. Roy. Soc. (London), Ser. B **122**, 3512 (1937).
90. MCNUTT, W. S.: The Exchange Between Free Purines and Pyrimidines and the Aglycones of Desoxyribosyl Purines and Desoxyribosyl Pyrimidines. Nature (London) **166**, 444 (1950).
91. — The Enzymically Catalyzed Transfer of the Desoxyribosyl Group from One Purine or Pyrimidine to Another. Biochemic. J. **30**, 384 (1952).
92. MCNUTT, W. S. and E. E. SNELL: Pyridoxal Phosphate and Pyridoxamine Phosphate as Growth Factors for Lactic Acid Bacteria. J. biol. Chemistry **182**, 557 (1950).
93. MANSON, L. A. and J. O. LAMPEN: Metabolism of Desoxyribosides in *Escherichia coli*. Federat. Proc. (Amer. Soc. exp. Biol.) **9**, 397 (1950).
94. MARSHAK, A.: The Purine and Pyrimidine Content of Three Strains of *Escherichia coli*. Science (New York) **113**, 181 (1951).
95. MICHELSON, A. M., W. DRELL and H. K. MITCHELL: A New Ribose Nucleoside from *Neurospora*: "Orotidine." Proc. nat. Acad. Sci. (USA) **37**, 396 (1951).

96. MITCHELL, H. K. and M. B. HOULAHAN: Adenine-Requiring Mutants of *Neurospora crassa*. Federat. Proc. (Amer. Soc. exp. Biol.) 5, 370 (1945).
97. — — Investigations on the Biosynthesis of Pyrimidine Nucleosides in *Neurospora*. Federat. Proc. (Amer. Soc. exp. Biol.) 6, 506 (1947).
98. MITCHELL, H. K., M. B. HOULAHAN and J. F. NYC: The Accumulation of Orotic Acid by a Pyrimidine-less Mutant of *Neurospora*. J. biol. Chemistry 172, 525 (1948).
98a. MITCHELL, J. H., Jr., H. E. SKIPPER and L. L. BENNETT: Investigation of Nucleic Acids of Viscera and Tumor Tissue from Animals Injected with Radioactive 8-Azaguanine. Cancer Res. 10, 647 (1950).
99. NAKAHARA, W., F. INUKAI and S. UGAMI: Studies on the Dietary Requirements for Lactation. XVIII. Chemical Nature of Vitamin L, as Evidenced by the Activity of Adenylthiomethylpentose. Sci. Pap. Inst. physic. chem. Res. (Japan) 40, 433 (1942/43).
99a. NOVELLI, G. D., R. M. FLYNN and F. LIPMANN: Coenzyme A as a Growth Factor for *Acetobacter suboxidans*. J. biol. Chemistry 177, 493 (1949).
99b. NOVELLI, G. D., J. D. GREGORY, R. M. FLYNN and F. J. SCHMETZ, Jr.: Structure of Coenzyme A. Federat. Proc. (Amer. Soc. exp. Biol.) 10, 229 (1951).
100. NYGAARD, A. P. and V. H. CHELDELIN: Nutritional Studies of Two Variants of *Lactobacillus gayonii*. J. Bacteriol. 61, 497 (1951).
101. OGINSKY, E. L., P. H. SMITH, N. E. TONKAZY and W. W. UMBREIT: The Influence of Vitamin B_{12} on Oxidation by a Mutant Strain of *Escherichia coli*. J. Bacteriol. 61, 581 (1951).
102. PAEGE, L. M. and F. SCHLENK: Pyrimidine Riboside Metabolism. Arch. Biochemistry 28, 348 (1950).
103. PAPPERHEIMER, A. M. and G. A. HOTTLE: The Effect of Certain Purines and CO_2 on the Growth of a Strain of Group A Hemolytic Streptococcus. Proc. Soc. exp. Biol. Med. 44, 645 (1940).
104. PAZUR, J. H. and D. FRENCH: The *trans*-Glucosidase of *Aspergillus oryzae*. J. Amer. chem. Soc. 73, 3536 (1951).
105. PEARSON, W. N.: A Purine-requiring Strain of *Photobacterium fischeri*. J. Bacteriol. 58, 653 (1949).
106. PENNINGTON, R. J.: A Heat-labile Vitamin B_{12} Complex in Faeces. Biochemic. J. 48, XVIII (1951).
107. PIERCE, J. G. and H. S. LORING: Growth Requirements of a Purine-deficient Strain of *Neurospora*. J. biol. Chemistry 160, 409 (1945).
108. — — Purine and Pyrimidine Antagonism in a Pyrimidine-deficient Mutant of *Neurospora*. J. biol. Chemistry 176, 1131 (1948).
109. PIERCE, J. V., A. C. PAGE, Jr., E. L. R. STOKSTAD and T. H. JUKES: Crystallization of Vitamin B_{12b}. J. Amer. chem. Soc. 71, 2952 (1949).
110. — — — — Studies of Some Characteristics of Vitamin B_{12b}. J. Amer. chem. Soc. 72, 2615 (1950).
111. POHLAND, A., E. H. FLYNN, R. G. JONES and W. SHIVE: A Proposed Structure for Folinic Acid-SF, a Growth Factor Derived from Pteroylglutamic Acid. J. Amer. chem. Soc. 73, 3247 (1951).
112. PRUSOFF, W. H., L. J. TEPLEY and C. G. KING: The Influence of Pteroylglutamic Acid on Nucleic Acid Synthesis in *Lactobacillus casei*. J. biol. Chemistry 176, 1309 (1948).
113. RABATÉ, M. J.: Contribution à l'étude biochimique des Salicacées. XI. Sur l'hydrolyse du salicoside par la poudre fermentaire de feuilles de *Salix purpurea* L. et sur quelques phénomènes qui en dérivent. Bull. Soc. Chim. biol. (Paris) 17, 572 (1935).

114. REGE, D. V. and A. SREENIVASAS: Folic Acid, Vitamin B_{12}, and Nucleic Acid Synthesis in *Lactobacillus casei*. Nature (London) 166, 1117 (1950).
115. REICHARD, P.: On the Turnover of Purines and Pyrimidines from Polynucleotides in the Rat Determined with N^{15}. Acta chem. Scand. 3, 422 (1949).
116. REICHARD, P. and S. BERGSTRÖM: Synthesis of Polynucleotides in Slices from Regenerating Liver. Acta chem. Scand. 5, 190 (1951).
117. REICHARD, P. and B. ESTBORN: Utilization of Desoxyribosides in the Synthesis of Polynucleotides. J. biol. Chemistry 188, 839 (1951).
118. ROBBINS, W. J., A. HERVEY and M. E. STEBBINS: Studies on *Euglena* and Vitamin B_{12}. Science (New York) 112, 455 (1950).
119. ROBBINS, W. J. and F. KAVANAGH: Hypoxanthine, a Growth Substance for *Phycomyces*. Proc. nat. Acad. Sci. (USA) 28, 65 (1942).
120. ROBERTS, M., T. K. FUKUHARA and D. W. VISSER: Biological Activity of Some Uridine Derivatives. Federat. Proc. (Amer. Soc. exp. Biol.) 9, 219 (1950).
121. ROBERTS, R. B. and M. SANDS: The Influence of Vitamin B_{12} on the Growth of Bacteriophage T_4 r. J. Bacteriol. 58, 710 (1949).
122. Ross, G. I.: Vitamin B_{12} Assay in Body Fluids. Nature (London) 166, 270 (1950).
123. SAUBERLICH, H. E.: The Purine and Pyrimidine Requirements of *Leuconostoc citrovorum* 8081. Arch. Biochemistry 24, 263 (1949).
124. — The Relationship of Folic Acid, Vitamin B_{12} and Thymidine in the Nutrition of *Leuconostoc citrovorum* 8081. Arch. Biochemistry 24, 224 (1949).
125. SAUBERLICH, H. E. and C. A. BAUMANN: Further Studies on the Factor Required by *Leuconostoc citrovorum* 8081. J. biol. Chemistry 181, 871 (1949).
126. SCHOPFER, W. H.: Sulfamidés et acides nucléiques. Experientia 2, 188 (1946).
127. SCHOPFER, W. H. and M. GUILLOUD: Antisulfonamide Action of Purines on a Microörganism *(Eremothecium ashbyii)*. Helv. Physiol. Pharmacol. Acta 4, 624 (1946) [Chem. Abstr. 40, 5794 (1946)].
128. SHIVE, W.: The Utilization of Antimetabolites in the Study of Biochemical Processes in Living Organisms. Ann. New York Acad. Sci. 52, 1212 (1950).
129. SHIVE, W., T. J. BARDOS, T. J. BOND and L. L. ROGERS: Synthetic Members of the Folinic Acid Group. J. Amer. chem. Soc. 72, 2817 (1950).
130. SHIVE, W., R. E. EAKIN, W. M. HARDING, J. M. RAVEL and J. E. SUTHERLAND: A Crystalline Factor Functionally Related to Folic Acid. J. Amer. chem. Soc. 70, 2299 (1948).
131. SHIVE, W., J. M. RAVEL and R. E. EAKIN: An Interrelationship of Thymidine and Vitamin B_{12}. J. Amer. chem. Soc. 70, 2614 (1948).
132. SHIVE, W., J. M. RAVEL and W. M. HARDING: An Interrelationship of Purines and Vitamin B_{12}. J. biol. Chemistry 176, 991 (1948).
133. SHIVE, W., M. E. SIBLEY and L. L. ROGERS: Replacement of Vitamin B_{12} by Desoxynucleotides in Promoting Growth of Certain Lactobacilli. J. Amer. chem. Soc. 73, 867 (1951).
134. SHULL, G. M. and W. H. PETERSON: The Nature of the "Sporogenes Vitamin" and Other Factors in the Nutrition of *Clostridium sporogenes*. Arch. Biochemistry 18, 69 (1948).
135. SKEGGS, H. R.: Microbiological Systems Involving Nucleic Acid Derivatives. J. cellular comparat. Physiol. 38, Suppl. 1, 227 (1951).
136. SKEGGS, H. R., L. D. WRIGHT, K. A. VALENTIK, H. NEPPLE and J. SPIZIZEN: Purine Ribose Nucleotide Inhibition of Desoxyribonucleic Acid Utilization by *Lactobacillus bifidus*. Federat. Proc. (Amer. Soc. exp. Biol.) 9, 228 (1950).
137. SNELL, E. E.: Nutrition of Microorganisms. Ann. Rev. Microbiol. 3, 97 (1949).

137a. SNELL, E. E., G. M. BROWN, V. J. PETERS, J. A. CRAIG, E. L. WITTLE, J. A. MOORE, V. M. MCGLOHON and O. D. BIRD: Chemical Nature and Synthesis of the *Lactobacillus bulgaricus* Factor. J. Amer. chem. Soc. **72**, 5349 (1950).
138. SNELL, E. E., E. KITAY and W. S. MCNUTT: Thymine Desoxyriboside as an Essential Growth Factor for Lactic Acid Bacteria. J. biol. Chemistry **175**, 473 (1948).
139. SNELL, E. E. and L. D. WRIGHT: The Water-soluble Vitamins. Annu. Rev. Biochem. **19**, 277 (1950).
140. SOARS, M. H. and D. HENDLIN: The Use of Potassium Cyanide in the *Lactobacillus leichmannii* Assay for Vitamin B_{12}. J. Bacteriol. **62**, 15 (1951).
141. STOKES, J. L.: Substitution of Thymine for "Folic Acid" in the Nutrition of Lactic Acid Bacteria. J. Bacteriol. **48**, 201 (1944).
142. STOKES, J. L., J. W. FOSTER and C. R. WOODWARD, Jr.: Synthesis of Pyridoxin by a "Pyridoxinless" X-Ray Mutant of *Neurospora sitophila*. Arch. Biochemistry **2**, 235 (1943).
143. STOKSTAD, E. L. R., T. H. JUKES, J. A. BROCKMAN, Jr., J. V. PIERCE and H. P. BROQUIST: Relation of Vitamin B_{12b} to Vitamin B_{12} and the Biological Activities of these Compounds. Federat. Proc. (Amer. Soc. exp. Biol.) **9**, 235 (1950).
144. TERNBERG, J. L. and R. E. EAKIN: Erythein and Apoerythein and Their Relation to the Antipernicious Anemia Principle. J. Amer. chem. Soc. **71**, 3858 (1949).
145. TOMARELLI, R. M., R. F. NORRIS and P. GYÖRGY: Inability of Vitamin B_{12} to Replace the Desoxyriboside Requirement of a *Lactobacillus bifidus*. J. biol. Chemistry **179**, 485 (1949).
146. VEER, W. L. C., J. H. EDELHAUSEN, H. G. WIJMENGA and J. LENS: Vitamin B_{12}. I. The Relation Between Vitamin B_{12} and B_{12b}. Biochem. Biophys. Acta **6**, 225 (1950).
147. VOLKIN, E., J. X. KHYM and W. E. COHN: The Preparation of Desoxynucleotides. J. Amer. chem. Soc. **73**, 1533 (1951).
148. WAJZER, J. et F. BARON: Réactions enzymatiques entre les esters phosphoriques du ribose et les bases puriques. Bull. Soc. Chim. biol. (Paris) **31**, 750 (1949).
149. WANG, T. P. and J. O. LAMPEN: The Cleavage of Adenosine, Cytidine and Xanthosine by *Lactobacillus pentosus*. J. biol. Chemistry **192**, 339 (1951).
150. WEED, L. L. and S. S. COHEN: The Utilization of Host Pyrimidines in the Synthesis of Bacterial Viruses. J. biol. Chemistry **192**, 693 (1951).
151. WEED, L. L. and D. W. WILSON: The Incorporation of C^{14}-Orotic Acid Into Nucleic Acid Pyrimidines *in vitro*. J. biol. Chemistry **189**, 435 (1951).
152. WEIL-MALHERBE, H.: The Enzymic Phosphorylation of Vitamin B_1. Biochemic. J. **33**, 1997 (1939).
153. WELCH, A. D. and M. F. WILSON: Mechanism of the Growth Promoting Effect of Ascorbic Acid on *Lactobacillus leichmannii* and the Reduction of Oxidation Products of Vitamin B_{12}. Arch. Biochemistry **22**, 486 (1949).
153a. WEYGAND, F., A. WACKER, H.-J. MANN u. E. ROWOLD: Mikrobiologisches Verhalten von hydrierten Pterinen. Z. Naturforsch. **6 b**, 174 (1951).
154. WEYGAND, F., A. WACKER, H.-J. MANN, E. ROWOLD u. H. LETTRÉ: Neue Wuchsstoffe für *Leuconostoc citrovorum* 8081 und ihre Wirkung beim Mäuse-Ascitestumor. Z. Naturforsch. **5 b**, 413 (1950).
155. WIJMENGA, H. G., W. L. C. VEER and J. LENS: Vitamin B_{12}. II. The Influence of HCN on Some Factors of the Vitamin B_{12} Group. Biochem. Biophys. Acta **6**, 229 (1950).

156. WILLIAMS, R. J., R. E. EAKIN, E. BEERSTECHER, Jr. and W. SHIVE: The Biochemistry of B Vitamins. New York: Reinhold Publ. Corp. 1950.
157. cf. Ref. No. *156*, p. 474.
158. WILSON, A. T.: Nucleic Acid Derivatives as Growth Factors for Certain Group A Hemolytic Streptococci. Proc. Soc. exp. Biol. Med. 58, 249 (1945).
159. WOODRUFF, H. B. and J. C. FOSTER: Analysis for Vitamin B_{12} and Vitamin B_{12a} by Paper Strip Chromatography. J. biol. Chemistry 183, 569 (1950).
160. WRIGHT, L. D., J. W. HUFF, H. R. SKEGGS, K. A. VALENTIK and D. K. BOSSHARDT: Orotic Acid, a Growth Factor for *Lactobacillus bulgaricus*. J. Amer. chem. Soc. 72, 2312 (1950).
161. WRIGHT, L. D., C. S. MILLER, H. R. SKEGGS, J. W. HUFF, L. L. WEED and D. W. WILSON: Biological Precursors of the Pyrimidines. J. Amer. chem. Soc. 73, 1898 (1951).
162. WRIGHT, L. D., H. R. SKEGGS and J. W. HUFF: The Ability of Thymidine to Replace Vitamin B_{12} as a Growth Factor for Certain Lactobacilli. J. biol. Chemistry 175, 475 (1948).
163. ZUCKER, T. and L. M. ZUCKER: "Animal Protein Factor' and Vitamin B_{12} in the Nutrition of Animals. Vitamins and Hormones 8, 1 (1950).

(Received, November 19, 1951.)

Some Current Concepts of the Chemical Nature of Antigens and Antibodies.

By Dan H. Campbell and N. Bulman, Pasadena, California.

With 3 Figures.

Contents.	Page
I. Introduction	443
II. Antigens and Haptens	445
1. Antigens	446
2. Haptens	449
III. Antibodies	451
1. Chemical Composition of Antibodies	451
2. Electrophoretic Properties of Antibodies	452
3. Shape and Size of Antibodies	453
4. Nature of Combining Sites	455
5. Purification of Antibodies	461
IV. The Physical Nature of Antigen-Antibody Reactions	463
1. The Properties of Specific Precipitates	463
a) Composition	463
b) Formation and Specificity	465
c) 'Ageing'	466
2. Thermodynamic Properties of Antigen-Antibody Reactions	466
a) The Free Energy and Heat Changes in Antigen-Antibody Reactions	467
b) Differences in Free Energies of Combination	468
3. Nature of the Forces Involved	471
4. Mathematical Interpretations of the Precipitin Reaction	475
5. A Note on the Use of Polyvalent Haptens	476
V. Conclusions	477
References	478

I. Introduction.

The basic principles of immunological phenomena have developed to a point where antigen-antibody reactions are an integral part of fundamental biochemistry and an important tool for the physical chemist.

This unique position of immunochemistry is due to the work of a large group of investigators during the past fifty years but particularly to LANDSTEINER for studies of specificity, HEIDELBERGER and colleagues for the development and emphasis of quantitative aspects of immunochemical reactions, and PAULING and his collaborators for the development and investigation of stimulating theoretical concepts. It is somewhat paradoxical that at present these studies have contributed as much if not more to our understanding of the physical nature of proteins than to an understanding of immunity mechanisms. Many of our present concepts may be erroneous or oversimplified but nevertheless furnish a useful basis for investigation of biological phenomena and reactions of proteins.

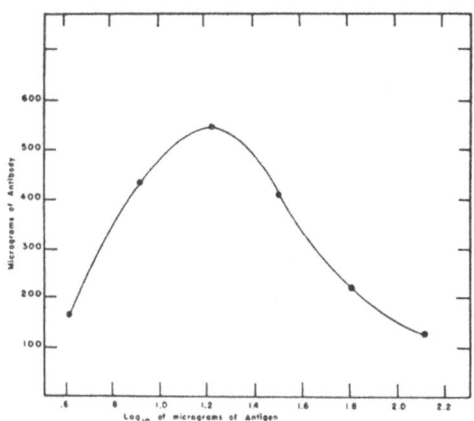

Fig. 1. A typical precipitation curve obtained by the reaction between a trivalent azophenylarsonic acid dye and serum from rabbits immunized with bovine globulin-azophenylarsonic acid.

These concepts have been developed to a large extent from studies of the in vitro reactions between antigens and antiserums and the physical properties of antigens and antibodies. Our survey will be limited to a consideration of the present concepts of the physical properties of antigens and antibodies and of their interactions. These concepts are based for the most part on specific precipitation; and since most of this review will be based on the specific precipitation reaction, a brief description of the fundamental experimental aspects will be useful for subsequent discussion. Figure 1 gives a typical precipitation curve obtained by adding 1.0 ml. of various antigen concentrations to 1.0 ml. of a rabbit antiserum. The antigen used in this particular test had the structure shown in the formula:

The serum was from rabbits which had been injected over a period of several months with bovine globulin-p-azophenylarsonic acid. The precipitates were analyzed for nitrogen by NESSLER's reaction as described by LANNI and CAMPBELL (76) and modified by LANNI (75). The general shape of the curve is basically the same for all precipitation reactions but may assume slightly different quantitative aspects. The maximum point usually contains all the antibody present and is used as a means of determining the concentration of precipitating antibody. The region of the curve in higher antigen concentration is referred to as antigen excess and the region in less antigen, as antibody excess. Precipitation may be reduced by a variety of conditions other than those indicated by the curve. For example, the amount of precipitate may be diminished by elevated temperatures, pH > 9 or < 4, high salt concentration (e. g. 10–15% NaCl) or non-precipitating or so-called inhibiting haptens. An example of the latter for this system would be p-aminophenylarsonic acid which combines with antibody to form a soluble compound and, consequently, prevents the antibody molecule from reacting with a polyvalent precipitating antigen or hapten.

There are many controversial problems regarding the chemistry of antigens and antibodies which cannot be discussed in this limited space. Such problems involve the question of antigenicity of polysaccharides, lipids and simple chemical substances, the possible antigenicity of autogenous substances (i. e. immunization against one's own proteins), the site and mechanism of antibody formation, applied immunochemistry and immunology, possible differences between serum and tissue antibodies, and, finally the biological significance of antibody-antigen reactions. Broader and more detailed features of these problems have been given within the past few years by LANDSTEINER (70), TREFFERS (133), STACEY (126), CAMPBELL (14, 15), KABAT (58), CAMPBELL and LANNI (20), GRABAR (35), DOERR's extensive work (27), KUZIN (66), BURNET and FENNER (11) as well as many others which will be found in a recent survey of the literature by GRABAR (36).

II. Antigens and Haptens.

For convenience, the specific reactants in immunochemical reactions will be designated as antigens, haptens and antibodies. *Antigens* will be considered as substances which have the ability of inducing the formation of *antibodies* in common laboratory animals and man; and *haptens* will be considered as those substances which react specifically with antibodies but lack the ability to induce the formation of antibodies unless combined with proteins. Such an uncompromising classification of substances as either antigens or haptens is based entirely on the fact that antigenic function and serological reactivities are totally different properties. With the development of more refined techniques many so-called haptens have been shown to have antigenic property but the basic distinction between compounds of different reaction capacities is still real and a useful designation.

1. Antigens.

The antigenic characterization of any substance depends upon the demonstration of its ability to induce the formation of specific antibodies. This often presents serious technical problems as in the case of insoluble materials (e. g. keratin), or when the antibody response is very weak (e. g. hemoglobin). Interpretation of results are also often difficult since biological reactions, such as antibody formation or the testing for the presence of antibody by biological methods, such as anaphylaxis or skin tests, are far more sensitive than most of our chemical analytical methods. Thus, an extremely small amount of a powerful antigenic material may confer an apparent antigenic property to a non-antigenic material. The detection of antibodies also offers a problem since relatively large amounts (at least a few micrograms per ml. of serum) must be present in the test serum in order to bring about the classical serological reactions upon which we still depend. A third question which often arises in studies of simple chemical compounds is whether the material is antigenic *per se* or whether it first reacts chemically with the test animal's own protein to give an essentially foreign antigenic complex.

Unfortunately the term antigen has no distinct significance from the point of chemical characterization; nevertheless, from the mass of contradictory data certain general properties seem to be predominantly significant for antigenic function. Thus, in order to induce antibody formation, an antigen must (*a*) be foreign or contain some foreign group, (*b*) have a molecular weight of at least 10000–15000, and (*c*) be metabolized or at least partially susceptible to the hydrolytic enzymes of the test animal.

Practically all known proteins have been tested for antigenicity at one time or another and except for only a few, have been found to be antigenic. The exceptions are particularly important since they may contain the clue to the phenomena of antigenicity. Of these exceptions, *gelatin* has received the most attention. The deficiency of aromatic groups in gelatin has led many authors to postulate the necessity of such groups for antigenicity. However, the addition of aromatic structures has consistently failed to produce a truly antigenic molecule but incorporation of aromatic amino acids into gelatin as part of a more natural polypeptide chain has not yet been done. Recent studies by WAKSMAN *et al.* (*136*) have also failed to demonstrate any antigenic properties for collagen. BOYD (*6*) has suggested that gelatin is eliminated too rapidly to have any influence on the antibody forming mechanism, but in view of recent work on the use of gelatin and modified gelatins (*19*) as substitutes for plasma there would be no doubt that such materials often stay in the blood stream for many days without inducing antibody formation. The question of the antigenicity of animal nucleoproteins is still unanswered due to the use of materials of questionable purity and to the fact that such nucleoproteins usually require conditions for solubility which are incompatible with the serological reactions. There seems to be no doubt that bacterial nucleoproteins are antigenic but MUDD et al. (*90*) found that the nucleic acid group apparently plays no part in the immunological reaction.

The antigenicity of a few chemically pure *polysaccharides* has been well established but their immunological importance lies chiefly in their role as haptens and specificity determinants of antigen-antibody reactions. [A brief review of some of the important immuno-polysaccharides has recently been given by HAYWORTH and STACEY (*43*)]. Polysaccharides which have been shown to possess antigenic properties are the acetyl derivatives obtained from the capsular material of pneumococci and some of the parasitic helminths. The chemical composition of the pneumococcus polysaccharides varies with specific type. Some types such as "II", "III", "VI", "VIII", "XVII", "XVIII", "XXII", "XXIII", and "XXXI" contain little or no nitrogen while the rest contain appreciable amounts of nitrogen. Only scattered observations have been made on the chemical compositions of the various polysaccharides except for type "III", but the evidence obtained so far indicates that they are long-chain polymers of glucose, glucuronic, galacturonic and aldobionic acids and in some instances of amino sugars. The recent report by RECORD and STACEY (*116*) on types "I", "II" and "III" pneumococcus polysaccharides gives molecular weight values of 171 000, 504 000, and 141 000, respectively, as determined by sedimentation data. The axial ratios $(x:y)$ of these materials were given as 60, 200, and 110. All three preparations were polydisperse. The ability of pneumococcus polysaccharides to induce antibody formation has been known for many years and studies of the quantitative aspects of antibody production in man by HEIDELBERGER et al. (*48*) showed that injection of about 0.10 mg. of either type "I", "II" or "V" would result in the production of as much as 1000 mg. of circulating antibody.

Antigenicity of the glycogen-like immuno-polysaccharides of helminths and a few other invertebrates has been well established by the work of CAMPBELL (*12, 13*), MELCHER (*87*), OLIVER-GONZALES (*92, 93*) and KUZIN et al. (*67*). The activity of these preparations differs from the pneumococcus polysaccharides in that they induce antibody formation in rabbits and sensitize guinea pigs. All polysaccharides from invertebrates which were tested were found to be antigenic for rabbits whereas similar preparations from vertebrates consistently failed to show antigenicity in rabbits or guinea pigs.

Other pure polysaccharides and more complex structures containing lipids and perhaps polypeptides have received considerable attention but antigenicity remained doubtful in most instances. Substances in this group which have been given the most attention are, agar, gum arabic, bacterial polysaccharides, blood group type A specific substance, and a few allergens. Of the bacterial polysaccharides, those produced by strains of *Leuconostoc* are receiving the most attention at present because of their use as an emergency substitute for human plasma [recently reviewed by STACEY and RICKETTS (*127*)]. The weight of evidence at present indicates that dextrans have little or no antigenic activity; however, occasional allergic-like reactions in patients suggest some haptenic activity. Although dextrans are branched-chain polyglucoses containing both 1 : 4 and 1 : 6 α-glucosidic linkages, they give remarkably strong cross reactions with types "II", "XII", and "XX" pneumococcus polysaccharides, which suggest similar spatial configurations with very little common chemical structure. The antigenicity of blood group A-substance has been fairly well established by antibody formation in rabbits by BROWN, BENNETT and NIEMANN (*9*), and a marked increase in anti-A titres of types "O" and "B" humans by KABAT et al. (*59*). The significance of such antigenicity is still open to question since these substances contain fairly large amounts of amino acids, probably in a polypeptide unit (*89*).

The antigenicity of *lipids* is still an unsolved problem and in no instance has there been a clear demonstration of antigenic power of a pure lipid compound. Antibodies will be formed in response to injections of crude lipids or mixtures of lipids and hog serum but the assumption that the lipids function as haptens only, is still justified.

The possible antigenicity of *low-molecular weight compounds* is an extremely interesting problem and is of considerable fundamental importance as well as of practical importance in allergy. The problem has recently been given stimulus by the studies of LOISELEUR (*80, 81*). Using viscosity as a tool to detect the presence of antibody-antigen reaction, he obtained 5–10% increases in viscosities when specific 'antigens' such as ethanol, raffinose, phloridzin etc. were added to serum of rabbits previously "immunized" with these various materials. In general, these simple substances would not be expected to combine chemically with tissue proteins to produce a foreign complex—an explanation which has been used so often to interpret the antigenicity of such reactive compounds as picryl chloride, mustard gases etc. Some of LOISELEUR's results are similar to those obtained by PAULING and CAMPBELL (*101*) which indicated that specific affinities could be imposed on protein molecules in vitro upon denaturation and renaturation of serum proteins in the presence of some dyes of known structure.

So far all attempts to destroy antigenic function has failed. In some instances the antigenic capacity has apparently been reduced by ultraviolet irradiation (49) but in general the principal effect of denaturing agents has been to change specificities without seriously affecting antigenic function.

2. Haptens.

The so-called haptenic compounds or structures are of the greatest importance in establishing concepts of the fundamental nature of immuno-

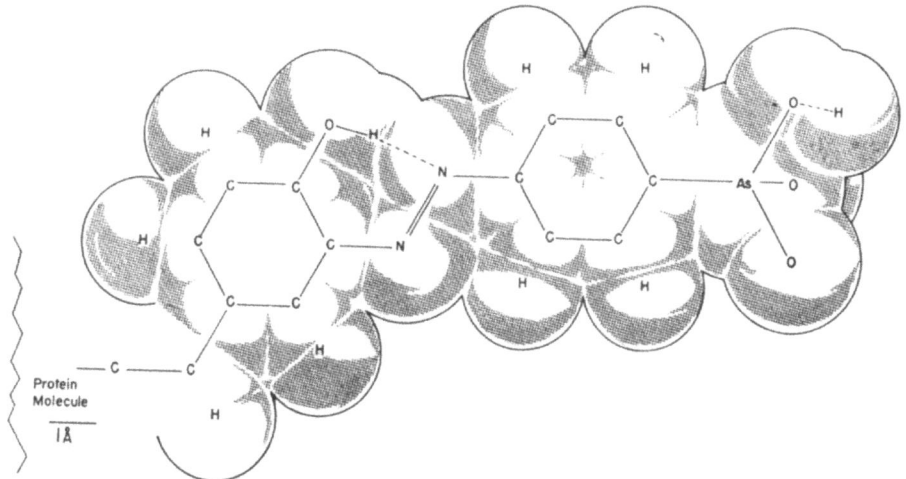

Fig. 2. Probable structure of the haptenic determinant p-azophenylarsonate when coupled to a tyrosine residue of a protein antigen. Scale drawing is based on VAN DER WAALS' radii and bond distances as given by PAULING (99).

chemical reactions. Two types of haptens are generally recognized. One type, because of its relatively large molecular size and number of specific combining sites, will combine with antibodies to give all the classical serological reactions. The other type, because of its relatively small molecular size and lack of more than a single specific reacting group, will combine with antibody but will not permit the formation of insoluble complexes. We have in this latter type the basic reacting group which determines the specificity of all immunochemical reactions, and the primary reacting structure.

An example of such a group that has been used extensively as a tool for studying immunochemical mechanisms is p-aminophenylarsonic acid (Figure 2). The antigen used to produce specific antibodies against this group is usually prepared by coupling the diazonium salt of p-aminophenylarsonate to a protein such as a foreign serum globulin or other serum protein or the stroma of erythrocytes. The antibodies produced are specific for the haptenic group and will react with it under the necessary physical conditions regardless as to whether it is present as the simple haptenic group itself or combined with large or small soluble or insoluble carrier structures. A large number of such artificial haptenic groups have been studied (70). If more than one haptenic group is present in the molecule such as was used in the precipitin test for Figure 1 (p. 444), the classical serological reactions can be obtained.

The concept of haptenic determinants suggests that native antigens could be looked upon as a mosaic of haptens and that small fragments of such molecules should have haptenic activity. Strangely enough there has been very little demonstration of this, due probably to lack of techniques which would break down large molecules without completely destroying the constituent haptenic structures. The most successful attempt so far was LANDSTEINER's who obtained preparations from an acid hydrolysate of silk fibroin (69) which behaved as simple haptens and combined with antifibroin antibodies. Studies by NEUBERGER and YUILL (91) indicated that the carbohydrate group of ovalbumin had no haptenic activity. BULL (10) studied the products obtained from egg albumin by peptic digestion and found that they would inhibit the reaction between ovalbumin and its antiserum. However, the hydrolysate also inhibited the reaction of a β-lactoglobulin system but to a lesser degree. KLEINSCHMIDT (65) investigated the inhibiting action of an imposing list of 61 simple compounds such as amino acids and similar materials on an ovalbumin-rabbit antiovalbumin precipitin system. Since only one antigen-antibody system was studied, the question of specificity arises but definite inhibitory activity was shown by proline, hydroxyproline, lysine, arginine, acetylated leucine, isoleucine, methionine, histidine, tyrosine, tryptophane, and cystine. The heterocyclic and aromatic carboxylic acids as a group were particularly effective and, of the amines, histamine was the most effective. GRABAR (36) has recently reported that glutathione will also inhibit the ovalbumin system. Studies of the fundamental haptenic structure of large molecules has been made chiefly of the pneumococcus polysaccharides. For example, the studies of GOEBEL (31) definitely established the haptenic significance of the cellobiuronic and gentiobiuronic acid portions of types "II", "III" and "VIII" polysaccharides.

One can reasonably assume that the *number of significant haptenic groups* on a natural antigenic molecule may be relatively large. The principal evidence for this is the high antibody/antigen ratios in specific precipitates which are formed in the presence of excess antibody. These ratios often approach those calculated for closest packing of the molecules involved and are dependent to some extent on the surface area of the antigen [see PAULING (*100*)]. Different specific haptenic configurations must also occur on a single antigen molecule but regardless of how logical this view may seem, there is little direct evidence to support it. Partial cross reactions between purified materials from different sources such as crystalline hen- and duck-ovalbumins suggest the presence of a common haptenic group as well as determinants which are specific for the species. This same sort of relation exists with the blood type specific A-substance in which the haptenic group responsible for the A type occurs on materials derived from a large number of sources. Other evidence for different haptenic groups on a single antigen molecule is suggested by denaturation studies in which treated proteins lose some of their specific reactivity with antibodies against the native protein but retain others; and that the groups destroyed may depend upon the physical or chemical agent to which the protein was exposed (*21*).

III. Antibodies.

Some characteristics of the physical properties of antibodies have been fairly well established for man, horse and rabbit. These data predominate our present concepts of the immunochemical nature of antibodies. No intensive study has been made of antibodies from other sources but it is quite likely that current immunochemical interest in antiserum of fowls (*33*) will produce detailed data for the physical properties of such antibodies. The only attempt to investigate the physical properties of antibody from invertebrates has been made by TYLER in his studies of the antibody-like fertilizin from sea urchins (*134*) and natural agglutinins in lobster blood (*135*).

One cannot help but think that perhaps antibody formation is an inherent property of all living systems but that we have been too fettered to the classical antigen-antibody systems to look for entirely different manifestations of antibody activity.

The general physical properties of antibodies which are described below represent in most instances the relative limits of these properties. These wide limits are due to such factors as a lack of knowledge of the theoretical and technical significance of certain data, the lability of material which is being studied and, finally, the insistence of some workers to study impure antibody preparations. In many instances, the actual amount of antibody in the preparation being characterized, is unknown.

1. Chemical Composition of Antibodies.

Most of the work reported so far on the chemical analysis of antibodies have actually been carried out on the γ-globulin fractions from serums of immunized animals. Amino acid and carbohydrate analyses of such globulins prepared by ethanol fractionation of serums from humans, horses and cattle have been made by SMITH et al. (*124*, *125*). The amino acid composition was essentially the same for the three animals (viz. 4.6% arginine, 2% histidine, 7% lysine, 3% isoleucine, 7–10% threonine, 3% cystine, 1% methionine, 2–3% hexoses, and 0.5% hexosamine). From an earlier report by RIMINGTON (*117*) one can assume that the hexose component probably consisted of mannose plus galactose. Unfortunately, the antibody content of SMITH's preparations was not determined and probably was quite low in most cases. More significant data have recently been reported by PORTER (*111*, *112*) on preparations of rabbit antibodies against egg albumin. γ-Globulin preparations containing 25–30% precipitable antibody protein were analyzed before and after removal of antibody as were the specific antigen-antibody precipitates. The analytical data for the amino acid composition of the terminal groups of antibody and normal globulin indicated a chain arrangement with alanine as the terminal group, followed by leucine, valine and aspartic acid, with glutamic acid as most likely in the fifth position. The end-group structure was the same for antibody and other γ-globulin.

Lipids probably play an important role in the chemical structure of antibodies but very little attention has been given to this problem. [A detailed discussion has been presented by PEDERSEN (*108*) and by MARRACK (*83*)]. The possible significance of lipids arose chiefly from the studies of HORSFALL and GOODNER (*56*) who found that removal of alcohol- and ether-soluble fractions from horse or rabbit antipneumococcus serum reduced its precipitability or agglutinability of antigen and, that the serological activity was restored when the lipids were replaced (lecithin for horse serum and cephalin for rabbit serum). KREUGER and HEIDELBERGER (*65 a*) recently repeated these experiments and concluded that serum lipids influence the rate but not the final equilibrium of the precipitation reaction. The amount of precipitate was the same for untreated and lipid-extracted serum but more time and greater centrifugal force were required for the latter serum. The role of such lipids is still a mystery but the ease with which they can be removed would certainly suggest they are present only as adsorbed materials influencing the secondary manifestations of antigen-antibody reactions, and are not integral parts of the antibody molecule combined by covalent bonds.

2. Electrophoretic Properties of Antibodies.

Detailed data on the electrophoretic properties of antibodies have been obtained in only a few instances but the general limiting values are fairly well established.

In most instances the mobility measurements give values characteristic of normal γ-globulin. These values center around -1.0×10^{-5} cm² sec^{-1} volt^{-1} at about pH 8.0 and ionic strengths of 0.15. On prolonged immunization of horses and rabbits, a faster moving component appears which often has been referred to as the T component and has a mobility between —1.0 and —2.0 × 10^{-5}. The recent studies of CANN et al. (24) reported the appearance of these two components in the electrophoretic range of γ-globulin of serums from rabbits immunized against bovine globulin-azophenylarsonate and that the faster moving component only occurred in high titre serums. Serums from rabbits which had been immunized for the same length of time, but gave only a low concentration of precipitating antibody, showed only a small increase in γ-globulin which was essentially electrophoretically homogeneous and of the normal slow moving type. These experiments were also of interest since they demonstrated for the first time that antibodies from the same animal, but having different specific combining affinities, could be separated on the basis of a difference in electrophoretic properties. Thus it was possible to effect a relative separation of the antibodies formed against the globulin portion of the antigen from those against the phenylarsonate group. These same investigations as well as others (22, 17) indicated that non-precipitating type antibodies (so-called "inhibiting" and "univalent" antibodies) which are present in anti-Rh* and allergic serums of humans, often occur in the globulins having mobilities much greater than γ-globulins. A detailed discussion of the electrophoretic properties of serums from normal and immunized animals and the problems involved has been given by WILLIAMS (137) in the present series.

The isoelectric points of antibodies also usually fall into the range characteristic for normal γ-globulins, namely p$_H$ 5.8–6.8. The one exception found so far is the isoelectric point of antibodies produced by horses, cattle and hogs against pneumococcus polysaccharide (61). These isoelectric points are significantly lower and range from p$_H$ 4.5 to 5.0.

3. Shape and Size of Antibodies.

Considerable study has been made of the shape and size of antibody molecules and, fortunately, many data have been obtained on purified preparations of antibodies isolated from specific antigen-antibody precipitates (57, 96, 109). Ultracentrifugal and diffusion data are in agreement with the assumption that γ-globulins and antibody molecules found in man, rabbits and horses are elongated in shape and have molecular weights varying from 150000 to 200000, except for the antibody produced in horses against pneumococcus polysaccharide which has a larger molecular weight of about 1000000. The recent work of CAMPBELL et al. (16) has

---

* The "Rh-factor" present in the blood cells of certain humans, is capable of inducing antibody formation in humans who lack this factor.

also shown that light scattering data on essentially pure preparations of rabbit antibody against p-azophenylarsonic acid gives a molecular weight of 160000. The studies by DEUTSCH et al. (26) on the isolation of a high-molecular weight antibody in human serum has recently been repeated by CANN et al. (23). Their results indicated that the relatively mild fractionation of fresh human serum by electrophoresis-convection failed to yield any high molecular weight antibody. One can reasonably assume that similar high molecular materials are actually complexes resulting from such rigorous denaturing procedures as alcohol precipitation or lyophilization.

Limiting values for molecular asymmetries can be calculated from frictional ratios. Using data supplied by KABAT and MAYER (61) and the values obtained by SVEDBERG and PEDERSEN (130) one obtains from PERRIN's equation, axial ratios for various antibodies as shown in Table 1. It is of interest that such dimensions are essentially in agreement with closest packing arrangements of antigen-antibody molecules in precipitates assuming a bivalent antibody and polyvalent antigen molecules [PAULING (100)]. It is also of interest that the minor axes of all antibody molecules studied so far are about the same. This is even true for the partially

Table 1. Asymmetry Values for Antibody Molecules.

Source of antibody [g]	Antibody against	Molecular weight	Frictional ratio	Dimensions in Å*
Horse	Diphtheria toxin [b]	184000	1.4	286/39
Horse	Diphtheria toxin [c]	113000	1.3	204/36
Horse	Diphtheria toxin [d]	90000	1.23	166/35
Horse	Pneumococcus [e]	920000	2.0	946/47
Man	Pneumococcus [e]	195000	1.5	338/37
Man (94)	Normal globulin [f]	156000	1.38	235/44
Monkey	Pneumococcus [e]	157000	1.5	312/34
Rabbit	Pneumococcus [e]	157000	1.4	272/37
Rabbit	Ovalbumin	165000	1.6	244/24

[a] Assuming an anhydrous partial specific volume of 0.73.

[b] The water-soluble fraction.

[c] The water-soluble fraction after treatment with pepsin.

[d] Crystallized preparation obtained by trypsin treatment of antitoxin-toxin precipitates.

[e] Antibody dissociated from specific precipitates by salt.

[f] Assuming a partial specific volume of 0.739.

[g] References for these sources are given by KABAT and MAYER (61).

* The authors are grateful to Dr. J. VINOGRAD for his helpful criticism and calculations of the dimensions listed in Table 1.

digested antitoxin molecules where the change in shape and size corresponds to a reduction of about one half the major axis of the original molecule.

4. Nature of the Combining Site.

Concepts of the nature of the antibody combining site involve such questions as the chemical composition of the site, the size or area, the total number on a single antibody molecule (valence), relative positions, degree of specificity with respect to a single determinant, the possibility that a single molecule contains sites of different specificities, and finally the stability of the site itself. These questions have been the subject of many publications and much speculation since EHRLICH's conception of the three dimensional "lock-and-key" hypothesis to account for the specific reaction between antigens and antibodies. A very good discussion of many of these problems has been presented by HOOKER (53). Some of these questions will be dealt with in more detail in the Section on antigen-antibody reactions (p. 463).

Until the recent work of PORTER (*111, 112*) no one had actually tackled the problem of the chemical composition of the antibody combining site. PORTER's work leaves many questions still unanswered but, as mentioned earlier, some of his data suggest for the first time the amino acid composition of the antibody reactive group. Antibody globulin was broken down by papain-HCN into fragments having one fourth of the size of the original molecule. The author concluded that only one of the fragments had antibody activity and that this fragment also gave the same end-group analysis as the original γ-globulin. There is no direct evidence that the end-group structure is the antibody combining site. However, the absence of aromatic groups in the antibody combining site is indirectly suggested by the fact that iodination of antibody has no marked effect on combining power. The fact that the composition of the terminal portion of antibody and normal γ-globulin is identical, lends support to PAULING's idea (*100*) that the chemical composition of all antibody molecules as well as of "normal" globulin is the same and that the immunological differences are the result of differences in specific spatial configurations assumed by the amino acids in the terminal portion of the polypeptide chains of antibody proteins.

Very little attention has been given so far to the exact size or area involved in an antibody combining site. The studies of LANDSTEINER and VAN DER SCHEER (*72*) however, would fix the upper limit at about 700 $Å^2$, since a substituted group such as represented in the formula, induced the formation of antibodies which reacted either with succinanilic or arsanilic acid but not with entire haptenic group. The lower limit of the size of determinant groups may be quite small, e. g. a fluorine atom can function as a specific determinant, but in such instances it is quite possible that a portion of the protein to which the substituent group is attached is also involved. HOOKER and BOYD's studies (54) showed that antibodies against protein-*p*-azophenylarsonate reacted

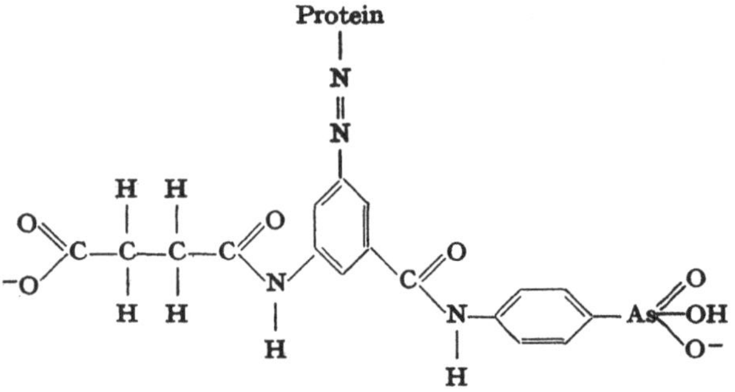

[Chemical structure diagram showing protein-N=N-phenyl group linked via amide bonds to a dicarboxylic acid derivative on one side and to *p*-aminophenylarsonate on the other.]

stronger with tyrosylazo-phenylarsonate than with *p*-aminophenylarsonate alone. These results, as well as others, described by LANDSTEINER (70) indicate that, in some instances at least, the antibody combining site may be directed to that portion of the protein carrier to which the substituent group is coupled.

The relative importance of the area of a protein carrier to which a haptenic group is attached should certainly be given more attention. It is entirely possible that the total area of antibody combining sites is essentially constant and the same for all cases. Thus, when artificial antigenic determinants are coupled to proteins for the purpose of inducing antibody formation against these groups, the spatial volume of such groups will determine whether portions of the protein carrier molecule will be involved in the template pattern for specific configuration of the antibody combining site. Fig. 2 (p. 449) is a representation of *p*-azophenylarsonic acid attached to the phenyl group of a tyrosine residue. The space occupied by such a structure would be approximately 800 Å^3 and the surface area of the corresponding antibody combining site would be about 400 Å^2, assuming a perfect fit. The idea that the antibody combining site is a closely fitting cavity on the surface of an antibody molecule is a useful and reasonable concept derived in part from experiments studying the reaction of haptens of varying structure with an antibody having a known specificity.

The number of combining sites (valencies) on antibody molecules is generally considered to be low (1, 2 or not more than 3). The principal controversy at present is concerned with the valence number of *precipitating* antibody molecules, with the weight of theoretical as well as *a priori* reasoning from experimental data favoring a valence of 2. (Some aspects of valence will be further discussed in the section on antigen-antibody reactions). The question of valency is extremely important for an understanding of the fundamental structure of antibody molecules as well as their reactions with antigens. A great deal of experimental effort as well as theoretical treatment has been given the problem which is too lengthy to discuss in detail here.

The chief proponents of the univalence of antibodies has been BOYD with his "occlusion" theory based on precipitation reactions with synthetic dye haptens (5), and TEORELL (*131*) with ideas based purely on theoretical grounds which fit much of the present experimental data. Regardless of what the theory of bivalent antibody is called, the simple framework structure of bivalent antibodies and multivalent antigens as described by MARRACK (*83*) has continued to be the most popular. Theoretical treatment has also shown that equations can be devised which correspond to experimental findings [KENDALL (*63*), PAULING, CAMPBELL and PRESSMAN (*102*)]. However, it is the experimental findings that land the greatest support to the idea of *bivalence*.

The following experiments of PAULING, PRESSMAN and CAMPBELL (*104*) have given such support to the bivalence of antibodies. A simple test antigen containing one carboxylic and one phenylarsonic acid coupled to H-acid (1-amino-8-hydroxy-naphthalene-3,6-disulfonic acid) failed to produce a precipitate when tested with either anti-phenylarsonic or anti-carboxylic serums alone but did produce a precipitate when a mixture of the serums was tested. Thus, a precipitating framework would only be possible if the respective antibody molecules were at least bivalent. This experiment (the so-called "RX" experiment) could be explained perhaps by a theory claiming that a complex of two univalent antibody molecules and one antigen molecule is insoluble and constitutes the precipitate. BOYD's "occlusion" theory is a theory of this type. However, the antigen-antibody ratios in the precipitates indicated a 1 : 1 mole ratio of antigen and antibody, and the precipitates could be dissolved to some extent by an excess of the mixed antiserum. These two facts would appear to be explained only by a "framework" theory involving bivalent antibody.

The chief criticism leveled against these experiments has been the possible polymerization of the test antigen and that instead of being a simple dye molecule it was actually a polymer containing many haptenic groups, and also that cross reactions occurred between arsonilic and carboxylic groups. That such dyes tend to polymerize and that weak cross reactions may occur is well recognized; however, the fact that the serums used in these studies failed to give precipitates when tested individually, under the conditions studied, tends to give such criticism very little justification.

Some of the most important data favoring the bivalence of antibodies have been derived from antigen-antibody systems in which the antigen was in great excess and hence all antibody valence sites would be combined with antigen (so-called "antigen excess") to give a soluble complex. PAPPENHEIMER, LUNDGREN and WILLIAMS (96) concluded from their electrophoretic and ultracentrifugal studies of such soluble complexes of diphtheria toxin and horse antitoxin that the limiting structure was 2 molecules of antigen combined with 1 antibody. More detailed electrophoretic and sedimentation experiments on purified soluble antigen-antibody complexes (bovine serum albumin and rabbit antibody) were made by SINGER and CAMPBELL (*122, 123*). Their data seem to be interpretable only on the basis of a complex containing a limiting antigen/antibody ratio of 2 molecules of antigen and 1 of antibody. MARRACK et al. (*84*) have also recently investigated the soluble complex formed by excess horse albumin antigen and its rabbit antibody by electrophoresis, and concluded that the antibody must be bivalent. Further evidence of the bivalence of antibody has been obtained by LERMAN (*77, 78*) as well as EISEN and KARUSH (*28*), by equilibrium studies of antibodies and simple univalent haptens separated by a cellophane membrane.

A recent report by BANKS et al. (*2*) concludes that antibodies are univalent as indicated by the studies of antigen-antibody complexes formed in the zone of maximum precipitation. Such complexes would be expected to contain some antibody molecules which have free antibody combining sites if such molecules are multivalent. The addition of hapten to the system failed to show any such hapten group in the precipitate and hence these authors concluded that antibody must be univalent. However, it is to be noted that in theoretical considerations of hapten inhibition (*105, 113, 106*) based on bivalency of antibody, the assumption is made that no hapten is present in the precipitate. As it is considered that antibody-hapten complexes are soluble, the system will have a higher ratio of antigen/antibody for precipitation than when no hapten is present, and possibly, the failure to bind hapten may be associated with this. Perhaps the strongest argument in favor of the univalence of antibody is the universal failure of all investigators to obtain the so-called "hetero-ligating" types with combining sites for more than a single specificity. For example, HAUROWITZ and SCHWERIN (*41*) carried out a series of experiments with antiserums against sheep serum coupled with various haptenic groups and found that the antiserums contained only antibodies which reacted with a single haptenic group. Thus, an antiserum prepared against a complex dihaptenic structure such as phenylarsonic and phenylsulfonic acids contained no single antibody molecule which would react with both haptenic groups.

The idea of the bivalence of precipitating antibodies has become so firmly entrenched in the minds of some immunologists that when antibodies are discovered which fail to form insoluble complexes with antigen, they are *a priori* described as univalent antibodies. Non-precipitating antibodies have been found in horses during the early stages of immunization against ovalbumin (95), and represent the chief antibodies ("reagin") of allergy (88) and the so-called Rh blocking antibody of man (115). Such antibodies might be referred to as inhibiting antibodies since, like inhibiting haptens, they combine with their counterpart and prevent the antigen from reacting with precipitating antibody. The recent studies of electrophoresis-convection fractions of antiserum from rabbits, immunized against protein-azophenylarsonic acid (24), would suggest that such antibodies are normally formed and that the precipitating power of antiserums depends upon the relative amounts of precipitating and non-precipitating antibodies. Although the designation of such antibodies as "univalent" is convenient and logical, other factors, e. g. variation in solubility of antibody molecules should be carefully investigated in each instance.

The fact that immunological reactions are characterized by a high degree of *specificity* may be considered the fundamental proposition of immunochemistry but this is not to be inferred as implying absolute specificity or homogeneity. Antibodies against a given protein molecule or even a specific haptenic group may show considerable variation. Such heterogeneities show up as differences in serological reaction capacity and are attributed to, (*a*) the number of combining sites per antibody molecule; (*b*) the specific haptenic portion of the antigen toward which the combining site is directed (e. g. antiserums against azoproteins contain antibodies against the azo group and against the protein carrier); (*c*) complementariness or preciseness of fit of the antibody combining site for its haptenic counterpart; and (*d*) the combining strength which involves variation in the forces affecting the antigen-antibody combination.

There is little doubt that some antibody molecules are extremely specific as evidenced by the reactions obtained by AVERY *et al.* (*1*) which distinguished between α- and β-glucosides. The studies of LANDSTEINER and VAN DER SCHEER (73) are of interest in this connection. They immunized rabbits against protein-*p*-azobenzenesulfonic acid and found that the resulting antiserum contained some antibodies which reacted specifically with the aminobenzenesulfonic acid group, and antibodies which would give cross reactions with aminobenzenearsonic acid, and others which would cross react with aminobenzoic acid. The quantitative immunochemical aspects of *heterogeneity* has been extensively studied by PAULING and his group and will be discussed later.

Heterogeneity is also suggested by qualitative serological and biological reactions and has been the subject of much controversy since EHRLICH's time through the period of the "unitarian theory of antibodies". With the development of improved methods for the fractionation of sera, the problem is once more gaining attention and reports are again being published which conclude that antibodies with different functional activities appear in a given serum. However, such complicated antigen systems (e. g. cells) are employed that interpretation is difficult. The studies of GOODNER and HORSFALL (*34*) are of interest in this connection since they were able to show that antipneumococcus antibodies, specifically precipitated with type-specific polysaccharide, could be resolved into fractions differing in their ability to confer immunity on mice.

Another type of heterogeneity which is often ignored arises from the fact that apparent chemical and physical homogeneity does not assure antigenic homogeneity. Thus, ovalbumin which has been crystallized many times may show several optimum points in a precipitation curve (cf. p. 444), each optimum being indicative of distinct antigen-antibody systems. As is well known, but so often forgotten, methods used for characterization of proteins are based on entirely different principles, and correlations are frequently purely fortuitous.

The *antibody combining site* itself is a relatively stable structure. Protein-denaturing agents may cause antiserums to lose their serological activity but the effects can usually be explained by complex formation of antibody with non-antibody serum proteins. Studies of denaturation of antibodies which have been made with heat by BAWDEN and KLECZKOWSKI (*3*) and by KLECZKOWSKI (*64*), guanidine-denaturation by ERICKSEN and NEURATH (*29*) and urea-denaturation according to CAMPBELL and CUSHING (*18*), all indicate that antibody can be inactivated by the covering of its reactive sites with non-antibody protein. The latter work was particularly significant since the antibody used was obtained from a specific antigen-antibody precipitate and was essentially pure. ROTHEN and LANDSTEINER (*120*) also showed that extended air-water interfacial films of antibody retained antibody activity but the completeness of the unfolding of each antibody molecule is not known.

The relative position of antibody combining sites with respect to each other and with respect to the asymmetric ellipsoid general pattern of the protein molecule is not known but it is an important problem in connection with the concepts of steric configuration for the framework theory of precipitation. It is generally assumed by those who support the bivalent theory of antibodies, that the two active sites are located at the terminal portions of the ellipsoid. There is no direct evidence for such an assumption but unimolecular film studies by HARKINS *et al.* (*37*) indicate that alternate layers of antigen and antibody can be built up on slides, indicating that combining sites must be some distance apart. Using the same techniques PORTER and PAPPENHEIMER (*110*) were also able to obtain multilayers of pneumococcus polysaccharide and horse antipneumococcus antibody.

5. Purification of Antibodies.

So many of the recent reviews have discussed the details for the common procedures used in antibody purification that there is little need to repeat them. These methods are based on the separation of certain antibody-containing serum protein components of immune serum by physical means (non-specific method) or by the removal of antibody from serum with a specific antigen (specific method).

Non-specific methods are in general directed toward removal of the γ-globulin fraction of serum. Since antibodies and normal γ-globulins have essentially the same physical properties, such methods obviously limit the purity of the final product to the antibody/γ-globulin ratio of the initial serum.

The common tools used for removal of γ-globulin are various precipitating agents such as ammonium or sodium sulfate, alcohol and acids or separation by means of electrophoresis. Although alcohol fractionation has been extensively employed as a result of the studies by the Harvard group, it requires such specialized equipment, techniques and experience, that salt precipitation is unquestionably the method of choice for most laboratories. In our own laboratory ammonium sulfate precipitation is used routinely for rabbit serum. The saturated salt solution is slowly added to give a final concentration of $1/_3$ saturation at room temperature and the pH adjusted to 7.8. The precipitate usually contains all of the γ-globulin present in the original serum and very little, if any, of the other serum proteins after two re-precipitations. Such protein fractions, obtained from pooled serums of rabbits immunized against bovine albumin or azoproteins usually contain 10–30% precipitable antibody protein. Further fractionation can be accomplished by dialysis against distilled water and consequent separation into water-soluble and water-insoluble components. The latter usually contains 2–3 times more antibody than the water-soluble fraction when high titre serum is used.

A new technique has recently been developed by KIRKWOOD and his group for fractionation of proteins in solution utilizing a combination of electrophoretic and convective transport of the components, and has been designated as the "electrophoresis-convection method". The extensive use of this method by this group for the fractionation of serum has proven its importance and resulted in some very significant studies [see the review by CANN and KIRKWOOD (25) and subsequent reports by these authors and their collaborators].

The apparatus consists of an upper and lower reservoir connected by a narrow rectangular channel constructed with cellophane membranes. A potential imposed across the channel produces a horizontal electrophoretic migration of the components. Final separation depends upon the vertical convective transport which is controlled by the density gradient in the channel. Thus, the faster moving electrophoretic components will tend to form a denser horizontal gradient and flow to the bottom reservoir by gravity. In the fractionation of serum at neutral pHs, albumin will become more concentrated in the lower compartment and the globulins more concentrated in the top. The apparatus developed so far is constructed to accommodate about 100 ml. of solution.

Electrophoresis-convection is undoubtedly a valuable tool for the fractionation of proteins but as for the isolation and purification of antibodies, it provokes the same basic objections to any electrophoretic method, namely, that the yield represents only a fraction of the total component, and, furthermore, that purity and yield of antibody bear an inverse relationship.

The so-called specific methods are based on the formation of insoluble antigen-antibody complexes with the subsequent dissociation of the complex and isolation of antibody. Dissociation of such complexes have been brought about by a variety of methods and in general are the same as those which inhibit precipitation such as increased salt concentration, high or low pH, increased temperatures, and use of inhibiting haptens, e. g. arsanilic acid in antiazophenylarsonic acid systems. Salt dissociation (around 15% NaCl) has been used rather extensively on specific precipitates of pneumococcus polysaccharides and was the method used by KABAT in his studies of the physical properties of antibodies. Unfortunately, salt dissociation effects are limited to polysaccharide systems and furthermore yield only a fraction of the total antibody, since dissociation is dependent upon establishing a higher antigen/antibody ratio for the insoluble state.

Other methods which have been described are dependent upon dissociation with dilute alkali or acid. The latter seems to be the method of choice since less denaturation occurs. If the antigen is insoluble at the pH used for dissociation, the procedure is fairly simple. For example, HAUROWITZ et al. (42) were able to obtain antibody by acid dissociation from specific precipitates using an acid-insoluble azoprotein. The yields were not stated, but in a few preparations the purity, as based on the percent of the total protein which would precipitate with antigen, was fairly high. The light scattering studies by CAMPBELL et al. (16) mentioned previously utilized an acid-insoluble azophenylarsonic acid dye for the removal of antiarsonilic acid antibodies from rabbit serum. Dissociation of the antigen-antibody complex was carried out at pH 3.5 by careful addition of dilute HCl. The yield obtained was close to the amount estimated to be present in the original serum. The final protein was about 100% precipitable with antigen but this value fell over a period of six months at 4° C. to approximately 80% where it remained constant. Antiovalbumin has been purified by LANNI (20) by acid dissociation of specific precipitates with yields of 50–60% and purities of 75–100%.

STERNBERGER and PRESSMAN (128) have recently reported a method for purification of antibody which they state will have general applicability. The method involves the coupling of p-aminobenzenarsonic or o-aminobenzoic acid to the precipitating antigen. In their studies they used bovine albumin and globulin antigens. The azoantigens were then dissociated from antibody with saturated calcium hydroxide solution and the azoprotein removed by precipitation with calcium aluminate. The yields were fairly low, being less than 20% but purity of the final preparations was about 100%.

Another method which has recently been reported by CAMPBELL et al. (21) also gives promise of wide applicability and depends upon the conversion of antigens into an insoluble state. This was first done with denatured ovalbumin in an attempt to secure an antigen that could be used to isolate the reagic antibodies* in humans allergic to ovalbumin. All of the preparations studied were either slightly soluble at the dissociation pH of 3.5 or too many native haptenic groups were destroyed during the preparation. The method finally developed, depended upon the chemical coupling of a protein to finely ground cellulose by means of an azobenzyl ether linkage. This was then used in the form of an adsorption column through which the specific antiserum was run. The column was then suspended and washed in cold saline and the antibody eluted with dilute HCl (final pH of the suspension was 3.5). The yields attained were approximately 100% as calculated for the original antiserum and the purity was also about 100%. This so-called "immunologic adsorbent" could be used over again after elution of antibody.

IV. The Physical Nature of Antigen-Antibody Reactions.

The following section of this review will be concerned chiefly with the fundamental nature of the reaction between molecules of antigens and antibodies. Such complicating factors as the role of lipids, complement, electrolytes etc. will be disregarded at the risk of oversimplification but it is desirable to do so here in order to clearly establish the basic concepts and problems of antigen-antibody reactions.

1. The Properties of Specific Precipitates.

a) Composition.

Since specific precipitates consist of both antigen and antibody, the ratio of these two components is of fundamental importance.

Analysis of the precipitate in the region of excess antibody and the so-called equivalence zone is relatively simple since in most instances all the antigen is precipitated. The precipitates are merely washed carefully and analyzed for nitrogen. The ratio of antibody to antigen can then be determined by subtracting the amount contributed by the antigen, which is the actual amount used for the test. When antigens contain little or no nitrogen as for example, specific dyes or polysaccharide antigens, the analysis becomes even simpler.

---

* "Reagins" are non-precipitating antibodies present in the serum of allergic individuals.

The portion of the precipitation curve which presents the greatest analytical difficulty is the region of antigen excess where soluble antigen-antibody complexes are formed. This problem has been solved in some instances by the use of tagged antigens which can be determined by chemical or physical means. For example, MALKIEL and BOYD (82) used the copper of a hemocyanin antigen, STOKINGER and HEIDELBERGER (129) used the iodine of thyroglobulin, MASOUREDIS et al. (86) added I^{131} to antibodies against human serum albumin and B. WÖSTMANN (to be published) tagged bovine serum albumin antigen with I^{131} and also the antibody with p-azophenylacetic ($C^{14}OOH$) acid. HEIDELBERGER and KENDALL (47) as well as KABAT and HEIDELBERGER (60) obtained composition data for untagged ovalbumin and horse serum albumin systems from immunochemical analysis for the unprecipitated antigen in the region of antigen excess. PAULING et al. (107) employed specific azophenylarsonic acid dyes which were easily determined by dissolving washed precipitates in dilute alkali and estimating dye concentration colorimetrically. The antibodies in these experiments were obtained from rabbits which had been immunized against sheep serum p-azophenylarsonic acid. The significant antibody/antigen ratios obtained by these various investigators have been listed in Table 2.

The antibody/antigen ratio will be seen to vary for all systems except the simple dye antigens. This would be expected since the proteins have a much greater surface and probably many specific combining sites.

Table 2. The Antibody/Antigen Mole Ratios of Specific Precipitations for Various Protein Systems Obtained at Different Regions of the Precipitation Curve.

Antigen	Molecular weight	Antibody/antigen mole ratios		
		Antibody excess	Equivalence zone	Antigen excess
Di- and trivalent azophenylarsonic acid dyes [b]	500—1200	0.9—0.7	0.9—0.7	0.9—0.7
Ovalbumin (47)	44000	5	4—3	2
Horse serum albumin (60)	70000	5	4—3	2
Human serum albumin (86)	70000	6—5	3—2	1.3—0.6
Bovine serum albumin [a]	70000	9	3.5	2
Diphtheria toxin (57)	74000	8	4—1.5	1
Thyroglobulin (129)	650000	40	14—10	2

[a] B. WÖSTMANN (unpublished). The value for antigen excess was obtained at a point where only about 0.5% of the total antigen present was precipitated. A significant value was obtained by using very large amounts of purified antibody in order to secure a valid sample.

[b] These are the range of values reported by PAULING et al. (103).

In fact, the data given for the large thyroglobulin molecules indicate that it will react with at least 40 molecules of antibody whereas albumins level off at 6 to 8 moles. It is of interest that in most instances the limiting value for an insoluble complex in antigen excess is about 2. The values of 1.3–0.6 for human serum albumin seem to be a little out of line and should be confirmed but the extensive analyses made on dye antigen and diphtheria toxin systems leave little doubt that these represent the true values. It will be recalled from the previous discussion on valence of antibody that antitoxin is peculiar in that its two combining sites are on the same terminal portion of an asymmetric molecule. The trivalent as well as the bivalent dyes are thought to be immunologically only bifunctional because of steric hindrances which prevent more than two antibody molecules from combining with one antigen. The rate of precipitation as well as the total amount may vary for the dye antigen systems but the ratio of the two components in the insoluble complex remains the same.

b) Formation and Specificity.

It is generally agreed that the initial reaction between molecules of antigen and antibody is highly specific and is based on a complementary stereochemical relation between portions of antigen and antibody molecules. The question which then arises is, whether the growth of these initial complexes is dependent upon the continued operation of a specific mechanism such as MARRACK and the majority of immunochemists visualize for a precipitating framework or, do these initial complexes aggregate to an insoluble state by non-specific mechanisms? The problem has been studied mostly by mixed antigen-antibody systems in which the antigens (e. g. cells) were morphologically distinguishable. The results were never very convincing although the weight of evidence indicated some specificity of small aggregates, e. g. a single clump would consist of only one species of antigen. HOOKER and BOYD (55) attacked the problem by determining the "particulation time" (i. e. the time between mixing of antigen + antiserum and the appearance of visible particles) of two non-cross reacting precipitation systems, when tested separately and when mixed. They noted that the particulation time was considerably reduced when the systems were mixed, suggesting a mutual non-specific interaction. LANNI (74) repeated these studies using a system containing partially purified rabbit antibodies against crystalline ovalbumin and type "III" pneumococcus polysaccharide. The floccules produced by these two systems differed sufficiently so that he was able to distinguish them in the dark-field microscope. Quantitative and qualitative studies of these systems clearly indicated that the reaction was specific to a point where microscopic particles (seromicrons) could be seen. Non-specific aggregation occurred only in the final stages of precipitation which is the stage commonly studied.

c) 'Ageing.'

The recent interest in practical procedures for isolation and purification of antibodies by specific methods has once more brought up the question of possible secondary changes in the nature of precipitates which might occur on standing. It has been our own experience with some systems that after 24 hours they become more difficult to dissolve by adding excess antigen or simple haptens. However, recent studies in our laboratory of precipitates of bovine serum albumin confirmed the previous ones of BOYD (4) that reversibility by addition of excess antigen was only slightly reduced after one week of standing at 4° C., provided that sufficient time (e. g. 24 hours at room temperature) was allowed for reversal. There is no doubt but that the rate of reversal of the precipitin reaction decreases with the ageing of the precipitate; however the final equilibrium may be eventually the same.

2. Thermodynamic Properties of Antigen-Antibody Reactions.

The nature of the forces involved in antigen-antibody reactions and the strength of the antigen-antibody bond are of fundamental importance. The complexity of the precipitin reaction has made the evaluation of an equilibrium constant or standard free energy of reaction difficult and the contribution of such information to the estimation of bonding forces is extremely small. The significant data have been obtained almost exclusively with non-precipitating systems involving simple haptens and their respective antibodies. LANDSTEINER and his collaborators demonstrated that antibodies could be prepared with specificities directed against groups of known chemical structure and that they would react with substances whose molecules contain the homologous or similar haptens. Studies first were made of the reaction of an antihapten serum with antigens prepared by coupling the hapten or simple related compound to a protein different from that used in immunization. Investigations with these simple chemical groups were extended to include the combination with the simple substances themselves. It was found that for a simple substance to form a specific precipitate with antibody it is necessary that its molecule contain two or more specific combining groups. While all such polyhaptenic substances do not necessarily form precipitates, no monohaptenic substances do. It has been demonstrated, however, that monohaptenic substances actually combine strongly with antibody and that the presence of monohaptenic substances specifically inhibits the precipitation of antibody by polyhaptenic azoproteins (68) or by simple polyvalent substances (71). This phenomena has been applied by LANDSTEINER and others (*70*) as well as by PAULING, PRESSMAN, CAMPBELL and their co-workers in extensive studies on specificity and its structural interpretation. MARRACK and SMITH (*85*) presented direct evidence for combination by equilibrium dialysis experiments which clearly showed that when a hapten was dialysed against its specific antibody, the hapten accumulated in the antibody solution. Their results were confirmed by HAUROWITZ and BREINL (*38*).

a) The Free Energy and Heat Changes in Antigen-Antibody Reactions.

Two different laboratories reported attempts to measure the free energy change (ΔF) of hapten-antibody reactions at about the same time. Both used essentially the same method which consisted of dialyzing solutions of highly purified antibody against monovalent hapten solutions of various concentrations. The concentration of free hapten and the amount of bound hapten was used to give an equilibrium constant for the combination.

EISEN and KARUSH (*28*) dialyzed *p*-(*p*-hydroxyphenylazo)-phenylarsonic acid against rabbit antibodies to bovine globulin-azophenylarsonic acid (at 29° C.). (These antibodies had been purified by specific precipitation.) They made a plot of $1/c$, the reciprocal of the free hapten concentration, against $1/r$ where r designates moles of hapten bound per mol of protein; and by extrapolation to $1/c = 0$ it was found that $r = 2$ which would indicate a valence of 2 for the precipitating antibody. By determining the equilibrium constant at the point where half the antibody sites are occupied, these authors concluded that the average free energy of combination was —7.7 kcal/mol hapten bound.

LERMAN (*77, 78*) conducted similar experiments using "H—acid—R" (prepared by coupling diazotized arsanilic acid to 8-amino-1-naphthol-3,6-disulfonic acid) as the hapten which also was dialyzed against highly purified antibody at 39° C. He obtained a value of —6.8 kcal/mol hapten bound for the binding of one hapten molecule per antibody molecule. LERMAN has discussed possible complicating factors in these experiments such as polymerization of the hapten and DONNAN equilibrium effects. However, he concluded that these seemed negligible at 39° C. and at physiological salt concentrations. EISEN and KARUSH (*28*) corrected for adsorption of the hapten by the cellophane membrane. Although these complications exist, it seems reasonable to conclude from these two investigations that the free energy change in such systems is of the magnitude of —10 kcal/mol of hapten bound.

More recently, HAUROWITZ *et al.* (*39*) have interpreted the solubility of labeled antigen-antibody precipitates as representing the following reaction, where G = antigen and B = antibody:

$$GB_n \rightarrow GB_{n-1} + B.$$

In these experiments azoproteins (arsanil-beef serum globulin and sulfanil-ovalbumin) were used to precipitate the homologous rabbit antibody and after washing, the precipitates were suspended in saline and their solubilities were determined at different temperatures. HAUROWITZ et al. state that the precipitates dissociated according to the above reaction and that the "solution contained mainly antibody". From these they determined the free energy change, heat change and entropy change of the reverse reaction. Their results were tabulated as follows:

Antigen	t_1	ΔF_1	t_2	ΔF_2	ΔH	ΔS
Arsanil-beef serum globulin	5° C.	−8.0	25° C.	−8.4	−2.0	21
	5° C.	−8.0	25° C.	−9.3	−2.1	21
Sulfanil-ovalbumin	4° C.	−8.5	29° C.	−9.0	−2.7	21
	4° C.	−8.7	29° C.	−9.2	−3.0	21

ΔF and ΔH are expressed in kcal/mol and ΔS in cal/mol/degree.

However, it would seem that great care would have to be taken to prove that the equation given above actually represents the reaction taking place and that a temperature variation in the solubility of the complexes themselves was not an influencing factor.

The only attempt so far to measure the heat of an antigen-antibody reaction by direct calorimetric means was made by BOYD et al. (8) who obtained values for the heat evolved at 31° C. when horse antihemocyanin serum and hemocyanin *ex Busycon cannaliculatum* were mixed. It was believed that the heat measured per mole of antigen was probably accurate to about 20%. They measured the heat involved in the region of antibody excess where no precipitate was formed and using 6.8×10^6 as the molecular weight of the antigen calculated a heat of 3300 kcal. per mol of antigen. The heat per mol of antibody is more interesting but the calculation of this quantity requires a knowledge of the ratio of antibody to antigen in the soluble complex. This ratio was estimated by extrapolating the results obtained in measuring the composition of specific precipitates. The value obtained for the heat change was 40 kcal per mole of antibody. BOYD and his colleagues assumed the reasonable value of −10 kcal for the free energy change per mol of antibody, and a value of about −100 cal/mol/degree for the entropy change was obtained. This is in contrast to the value given by HAUROWITZ of 21 cal/mol/degree.

The limited amount of data that is available indicates that the heat change involved in the combination of an antigen with its homologous antibody site is much less than the heat of an ordinary chemical reaction.

b) *Differences in Free Energies of Combination.*

PAULING and collaborators (*105, 113, 106*) have developed theoretical aspects of the reaction involving precipitation as well as hapten inhibition and these have been used to explain in part the free-energy relationship of antigen-antibody combinations. The essential parts of their assumptions and theory are as follows.

For the involved mathematical portions of the work the reader is referred to the original publications.

It is assumed that in an ideal antigen-antibody system, the solution contains antigen molecules A, antibody molecules B, soluble complex molecules A_2B, and molecules AB in equilibrium with a precipitate AB. It is also assumed that the two bonds in $A-B-A$ are equal in strength and hence the equilibrium constants for the two reactions

$$A + B \rightleftarrows AB; \quad A + AB \rightleftarrows A_2B$$

differ only by the entropy factor 4 being $4K$ and K respectively. Extending this to include the soluble molecular species H (hapten) BH, BH_2, and ABH and making a similar assumption about the B—H bond strength and the equilibrium constants, the authors derived an expression for the variation of the amount of precipitate with hapten concentration in any given antigen-antibody system at the equivalence zone. However, the expression developed provided only partial agreement with the experimental data. Deviation of the experimental results from the theoretical curve was attributed to the heterogeneity of the antiserum which appeared to contain antibodies of greatly varying combining powers. They then extended the theory by assuming that the heterogeneity of the antiserum was such that it could be described by a probability distribution function which is an error function in the effective free energy of combination of hapten and antibody (in competition with antigen); that is, in the quantity $\ln (K'/K_0')$, where K' is the effective hapten inhibition constant (or the equilibrium constant for the combination of hapten and complementary region of the antibody) of the particular antibody molecule under consideration and K_0' is an average effective hapten inhibition constant. Theoretical curves were then calculated for various arbitrary values of the heterogeneity index[*] in which the amount of precipitate was plotted against the logarithm of the amount of hapten.

The shapes of such curves are determined entirely by the values selected for σ and were used to determine the σ for experimental data by merely selecting the theoretical curve which showed the greatest

---

[*] The normalized distribution function used was $\dfrac{1}{\sqrt{\pi}\,\sigma} e^{-[\ln (K'/K_0')]^2/\sigma^2}$ and the fractional number of antibody molecules with given value of K' in a differential region was $\dfrac{dN}{N} = \dfrac{1}{\sqrt{\pi}\,\sigma} e^{-[\ln (K'/K_0')]^2/\sigma^2} d \ln (K'/K_0')$.

degree of fit. For a given antigen-antibody system the effect of a different K_0' (corresponding to different haptens) is merely to shift the curve along the log concentration of hapten axis; and hence the position of the curve will be dependent upon the K_0' which is a measure of the free energy of combination of the hapten and antibody. The ratio of any two K_0''s can be used to evaluate the difference in the free energy of combination of the haptens with the antibody. Usually PAULING et al. selected a given hapten or group of haptens as having $K_0' = 1$ and evaluated the other hapten constants relative to this standard. As an example, two haptens may be considered which give inhibition curves of the same shape (i. e. same σ) and give 50% inhibition with 5×10^{-7} and 50×10^{-7} moles of hapten respectively. The difference of their free energies of combination with antibody is, RT ($\ln 50 \times 10^{-7} - \ln 5 \times 10^{-7}$) $\simeq$ $\simeq RT \ln 10 \simeq 1.4$ kcal.

It should be emphasized that such analyses only give the difference in free energies of combination and not the absolute free energy. Although PAULING and collaborators have used the theory outlined to arrive at these results they point out (103) that "the evaluation and discussion of differences in the standard free energy of combination of different haptens with antibody is not dependent on the assumption of any particular theories of the precipitation of antibody and antigen. We may consider that the precipitation with antigen is used as a method for fixing a standard concentration of free antibody (for example, that which corresponds under the conditions of the experiment to the formation of one-half as much precipitate as is formed in the absence of hapten); the concentration of hapten necessary to reduce, by combination with part of the antibody, the concentration of free antibody to the standard value is determined experimentally and the ratio of concentrations for two haptens may then be introduced in the well-known equation relating standard free energy change and equilibrium constant to give a value of the difference in the standard free energy of combination of the two haptens with antibody. The only assumption underlying our treatment of hapten inhibition data is that inhibition of precipitation is the result of combination of the antibody with molecules of the hapten."

Antigen-antibody ratios have also been obtained in these systems but these experiments may be complicated by the binding of the simple substances by albumin present in the serum and by the polymerization of the haptens. Hapten polymers in the precipitates would have a marked effect upon the antigen-antibody ratios. However, PARDEE and PAULING (97) have concluded that this effect is only significant when purified antibody is used instead of whole serum and hence maintain that the assumption of approximately equal molal quantities of antibody and polyhaptenic simple substance in the precipitates is justified.

3. Nature of the Forces Involved.

It is now generally accepted that complementary surfaces are required for antigen-antibody reactions and that the bond strengths involved are weaker than those of ordinary chemical bonds. Consequently, it is believed that primary valence forces play no role in these reactions and instead VAN DER WAALS forces, hydrogen bonding, and coulombic attraction of oppositely charged groups, are the important factors. This has been discussed by PAULING et al. (*102*) and again by PAULING (*70*). While these forces are themselves non-specific, the specificity results from the complementariness of antigen and antibody.

A theoretical discussion of VAN DER WAALS forces by LONDON (*79*) showed that the forces involved were inversely proportional to the seventh power of the distance between the atoms involved. VAN DER WAALS attraction between two molecules on contact is thus due largely to interactions of pairs of atoms in the two molecules which are themselves in contact; and the magnitude of the attraction is determined by the number of pairs of atoms which can be brought into contact. In consequence, two molecules which can bring large portions of their surfaces into close contact will in general show much stronger mutual attraction than molecules with less extensive complementariness of surface topology.

Steric restrictions and the limitation of the class of atoms forming good hydrogen bonds gives to the hydrogen bond even somewhat greater stereochemical significance than is shown by electronic VAN DER WAALS attraction.

The third type of attractive force considered, coulombic attraction of oppositely charged groups also depends upon the charged groups being quite close together.

It indeed seems that the specificity of antigen-antibody reactions can be explained in terms of these well known non-specific *short-range forces* with the specificity resulting from the complementariness of surface required to give a stable bond.

Recently, there has been considerable discussion about the possibility of specific *long-range forces* being of importance in immunological and enzymatic reactions.

Experiments with thin films of antigen and antibody separated by inert "barriers" led ROTHEN (*118, 119*) to postulate that long range forces may be involved in the reaction of antigen with antibody and enzyme with its substrate. ROTHEN spread films of antigen (such as egg albumin or bovine albumin) on an air-water interface and transferred them onto metal slides. Rabbit antibodies would react with these films and form a layer of antibody even when the antigen film was covered with a film of barium stearate, octadecylamine or "Formvar" (an artificial high polymer preparation). He believed that the results could not be accounted for by holes in the barrier since, (a) the rate of the reaction was not affected by the films; (b) barium stearate and Formvar films prevented the reaction of insulin with films of protamine clupein sulfate; and (c) antigen-antibody films covered with Formvar were not washed off with sodium chloride solution as were unprotected films. He concluded that antigen-antibody forces extended over 200 Å, a distance many times greater than that calculated for interactions of the type discussed above. He has postulated that antigen and antibody "interact as two resonating oscillators—perhaps involving a characteristic frequency" but he has not attempted to account for the specificity of the reactions.

Electron microscope studies have indicated that these films are not regular (62, 121), and it has been proposed that the reactions do indeed occur through holes in the barrier. It has been pointed out that the method of measuring the thickness of these films (ellipsometer readings) is unable to distinguish between a smooth film and an irregular one of about the same average thickness and hence clumps of antibody over holes in the barrier could be mistaken for a uniform film of antibody. In particular SINGER (121) was able to carry out experiments like those of ROTHEN and at the same time demonstrated imperfections in the barrier films. SINGER discussed ROTHEN's arguments against the possibility of the reaction occurring through holes in the barrier and concluded these were not sufficient to eliminate this possibility. He has calculated that the effect on the rate of the reaction would not be detected and also states that the salt dissociation of the antigen-antibody complex may only produce relatively large soluble aggregates of antigen and antibody which are unable to diffuse through the holes in the Formvar films. SINGER has discussed the protamine-insulin-reaction and believes that the barium stearate films themselves react with the protamine film; and he is of the opinion that the fact that Formvar films permitted some insulin adsorption is significant.

SINGER also conducted experiments with films of certain cellulose derivatives and found that these films did not permit nearly as much adsorption of antibody as the Formvar and barium stearate films. He also studied evaporated films of silica as inert barriers; and films 18 Å in thickness were found to completely inhibit the subsequent adsorption of antibody. This was interpreted as meaning that these films were more perfect than the others. SINGER concluded: "it appears unlikely that specific long rangs forces operate in these reactions." As most of the experimental evidence in immunochemistry is in accord with the well established principles involving short-range forces, at present there is not sufficient evidence to warrant the acceptance of specific long-range forces.

The present concept of antigen-antibody reactions occurring because of the existence of complementary sites which allow short-range forces to produce a stable bond receives considerable support from the hapten inhibition studies previously described. These experiments provide relative equilibrium constants which presumably represent differences in the binding of monohaptenic substances. These relative differences in the free energy of combination have been interpreted as in the following ways.

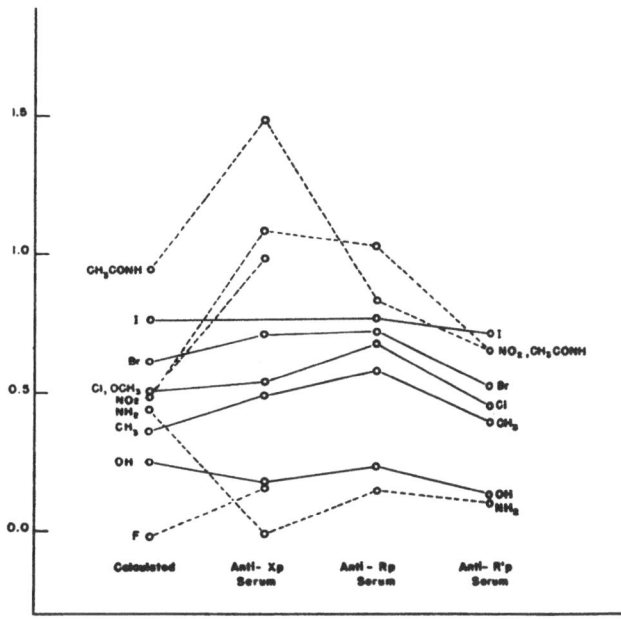

Fig. 3. Experimental values of log K_0' for para-substituted haptens in three different antibody systems and the theoretical K_0' values as calculated from VAN DER WAALS attraction energies. [From: Journal of the American Chemical Society 67, 1003 (1945)].

Using the approximate expression developed by LONDON for the interaction energy of two atoms, groups, or molecules (79) and making certain other assumptions, PAULING and PRESSMAN (103) developed an expression for the change in free energy of combination of hapten and antibody when certain modifications are made in the hapten structure. This expression was used to evaluate theoretical inhibition constants for several series of haptens and these values were compared with the experimental value.

Antiserums were obtained from rabbits which were immunized against sheepserum-p-azobenzoic acid (anti-X_p serum), -p-azophenylarsonic acid (anti-R_p serum), and -p-(p-azophenyl)-azophenylarsonic acid (anti-R_p' serum). Hapten inhibition experiments were carried out with these serums using as haptens various para-substituted homologs of the immunizing hapten. These para substituents occupy the position of the azo linkage in the immunizing antigen and hence steric effects would not be expected to be too significant. In each series the hapten inhibition constants were evaluated relative to the specific unsubstituted hapten, viz., benzoic acid, phenylarsonic acid, and *p*-phenylazophenyl-arsonic acid, respectively. Figure 3 which is taken from one of their papers (*103*) shows a comparison of the experimental values in the three systems with the calculated value.

The somewhat low values obtained with anti-R_p serum were attributed to a poorer fit of these antibodies. This poorer fit had previously been postulated because of the much smaller effect of position of substituents in the benzene ring of a hapten for this system as compared with the anti-R_p system. It can be seen that for the iodo, bromo, chloro, methyl, and hydroxy groups that the correlation between calculated and observed values is good. It seemed likely that the values for the haptens containing the acetoamino, nitro, methoxy, and fluoro groups were increased by hydrogen bonding with the antibody providing the proton for the bond which in combination with the immunizing antigen would be directed to the azo linkage. This effect does not occur with the amino group because the unshared electron pair is used in conjugation with the benzene ring. The formation of hydrogen bonds with the hydrogen atoms of the amino group may not be effective because of a lack of a suitable electronegative group in the antibody combining site, whereas hydrogen bonding to water molecules would decrease the value of K_0'. The phenolic hydroxyl group and the acetoamino group can form hydrogen bonds of both kinds and the two resultant opposed effects on the value of K_0' appear largely to cancel each other in the case of the hydroxyl group; and for the acetoamino group the first effect is the more important.

An interesting calculation has been made by PRESSMAN et al. (*114*) who found that two haptens which were identical except that one carried a $N(CH_3)_3^+$ group where the other had a $C(CH_3)_3$ group, gave a ratio of 15.5 for their hapten inhibition constants. This corresponds to a difference in standard free energy of combination with antibody of 1.5 kcal. Identifying this free energy difference as being coulomb interaction energy between the $N(CH_3)_3^+$ group and the antibody, they calculated that the positive charge of the hapten must be 7 Å apart from a negative charge on the antibody which appeared to be reasonable.

In general it can be concluded that the experimental observations in hapten inhibition studies can be explained in terms of steric considerations, VAN DER WAALS forces, hydrogen bonding, and coulombic attraction.

4. Mathematical Interpretations of the Precipitin Reaction.

Numerous equations have been proposed and derived to quantitatively interpret the precipitin reaction. These equations are all derived on the basis of certain more or less justified simplifying assumptions and usually the final expressions contain constants which are adjusted to make the equations fit the experimental data. The limited experimental evidence on antigen-antibody ratios makes it difficult to evaluate the adequateness of the relationships and in many cases it is still not possible to validate the initial assumptions.

Perhaps the most widely used of these equations has been the one proposed by Heidelberger and Kendall (*44*, *45*, *46*). Kendall (*63*) re-derived this equation which was restricted to the zone of antibody excess and assumed that all antibodies were divalent, antigens multivalent and that all combined antibody molecules precipitate. Thus, antibodies with either one or both combining sites attached to antigen would be in the precipitate. The assumption that all antigen present precipitated was also made. The equation derived was of the form,

$$p = 2Rx - \left(\frac{R^2 x^2}{q}\right)$$

where p is milligrams of antibody precipitated, x milligrams of antigen added, q milligrams of antibody added (a constant), and R is a constant dependent upon the molecular weights of antigen and antibody and their valences. This equation is rewritten in the form, $r = a - bx$, where r is the weight ratio of antibody to antigen in the precipitate. The usual procedure in applying this equation is to express the amounts of antigen and antibody in milligrams of nitrogen and then plot r against x and obtain the values of the constants a and b from the graph. Heidelberger and Kendall found that an empirical equation of the form $r = a - bx^{1/2}$ gave better agreement with some of their experimental data.

Ghosh (*30*) using similar considerations derived an equation similar to the adsorption isotherm expression of Langmuir. An equation of this type has also been proposed by Boyd (*6*).

Hershey (*50*, *51*) developed a theory which described a system of complexes prior to aggregation. These complexes were the results of reactions in which antigen is multivalent and antibody acts as effectively univalent. It was necessary to either assume that precipitation occurred without disturbing this equilibrium or to introduce non-equilibrium concepts in order to arrive at a final solution. Hershey favored a specific type of aggregation with most satisfactory results obtained by considering antibody bivalent. His treatment is based in part on probability considerations. By assuming ovalbumin to be an antigen with a maximum valence of 5, satisfactory agreement with experiment could be obtained for the ovalbumin-rabbit antiovalbumin system. HERSHEY later developed a theory for bivalent antigen and bivalent antibody in a system thus restricted to the formation of linear chains (*52*).

As outlined previously, PAULING *et al.* (*105*) have derived an expression on application of the principles of chemical equilibrium which was applied to their studies on the serological properties of simple substances. The equation developed accounted for the observed properties of their systems.

TEORELL (*131, 132*) predicted antibody-antigen relationships, assuming antibody to be univalent and antigen multivalent. His treatment was patterned after the treatment of polybasic acid equilibria. These considerations are very similar to those used by HERSHEY in setting up the initial conditions before aggregation occurred. TEORELL considers certain of his aggregates insoluble whereas HERSHEY has considered the specific aggregation of these complexes. TEORELL was able to obtain good agreement with various sets of data over the entire range.

More recently, GOLDBERG (*32*) developed a theory for the reactions of multivalent antigen molecules with bivalent and univalent antibody molecules. This statistical mechanical theory requires the assumption that antigen-antibody aggregates are three-dimensional networks with no cyclical structures; and also that any unreacted site is as reactive as any other site regardless of the size or shape of the aggregate to which it is attached. Using this theory GOLDBERG was able to place narrow limits upon antibody-antigen ratios in precipitates which were in very good agreement with experimental data.

5. A Note on the Use of Polyvalent Haptens.

Considerable criticism has been leveled against the use of many of the dyes used for specific precipitation because of their tendency to aggregate in solution. In fact, LANDSTEINER (*70*) was the first to suggest that the precipitating power of these simple antigens might be connected with their partial aggregation in solution. PAULING and his co-workers were aware of this possibility as suggested by diffusion and osmotic pressure studies with simple antigens but did not consider association important in most instances. BOYD and BEHNKE (*7*) have again brought up the question of polymerization of these simple compounds. These workers and others have seriously questioned the interpretations that PAULING *et al.* have made of their results because of this phenomenon. Their criticism has led PAULING and his associates to investigate the precipitation of purified antibody by simple substances and to consider polymerization of precipitating haptens (*97, 98*).

In an earlier part of the present review the statement was made that monohaptenic substances do not cause precipitation. This statement must now be qualified because if the monohaptenic substance has a structure that favors the formation of polymers in saline solution, precipitation with purified antibody is observed to occur as the polymer is effectively polyhaptenic. However, it has been shown that precipitation does not occur with whole serum. PARDEE and PAULING (97) believe that in serum the polymerization equilibrium is shifted toward the monomeric form since the concentration of free hapten in solution is greatly reduced by its reversible combination with serum albumin. They assume that this nonspecific binding effectively keeps the concentration of free hapten low and prevents the formation of aggregates. They believe that the conclusions drawn from their earlier experiments are valid.

Some of the simple antigens used by PAULING and his group (97) were colorless and hence almost certainly not associated; however, antigen-antibody ratios in these precipitates were not determined. The ratio antigen/antibody in the studies with purified antibody were extremely high due to the presence of aggregates; however, as the concentration of antigen was decreased the antigen/antibody ratios approached unity in some cases. PARDEE and PAULING believe that these observations support their earlier conclusions. They also state that aggregation has no influence on the hapten inhibition data because of the use of whole serum and the use of a standard amount of inhibition in comparing the inhibitory powers. Most of the haptens do not aggregate and the values would be expected to be the same whether the precipitating antigen was aggregated or not. However, the role of albumin may present an additional complication.

Perhaps one of the strongest indications that aggregation plays no part in the original experiments was that a simple substance containing one haptenic group against one antiserum and another against another antiserum did not precipitate with either antiserum separately (as it should if it were polymerized) but did precipitate with the mixed antiserums. This antigen did precipitate with purified antibody from one of the serums. This experiment (the so-called "RX" experiment) is one of the strongest supports for the framework theory and its original interpretation still seems valid (cf. p. 457).

Nevertheless, as PARDEE and PAULING have stated, it would "be desirable, in order to carry conviction for additional experiments to be made, with the use of nonaggregating polyhaptenic precipitating antigens". The studies using purified antibodies are certainly of great importance.

V. Conclusions.

An attempt has been made in the foregoing discussion to present up-to-date concepts of the physical nature of antigens and antibodies and certain aspects of their interactions. It has been assumed that some of the readers will have had only slight contact with the problems of immunochemistry and hence some of the discussion may have seemed unnecessarily simplified and repititious to those more familar with the subject. Some of the problems are obviously controversial and interpretation of available data must depend in part upon speculative extrapolation. However, from this mass of contradictory data and speculation emerge tangible concepts which are useful to both chemists and biologists.

The concepts of the physical nature of antigens and antibodies which may be considered as being generally accepted at the present time may be summarized as follows: (a) Antibody molecules are proteins with physical properties very similar to any other serum protein; (b) antigens are fairly large molecular weight materials which must have certain chemical properties in order to induce antibody formation; (c) the haptenic determinant of immunochemical specificity may be only a single atom but larger structures containing two or more such groups are necessary to form insoluble complexes with antibodies; (d) antibodies formed against a single molecular species may show great heterogeneity with respect to number of combining sites, specificities, and strength of reaction; (e) the basis of specificity is the spatial configuration of the antibody-combining site and the specific determinant on the antigen; and (f) the forces which are responsible for the combination of antibody and antigen molecules are the relatively weak forces of VAN DER WAALS interaction, hydrogen bonding, and coulombic attraction which can be effective in forming a stable bond when suitable complementariness of surface exists.

References.

1. AVERY, O. T. and W. F. GOEBEL: Immunochemical Specificity of Synthetic Sugar-Protein Antigens. J. exp. Medicine 50, 533, 551 (1929).
2. BANKS, T. E., G. E. FRANCIS, W. MULLIGAN and A. WORMALL: The Use of Radioactive Isotopes in Immunological Investigations. Biochemic. J. 48, 371 (1951).
3. BAWDEN, F. C. and A. KLECZKOWSKI: The Effects of Heat on the Serological Reactions of Antisera. Brit. J. exp. Pathol. 23, 178 (1942).
4. BOYD, W. C.: The Time Factor in Solubility of Precipitates in Excess of Antigen. J. Immunology 38, 143 (1940).
5. — On the Mechanism of Specific Precipitation. J. exp. Med. 75, 407 (1942).
6. — Fundamentals of Immunology. 2nd ed. New York: Interscience. 1947.
7. BOYD, W. C. and J. BEHNKE: Aggregation in Solution of a Synthetic Hapten. Science (New York) 100, 13 (1944).
8. BOYD, W.C., J.B.CONN, D.C. GREGG, G.B. KISTIAKOWSKY and R.M. ROBERTS: The Heat of an Antibody-Antigen Reaction. J. biol. Chemistry 139, 787 (1941).
9. BROWN, D. H., E. L. BENNETT and C. NIEMANN: The Appearance of Sheep Cell Lysins and Human A Cell Agglutinin in a Rabbit Immunized with a Partially Purified Blood Group A-Specific Substance from Hog Gastric Mucin. J. Immunology 56, 1 (1947).

10. BULL, H.: On the Haptenic Activity of Protein Hydrolysates. (Personal communication.)
11. BURNET, F. M. and F. FENNER: The Production of Antibodies. London: Macmillan & Co. 1949.
12. CAMPBELL, D. H.: An Antigenic Polysaccharide *Ascaris Suum*. J. infect Diseases **59**, 266 (1936).
13. — The Immunological Specificity of Polysaccharides from Some Common Parasitic Helminths. J. Parasitology **23**, 348 (1937).
14. — Immunochemical Aspects of Allergy. J. Allergy **19**, 151 (1948).
15. — The Nature of Antibodies. Ann. Rev. Microbiol. **2**, 269 (1948).
16. CAMPBELL, D. H., R. H. BLAKER and A. B. PARDEE: The Purification and Properties of Rabbit Antibody Against p-Azophenylarsonic Acid with Particular Reference to Molecular Weight Determination from Light Scattering Data. J. Amer. chem. Soc. **70**, 1293 (1948).
17. CAMPBELL, D. H., J. R. CANN, T. B. FRIEDMAN and R. A. BROWN: Reagic Serum Fractions Obtained by the Electrophoresis-Convection Method. J. Allergy **21**, 522 (1951).
18. CAMPBELL, D. H. and J. CUSHING, Jr.: The Effect of Concentrated Urea Solutions on the Precipitating Power of Antiovalbumin. Significance of the Formation of Protein Complexes. Science (New York) **102**, 564 (1945).
19. CAMPBELL, D. H., J. B. KOEPFLI, L. PAULING, N. ABRAHAMSEN, W. DANDLIKER, G. FEIGEN, F. LANNI and A. L: LEROSEN: The Preparation and Properties of A Modified Gelatin (Oxypolygelatin) as an Oncotic Substitute for Serum Albumin. Texas Reports on Biol. and Med. **9**, 235 (1951).
20. CAMPBELL, D. H. and F. LANNI: The Chemistry of Antibodies. In: Amino Acids and Proteins, ed. D. GREENBERG. Chapter XI, p. 649. Springfield, Ill.: Charles C. Thomas. 1951.
21. CAMPBELL, D. H., E. LUESCHER and L. S. LERMAN: Immunologic Adsorbents. I. Isolation of Antibody by Means of a Cellulose-Protein Antigen. Proc. nat. Acad. Sci. USA **37**, 575 (1951).
22. CANN, J. R., R. A. BROWN, D. C. GAJDUSEK, J. G. KIRKWOOD and P. STURGEON: Fractionation of Rh Antiserum by Electrophoresis-Convection. J. Immunology **66**, 137 (1951).
23. CANN, J. R., R. A. BROWN, S. J. SINGER, J. B. SHUMAKER, Jr. and J. G. KIRKWOOD: Ultracentrifugal Studies of γ-Globulins Prepared by Electrophoresis-Convection. Science (New York) **114**, 30 (1951).
24. CANN, J. R., D. H. CAMPBELL, R. A. BROWN and J. G. KIRKWOOD: Fractionation of Rabbit Antiserum (Antiphenylarsonicazo-Bovine Globulin) by Electrophoresis-Convection. J. Amer. chem. Soc. **73**, 4611 (1951).
25. CANN, J. R. and J. G. KIRKWOOD: The Fractionation of Proteins by Electrophoresis-Convection. Cold Spring Harbor Sympos. quantitat. Biol. **14**, 9 (1950).
26. DEUTSCH, H. F., R. A. ALBERTY, L. J. GOSTING and J. W. WILLIAMS: Biophysical Studies of Blood Plasma Proteins. J. Immunology **56**, 183 (1947).
27. DOERR, R.: Die Immunitätsforschung (7 Bände). Wien: Springer-Verlag. 1947 — 1951.
28. EISEN, H. W. and F. KARUSH: The Interaction of Purified Antibody with Homologous Hapten. Antibody Valence and Binding Constant. J. Amer. chem. Soc. **71**, 363 (1949).
29. ERICKSEN, J. O. and H. NEURATH: Immunochemical Properties of Native and Denatured Horse Serum Globulins. J. gen. Physiol. **28**, 421 (1945).
30. GHOSH, B. N.: The Adsorption of Antigens by Antibodies or vice versa. I. Theoretical. Indian J. med. Res. **23**, 285 (1935).

31. GOEBEL, W. F.: Studies on Anti-bacterial Immunity Induced by Artificial Antigens. J. exp. Medicine **72**, 33 (1940).
32. GOLDBERG, R. J.: A Theory for Antigen-Antibody Reactions and Ph. D. Thesis, California Institute of Technology (1952). Federat. Proc. (Amer. Soc. exp. Biol.) **10**, 409 (1951).
33. GOODMAN, M., H. R. WOLFE and S. NORTON: Precipitin Production in Chickens. J. Immunology **66**, 225 (1951).
34. GOODNER, K. and F. L. HORSFALL: Properties of Type Specific Proteins of Type I Antipneumococcal Horse and Rabbit Serums. J. exp. Medicine **66**, 425 (1937).
35. GRABAR, P.: Les Globulines du Serum Sanguin. Liège, Belgique: Disoer. 1947.
36. — Immunochemistry. Annu. Rev. Biochem. **19**, 453 (1950).
37. HARKINS, W. D., L. FOURT and P. C. FOURT: Immunochemistry of Catalase. II. Activity of Multilayers. J. biol. Chemistry **132**, 111 (1940).
38. HAUROWITZ, F. und F. BREINL: Chemische Untersuchung der spezifischen Bindung von Arsanil-Eiweiß und Arsanilsäure an Immunserum. Hoppe-Seyler's Z. physiol. Chem. **214**, 111 (1933).
39. HAUROWITZ, F., C. F. CRAMPTON and R. SOWINSKI: Immunochemical Studies with Labelled Antigens. Federat. Proc. (Amer. Soc. exp. Biol.) **10**, 560 (1951).
40. HAUROWITZ, F. and P. SCHWERIN: The Valence of Antibodies and the Structure of Antigen-Antibody Precipitates. Brit. J. exp. Pathol. **23**, 146 (1942).
41. — — The Specificity of Antibodies to Antigens Containing Two Different Determinant Groups. J. Immunology **47**, 111 (1943).
42. HAUROWITZ, F., SH. TEKMAN, M. BILEN and P. SCHWERIN: The Purification of Azoprotein Antibodies by Dissociation of Specific Precipitates. Biochemic. J. **41**, 304 (1947).
43. HAYWORTH, N. and M. STACEY: The Chemistry of Immunopolysaccharides. Annu. Rev. Biochem. **17**, 97 (1948).
44. HEIDELBERGER, M. and F. E. KENDALL: A Quantitative Theory of the Precipitin Reaction. III. The Reaction Between Crystalline Egg Albumin and its Homologous Antibody. J. exp. Medicine **62**, 697 (1935).
45. — — A Quantitative Theory of the Precipitin Reaction. II. A Study of an Azoprotein-Antibody System. J. exp. Medicine **62**, 467 (1935).
46. — — The Precipitin Reaction Between Type III Pneumococcus Polysaccharide and Homologous Antibody. II. Conditions for Quantitative Precipitation of Antibody in Horse Sera. III. A Quantitative Study and a Theory of the Reaction Mechanism. J. exp. Medicine **61**, 559 (1935).
47. — — The Reaction Between Crystalline Egg Albumin and its Homologous Antibody. J. exp. Medicine **62**, 697 (1935).
48. HEIDELBERGER, M., C. M. MACLEOD, S. J. KAISER and B. ROBINSON: Antibody Formation in Volunteers Following Injection of Pneumococci or their Type-Specific Polysaccharides. J. exp. Medicine **83**, 303 (1946).
49. HENRY, J. P.: Quantitative Studies of the Photochemical Despeciation of Horse Serum. J. exp. Medicine **76**, 451 (1942).
50. HERSHEY, A. D.: A Descriptive Theory of Specific Precipitation. I. The Theory. J. Immunology **42**, 455 (1941).
51. — Specific Precipitation IV. Quantitative Application of the Restricted Theory. J. Immunology **45**, 39 (1942).
52. — Specific Precipitation. VI. The Restricted System Bivalent Antigen, Bivalent Antibody, as an Example of Reversible Bifunctional Polymerization. J. Immunology **48**, 381 (1944).
53. HOOKER, S. B.: The Nature of Antibodies. J. Immunology **33**, 57 (1937).

54. HOOKER, S. B. and W. C. BOYD: Tyrosin- and Histidin-diazo-arsonilic Acids as Haptens. J. Immunology 25, 61 (1933).
55. — — The Nonspecificity of the Flocculative Phase of Serologic Aggregation. J. Immunology 33, 337 (1937).
56. HORSFALL, F. L. and K. GOODNER: Lipoids and Immunological Reactions. J. exp. Medicine 62, 485 (1935).
57. KABAT, E. A.: The Molecular Weight of Antibodies. J. exp. Medicine 69, 103 (1939).
58. — Immunochemistry. Progress Sci. Allergy 2, 11 (1949).
59. KABAT, E. A., A. BENDISCH, A. E. BEZER and S. M. BEISER: Immunochemical Studies of Blood Groups. J. exp. Medicine 85, 685 (1947).
60. KABAT, E. A. and M. HEIDELBERGER: The Reaction Between Crystalline Horse Serum Albumin and Antibody Formed in the Rabbit. J. exp. Medicine 66, 229 (1937).
61. KABAT, E. A. and M. M. MAYER: Experimental Immunochemistry, p. 168. Springfield, Ill.: Charles C. Thomas. 1948.
62. KARUSH, F. and B. M. SIEGEL: The Structure of Antigen Films and Long-Range Forces. Science (New York) 108, 107 (1948).
63. KENDALL, F. E.: The Quantitative Relationship Between Antigen and Antibody in the Precipitin Reaction. Ann. New York Acad. Sci. 43, 85 (1942).
64. KLECZKOWSKI, A.: Effect of Heat on Flocculating Antibodies of Rabbit Antisera. Brit. J. exp. Pathol. 22, 192 (1941).
65. KLEINSCHMIDT, W. J.: Inhibition of Eggalbumin-Antieggalbumin Reaction by Amino Acids and Other Organic Residues. Thesis, University of Minnesota, Minneapolis. 1950.
65 a. KREUGER, R. C. and M. HEIDELBERGER: Effect of the Removal ot Lipids on Specitic Precipitation: J. exper. Med. 92, 382 (1950).
66. KUZIN, A. M.: Chemistry of Antigens. Zhur. Mikrobiol. Epidemiol. Immunobiol. 11, 52 (1947) [Chem. Abstr. 43, 7126 (1949)].
67. KUZIN, A. M., S. N. BABADZHANOV and O. I. POLYAKOVA: The Chemical Nature of the Antigens of Some Helminths. Biokhimiya 14, 65 (1949) [Chem. Abstr. 43, 5110 (1949)].
68. LANDSTEINER, K.: Spezifische Serumreaktionen mit einfach zusammengesetzten Substanzen von bekannter Konstitution (organischen Säuren). XIV. Mitt. über Antigene und serologische Spezifität. Biochem. Z. 104, 280 (1920).
69. — Serological Reactivity of Hydrolytic Products from Silk. J. exp. Medicine 75, 269 (1942).
70. — The Specificity of Serological Reactions. Rev. ed. Harvard University Press, Cambridge, Mass. (1945).
71. LANDSTEINER, K. and J. VAN DER SCHEER: Serological Reactions with Simple Chemical Compounds (Precipitin Reactions). J. exp. Medicine 56, 399 (1932).
72. — — Antigens with Azo-Components Containing Two Determinant Groups. J. exp. Medicine 67, 709 (1938).
73. — — On Cross Reactions of Immune Sera to Azoprotein. J. exp. Medicine 63, 325 (1939).
74. LANNI, F.: The Specificity of Serological Precipitation. J. exp. Medicine 84, 167 (1946).
75. — Determinations of Small Quantities of Nitrogen in Serological Precipitates and Other Biological Materials. Proc. Soc. exp. Biol. Med. 74, 4 (1950).
76. LANNI, F. and D. H. CAMPBELL: A Search for Heteroligating Antibody and the Significance of the Results to the Mechanism of Antibody Formation. Stanford med. Bull. 6, 97 (1948).

77. LERMAN, L. S.: Equilibrium Studies on the Reaction of Hapten and Antibody. Federat. Proc. (Amer. Soc. exp. Biol.) **8**, 406 (1949).
78. — Studies on the Reaction of Antibody with Simple Substances. Thesis, California Institute of Technology. 1949.
79. LONDON, F.: Zur Theorie und Systematik der Molekularkräfte. Z. Physik **63**, 245 (1930).
80. LOISELEUR, J.: Sur le mécanism de l'adaptation in vitro d'un protéide à un antigène. C. R. hebd. Séances Acad. Sci. **224**, 687 (1947).
81. LOISELEUR, J. et M. LÉVY: La specificité des anticorps consécutifs à l'injection directe de molécules organiques de faible poids moléculaire. Ann. Inst. Pasteur **73**, 116 (1947).
82. MALKIEL, S. and W. C. BOYD: The Composition of Specific Precipitates in the Region of Antigen Excess. J. exp. Medicine **66**, 383 (1937).
83. MARRACK, J. R.: Chemistry of Antigens and Antibodies. His Majesty's Stationary Office, London, p. 154 (1938).
84. MARRACK, J. R., H. HOCH and R. G. S. JOHNS: Complexes Formed by Antibodies with Excess of Antigen and the Valence of Antibodies. J. gen. Microbiol. **3**, XXVIII (1949).
85. MARRACK, J. R. and F. C. SMITH: Quantitative Aspects of Immunity Reactions: The Combination of Antibodies with Simple Haptens. Brit. J. exp. Pathol. **13**, 394 (1932).
86. MASOUREDIS, S. P., L. R. MELCHER and D. C. KOBLICK: Specificity of Radio-iodinated (I^{131}) Immune Globulin as Determined by Quantitative Precipitin Reaction. J. Immunology **66**, 297 (1951).
87. MELCHER, L. R.: An Antigenic Analysis of *Trichinella spiralis*. J. infect. Diseases **73**, 31 (1943).
88. MILLER, H. and D. H. CAMPBELL: Immunochemical Studies of Reagins. Ann. Allergy **5**, 236 (1947).
89. MORGAN, W. T. J. and W. M. WATKINS: Amino Acid Composition of A-Substance from Hog Mucin. Brit. J exp. Pathol. **25**, 221 (1944).
90. MUDD, S. and D. B. LACKMAN: The Antigenic Structure of Hemolytic Streptococci of the Lancefield Group A. VI. The Immunological Behavior of Neucleoproteins Extracted at Neutrality. J. Immunology **39**, 495 (1940).
91. NEUBERGER, A. and M. E. YUILL: A Note on the Possible Relationship of the Carbohydrate Component of Proteins to their Antigenic Properties. Biochemic. J. **34**, 109 (1940).
92. OLIVER-GONZALES, J.: The Inhibition of Human Isoagglutinins by a Polysaccharide from *Ascaris Suum*. J. infect. Diseases **74**, 81 (1944).
93. — Functional Antigens in Helminths. J. infect. Diseases **78**, 232 (1946).
94. ONCLEY, J. L., G. SCATCHARD and A. BROWN: Physical-Chemical Characteristics of Certain of the Proteins of Normal Human Plasma. J. Phys. Colloid Chem. **51**, 184 (1947).
95. PAPPENHEIMER, A. M.: Anti-Egg Albumin Antibody in the Horse. J. exp. Medicine **71**, 263 (1940).
96. PAPPENHEIMER, A. M., H. P. LUNDGREN and J. W. WILLIAMS: Studies on the Molecular Weight of Diphtheria Toxin, Antitoxin and their Reaction Products. J. exp. Medicine **71**, 247 (1940).
97. PARDEE, A. B. and L. PAULING: The Serological Properties of Simple Substances. XIV. The Reaction of Simple Antigens with Purified Antibody. J. Amer. chem. Soc. **71**, 143 (1949).
98. PARDEE, A. B. and S. M. SWINGLE: The Degree of Association of Some Simple Antigens. J. Amer. chem. Soc. **71**, 148 (1949).

99. PAULING, L.: The Nature of the Chemical Bond, 2nd ed. Ithaca: Cornell University Press. 1940.
100. — A Theory of the Structure and Process of Formation of Antibodies. J. Amer. chem. Soc. **62**, 2643 (1940).
101. PAULING, L. and D. H. CAMPBELL: The Manufacture of Antibodies in Vitro. J. exp. Medicine **76**, 211 (1942).
102. PAULING, L., D. H. CAMPBELL and D. PRESSMAN: The Nature of Forces Between Antigen and Antibody and of the Precipitin Reaction. Physiologic. Rev. **23**, 203 (1943).
103. PAULING, L. and D. PRESSMAN: The Serological Properties of Simple Substances. IX. Hapten Inhibition of Precipitation of Antisera Homologous to the o-, m-, and p-Azophenylarsonic Acid Groups. J. Amer. chem. Soc. **67**, 1003 (1945).
104. PAULING, L., D. PRESSMAN and D. H. CAMPBELL: The Precipitation of a Mixture of Two Specific Antisera by a Dihaptenic Substance Containing the Two Corresponding Haptenic Groups; Evidence for the Framework Theory of Serological Precipitation. J. Amer. chem. Soc. **66**, 330 (1944).
105. PAULING, L., D. PRESSMAN, D. H. CAMPBELL and C. IKEDA: The Serological Properties of Simple Substances. II. The Effects of Changed Conditions and of Added Haptens on Precipitation Reactions of Polyhaptenic Simple Substances. J. Amer. chem. Soc. **64**, 3003 (1942).
106. PAULING, L., D. PRESSMAN and A. L. GROSSBERG: The Serological Properties of Simple Substances. VII. A Quantitative Theory of the Inhibition by Haptens of the Precipitation of Heterogeneous Antisera with Antigens and Comparison with Experimental Results for Polyhaptenic Simple Substances and for Azoproteins. J. Amer. chem. Soc. **66**, 784 (1944).
107. PAULING, L., D. PRESSMAN and C. IKEDA: The Serological Properties of Simple Substances. III. The Composition of Precipitates of Antibodies and Polyhaptenic Simple Substances; the Valence of Antibodies. J. Amer. chem. Soc. **64**, 3010 (1942).
108. PEDERSEN, K. O.: Ultracentrifugal Studies on Serum and Serum Fractions. Upsala: Almquist & Wiksell. 1945.
109. PETERMAN, M. L. and A. M. PAPPENHEIMER: The Ultracentrifugal Analysis of Diphtheria Proteins. J. physic. Chem. **45**, 1 (1941).
110. PORTER, E. F. and A. M. PAPPENHEIMER: Antigen-Antibody Reactions Between Layers Adsorbed on Built-up Stearate Films. J. exp. Medicine **69**, 755 (1939).
111. PORTER, R. R.: A Chemical Study of Rabbit Antiovalbumin. Biochemic. J. **46**, 473 (1950).
112. — The Formation of Specific Inhibitor by Hydrolysis of Rabbit Antiovalbumin. Biochemic. J. **46**, 479 (1950).
113. PRESSMAN, D., D. H. BROWN and L. PAULING: The Serological Properties of Simple Substances. IV. Hapten Inhibition of Precipitation of Antibodies and Polyhaptenic Simple Substances. J. Amer. chem. Soc. **64**, 3015 (1942).
114. PRESSMAN, D., A. L. GROSSBERG, L. H. PENCE and L. PAULING: The Serological Proporties of Simple Substances. XII. The Reactions of Antiserum Homologous to the p-Azophenyltrimethylammonium Group. J. Amer. chem. Soc. **68**, 250 (1946).
115. RACE, R. R.: An "Incomplete" Antibody in Human Serum. Nature (London) **153**, 771 (1944).
116. RECORD, B. R. and M. STACEY: Some Physical Properties of the Specific Polysaccharides from the Types I, II and III Pneumococcus. J. chem. Soc. (London) **1948**, 1561.

117. RIMINGTON, C.: The Chemistry of the Proteins and Amino Acids. Annu. Rev. Biochem. 5, 138 (1936).
118. ROTHEN, A.: Immunological Reactions Between Films of Antigen and Antibody Molecules. J. biol. Chemistry **168**, 75 (1947).
119. — Long Range Enzymatic Action on Films of Antigen. J. Amer. chem. Soc. **70**, 2732 (1948).
120. ROTHEN, A. and K. LANDSTEINER: Serological Reactions of Protein Films and Denatured Proteins. J. exp. Medicine **76**, 437 (1942).
121. SINGER, S. J.: On the Unlikelihood of Specific Long Range Forces in Immunologic and Enzymatic Reactions. J. biol. Chemistry **182**, 189 (1950).
122. SINGER, S. J. and D. H. CAMPBELL: The Valence of Precipitating Antibody. J. Amer. chem. Soc. **73**, 3543 (1951).
123. — — Physical Chemical Studies of Antigen-Antibody Complexes. I. The Valence of Precipitating Rabbit Antibody. J. Amer. chem. Soc. **74**, 1794 (1952).
124. SMITH, E. L. and R. D. GREENE: Further Studies on the Amino Acid Composition of Immune Proteins. J. biol. Chemistry **171**, 355 (1947).
125. SMITH, E. L., R. D. GREENE and E. BARTNER: Amino Acid and Carbohydrate Analysis of some Immune Proteins. J. biol. Chemistry **164**, 359 (1949).
126. STACEY, M.: Aspects of Immunochemistry. Quart. Revs. (London) **1**, 179, 213 (1947).
127. STACEY, M. and C. R. RICKETTS: Bacterial Dextrans. Fortschr. Chem. organ. Naturstoffe **8**, 28 (1951).
128. STERNBERGER, L. A. and D. PRESSMAN: A General Method for the Specific Purification of Antiprotein Antibodies. J. Immunology **65**, 65 (1950).
129. STOKINGER, H. E. and M. HEIDELBERGER: The Reaction Between Mammalian Thyroglobulin and Antibodies to Homologous and Heterologous Preparations. J. exp. Medicine **66**, 251 (1937).
130. SVEDBERG, T. and K. O. PEDERSEN: The Ultracentrifuge, p. 41, Oxford University Press. 1940.
131. TEORELL, T.: Quantitative Aspects of Antigen-Antibody Reactions. J. Hygiene **44**, 227 (1946).
132. — A Quantitative Theory of the Precipitin Reaction. Nature (London) **151**, 696 (1943).
133. TREFFERS, H. P.: Immunochemistry. Ann. Rev. Microbiol. **1**, 263 (1947).
134. TYLER, A.: Properties of Fertilizin and Related Substances of Eggs and Sperm of Marine Animals. Amer. Naturalist **83**, 195 (1949).
135. TYLER, A. and C. B. METZ: Natural Heteroagglutinins in the Serum of the Spiny Lobster. J. exp. Zoology **100**, 387 (1945).
136. WAKSMAN, B. H. and H. L. MASON: The Antigenicity of Collagen. J. Immunology **63**, 427 (1949).
137. WILLIAMS, J. W.: Some Recent Developments in the Chemistry of Antibodies. Fortschr. Chem. organ. Naturstoffe **7**, 270 (1950).

(Received, January 31, 1952.)

Namenverzeichnis. Index of Names. Index des Auteurs.

ABDOU, I. A. 298, 352.
ABE, N. 194, 213, 220.
ABRAHAMSEN, N. 447, 449.
ABRAMS, A. 376, 382, 395.
ABRAMS, R. 310, 352, 393, 394, 395.
ACKERMANN, W. W. 100, 112, 389, 400.
ACREE, F., Jr. 254, 286.
ADAMS, R. 282, 285.
AKABORI, S. 119, 170.
ALBERTY, R. A. 453, 479.
ALDAG, J. 12, 19, 38, 39.
ALLBRIGHT, F., Jr. 306, 358.
ALLFREY, V. 367, 400.
AMES, S. R. 66, 70, 82.
ANDREAE, W. A. 245, 285.
ANDREWS, L. J. 46, 48, 87.
ANET, L. 250, 281, 285.
ANFINSEN, C. B. 294, 299, 310, 312, 345, 348, 349, 350, 352, 358.
ANGIER, R. B. 338, 352.
ANSCHÜTZ, R. 257.
ANTOPOL, W. 95, 113.
AQUILONIUS, L. 305, 315, 353.
ARENS, J. F. 43, 52, 53, 54, 56, 57, 65, 71, 72, 73, 75, 79, 80, 81, 82, 87.
ARIMA, J. 272, 285.
ARMITAGE, J. B. 12, 38.
ARNDT, F. 234, 285.
ARVIDSON, H. 392, 395, 416, 433.
ASADA, T. 190, 222.
ASAHINA, Y. 209, 213, 214.
ASAI, T. 241, 285.
ASODA, A. 213, 220.
AUHAGEN, E. 102, 107.
AVAKIAN, S. 96, 98, 110.
AVERY, O. T. 459, 478.
AXELROD, A. E. 100, 109.
AYÇA, E. 234, 285.

BABADZHANOV, S. N. 447, 481.
BACHARACH, A. L. 42, 81.
BADDILEY, J. 422, 434.
BADIN, J. 106, 107, 109.
BAETCKE, E. 262, 291.

BAKER, B. R. 282, 285.
BALAIAH, V. 237, 285.
BALIS, M. E. 392, 395, 399, 414, 434.
BALL, E. G. 366, 395.
BALL, S. 43, 52, 56, 71, 75, 81.
BANKS, M. G. 414, 423, 424, 434.
BANKS, T. E. 458, 478.
BARANOVSKI, T. 390, 399.
BARDOS, T. J. 420, 431, 432, 434, 436, 440.
BARGELLINI, G. 231, 248, 285.
BARGER, G. 117, 118, 119, 121, 134, 136, 144, 156, 157, 170.
BARKER, H. A. 339, 352, 370, 397, 425, 435.
BARKER, S. A. 422, 434.
BARON, F. 384, 400, 422, 441.
BARRON, E. S. G. 310, 352.
BARRY, J. M. 414, 423, 424, 434.
BARTH, L. 243.
BARTLETT, M. N. 306, 357.
BARTNER, E. 452, 484.
BARTRAM, K. 19, 22, 38, 54, 83.
BARUA, R. K. 73, 81.
BATTY, J. W. 35, 36, 38.
BAUHIN 116.
BAUMANN, C. A. 297, 314, 359, 420, 431, 440.
BAUMGARTEN, P. 90, 107.
BAWDEN, F. C. 460, 478.
BAXTER, J. G. 38, 41, 42, 43, 46, 49, 51, 53, 66, 70, 73, 74, 81, 86.
BAYERLE, H. 242, 285.
BEALL, B. 306, 357.
BEAVEN, G. H. 419, 420, 434, 435.
BECKER, B. 135, 136, 144, 173.
BEERSTECHER, E., Jr. 417, 421, 442.
BEHNKE, J. 476, 478.
BEHRE, J. A. 119, 170.
BEHRENS, O. K. 326, 352.
BEILER, J. M. 97, 107, 420, 434.
BEIN, M. L. 91, 112.
BEISER, S. M. 448, 481.
BELIČ, I. 19, 22, 40, 80, 86.

BELL, J. C. 277, 278, 285.
BELOFF, A. 310, 311, 352.
BELT, M. 418, 437.
BENDICH, A. 413, 414, 434, 435, 448, 481.
BENEDICT, S. R. 119, 172, 365, 370, 396.
BEN-ISHAI, R. 415, 434.
BENNETT, E. L. 448, 478.
BENNETT, L. L. 415, 439.
BENZ, J. 72, 78, 84.
BERGMAN, E. D. 415, 434.
BERGMANN, M. 136, 138, 170, 309, 322, 323, 325, 326, 328, 333, 334, 352.
BERGSTRÖM, S. 392, 395, 416, 433, 440.
BERT, L. 236, 285.
BESANCON, A. P. 41.
BEST, R. J. 240, 245, 285.
BEZER, A. E. 448, 481.
BICK, I. R. C. 197, 198, 205, 214.
BIELSCHOWSKY, F. 365, 395.
BIGINELLI, P. 247.
BILEN, M. 462, 480.
BILLET, D. 89, 110.
BIRD, O. D. 409, 420, 434, 441.
BLAKER, R. H. 453, 462, 479.
BLANCHARD, M. 358.
BLANKS, F. R. 250, 285.
BLOCH, E. 43, 46, 87.
BLOCH, H. 103, 108.
BLOCH, K. 332, 341, 342, 343, 352, 356.
BLOOM, S. 305, 355.
BLUM, W. P. 41, 60.
BODANSKY, O. 297, 352.
BOEHM, R. 200, 202, 214.
BOGGIANO, E. 410, 437.
BOHLMANN, F. 12, 13, 19, 20, 22, 38, 43, 54, 56, 57, 80, 83.
BOHLMANN, M. 56, 57, 83.
BÖHME, H. 252, 285.
BOLDINGH, J. 74, 81.
BOND, A. C., Jr. 43, 82.
BOND, T. J. 420, 431, 432, 434, 436, 440.
BOOTHE, J. H. 338, 352.
BOREK, E. 335, 345, 355, 361.
BORK, S. 12, 38.
BORSOOK, H. 292, 294, 295, 296, 298, 299, 303, 304, 305, 308, 310, 311, 313, 314, 315, 316, 317, 318, 319, 320, 321, 327, 333, 337, 338, 343, 344, 345, 346, 348, 351, 353, 361.
BOSE, P. K. 242, 245, 271, 278, 280, 285, 286, 288.
Boss (Mlle.) 98, 107, 112.

BOSSHARDT, D. K. 412, 417, 442.
BOTTOMLEY, W. 254, 278, 286.
BOURNE, E. J. 422, 434.
BOVET, D. 100, 112.
BOYD, J. 237, 286.
BOYD, W. C. 446, 456, 457, 464, 465, 466, 468, 475, 476, 478, 481, 482.
BRADBURY, T. 74, 81.
BRADY, T. G. 365, 379, 396.
BRACHET, J. 305, 315, 353.
BRAGG, W. L. 138, 170.
BRANDT, B. J. 52, 65, 81.
BRANDT, K. 305, 353.
BRAUDE, R. 74, 81.
BREINL, F. 467, 480.
BRENNER, M. 333, 334, 353.
BRETSCHER, E. 76, 84.
BREWER, C. R. 406, 434.
BRIDGE, W. 277, 285.
BRINK, N. G. 417, 419, 434, 436.
BRINKMAN, J. H. 66, 70, 71, 76, 82, 87.
BROCKMAN, J. A., Jr. 101, 108, 419, 420, 431, 434, 441.
BROQUIST, H. P. 101, 108, 419, 420, 424, 431, 432, 434, 435, 437, 441.
BROWN, A. 454, 482.
BROWN, D. H. 448, 458, 468, 478, 483.
BROWN, G. B. 98, 110, 379, 391, 392, 394, 395, 396, 397, 398, 399, 408, 413, 414, 434, 435, 437.
BROWN, G. M. 409, 441.
BROWN, R. A. 453, 454, 459, 479.
BRUCHHAUSEN, F. V. 177, 181, 182, 193, 194, 196, 197, 207, 208, 209, 214, 215, 274, 286.
BRUCK, J. 254, 288.
BUCHANAN, J. M. 391, 396.
BUCHANAN, J. G. 417, 435.
BUHS, R. P. 419, 420, 435.
BULL, H. 450, 479.
BULMAN, N. 443.
BURAWOY, A. 35, 36, 38.
BURCKHARDT, E. 118, 120, 122, 149, 151, 159, 173.
BURNET, F. M. 445, 479.
BURROWS, H. 245, 291.
BUSHRA, A. 261, 286.
BUTENANDT, A. 254, 267, 286.
BUU-HOÏ, NG. PH. 105, 110.

CAGNIANT, P. 105, 110.
CAHILL, J. J. 417, 434.

CAHNMANN, H. J. 48, 49, 50, 56, 57, 86.
CALDWELL, A. G. 249, 286.
CAMA, H. R. 74, 81.
CAMPBELL, D. H. 443, 445, 447, 448, 451, 453, 457, 458, 459, 460, 462, 463, 467, 468, 471, 476, 479, 481, 482, 483, 484.
CAMPBELL, H. A. 242, 257, 286.
CAMPBELL, M. A. 58, 62, 86.
CANN, J. R. 453, 454, 459, 461, 479.
CANNON, P. R. 316, 353.
CANTER, F. W. 229, 286.
CAPELL, L. T. 147, 172.
CAPUTTO, R. 376, 390, 396.
CARDINI, C. E. 376, 396.
CARLO, F. J. DI 408, 435.
CARLSON, W. E. 92, 109.
CARR, F. H. 117, 118, 121, 156, 157, 170.
CARRÉ-CHENAVIER, P. 104, 108.
CARTER, C. E. 375, 396, 411, 435.
CASPARIS, P. 256.
CASPERRSON, T. 305, 315, 353, 354.
CAVALIERI, L. F. 391, 396.
CAWLEY, J. D. 41, 43, 46, 50, 51, 53, 56, 70, 71, 72, 75, 76, 77, 78, 79, 81, 82, 87.
CERECEDO, L. R. 91, 112.
CHANG, K. J. 207, 214.
CHANTRENNE, H. 339, 354.
CHARAUX, C. 243.
CHARGAFF, E. 380, 396.
CHARKOFF, I. L. 308, 354.
CHASSAR MOIR, J. 118, 120, 122, 149, 170, 171.
CHATAIN, H. 70, 73, 81.
CHATTERJEE, A. 272, 286.
CHATTERJEE, R. 196, 214.
CHAUDHRY, G. R. 207, 214.
CHAUDHURY, D. N. 245, 286.
CHAUDHURY, J. C. 271, 285.
CHECHAK, A. J. 41.
CHEESEMAN, G. W. H. 60, 61, 70, 71, 81.
CHELDELIN, V. H. 409, 411, 438, 439.
CHENAVIER, P. 103, 104, 108, 110.
CHEVREUIL, M. E. 283.
CHOUKROUN, J. 106, 109.
CHRISTIAN, W. 377, 400.
CHRISTIANI, A. v. 247, 267, 268, 286, 288.
CHUANG, C. K. 207, 214.
CLAPP, R. C. 99, 108.
CLARK, D. A. 379, 396.
CLARK, I. 96, 97, 111.
CLARK, J. H. 282, 285.
CLARK V. M. 388, 395, 396.

CLOUGH, F. B. 41, 46, 70, 73, 81, 87.
COBURN, A. F. 358.
COHEN, P. P. 338, 339, 345, 354, 357.
COHEN, S. S. 315, 354, 375, 377, 396, 399, 423, 441.
COHN, M. 376, 400.
COHN, W. E. 411, 424, 429, 435, 441.
COLE, Q. P. 99, 108.
COLLIER, H. B. 327, 354.
COLLINS, F. D. 74, 81.
COLOWICK, S. P. 376, 382, 396, 408, 409, 435.
CONN, J. B. 420, 435, 468, 478.
CONGER, T. W. 75, 82.
CONSULICH, D. B. 101, 108, 338, 352, 420, 431, 434.
COOK, E. S. 405, 435.
COOLEY, G. 419, 420, 435.
COOMBES, A. I. 93, 112.
COOPERMAN, J. M. 420, 435.
CORBET, R. E. 42, 83.
CORDDRY, E. I. 96, 109.
CORI, C. F. 367, 376, 382, 396, 400.
CORI, G. T. 367, 376, 382, 396.
CORMIER, M. 103, 108.
COTZIAS, G. C. 305, 354.
COVO, G. A. 346, 354.
CRAIG, J. A. 409, 441.
CRAIG, L. C. 118, 122, 123, 124, 134, 136, 137, 151, 167, 168, 170, 171.
CRAIG, L. E. 214.
CRAMPTON, C. F. 467, 480.
CRAVENS, W. W. 97, 108.
CRISTOL, S. J. 46, 48, 87.
CRONIN, S. A. G. 405, 435.
CROSS, R. J. 346, 354.
CUNNINGHAM, L. 298, 305, 314, 354, 355.
CUSHING, J., Jr. 460, 479.
CUTHBERTSON, W. F. J. 430, 435.
CYMERMAN, J. 17, 38.

DAKIN, H. D. 119, 172.
DALE, H. 119, 149, 170.
DALGLIESH, C. E. 89, 108.
DAM, H. 104, 106, 108.
DAMODARAN, M. 316, 362.
DANDLIKER, W. 447, 479.
DAVIDSON, J. N. 311, 315, 354.
DAVIES, W. H. 52, 81.
DAVIS, A. R. 365, 370, 396.
DAVIS, B. D. 101, 108, 418, 419, 429, 435.
DAVIS, P. 19, 22, 40, 80, 86.
DAVOLL, J. 365, 379, 396.

Day, H. G. 91, 108.
Day, P. L. 311, 360.
Dean, F. M. 225.
Deasy, C. L. 298, 303, 304, 305, 308, 310, 311, 313, 314, 315, 316, 317, 318, 333, 343, 344, 345, 346, 348, 351, 353.
Debodard, M. 70, 73, 81.
Delis, D. 136, 170.
Dellweg, H. 365, 400.
Delor, J. 105, 109.
Delwiche, C. C. 335, 354.
Demange, G. 63, 65, 82.
Demmer, E. 247, 248, 291.
Dempster, E. 358.
Denis, W. 294, 354.
Denkewalter, R. G. 420, 437.
Deuel, H. J., Jr. 26, 33, 40, 82.
Deutsch, H. F. 454, 479.
Deutsch, W. 372, 396.
Devine, J. 82.
Dewey, V. C. 411, 415, 416, 421, 437.
Dey, B. B. 251, 286, 288.
Dialer, K. 58, 60, 61, 71, 74, 75, 84.
Diamant, E. 257, 288.
Di Carlo, F. J. 408, 435.
Dickens, F. 377, 396.
Diels, O. 147, 170.
Dieterle, H. 280, 286.
Dietrich, H. 199, 215.
Dijk, G. van 81, 52, 65.
Dimroth, K. 365, 396.
Dinjaški, K. 291.
Dische, Z. 365, 375, 376, 377, 396.
Dittmer, K. 98, 99, 108, 110, 413, 435.
Doblhammer, F. 199, 202, 215.
Dobrovolny, E. 248, 266, 280, 288, 289.
Dobry, A. 331.
Doerr, R. 445, 479.
Doi, K. 199, 223.
Dole, V. P. 305, 354.
Dolphin, J. 306, 357.
Domagk, G. 100, 108.
Dornow, A. 90, 103.
Dorp, D. A. van 43, 52, 53, 54, 56, 57, 65, 71, 72, 73, 75, 79, 80, 81, 82, 87.
Doudoroff, M. 367, 370, 396, 397, 425, 435.
Drell, W. 392, 399, 409, 410, 416, 435, 438.
Dreyfus, J.-C. 294, 356.
Drucker, R. 420, 435.
Drummond, J. C. 42, 82.

Drury, A. N. 378, 397.
Dubnoff, J. W. 295, 296, 320, 321, 337, 338, 345, 358.
Dudley, H. W. 118, 120, 122, 149, 171.
Dulou, R. 78, 85.
Dutcher, J. D. 200, 202, 203, 215.
Dutta, P. 252, 286.
Du Vigneaud, V. 98, 99, 108, 110.

Eagles, B. A. 119, 171.
Eakin, R. E. 100, 112, 389, 400, 417, 420, 421, 440, 441, 442.
Ebnöther, A. 274, 288.
Ecker, P. G. E. 327, 354.
Eddinger, C. C. 41, 52, 86.
Edelhausen, J. H. 420, 441.
Edgerton, R. O. 58, 86.
Edwards, R. R. 311, 360.
Ehrlich, H. 240.
Ehrlich, P. 455, 459.
Eijkmann, J. F. 243.
Eisen, H. W. 458, 467, 479.
Eiter, K. 227, 278, 289.
Elderfield, R. C. 90, 112.
Eliasson, N. A. 392, 395, 416, 433.
Elion, G. B. 395, 407, 413, 414, 434, 435, 437.
Elliott, D. F. 300, 337, 354.
Elliott, W. H. 305, 354.
Ellis, B. 419, 420, 434, 435.
El Ridi, M. S. 5.
Elvehjem, C. A. 93, 112, 297, 299, 314, 357, 359, 361.
Embree, N. D. 41, 43, 46, 53, 70, 72, 73, 77, 78, 79, 81, 82, 87.
Emerson, G. 417, 419, 436.
Emerson, G. A. 96, 108.
Engeland, R. 119, 171.
English, J. P. 99, 108.
Entenman, C. 308, 354.
Ericksen, J. O. 460, 479.
Estborn, B. 393, 399, 402, 423, 440.
Eugster, C. H. 19, 28, 31, 35, 39, 80, 84.
Euler, H. v. 72, 78, 82.
Evans, C. A. 92, 109.
Evans, H. M. 97, 111, 357.
Ewen, E. S. 198, 214.
Ewing, J. 250, 286.
Ewins, A. J. 119, 134, 144, 170, 171.

Fahrenbach, M. J. 101, 108, 338, 352, 420, 431, 434.
Fairley, J. L., Jr. 407, 436.

FALCO, E. A. 413, 437.
FALCONER, R. 365, 397.
FALTIS, F. 176, 177, 181, 195, 196, 199, 202, 203, 209, 215.
FARBER, E. 348, 359.
FARRAR, K. R. 71, 76, 78, 82.
FAUST, M. 35, 39.
FEIGELSON, F. 299, 357.
FEIGEN, G. 447, 479.
FELDHAUS, A. 182, 193, 194, 196, 207, 214.
FELIX, E. L. 106, 112.
FENNER, F. 445, 479.
FERRANDO, R. 71, 75, 76, 85, 93, 102, 103, 108, 110.
FIESER, L. F. 76, 82.
FINHOLT, A. E. 43, 82.
FINKELSTEIN, J. 187, 215.
FISCHER, R. 240.
FLAVIN, M. 414, 436.
FLEŠ, D. A. 19, 22, 40, 80, 86.
FLYNN, E. H. 420, 429, 431, 436, 439.
FLYNN, R. M. 408, 409, 439.
FOLIN, O. 294, 354.
FOLKERS, K. 417, 419, 420, 434, 436, 437.
FOLLEY, S. J. 327, 354.
FOOT, A. S. 74, 81.
FOOTE, M. W. 105, 110.
FORKER, L. L. 308, 354.
FORREST, H. S. 384, 397.
FOSTER, C. 91, 109.
FOSTER, J. C. 419, 442.
FOSTER, J. W. 441.
FOSTER, R. T. 261, 286.
FOURT, L. 460, 480.
FOURT, P. C. 460, 480.
FOX, C. L., Jr. 389, 400.
FOX, J. J. 413, 436.
FRAENKEL-CONRAT, H. 323, 325, 328, 352.
FRANCIS, G. E. 458, 478.
FRANKEL, S. 97, 111.
FRÄNKEL, S. 119, 171.
FREED, A. M. 93, 94, 113.
FRENCH, D. 422, 425, 439.
FRESCO, J. R. 423, 424, 436.
FRIEDBERG, F. 304, 305, 306, 307, 310, 345, 350, 355, 361.
FRIEDKIN, M. 367, 369, 370, 372, 382, 383, 385, 386, 397, 415, 422, 436.
FRIEDMAN, T. B. 453, 479.
FRIES, N. 407, 408, 409, 411, 413, 415, 436.
FROHMAN, C. E. 91, 108.

FRUTON, J. S. 325, 326, 329, 330, 331, 332, 333, 334, 341, 343, 352, 355, 356, 359.
FUJII, M. 207, 217.
FUJITA, E. 184, 193, 194, 196, 198, 199, 200, 208, 215, 222.
FUKUHARA, T. K. 412, 417, 436, 440.
FULD, M. 340, 359.
FURST, S. S. 392, 397, 413, 414, 434.
FUSE, S. 213, 220.
FRANKLIN, A. L. 418, 437.
FRANTZ, I. D., Jr. 305, 311, 315, 332, 354, 355, 362.
FRAUENDORFER, H. 181, 196, 199, 215.
FRAZIER, L. J. 316, 353.

GADAMER, J. 196, 215.
GAIND, K. N. 274, 286.
GAJDUSEK, D. C. 453, 479.
GALINOVSKY, F. 230, 289.
GANDINI, A. 247, 286.
GANGULY, J. 26, 33, 40.
GARZA, H. M. 193, 220.
GATCHELL, H. D. 316, 355.
GAVARRET, J. 63, 65, 82.
GEIGER, A. 76, 84.
GEIGER, E. 316, 355, 359.
GELLHORN, A. 96, 109.
GERICKE, P. H. 182, 193, 196, 214.
GETZENDANER, M. E. 100, 112, 389, 400.
GEYR, J. 240.
GHOSH, B. N. 475, 479.
GILLAM, A. E. 5, 43, 76, 82.
GINGRICH, W. 405, 436.
GLADDING, J. K. 58, 62, 86.
GLADSTONE, G. P. 406, 436.
GLEY, P. 105, 109.
GLICK, D. 46, 87.
GLOVER, J. 56, 75, 82.
GOEBEL, W. F. 450, 459, 478, 480.
GOLDBERG, R. J. 476, 480.
GOLDINGER, J. M. 310, 352, 393, 395.
GOLDWASSER, E. 394, 397.
GOLSE, R. 63, 65, 82.
GOODMAN, I. 413, 435, 436.
GOODMAN, M. 451, 480.
GOODNER, K. 452, 460, 480, 481.
GOODWIN, R. H. 240, 241, 245, 286.
GOODWIN, T. W. 43, 52, 54, 56, 71, 75, 81, 82.
GORDON, E. 106, 109.
GORDON, M. 100, 112, 389, 400.
GOSS, G. C. 80, 82.

GOSTING, L. J. 454, 479.
GOTO, K. 176, 188, 189, 191, 207, 215, 216.
GOULD, R. G., Jr. 50, 82, 118, 124, 131, 170, 171.
GOUREVITCH, M. 71, 72, 79, 85.
GRABAR, P. 445, 450, 480.
GRAFF, S. 414, 436.
GRAHAM, W. 54, 56, 57, 71, 73, 75, 82.
GRATZL, K. 107, 109.
GRAY, E. LeB. 76, 82.
GREGG, D. C. 468, 478.
GREGORY, J. D. 408, 409, 439.
GREEN, A. A. 367, 396.
GREEN, D. E. 340, 346, 354, 358, 359.
GREEN, R. G. 92, 109.
GREENBERG, D. M. 304, 305, 306, 307, 310, 311, 316, 345, 350, 355, 358, 359, 361, 362.
GREENBERG, G. R. 384, 391, 397, 415, 436.
GREENBERG, S. M. 26, 33, 40.
GREENE, C. H. 296, 355.
GREENE, R. D. 452, 484.
GRIDGEMAN, N. T. 74, 81.
GRIFFIN, A. C. 298, 305, 314, 354, 355.
GRIFFIN, P. J. 406, 436.
GROB, E. C. 107, 112.
GROSSBERG, A. L. 458, 468, 474, 483.
GROSSI, F. X. 58, 60, 62, 86.
GROSSOWICZ, N. 335, 354, 355, 359, 361.
GRUBER, W. 229, 286.
GUÉRRILLOT-VINET, A. 71, 72, 79, 85, 106, 107, 109.
GUEX, W. 58, 60, 61, 71, 74, 75, 84.
GUHA, M. P. 196, 214.
GUHA, N. C. 278, 285, 288.
GUILLOUD, M. 100, 112, 410, 440.
GULLAND, J. M. 365, 397.
GUPTA, I. S. 274, 286.
GUTHNECK, B. T. 298, 359.
GYÖRGY, P. 99, 108, 429, 441.

HAAG, J. R. 93, 94, 113.
HAAGEN-SMIT, A. J. 298, 303, 304, 305, 308, 310, 311, 313, 314, 315, 316, 317, 318, 333, 343, 344, 345, 346, 348, 351, 353.
HAAN, P. G. DE 100, 113.
HAGEMANN, W. 245, 287.
HAGERTY, E. B. 316, 355.
HAKALA, M. 327, 361.
HALLER, H. L. 254, 286.
HALVERSTADT, I. F. 99, 108.

HAMADA, Z. 213, 220.
HAMANO, S. 42, 82.
HAMLET, J. C. 71, 76, 78, 82.
HAMM, J. H. 306, 357.
HAMMARSTEN, E. 392, 393, 394, 397, 402, 414, 416, 422, 433, 436.
HANES, C. S. 331, 332, 333, 336, 343, 355, 382, 397.
HANZE, A. R. 75, 82.
HARDING, W. M. 407, 420, 440.
HARINGTON, C. R. 119, 171.
HARKINS, W. D. 460, 480.
HARKNESS, D. M. 298, 314, 358, 359.
HARPER, S. H. 35, 36, 38, 43, 48, 52, 63, 66, 82.
HARRINGTON, T. M. 43, 86.
HARRIS, P. L. 41, 66, 70, 82.
HARRIS, R. S. 316, 356.
HARRIS, S. A. 90, 113.
HARRISON, D. C. 119, 172.
HARTE, R. A. 316, 355.
HASEGAWA, S. 213, 216.
HASHIMOTO, T. 213, 216.
HASSID, W. Z. 370, 397, 425, 435.
HASTINGS, A. B. 310, 311, 352.
HATTORI, S. 246, 286.
HAUGAARD, G. 321, 355.
HAUROWITZ, F. 138, 171, 458, 462, 467, 468, 480.
HAYWORTH, N. 447, 480.
HEAD, F. S. H. 244, 245, 287.
HECZKO, T. 199, 215.
HEIDELBERGER, C. 305, 357, 360, 392, 398.
HEIDELBERGER, M. 302, 359, 444, 447, 452, 464, 475, 480, 481, 484.
HEILBRON, I. 1, 2, 10, 17, 35, 36, 38, 42, 43, 52, 53, 54, 56, 58, 60, 61, 70, 71, 72, 74, 76, 79, 81, 82, 83.
HEINRICH, M. R. 415, 416, 421, 437.
HEINZEL, D. 365, 396.
HENBEST, H. B. 71, 76, 77, 78, 82, 83.
HENDERSON, R. 316, 356.
HENDLIN, D. 419, 420, 427, 437, 438, 441.
HENLE, W. 91, 109.
HENRY, J. P. 449, 480.
HENRY, K. M. 74, 81.
HENRY, T. A. 178, 216.
HEPPEL, L. A. 365, 375, 397.
HERBST, E. J. 406, 409, 411, 423, 434, 437.
HERBST, R. M. 339, 356.
HERRING, D. A. 214, 219.

HERSHEY, A. D. 475, 476, 480.
HERVEY, A. 420, 440.
HERZIG, J. 284, 287.
HERZIG, L. 243.
HERZOG, J. 253, 267.
HESLOP, R. N. 42, 82.
HESSE, O. 193, 196, 216.
HEYES, R. G. 246, 287.
HEYL, D. 417, 436.
HICKMAN, K. C. D. 42, 83.
HILLEL, R. 277, 278, 289.
HILLS, G. M. 406, 437.
HILMOE, R. J. 365, 397.
HINDLEY, N. C. 58, 60, 61, 71, 74, 75, 84.
HIRAOKA, S. 213, 220.
HIRD, F. J. R. 331, 332, 333, 336, 355.
HITCHCOCK, C. R. 356.
HITCHCOCK, D. 322, 356.
HITCHINGS, G. H. 395, 407, 413, 414, 434, 435, 437.
HOBERMAN, H. D. 308, 356.
HOCH, H. 458, 471, 473, 474, 476, 482.
HOCHBERG, M. 93, 110.
HOEVEL, B. 420, 434.
HOFF-JÖRGENSEN, E. 370, 380, 386, 388, 394, 397, 398, 420, 423, 424, 425, 437.
HOFFMANN, C. 247, 286.
HOFFMANN, C. E. 371, 397, 418, 437.
HOFFMANN, H. 274, 286.
HOFFMANN-OSTENHOF, O. 107, 109.
HOFMANN, A. 117, 118, 119, 120, 122, 123, 126, 128, 131, 134, 135, 136, 137, 139, 142, 144, 149, 152, 153, 155, 157, 161, 162, 163, 164, 166, 167, 168, 169, 173.
HOFMANN, K. 98, 100, 109, 110.
HOGEBOOM, G. H. 305, 356.
HOLIDAY, E. R. 419, 420, 434, 435.
HOLLAND, A. 420, 437.
HOLLAND, R. A. 245, 286.
HOLLOWAY, B. J. 305, 315, 356.
HOLLY, F. W. 417, 434, 436.
HOLMES, H. N. 42, 83.
HOLZEN, H. 256, 267, 268, 269, 289.
HOLZINGER, L. 199, 202, 203, 215.
HONDA, I. 213, 220.
HOOKER, S. B. 455, 456, 465, 480, 481.
HOOPS, L. 257, 287.
HORECKER, B. L. 375, 377, 378, 397.
HORIUCHI, K. 213, 216.
HORNING, E. C. 261, 287.
HORSFALL, F. L. 452, 460, 480, 481.
HORST, W. P. TER 106, 112.

HOTTINGER, A. 103, 108.
HOTTLE, G. A. 406, 439.
HOULAHAN, M. B. 414, 415, 439.
HOWE, A. F. 406, 434.
HOWELL, W. N. 262, 287.
HSING, C. Y. 207, 214.
HUBBARD, R. 43, 56, 83, 87.
HUBER, W. 14, 17, 22, 39, 43, 58, 60, 61, 64, 71, 74, 75, 84.
HUEBNER, C. F. 104, 112, 242, 257, 287, 291.
HUFF, J. W. 412, 417, 418, 420, 421, 442.
HUFFMAN, H. M. 299, 353.
HUGHES, G. K. 250, 281, 285, 286.
HUISMAN, H. O. 48, 63, 66, 83.
HULTQUIST, M. E. 101, 108, 338, 352, 420, 431, 434.
HUMPHREYS, J. 420, 431, 432, 434.
HUNTER, G. 119, 171.
HUNTER, R. F. 43, 80, 82, 83.
HUNTER, S. W. 356.
HURLBERT, R. B. 305, 358.
HUTCHINGS, B. L. 338, 352, 410, 437.
HUTNER, S. H. 418, 437.
HYDÉN, H. 305, 315, 356.

IIDA, G. 241, 287.
IKEDA, C. 458, 464, 468, 476, 483.
IKEDA, T. 193, 219.
IMAIDA, M. 243, 288.
INABA, R. 189, 216.
INHOFFEN, H. H. 1, 11, 12, 13, 16, 19, 20, 22, 26, 33, 38, 39, 40, 43, 54, 56, 57, 80, 83, 84.
INUBUSHI, Y. 184, 185, 187, 193, 208, 209, 216, 222, 223.
INUKAI, F. 408, 439.
ISHERWOOD, F. A. 331, 332, 333, 336, 355, 382, 397.
ISHIKAWA, S. 58, 60, 84.
ISHIWARI, N. 176, 188, 213, 216.
ISLER, O. 11, 14, 17, 22, 39, 43, 58, 60, 61, 62, 63, 64, 65, 71, 74, 75, 81, 84.
ITA, P. 199, 202, 203, 215.
ITALLIE, L. VAN 207, 216.
ITO, M. 213, 221.

JACOBS, W. A. 118, 122, 123, 124, 125, 126, 127, 131, 134, 136, 137, 151, 167, 168, 170, 171, 174.
JAENICKE, L. 365, 396.
JAQUES, L. B. 106, 109.

JERZMANOWSKA-SIENKIEWICZOWA, Z. 248, 249, 289.
JOHNS, R. G. S. 458, 482.
JOHNSON, A. W. 13, 39, 43, 58, 60, 83, 84, 417, 435.
JOHNSON, E. A. 419, 420, 434, 435.
JOHNSON, T. B. 119, 171.
JOHNSTON, R. B. 329, 330, 332, 333, 341, 342, 343, 356.
JOINER, R. R. 90, 112.
JOIS, H. S. 260, 290.
JONES, E. R. H. 10, 12, 17, 38, 52, 53, 54, 56, 58, 60, 61, 70, 71, 72, 76, 78, 81, 82, 83, 249, 286.
JONES, H. B. 305, 338, 358.
JONES, J. H. 91, 109.
JONES, L. O. 96, 109.
JONES, R. G. 420, 429, 431, 439.
JONES, W. E. 35, 36, 38, 43, 52, 53, 56, 76, 81, 82, 83.
JOUANNETEAU, J. 56, 71, 72, 73, 75, 77, 79, 80, 85, 102, 103, 110, 111.
JUCKER, E. 1, 39, 56, 73, 84, 167, 173.
JUKES, T. H. 101, 108, 418, 419, 420, 424, 431, 432, 434, 435, 437, 439, 441.

KABAT, E. A. 445, 448, 453, 454, 462, 464, 481.
KACZKA, E. A. 419, 420, 437.
KADIERA, K. 199, 202, 215.
KAHOVEC, L. 266, 289.
KAINRATH, P. 256, 269, 272, 289.
KAISER, S. J. 447, 480.
KALCKAR, H. M. 363, 366, 367, 368, 370, 371, 374, 375, 379, 380, 385, 386, 388, 392, 395, 396, 397, 398, 399, 402, 413, 422, 424, 425, 430, 436, 437.
KALLAB, F. 264, 266, 291.
KAMEN, M. D. 305, 315, 346, 360.
KAO, Y. S. 207, 214.
KAPLAN, N. O. 408, 409, 435.
KARMAS, G. 62, 64, 86.
KARNOFSKY, D. A. 97, 109.
KARRER, P. 1, 2, 3, 19, 28, 31, 34, 35, 38, 39, 42, 43, 46, 50, 56, 72, 73, 76, 78, 80, 82, 84, 85, 91, 94, 109.
KARUSH, F. 458, 467, 472, 479, 481.
KASCHER, H. M. 46, 50, 70, 73, 81, 87.
KAUFMAN, S. 334, 356, 359.
KAVANAGH, F. 240, 241, 245, 286, 408, 440.
KAWANAMI, M. 269, 270, 287.

KEIGHLEY, G. 298, 303, 304, 305, 308, 310, 311, 313, 314, 315, 316, 317, 318, 333, 343, 344, 345, 346, 348, 351, 353.
KEIGHLEY, G. L. 294, 353.
KEIMATSU, I. 194, 195, 217.
KELLER, E. B. 298, 305, 356.
KELLEY, B. 311, 360.
KEMEN, A. J. 356.
KENDALL, F. E. 457, 464, 475, 480, 481.
KENNER, G. W. 364, 398, 402, 437.
KERKKONEN, H. K. 327, 361.
KERR, S. E. 392, 394, 398, 408, 414, 437.
KESTON, A. 294, 356.
KESTON, A. S. 297, 301, 358.
KHARASCH, M. S. 118, 120, 122, 149, 171, 172.
KHYM, J. X. 424, 429, 441.
KIDDER, G. W. 411, 415, 416, 421, 437.
KIELLEY, R. K. 305, 340, 356.
KIKUCHI, K. 213, 221.
KING, C. G. 421, 439.
KING, H. 149, 171, 177, 180, 181, 187, 199, 200, 201, 202, 207, 209, 213, 216, 217.
KING, T. E. 408, 409, 438.
KIPPING, F. B. 65, 85.
KIRKWOOD, J. G. 453, 454, 459, 461, 479.
KISTIAKOWSKY, G. B. 468, 478.
KITAGISHI, H. 193, 223.
KITASATO, Z. 191, 216.
KITAY, E. 418, 420, 424, 426, 428, 429, 430, 431, 432, 438, 441.
KLAGER, K. 253, 258, 267, 273, 289.
KLECZKOWSKI, A. 460, 478, 481.
KLEIN, W. 365, 372, 395, 398.
KLEINSCHMIDT, W. J. 450, 481.
KLENOW, H. 375, 376, 382, 395, 398.
KLIOZE, O. 311, 357.
KLOTZ, I. M. 311, 346, 357.
KNOX, W. E. 89, 108.
KOBERT, R. 116, 119, 172.
KOBLICK, D. C. 464, 482.
KOCH, A. L. 414, 423, 424, 434, 438.
KOCH, O. 199, 221.
KOCHAKIAN, C. D. 306, 356, 357.
KOCHER, V. 420, 424, 438.
KODITSCHEK, L. K. 427, 438.
KOEPFLI, J. B. 447, 479.
KOFLER, A. 159, 172.
KOFLER, L. 159, 172.
KOFLER, M. 14, 17, 22, 39, 43, 58, 60, 61, 64, 71, 74, 75, 81, 84, 240.

Kolbe, A. 196, 222.
Kon, S. K. 74, 81.
Kondo, H. 176, 177, 178, 181, 186, 187, 188, 189, 190, 191, 192, 193, 194, 195, 196, 198, 199, 200, 203, 204, 205, 207, 209, 217, 218, 219.
Kondo, T. 176, 186, 187, 217, 219.
Koniuszy, F. 417, 436.
Kornberg, A. 368, 369, 370, 380, 382, 384, 391, 398, 399.
Körner, G. 247.
Kossel, A. 119, 172.
Kraft, F. 117, 118, 119, 122, 156, 157, 172.
Kranzfelder, A. L. 185, 219.
Kream, J. 380, 396.
Krehl, W. A. 103, 110.
Kreke, C. W. 405, 435.
Kreuger, R. C. 452, 481.
Krishnaswamy, B. 259, 287.
Kritskiĭ, G. A. 426, 438.
Krohn, D. 253, 267.
Krukovsky, V. N. 103, 113.
Kruta, E. 280, 286.
Kubiczek, G. 262, 263, 289, 290, 291.
Kuehas, E. 195, 199, 215.
Kuehl, F. A., Jr. 419, 420, 434, 437.
Kuffner, F. 202, 222, 241, 260, 289, 290.
Kuh, E. 338, 352.
Kühn, K. 323, 361.
Kuhn, R. 12, 34, 39, 43, 52, 85, 94, 100, 101, 109, 147, 172, 251.
Kun, C. P. 284, 288.
Kunz, K. 257, 287.
Küssner, W. 159, 172.
Kusuda, F. 187, 209, 222, 223.
Kutscher, Fr. 119, 171, 172.
Kuzin, A. M. 445, 447, 481.

Laaksonen, T. 327, 361.
Lackman, D. B. 447, 482.
Ladeck, F. 201, 222.
Lajtha, A. 335, 359.
Lampen, J. O. 99, 108, 369, 372, 376, 378, 380, 399, 400, 412, 422, 430, 438, 441.
Landsteiner, K. 444, 445, 450, 455, 456, 459, 460, 466, 476, 481, 484.
Landström-Hydén, H. 305, 315, 353.
Lankford, C. E. 406, 438.
Lanner, K. 257, 288.
Lanni, F. 445, 447, 462, 465, 479, 481.
Laser, R. 372, 396.

Lavollay, J. 101, 109.
Lederer, E. 43, 76, 82, 85, 228, 239, 282, 284, 287.
Ledingham, A. E. 209, 219.
Lee, S. W. 58, 60, 86.
Leemann, J. 26, 33, 40.
Lefwich, W. B. 96, 109
Legault, R. R. 118, 120, 122, 149, 172.
Leibner, G. 12, 38.
Leithe, W. 191, 201, 219, 222.
Leloir, L. F. 364, 376, 382, 384, 396, 398.
Lens, J. 420, 441.
Leone, P. 246.
Le Page, G. A. 305, 357, 360, 392, 398.
Lepp, E. 106, 109.
Lerman, L. S. 451, 458, 463, 467, 479, 482.
Le Rosen, A. L. 5, 8, 33, 40, 447, 479.
Lettré, H. 431, 441.
Levaditi, C. 101, 109.
Levine, M. 304, 307, 308, 350, 357.
Lévy, M. 448, 482.
Levy-Solal, E. 106, 109.
Lewis, A. D. 48, 49, 50, 56, 57, 86.
Lewis, D. G. 72, 83.
Li, C. H. 357.
Limaye, D. B. 287.
Link, K. P. 104, 112, 240, 242, 257, 286, 287, 288, 291.
Linnell, W. H. 74, 85.
Lintner, J. 256, 290.
Lipmann, F. 346, 357, 377, 398, 409, 439.
Little, J. E. 105, 110.
Litwack, G. 299, 357.
Lloyd, J. T. 430, 435.
Locher, L. M. 409, 438.
Lochhead, A. G. 420, 438.
Loewe, L. 234, 285.
Loftfield, R. B. 305, 311, 332, 345, 354, 362.
Loiseleur, J. 448, 482.
London, F. 471, 473, 482.
London, J. M. 311, 359.
Lonicerus 118.
Lonitzer, A. 116, 118.
Loofbourow, J. R. 405, 438.
Loomis, W. D. 335, 345, 346, 354, 357, 360.
Loosli, J. K. 103, 113.
Lopez, J. A. 367, 382, 385, 398.
Loring, H. S. 402, 407, 408, 409, 411, 412, 415, 417, 436, 438, 439.

LOTSPEICH, W. D. 307, 357.
LOWE, A. 52, 53, 56, 81, 83.
LOWRY, O. H. 367, 379, 382, 385, 398.
LOWY, P. H. 298, 303, 304, 305, 308, 310, 311, 313, 314, 315, 316, 317, 318, 333, 343, 344, 345, 346, 348, 351, 353.
LUCK, J. M. 298, 305, 314, 354, 355.
LUESCHER, E. 451, 463, 479.
LUNDGREN, H. P. 453, 458, 482.
LUTWAK-MANN, C. 376, 398.
LUTZMANN, H. 243, 287.
LWOFF, A. 405, 438.
LWOFF, M. 405, 438.
LYTHGOE, B. 365, 396, 398.

MCCOMBIE, J. T. 10, 38.
MCCULLOUGH, W. G. 406, 434.
MACDONALD, N. S. 58, 60, 86.
MCELROY, W. D. 381, 399.
MCFARLANE, W. D. 80, 82.
MCGILVERY, R. W. 338, 339, 345, 354, 357.
MCGLOHON, V. M. 409, 441.
MACLEOD, C. M. 447, 480.
MCMANUS, D. K. 408, 435.
MCNAIR SCOTT, D. B. 375, 377, 396, 399.
MCNULTY, H. P. 413, 435.
MCNUTT, W. S. 364, 372, 373, 380, 387, 388, 389, 390, 395, 398, 399, 401, 411, 418, 420, 422, 424, 425, 426, 428, 429, 431, 432, 433, 437, 438, 441.
MALKIEL, S. 464, 482.
MALLEIN, R. 71, 75, 85, 103, 111.
MANJUNATH, B. L. 260, 290.
MANN, H.-J. 431, 441.
MANSKE, R. H. F. 184, 192, 196, 209, 219.
MANSON, L. A. 371, 372, 376, 378, 397, 399, 422, 430, 438.
MARNAY, CH. 102, 107, 110.
MARRACK, J. R. 452, 457, 458, 467, 482.
MARRIAN, D. H. 392, 399.
MARSH, D. F. 214, 219.
MARSHAK, A. 423, 424, 430, 436, 438.
MARTEN, A. 254, 267, 286.
MARTIN, C. J. 316, 357.
MARTIN, G. J. 96, 97, 98, 107, 110, 420, 434.
MARX, R. 242, 285.
MASON, H. L. 446, 484.
MASOUREDIS, S. P. 464, 482.
MATSUMOTO, S. 213, 220.
MATSUNO, T. 191, 217, 220.
MATSUOKA, M. 213, 220.

MATSUURA, T. 58, 60, 84.
MATZKE, J. 278, 288.
MAUTHNER, F. 243.
MAYER, J. 103, 110.
MAYER, M. M. 453, 454, 481.
MEAD, T. H. 42, 74, 81, 85.
MEBANE, A. D. 62, 64, 86.
MEIJER, TH. M. 249, 287.
MEINHARD, T. 227, 278, 289.
MELCHER, L. R. 447, 464, 482.
MELCHIOR, J. B. 311, 346, 350, 357.
MELLANBY, E. 119, 172.
MELLODY, M. 311, 346, 357.
MELNICK, D. 93, 110.
MELVILLE, D. B. 98, 110.
MENTZER, C. 89, 101, 104, 105, 106, 107, 109, 110.
MERZ, K. W. 244, 245, 287.
METH, E. G. 11, 16, 19, 20, 39.
METZ, C. B. 451, 484.
MEUNIER, P. 56, 71, 72, 73, 75, 76, 77, 78, 79, 80, 85, 88, 89, 101, 102, 103, 104, 105, 106, 107, 108, 110, 111.
MEYER, G. M. 296, 360.
MEYER, H. 65, 87.
MICHEL, I. 256.
MICHELSON, A. M. 392, 399, 416, 438.
MICHELSON, C. 335, 345, 360.
MIEKELEY, A. 136, 170.
MILAS, N. A. 19, 22, 40, 43, 52, 56, 58, 60, 62, 64, 80, 85, 86.
MILLER, C. S. 417, 442.
MILLER, E. C. 298, 314, 358, 359.
MILLER, H. 459, 482.
MILLER, J. A. 298, 314, 358, 359.
MILLER, L. L. 298, 299, 357.
MILLER, W. W. 311, 345, 354.
MILLS, J. A. 417, 435.
MILLS, R. C. 406, 434.
MIMURA, T. 213, 220.
MINGIOLI, E. S. 418, 419, 429, 435.
MIRICK, G. S. 96, 109.
MIRSKY, A. E. 367, 400.
MITCHELL, H. K. 300, 360, 381, 390, 392, 399, 414, 415, 416, 438, 439.
MITCHELL, J. H., Jr. 415, 439.
MITRA, S. S. 272, 286.
MIWA, T. 199, 223.
MIYAJIMA, S. 254, 290, 291.
MIYAZAWA, Y. 213, 221.
MOČNIK, W. 277, 290.
MOE, J. G. 306, 357.

Molho, D. 89, 105, 106, 107, 110, 111.
Molho-Lacroix, L. 89, 101, 107, 110, 111.
Möller, E. F. 94, 109.
Monti, L. 231, 285.
Mookerjee, A. 245, 280, 286, 288.
Moore, G. E. 356.
Moore, J. A. 409, 441.
Moraux, J. 105, 106, 107, 111.
Morf, R. 42, 46, 50, 84.
Morgan, W. T. J. 448, 482.
Morris, C. J. O. R. 43 52 85.
Morton, R. A. 42, 43, 52, 54, 56, 71, 73, 74, 75, 76, 77, 81, 82, 86.
Mosettig, E. 191, 222.
Moss, J. 96, 98, 110.
Moss, J. N. 420, 434.
Moura Campos, F. A., de 241.
Mowat, J. H. 338, 352.
Mudd, S. 447, 482.
Mueller, J. F. 97, 111.
Müller, H. R. 333, 353.
Mulligan, W. 458, 478.
Muntwyler, E. 298, 314, 358, 359.
Murai, F. 184, 193, 194, 196, 199, 200, 208, 215, 222.
Murakami, M. 192, 194, 196, 217.
Murphy, M. K. 97, 113.
Mycek, M. J. 329, 330, 332, 333, 356.

Nakaguchi, K. 187, 223.
Nakahara, W. 408, 439.
Nakajima, T. 207, 217, 218.
Nakamura, T. 193, 207, 217, 222.
Nakano, T. 188, 223.
Nakazato, T. 203, 204, 217, 220.
Narita, Z. 192, 194, 196, 217.
Navez, A. 243.
Nelson, M. M. 97, 111.
Nepple, H. 410, 440.
Neuberger, A. 89, 108, 300, 337, 354, 450, 482.
Neufeld, O. 243, 278, 290.
Neuman, J. 101, 109.
Neumann, F. 181, 199, 215.
Neurad, K. 257, 288.
Neurath, H. 334, 356, 359, 460, 479.
Newstead, E. G. 419, 420, 435.
Newton, E. B. 119, 172, 365, 370, 396.
Niemann, C. 138, 170, 172, 448, 478.
Nierenstein, M. 283, 288.
Nitti, F. 100, 112.
Niwa, H. 184, 185, 208, 222, 223.

Noguchi, T. 269, 270, 287.
Norman, S. L. 420, 435.
Norris, R. F. 429, 441.
Northey, E. H. 338, 352.
Northrop, J. 327, 358.
Norton, S. 451, 480.
Novelli, G. D. 408, 409, 439.
Nozaki, H. 189, 216.
Nozoe, T. 199, 218.
Nyc, J. F. 415, 439.
Nygaard, A. P. 411, 439.

Oberembt, H. 182, 193, 194, 196, 207, 214.
Ochiai, E. 176, 188, 207, 218, 220.
Odera, T. 207, 218.
Oginsky, E. L. 419, 439.
Ohta, T. 245, 287.
Okada, M. 213, 220.
Okahara, K. 260, 290.
Okamoto, Y. 213, 221.
Oliver-Gonzales, J. 447, 482.
Oncley, J. L. 454, 482.
Ono, M. 254, 291.
Opitz, H. 119, 174.
Ordway, G. L. 409, 411, 412, 438.
Orechov, A. P. 181, 192, 193, 221.
Orla-Jensen, A. D. 242, 287.
Oroshnik, W. 62, 64, 86.
Osan 268.
Oser, B. L. 93, 110.
Ostern, P. 390, 399.
O'Sullivan, D. G. 53, 54, 56, 83.
Ott, W. H. 96, 97, 111.
Oughton, J. F. 43, 48, 52, 63, 66, 82.
Overhoff, J. 119, 171.
Owades, P. 335, 361.

Paege, L. M. 416, 439.
Page, A. C., Jr. 419, 420, 439.
Pailer, M. 260, 261, 264, 290.
Paladini, A. C. 376, 396.
Pallares, E. S. 193, 220.
Pappenheimer, A. M. 406, 439, 453, 458, 459, 460, 482, 483.
Pardee, A. B. 453, 462, 470, 476, 477, 479, 482.
Parikh, R. J. 231, 288.
Parker, R. P. 101, 108, 420, 431, 434.
Parks, R. E., Jr., 415, 416, 421, 437.
Patterson, A. M. 147, 172.
Patterson, L. P. 97, 109.

Namenverzeichnis. Index of Names. Index des Auteurs.

PAULING, L. 5, 7, 8, 33, 40, 51, 138, 172, 444, 447, 448, 449, 451, 454, 455, 457, 458, 464, 467, 468, 470, 471, 473, 474, 476, 477, 479, 482, 483.
PAVOLINO, T. 228.
PAZUR, J. H. 422, 425, 439.
PEARSON, O. H. 327, 358.
PEARSON, W. N. 407, 439.
PEASE, D. C. 282, 285.
PEAT, S. 422, 434.
PEDERSEN, K. O. 452, 454, 483, 484.
PEEL, E. W. 417, 434.
PENCE, L. H. 474, 483.
PENNINGTON, R. J. 420, 439.
PEREIRA, A. 261, 291.
PERKIN, A. G. 283, 288.
PERRAULT, R. 101, 109.
PESTA, O. 254, 261, 290.
PETERMAN, M. L. 453, 483.
PETERS, T., Jr. 294, 310, 312, 345, 348, 358.
PETERS, V. J. 409, 441.
PETERSON, E. A. 310, 311, 350, 358, 362.
PETERSON, W. H. 407, 440.
PETROW, V. 65, 86, 419, 420, 434, 435.
PETRŮ, F. 257, 288.
PETRZILKA, T. 117, 119, 131, 134, 135, 136, 137, 139, 142, 144, 148, 155, 161, 162, 163, 167, 169, 173.
PFISTER, R. W. 333, 334, 353.
PHILIPS, F. S. 379, 396.
PHIPSON, T. L. 243.
PIERCE, J. G. 402, 407, 408, 409, 411, 412, 415, 417, 438, 439.
PIERCE, J. V. 419, 420, 434, 439, 441.
PIETSCH, G. 252, 285.
PIKL, J. 181, 196, 222.
PILLAY, P. P. 251, 286.
PILLOW, A. 147, 170.
PINCKARD, J. H. 5, 40.
PIRIE, N. W. 119, 172.
PLAICHINGER, I. 233, 291.
PLATI, J. T. 58, 62, 86.
PLATZER, N. 275, 290.
PLENTL, A. A. 391, 393, 396, 399.
POHLAND, A. 420, 429, 431, 439.
POLAK, J. 284, 287.
POLEX 195.
POLGÁR, A. 4, 5, 8, 16, 26, 33, 40.
POLONSKY, J. 282.
POLYAKOVA, O. I. 447, 481.
POMERANZ, C. 262.

POMMER, H. 11, 16, 19, 20, 22, 38, 39, 80, 83, 84.
POMMEREHNE, H. 193, 196, 221.
PORTER, C. C. 96, 97, 111.
PORTER, E. F. 460, 483.
PORTER, R. R. 351, 358, 452, 455, 483.
POSTERNAK, T. 376, 400.
POTTER, V. R. 305, 358.
PRANGE, I. 104, 108.
PREOBRAZHENSKII, N. A. 52, 86.
PRESSMANN, D. 457, 458, 462, 464, 467, 470, 483, 484.
PRICE, J. M. 298, 314, 358, 359.
PRICE, V. 367, 368, 399.
PRICER, W. E., Jr. 380, 391, 398.
PRIESS, H. 241, 264.
PROSKOURNINA, N. F. 181, 192, 193, 221.
PROVASOLI, L. 418, 437.
PRUSOFF, W. H. 421, 439.
PYMAN, F. L. 119, 172, 198, 205, 221.

QUIBILAN, G. Q. 207, 221.

RABATÉ, M. J. 422, 439.
RACE, R. R. 459, 483.
RACKER, E. 376, 377, 378, 399.
RAFFAUF, R. F. 90, 111.
RAINER, J. 119, 171.
RATNER, S. 298, 300, 302, 307, 347, 358, 359.
RAUDNITZ, H. 257, 288.
RAVEL, J. M. 407, 420, 440.
RAY, J. N. 274, 286.
REA, J. L. 42, 82.
RECKNAGEL, R. O. 305, 358.
RECORD, B. R. 447, 483.
REESE, J. W. 305, 355.
REGE, D. V. 421, 440.
REICHARD, P. 392, 393, 395, 397, 399, 402, 414, 416, 422, 423, 433, 436, 440.
REICHSTEIN, T. 90.
REID, J. C. 305, 338, 358.
REIFENSTEIN, E. C. 306, 358.
REINHARDT, W. O. 306, 307, 360.
REISNER, D. B. 261, 287.
RENZ, J. 261, 291.
REYCHLER, A. 229, 288.
RICHARDSON, R. W. 72, 83.
RICHERT, D. A. 299, 361.
RICKETTS, C. R. 448, 484.
RIDGWAY, L. P. 97, 109.
RIMINGTON, C. 452, 484.
RINDL, M. 247, 288.

Ripley, S. H. 305, 315, 356.
Ritchie, E. 250, 281, 285, 286.
Rittenberg, D. 297, 298, 300, 301, 302, 303, 305, 307, 311, 332, 338, 340, 347, 358, 359, 360, 361.
Rivers, J. T. 58, 62, 86.
Robbins, W. J. 90, 111, 408, 420, 440.
Roberts, D. 372, 397.
Roberts, E. 97, 111.
Roberts, E. C. 100, 112.
Roberts, M. 412, 440.
Roberts, R. B. 419, 440.
Roberts, R. M. 321, 355, 468, 478.
Roberts, W. L. 240, 288.
Robertson, A. 225, 229, 237, 244, 245, 246, 261, 262, 277, 278, 279, 281, 285, 286, 287, 288.
Robeson, C. D. 41, 42, 43, 46, 49, 50, 51, 52, 53, 66, 70, 73, 74, 81, 86, 87.
Robinson, B. 447, 480.
Robinson, R. 228.
Robison, R. 316, 357.
Roblin, R. O., Jr. 89, 99, 108, 111.
Roessler, W. G. 406, 434.
Rogers, L. L. 99, 112, 420, 424, 429, 431, 440.
Roll, P. M. 391, 392, 396, 397, 399, 413, 414, 435.
Ronco, A. 14, 17, 22, 39, 43, 58, 60, 61, 64, 71, 74, 75, 84.
Rose, C. S. 99, 108.
Rosenblum, C. 54, 75, 80, 87.
Ross, G. I. 420, 440.
Rost 241.
Roth, B. 101, 108, 420, 431, 434.
Roth, H. 147, 172, 251.
Rothen, A. 460, 471, 472, 484.
Rothlin, E. 118, 121, 157.
Rothman, F. 342, 360.
Rowen, J. W. 368, 369, 370, 382, 384, 391, 399.
Rowley, D. A. 316, 353.
Rowold, E. 431, 441.
Rozanova, V. A. 76, 85.
Rubstov, I. A. 52, 86.
Ruckstuhl, H. 94, 109.
Ruedel, C. 193, 196, 221.
Ruegger, A. 19.
Rüfenacht, K. 334, 353.
Rummert, G. 13, 22, 38.
Rummert, S. 22, 38.
Russell, P. B. 413, 437.

Rutman, R. 358.
Rutschmann, J. 34, 39, 119, 131, 133, 167, 173.
Rydon, H. N. 91, 112.
Sable, H. Z. 375, 380, 399, 400.
Saetren, H. 367, 400.
Sailer, E. 334, 353.
Sakal, E. 58, 60, 62, 86.
Salah, M. K. 71, 76, 77, 86.
Salm-Horstmar, W. F. 247.
Salomon, H. 46, 50, 84.
Salter, W. T. 327, 358.
Saluste, E. 392, 393, 397, 402, 416, 422, 436.
Sanada, T. 189, 199, 207, 218, 221.
Sanadi, D. R. 304, 306, 307, 316, 359.
Sands, M. 419, 440.
Sanger, F. 351, 358.
Santos, A. C. 193, 195, 207, 221.
Sareen, K. N. 274, 286.
Sarich, P. 316, 355.
Sarkar, N. 340, 359.
Sartoretto, P. A. 185, 221.
Sasaki, T. 187, 221.
Sato, T. 207, 217.
Satomi, M. 193, 207, 218, 219.
Sauberlich, H. E. 101, 112, 297, 314, 359, 408, 413, 420, 429, 431, 440.
Sawa, S. 199, 223.
Scatchard, G. 454, 482.
Schaedel, M. L. 370, 379, 399.
Schaeffer, A. J. 316, 359.
Scheer, J. van der 455, 459, 466, 481.
Schick, E. 56, 84.
Schiff, H. 283.
Schindler, O. 420, 424, 438.
Schläger, J. 245, 288.
Schlatter, C. H. 273.
Schlenk, F. 370, 375, 378, 379, 380, 399, 405, 416, 436, 439.
Schlesinger, H. I. 43, 82.
Schlientz, W. 119, 131, 133, 173.
Schlösser, C. 273, 289.
Schmelkes, F. C. 90, 112.
Schmetz, F. J., Jr. 408, 409, 439.
Schmid, H. 249, 265, 274, 275, 276, 280, 287, 288, 290.
Schmidt, C. L. A. 298, 307, 317, 360.
Schmidt, E. 245, 288.
Schmidt, G. 366, 379, 399.
Schneider, E. 252, 285.
Schneider, P. 76, 84.

SCHNEIDER, W. C. 305, 315, 340, 356, 359.
SCHOELLER, M. 91, 109.
SCHOENHEIMER, R. 297, 298, 299, 300, 301, 302, 303, 306, 307, 347, 358, 359, 361, 393, 399.
SCHOLTZ, M. 176, 199, 201, 202, 221.
SCHOPFER, W. H. 90, 91, 94, 98, 100, 107, 112, 410, 440.
SCHÖPP, K. 42, 84.
SCHOU, M. 335, 359, 361.
SCHROEDER, W. A. 5, 8, 33, 40.
SCHUERCH, C., Jr. 58, 86.
SCHULMAN, M. P. 310, 355.
SCHULTZ, A. S. 408, 435.
SCHULTZ, J. 305, 354.
SCHULTZE, H. 181, 196, 197, 214.
SCHWARZ, R. 199, 202, 203, 215.
SCHWARZKOPF, O. 48, 49, 50, 56, 57, 86.
SCHWEET, R. S. 348, 359.
SCHWEIGERT, B. S. 297, 298, 314, 359.
SCHWERIN, P. 458, 462, 480.
SCHWERT, G. W. 334, 356, 359.
SCHWYZER, R. 78, 85.
SEEGER, D. R. 338, 352.
SEIDEL, G. R. 41, 50, 51.
SEIFTER, S. 298, 314, 358, 359.
SEMB, J. 338, 352.
SEN, P. B. 242, 286.
SERAIDARIAN, K. 392, 394, 398, 408, 414, 437.
SESHADRI, T. R. 237, 259, 285, 287.
SETHNA, S. 231, 288.
SHAH, N. M. 237, 291.
SHANTZ, E. M. 41, 43, 46, 50, 53, 70, 71, 72, 73, 76, 77, 78, 79, 81, 82, 87.
SHAW, E. 389, 399.
SHEDLOVSKY, T. 118, 124, 170.
SHEMIN, D. 298, 301, 302, 305, 311, 338, 339, 340, 356, 359, 394, 395.
SHEN, C. C. 74, 85.
SHERWOOD, M. B. 413, 437.
SHIINA, S. 213, 220.
SHIMURA, J. 213, 221.
SHINODA, J. 243, 284, 288.
SHINOZUKA, T. 213, 221.
SHIRAI, H. 189, 190, 200, 221, 222, 223.
SHISHIDO, H. 189, 216.
SHIVE, W. 99, 100, 112, 389, 400, 407, 410, 417, 419, 420, 421, 424, 429, 431, 432, 434, 436, 439, 440, 442.
SHULL, G. M. 407, 440.
SHUMAKER, J. B., Jr. 454, 479.

SHUNK, C. H. 417, 434, 436.
SIBLEY, M. E. 424, 429, 431, 434.
SICKELS, J. P. 338, 352.
SIDDIQUI, S. 207, 214.
SIEBURG, E. 241.
SIEGEL, B. M. 472, 481.
SIEKEVITZ, P. 314, 359.
SIEMER, H. 1, 80, 84.
SILBER, R. H. 96, 97, 111.
SIMMONDS, S. 341, 359.
SIMON, A. F. J. 256, 290.
SIMPSON, M. V. 348, 359.
SINGER, S. J. 454, 458, 472, 479, 484.
SKAGGS, P. K. 406, 438.
SKEGGS, H. R. 410, 412, 416, 418, 420, 421, 440, 442.
SKIPPER, H. E. 415, 439.
SLATES, H. L. 43, 48, 51, 52, 54, 56, 87.
SLEETH, C. K. 214, 219.
SLOANE, N. H. 410, 437.
SLYKE, D. D. VAN 296, 360.
SMITH, E. L. 42, 81, 333, 359, 452, 484.
SMITH, F. C. 467, 482.
SMITH, J. M., Jr. 101, 108, 338, 352, 420, 431, 434.
SMITH, P. H. 419, 439.
SMITH, S. 118, 120, 121, 122, 144, 155, 157, 159, 172.
SMYRNIOTIS, P. Z. 375, 377, 378, 397.
SNELL, E. E. 97, 108, 409, 411, 417, 418, 420, 423, 424, 426, 428, 429, 430, 431, 432, 437, 438, 440, 441.
SNOKE, J. E. 334, 342, 359, 360.
SOARS, M. H. 419, 420, 437, 441.
SOBOTKA, H. 43, 46, 87.
SOCIAS, L. 263, 290.
SOLOMON, A. K. 310, 311, 352.
SÖNDERGAARD, E. 104, 106, 108.
SONDHEIMER, F. 60, 61, 70, 71, 81.
SOODAC, M. 91, 112.
SOWA, F. J. 185, 219, 221, 224.
SOWINSKI, R. 467, 480.
SPÄTH, E. 177, 181, 191, 196, 201, 202, 222, 227, 228, 230, 235, 239, 241, 243, 245, 248, 249, 251, 253, 254, 256, 258, 260, 261, 262, 263, 264, 265, 266, 267, 268, 269, 272, 273, 275, 276, 277, 278, 280, 288, 289, 290, 291.
SPECK, J. F. 337, 360.
SPECTOR, H. 300, 360.
SPENCER, E. Y. 232, 291.
SPICER, V. L. 392, 399.

SPIEGELMAN, S. 305, 315, 346, 360.
SPINKS, A. 58, 60, 83.
SPIRO 121.
SPITZER, E. H. 93, 112.
SPIZIZEN, J. 410, 440.
SPRINSON, D. B. 297, 301, 302, 303, 305, 360.
SPROSTON, T. J. 105, 110.
SREENIVASAS, A. 421, 440.
STACEY, M. 445, 447, 448, 480, 483, 484.
STAHMANN, M. A. 104, 112, 242, 257, 291.
STEARNS, J. 116, 118, 172.
STEBBINS, M. E. 420, 440.
STEENHAUER, A. J. 207, 216.
STEFFEE, C. H. 316, 353.
STEINBERG, D. 348, 349, 350, 352.
STEPHENSON, M. 376, 388, 400.
STEPHENSON, M. L. 305, 311, 355, 362.
STEPHENSON, O. 65, 86.
STEPTO, P. C. 316, 353.
STERN, H. 367, 400.
STERN, M. H. 41.
STERNBERGER, L. A. 462, 484.
STETTEN, M. R. 389, 400.
STOCK, C. C. 97, 109.
STOERK, H. C. 96, 112.
STOKES, J. L. 421, 430, 441.
STOKINGER, H. E. 464, 484.
STOKSTAD, E. L. R. 101, 108, 338, 352, 418, 419, 420, 424, 431, 432, 434, 435, 437, 439, 441.
STOLL, A. 114, 117, 118, 120, 121, 122, 123, 126, 128, 131, 133, 134, 135, 136, 137, 139, 142, 144, 148, 149, 151, 152, 153, 154, 155, 157, 159, 161, 162, 163, 164, 166, 167, 168, 169, 171, 172, 173, 261, 291.
STRONG, F. M. 408, 409, 438.
STUBBS, A. L. 76, 86.
STUMPF, P. K. 334, 335, 345, 354, 360.
STURGEON, P. 453, 479.
STURM, K. 246, 291.
STURTEVANT, J. M. 331.
SUBBAROW, Y. 338, 352.
SUBRAMANIAM, T. S. 278, 279, 281, 285, 288.
SUDZUKI, H. 188, 191, 207, 216.
SUGIURA, K. 413, 414, 435.
SURIE, E. 119, 172.
SUTHERLAND, E. W. 376, 400.
SUTHERLAND, J. E. 420, 440.
SVEDBERG, T. 454, 484.

SWIDINSKY, J. 48, 49, 50, 56, 57, 86.
SWINGLE, S. M. 476, 482.
SZENT-GYÖRGYI, A. 378, 397.

TABENKIN, B. 420, 435.
TAGGART, J. V. 346, 354.
TAKAGI, S. 187, 223.
TAKAHASHI, K. 213, 216, 221.
TAKEI, S. 254, 290, 291.
TANAKA, K. 199, 218, 222.
TANI, T. 190, 203, 204, 222, 223.
TANRET, C. 117, 118, 119, 156, 173, 174.
TARVER, H. 298, 304, 306, 307, 308, 311, 317, 348, 350, 352, 354, 355, 357, 358, 359, 360.
TATUM, E. L. 341, 359.
TAUBER, H. T. 327, 360.
TAYLOR, R. J. 74, 81.
TEKMAN, SH. 462, 480.
TEORELL, T. 457, 476, 484.
TEPLEY, L. J. 421, 439.
TERESI, J. D. 305, 355.
TER HORST, W. P. 106, 112.
TERNBERG, J. L. 420, 441.
TERSZACOVEĆ, J. 390, 399.
THANNHAUSER, S. J. 365, 379, 398, 399.
THAKOR, V. M. 237, 291.
THEXTON, R. H. 420, 438.
THOMPSON, A. F., Jr. 50, 82.
THOMPSON, M. R. 118, 120, 122, 149, 171, 174.
THOMPSON, S. Y. 74, 81.
THOMS, H. 262, 264, 291.
THORELL, B. 305, 315, 354, 360.
TILDEN, W. A. 245, 291.
TIMMIS, G. M. 118, 120, 121, 122, 144, 155, 157, 159, 172.
TISHLER, M. 43, 48, 51, 52, 54, 56, 75, 80, 87.
TOBLER, E. 28, 39.
TODD, A. R. 177, 197, 198, 205, 214, 365, 370, 384, 396, 397, 398, 417, 435.
TOMARELLI, R. M. 429, 441.
TOMITA, M. 175, 177, 179, 184, 185, 187, 188, 189, 190, 193, 194, 195, 196, 197, 198, 199, 200, 203, 204, 205, 208, 209, 212, 218, 219, 222, 223.
TONKAZY, N. E. 419, 439.
TOTTER, J. R. 311, 360.
TRACY, A. H. 90, 112.
TRAEGER, W. 95, 112.
TRAVERS, J. J. 316, 355.
TREFFERS, H. P. 445, 484.

TRÉFOUEL, J. 100, 112.
TRÉFOUEL (Mme.) 100, 112.
TRENNER, N. R. 43, 48, 51, 52, 54, 56, 87, 419, 420, 435.
TRIM, A. R. 376, 388, 400.
TROLLER, A. 181, 199, 215.
TROXLER, F. 118, 126, 128, 167, 173.
TRUCCO, R. E. 376, 396.
TSURUTA, S. 213, 223.
TURK, K. L. 103, 113.
TYLER, A. 451, 484.
TYNER, E. P. 305, 360.
TYRAY, E. 251, 272, 288, 290.

UBISCH, H. V. 392, 395, 416, 433.
UGAMI, S. 408, 439.
UHLE, F. C. 118, 124, 125, 174.
UMBREIT, W. W. 97, 113, 419, 439.
ÜN, R. 234, 285.
UNNA, K. 95, 113.
UYEO, S. 192, 199, 200, 205, 208, 217, 219, 223.

VAHLEN, E. 119, 174.
VALENTIK, K. A. 410, 412, 417, 440, 442.
VALLANOS 207, 221.
VAN DER SCHEER, J. 455, 459, 466, 481.
VANDER WERFF, H. 395, 413, 414, 434, 437.
VAN DORP, D. A. 43, 52, 53, 54, 56, 57, 65, 71, 72, 73, 75, 79, 80, 81, 82, 87.
VAN SLYKE, D. D. 296, 360.
VAUQUELIN, L. N. 117, 118, 174.
VEER, W. L. C. 420, 441.
VENDRELY, C. 298, 314, 361.
VENDRELY, R. 298, 314, 361.
VENKATESWARLU, V. 237, 285.
VERBANC, J. J. 185, 219.
VIERHAPPER, F. 251, 265, 290.
VIERORDT, H. 300, 361.
VIGNEAUD, V. DU 98, 99, 108, 110.
VILTER, R. W. 97, 111.
VINET, A. 78, 85, 103, 111.
VINOGRAD, J. 454.
VIRTANEN, A. I. 327, 361.
VISSER, D. 413, 435.
VISSER, D. W. 412, 417, 436, 440.
VOGT, W. 119, 174.
VOLCANI, B. 415, 434.
VOLKIN, E. 424, 429, 441.

WACKER, A. 365, 400, 431, 441.
WADDELL, J. G. 97, 113.

WAELSCH, H. 332, 334, 335, 336, 345, 355, 359, 361.
WAINFAN, E. 335, 345, 355.
WAJZER, J. 384, 400, 422, 441.
WAKSMAN, B. H. 446, 484.
WALD, G. 42, 43, 56, 75, 76, 83, 87.
WALDSCHMIDT-LEITZ, E. 323, 361.
WALDVOGEL, M. J. 370, 375, 378, 379, 399.
WALKER, O. 46, 50, 84.
WALLENFELS, K. 12, 39.
WALLER, C. W. 338, 352.
WALPOLE, G. S. 119, 170.
WALSH, S. T. M. 405, 435.
WATANABE, Y. 190, 222.
WANG, T. P. 369, 380, 400, 412, 441.
WARBURG, O. 377, 400.
WARTMAN, T. G. 420, 435.
WASICKY, R. 241.
WASTENEYS, H. 327, 361.
WATANABE, S. 213, 220.
WATANABE, T. 219.
WATKINS, W. M. 448, 482.
WATT, G. W. 185, 224.
WEBER, F. C. 185, 224.
WEBER, G. M. 298, 314, 358.
WEBSTER, E. T. 42, 82.
WEED, L. L. 416, 417, 423, 441, 442.
WEEDON, B. C. L. 10, 38, 52, 60, 61, 70, 71, 72, 74, 81, 83, 87.
WEHMER, C. 178, 196, 224.
WEIL-MALHERBE, H. 406, 441.
WEISBLAT, D. I. 75, 82.
WEISLER, L. 41, 43, 46, 53, 70, 81.
WEISS, Z. 58, 62, 86.
WEISSMAN, N. 300, 361.
WELCH, A. D. 427, 441.
WELIKY, I. 392, 399.
WELLAND, A. S. 74, 81.
WELLS, S. L. 306, 358.
WENDLER, N. L. 43, 48, 51, 52, 54, 56, 75, 80, 87.
WERFF, VANDER, H. 395, 413, 414, 434, 437.
WERTHEIMER, P. 107, 109.
WESSELY, F. 233, 246, 247, 248, 262, 264, 265, 266, 290, 291.
WESTERFIELD, W. W. 299, 361.
WESTPHAL, F. 19, 39, 80, 84.
WESWIG, P. H. 93, 94, 113.
WEYGAND, F. 94, 109, 365, 400, 431, 441.
WHALLEY, T. G. 197, 198, 207, 214.

WHALLEY, W. B. 279, 288.
WHITE, A. G. C. 90, 113.
WHITE, D. E. 254, 278, 286.
WHITING, F. 103, 113.
WHITING, M. C. 12, 38.
WIJMENGA, H. G. 420, 441.
WILD, F. 65, 85.
WILKINSON, I. A. 422, 434.
WILLIAMS, J. N., Jr. 299, 357, 361.
WILLIAMS, J. W. 453, 454, 458, 479, 482, 484.
WILLIAMS, N. E. 80, 82, 83.
WILLIAMS, R. J. 417, 421, 442.
WILSON, A. N. 90, 113.
WILSON, A. T. 406, 442.
WILSON, D. W. 416, 417, 441, 442.
WILSON, M. F. 427, 441.
WINDAUS, A. 119, 174.
WINKELMANN, K. 16, 19, 39.
WINKLER, K. C. 100, 113.
WINNICK, T. 99, 109, 304, 305, 306, 307, 310, 311, 316, 345, 346, 350, 355, 358, 361, 362.
WISE, E. C. 75, 82.
WISNICKY, W. 93, 112.
WITTLE, E. L. 409, 441.
WOHLERS, H. C. 58, 60, 86.
WOLF, D. E. 419, 420, 437.
WOLF, E. 159, 170.
WOLFE, H. R. 451, 480.
WOODRUFF, H. B. 419, 427, 438, 442.
WOODS, D. D. 100, 102, 113.
WOODS, R. J. 70, 74, 87.
WOODWARD, C. R., Jr. 441.
WOODWARD, G. E. 306, 362.
WOOLEY, J. G. 97, 113.
WOOLLEY, D. W. 89, 90, 93, 94, 95, 98, 100, 102, 106, 113, 389, 399.

WORK, E. 309, 362.
WORK, T. S. 309, 362.
WORMALL, A. 458, 478.
WÖSTMANN, B. 464.
WRANN, S. 195, 199, 215.
WRIGHT, G. F. 232, 291.
WRIGHT, H. F. 58, 60, 62, 86.
WRIGHT, H. R. 53, 56, 83.
WRIGHT, L. D. 410, 412, 417, 418, 420, 421, 440, 441, 442.
WRINCH, D. M. 138, 148, 174.
WÜEST, H. M. 48, 49, 50, 56, 57, 86.

YAMAGATA, M. 187, 223.
YAMASHITA, M. 256, 291.
YAMASHITA, Y. 195, 217.
YANO, K. 194, 205, 219, 224.
YESHODA, K. M. 316, 362.
YOSHINO, M. 213, 220.
YOUNG, W. G. 46, 48, 87.
YOUNGER, F. 97, 111.
YUILL, M. E. 450, 482.
YUNUSOV, S. 179, 224.

ZAMECNIK, P. C. 305, 311, 314, 315, 355, 359, 362.
ZECHMEISTER, L. 4, 5, 8, 16, 26, 27, 33, 40, 51, 87.
ZEILE, K. 65, 87.
ZEUTHEN, E. 394, 400.
ZIEGLER, K. 53.
ZUCKER, L. M. 417, 442.
ZUCKER, T. 417, 442.
ZWERINA, K. 181, 199, 215.
ZWINGELSTEIN, G. 56, 71, 75, 77, 80, 85, 103, 111.

Sachverzeichnis. Index of Subjects. Index des Matières.

Acetate of C_{30}-diol 61.
Acétate de tocophérol 104.
Acetobacter suboxidans 409.
3-Acetoxy-coumaran 260.
2-Acétoxy-3,4-diacétyl-méthyltoluène 96.
Acetyl, antigenicity 447.
N-Acetylated amino acids, peptide and protein biosynthesis 340.
Acetylated leucine, inhibitory activity 450.
8-Acetyl-daphnetin 247.
Acetylen-dimagnesium-bromid 9, 15, 24.
C_{38}-Acetylen-diol 16.
β-C_{38}-Acetylen-diol 15.
Acetylene 22, 58, 62.
C_{20}-Acetylenic diol 61.
Acetylen-Lithium 10.
N-Acetyl-glycine 338.
Acetyl group 324.
Acetyl-phenylalanine 341.
Acetyl-*L*-phenylalanine 341.
Acetyl-*L*-phenylalanyl-*L*-glutamylanilide 324.
Acetyl-phenylalanyl-glycine 326, 327.
Acetyl-*DL*-phenylalanyl-glycine 326.
Acetyl-*L*-phenylalanyl-glycine anilide 326.
Acetyl-phenylalanyl-glycyl-glycyl-leucine 326, 327.
Acétyl-pyridine 98.
3-Acétyl-pyridine 98.
Acetyl-tyrosine 341.
Acétyl-*L*-tyrosine 341.
C_{16}-Acid 72, 79.
C_{17}-Acid 72, 79.
C_{18}-Acid 72, 79.
C_{19}-Acid 72, 79.
C_{20}-Acid 56.
Acide *p*-aminobenzoïque 100.
Acide *p*-aminobenzoïque, activité antisulfamide et antisulfone 101.
Acide *p*-aminobenzoïque et sulfamidés, antagonisme 101.
Acide folique 100.

Acide *p*-hydroxybenzoïque 101.
Acide nicotinique, antagonistes 98.
Acide *p*-oxybenzoïque 101.
Acide pantothénique 95.
Acide pantothénique, antagonistes 95.
Acide xanthurénique 96.
Acrodynie 96.
Actinomycetes and vitamin B_{12} 420.
Activated carbonyl 330.
Activité antifongique, dichloro-2,3-naphtoquinone 106.
Activité antisulfamide et antisulfone de l'acide *p*-aminobenzoïque 101.
Aculeatin 252.
Aculeatin hydrate 251.
Acutumin 177, 207.
N-Acylated amino acids in peptide and protein biosynthesis 340.
Acyl group 324.
8-Acyl-7-hydroxy-coumarins 232, 233.
2-Acyl-resorcinol 233.
Adenine 324, 380, 390, 392, 394, 407.
Adenine, deamination 393.
Adenine, labeled 414.
Adenine compounds, deamination to hypoxanthine compounds 379.
Adenine-cytosine dinucleotide 410.
Adenine desoxyribonucleotide 424.
Adenine desoxyriboside 404, 432.
Adenine nucleosides 410.
Adenine nucleosides, estimation 366.
Adenine nucleotides 410.
Adenine thiomethyl-pentoside 408.
Adenine thiomethyl-riboside 370.
Adenine triphosphate 391.
Adenosine 375, 376, 379, 380, 391, 394, 404, 405, 407, 410.
Adenosine, estimation 366.
Adenosine, synthesis 365.
Adenosine compounds, pharmacological action 378.
Adenosine deaminase 371, 378.
Adenosine deaminase, bacterial 380.
Adenosine deaminase, intestinal 379.

Adenosine diphosphate 334, 337.
Adenosine-5'-diphosphate 403.
Adenosine, growth inhibitor 409.
Adenosinekinase 391.
Adenosine monophosphate 337, 339.
Adenosine-monophosphate in glutathione synthesis 342.
Adenosine-3'-phosphate 405, 407, 408, 410.
Adenosine-3'-phosphate, growth factor 406, 410, 411.
Adenosine-3'-phosphate, growth inhibitor 309, 310.
Adenosine-5-phosphate 391.
Adenosine-5'-phosphate 405, 408, 410.
Adenosine-5'-phosphate, growth factor 411.
Adenosine-5'-phosphate, growth inhibitor 410.
Adenosine polyphosphate 364, 379.
Adenosine triphosphate 314, 334, 336, 337, 341, 405.
Adenosine-5'-triphosphate 403.
Adenosine triphosphate, growth factor 405, 406.
Adenosine triphosphate in glutathione synthesis 342.
Adenosine triphosphate requirement, *Hemophilus piscium* 406.
Adenylic acids 404.
5-Adenylic acid 379, 391.
Adenylic acid a, growth inhibitor 410.
Adenylic acid b, growth inhibitor 410.
Adenylic acid, labeled 392.
ADP 403.
Aegle marmelos 243, 272.
Aesculetin 226, 244.
Aesculetin dimethylether 245.
Aesculetin glucosides 244.
Aesculetin 6-methyl ether 245.
Aesculin 244.
Aesculin, use for identification of bacteria 242.
Aesculus hippocastanum 244.
Aesculus turbinata 244, 247.
Ageing, antigen-antibody precipitates 466.
Agglutinins in lobster blood 451.
Aggregation, polyvalent haptens 476.
Agmatine 119.
L-Alanine 166.
DL-Alanine 321.

DL-Alanine, free energy of formation 320.
Alanine, labeled 311, 313.
Alanine methyl ester 141.
Alanyl-alanine 349.
Alanyl-glycine 321.
DL-Alanyl-glycine 321.
DL-Alanyl-glycine, free energy of formation 320.
Alanyl-glycine, synthesis 322.
Alanyl group 324.
Albertisia papuana 178.
C_{17}-Alcohol 70, 74.
C_{14}-Aldehyde 58, 60, 61, 62.
C_{14}-Aldehyde, infrared spectrum 60.
C_{14}-Aldehyde, RAMAN spectrum 60.
β-C_{18}-Aldehyd 15.
Aldéhydodéhydrase 94.
Alkaloide (Menispermaceae) 175, 177.
Allo-bergapten 262, 263.
Allo-imperatorin 268, 269.
Alloxan 307.
Allo-xanthoxyletin 281.
All-*trans*-β-carotin 4, 26.
All-*trans*-16,16'-homo-β-carotin 32.
All-*trans*-Konfiguration 9.
Allylic rearrangement 60, 65.
7-Allyl-oxycoumarin 232.
8-Allyl-umbelliferone 232.
Allyl-Umlagerung 9, 11, 15, 17.
Aluminiumbutylat, tert. 35.
Ambalin 178, 207.
Ambalinin 178, 207.
Amide bonds, enolization 331.
Amino acids 166.
Amino acid concentration in blood 297.
Amino acid esters in peptide synthesis 333.
Amino acid exchange 298.
Amino acids, free energy of formation 319.
Amino acids (foreign), incorporation 308.
Amino acid incorporation, bone marrow cells 343.
Amino acid incorporation, heat-stable cofactors 346.
Amino acid incorporation, inhibitors 343.
Amino acid incorporation, mechanism 343.
Amino acid incorporation and phosphorylation 346.
Amino acid incorporation, proteins 343.
Amino acid incorporation, *de novo* protein synthesis 347.

Amino acids, incorporation *in vivo* and *in vitro* 313.
Amino acid, labeled 300.
Amino acids, labeled, incorporation *in vitro* 317.
Amino acids in plasma 297.
Amino acid turnover, glutathione 332.
Amino acid turnover, proteins 302.
Aminoacyl group 324.
2-Amino-adenine 384.
2-Amino-adenosine 379, 391.
2-Amino-adenosine triphosphate 391.
α-Amino-adipic acid, labeled 308.
p-Amino-benzenarsonic acid 462.
o-Amino-benzoic acid 324, 462.
p-Amino-benzoic acid 338, 340.
p-Amino-benzoic acid, condensation with hippuric acid 324.
p-Amino-benzoic acid, reversion of growth inhibition 410.
3'-Amino-5 : 6-benzoquinoline-3 : 7-dicarboxylic acid lactam 125.
3'-Amino-5 : 6-benzoquinolone(4)-3 : 7-dicarboxylic acid ethyl ester 132.
3'-Amino-5 : 6-benzoquinolone(4)-3 : 7-dicarboxylic acid lactam ethyl ester 132.
Amino-bergapten 262.
α-Amino-butyric acid 166.
Amino-cobalamin 420.
p-Amino-hippuric acid 345.
p-Amino-hippuric acid, enzymatic synthesis 338, 340.
p-Amino-hippuric acid, synthesis 338.
1-Amino-8-hydroxy-naphthalene-3,6-disulfonic acid 457.
4-Amino-5-imidazole 390.
4-Amino-5-imidazole carboxamide 415.
4-Amino-imidazole-5-carboxamide-desoxyriboside 389.
4-Amino-imidazole-5-carboxamide, precursor of purines 389.
3'-Amino-1-methyl-1 : 2 : 3 : 4-tetrahydro-5 : 6-benzoquinoline-3 : 7-dicarboxylic acid lactam 125.
8-Amino-1-naphthol-3,6-disulfonic acid 467.
4-Amino-naphthostyril 131, 132.
Amino nitrogen in blood 296.
Amino nitrogen in muscle 296.
Amino nitrogen in pea seedlings 296.
Amino nitrogen in salmon 296.

Amino nitrogen in tissues 296.
p-Amino-L-ornithuric acid 340.
p'-Amino-L-ornithuric acid 340.
o-Amino-phenol, condensation with hippuric acid 324.
p-Amino-phenol, condensation with hippuric acid 324.
p-Amino-phenylarsonate 456.
p-Amino-phenylarsonate, coupled to protein 449, 450.
p-Amino-phenylarsonic acid 449.
p-Amino-phénylsulfamide 100.
D-2-Amino-propanol 151.
L-2-Amino-propanol 151.
L-2-Amino-propanol-(1) 152.
Amino-ptérine 101.
4-Amino-pteroylglutamic acid 431.
6-Amino-pteroylglutamic acid 431.
α-Amino-pyridine 324.
Ammidin 268.
Ammi majus 263, 265.
Ammoidin 264, 265.
Ammonia 134, 324.
Ammonia, labeled 393.
Ammoniacum 243.
Ammonium molybdate (inhibitor) 344.
Ammoresinol 226, 227, 240, 256.
Ammoresinol diacetate 256.
Amphibian egg 394.
Anaerobiosis (inhibitor) 343, 344.
Anaerobiosis (inhibitor of amino acid incorporation) 345.
Anamirta cocculus 178.
Anamirta paniculata 178, 207.
Aneurine 92.
Angelic acid 268.
Angelica archangelica 261, 265.
Angelica glabra 269, 270.
Angelicin 259, 261.
Anhydro-byakangelicin 271.
Anhydro-nodakenetin 272.
Anhydro-vitamin A 43, 61, 72, 77, 78.
Anhydro-vitamin A_1 78.
Anhydro-vitamin A_2 72, 77, 78.
Aniline 324, 328.
Aniline, condensation with hippuric acid 324.
4-Anilino-coumarin 233, 257.
Animal nucleoproteins, antigenicity 447.
o-Anisidin, condensation with hippuric acid 324.
Anonaceae, Biscoclaurin-Basen 179.

Anorexie 92.
Antagonistes de l'acide nicotinique 98.
Antagonistes de l'acide panthothénique 95.
Antagonistes de la biotine 98.
Antagonistes des flavines 94.
Antagonistes de la pyridoxine 95.
Antagonistes de la riboflavine 94.
Antagonistes de la thiamine 90.
Antagonistes de la thiamine de nature enzymatique 92.
Antagonistes de la vitamine B_{12} 94.
Antagonistes des vitamines hydrosolubles 90.
Antagonistes des vitamines liposolubles 102.
Anthrosolen polycephalus 247.
Antiacides foliques 100.
Antibodies 445.
Antibody, acid dissociation 462.
Antibody activity, globulin fragments 455.
Antibodies, amino acid composition 451.
Antibody/antigen ratio 464.
Antibody, bivalence 457, 458.
Antibodies, chemical nature 443.
Antibody combining site 455, 460.
Antibody combining site, area 455.
Antibody, denaturation 460.
Antibody, electrophoretic properties 452.
Antibody, framework theory 457.
Antibody/γ-globulin ratio 461.
Antibody, heterogeneity 459.
Antibodies, isoelectric points 453.
Antibodies, molecular weight 454.
Antibody molecules, asymmetry values 454.
Antibody molecules, shape and size 453.
Antibody molecules, valency 456.
Antibodies, physical nature 478.
Antibodies, properties 451.
Antibody proteins 298.
Antibodies, purification 461.
Antibody response 446.
Antibodies, unitarian theory 459.
Antibodies, univalence 457, 458.
Antigen-antibody combinations, free-energies 468, 469.
Antigen-antibody complexes 462.
Antigen-antibody, initial reaction 465.
Antigen-antibody molecules in precipitates 454.
Antigen-antibody precipitates 463.
Antigen-antibody precipitates, ageing 466.
Antigen-antibody precipitates, labeled 467.
Antigen/antibody ratio 470.
Antigen-antibody reactions 443.
Antigen-antibody reactions, coulombic attraction 471.
Antigen-antibody reactions, forces involved 471.
Antigen-antibody reactions, free energy and heat changes 467, 468.
Antigen-antibody reactions, hydrogen bonding 471.
Antigen-antibody reactions, physical nature 463.
Antigen-antibody reactions, short-range forces 471.
Antigen-antibody reactions, thermodynamic properties 466.
Antigen-antibody reactions, VAN DER WAALS forces 471.
Antigen and antibody separated by barriers 471.
Antigens 445, 446.
Antigens, chemical nature 443.
Antigens, conversion into insoluble state 463.
Antigens, general properties 446.
Antigens, physical nature 478.
Antigens, tagged 464.
Antigenicity 445.
Antigenicity, acetyl derivatives 447.
Antigenicity, animal nucleoproteins 447.
Antigenicity, invertebrate polysaccharides 447.
Antigenicity, lipids 448.
Antigenicity, low-molecular weight compounds 448.
Antigenicity, polysaccharides 447.
Antiovalbumin 462.
Antipneumococcus antibodies 460.
Antipneumococcus serum 452.
Anti-R_p serum 473.
Anti-R_p' serum 473, 474.
Anti-X_p serum 473.
Antisulfamides 100.
Antithiamines des fougères 93.
Antivitamines 88.
Antivitamine A 102.
Antivitamine B_1 90, 91, 93.

Antivitamine B_6 97.
Antivitamine B_6, constitution 96.
Antivitamine E 102, 103.
Antivitamine K 102, 104, 106.
Antivitamin K effect, coumarins 242.
Apionol-1-methyl ether 248.
Aporphin-Typus 180.
Apoxanthoxyletin 262, 279.
Arbacia 394.
Archangelica officinalis 261, 265.
Archangelisia flava 177.
Archangelisia lemniscata 178.
Arginase 298.
Arginine 452.
Arginine, inhibitory activity 450.
Aromolin 179, 180, 198.
Arsanil-beef serum globulin 468.
Arsanilic acid 467.
Arsenate (activator) 365.
Arsenate (inhibitor) 343, 344.
Arsenite (inhibitor) 343, 344.
Artemisia capillaris 245.
Artificial haptenic groups 450.
Asafaetida 243.
Ascorbic acid 411, 428, 429, 433.
Ascorbic acid, growth promoter 427.
Ash 246.
Aspartic acid 349.
Aspartic acid, labeled 310, 312.
Aspergillus niger 101, 107.
Aspergillus oryzae 422, 425.
Asparto-amide transferase 336.
Asparto-transferase 334.
Asymmetry values, antibody molecules 454.
Athamanta oreoselinum 275.
Athamantin 275, 276.
α-C_{16}-Äthinyl-carbinol 30.
β-C_{16}-Äthinyl-carbinol 10, 19.
C_{21}-Äthinyl-carbinol 35.
ATP 403.
Auraptene 252, 253.
Avena roots (scopoletin) 241.
Avitaminose B_1 90, 92.
Axerophthene 72, 78.
Axerophthol 42.
Ayapanin 243.
Ayapanin, haemostatic properties 242.
Ayapin 245.
Azaguanine 369.
8-Azaguanine 404.

8-Azaguanine, antagonism against guanine 415.
8-Azaguanine nucleoside 415.
Azide (inhibitor) 344, 346.
Azin des C_{19}-Carotinoid-Aldehyds 24.
Azo-antigens 462.
p-Azophenylarsonic acid 454.
p-Azophenylacetic acid, labeled 464.
Azophenylarsonic acid dyes 444, 464.
Azophenylarsonic acid dyes, divalent 464.
Azophenylarsonic acid dyes, trivalent 464.
Azoprotein 459, 462, 468.
Aztequin 179, 180, 193.

Bacillus anthracis, adenosine requirement 406.
Bacillus anthracis, germination 406.
Bacillus lactis acidi 94.
Bacillus megatherium 95.
Bacterial adenosine deaminases 380.
Bacterial growth, desoxyribosides 424.
Bacterial polysaccharides 448.
Bacterial trans-N-glycosidase 388.
Bacterial viruses 423.
Bacteriophage T_4 and vitamin B_{12} 419.
Bacteriophage T_6 423.
T_2-Bactériophage du colibacille 96, 97.
Barium stearate 471, 472.
Barrier films, imperfections 472.
VIII-Base 177, 207.
Beaver 282.
Bebeeria 201.
Bebeerin 201, 212.
D-Bebeerin 178, 180, 201, 202.
L-Bebeerin 178, 180, 201.
Bebeerin, Biosynthese 212.
Bebeerinum purum 201.
Benzimidazole nucleotide 417.
α-Benzimidazole riboside 417.
Benzoate ion 321.
Benzoate ion, free energy of formation 320.
Benzochinolizin-Typus 180, 191.
3 : 4-Benzocoumarin 228, 282, 283.
3 : 4-Benzocoumarin (beaver) 239.
Benzoic acid 321.
Benzoic acid, free energy of formation 320.
Benzoyl-arginine amide 329.
Benzoyl-L-arginine amide 329, 330.
Benzoyl-glycine amide 329.
Benzoyl-glycine amide, digestion with papain 329.
Benzoyl group 324.
Benzoyl-L-leucine 325, 328.

Benzoyl-L-leucine anilide 328.
Benzoyl-L-leucyl 328.
Benzoyl-leucyl-leucine anilide 325.
Benzoyl-L-leucyl-leucine anilide 328,
Benzoyl-L-leucyl-L-leucine anilide 325.
α-Benzoyl-L-ornithine 339, 340.
δ-Benzoyl-L-ornithine 339, 340.
Benzoyl-phenylalanine 325.
Benzoyl-phenylalanyl-leucine anilide 325.
Benzoyl-L-tyrosine 325, 331.
N-Benzoyl-tyrosine 331.
Benzoyl-tyrosyl-glycine amide 330.
Benzoyl-1-tyrosyl-glycine amide 330.
Benzoyl-L-tyrosyl-glycine amide 325.
N-Benzoyl-tyrosyl-glycine amide 331.
Benzoyl-tyrosyl-glycine anilide 325.
Benzoyl-L-tyrosyl-glycine anilide 325.
Benzoyl-tyrosyl-leucine anilide 325.
5 : 6-Benz-α-pyrone 229.
Benzyl-amine 324.
N-Benzyl-ergobasine 165.
Benzyl-isochinolin-Typus 180, 186.
Berbamin 177, 179, 180, 181, 182, 185, 186, 193, 194, 196, 210.
Berbamin, Biosynthese 210.
Berberidaceae, Biscoclaurin-Basen 179.
Berberin 177, 178, 190, 191.
Berberin-Typus 180, 190.
Berberis aquifolium 179.
Berberis heteropoda 179.
Berberis repens 179.
Berberis Thunbergii 179, 196.
Berberis vulgaris 179, 193.
Berberitze 179, 181, 195.
Bergamot camphor 262.
Bergamot oil 263.
Bergamottin 269.
Bergapten 262, 263, 264.
Bergapten-quinone 262, 264, 265, 269.
Bergaptin 269.
Bergaptol 263, 264, 266, 267, 268.
Bergaptol-y : y-dimethylallyl ether 266.
Bergaptol geranyl ether 269.
L-C^{14}-Berine, incorporation 307.
BERT's method 244.
Betaine 119.
Bezoar stones 283.
Biogenese, Biscoclaurin-Basen 209.
Biosynthese, Bebeerin 212.
Biosynthese, Berbamin 210.
Biosynthese, Cepharanthin 211.
Biosynthese, Cycleanin 212.

Biosynthese, Daphnolin 210.
Biosynthese, Dauricin 210
Biosynthesis, desoxyribosides 421, 432.
Biosynthese, Epistephanin 213.
Biosynthese, Insularin 212.
Biosynthese, Isochondodendrin 212.
Biosynthese, Isotetrandrin 210.
Biosynthese, Isotrilobin 210.
Biosynthese, Magnolin 210.
Biosynthese, Menisarin 211.
Biosynthese, Micranthin 211.
Biosynthese, Oxyacanthin 210, 213.
Biosynthesis, peptides 292.
Biosynthese, Phaeanthin 210.
Biosynthesis, proteins 292.
Biosynthesis, ribonucleotides 413.
Biosynthesis, ribosides 413.
Biosynthese, Tetrandrin 210.
Biosynthese, Trilobamin 210.
Biosynthese, Trilobin 211.
Biotine 98.
Biotine, analogues guanidinés 100.
Biotine, antagonistes 98.
Biotine-sulfone 99.
Bis-benzylisochinolin-Alkaloid 180.
Biscoclaurin-Alkaloid 177, 180.
Biscoclaurin-Alkaloide, Abbau 181.
Biscoclaurin-Alkaloide in Curare 179.
Biscoclaurin-Alkaloide, Ozon-Spaltung von Methinbasen 181.
Biscoclaurin-Alkaloide, Permanganat-Oxydation 181.
Biscoclaurin-Basen, Asymmetrie-Zentren 207, 208.
Biscoclaurin-Basen, Biogenese 209.
Biscoclaurin-Basen, Charakterisierung 209.
Biscoclaurin-Basen, Diphenyläther-Struktur 209.
Biscoclaurin-Basen, Diphenylendioxyd-Gruppierung 209.
Biscoclaurin-Basen, optische Isomerie 207.
Biscoclaurin-Basen, Pharmakologie 213.
Biscoclaurin-Basen, Spaltung durch Natrium in flüssigem Ammoniak 184.
Biscoclaurin-Basen, Systematik 180.
Biscoclaurin-Basen in Anonaceae 179.
Biscoclaurin-Basen in Berberidaceae 179.
Biscoclaurin-Basen in Magnoliaceae 179.
Biscoclaurin-Basen in Monimiaceae 179.
Biscoclaurin-Typus 180, 192.

Bis-nor-methyl-β-carotin 13.
13,13'-Bis-nor-methyl-β-carotin 15, 16.
13,13'-Bis-nor-methyl-β-carotinin 15.
13,13'-Bis-nor-methyl-15,15'-dehydro-β-carotin 15.
Bivalence, antibodies 457, 458.
Bivalent antibodies 454.
Bleu de méthylène 104.
Blood, amino acid concentration 297.
Blood, free amino nitrogen 296.
Blow-fly 240.
Bone marrow cells, incorporation of amino acids 317, 343.
Bone marrow cells and labeled acetate 310.
Bound pantothenic acid 409.
Bovine serum albumin 458, 464.
Bovine serum albumin antigen, labeled 464.
Brayleyanin 250, 251.
Braylin 250, 281, 282.
Brochet 93.
1-Brom-3-methyl-penten-2-in-4 16.
Bromoacetal 124.
2-Bromo-5-hydroxybenzoic acid 282.
γ-Bromosenecioate 63, 66.
N-Bromosuccinimide 77.
5-Bromouridine, growth inhibitor 412.
BTGA 330.
Busycon cannaliculatum 468.
t-Butyl hypochlorite 64.
Byakangelicin 227, 270, 271.
Byakangelicin diacetate 271.
Byakangelicol 270, 271, 272.

Cadaverine 119.
Carbobenzoxy-*iso*asparagine 330.
Carbobenzoxy-*L*-*iso*asparagine 329.
Carbobenzoxy-*iso*glutamine 330.
Carbobenzoxy-*L*-*iso*glutamine 329.
Carbobenzoxyl group 324.
Carbobenzoxy-methionine amide 330.
Carbobenzoxy-*D*-methionine amide 329.
Carbobenzoxy-*L*-methionine amide 329.
Carbobenzoxy-phenylalanyl-glycine 325, 326.
Carbobenzoxy-*L*-phenylalanyl-glycine 327.
Carbobenzoxy-phenylalanyl-glycyl-glycine anilide 326.
Carbobenzoxy-phenylalanyl-glycyl-tyrosine amide 325.
Carbobenzoxy-serine amide 330.

Carbobenzoxy-*L*-serine amide 329.
Carbon dioxide incorporation into amino acids 310.
Carbon dioxide, labeled 310, 312, 351.
Carboxamide chlorohydrate 390.
Carboxy-peptidase 332, 351.
4-Carboxy-uracil 403.
4-Carboxy-uracil, growth factor 412.
4-Carboxy-vitamin A acid 65.
cis-4-Carboxy-vitamin A acid 65.
trans-4-Carboxy-vitamin A acid 65.
4-Carboxyvitamin A acid, photomicrograph 69.
DL-α-Carotin 31.
β-Carotin 21, 28, 31.
γ-Carotin 2, 3.
ε_1-Carotin 31.
β-Carotin-15,15'-in 25.
all-*trans*-β-Carotin 4, 26.
15,15'-mono-*cis*-β-Carotin 26.
15,15'-mono-*cis*-β-Carotin, Bioeffekt 26.
15,15'-mono-*cis*-β-Carotin, Stereoisomerisierung 26.
neo-β-Carotine 4.
β-Carotin, Bezifferung der C-Atome 4.
β-Carotène, oxydation 102.
β-Carotin, spontane Isomerisierung 5.
β-Carotin, Synthesen 19.
β-Carotin-Synthesen $C_{16} + C_8 + C_{16}$ 19.
β-Carotin-Synthese $C_{18} + C_4 + C_{18}$ 27.
β-Carotin-Synthese $C_{19} + C_2 + C_{19}$ 22.
γ-Carotin, Bezifferung der C-Atome 4.
ε_1-Carotin, CARR-PRICE-Reaktion 31.
ε_1-Carotin, *cis-trans*-Isomere 30.
ε_1-Carotin, Synthese 30.
Carotenes and vitamin A, relationship 80.
β-Carotinin 25.
Carotinoid-Aldehyd 24.
Carotinoid-Aldehyd, Azin 24.
Carotinoide, Bezifferung 3.
all-*trans*-Carotinoide, Eigenschaften 6.
Carotinoide, *cis-trans*-Isomerie 4.
Carotinoide, Isomerisierung durch Jod 6.
Carotinoide, Isomerisierung mit Säure 6.
Carotinoide, Isomerisierung durch Schmelzen 6.
Carotinoide, Jod-Katalyse 5.
Carotinoide, Methoden der Stereoisomerisierung 5.
Carotinoide, Nomenklatur 2.
Carotinoide, Photo-isomerisierung 6.
Carotinoid-epoxyde 2.

Carotinoide, spektrale Veränderungen durch Stereoisomerisierung 8.
Carotinoide, Spektrum 9.
Carotinoide, Stereochemie 4.
Carotinoide, sterische Hinderung 7.
Carotinoid-Synthesen mit Diacetylen 12.
Carotinoide, synthetische Chemie 1.
Carpe 92.
Castoreum 282.
Catalase 298.
Cathepsin 298.
Cell fractions, incorporation of amino acids 313.
Cell nuclei, incorporation of amino acids 313, 314.
Cellobiuronic acid 450.
Cepharanthin 177, 180, 195, 211, 213.
Cepharanthin, Biosynthese 211.
Chemical equilibrium 476.
Chicken liver slices (protein synthesis) 312.
2-Chloro-adenosine 379.
Chloro-cobalamin 420.
1-Chloro-2-methyl-4-methoxy-2-butene 64.
Chloro-2-naphtoquinone-1,4 107.
5-Chloro-uridine 417.
5-Chloro-uridine, growth inhibitor 412.
Choline 119.
Chondocurin 178, 202, 203.
D-Chondocurin 180.
Chondrodendrin 201.
Chondrodendron candicans 178, 201.
Chondrodendron microphyllum 178, 201.
Chondrodendron platyphyllum 178, 201, 202.
Chondrodendron tomentosum 178, 179, 200, 202, 207.
Chondrofolin 178, 180, 202.
Chromeno-α-pyrones 276.
545-Chromogen 75.
Chromones 235.
Chrysatropic acid 245.
Ch'uan sen 239.
Chymotrypsin 325, 328, 331, 333, 351.
Chymotrypsin benzoyl-L-tyrosine 325.
Cichorigenin 244.
Cichoriin 244.
Cissampelos insularis 177.
Cissampelos ochiaiana 177.
Citidylic acid 403.
Citropten 226, 245, 246, 250.
Citrovorum factor 417, 420, 422, 431, 432.

Citrovorum-factor in growth of microorganisms 426.
Citrus acida 263, 277.
Citrus Medica 263, 277.
CLAISEN migration 232.
CLAISEN reaction 237.
Claviceps purpurea 114, 119.
Clavine 119.
Clostridium sporogenes 407.
Cobalamin 419.
Cocarboxylase 338, 339.
Cocculidin 179.
Cocculin 179.
Cocculus bakis 178.
Cocculus crispus 178.
Cocculus laurifolius 176, 177, 179, 187, 213.
Cocculus ovalifolius 178.
Cocculus peltata 177.
Cocculus sarmentosus 177, 203, 204.
Cocculus trilobus 177, 178, 198, 203, 204, 213.
Cocculus umbellatus 178.
Coclaurin 176, 177, 180, 186, 187, 213.
Coclaurin, Biosynthese 211.
Coenzyme I 338, 339, 403, 405, 407.
Coenzyme II 339, 403, 405.
Coenzyme A 408.
Collagen, lack of antigenicity 446.
Collinin 250.
Columbamin 177, 178, 190, 191.
Columbo-Wurzel 178, 213.
Combining site, antibodies 455, 460.
Compound „Y" 71, 75.
Compound „Z" 71, 75.
Convulsive ergotism 115.
Corn pegs 116.
Coronilla glauca 261.
Corps „Z" 103, 102.
Corydaldin 181.
Coscinium blumeanum 177, 190.
Coscinium fenestratum 177.
Coulombic attraction, antigen-antibody reactions 471.
Coumaric acid 228, 229, 230.
Coumaric acid, isomerization 229.
Coumaric acid, reduction 230.
Coumarilic acid 232.
Coumarin 225, 226, 230, 232, 242.
Coumarins, antivitamin K effect 242.
Coumarins, biochemical properties 240.
Coumarins, biogenesis 228.

Coumarins and carbonyl reagents 231.
Coumarin-3-carboxylic acid 235.
Coumarin-3-carboxylic acid, decarboxylation 235.
Coumarins, color reactions 240.
Coumarins, degradation by alkali 231.
Coumarins, determination 239.
Coumarin, dimer 233.
Coumarins, dimerization 233.
Coumarin, dipole moment 237.
Coumarins, electronic interpretation 237.
Coumarins, estimation 240.
Coumarin glycoside 239.
Coumarin as growth inhibitor 240.
Coumarins, isolation 239.
Coumarins, isoprenoid units 226.
Coumarins, methylation 232.
Coumarin as narcotic 241.
Coumarin, nitration 232.
Coumarins, occurence 239.
Coumarins, odor 240.
Coumarins, oxydation 230.
Coumarin, RAMAN spectrum 239.
Coumarins, structure 226.
Coumarins in sunburn prevention 241.
Coumarins, synthesis 235.
Coumarin system, conversions 229.
Coumarin system, degradation 229.
Coumarins, toxicity 241.
Coumarinic acid 228, 229, 230.
Coumarinic acid, reduction 230.
Crategin 244.
Crebanin 177, 180, 189, 190.
Crocetin 3.
Cularin 184, 209.
Curare 199.
Curare (Biscoclaurin-Alkaloide in) 179.
Curare-Alkaloide 179.
Curin 180.
L-Curin 178, 201, 203.
Cyanide (inhibitor) 345.
Cyano-acetal 124.
Cyano-cobalamin 419, 420.
Cyano-cobaltamine (synthesis of desoxyribosides) 395.
2-Cyano-2-formylethylidene-4-aminonaphthostyril 124, 125.
Cyano-malonaldehyde 124.
6-Cyano-7-methoxycoumarin 254.
Cyclea barbata 178.
Cyclea insularis 177, 200, 205.
Cyclea peltata 177, 207.

Cycleanin 177, 180, 200, 212.
Cycleanin, Biosynthese 212.
Cyclein 177, 207.
Cyclic peptides 327.
Cyclized peptides 138.
β-Cyclocitral 63, 66.
Cyclohexamine 324.
Cyclol structure 146.
Cyclol theory of proteins 138.
Cynomyia cadaverina 240.
Cystine 452.
Cystine, inhibitory activity 450.
Cytidine 375, 380, 403, 409, 412.
Cytidine, conversion into cytosine desoxyriboside 423.
Cytidine deaminase 380.
Cytidine, deamination 417.
Cytidine + uridine, microbiological assay 412.
Cytidylic acid 380, 409, 412.
Cytidylic acid, growth factor 410, 411.
Cytosine 380, 412, 415.
Cytosine deaminase 423.
Cytosine desoxyribonucleotide, growth effect 424.
Cytosine desoxyriboside 380, 403, 423, 428, 429, 432.
Cytosine desoxyriboside deaminase 423.
Cytosine desoxyriboside, growth effect 424.
Cytosine desoxyriboside, growth factor 412.

DAKIN's reaction 265.
Daphnandra aromatica 179, 198.
Daphnandra Dielsii 179, 207.
Daphnandra micrantha 179, 198, 205.
Daphnandra repandula 179, 197, 207.
Daphnandrin 179, 180, 198.
Daphne odora 241.
Daphnetin 226, 246.
Daphnetin diethylether 247.
Daphnetin-8-methyl ether 250.
Daphnin 241, 246, 247.
Daphnolin 179, 180, 196, 198, 210.
Daphnolin, Biosynthese 210.
Dauricin 177, 180, 192, 210.
Dauricin, Biosynthese 210.
Deaminase (adenosine) 379.
Deaminase, bacterial 380.
Deaminase (cytidine) 380.
Deaminase, intestinal 366, 379.
Deaminase (purines) 381.

Deaminase (pyrimidines) 381.
Decapreno-ε_1-carotin, CARR-PRICE-Reaktion 35.
Decapreno-β-carotin, Synthese 34.
Decapreno-ε_1-carotin, Synthese 35.
Décarboxylase 97.
Dehydro-β-carotin 28.
15,15'-Dehydro-β-carotin 25.
Dehydrohalogenation 77.
3-Dehydro-β-ionone 77.
Denatured ovalbumin 463.
Denaturated proteins 451.
De novo protein synthesis by amino acid incorporation 347, 348.
Deoxy- = Desoxy-
Depsidan 209.
Depsidon 209.
Desamino-codehydrogenase I 405.
Des-Base von O,O-Đimethyl-isochondodendrin 181.
Desmethyl-axerophthene 72.
Desoxy-apoxanthoxyletin 279.
Desoxydihydro-oreoselone 272, 273, 274.
Désoxypyridoxine 96, 97.
Désoxypyridoxine, activité antivitaminique B_6 96.
Desoxyribocytidine 393.
Desoxyribonucleic acid 422.
Desoxyribonucleic acid in bacterial cells, estimation 425.
Desoxyribonucleic acid, biosynthesis 416, 421.
Desoxyribonucleic acid, release from tissues 425.
Desoxyribonucleotides, growth factors 411, 424.
β-D-2-Desoxyribose 364.
Desoxyribose-1-arsenate 378.
Desoxyribose formation 390.
Desoxyribosenucleic acid 315.
Desoxyribose nucleosides 420.
Desoxyribose nucleosides, trans-N-glycosidic reactions 387.
Desoxyribose nucleotides 420.
Desoxyribose-1-phosphate 372, 385, 386, 387, 418.
Desoxyribose-5-phosphate 377.
2-Desoxy-D-ribose-1-phosphate 385.
Desoxyribose-1-phosphate, biosynthesis 418.
Deoxyribose-1-phosphate in trans-N-glycosidic reactions 387.

Desoxyribosides 373.
Desoxyribosides in bacterial growth 424.
Desoxyribosides, biological function 364.
Desoxyribosides, biosynthesis 421, 432.
Desoxyribosides, enzymatic formation 388.
Desoxyribosides, enzymatic synthesis 389.
Desoxyribosides in frog eggs 394.
Desoxyribosides in growth of microorganisms 426.
Desoxyribosides, growth-promoting activity 424.
Desoxyribosides, microbiological assay 424.
Desoxyribosides, non-specificity in bacterial growth 425.
Desoxyribosides, paper chromatography 425.
Desoxyriboside phosphorylase and riboside phosphorylase, relationship 370.
Desoxyribosidic linkage, formation 422.
Desoxyribosyl group, enzymatic transfer 425.
Desthiobiotine 98.
Desthiobiotine, dérivés 99.
Dextrans, haptenic activity 448.
Dextrorotatory ergot alkaloids, dihydro derivatives 168.
3:6-Diacetoxy-coumaran 260.
Diacetylen 12, 31.
Diacetylen in Carotinoid-Synthesen 12.
Diacetylen-dimagnesiumbromid 12.
2,6-Diamino-purine 384, 388, 390, 404.
2,6-Diamino-purine, growth-inhibitor 410.
2,6-Diaminopurine, incorporation 414.
2,6-Diaminopurine, labeled 391.
2,6-Diaminopurine, precursor of guanine 413.
2,6-Diamino-9-β-D-ribofuranosyl-purine 391.
Diaphragm, incorporation of amino acids 317.
Di-äthinyl-Kohlenwasserstoff 18.
Diazomethane and 4-hydroxycoumarins 234.
Dicentrin 177, 190.
D-Dicentrin 190.
1,2-Dichloro-4,5-diaminobenzène 94.
Dichloro-2,3-naphtoquinone, activité antifongique 106.
1:3-Dichloro-propylene 236.
Dichrin A 243.

Dichroa febrifuga 243.
Dicoumarol 105, 106, 226, 242, 257.
Dicoumarol, antogonists 242.
Di-o-crésyle 104.
C_{40}-Dien-diol 27.
3 : 4-Diethoxy-2 : 5-dimethoxy-benzoic acid 247, 248.
2 : 3-Diethoxy-4-methoxy-benzoic acid 246.
Diethyl-dithiocarbamate (inhibitor) 344.
Diethyl-ethoxymethylene-malonate 132.
Differential enzymatic spectrophotometry 266.
Diglycyl-glycine 166.
C_{30}-Diin-diol 12.
C_{32}-Diin-diol 13.
C_{40}-Diin-diol 18.
C_{42}-Diin-diol 31.
Dihydro-alloxanthoxyletin 281.
Dihydro-auraptenic acid 253.
Dihydro-braylin 281.
Dihydro-β-carotin 19, 22.
7,7'-Dihydro-β-carotin 13, 16.
Dihydro-coumarin 243.
Dihydro-coumarin-4-sulfonic acid 231.
Dichloroflavine 94.
Dihydro-ergobasine 143, 167.
Dihydro-ergobasinineI 168.
Dihydro-ergobasinineII 168.
Dihydro-ergocornine 118, 167.
Dihydro-ergocorninineI 168.
Dihydro-ergocorninineII 168.
Dihydro-ergocristine 118, 135, 144, 167.
Dihydro-ergocristinineI 168.
Dihydro-ergocristinineII 168.
Dihydro-ergokryptine 118, 135, 144, 167.
Dihydro-ergokryptine, cleavage to polyamines 142.
Dihydro-ergokryptinineI 168.
Dihydro-ergokryptinineII 168.
Dihydro-ergonovine 143.
Dihydro-ergosine 167.
Dihydro-ergosine, cleavage to polyamines 142.
Dihydro-ergosinineI 168.
Dihydro-ergosinineII 168.
Dihydro ergot alkaloids in medicine 169.
Dihydro-ergotamine 135, 167.
Dihydro-ergotaminineI 168.
Dihydro-ergotaminineII 168.
Dihydro-isolysergic acid 118, 131, 133, 168.

Dihydro-isolysergic acid I 130.
Dihydro-isolysergic acid II 130.
Dihydro-isolysergic acid amide 133.
Dihydro-isolysergic acid azide 133.
Dihydro-isolysergic acidI, derivatives 133.
Dihydro-isolysergic acidII, derivatives 133.
Dihydro-isolysergic acid hydrazide 133.
Dihydro-isolysergic acid methyl ester 133.
Dihydro-luvangetin 280.
Dihydro-lysergic acid 118, 125, 131, 132, 133, 168.
Dihydro-lysergic acid amide 133, 144, 145.
Dihydro-lysergic acid azide 133, 141.
Dihydro-lysergic acid, derivatives 133.
Dihydro-lysergic acid hydrazide 133.
Dihydro-lysergic acid-D-isopropanolamide 143.
Dihydro-lysergic acid-L-isopropanolamide 143.
Dihydro-lysergic acid methyl ester 133.
Dihydro-lysergic acid, peptide-like derivatives 166.
Dihydro-lysergic acid, spectrum 127.
Dihydro-lysergic acid-D-valine methylester 143.
Dihydro-lysergic acid-L-valine methylester 143.
Dihydro-lysergyl-amino acid azide 141, 142.
Dihydro-lysergyl-amino acid ester 141, 142.
Dihydro-nor-isolysergic acid 132.
Dihydro-nor-lysergic acid 132.
Dihydro-oreoselone 273.
Dihydro-oroselone 275.
Dihydro-osthol 254, 255.
Dihydro-psoralene 260.
3' : 4'-Dihydro-α-pyronocoumaran 258.
3' : 4'-Dihydro-α-pyronocoumarone 258.
Dihydro-seselin 254, 278.
Dihydro-thymine 390.
Dihydro-xanthotoxol 265.
Dihydro-xanthoxyletin 280.
Dihydro-xanthyletin 277.
4 : 6-Dihydroxy-coumaran 262.
6 : 7-Dihydroxy-coumaran 264.
5 : 7-Dihydroxy-coumarin 246.
2' : 3''-Dihydroxy-dibenz-α-pyrone 282.
2 : 5-Dihydroxy-4 : 6-dimethoxy-benzaldehyde 249.

4 : 4'-Dihydroxy-6 : 6'-dimethoxy-diphenic acid dilactone 284.
4 : 6 : 4' : 6'-Dihydroxy-diphenic acid dilactone 282, 283.
2 : 4-Dihydroxy-*iso*phthalaldehyde 275, 278.
2 : 3-Dihydroxy-4-methoxy-benzaldehyde 248.
4 : 7-Dihydroxy-3-methyl-coumarin 240, 256.
Dihydroxy-5,6-vitamin A aldehyde 71.
5,6-Dihydroxy-vitamin A aldehyde 75, 76.
Diketopiperazine 137.
2 : 4-Dimethoxy-benzene-1 : 3-dicarboxylic acid 253.
5 : 7-Dimethoxy-2 : 2-dimethyl-chroman 280.
Dimethoxy-diphenyläther-tricarbonsäure 181.
2,3-Dimethoxy-diphenyläther-5,6,4'-tricarbonsäure 181, 182, 205, 206.
6 : 7-Dimethoxy-8-ethoxy-coumarin 248.
2 : 6-Dimethoxy-4-ethoxy-3-methylcinnamic acid 279.
Dimethoxy-furanocoumarin 266.
6 : 7-Dimethoxy-8-hydroxy-coumarin 248.
2 : 4-Dimethoxy-*iso*phthalic acid 261.
5 : 7-Dimethoxy-8-methyl-coumarin 279.
2,3-Dimethoxy-2'-methyl-diphenyläther-5,4'-dicarbonsäure 205, 206.
γ : γ-Dimethylallyl bromide 250, 266.
6-(γ : γ-Dimethylallyl)-7-methoxy-coumarin 250.
o-(3 : 3-Dimethylallyl)-phenol 277.
5,6-Dimethyl-benzimidazole 417.
D-Dimethyl-chonaocurin-dijodmethylat 203.
2 : 2-Dimethyl-chroman 276.
2 : 2-Dimethyl-chromene 276.
2 : 2-Dimethyl-Δ^3-chromene 276.
1,2-Diméthyl-4,5-diaminobenzène 94.
2,4-Diméthyl-3-hydroxy-5-hydroxyméthyl-pyridine 96.
D-Dimethyl-isochondodendrin 178.
O,O-Dimethyl-isochondodendrin 181, 182, 200.
O,O-Dimethyl-isochondodendrin, Des-Base 181.
O,O-Dimethyl-N-methyl-micranthin-methylmethin 205.
4 : 8-Dimethylnonan-1-oic acid 253.
Dimethyl-pyruvic acid 134.

Dimethyl-pyruvic acid amide 134.
D-Dimethyl-tubocurarin-jodid 203.
C_{20}-Diol, acetate 61.
Diospyros maritima 245.
1 : 3-Dioxy-cyclo-butane 147.
3,4-Dioxy-5-methoxy-phthalsäure-dimethylester 205.
Dipeptides 166.
Diphenyläther 209.
Diphenylendioxyd 105, 209.
Diphenylenoxyd 209.
Diphenyl-triacontapentadecaen 34.
Diphosphopyridine nucleotide 403.
Diphtheria toxin 454, 458, 464.
Diphtiocol 105, 106.
Diploclisia Kunstleri 178.
$\alpha\alpha'$-Dipyridyl (inhibitor) 344.
Disinomenin 177, 188, 189.
ψ-Disinomenin 188.
Di-(p-toluyl)-D-tartaric acid 151.
Di-(p-toluyl)-L-tartaric acid 151, 157.
Divalent azophenylarsonic acid dyes 464.
Diversin 177, 207.
Dodecapreno-β-carotin 37, 38.
Dodecapreno-β-carotin, Synthese 35.
Dowex-1 378.
DPN 403.
DUFF reaction 232, 265.

Echinoderm eggs 394.
Effet antisulfamide 101.
Egg albumin, labeled 350.
Electrophoresis-convection method 461, 462.
Electrophoretic properties (antibodies) 452.
Ellagic acid 228, 283, 284.
Ellagic acid tetraacetate 283.
Ellagic acid tetramethyl ether 283.
Ellagitannin 283.
Endogenous protein metabolism 294.
Enzymatic formation of desoxyribosides 388.
Enzymatic phosphorylation of nucleosides 390.
Enzymatic reactions, long- range forces 471.
Enzymatic synthesis of p-aminohippuric acid 338, 340.
Enzymatic synthesis of glutamine 336.
Enzymatic synthesis of gluthathione 341.
Enzymatic synthesis of hippuric acid 337, 340.

Enzymatic synthesis of hypoxanthine desoxy-riboside 386.
Enzymatic synthesis of ornithuric acids 339.
Enzymatic synthesis of protein 293.
Enzymatic synthesis of ribosides 382.
Enzymes of nucleoside metabolism 363, 365.
Enzyme proteins, lability 298.
Epistephanin 177, 180, 196, 199, 213.
Epistephanin, Biosynthese 212, 213.
ψ-Epistephanin 199.
Eremothecium Ashbyii 94, 410.
Erepsin 351.
Ergine 118.
Ergobasine 118, 120, 122, 149, 159, 164, 167, 168.
D-Ergobasine 151, 152.
L-Ergobasine 151, 152.
D-Ergobasine, crystal form 153.
L-Ergobasine, crystal form 153.
Ergobasine derivatives 165.
Ergobasine, isomers 152.
Ergobasine, formula 152.
Ergobasine group 167.
Ergobasine homologs 164.
Ergobasine, natural 151.
Ergobasinine 120, 122, 149, 164.
D-Ergobasinine 151, 152.
L-Ergobasinine 151, 152.
D-Ergobasinine, crystal form 153.
L-Ergobasinine, crystal form 153.
Ergobasinine homologs 164.
Ergobasinine, natural 151.
Ergoclavine 159.
Ergocornine 118, 120, 121, 134, 135, 137, 139, 145, 147, 157, 158, 162, 163.
Ergocornine, crystal form 163.
Ergocornine, formula 162.
Ergocornine, thermal cleavage 144.
Ergocorninine 120, 121, 162.
Ergocorninine, crystal form 163.
Ergocristine 118, 120, 121, 134, 137, 139, 145, 147, 157, 158, 159.
Ergocristine, crystal form 160.
Ergocristine, formula 160.
Ergocristinine 120, 121, 159, 161.
Ergocristinine, crystal form 160.
Ergokryptine 118, 120, 121, 134, 137, 139, 145, 147, 157, 158, 161.
Ergokryptine, crystal form 161.
Ergokryptine, formula 162.

Ergokryptine, thermal cleavage 144.
Ergokryptinine 120, 121, 161.
Ergokryptinine, crystal form 161.
Ergometrine 118, 122, 149.
D-Ergometrine 151.
D-Ergometrine, natural 151.
Ergometrinine 149.
D-Ergometrinine 151.
Ergometrinine, natural 151.
Ergonovine 118, 122, 149.
D-Ergonovine, crystal form 153.
L-Ergonovine, crystal form 153.
Ergosine 118, 120, 121, 137, 139, 145, 147, 155, 159.
Ergosine, crystal form 156.
Ergosine, formula 156.
Ergosinine 120, 121, 155, 159.
Ergosinine, crystal form 156.
Ergosinine, formula 156.
Ergosterine 119.
Ergot, ascospores 115.
Ergot, history 114.
Ergot, hyphae 115.
Ergot in midwifery 116.
Ergot, mixed alkaloids 159.
Ergot, non-specific compounds 119.
Ergot, perithecia 115.
Ergot research, history 118.
Ergot of rye 114.
Ergot sclerotium 115, 150.
Ergot stroma heads 115.
Ergot alkaloids 114, 117.
Ergot alkaloids, cleavage with hydrazine hydrate 122.
Ergot alkaloids, connecting link 148.
Ergot alkaloids, cyclized peptide structure 138.
Ergot alkaloids, cyclol grouping 148.
Ergot alkaloids, dextrorotatory isomers 120.
Ergot alkaloids, dihydro derivatives 167.
Ergot alkaloids, general formula 147.
Ergot alkaloids, hydrogenated derivatives 163.
Ergot alkaloids, infrared spectra of cleavage products 146, 147.
Ergot alkaloids, peptide fragment 144.
Ergot alkaloids, peptide section 134.
Ergot alkaloids, reduction with LiAlH$_4$ 142, 143.
Ergot alkaloids, stereoisomeric pairs 120, 121.

Ergot alkaloids, structure 119.
Ergot alkaloids, synthetic derivatives 163.
Ergot alkaloids, thermal cleavage 144.
Ergotamine 117, 118, 120, 134, 136, 137, 139, 145, 147, 153, 159.
Ergotamine, crystal form 154.
Ergotamine, formula 155.
Ergotamine group 167.
Ergotamine, pharmacology 121.
Ergotamine, rearrangement 154.
Ergotamine, thermal cleavage 144.
Ergotamine, thermal cleavage product, infrared spectrum 146.
Ergotaminine 120, 153, 159.
Ergotaminine, crystal form 154.
Ergothioneine 119.
Ergotinic acid 119.
Ergotinine 118, 123.
Ergotinine cristallisée 117, 156.
Ergotism 115.
Ergotism, convulsive 115.
Ergotism epidemics 116.
Ergotoxine 118, 122, 159.
Ergotoxine group 120, 121, 156, 158, 167.
Ergotoxine group, dextrorotatory isomers 158.
Ergotoxine, pharmacology 157.
Ergotoxine preparations 157.
Erythrotin 419.
Escherichia coli 102, 344, 346, 371, 374, 376, 378, 380, 381, 393, 410, 413, 414, 416, 418, 419, 423, 430.
Escherichia coli, action on purines, pyrimidines 380.
Escherichia coli, nucleoside phosphorylase 416, 430.
Escherichia coli, response to desoxyribosides 429.
Escherichia coli, response to reducing agents 429.
Escherichia coli, response to vitamin B_{12} 429.
Escherichia coli, sulfanilamide inhibition 419.
Ethionine, incorporation into proteins 308.
Ethionine (inhibitor) 348.
Ethionine, labeled 308.
Ethoxyacetylene-magnesium bromide 52, 53.
2-Ethoxy-4 : 6-dimethoxybenzene-1 : 3-dicarboxylic acid 252.

8-Ethoxy-7-methoxy-coumarin 246, 248.
Ethoxy-methylene-malonic acid diethyl ester 131.
Ethoxyvinyl-carbinol 53.
2-Ethyl-3-amino-4-éthoxyméthyl-5-amino-méthylpyridine 96, 97, 98.
Ethyl carbonate 237.
Ethyl chloroacetate 58.
Ethyl ester of β-ionolacetic acid 48.
Ethyl β-ionylideneacetate 46, 47, 48, 50.
Ethyl β-ionylideneacetate, spectrum 44.
Ethyl β-ionylideneacetate, isomer 48.
Ethyl β-ionylideneacetate isomer, spectrum 44.
α-Ethynyl-β-ionol 64.
Eugenin 249.
Euglena 420.
Euglena gracilis var. *bacillaris*, growth factor 418.
Eupatorium ayapana 243, 245.
Eupatorium triplinerve 243, 245.
Euphorbia formosanum 284.
Euphorbia lathyris 244, 246.
Exchange reaction (purines) 390.
Exogenous protein metabolism 294.

Fabiana imbricata 245.
Fabiatrin 245.
Factor for incorporation of amino acids 347.
Fagara zanthoxyloides 264.
Fangchinin 178.
Fangchinolin 178, 207.
Fertilizin 451.
Ferula alliacea 271.
Ferulin 271, 272.
Ferulinic acid 271.
Feu sacré 116.
Fibraurea chloroleuca 177, 190.
Fibraurea tinctoria 177.
Ficus carica 260.
Ficusin 260.
Fish liver oils 43.
Flavines, antagonistes 94.
Flindersia brayleyana 281.
Flindersia collina 250.
Fluorene 282, 283, 284.
Fluoride (inhibitor) 344, 345, 346.
Foetal tissues, incorporation of amino acids 315.
Folic acid 395.
Folinic acids 431.
FOLIN's theory, protein metabolism 294.

Foreign amino acids, incorporation 308.
Formamido-malonamamidine 390.
Formvar films 471, 472.
Formvar films, imperfections 472.
Formylacetic acid 235.
Formylacetic ester 235.
5-Formyl-6-hydroxy-coumaran 261.
6-Formyl-7-hydroxy-5-methoxy-coumarin 262.
8-Formyl-7-hydroxy-5-methoxy-2 : 2- dimethyl-chroman 281.
6-Formyl-8-methoxy-umbelliferone 280.
5-Formyl-5,6,7,8-tetrahydro-pteroyl-glutamic acid 429, 431.
Formyl-4,5,6-triamino-pyrimidine 390.
8-Formyl-umbelliferone 232, 259, 261, 275, 277, 278.
Fougères, antithiamines 93.
Framework theory, antibodies 457.
Fraxetin 226, 247, 248, 265.
Fraxidin 248, 249.
Fraxin 247, 248.
Fraxinol 249.
Fraxinus species 244.
Free amino acids in plasma 297.
Free amino acids, *de novo* protein synthesis 348.
Free amino nitrogen in blood 296.
Free amino nitrogen in muscle 296.
Free amino nitrogen in pea seedlings 296.
Free amino nitrogen in salmon 296.
Free amino nitrogen in tissues 296.
FRIEDEL-CRAFTS reaction 236.
FRIES rearrangement 232.
Frog eggs (desoxyribosides) 394.
Fryxin, pressor effect 241.
Furan-2 : 3-dicarboxylic acid 258, 259, 260, 261, 265, 267, 268, 269.
Furanocoumarins 257.

Galbanum 243.
Gallotannin 283.
Gangrenous ergotism 115.
GATTERMANN reaction 280.
GATTERMANN synthesis 259.
Gelatin, lack of antigenicity 446.
Gelseminic acid 245.
Gentiobiuronic acid 450.
5-Geranoxy-7-methoxycoumarin 249.
Gindaricin 178, 207.
Gindarin 178, 207.
Gindarinin 178, 207.
Glaucoma piriformis 107.

γ-Globulin 451, 452, 453, 461.
Globulin-azophenylarsonic acid 444, 467.
Globulin-p-azophenylarsonic acid 445.
Globulin fragments, antibody activity 455.
Globulin, mobility measurements 452.
Globulin, normal 454.
Glucose, conversion into ribose 377.
α-Glucose-1,6-diphosphate 376.
Glucose-6-phosphate 377.
α-Glucosides 459.
β-Glucosides 459.
7-Glucosido-daphnetin 247.
8-Glucosido-daphnetin 246, 247.
8-Glucosido-fraxetin 248.
Glutamic acid 336.
Glutamic acid, condensation of its γ-carbonyl 332.
Glutamic acid, labeled 310, 312.
Glutamine 336.
Glutamine, enzymatic synthesis 336, 337.
Glutamo-amide transferase 336.
Glutamo-transferase 334, 342, 345.
γ-L-Glutamyl-L-cysteine 342.
γ-Glutamyl-cysteinyl-glycine 332.
Glutathione 331, 332.
Glutathione, amino acid turnover 332.
Glutathione, enzymatic synthesis 341.
S^{35}-Glutathione, incorporation 306.
Glycine 321.
Glycine amide 330, 331.
Glycine anilide 325, 326, 328.
Glycine, free energy of formation 320.
Glycine, incorporation 348.
C^{14}-Glycine, incorporation 306.
N^{15}-Glycine, incorporation 306.
Glycine, incorporation into nucleic acid purines 394.
Glycine, labeled 301, 304, 305, 311, 317, 318, 351, 392.
Glycyl-glycine 166, 321.
Glycyl-glycine, free energy of formation 320.
Glycyl-glycine, hydrolysis in the presence of labeled glycine 332.
Glycyl-leucine 326, 327.
Glycyl-L-leucine 166.
Glyoxalate 338.
Gnidia polycephala 247.
GRIESSMAYER reaction 282, 284.
Growth factor, adenosine-3'-phosphate 406, 410, 411.

Growth factor, adenosine-5'-phosphate 411.
Growth factor, adenosine triphosphate 406.
Growth factor, 4-carboxy-uracil 412.
Growth factor, cytidylic acid 410, 411.
Growth factor, cytosine desoxyriboside 412.
Growth factor, desoxyribonucleotides 411.
Growth factor in *Euglena gracilis* var. *bacillaris* 418.
Growth factors in guanine-less microorganisms 407.
Growth factor, guanylic acid 410.
Growth factors, hemolytic streptococci 408.
Growth factor in *Hemophilus influenzae* 405.
Growth factor in *Hemophilus parainfluenzae* 405.
Growth factor in *Lactobacillus lactis* 407.
Growth factor in *Lactobacillus pentosis* 412.
Growth factor in *Neisseria gonorrhoeae* 406.
Growth factor in *Neurospora* 407.
Growth factor, orotic acid 412.
Growth factor in *Phycomyces* 408.
Growth factors in purine-less microorganisms 407.
Growth factor in *Tetrahymena geleii* 407.
Growth factor, uracil 411, 412.
Growth factor, uridine 412.
Growth factor, uridylic acid 410.
Growth factor ,,V" 405.
Growth inhibition, adenosine 409.
Growth inhibition, adenosine-3'-phosphate 409.
Growth inhibition, 5-bromo-uridine 412.
Growth inhibition, 5-chloro-uridine 412.
Growth inhibition, guanosine 409.
Growth inhibition, guanylic acid 409.
Growth-inhibition in microorganisms 409.
Growth-inhibition, pyrimidine nucleosides 412.
Growth-inhibition, pyrimidine nucleotides 412.
Growth inhibitors 410.
Growth-promoting activity of desoxyribosides 424.
Growth-promoting activity of desoxyribonucleotides 424.

Growth-promoting activity, nucleosides 411.
Growth-promoting activity, nucleotides 405, 411.
Growth-promoting activity, purine nucleosides 405.
Growth-promoting activity, purine nucleotides 405.
Growth-promoting activity, pyrimidine nucleosides 411.
Growth-promoting activity, pyrimidine nucleotides 411.
Growth-requirement, lactic acid bacteria 424.
Growth requirement, *Tetrahymena geleii* 411.
Guanase 371.
D-Guanidyl-butylamine 119.
Guanine 380, 390.
Guanine, conversion into adenine 395.
Guanine, labeled 395.
Guanine, sparing action on adenine 407.
Guanine desoxyriboside 385, 387, 404, 432.
Guanine des oxyribosides, phosphorolysis 370.
Guanine-less microorganisms, growth factors 407.
Guanine ribosides, phosphorolysis 369.
Guanosine 375, 380, 404, 405.
Guanosine, estimation 366.
Guanosine, growth inhibitor 409.
Guanylic acid 404.
Guanylic acid, growth factors 410.
Guanylic acid, growth inhibitor 409, 410.
Guinea pig liver homogenate, incorporation of amino acids 344.
Gum ammoniacum 256.

Han-fang-chi 178, 207, 213.
Hapten-antibody reactions, free energy 467.
Hapten bond, free energy 467.
Hapten inhibition 458, 473, 474.
Haptenic compounds 449.
Haptenic determinants 450.
Haptenic determinant, p-azophenylarsonate 449.
Haptenic groups, number 450.
Hapten polymers 470.
Haptens 445, 449.
Haptens, inhibiting 445.
Hasubanin 177.

Hasubanonin 177, 207.
Heat-stable co-factors, amino acid incorporation 346.
Helicin 243.
Helminths, immuno-polysaccharides 447.
Hemocyanin 468.
Hemocyanin antigen, labeled 464.
Hemoglobin 446.
Hemolytic streptococci 406.
Hemolytic streptococci, growth factors 408.
Hemophilus influenzae, growth factor 405.
Hemophilus parainfluenzae, growth factor 405, 423.
Hemophilus parainfluenzae, pyrimidine requirement 411.
Hemophilus piscium, adenosine triphosphate requirement 406.
Heracleum nepalanese 263, 270.
Heracleum sphondylium 263, 264, 265.
Heraclin 262, 263.
Herniania hirsuta 243.
Herniarin 243.
Herniarin, dimerization 233.
Heterogeneity, antibodies 459.
Heteroligating types 458.
Hétérovitamine B_1 90.
Hexahydroxy-diphenyl 283.
Hexahydro-oroselone 275.
Hexahydro-m-tolylsäure-methylester 205.
Hexapeptide 349.
Hexosamine 452.
Hexose-diphosphate 338.
Hexose-6-phosphate 375.
Hippurate ion 321.
Hippurate ion, free energy of formation 320.
Hippuric acid 321, 323, 345.
Hippuric acid anilide 328.
Hippuric acid, enzymatic synthesis 337, 340.
Hippuric acid, free energy of formation 320.
Hippuryl-anilide, enzymatic synthesis 323.
Hippuryl-anilide, free energy of formation 324.
Hippuryl-glycine 321.
Histamine 119.
Histidine 119, 452.
Histidine, inhibitory effect 450.
Histidine, labeled 311, 318.

L-Histidine, labeled 305, 318, 351.
Holy Fire 116.
16,16'-Homo-β-carotin 2.
all-*trans*-16,16'-Homo-β-carotin 32.
mono-*cis*-16,16'-Homo-β-carotin 32.
di-*cis*-16,16'-Homo-β-carotin 32.
16,16'-Homo-β-carotin, *cis-trans* Isomere 31, 32.
16,16'-Homo-β-carotin, Synthese 31.
Homocysteine 418.
Homostephanolin 177, 207.
Homothiamine 92.
Homothiamine-glycol 91, 92.
Homotrilobin 204.
Hormonal effects on incorporation of labeled amino acids 306.
Horse antitoxin 458.
Horse serum albumin 464.
Huile de foie de morue 103.
Huiles de foies de poissons, action antivitaminique 103.
Human plasma substitute 448.
Human serum albumin 464.
Humans, nitrogen pool 302.
Humans, protein turnover 303.
Hydergine 118, 169.
Hydrocoumarin 226, 228, 243.
Hydroergotinine 117, 118, 156.
Hydrogen bonding, antigen-antibody reactions 471.
α-Hydro-β-hydroxy-vitamin A acid ethyl ester 46, 47, 48.
α-Hydro-β-hydroxy-vitamin A acid ethyl ester, spectrum 45.
Hydroxamic acid 231, 329.
Hydroxo-cobalamin 419, 420.
o-Hydroxy-acetophenone 237.
2-Hydroxy-6-aminopurine 404.
2-Hydroxy-6-aminopurine-D-riboside 379.
1 : 3-*bis*(2-Hydroxybenzoyl)-ethane 257.
2-Hydroxy-chromones 234.
cis-o-Hydroxy-cinnamic acid 229.
trans-o-Hydroxy-cinnamic acid 229.
o-Hydroxy-cinnamyl alcohol 230.
3-Hydroxy-coumaran 272.
6-Hydroxy-coumaran 259, 260, 261.
4-Hydroxy-coumarin 233, 237.
4-Hydroxy-coumarins and diazomethane 234.
5-Hydroxy-coumarin 238.
6-Hydroxy-coumarin 238.
7-Hydroxy-coumarin 233, 238.

Hydroxy-coumarins, fluorescence 238.
Hydroxy-coumarins, glycosides 228.
6-Hydroxy-coumarone 259.
4-Hydroxy-dihydrocoumarin 234.
6-Hydroxy-7 : 8-dimethoxy-coumarin 249.
7-Hydroxy-8-(y : y-dimethylallyl)-coumarin 254.
7-Hydroxy-2 : 2-dimethyl-chroman 277.
7-Hydroxy-6-(1 : 2-diphenylethyl)-coumarin 259.
Hydroxy-ergobasine 165.
Hydroxy-ergobasinine 165.
C_{20}-Hydroxy ester 55.
α-Hydroxy-*iso*butyric acid 268, 270, 273, 276, 278, 280.
α-Hydroxy-*iso*propylacetylene 227, 276, 277, 278, 281.
Hydroxylamine (inhibitor) 344.
Hydroxylated vitamin A aldehydes 71.
Hydroxy-maleic anhydride 52.
2-Hydroxy-4-methoxy-benzaldehyde 256.
2-Hydroxy-4-methoxy-benzoic acid 254.
5-Hydroxy-7-methoxy-coumarin 249, 250.
7-Hydroxy-5-methoxy-coumarin 249.
5-Hydroxy-7-methoxy-2-methylchromone 249.
7-Hydroxy-5-methoxy-4-methylcoumarin 249.
8-Hydroxy-5-methoxy-psoralene 269, 270.
8-Hydroxy-5-methoxy-psoralene y : y-dimethylallyl ether 270.
1-Hydroxy-3-methyl-2-pentene-4-yne 58, 61.
Hydroxy-peucedanin 267.
p-(p-Hydroxyphenylazo)-phenylarsonic acid 467.
3-[2'-Hydroxyphenyl]-propanol 230.
Hydroxy-proline, inhibitory activity 450.
Hydroxy-quinol 226.
o-Hydroxy-stilbene 230.
Hypoepistephanin 177, 180, 196, 199.
Hypoxanthine 376, 383, 386, 388, 390, 392, 394, 404, 407.
Hypoxanthine desoxyriboside 373, 374, 385, 386, 388, 393, 432.
Hypoxanthine desoxyriboside (bacterial extracts) 374.
Hypoxanthine desoxyriboside, enzymatic synthesis 386.
Hypoxanthin desoxyriboside, estimation 366.

Hypoxanthine desoxyribosides, phosphorolysis 370.
Hypoxanthine riboside 367.
Hypoxanthine riboside, estimation 366.
Hypoxanthine ribosides, phosphorolysis 368.
Hypserpa cuspidata 178.
Ignis sacer 116.
Immunochemistry 444.
Immunological reactions, long-range forces 471.
Immunological reactions, specificity 459.
Immuno-polysaccharides 447.
Immuno-polysaccharides (helminths) 447.
Imperatoria ostruthium 266.
Imperatoria rhizome 253, 254.
Imperatorin 268, 269.
Incorporated amino acids, linkage 349, 351.
Incorporation of amino acids, effect of concentration 315.
Incorporation of amino acids, into foetal tissues 315.
Incorporation of amino acid and presence of others 316.
Incorporation of amino acids, into tumor slices 315.
Incorporation of foreign amino acids 308.
Incorporation of labeled amino acids 305.
Incorporation *in vitro* of labeled amino acids 317.
Incorporation *in vitro* of labeled glycine 318.
Incorporation *in vitro* of labeled L-histidine 318.
Incorporation *in vitro* of labeled L-lysine 318.
Incorporation of amino acids into proteins 343.
Incorporation of purines into nucleic acids 391.
Incorporation of pyrimidines into nucleic acids 391.
Incorporation rates, labeled amino acids 306, 307.
Infra-red spectrum, ergotamine cleavage products 146.
Inhibiteurs 89.
Inhibiting antibodies 453, 459.
Inhibiting haptens 445.
Inhibition of precipitation 470.
Inhibitors, amino acid incorporation 343.

Inhibitory activity, acetylated leucine 450.
Inhibitory activity, arginine 450.
Inhibitory activity, cystine 450.
Inhibitory activity, histidine 450.
Inhibitory activity, hydroxyproline 450.
Inhibitory activity, isoleucine 450.
Inhibitory activity, lysine 450.
Inhibitory activity, methionine 450.
Inhibitory activity, proline 450.
Inhibitory activity, tryptophane 450.
Inhibitory activity, tyrosine 450.
Inosine 375, 376, 383, 392, 394, 404.
Inosine, estimation 366.
Inosinic acid 404.
Insularin 177, 180, 205, 206, 208, 212.
Insularin, Biosynthese 212.
Insularinsäure 205, 206.
Insulin 307.
Intestinal adenosine deaminase 379.
Intestinal deaminase 366.
Invertebrate polysaccharides, antigenicity 447.
Iodoacetate (inhibitor) 344, 345.
β-Ionolacetic acid 46.
β-Ionolacetic acid ethyl ester, spectrum 44.
β-Ionone 43, 47, 53, 55, 58, 62, 63.
β-Ionone, propargyl bromide 63.
Ionylidene-acetaldehyde 63.
β-Ionylidene-acetaldehyde 48, 52, 63, 66.
cis-β-Ionylidene-acetaldehyde 48, 56.
trans-β-Ionylidene-acetaldehyde 48.
β-Ionylidene-acetaldehyde, cis-trans isomers 52.
cis-β-Ionylidene-acetaldehyde semicarbazone 48.
trans-β-Ionylidene-acetaldehyde semicarbazone 48.
β-Ionylidene-acetaldehyde, spectrum 44.
β-Ionylidene acetate 51.
β-Ionylidene-acetic acid 43, 48, 50.
β-Ionylidene-crotonic acid 53, 55.
cis-β-Ionylidene-crotonic acid 54, 56.
trans-β-Ionylidene-crotonic acid 54.
β-Ionylidene-crotonic acid, isomer 54.
β-Ionylidene-ethanol 46, 47, 48, 50, 52.
β-Ionylidene-ethanol, isomer 48.
β-Ionylidene-ethanol isomer, spectrum 44.
β-Ionylidene-ethanol, spectrum 44.
β-Ionylidene-pyruvic acid 52.
Iron 2.

Isoamylalcohol 266, 268.
Isoamylamine 119.
6-Isoamyl-7-methoxycoumarin 256.
Isoanhydrovitamin A 43, 61, 72, 79.
Isoauraptene 252, 253.
Isobebeerin 178, 180, 199.
Isobergapten 263, 264.
Isobutyric acid 267, 271, 273, 275.
Isobyakangelicolic acid 270, 272.
Isocaproic acid 269.
Isochondodendrin 178, 180, 199, 200, 212.
D-Isochondodendrin 178, 203.
Isochondodendrin, Biosynthese 212.
D-Isochondodendrin-dimethyläther 203.
Isochondodendrin-Typus 180.
Isococlaurin 178, 187.
Isocytosine 380.
Isofraxidin 249.
Isoguanine 404, 407.
Isoguanosine 370, 379, 404.
Isoimperatorin 226, 227, 266, 267, 268.
Isoleucine 452.
Isoleucine, inhibitory activity 450.
Isolysergic acid 121.
D-Isolysergic acid 129.
L-Isolysergic acid 129.
Isolysergic acid, configuration 129.
Isolysergic acid, condensation with 2-aminopropanol-(1) 151.
Isolysergic acid, decarboxylation product, spectrum 127.
Isolysergic acid, enol form 130.
Isolysergic acid, formula 124, 128.
Isolysergic acid, peptide-like derivatives 166.
Isolysergic acid, position of the double bond 126, 127.
Isolysergic acid, spectrum 127.
Isolysergic acid, synthesis 125.
Isolysergic acid azide 152.
D-Isolysergic acid azide 151.
L-Isolysergic acid azide 151.
D-Isolysergic acid-(+)-butanolamide-(2) 164.
D-Isolysergic acid diethylamide 165.
D-Isolysergic acid-2-diethylamino-ethylamide 165.
D-Isolysergic acid-1 : 3-dihydroxypropaneamide-(2) 165.
D-Isolysergic acid-D-nor-ephedride 166.
D-Isolysergic acid-D-nor-ψ-ephedride 166.
D-Isolysergic acid-L-nor-ephedride 166.

Sachverzeichnis. Index of Subjects. Index des Matières. 521

L-Isolysergic acid-L-nor-ephedride 166.
D-Isolysergic acid ethanolamide 164.
Isolysergic acid hydrazide 122, 123, 164.
D-Isolysergic acid hydrazide 123, 151.
L-Isolysergic acid hydrazide 123, 151.
D,L-Isolysergic acid hydrazide 151.
rac. Isolysergic acid hydrazide 123.
Isolysergic acid lactam, spectrum 127.
D-Isolysergic acid-L-(+)-4-methylpentanolamide-(2) 164.
D-Isolysergic acid-D-propanolamide-(2) 151, 152.
D-Isolysergic acid-L-propanolamide-(2) 151, 152.
D-Isolysergic acid-L-(+)-propanolamide-(2) 164.
L-Isolysergic acid-D-propanolamide-(2) 151, 152.
L-Isolysergic acid-L-propanolamide-(2) 151, 152.
D-Isolysergic acid-D-propanolamide-(2), crystal form 153.
D-Isolysergic acid-L-propanolamide-(2), crystal form 153.
L-Isolysergic acid-D-propanolamide-(2), crystal form 153.
L-Isolysergic acid-L-propanolamide-(2), crystal form 153.
Isomaltose 422.
cis-trans Isomerism, β-ionylidene-acetaldehyde 52.
cis-trans Isomerism, vitamin A 51.
Isooxypeucedanin 267, 268.
Isopimpinellin 265, 266.
Isoprene hydrobromide 269.
Isoprene, reaction with umbelliferone 227.
Isoprene system in coumarins 226.
α-Isopropyl-coumaran 277.
Isopropyl-ergobasine 164.
Isopropyl-ergobasinine 164.
α-Isopropyl-furan 276.
α-Isopropylidene-coumaran 276.
Isopsoralene 261.
Isotetrandrin 177, 179, 180, 184, 185, 194, 207, 210.
Isotetrandrin, Biosynthese 210.
Isotopic tracer studies (biosynthesis of proteins and peptides) 292.
Isotrilobin 177, 180, 182, 204, 211.
Isotrilobin, Biosynthese 211.
Isovaleric acid 275.
Isovaleryl-L-leucyl-L-proline 135.

Isovaleryl-L-phenylalanyl-L-proline 135.
Isovaleryl-L-valyl-L-proline 135.
Isovitamin A acetate 71, 74.
Isovitamin A methyl ether 71, 75.

January Bossie 247.
Jatrorrhiza columba 178.
Jatrorrhiza palmata 178.
Jatrorrhizin 177, 178, 190, 191.
Jod-Katalyse (Carotinoide) 5.
β-Jonon 2, 10, 12.
Jonyliden-aceton 2.

Keratin 446.
α-Ketoglutarate 314.
C_{18}-Ketone 46, 48, 50, 53, 54.
cis-C_{18}-Ketone 54, 56.
C_{18}-Ketone, semicarbazone 56.
cis-C_{18}-Ketone, semicarbazone 54.
trans-C_{18}-Ketone, semicarbazone 54.
C_{18}-Ketone, spectrum 45.
2-Keto-6-phospho-gluconate 377, 378.
Keuchhusten 213.
Kitol 43, 70, 73.
Kitol diacetate 70, 73.
Kohlenwasserstoff $C_{30}H_{42}$ 9.
C_{30}- und C_{32}-Kohlenwasserstoffe, Synthese 9.
all-trans-Konfiguration 9.
poly-cis-Konfiguration 9.
Kreuterbuch 116, 118.

Labeled adenine 388, 414.
Labeled adenylic acid 392.
Labeled alanine 311, 313.
Labeled amino acid 300.
N^{15}-Labeled amino acids 297.
Labeled amino-acids, incorporation 303, 304, 305.
Labeled amino acids, incorporation in vivo 300.
Labeled amino acids, incorporation in vitro 309, 317.
Labeled amino acids, incorporation into tissue proteins 310, 311.
Labeled amino acids, rabbit liver 298.
Labeled amino acids, relative rates of incorporation 306, 307.
C^{14}-Labeled amino acids as tracers 303.
N^{15}-Labeled amino acids as tracers 300.
S^{35}-Labeled amino acids as tracers 303.
Labeled amino acids in visceral and plasma proteins 304.
Labeled α-aminoadipic acid 308.

Labeled ammonia 393.
Labeled antigen-antibody precipitates 467.
Labeled aspartic acid 312.
Labeled aspartic acid, incorporation 310.
Labeled p-azophenylacetic acid 464.
Labeled bovine serum albumin antigen 464.
Labeled carbon dioxide 310, 312, 351.
Labeled 2,6-diaminopurine 391.
Labeled egg albumin 350.
Labeled ethionine 308.
Labeled glycine 301, 304, 305, 311, 317, 318, 351, 392.
Labeled glycine, incorporation 394.
Labeled glutamic acid 312.
Labeled glutamic acid, incorporation 310.
Labeled guanine 395.
Labeled hemocyanin antigen 464.
Labeled histidine 304, 311, 318.
Labeled L-histidine 305, 318, 351.
Labeled leucine 301, 304, 311, 318.
Labeled L-leucine 298, 304, 305, 317, 318, 351.
Labeled lysine 301, 304, 311, 318.
Labeled L-lysine 305, 317, 318, 351.
Labeled methionine 311.
S^{35}-Labeled methionine 307.
Labeled nucleic acid purines 392.
Labeled purines 391.
Labeled purines, *in vitro* studies 393.
Labeled purine nucleosides 392.
Labeled pyrimidines 392.
Labeled pyrimidine desoxyribosides 393.
Labeled ribosides 423.
Labeled L-serine 304.
Labeled threonine 301.
Labeled thyroglobulin 464.
Labeled tryptophane 316.
Labeled L-tryptophane 304.
Labeled tyrosine 301.
Labeled DL-tyrosine 304.
Lactam from *iso*lysergic acid 126.
Lactam from lysergic acid 126.
Lactic acid bacteria, growth requirement 424.
Lactobacillus acidophilus 424, 426, 427, 430.
Lactobacillus acidophilus, response to desoxyribosides 428, 429.
Lactobacillus acidophilus, response to reducing agents 428, 429.

Lactobacillus acidophilus, response to vitamin B_{12} 428, 429.
Lactobacillus arabinosus 409, 424.
Lactobacillus bifidus 427.
Lactobacillus bifidus, response to desoxyribosides 429.
Lactobacillus bifidus, response to reducing agents 429.
Lactobacillus bifidus, response to vitamin B_{12} 429.
Lactobacillus bifidus, utilization of desoxyribonuc eic acid 410.
Lactobacillus bulgaricus 409, 412.
Lactobacillus bulgaricus, orotic acid requirement 417.
Lactobacillus casei 98, 99, 100, 371, 374, 380, 395, 407, 413, 414, 421, 431.
Lactobacillus casei, incorporation of adenine 395.
Lactobacillus delbrückii 371, 373, 374, 387, 424, 427, 430, 433.
Lactobacillus delbrückii, response to desoxyribosides 429.
Lactobacillus delbrückii, response to reducing agents 429.
Lactobacillus delbrückii, response to vitamin B_{12} 429.
Lactobacillus gayonii 411, 433.
Lactobacillus gayonii, nucleotides in the nutrition 410.
Lactobacillus gayonii, stimulation of growth by nucleotides 410.
Lactobacillus helveticus 371, 372, 373, 374, 380, 387, 389, 411, 425, 426, 427, 430, 433.
Lactobacillus helveticus, response to desoxyribosides 428.
Lactobacillus helveticus, response to reducing agents 428.
Lactobacillus helveticus, response to vitamin B_{12} 428.
Lactobacillus helveticus, enzyme 387.
Lactobacillus lactis 417, 420, 424, 430.
Lactobacillus lactis, growth factors 407.
Lactobacillus lactis, vitamin B_{12} requirement 427.
Lactobacillus leichmannii 374, 419, 420, 424, 427, 432.
Lactobacillus leichmannii, nutritional behavior 432.
Lactobacillus leichmannii, response to desoxyribosides 428.

Lactobacillus leichmannii, response to reducing agents 428.
Lactobacillus leichmannii, response to vitamin B_{12} 428.
Lactobacillus pentosus 369.
Lactobacillus pentosus, growth factor 412.
Lactoflavine 94.
β-Lactoglobulin 450.
Laserpitium latifolium 242.
Lavandula delphinensis 242.
Lavandula spica 242.
Lavender oil 243.
Lepra 213.
Leucine 331.
DL-Leucine 321.
L-Leucine 134, 166.
Leucine anilide 325.
L-Leucine anilide 325, 328.
DL-Leucine, free energy of formation 320.
L-S^{35}-Leucine, incorporation 307.
Leucine, labeled 301, 304, 311, 318.
L-Leucine, labeled 298, 304, 305, 317, 318, 351.
Leuconostoc 448.
Leuconostoc citrovorum 101, 373, 389, 408, 413, 420, 426, 431, 432, 433.
Leuconostoc citrovorum, nutritional behavior 432.
Leuconostoc citrovorum, response to desoxyribosides 429.
Leuconostoc citrovorum, response to reducing agents 429.
Leuconostoc citrovorum, response to vitamin B_{12} 429.
Leuconostoc citrovorum, thymidine requirement 424.
Leuconostoc mesenteroides 420, 421, 431, 433.
Leucyl-glycine 321.
DL-Leucyl-glycine 321.
DL-Leucyl-glycine, free energy of formation 320.
Leucyl group 324.
L-Leucyl-*L*-proline lactam, reduction 144.
L-Leucyl-*D*-proline lactam, thermal cleavage 144.
Levorotatory ergot alkaloids, dihydro derivatives 167.
Levures, action des quinones 107.
Ligusticum acutilobum 263.
Limacia cuspidata 178.
Limacia obrenga 178.
Limacia veltina 178.
Lime oil 266.
Lime oil deposits 249.
Limettin 245.
Linkage of incorporated amino acids 349, 351.
Lipids in antibodies 452.
Lipids, antigenicity 448.
Lithium-acetylid 22.
Lithium aluminum hydride 43, 230.
Liver mitochondira, incorporation of amino acids 317.
Liver nucleosidase 366.
Liver nucleoside phosphorylase 366.
Lobster blood (agglutinins) 451.
Lock-and-key hypothesis (antigens-antibodies) 455.
Long-range forces, enzymatic reactions 471.
Long-range forces, immunological reactions 471.
Low-molecular weight compounds, antigenicity 448.
Luvanga scandens 265, 277, 280.
Luvangetin 280, 281.
Lycopin, Synthese 28.
Lymphosarcomes de la souris 96.
Lysergic acid 120, 134.
D-Lysergic acid 129.
L-Lysergic acid 129.
Lysergic acid, condensation with 2-amino-propanol-(1) 151.
Lysergic acid, configuration 129.
Lysergic acid, decarboxylation product, spectrum 127.
Lysergic acid, enol form 130.
Lysergic acid, formula 122, 124, 128.
Lysergic acid, hydrogenation 131.
Lysergic acid, partially synthetic derivatives 164.
Lysergic acid, peptide-like derivatives 166.
Lysergic acid, position of the double bond 126.
Lysergic acid residue, connecting link 136.
Lysergic acid, spectrum 127.
Lysergic acid, stereoisomerization 122.
Lysergic acid, synthesis 125.
Lysergic acid amide 134, 136.
D-Lysergic acid-*L*-*N*-benzylpropanolamide-(2) 165.
D-Lysergic acid-(+)-butanolamide-(2) 164.

Lysergic acid diethylamide 165.
D-Lysergic acid diethylamide 165, 166.
D-Lysergic acid-2-diethylamino-ethylamide 165.
D-Lysergic acid-1 : 3-dihydroxypropaneamide-(2) 165.
D-Lysergic acid-L-ephedride 165.
D-Lysergic acid-D-nor-ephedride 166.
D-Lysergic acid-D-nor-ψ-ephedride 166.
D-Lysergic acid-L-nor-ephedride 166.
L-Lysergic acid-L-nor-ephedride 166.
D-Lysergic acid-ethanolamide 164.
Lysergic acid hydrazide 123, 152.
D-Lysergic acid hydrazide 123.
L-Lysergic acid hydrazide 123.
rac. Lysergic acid hydrazide 123.
Lysergic acid lactam, spectrum 127.
D-Lysergic acid-L-(+)-4-methylpentanolamide-(2) 164.
D-Lysergic acid-D-propanolamide-(2) 151, 152.
D-Lysergic acid-D-propanolamide-(2), crystal form 153.
D-Lysergic acid-L-propanolamide-(2) 151, 152.
D-Lysergic acid-L-(+)-propanolamide-(2) 164.
D-Lysergic acid-L-propanolamide-(2), crystal form 153.
L-Lysergic acid-D-propanolamide-(2) 151, 152.
L-Lysergic acid-D-propanolamide-(2), crystal form 153.
L-Lysergic acid-L-propanolamide-(2) 151, 152.
L-Lysergic acid-L-propanolamide-(2), crystal form 153.
Lysine 452.
D-Lysine, incorporation 313.
Lysine, inhibitory activity 450.
Lysine, labeled 301, 304, 311, 318.
L-Lysine, labeled 305, 317, 318, 351.

Magnesium-ion (activator) 342, 345.
Magnocurarin 179, 187.
Magnolamin 179, 180, 192.
Magnolia fuscata 179, 192.
Magnolia Kobus 179, 188.
Magnolia obovata 179, 187.
Magnolia salicifolia 179, 188.
Magnoliaceae, Biscoclaurin-Basen 179.
Magnolin 179, 180, 192, 210.
Magnolin, Biosynthese 210.

Mahonia acanthifolia 179.
Mahonia borealis 179.
Mahonia Griffithii 179.
Mahonia japonica 179, 194.
Mahonia leschenaultii 179.
Mahonia manipurensis 179.
Mahonia sikkimensis 179.
Mahonia Simonsii 179.
Mahonia Swaseyi 179.
Majudin 262, 263.
Malaria 95.
Malate 314.
Malic acid 261.
Malonate (inhibitor) 346.
Maltose 422.
Mammalian purine nucleoside phosphorylase 367.
Mangano-ion (activator) 335, 337, 342, 345.
Maquereau 93.
Marmelosin 268.
Marmesin 272.
Matricaria chamonilla 243.
Mechanism, amino acid incorporation 343.
Melilotic anhydride 243.
Melilotol 243.
Melilotoside 242, 243.
Melilotus altissima 242.
Melilotus arvensis 243.
Melilotus dentata 242.
Melilotus officinalis 242, 243.
Menisarin 177, 180, 204, 211.
Menisarin, Biosynthese 211.
Menisin 178.
Menisidin 178.
Menispermaceae-Alkaloide 175, 177
Menispermaceae-Alkaloide, Benzochinolizin-Typus 180, 191.
Menispermaceae-Alkaloide, Benzylisochinolin-Typus 180, 186.
Menispermaceae-Alkaloide, Berberin-Typus 180, 190.
Menispermaceae-Alkaloide, Biscoclaurin-Typus 180, 192.
Menispermaceae-Alkaloide, Klassifizierung 180.
Menispermaceae-Alkaloide, medizinische Anwendungen 213.
Menispermaceae-Alkaloide, Phenanthropyridin-Typus 180, 188.
Menispermin 178, 207.
Menispermum canadense 177, 192, 196.

Sachverzeichnis. Index of Subjects. Index des Matières. 525

Menispermum cocculus 178.
Menispermum dauricum 177, 178, 192, 194.
β-Mercaptoethylamino derivative of pantothenic acid 409.
Méristème radiculaire de Pisum 92.
Merlan 93.
Metaphanin 177, 207.
Methionine 418, 419, 429, 430, 452.
DL-Methionine 333.
S^{35}-Methionine 350.
Methionine incorporation 348.
S^{35}-Methionine, incorporation 306, 307, 308, 346.
Methionine, inhibitory activity 450.
DL-Methionine-isopropyl ester 333, 334.
Methionine, labeled 307, 311.
Methionine peptides 333.
L-Methionyl-*L*-methionine 334.
Methochloride 125.
o-Methoxy-benzoylacetic ester 257.
L-1-(4'-Methoxybenzyl)-6,7-dimethoxy-N-methyl-1,2,3,4-tetrahydroisochinolin 184, 185, 186.
2-Methoxy-chromones 234.
o-Methoxy-cinnamic acid 229, 230.
o-Methoxy-cinnamic acid, *cis-trans* isomerism 229.
4-Methoxy-coumarin 234.
7-Methoxy-coumarin 233, 243.
7-Methoxy-coumarin-6-aldehyde 250.
7-Methoxy-coumarin-6-carboxylic acid 250.
7-Methoxy-8-coumarinylacetic acid 252.
2-Methoxy-diphenyläther-5,4'-dialdehyd 182, 183.
2-Methoxy-diphenyläther-5,4'-dicarbonsäure 181, 182.
5-Methoxy-7-ethoxy-8-methylcoumarin (synthetic) 279.
Methoxy-furanocoumarins 259.
7-Methoxy-8-methylcoumarin 252, 254.
1-Methoxy-5,6-methylendioxy-apomorphin 189.
Méthoxy-2-naphtoquinone-1,4 105, 107.
Méthoxy-pyridoxine 96, 97.
Méthoxy-pyridoxine, action inhibitrice 97.
N-Methyl-aniline 324.
O-Methyl-berbamin 184, 194, 207.
Methyl-γ-bromocrotonate 53.
Methyl-α-bromo*iso*butyrate 278.
Methyl-γ-bromosenecioate 63, 66.

5-Methyl-cytosine 389, 390.
4-Methyl-cytosine desoxyriboside 424, 425.
5-Methyl-cytosine desoxyriboside 403.
8-O-Methyl-daphnetin 281.
N-Methyl-dihydro-epistephanin-A 208.
N-Methyl-dihydro-menisarin-methylmethin 205.
3 : 3'-Methylene-*bis*-(4-hydroxycoumarin) 105, 257.
Methylene-*bis*-[4-hydroxy-3-coumarinyl] 242.
Méthylène-3 : 3'-*bis*-oxy-2-naphtoquinone 105.
Methyl-ergobasine 164.
Methyl-ergobasinine 164.
β-Methyl-glutaconate 63, 65.
β-Methyl-glutaconic anhydride 65.
Methyl-heptenone 256, 257.
N-Methyl-hydroepistephanin-A 199.
2-Méthyl-3-hydroxy-4-diméthylaminométhyl-pyridine 96.
2-Méthyl-3-hydroxy-4-hydroxyméthyl-pyridine 96, 97.
Methyl-lithium 53.
Méthyl-2-naphtoquinone 98, 105.
Méthyl-2-naphtoquinone-1,4 105, 107.
O-Methyl-oxyacanthin 197, 199, 208.
o-Methyl-oxyacanthin-methylmethin 182, 183.
N-Methyl-phenyl-ergobasine 165.
N-Methyl-quinoline-betaine-tricarboxylic acid 123, 124.
4-Methoxy-resorcinol 245.
1-(α,β)-Methyl-*D*-ribopyranoside-3-phosphate 411.
O-Methyl-tetrahydro-braylinic acid 281.
Methyl thiouracil 389.
6-Methyl-umbelliferone 277.
Methyl-vinyl-ketone 58, 62.
MICHAEL condensation 233.
Micranthin 179, 180, 198, 205, 211.
Micranthin, Biosynthese 211.
Microbiological assay of cytidine + uridine 412.
Microbiological assay, desoxyribosides 424.
Microbiological functions of vitamin B_{12} 418.
Microorganisms, growth-inhibition 409.
Micro-organisms, incorporation of purines 394.

Microorganisms, purine-less 409.
Microorganisms, purine precursors 413.
Microorganisms, pyrimidine-less 409.
Microsomes, amino acid incorporation 314.
Microsomes, incorporation of amino acids 313.
Migraine 169.
Mitochondria, incorporation of amino acids 313, 314.
Monimiaceae, Biscoclaurin-Basen 179.
Mu-fang-chi 178, 213.
Mufangchin 178.
Multivalent antigen molecules, reactions 476.
Murraya exotica 245.
Muscle, free amino nitrogen 296.
Mycobacterium tuberculosis 412.

Natrium-acetylid 17.
Natural ergobasine 151.
Natural ergobasinine 151.
Natural *D*-ergometrine 151.
Natural ergometrinine 151.
Neisseria gonorrhoeae, growth factors 406.
Neoprotocuridin 179, 180, 201.
Néopyrithiamine 90, 91.
Neo-β-carotin 4.
Neovitamin A 42, 46, 49, 70.
Neovitamin A, biopoteny 66.
Neovitamin A, photomicrograph 67.
Neovitamin A acid, photomicrograph 68.
Neovitamin A anthraquinone-β-carboxylate 49.
Neovitamin A *p*-phenylazobenzoate 49.
Neuralgie 213.
Neurospora 392, 402, 408, 409, 410, 412, 414, 415, 416.
Neurospora, growth factors 407.
Neurospora, pyrimidine-less mutant strains 412.
Neurospora sitophila 430.
Niacine 98.
Nicotinamide 98, 338, 369, 384.
Nicotinamide riboside 369, 384, 405.
NIDHONE synthesis 233.
Nitrobergapten 262.
Nitrogen balance 295.
Nitrogen pool in humans 302.
Nitrogen requirement, *Torula utilis* 408.
Nitrogen turnover 295.
Nitroso-cobalamin 420.
Nodakenetin 227, 272.

Nodakenin 272.
Nodakenin tetraacetate 272.
Non-precipitating antibodies 459.
Nor-bixin 3.
Nor-dihydrolysergic acid 132.
Nor-dihydrolysergic acid methyl ester 132.
L-Nor-ephedrine 122.
Nor-ergobasine 164.
Nor-ergobasinine 164.
Normal globulin 454.
Nor-menisarin 177, 180, 204.
Nor-methyl-vitamin A 2, 3.
Nor-vitamin A acetate 70, 74.
Nor-vitamin A methyl ether 71, 75.
Nutritional behavior, *Lactobacillus leichmannii* 432.
Nutritional behavior, *Leuconostoc citrovorum* 432.
Nucleic acid biosynthesis, purine precursors 413.
Nucleic acids, cleavage to nucleosides 364.
Nucleic acid and protein synthesis 315.
Nucleic acid purines, formation 391.
Nucleic acid purines, incorporation of glycine 394.
Nucleic acid purines, labeled 392.
Nucleoproteins (animal), antigenicity 447.
Nucleosidases 365.
Nucleoside deaminase 378.
Nucleosides of desoxyribose 420.
Nucleosides as growth factors 401.
Nucleoside metabolism 364.
Nucleoside metabolism, enzymes 363, 365.
Nucleoside phosphorylase 366, 368, 374, 381, 385, 392, 426, 430.
Nucleoside phosphorylase, *Escherichia coli* 416.
Nucleoside phosphorylases, function 393.
Nucleosides, phosphorylation 390.
Nucleosides, preparation 364.
Nucleosides, synthetic 413.
5-Nucleotidase 365.
Nucleotides of desoxyribose 420.
Nucleotides, growth-promoting activity 401, 405, 411.

Oak galls 283.
Occlusion theory (antigens-antibodies) 457.
Octadecylamine 471.
Octendion 20, 28, 30.
Ophiostoma 408, 409, 413, 416.

Ophiostoma, guanine-less strains 415.
Ophiostoma, pyrimidine-less mutant strains 412.
Ophiostoma multiannulatum 407.
Orange-peel oil 252.
Oreoselone 272, 273, 274, 275.
Oreoselone acetate 273.
Ornithuric acids, enzymatic synthesis 339.
Ornithuric acids, synthesis 340.
Oroselone 258, 275.
Orotic acid 403, 415, 416.
Orotic acid, growth factor 412.
Orotic acid, precursor, of pyrimidine 417.
Orotic acid, precursor of uridine and cytidine 416.
Orotic acid decarboxylase 416.
Orotic acid riboside 392, 416.
Osthenol 254, 255, 278.
Osthol 253, 254, 255, 256.
Ostholic acid 252, 253, 255.
Ostruthin 226, 253, 254.
Ostruthin, toxicity 241.
Ostruthol 267, 268.
Ovalbumin 348, 450, 454, 464, 465.
Ovalbumin, precipitation curve 460.
Ovalbumin-rabbit antiovalbumin precipitin 450.
Oxalacetic acid 415.
Oxalic acid 275.
Oxalsäure, wasserfreie 11.
Oxidation inhibitors of amino acid incorporation 345, 346.
Oxyacanthin 179, 180, 181, 182, 195, 196, 210, 213.
Oxyacanthin, Abbau 182.
Oxyacanthin, Biosynthese 210, 213.
D-1-(4'-Oxybenzyl)-6,7-dimethoxy-N-methyl-1,2,3,4-tetrahydroisochinolin 185, 186.
L-1-(4'-Oxybenzyl)-6,7-dimethoxy-N-methyl-1,2,3,4-tetrahydroisochinolin 200.
D-1-(4'-Oxybenzyl)-6-methoxy-7-oxy-N-methyl-1,2,3,4-tetrahydroisochinolin 184, 185.
Oxybiotine 99.
Oxybiotine, analogues guanidinés 100.
Oxydase de D-amino-acides 94.
Oxyde de vanadium 102.
3-Oxy-5,6-dimethoxy-N-*nor*-apomorphin 189.
2-Oxy-diphenyläther 185.

4,4'-*p*-Oxydiphénylsulfone 101.
C_{18}-Oxy-enoläther 13.
Oxyimperatorin 269.
1-Oxy-3-methylpenten-2-in-4 17.
Oxypeucedanin 227, 267.
Oxypeucedanin hydrate 267, 268.
Oxypeucedaninic acid 267.
Oxythiamine 91.

Palmatin 177, 178, 190, 191.
Pantothenic acid, β-mercaptoethylamino derivative 409.
Pantoyltaurine 95.
Papain 323, 325, 327, 328.
Papain-HCN 326.
Paper chromatography, desoxyribosides 425.
Parabaena hirsuta 177, 190.
Paralysie de CHASTEK 92, 93.
Paramenispermin 178, 207.
Paraplégie 92.
Pareira-Basen 179.
Pareira-Wurzel 178, 199, 201, 213.
Particulation time 465.
Pastinaca sativa 242.
Pea seedlings, free amino nitrogen 296.
cis-Peak 8, 9, 16, 33.
PECHMANN reaction 244.
Pelosin 178, 201.
Penicillin 344.
β-Pentaacetyl-D-glucose 272.
Pentose diphosphate 382.
Pepsin 327.
Peptidases, role in peptide synthesis 322.
Peptidases, role in protein synthesis 322.
Peptides, biosynthesis 292.
Peptide bond biosynthesis, mechanism 319.
Peptide bonds, enolization 331.
Peptide bonds, equilibrium constants of formation 321.
Peptide bonds, free energy of formation 321.
Peptide chain, lengthening by amino acid esters 334.
Peptide formation, effect of p_H on the free energy change 321.
Peptide fragment (ergot alkaloids) 144.
Peptides, free energy of formation 319, 320.
Peptide hydrolysis, activation of carbonyl 330.

Peptides as intermediates in protein synthesis 348.
Peptides, less soluble ones 326.
Peptide synthesis from amino acid esters 333.
Peptide synthesis in exchange reactions 328.
Peptide type ergot alkaloids, cleavage products 134.
Pericampylus glaucus 178.
Pericampylus incanus 178.
Permanganat-Oxydation (Biscoclaurin-Alkaloide) 181.
Petroselinum sativum 242.
Peucedanin 227, 228, 260, 272, 273, 274.
Peucedanin, toxicity 241.
Peucedanum decursivum 272.
Peucedanum officinale 273.
Peucedanum oreoselinum 275.
Peucedanum ostruthium 266.
Phaeanthin 179, 180, 195, 207, 210.
Phaeanthin, Biosynthese 210.
Phaeantus ebracteoratus 179, 195.
Phanostenin 177, 190.
Pharmacological action of adenosine compounds 378.
Phellopterin 269.
Phellopterin epoxide 270.
Phellapterus littoralis 269.
Phenanthropyridin-Typus 180, 188.
Phenoxy-dimethylacetic acid 276.
Phenyl-alanine 331.
L-Phenyl-alanine 134, 166, 341.
Phenyl-alanine esters 334.
Phenyl-alanyl group 324.
L-Phenyl-alanyl-D-proline lactam 135.
o-Phenylene-diamine, condensation with hippuric acid 324.
p-Phenylene-diamine, condensation with hippuric acid 324.
Phenyl-ergobasine, stereoisomers 166.
Phenyl-ergobasinine, stereoisomers 166.
Phényl-indanedione 105, 106.
Phenyl-hydrazine 324.
N-Phenyl-morpholine 77.
Phloroglucinaldehyde 246.
Phloroglucinol 226, 245, 267.
Phosphatase 298, 371, 380.
Phosphate (activator) 365.
Phosphate de tri-o-crésyle 103, 104.
1-Phospho-desoxyriboside 386.
1-Phospho-furano-riboside 374.

Phosphoglucomutase 376.
6-Phospho-gluconic acid 375, 377.
6-Phospho-gluconic acid, enzymatic oxidation 375, 377.
3-Phospho-pyruvate 391.
Phosphordijodid 12, 27.
Phosphoribomutase 369, 372, 375, 376, 381, 382.
Phosphoribomutase in muscle extracts 375.
1-Phospho-ribose 367.
Phospho-ribosides 381.
1-Phospho-riboside 384.
Phosphorylation and amino acid incorporation 346.
Phosphorylation, inhibitor of amino acid incorporation 345, 346.
Phosphorylation of nucleosides 390.
Photobacterium fischeri 407.
Phtiocol 105.
Pimpinella saxifraga 264, 265.
Pimpinellin 266.
Pimpinellin, toxicity 241.
Pituitary hormones 307.
Plakalbumin 348.
Plasma, free amino acids 297.
Plasmodium lophurrae 95.
Plastein, cyclic structure 327.
Plastein formation 327.
Pneumococcus 454.
Pneumococcus polysaccharide 447, 453, 465.
Phycomyces, growth factor 408.
Physical nature, antibodies 478.
Physical nature, antigens 478.
Poisson cru, effets pathologiques 93.
Polyamines, from ergot alkaloids 139.
Polyamines, synthesis 141.
Polyencéphalite hémorragique 92.
Polyhaptenic azoproteins 466.
Poly-cis-Konfiguration 9.
Polymerization, precipitating haptens 476.
Polypeptide formation 339.
Polysaccharides, antigenicity 447.
Polyvalent antigen molecules 454.
Polyvalent haptens, aggregation 476.
Prangos patularia 267.
Precipitating haptens, polymerization 476.
Precipitation curve (ovalbumin) 460.
Precipitation curve (rabbit serum) 444.

Precipitin reaction, mathematical interpretations 475.
Precipitin test 450.
Prickly ash bark 279.
Proline 148.
D-Proline 134.
Proline, inhibitory activity 450.
Prontosil rubrum 100.
Propargyl bromide 10, 28, 35, 36, 62, 63, 65.
Propionyl-L-phenylalanyl-L-proline 135.
Proteases, role in peptide synthesis 322.
Proteases, role in protein synthesis 322.
Protein-*p*-azobenzenesulfonic acid 459.
Protein-*p*-azophenylarsonate 456.
Protein-azophenylarsonic acid 459.
Proteins, biosynthesis 292.
Protein carrier combining area 456.
Protein catabolism 294.
Proteins, cyclol theory 138.
Proteins, denaturated 451.
Proteins, enzymatic synthesis 293.
Proteins, half-life 303.
Proteins, high lability in the cell 319.
Protein hydrolysis 328.
Protein hydrolysis, formation of new peptides 333.
Proteins, maintenance of the amino acid pattern 314.
Protein metabolism as a dynamic steady state 294.
Protein metabolism, endogenous 294.
Protein metabolism, exogenous 294.
Proteins, net synthesis *in vitro* 312.
Proteins *ex* labeled amino acids 351.
Protein synthesis, chicken liver slices 312.
Protein synthesis and nucleic acid 315.
Protein synthesis, peptides as intermediates 348.
Protein synthesis, rate 302.
Protein turnover 299.
Proteus vulgaris 345.
Prothrombin 106, 242.
Protocuridin 179, 180, 200, 201.
Protostephanin 177, 207.
Pseudo-jonen 28.
Pseudomonas saccharophila 425.
Pseudo-parenchyma 115.
Pseudo-pyridoxine 95.
Psoralea corylifolia 260.
Psoralene 259, 260, 261.
Psoralene-glycoside 261.

Pteris aquilina 93.
Ptéroyl-glutamate 100.
Pteroyl-glutamic acid 414, 416, 417, 418, 419, 420, 421, 429, 431, 432.
Pumpkin seedling enzyme 335.
Pyridoxal 95, 96, 97.
Pyridoxamine 95, 96, 97.
Pyridoxine 95, 96, 97
Purines 363.
Purine deaminase 381.
Purines, estimation 365.
Purines, exchange reactions 390, 392.
Purines, incorporation 391, 414.
Purines, incorporation into nucleic acids 391.
Purines, labeled 391.
Purine-less microorganisms 409.
Purine-less microorganisms (growth factors) 407.
Purine nucleosides, growth-promoting activity 405.
Purine nucleosides, labeled 392.
Purine nucleoside phosphorylase 367, 369, 372, 381.
Purine nucleoside phosphorylase, assay 371.
Purine nucleoside phosphorylase, mammalian 367.
Purine nucleoside ribosidase 375.
Purine-nucleosides, utilization 414.
Purine mononucleotides, growth inhibitors 410.
Purine-nucleotides, growth-promoting activity 405.
Purine precursors in microorganisms 413.
Purine precursors in nucleic acid biosynthesis 413.
Purines, purine-nucleosides and nucleotides, comparative growth effects 407.
Purines and pyrimidines, exchange 372, 373.
Purine-ribosides, antagonism 417.
Purine ribosides, cleavage 374.
Purine ribosides, phosphorolysis 371.
Purines, spectra 365.
Putrescine 119.
Pycnarrhena manillensis 178, 207.
Pyrazoline 232.
Pyridoxal phosphate 339.
Pyridoxine 430.
Pyridoxine, antagonistes 95.
Pyrimidines 363.

Pyrimidine deaminase 381.
Pyrimidine desoxynucleoside 393.
Pyrimidine desoxyriboside 387, 402.
Pyrimidine desoxyriboside, labeled 393.
Pyrimidines, estimation 365.
Pyrimidines, incorporaton 392.
Pyrimidines, incorporation into nucleic acids 391.
Pyrimidines, labeled 392.
Pyrimidine-less microorganisms 409.
Pyrimidine-nucleoside biosynthesis 415.
Pyrimidine nucleosides, growth-inhibiting effect 412.
Pyrimidine-nucleosides, growth-promoting effect 411.
Pyrimidine nucleosides, incorporation 393.
Pyrimidine nucleosides, metabolism 372.
Pyrimidine nucleoside phosphorylase 372.
Pyrimidine nucleotides, growth-inhibiting effect 412.
Pyrimidines and purines, exchange 372, 373.
Pyrimidine requirement, *Hemophilus parainfluenzae* 411.
Pyrimidine ribosides 370, 402, 417.
Pyrimidine ribosides, incorporation 393.
Pyrimidine ribosides as a source of pyrimidines 412.
Pyrithiamine 90.
Pyrogallol 226.
Pyruvic acid 134, 137.
Pyruvic acid amide 144.
Pyruvyl-L-phenylalanyl-L-proline 136

Q-Enzyme 422.
Quinones, action sur la levure 107.

Rabbit antibody 458.
Rabbit liver, labeled amino acids 298.
Rabbit serum, precipitation curve 444.
Radix Pareira brava 187.
Rat liver nucleoside phosphorylase 368.
Reagic antibodies 463.
Reagin 463.
Reducing agents in growth of microorganisms 426.
Reductive amination 339.
Rehydrovitamin A 70, 73.
Repandin 179, 180, 197.
Repandulin 179, 207.
Reserve proteins 298.
Resodicarboxylic acid 260.

α-**Resodicarboxylic acid 273.**
Reticulocytes, amino acid incorporation 317.
Reticulocytes, amino acid incorporation, inhibitors 343.
Reticulocytosis 315.
Retinene$_1$ 42, 56, 75.
Rétinène, oxydation 102.
Retro-ionylidene structure 65.
Rh blocking antibody 459.
Rh-factor 453.
Rheumatismus 213.
Rhodopsin 42, 43, 56.
cis-Rhodopsin 43.
Riboflavine, antagoniste 94.
1-α-D-Ribofuranoside 417.
1-α-D-Ribofuranosido-5,6-dimethylbenzimidazole 404.
9-β-D-Ribofuranosido-isoguanine 404.
Ribonuclease, radioactive 299.
Ribonucleic acid 407, 410.
Ribonucleic acid, biosynthesis 416.
Ribonucleic acid, growth inhibitor 410.
Ribonucleic acid mononucleotides, two forms 411.
Ribonucleotides, biosynthesis 413.
β-D-**Ribose 364.**
Ribose-1-5-diphosphate 376.
Ribose-1-hypoxanthine 383.
Ribosenucleic acids 315.
Ribose-nucleosides 405.
Ribose-nucleotides 405.
Ribose-1-phosphate 369, 370, 381, 382, 383, 411.
Ribose-1-phosphate, cyclo-hexylamine salt 382.
Ribose-1-phosphate, enzymatic preparation 381.
Ribose-1-phosphate and inorganic phosphate, exchange 370.
Ribose-1-phosphate in nucleotide synthesis 384.
Ribose-1-phosphate, properties 382.
β-D-Ribose-1-phosphate 369.
Ribose-3-phosphate 375, 384.
Ribose-5-phosphate 375, 378, 411.
Ribose-5-phosphate, conversion into inosine 375.
β-D-Ribose-5-phosphate 377.
Ribose-3-phosphoric acid 422.
Ribose phosphoric esters, degradation and synthesis 376.

Ribosidase 374.
Ribosides, biological function 364.
Ribosides, biosynthesis 413.
Ribosides, conversion into desoxyribosides 393, 423.
Ribosides, enzymatic synthesis 382.
Riboside formation from hexoses 377.
Ribosides, labeled 423.
Riboside phosphorylase and desoxyriboside phosphorylase, relationship 370.
Ribosyl group, enzymatic transfer 426.
Ribulose phosphate 378.
Rotundin 177, 180, 191.
RX experiment 457, 477.
Rye, infected by ergot 150.

Saccharomyces cerevisiae 96, 98, 99, 100, 394, 395.
Saccharomyces cerevisiae, growth requirement 405.
Salicifolin 179, 213.
Salicifolinchlorid 188.
Salicylaldehyde D-glucoside 243.
Salicylic acid (from coumarins) 242.
Salix purpurea 422.
Salmon, free amino nitrogen 296.
Sangolin 178, 207.
Sarcopetalum Harveyanum 178.
Sauerdorn 195.
Scopoletin 240, 244, 245, 250, 251, 265.
Scopoletin-β-primeveroside 245.
Scopolia japonica 245.
Scopolin 244, 245.
Sea urchin 451.
Secale cornutum 114.
Semen angelicae 239, 263, 265, 266.
Sensibamine 159.
Serine 419, 429.
L-Serine, labeled 304.
Serological properties 476.
Seromicrons 465.
Serum lipids 452.
L-Seryl-L-leucine 166.
L-Seryl-L-leucine-D-proline 166.
Seseli indicum 278.
Seselin 227, 228, 276, 278.
Sheepserum-p-azobenzoic acid 473.
Sheepserum-p-azophenylarsonic acid 473.
Sheepserum-p-(p-azophenyl)-azophenylarsonic acid 473.
Short-range forces, antigen-antibody reactions 471.

Silk fibroin hydrolysate 450.
Simple coumarins 242.
Sinactin 177, 191.
Sinomenin 176, 177, 180, 188.
Sinomenium acutum 176, 177, 178, 188, 189, 191, 207, 213.
Sinomenium diversifolium 177.
Sitosterol 261.
Skimmetin 243.
Skimmia japonica 243, 278.
Soil bacteria and vitamin B_{12} 420.
Souche de LANSING 91.
Specific haptenic portion 459.
Specific precipitates 463.
Sphondin 265.
Sphondylin 265.
St. Anthony's Fire 116.
Staphylococcus aureus 94, 337.
Stephania capitata 177, 189, 190, 200.
Stephania cepharantha 177, 194, 195, 200.
Stephania glabra 178, 207.
Stephania japonica 177, 178, 189, 199, 207, 213.
Stephania rotunda 177, 191.
Stephania Sasakii 177, 189, 190, 195.
Stephania tetrandra 177, 178, 194.
Stephania venosa 178.
Stephanin 177, 189, 213.
Stephanolin 177, 207.
Stereochemie der Carotinoide 4.
Stereoisomere Carotinoide, Anzahl 6.
Stereoisomerisierung, 15,15'-mono-cis-β-Carotin 26.
Stereoisomerisierung, 16,16'-Homo-β-carotin 31, 32.
Storage protein 298.
Streptobacterium plantarum 94, 102.
Streptococcus faecalis 97, 421.
Streptococcus lactis 95.
Streptomyces aureofaciens 419.
Suberosin 250.
Subvitamin A 43, 70, 73.
Succinate 314.
Succinate de di-o-crésyle 104.
Succinic acid 273, 275, 280.
Sulfanilamide, condensation with hippuric acid 324.
Sulfanilamide growth-inhibition and adenine 410.
Sulfanilic acid 324.
Sulfanil-ovalbumin 468.
Sulfato-cobalamin 420.

Sulfonamide growth-inhibition, reversion 410.
Sweet clover 257.
Sweet clover disease 241.
Synthese, β-Carotin 19.
Synthese, ε_1-Carotins 30.
Synthese, Decapreno-β-carotin 34.
Synthese, Decapreno-ε_1-carotin 35.
Synthese, Dodecapreno-β-carotin 35.
Synthese höherer Carotin-Homologe 31.
Synthese, Lycopin 28.
Synthesis of ornithuric acids 340.
Synthesis of proteins *in vitro* 312.
Synthesis of vitamin A *via* "C_{14}-aldehyde" 58, 59.
Synthesis of vitamin A *via* C_{14}-aldehyde, extinction values of products 61.
Synthesis of vitamin A, *via* β-ionylideneacetic acid esters 43, 47.
Synthesis of vitamin A, *via* β-ionylidenecrotonic acid esters 52, 54, 55.
Synthoxylin N 279.

Tagged antigens 464.
Talauma mexicana 179, 193.
Tert. Albuminiumbutylat 35.
O-Tetraacetyl-α-glucosidyl bromide 246, 247.
Tetrahydro-braylin 281.
L-Tetrahydro-epiberberin 191.
Tetrahydro-luvangetin 280.
Tetrahydro-oroselone 275.
Tetrahydro-osthol 255.
Tetrahydro-peucedanin 273.
Tetrahydro-pteroylglutamic acid 431.
Tetrahydro-seselin 278.
Tetrahydro-tubaic acid-4-methyl ether 254.
1 : 2 : 3 : 4-Tetrahydroxy-benzene 226.
Tetrahymena geleii 414, 415, 421.
Tetrahymena gelei, growth factors 407, 411.
Tetrahymena geleii, uridine formation 416.
2 : 3 : 4 : 5-Tetramethoxy-benzoic acid 247, 248.
O-Tetramethyl-ellagic acid 284.
Tetrandrin 177, 178, 180, 194, 195, 207, 210.
Tetrandrin, Biosynthese 210.
Tetrandrin-Typus 180.
Tetrapeptide 348.
Theatrum botanicum 116.

Thermobacterium acidophilus 373, 374, 386, 387, 389, 424, 427.
Thermobacterium lactis 374.
Thiaminase 92, 93.
Thiamine 90, 93.
Thiamine, antagonistes 90, 92.
Thiamine, destruction *in vivo* 93.
Thiamine, isostère pyridinique 90.
Thiocyanate analog 420.
Thioglycolic acid 350.
5-Thiomethyl-*D*-ribose 379.
Threonine 452.
DL-Threonine 333.
DL-Threonine-isopropylester 334.
Threonine, labeled 301.
L-Threonyl-*L*-threonine 334.
Thymidine 372, 375, 388, 393, 403, 411, 413, 418, 421, 426, 428, 429, 431, 432.
Thymidine, biosynthesis 418, 419.
Thymidylic acid 424.
Thymine 372, 373, 387, 388, 393, 412, 418, 419, 421, 429.
Thymine-desoxyriboside 388.
Thyroglobulin 464.
Thyroglobulin, labeled 464.
Tiliacora acuminata 178.
Tiliacora racemosa 178, 207.
Tiliacorin 178, 207.
Tinospora bakis 178, 207.
Tinospora crispa 178.
Tinospora Rumphii 178.
Tissues, amino nitrogen 296.
Tocophérol 103.
Toddalia aculeata 251.
Toddalolactone 251, 252.
p-Toluene-sulfonic acid 57.
p-Toluenesulfonyl-glycine-anilide 324.
m-Toluidine, condensation with hippuric acid 324.
o-Toluidine, condensation with hippuric acid 324.
p-Toluidine, condensation with hippuric acid 324.
p-Toluolsulfosäure 9, 10, 15, 23, 25, 28, 35.
D-Tomentocurin 178, 207.
Topf-curare 179.
Torresea ceariensis 242.
Torulopsis utilis 408, 414.
Torula utilis, nitrogen requirement 408.
TPN 403.
Transamidation 328, 334, 335.
Transamination 339.

Trans-desoxyribosidase 425.
Trans-desoxyribosidation 433.
Transferase 336.
Trans-glucosidase 422:
Trans-N-glycosidase 372, 373, 374, 425.
Trans-N-glycosidase, bacterial 388.
Trans-N-glycosidase, spectroscopic demonstration 373.
Trans-O-glycosidases 425.
Trans-N-glycosidation 387, 422.
Trans-N-glycosidic reactions (desoxyribose nucleosides) 387.
Transmutase 372.
Transpeptidation 328, 331.
Trans-phosphorylase 411.
3 : 4 : 6-Triacetoxy-coumaran 262.
3 : 4 : 6-Triacetoxy-coumarone 262.
4,5,6-Triamino-pyrimidine 389.
Trichloroacetic acid 350.
Triclisia Gilletii 178.
Tri-*o*-crésyle 103.
2 : 3 : 4-Trihydroxy-benzoic acid 264.
6 : 7 : 8-Trihydroxy coumarin 6-methylether 248.
Trilobamin 177, 180, 196, 198, 210.
Trilobamin, Biosynthese 210.
Trilobin 177, 180, 182, 203, 204, 211.
Trilobin, Biosynthese 211.
2 : 3 : 4-Trimethoxy-benzoic acid 280.
5 : 6 : 7-Trimethoxy-coumarin 249.
Trimethoxy-diäthyl-dimethyl-diphenyläther 182, 183.
Trimethoxy-diäthyl-diphenyläther-dialdehyd 182, 183.
2 : 4 : 6-Trimethoxy-3-methyl-cinnamic acid 279.
Trimethylamine 119.
$L:L$-1 : 2-Trimethylene-5-benzyl-piperazine 144.
$L:L$-1 : 2-Trimethylene-5-isobutyl-piperazine 144.
2 : 6 : 10-Trimethyl-tetradecan-14-oic acid 256.
Tripeptides 166.
Triphosphopyridine nucleotide 403.
Trivalent azophenylarsonic acid dyes 464.
Truxillic acid 233.
Trypsin 351.
Tryptophane 96, 119.
L-Tryptophane 166.
Tryptophane esters 334.
L-C^{14}-Tryptophane, incorporation 307.

Tryptophane, inhibitory activity 450.
Tryptophane, labeled 316.
L-Tryptophane, labeled 304.
Tubocurare 179, 200, 201.
Tubocurarinchlorid 180, 202, 213.
D-Tubocurarinchlorid 178, 203.
L-Tubocurarinchlorid 178.
Tuduranin 177, 189.
Tumors 305.
Tumor slices, incorporation of amino acids 315.
Tyramine 119.
Tyrosine 119.
L-Tyrosine 341.
Tyrosine amide 325.
Tyrosine esters 334.
DL-C^{14}-Tyrosine, incorporation 307.
Tyrosine, inhibitory activity 450.
Tyrosine, labeled 301.
DL-Tyrosine, labeled 304.
Tyrosylazo-phenylarsonate 456.

Umbelliferone 226, 228, 232, 240, 243, 276, 278.
Umbelliferone-6-carboxylic acid 272, 274.
Umbelliferone-8-carboxylic acid 261.
Umbelliferone, electronic structure 238.
Umbelliferone glucoside 243.
Umbelliferone, reaction with isoprene 227.
Umbelliprenin 251.
Univalent antibodies 453, 457, 458.
Unnatural amino acids, incorporation into proteins 309.
Uracil 380, 387, 415.
Uracil desoxyriboside 389, 403, 423, 424.
Uracil, growth factor 411, 412.
Uric acid 388, 390.
Uric acid riboside 365, 370.
Uridine 380, 403, 409, 412, 416.
Uridine formation *(Tetrahymena geleii)* 416.
Uridine, growth factor 412.
Uridine nucleosidase 375.
Uridylic acid 403, 409.
Uridylic acid, growth factor 410.

Valence number of precipitating antibody molecules 456.
Valine 331, 332.
L-Valine 134.
Valine methyl ester 141.
Vanadium tetroxide 75.

VAN DER WAALS forces, antigen-antibody reactions 471.
Velleia discophora 254.
Verbascum thapsus 242.
Vinylmethylketon 17.
Virus de la pneumonie 96.
Viscera 302.
Vitamines hydrosolubles, antagonistes 90.
Vitamines liposolubles, antagonistes 102.
Vitamin A 2, 41, 42, 49, 55, 61.
Vitamin A_1 42.
Vitamin A_2 43, 71, 76, 77.
Vitamin A_3 57, 58, 73.
Δ^2-cis-Δ^6-trans-Vitamin A 51.
Δ^2-trans-Δ^6-cis-Vitamin A 54, 57, 73.
Δ^2-trans-Δ^6-trans-Vitamin A 51.
Vitamin A acetate 35, 42, 49, 61.
all-trans-Vitamin A acetate 57.
Vitamin A acid 43, 55, 56, 72, 79.
all-trans-Vitamin A acid 52, 54.
all-trans-Vitamin A acid, photomicrograph 68.
Δ^2-cis-Δ^6-trans-Vitamin A acid 54, 57.
Δ^2-cis-Δ^6-trans-Vitamin A acid, photomicrograph 68.
Vitamin A_2 acid ester 57.
Vitamin A acid ethyl ester 47, 49, 50.
Vitamin A acid ethyl ester, isomer 49, 50.
Vitamin A acid ethyl ester, spectrum 45.
Vitamin A acid ethyl ester isomer, spectrum 45.
Vitamin A acid methyl ester 49.
Vitamin A_2 acid methyl ester 77.
Vitamin A alcohol 42.
Vitamin A_2 alcohol 71.
Vitamin A alcohol, spectrum 45.
Vitamin A alcohol isomer, spectrum 45.
Vitamin A aldehyde 43, 55, 56, 71, 75, 80.
Vitamin A_2 aldehyde 71, 77.
all-trans-Vitamin A aldehyde 54, 56.
cis-Vitamin A aldehyde 75.
Δ^2-trans-Δ^6-cis-Vitamin A aldehyde 54, 56, 71, 75.
all-trans-Vitamin A aldehyde, photomicrograph 69.
all-trans-Vitamin A aldehyde, semicarbazone 54.
Vitamin A anthraquinone-β-carboxylate 42, 53, 54, 61.
Vitamin A, benzene analog 70, 74.
Vitamin A and carotenes, relationship 80.
Vitamin A, compounds related to 70.

Vitamin A esters 64, 74.
Vitamin A ethers 74.
Vitamin A isomer 49, 51.
Vitamin A isomers, growth-promoting effect 73.
Vitamin A, cis-trans-isomerism 51.
Vitamin A methyl ether 43, 60, 61, 64, 71, 75.
Vitamin A β-naphthoate 42, 53, 54, 61.
Vitamin A and neovitamin A, relative potencies 66.
Vitamin A, oxidation 56.
Vitamin A palmitate 61.
Vitamin A p-phenylazobenzoate 49, 61.
Vitamin A phenyl ether 60, 61, 71, 75.
Vitamin A, photomicrograph 67.
Vitamin A succinate 42.
Vitamin A synthesis 41, 43.
Vitamin A synthesis via "C_{14}-aldehyde" 58, 59.
Vitamin A synthesis via "C_{14}-aldehyde", extinction values of products 61.
Vitamin A synthesis via β-ionylideneacetic acid esters 43, 47.
Vitamin A synthesis via ionylidenecrotonic acid esters 52, 54, 55.
Vitamin A synthesis, intermediates biopotencies 62.
Vitamin A synthesis, spectra of intermediates 48, 54.
Vitamin A tetrahydropyranyl ether 60.
Vitamine B_1 93.
Vitamine B_6 95, 96.
Vitamin B_{12} 95, 404, 411, 417, 418, 419, 421, 426, 428, 429, 432.
Vitamin B_{12b} 420.
Vitamin B_{12} in animal nutrition 419.
Vitamine B_{12}, antagonistes 94.
Vitamin B_{12}, different forms 419.
Vitamin B_{12} in growth of microorganisms 426.
Vitamin B_{12}, microbiological functions 418.
Vitamin B_{12} requirement and reducing agents 427.
Vitamin D 119.
Vitamin K 242.

Water, free energy of formation 320.
Whitening agents 241.
Wild cloves 249.

Xanthine 390.
Xanthine déhydrase 94.
Xanthine dehydrogenase 298.
Xanthine desoxyriboside 389.
Xanthine oxidase 366, 373, 381.
Xanthine oxidase in liver 299.
Xanthotoxin 264, 265, 269.
Xanthotoxol y : y-dimethylallyl ether 268.
Xanthoxyletin 277, 278, 279, 280.

Xanthoxylum senegalense 264.
Xanthotoxol 264, 265.
Xanthyletin 227, 277.
Xynthosine 404.
Xynthotoxol 268.

Yoloxochitl 179.

Zanthoxyl suberosum 250.
Zanthoxylum americanum 277, 278, 281.